HANDBOOK OF
Neurolinguistics

HANDBOOK OF
Neurolinguistics

Edited by

BRIGITTE STEMMER
Centre de Recherche du Centre Hospitalier
Côtes-des-Neiges, Montréal, Canada
and
Lurija Institute for Rehabilitation and Health Sciences at the University of
Konstanz, Kliniken Schmieder, Allensbach, Germany

HARRY A. WHITAKER
Department of Psychology
Northern Michigan University
Marquette, Michigan

ACADEMIC PRESS

San Diego   London   New York   Boston   Sydney   Tokyo   Toronto

Academic Press
a division of Harcourt Brace & Company
525 B Street, Suite 1900, San Diego, California 92101-4495, USA
http://www.apnet.com

Academic Press Limited
24-28 Oval Road, London NW1 7DX, UK
http://www.hbuk.co.uk/ap/

Library of Congress Card Catalog Number: 97-80297

International Standard Book Number: 0-12-666055-7

PRINTED IN THE UNITED STATES OF AMERICA
97 98 99 00 01 02 EB 9 8 7 6 5 4 3 2 1

CONTENTS

PART II   Clinical and Experimental Methods in Neurolinguistics

2. Methodological and Statistical Considerations in Cognitive Neurolinguistics
Klaus Willmes

3. Clinical Assessment Strategies: Evaluation of Language Comprehension and Production by Formal Test Batteries
Jean Neils-Strunjaš

4. Research Strategies: Psychological and Psycholinguistic Methods in Neurolinguistics
Chris Westbury

PART III Experimental Neurolinguistics

A. Levels of Language Representation and Processing: Linguistic and Psychological Aspects

PART IV   **Clinical Neurolinguistics**

A. *Language and Communication in Special Populations*
and in Various Disease Processes

CONTRIBUTORS

Numbers in parentheses indicate the pages on which the authors' contributions begin.

Francisco Aboitiz (393), Programa de Morfología, Instituto de Ciencias Biomedicas, Facultad de Medicina, Universidad de Chile, Santiago, Chile

Luis F. H. Basile (143), Department of Neurosurgery, University of Texas—Houston, Health Science Center, Houston, Texas 77030

D. Frank Benson (447), Department of Neurology, University of Southern California at Los Angeles School of Medicine, Los Angeles, California 90024

Leo Blomert (547), Department of Psychology, University of Maastricht, Maastricht, The Netherlands

Stefano F. Cappa (535), Laboratorio di Neuropsicologia, Università di Brescia, Brescia, and Dipartimento di Scienze Cognitive, Scientific Institute H S. Raffaele, Milan, Italy

Paulo Caramelli (463), Cognitive and Behavioral Neurology Unit, Neurology Division, Hospital das Clínicas of the University of São Paulo School of Medicine, São Paulo, Brazil

Dominique Cardebat (261), INSERM U455, Hopital Purpan, Toulouse, France

Yves Chantraine (261), Centre de Recherche de l'Institut Universitaire de Gériatrie de Montréal, and École d'Orthophonie et d'Audiologie, Faculté de Médecine, Université de Montréal, Montréal, Québec, Canada, H3W 1W5

Monique M. Cherrier (447), Department of Psychiatry and Biobehavioral Sciences, University of California at Los Angeles School of Medicine, Los Angeles, California 90024 and West Los Angeles Veterans Affairs Medical Center, Los Angeles, California 90073

Hélène Chevalier (95), Department of Psychology, Brock University, St. Catharines, Ontario, Canada L25 3A1

Henri Cohen (475), Laboratoire de Neuroscience de la Cognition et Département de Psychologie, Université du Québec à Montréal, Montréal, Québec, Canada H3C 3P8

Laurent Cohen (331), Service de Neurologie, Hôpital de la Salpêtrière, Paris, France

David Corina (313), Department of Psychology, University of Washington, Seattle, Washington 98195

Bruce Crosson (431), Department of Clinical and Health Psychology, University of Florida Health Science Center, Gainesville, Florida 32610

Jeffrey L. Cummings (447), Department of Psychiatry and Biobehavioral Sciences, University of California at Los Angeles School of Medicine and West Los Angeles Veterans Affairs Medical Center and Department of Neurology, University of California School of Medicine, Los Angeles, California 90024

Stanislas Dehaene (331), INSERM, CNRS, and EHESS, Laboratoire de Sciences Cognitives et Psycholinguistique, Paris, France

Jean-François Démonet (131), INSERM U455, Service de Neurologie, Hôpital Purpan, Toulouse Cedex, France

Gianfranco Denes (507), Department of Neurological and Psychiatric Sciences, University of Padova, Padova, Italy

Nina F. Dronkers (173), Veterans Affairs Northern California Health Care System, Martinez, California 94553 and University of California, Davis, California 95616

Zohar Eviatar (275), Psychology Department, University of Haifa, Haifa, Israel

Julie A. Fields (189), Department of Neurology, University of Kansas Medical Center, Kansas City, Kansas 66160

Brigitta Gahl (617), Kliniken Schmieder, Allensbach, Germany

Guido Gainotti (3), Institute of Neurology of the Catholic University of Rome, Rome, Italy

Jackson T. Gandour (207), Department of Audiology and Speech Sciences, Purdue University, West Lafayette, Indiana 47907

Harold Goodglass (13), Aphasia Research Center, Department of Neurology, Boston University School of Medicine, Boston, Massachusetts 02130

Peter Hagoort (235), Max Planck Institute for Psycholinguistics, Nijmegen, The Netherlands

Francesca Happé (525), Social, Genetic and Developmental Psychiatry Research Centre, Institute of Psychiatry, London, UK

Anthony E. Harris (343), Intelligent Systems Program, Center for the Neural Basis of Cognition, and the Department of Neurology, University of Pittsburgh, Pittsburgh, Pennsylvania 15261

Joseph B. Hellige (405), Department of Psychology, University of Southern California, Los Angeles, California 90089

Manfred Hild (665), Kliniken Schmieder, Allensbach, Germany

Merrill Hiscock (357), Department of Psychology, University of Houston, Houston, Texas 77204

Andrés Ide (393), Programa de Morfología, Instituto de Ciencias Biomedicas, Facultad de Medicina, Universidad de Chile, Santiago, Chile

Gonia Jarema (221), Department of Linguistics, Université de Montréal, Montréal, Québec, Canada, and Centre de Rêcherche de l'Institute Universitaire de Gériatrie de Montréal, Montréal, Québec, Canada H3W 1W5

Yves Joanette (261), Centre de Recherche de l'Institute Universitaire de Gériatrie de Montréal, and École d'Orthophonie et d'Audiologie, Faculté de Médecine, Université de Montréal, Montréal, Québec, Canada H3W 1W5

Richard C. Katz (585), Department of Veterans Affairs Medical Center, Phoenix, Arizona 85012 and Arizona State University, Tempe, Arizona 85287

Marcel Kinsbourne (385), New School for Social Research, New York, New York 10011

Herman Kolk (249), Nijmegen Institute for Cognition and Information, University of Nijmegen, Nijmegen, The Netherlands

Sieglinde Lacher (641), Kliniken Schmieder, Allensbach, Germany

André-Roch Lecours (17), Centre de Recherche du Centre Hospitalier Côte-des-Neiges, and Faculté de Médecine, Université de Montréal, Montréal, Québec, Canada H3W 1W5

Leticia Lessa Mansur (463), Department of Internal Medicine (Speech Pathology), University of São Paulo School of Medicine, São Paulo, Brazil

Carl A. Ludy (173), Veterans Affairs Northern California Health Care System, Martinez, California 94553

Phan Luu (159), Department of Psychology, University of Oregon, Eugene, Oregon 97403

Brian MacWhinney (599), Department of Psychology, Carnegie Mellon University, Pittsburgh, Pennsylvania 15213

Nadine Martin (559), Center for Cognitive Neuroscience, Temple University School of Medicine, Philadelphia, Pennsylvania 19140

Skye McDonald (485), School of Psychology, University of New South Wales, Sydney, Australia

Mario F. Mendez (447), West Los Angeles Veterans Affairs Medical Center, Los Angeles, California 90073 and Department of Neurology, University of California at Los Angeles School of Medicine, Los Angeles, California 90024

Dennis L. Molfese (515), Department of Psychology, Southern Illinois University, Carbondale, Illinois 62901

Stephen E. Nadeau (431), Department of Clinical and Health Psychology, University of Florida Health Science Center, Gainesville, Florida 32610

Jean Neils-Strunjaš (71), Department of Communication Sciences and Disorders, University of Cincinnati, Cincinnati, Ohio 45221

Ricardo Nitrini (463), Cognitive and Behavioral Neurology Unit, Neurology Division, Hospital das Clinicas of the University of São Paulo School of Medicine, São Paulo, Brazil

Nancy A. Pachana (301), Department of Psychology, Massey University, Palmerston North, New Zealand

Andrew C. Papanicolaou (143), Department of Neurosurgery, University of Texas-Houston Health Science Center, Houston, Texas 77030

Michel Paradis (417), Department of Linguistics, McGill University, Montréal, Québec, Canada H3A 1G5

Martine Poncelet (289), Neuropsychology Unit, University of Liège, Liège, Belgium

Volkbert M. Roth (585), Kliniken Schmieder, Allensbach, Germany, and Universität Konstanz, Germany

Shirin Sarkari (515), Department of Psychology, Southern Illinois University, Carbondale, Illinois 62901

Sidney J. Segalowitz (95), Department of Psychology, Brock University, St. Catharines, Ontario, Canada L2S 3A1

Philip H. K. Seymour (573), Department of Psychology, The University of Dundee, Dundee, Scotland

Panagiotis G. Simos (143), Department of Neurosurgery, University of Texas-Houston Health Science Center, Houston, Texas 77030

Martine Simard (17), McGill Centre for Studies in Aging, Douglas Hospital, Verdun, Québec, Canada H3W 1W5

Steven L. Small (343), Intelligent Systems Program, Center for the Neural Basis of Cognition, and the Department of Neurology, University of Pittsburgh, Pittsburgh, Pennsylvania, 15261

Brigitte Stemmer (617), Centre de Recherche du Centre Hospitalier Côte-des-Neiges, Montréal, Québec, Canada H3W 1W5 and Lurija Institute for Rehabilitation and Health Sciences at the University of Konstanz, Kliniken Schmieder, 78476 Allensbach, Germany. E-mail: Brigitte.Stemmer@uni-konstanz.de

Arlene A. Tan (515), Department of Psychology, Southern Illinois University, Carbondale, Illinois 62901

Joseph I. Tracy (495), Allegheny University of the Health Sciences, Medical College of Pennsylvania/Hahnemann School of Medicine and the Norristown State Hospital Clinical Research Center, Philadelphia, Pennsylvania 19129

Alexander I. Tröster (189), Department of Neurology, University of Kansas Medical Center, Kansas City, Kansas 66160

Don M. Tucker (159), Department of Psychology, University of Oregon, Eugene, Oregon 97403 and Electrical Geodesics, Inc., Eugene, Oregon 97403

Martial Van der Linden (289), Neuropsychology Unit, University of Liège, Liège, Belgium

Diana Van Lancker (301), Department of Neurology, University of Southern California Medical Center, Los Angeles, California 90033 and Veterans Affairs Outpatient Clinic, Los Angeles, California 90012

Chris Westbury (83), Center for Cognitive Studies, Tufts University, Medford, Massachusetts 02155

Harry A. Whitaker (27), Professor and Head of the Department of Psychology, Northern Michigan University, Marquette, Michigan 49855. E-mail: hwhitake@nmu.edu

Klaus Willmes (57), Neurologische Klinik—Neuropsychologie, Universitätsklinikum RWTH Aachen, Aachen, Germany

Eran Zaidel (369), Department of Psychology, University of California at Los Angeles, Los Angeles, California 90095

PREFACE

This handbook is intended as a state-of-the-art reference and resource book describing current research and theory in the many subfields of neurolinguistics and their clinical applications. The handbook aims at reaching the newcomer to the field, as well as the expert searching for the latest developments in neurolinguistics.

The contents of this handbook were shaped while the editors were driving from Montréal to Vancouver. The length of the trip and the beauty of the scenery contributed to the number and creative combination of topics. In order to cover the many subfields and, at the same time, keep the handbook manageable, we opted to include short articles on a variety of different topics instead of longer articles. To keep with this goal and the goal of the handbook, the authors were asked—not always to their delight—to focus on the most recent advances in the field and to point the reader to older literature by referring to review articles. Despite these limitations, we think the authors did a remarkable job, and the many areas covered attest to the broadness and interdisciplinarity of the field.

The book starts out with three prologues; the first written by Guido Gainotti, the second by Harold Goodglass, and the third by André-Roch Lecours and Martine Simard. Gainotti discusses category-specific disorders for nouns and verbs—a topic that was raised by Goodglass and his co-workers in 1966 and which does not seem to have lost topicality 30 years later. Goodglass surveys the advances that have been made in neurolinguistic research, and Lecours and Simard discuss the cerebral substrate of language from the beginning to the end of life and the impact of environmental and biological factors.

The book is organized in five major sections, with each section subsuming chapters that share a common framework but that may differ widely in the topics they cover.

Part I covers the latest research in and provides a survey of the history of neurolinguistics. Part II contains chapters dealing with clinical and experimental methods in neurolinguistics and covers methodological and statistical considerations; clinical assessment and research strategies; neurophysiological and imaging techniques; brain lesion analysis in clinical research; the sodium amytal test; and a theoretical outline of neurolinguistics that tries to integrate neural organization across brainstem, limbic, and cortical levels. In Part III experimental aspects of neurolinguistics are pursued, such as current views on linguistic and psychological aspects of various levels of language representation and processing, and the issue of lateralization of language and communication is investigated from various perspectives. In Part IV the clinical aspect of neurolinguistics is emphasized. Language and communication in multilinguals, in people affected with various disease processes, such as dementia, Parkinson's disease, traumatic brain injury, and psychosis, and in people with developmental disorders is discussed. This section of the handbook ends with a collection of chapters devoted to the important issue of recovery from and rehabilitation of language and communication disorders. The final section of the handbook, Part V, provides information and techniques that should help the clinician and the researcher in their daily work: A methodology for eliciting, recording, transcribing, and analyzing language and communication data is described by Brian MacWhinney. Brigitte Stemmer and her colleagues provide a list of neurolinguistic assessment and rehabilitation software and journal and book resources relevant to neurolinguistics and related fields. The final chapter of the book is a compilation of useful resources on the internet.

We not only hope that this book will be a practical and useful resource for the newcomer and expert but will also ignite constructive criticism from our readers. Despite the inspiring environment that helped create the contents and the efforts put into the book by the contributing authors, some topics may have slipped the editors' attention, or some readers may wish that certain topics had been covered or presented in a different way. As we hope that the book will create enough interest for the printing of a revised edition, we would like to encourage all readers to provide us with feedback, suggestions, and information to help make a revised edition even better.

We would like to thank the authors of the handbook for dedicating their valuable time to the project, for their cooperation, for their willingness to keep within the page limits, and for their patience with the editors throughout the entire project. The first author is grateful to the Centre de Recherche du Centre Hospitalier Côte-des-Neiges, Montréal, Canada for the financial and accommodating support. Finally, the first author would like to thank her colleagues and friends for their moral support and their tolerance of her antisocial behavior during seemingly endless times of manuscript battles.

Brigitte Stemmer
Harry A. Whitaker

Prologue

Category-Specific Disorders for Nouns and Verbs

A Very Old and Very New Problem

Guido Gainotti

Institute of Neurology of the Catholic University of Rome, Rome, Italy

Several clinical and experimental studies have shown that nouns and verbs can be independently disrupted by brain damage.

A prevalent impairment in naming actions (producing verbs) is usually observed in nonfluent/agrammatic patients, with a lesion involving the frontal lobes, whereas a selective defect in naming objects (producing nouns) is usually observed in anomic patients, with lesions involving the temporal lobe and the posterior association areas. The speculation is offered that action schemata, subserved by the frontal lobes, may play a crucial role in the semantic representations of verbs and that schemata of sensory integration, subserved by the posterior association areas, may play a critical role in the semantic representations of objects/nouns.

The possibility that nouns and verbs may be selectively and independently disrupted by brain damage is considered by many authors to be one of the most recent and intriguing discoveries in the field of neurolinguistics. As a matter of fact, the problem has been raised in the modern neuropsychological literature in a paper by Goodglass, Klein, Carey, and Jones (1966) dealing with specific semantic word categories in aphasia, but the interest for disorders selectively affecting nouns and verbs is at the same time older and younger than their paper. It is older because, for at least two centuries, the selective impairment of nouns and verbs in patients with brain damage had attracted the attention of literary men, philosophers, and neurologists. It is younger because the Gloodglass et al.'s paper did not immediately orient the work of other researchers, but rather became influential some years later, when the seminal work of Warrington and coworkers (Baxter & Warrington, 1985; McCarthy & Warrington, 1985; McKenna & Warrington, 1978; Warrington, 1975, 1981b; Warrington & McCarthy, 1983, 1987; Warrington & Shallice, 1984b) convincingly showed that different categories of knowledge can be selectively disorganized by brain damage.

Subsequently, several investigations have confirmed that nouns and verbs can be selectively disrupted by brain injury and have shown that this selective impairment can be predicted on the basis of clinical (syndromic and anatomical) variables. Furthermore, some interpretations aiming to clarify the meaning of these category-specific disorders have been advanced. This essay will focus on *(a)* recalling the pioneer work of authors who first called attention to disorders specifically affecting nouns and verbs, *(b)* summarizing results of clinical investigations that have given a detailed description of these disturbances and trying to elucidate the underlying pathophysiological mechanisms from the cognitive and the anatomical point of view, and *(c)* giving a short account of some recent experimental studies that used neurophysiological or neuroimaging procedures to investigate these questions.

THE PIONEERS

A curious phenomenon such as a selective loss of nouns and verbs could not escape the attention of literary persons. Indeed, Denes and Dalla Barba (1995) showed that the Italian philosopher Giovanni Battista Vico reported, in the third edition of his *Principi di Scienza Nuova,* the observation of a patient affected by a severe stroke who could remember nouns but had forgotten verbs. According to Vico (1744), this selective impairment of verbs was due to the greater fragility of this part of speech, since "nouns elicit ideas that leave lasting traces, whereas verbs denote movements, which refer to transient fractions of time and space." If Vico's observation was anecdotal and important only from the historical point of view, much more relevant from the viewpoint of contemporary research is the contribution of Pitres (1898a, 1898b) in his papers on amnesic aphasia.[1]

In these papers, Pitres offers two important contributions: (1) he anticipates our present assumptions that an important component of aphasia (namely, the word-finding impairment) may result from disruption of a component of the memory system (namely, of semantic memory); (2) he also describes a clinical variety of amnesic aphasia, which he calls "antonomasie," in which the defect selectively concerns nouns. In his discussion of the pathophysiological mechanisms underlying this category-specific impairment for nouns, Pitres firmly discards the possibility that this

---

[1]I want to express my gratitude to Prof. M. Poncet (Marseille), who called my attention to the Pitres papers.

selective defect may result from disruption of a brain "center" that specifically stores the representation of nouns. Analogous to Vico, he suggests that the selective impairment for nouns may be due to a different rate of memory decay of the various parts of speech. However, in contrast to Vico, and drawing on a model proposed by Ribot (1904) to account for various aspects of memory disorders, Pitres claims that words denoting qualities and actions (such as adjectives and verbs) are more resistant to memory decay than words denoting concrete entities, such as nouns. But more important than these speculations is the fact that Pitres also reviews the anatomoclinical cases of patients who had shown a category-specific impairment for nouns. Results of this review clearly demonstrate, in contrast with Pitres's theoretical positions, that the large majority of patients with a selective impairment for nouns had a well-localized brain lesion, encroaching upon the left temporal lobe and, in particular, upon the left angular gyrus. This finding, as we will see in section 4 of this essay dealing with the anatomical correlates of category-specific disorders for nouns and verbs, can have a certain theoretical interest.

THE CLINICAL STUDIES

As I noted at the beginning of this essay, the problem of disorders selectively concerning nouns or verbs was raised by Goodglass et al. (1966), who were able to demonstrate an opposite pattern of performance in Broca's aphasics and in patients with fluent aphasia on naming tasks using pictures of objects and of actions as stimuli. Broca's aphasics were mainly impaired in naming actions (producing verbs), whereas fluent aphasics showed a prevalent impairment in naming objects (producing nouns).

This double dissociation was subsequently confirmed by authors who focused attention on agrammatic patients (within Broca's aphasics) and on anomic patients within fluent aphasics. Thus, Myerson and Goodglass (1972), Marin, Saffran, and Schwartz (1976), and Miceli, Mazzucchi, Menn, and Goodglass (1983) noticed that agrammatic patients tend to omit main (root) verbs, whereas Benson (1979) stressed the selective difficulty met by anomic patients in the selection of nouns. Other facets of the problem have been investigated by means of both group studies and single-case studies. With the first methodology, Miceli and coworkers have shown (a) that agrammatic patients are more impaired in naming actions than in naming objects, whereas anomic patients show the reverse pattern of impairment (Miceli, Silveri, Villa, & Caramazza, 1984) and (b) that the same kind of dissociation can be observed in comprehension of nouns and verbs (Miceli, Silveri, Nocentini, & Caramazza, 1988). With the second methodology, McCarthy and Warrington (1985) have observed a selective impairment both in comprehending and in naming verbs in an agrammatic patient and have attributed this defect to a degradation of the semantic representation of actions, whereas Zingeser and Berndt (1988) have described an anomic patient who presented a selective difficulty in producing nouns both in the oral and in the written modality. An even more specific defect in verb production has been reported by Caramazza and Hillis (1991), broadening the field of investigation beyond the syndromes of agrammatism and anomia. One of their patients (H.W.), who showed a phonological working memory disorder, was unable to produce verbs in the oral, but not in the written, modality, whereas a second patient (S.J.D.), who showed a fluent but paraphasic speech, had a selective difficulty producing verbs in the written, but not in the oral, modality. Finally, some authors have studied category-specific disorders for nouns and for verbs in patients with focal progressive degenerative diseases of the brain, showing fluent or nonfluent forms of primary progressive aphasia. A category-specific impairment for verbs has usually been found in patients with a nonfluent form of primary progressive aphasia (e.g., patient R.O.X. of McCarthy & Warrington, 1985, and patient R.A. of Daniele, Giustolisi, Silveri, Colosimo, & Gainotti, 1994), whereas a selective impairment for nouns has been observed in patients with a fluent form of primary progressive aphasia (e.g., patient P.G. of Daniele et al., 1994, and patient D.A. of Silveri & Di Betta, in press). As with all the clinical data, these clinical findings have a statistical rather than an absolute value. Thus, Williams and Carter (1987) have described anomic patients who were more impaired in naming actions than in naming objects, whereas Miceli et al. (1984) have described a classical agrammatic patient who was equally impaired both with nouns and with verbs. Even with these reservations, however, the prevalent impairment of nouns in anomic aphasia and of verbs in agrammatism remains a well-established clinical fact that requires an appropriate explanation. Both the intrinsic structure of the linguistic disorganization and the location of the underlying brain lesion have been used to clarify this issue. In the present survey, I will summarize first the interpretations framed in cognitive terms and then those based on anatomical considerations.

THE NEUROPSYCHOLOGICAL LOCUS OF COGNITIVE IMPAIRMENT

Some controversies exist about the neuropsychological locus of cognitive impairment leading to a selective defect for nouns or verbs. According to some authors (e.g., Miceli et al., 1984, 1988; Caramazza & Hillis, 1991), this impairment should affect the lexical representations (and in particular the phonological or the graphemic output lexicon), whereas, according to other authors (e.g., McCarthy & Warrington, 1985; Gainotti, Silveri, Daniele, & Giustolisi, 1995), a disruption of the semantic representations of objects and actions might underpin the selective disorders for nouns and verbs. Both the lexical and the semantic defect hypothesis make reference to current information-processing models, assuming that the functional architecture of the semantic-lexical system may be based on five independent representations: two modality-specific (visual and auditory) input structures addressing a central semantic system and two response-specific (phonological and orthographic) output stores (Caramazza & Hillis, 1990b; Hillis & Caramazza, 1991). According to this model, a central semantic disorder should express itself both at the expressive and at the receptive level in any lexical task requiring a consultation of the semantic system (Gainotti, 1976;

Butterworth, Howard, & McLoughlin, 1984; Hillis, Rapp, Romani, & Caramazza, 1990), whereas disruption of a single lexical entry should affect only tasks relying on the damaged lexical modality. The hypothesis that the selective impairment of nouns and verbs may be due to purely lexical defects is supported by two sets of data: *(a)* the observation of Miceli et al. (1988) that some agrammatic patients have a category-specific impairment for verbs (and some anomics for nouns) circumscribed to the production stage, in the absence of comprehension disorders affecting the same categories; *(b)* the evidence of an even more specific defect in verb production reported by Caramazza and Hillis (1991), whose patient H.W. was unable to produce verbs in the oral modality only, in contrast with the other patient, S.J.D., who showed a selective deficit in verb production circumscribed to the written modality.

On the other hand, the hypothesis that, at least in some patients, the defect may be located at the semantic level is supported by two equally convincing sets of clinical data: *(a)* the observation of McCarthy and Warrington (1985) and of Miceli et al. (1988) that some anomics and some agrammatic patients show the same category-specific defect (for nouns or for verbs) both in production and in comprehension; *(b)* longitudinal data obtained by Daniele et al. (1994) in patients with category-specific disorders for nouns or for verbs arising in the context of degenerative diseases. These authors have shown, in fact, that in their patients category-specific disorders for nouns or for verbs were usually observed in a single output lexical modality in the first stages of the disease (a pattern of impairment considered by Caramazza and Hillis [1991] as being due to a disruption of the part of the corresponding lexical story where the affected word class is specified). With the disease progression, however, the same category-specific impairment tended to generalize to the other lexical representations, affecting first the other output lexical modality and, finally, also the level of lexical comprehension. Now, since it is very unlikely that with the disease progression the same pattern of category-specific impairment may develop independently in different lexical modalities, these data suggest that even a deficit apparently circumscribed to a single lexical modality might, indeed, be due to a semantic impairment. The order of progression of the defect across the different output and input lexical modalities could, in this case, be explained by assuming that the same semantic impairment may have a different clinical expression in different lexical modalities (or in different lexical tasks) according to the attentional demand of the task. According to this interpretation, confrontation naming tasks might be affected first because they require a more active and effortful search within the mental lexicon, whereas comprehension (matching-to-sample) tasks might be affected last because they are more simple and less demanding.

From the theoretical point of view, the lexical defect hypothesis makes the rather strong assumption that not only the semantic system, but also the modality-specific lexical entries may have a categorical organization. According to Rapp, Benzing, & Caramazza (1995), this redundant representation of the word class distinction could be due to the fact that these categorical effects are relevant at multiple levels of speech production (Garret, 1980):

1. at the level of the semantic/syntactic representation of lexical items;
2. in the application of syntactic rules for generating surface phrasal structures;
3. in the procedures responsible for assigning a phonological content to planning frames used in speech production.

The semantic defect hypothesis, on the other hand, has the advantage of a greater parsimony in that it does not assume a redundant representation of categorical information both at the semantic and at the lexical level. Furthermore, it gives a better account of the fact that disorders specifically concerning nouns and verbs were not observed during connected speech but during tasks exploring the knowledge of individual words (namely, confrontation naming and word-picture-matching tasks) using pictures of objects or of actions as stimuli. This hypothesis, however, oversimplifies the complex interaction of semantic and syntactic aspects of verb representation, since it focuses on the central meaning of verbs (actions), paying only marginal attention to the roles played by a verb's arguments (thematic roles) and ignoring syntactic aspects, such as the subcategorization frame and the argument structure of the verb (Breedin & Martin, 1996).

THE ANATOMICAL CORRELATES OF SELECTIVE DISORDERS FOR NOUNS OR VERBS

As I said in section 1, the majority of patients reviewed a century ago by Pitres (1898) for a disorder selectively affecting nouns had a lesion involving the left temporal lobe and extending toward the angular gyrus. This finding has been confirmed by recent investigations, which have studied lesion location in patients with a category-specific impairment for nouns. As a matter of fact, all patients reported for this defect by Miceli et al. (1988), Zingeser and Berndt (1988), Damasio and Tranel (1993), Daniele et al. (1994), Miozzo, Soardi, and Cappa (1994), and Silveri and Di Betta (in press) had a lesion of the left temporal lobe, extending in some cases toward the posterior association areas. It must also be noted that these findings are consistent with the standard lesion location in anomia, that is, in the aphasic syndrome where a selective impairment for nouns is usually observed. Less clear are the anatomical correlates of disorders specifically concerning verbs, since in this case the lesions are more scattered. In the majority of these patients, however, a definite pattern of lesion location, extending from the antero-superior parts of the temporal lobe toward the infero-posterior parts of the left frontal lobe can be observed (see Gainotti et al., 1995, for a detailed discussion of this issue).

Even in this case, the localization of lesions provoking a selective impairment of verbs is consistent with the usual anatomical correlates of agrammatism (where a category-specific disorder for verbs is generally observed). Taken together, anatomical data obtained in patients with a selective disorder for nouns or for verbs therefore suggest that a temporal lobe lesion is in any case necessary to produce a defect of lexical retrieval. However, the lesion extends from the temporal toward the frontal lobe when the word-retrieval defect concerns the action names and toward the posterior association areas when the object names are more difficult to produce and to comprehend.

At present, only speculation can be offered to explain why the word-finding disorder may mainly concern action names when the lesion extends toward the frontal lobe and object names when it extends toward the posterior association areas. My own speculation (Gainotti, 1990; Gainotti et al., 1995) is based on three main points: (1) the distinction between operations of action planning and of motor execution subserved by the frontal cortical areas and the operations of sensory analysis and of perceptual integration subserved by the posterior association areas; (2) the assumption made by contemporary connectionistic models (e.g., Ballard, 1986; Farah & Mc-Clelland, 1991) that information processing and storage are not separated, but closely interwined in a network, since information is stored as a pattern of activity in the connections between the units of the net processing the same information; according to this assumption, the representations of different semantic categories should be closely interwined with the neurophysiological mechanisms that have critically contributed to their acquisition; (3) the hypothesis that the frontal lobe, housing the neurophysiological mechanisms involved in action planning and execution, may play a greater role in the acquisition of the semantic representations of actions and that the posterior association areas, subserving operations of analysis and integration of sensory information, may play a critical role in the acquisition of the semantic representations of objects.

A rather similar "neurological model" of naming was proposed some years ago by Geschwind (1967). According to this author, aphasic anomia results from damage to the left angular gyrus because a large variety of sensory information converges in this region, supplying the subject with the "attributes" necessary to arouse the verbal label corresponding to the semantic representation of the stimulus object. In my opinion, the Geschwind model, stressing the role of sensory attributes in establishing the semantic representation of objects, could explain why a lesion of their convergence zone, in the posterior association areas, produces a selective inability to name objects. A similar model, stressing the importance of action schemata in constructing the semantic representations of actions, could explain why agrammatic patients, with a left frontal lesion, are particularly impaired in producing and comprehending verbs (action names).

Although certainly speculative, this interpretation is consistent with the view, originally proposed by Warrington and coworkers (Warrington & Shallice, 1984b; Warrington & McCarthy, 1983, 1987) and now accepted by many other authors (see Hagoort, this volume), of a relationship between categories of semantic knowledge and brain systems for perception and action. According to this view, various kinds of perceptual and functional information usually converge in the acquisition of different semantic categories, but the relative salience of each source of information may be different for different semantic categories. Disruption of a particular source might, therefore, lead to a prevalent impairment of the semantic categories that are more heavily based on the disrupted brain mechanism.

RECENT NEUROPHYSIOLOGICAL AND NEUROIMAGING INVESTIGATIONS

The hypothesis that nouns and verbs may be stored in different brain areas was recently tested in normal subjects by Warburton et al. (1996) with a PET study and by Preissl, Pulvermüller, Lutzenberger, & Birbaumer (1995) with an Evoked Potentials experiment. These two investigations did not give consistent results. The PET study showed only a quantitative, but no qualitative, difference between the pattern of cerebral activation produced by tasks of verb retrieval and noun retrieval, since the rCBF increase was greater during the verb retrieval task (which was, indeed, more difficult) but the distribution of activation was similar in the frontal, parietal, and temporal regions of the left hemisphere.

On the other hand, the Evoked Potentials study demonstrated a significant interaction between word class and electrode site, with a greater representation over the frontal regions of a positive wave form generated post stimulus onset. Because, in a preexperiment aiming to assess the motor and visual associations elicited by the same stimuli, motor associations had been much stronger for verbs than for nouns, the authors conclude that the greater amplitude of the Evoked Potentials recorded from the frontal lobes has its counterpart in the stronger motor association elicited by verbs.

CONCLUDING REMARKS

Even if the observation of disorders selectively affecting nouns or verbs is a very old one, the systematic study of these disturbances with well-controlled experimental designs based on sophisticated linguistic or cognitive models is a very recent enterprise. It is chiefly for this reason that I have focused attention in this essay on a rather simple and clear line of investigation, rather than discussing in detail other more complex and controversial facets of the problem. We cannot, however, ignore some of the most puzzling aspects of this issue, aspects that mainly concern category-specific disorders for verbs. Both theoretical and factual reasons can probably account for the greater uncertainties concerning the clinical features, the pathophysiological mechanisms, and the anatomical correlates of disorders selectively affecting verbs. The theoretical reasons derive from the controversies existing about the relationships between semantic and syntactic aspects of verb representation (Pinker, 1989; Jackendoff, 1990), whereas the factual reasons refer to the observation that a very heterogeneous set of disturbances is probably grouped under the heading "category-specific impairments for verbs."

As a matter of fact, if we look at patients reported in some detail by various authors, the following points emerge: (1) some patients, as I have noted, show a selective impairment for verbs limited to a single lexical modality, whereas in other patients this defect can be observed at both the expressive and the receptive level; (2) in some of these subjects, the capacity to select the appropriate abstract lexical item (the "lemma," according to Levelt, 1992) seems relatively spared, whereas the defect mainly concerns postselection aspects of the verb processing. I refer here, for example, to patient K.S.R. (Rapp et al., 1995), who showed phonemic distortions and neologisms in oral production of verbs and misspellings in written production of nouns, or to patient S.M. (Silveri and Di Betta, in press), who showed morphological (inflexional) errors with verbs, both in single word production and in (matching-to-sample) lexical comprehension tasks.

The heterogeneity of category-specific disorders for verbs has been confirmed in an interesting paper by Breedin and Martin (1996), who have shown that different patterns of verb disruption (involving theoretically relevant aspects of the semantic and syntactic features of verb representation) can be observed in patients showing a selective loss of verbs. A careful quantitative analysis of errors shown on production tasks and of the pattern of impairment shown on theoretically relevant tasks should, therefore, constitute the prerequisite necessary to gather the database required to elaborate more complex and detailed interpretations of category-specific impairments for verbs.

Advances in Neurolinguistic Research

Harold Goodglass

Aphasia Research Center, Department of Neurology, Boston University School of Medicine, Boston, Massachusetts 02130

The last decade of the 20th century is an auspicious time to survey the advances that have taken place in neurolinguistics in 50 years—much of it since the mid-1980s. This progress has taken place through a new appreciation of cognitive-linguistic factors in the definition of aphasic symptoms, through sophisticated, on-line methods of behavioral analysis, through continued research with the tried and true methods of clinico-anatomic correlations, and through the introduction of CT and MRI scans, permitting superb *in vivo* lesion studies, followed by methods for functional imaging of the normal brain in action. A particularly gratifying development is the growth of interdisciplinary research carried out increasingly by individuals who are schooled in multiple areas, such as advanced imaging technology, along with techniques of cognitively sophisticated experimental design, and who also share in the accumulated knowledge from the clinico-anatomical investigation of aphasia.

Looking back to the period immediately after World War II, it is fair to say that concepts of brain organization for language and methods of study were largely unchanged from those at the end of the 19th century. The dominant models conceived of language as a set of faculties, corresponding to the input and output channels of auditory comprehension, speech output, reading, and writing, with a storage center for each and postulated connections between them. Only in the case of interhemispheric disconnection phenomena such as pure alexia and unilateral agraphia and apraxia had the role of connecting pathways been concretely demonstrated. To be sure, the clinician-scholars of the 19th and early 20th centuries had been well aware of linguistically defined symptoms such as anomia and agrammatism that cut across modalities. These features were noted descriptively but did not influence the structure of the anatomical models.

However simplistic the classical anatomic-connectionist models might seem, they served the goal of clinical communication and lesion localization reasonably well. The only competition that they had came from the theorists of the noetic or mentalistic school, such as Hughlings Jackson, Finkelnburg, Goldstein, and Head.

The approach that these writers had in common was to reduce aphasia to the loss of a general symbolizing capacity. Jackson's concepts of automaticity versus propositionality of language use and Goldstein's notion of abstract and concrete behavior have been incorporated into the thinking of contemporary neurolinguists. However, as an explanatory model, the concept of a general symbolic capacity has been completely overshadowed by the accumulation of clinico-pathological evidence concerning the specific lesion-to-symptom relationships.

A major change in the focus of aphasia research occurred in the 1950s, with the influx of psycholinguists and cognitive psychologists. The very factors that had been neglected by the classical theorists moved into the foreground, along with the research methodology of experimental psychology: controlled comparisons between contrasted groups or contrasted experimental conditions and the use of statistical tests of hypotheses. But it was not merely the trappings of experimental methodology that were new. The questions asked began to probe the underlying rules that might explain the mechanisms that gave rise to linguistic symptoms. Hypotheses were imported from linguistic theory and from the psycholinguistics of normal language. For example, Goodglass and Hunt (1958) compared the accessibility of the plural and possessive final *s* inflection for aphasics, to see if they conformed to predictions from Jakobson's (1956) theory. Wepman, Bock, Jones, and Van Pelt (1956) examined the frequency relationships among parts of speech used by anomic patients, to see if they could be accounted for by a frequency-based account.

Two studies in the early 1970s greatly influenced the direction of subsequent studies of syntactic processing and of reading disorders in aphasia. Zurif, Caramazza, and Myerson (1972), by revealing that agrammatic Broca's aphasics had difficulties in syntactic decoding, opened the door to research guided by the "central syntactic deficit" hypothesis. Marshall and Newcombe (1973), by their analysis of the basic disorder of what came to be called "deep dyslexia" and "surface dyslexia," opened the door to the cognitive neuropsychology of written language.

Whereas classical anatomic connectionism had as its goal the construction of an anatomically based architecture of language processing, the often-reiterated goal of cognitive neuropsychology is the determination of the "cognitive architecture" of language. These hypothetical architectures are commonly displayed as flow-chart diagrams of sequential operational stages, subject to injury either of a processing component or its connection to the next component in the sequence. They differ from the anatomical diagrams mainly in that they make no claim to a structural basis.

Much of the work in cognitive neuropsychology has relied on in-depth studies of single cases that displayed linguistically defined dissociations of particular theoretical interest. Unlike the psycholinguistic studies in aphasia of the 1950s and 1960s, the objectives of cognitive neuropsychology are not so much concerned with the clinical phenomenology of aphasia as with the elucidation of normal language processes through data gleaned from the damaged system. The movement to single case studies has been motivated by several considerations. One is that selection of research subjects on the basis of syndrome labels is too unreliable a predictor of how a linguistic process of interest is affected in any patient (cf. Schwartz, 1984). Another is the doctrine of "transparency," expressed in its strongest form by Caramazza (1984). This is the presumption that the normal cognitive architecture for language is defined by a fixed set of basic components, any of which may be injured in aphasia, resulting in a deficit symptom. The transparency doctrine postulates that brain injury cannot bring about the appearance of processing mechanisms that were not available premorbidly; hence there is a one-to-one relationship between any symptom of deficit and a preexisting component.

Despite criticisms leveled at the assumptions of the transparency hypothesis and of a universal normal cognitive architecture (cf. Goodglass, 1993), single case analyses by cognitive neuropsychologists have been extremely fruitful in uncovering new syndromes. They have also provided new ways of conceptualizing relationships between previously described symptoms. For example, the syndrome of deep dyslexia and its cognitive underpinnings, as well as the analysis and interpretation of category-specific and modality-specific lexical dissociations, are products of cognitive neuropsychological analyses of single cases. Thus, the application of the method has added to awareness of the phenomena to be accounted for in the cognitive-linguistic domain.

Methods of behavioral analysis in neurolinguistics have expanded to include an array of on-line methods. Until recently, the study of linguistic operations was confined to the recording of the end product of what might be a multicomponent process, parts of which were outside of awareness. The last decade has seen the introduction of an array of on-line techniques that reveal the unfolding of language events in real time. They include the use of lexical decision probes, event-related potentials, autonomic measures, and functional brain imaging.

The introduction of computed tomography (CT), soon followed by magnetic resonance imaging (MRI), has accelerated the pace of new clinico-anatomic correlations beyond anything that could have been imagined four decades ago. Freed from the slow and haphazard accumulation of postmortem lesion information, every type of structure-to-function correlation is virtually universally available. But a second revolution came into being with the introduction of functional brain imaging—techniques for identifying the sites of increased local cerebral blood flow or glucose metabolism that signal the brain regions that are activated by an imposed cognitive task. Now it becomes possible to examine the normal brain in action, checking out hypotheses derived from lesion data as to the regions that may be vital for particular cognitive or linguistic operations.

The problem of category-specific lexical dissociations is perhaps the most dramatic instance in which cognitive theory and investigative techniques have converged with advanced techniques in brain anatomy and physiology. Over a relatively short span of years, patients have been studied as a result of their presenting circumscribed disorders in retrieving or in identifying objects of particular categories. Among the affected categories are animals, fruits and vegetables, and flowers—that is, objects of nature that are known largely by their visual properties (Warrington & McCarthy, 1987). These deficits appear most commonly in recovered herpes encephalitis. A reversal of this pattern has been seen in a number of aphasic patients, who performed well on animate objects, but poorly on man-made ones (Warrington & McCarthy, 1983, 1987). Clues to the mechanism for category-specific dissociations have come from cognitive analyses of cases (Hart & Gordon, 1992; Sartori, Job, Miozzo, Zago, & Marchiori, 1993; Warrington & McCarthy, 1987), group studies of clinico-anatomic data (Damasio, Grabowski, Tranel, Hichwa, & Damasio, 1996), and functional imaging based on normal subjects performing lexical retrieval of items from contrasting categories (Damasio et al., 1996; Martin, Wiggs, Ungerleider, & Haxby, 1996). Not only do the data from these sources reinforce each other, they also converge with recent insights from primate research on the role of the inferior temporal lobe in visual object recognition (Ungerleider & Mishkin, 1982).

One of the by-products of functional imaging has been the general consensus that the simple anatomic serial processing models probably have to give way to networks of interacting units. In the study of brain and language, insight has had the habit of arriving in small steps. Although functional imaging has proven to be an exciting new channel of investigation, it has thus far brought additional bits of confirmatory data or new observations to be explained and reconciled with others. Thus, the outlook for the upcoming century is for accelerating progress with new tools in the areas of technology and knowledge, while the goal of understanding remains a daunting challenge.

Acknowledgments

This essay is based on ideas developed in collaboration with Dr. Arthur Wingfield, which are to be published at greater length elsewhere. The preparation of this paper was supported in part by USPHS Grant DC 00081.

Cerebral Substrate of Language

Ontogenesis, Senescence, Aphasia, and Recoveries

André-Roch Lecours[1] and Martine Simard[2]

[1]Centre de Recherche du Centre Hospitalier Côte-des-Neiges and Faculté de Médecine, Université de Montréal, Montréal, Québec, Canada H3W 1W5; [2]McGill Centre for Studies in Aging, Douglas Hospital, Verdun, Québec, Canada H3W 1W5

The "aging" notion covers a time span that begins in utero and ends with death. This is also true for the development of language. Indeed, we shall verify this assertion with a discussion on the embryogenesis and the ontogenesis of the cerebral substrate of language. Later, we will demonstrate the continuing process of the ontogenesis of language with a discussion on sociocultural factors molding the evolution of oral and written language and the biology of cerebral language areas. These interactions between environment and biology, lasting until death, insofar as the brain is not destroyed, will affect the capacity to exploit language during childhood, adulthood, and senescence and will also determine the ability to recover following a brain injury.

CEREBRAL ONTOGENESIS

The nervous system has its origin in the most external layer of the human embryo: the "ectodermal" layer, which is already present on the fifth day of gestation. At this time, the rostral extremity of the neural tube resembles a median holosphere with two *anlagen* distributed symmetrically, on its right and left sides. These *anlagen* represent the subcortical structures of the brain, and the cerebellar hemispheres will be one of the results of their development. In this area of the brain, the left and right parts grow equally, with no laterality effect on the rest of the body, realizing in this way what Yakovlev (1968) has defined as the "primary parity." Later, in the fourth fetal week, the telencephalon is represented by a holosphere that grows from the median fountainhead (located in the lamina terminalis) rostrally to the optic vesicle. This holosphere is the *anlagen* of the septum, the paraolfactory areas, the olfactory bulbs, and the hippocampi. Yakovlev has labeled it the "telencephalon impar" or "rhinic brain." The projections of these new structures are ablateral. The genetic program continues its course and around the sixth fetal week, laterally to the "telencephalon impar," a first pair of paramedian vesicles appears and constitutes what Yakovlev has called the "telencephalon semipar" or "limbic brain" (the limbic lobe of Broca). These evaginations are the *anlagen* of the cingular gyrus, the isthmus, the parahippocampal gyrus, the limen, and the insula. The duality of the limbic brain is not complete because ventrally the vesicles are reunited in the midline (or in the medial band). This fact justifies the term "semipar" used by Yakovlev. The projections of the "telencephalon semipar" are ambilateral. The formation of the limbic brain is followed, in the ninth gestational week, by the emergence of a second pair of left and right vesicles, dorsolaterally to the first ones: it represents the supralimbic structures and Yakovlev has named it the "telencephalon totopar." "Totopar" means a reinstallation of the parity between the left and the right, in other words, a "secondary parity." This parity is nevertheless different from the primary parity of the rhinic brain: the projections of the supralimbic brain are mostly controlateral. The evaginations of the "telencephalon totopar" are the *anlagen* of the frontal, parietal, temporal, and occipital lobes (Yakovlev, 1968).

Yakovlev considered that each of the "three brains" corresponds to a different type of motility (see also MacLean, 1987). According to this point of view, the rhinic brain is devoted to the "endokinesis" or the "cell-bound movement" related to visceral motility; the limbic brain plays a role in the "ereismokinesis" or the "body-bound movement" dedicated to the expression of emotions; and finally, the supralimbic brain is attached to the "telokinesis" or the "object-bound movement" oriented to the external world. Expressive language is a good example of the "telokinetic" role played by the supralimbic structures (Yakovlev, 1948, 1963, 1970).

Asymmetries appear soon, between the left and right hemispheres of the supralimbic structures, during the uterine period, by way of cell differentiation, migration, and maturation and the synaptic competition in regions later devoted to language. The maturation phase is characterized by the development of dendrites, and the creation of synapses and will continue even during the senescence. In the uterus, the genetic program already decides that there will be more cells in some regions of one hemisphere compared to the homolateral regions of the other hemisphere. The best-known assymmetry is the one of the temporal planum, which is bigger in the left hemisphere than it is in the right (Geschwind & Levitsky, 1968).

Some authors believe that the right hemisphere completes its maturation sooner than the left hemisphere (Chi, Dooling, & Gilles, 1977). Slower in its actualization, the genetic program of the left hemisphere would otherwise lead to a more advanced stage of maturity (Galaburda, 1984). In addition to this difference of tempo between the two cerebral hemispheres, one might also describe the opposition between the left and the right in qualitative terms instead of quantitative terms. According to this point of view, the genetic program heading the cortical maturation would not be the same for the left and the right hemispheres.

There is also a difference of tempo in the myelinogenetic maturation of ipsilateral structures. Yakovlev and one of us (Lecours) have demonstrated that the maturation of the "rhinic" and "limbic" brains is completed around puberty. However, the maturation of the "supralimbic" brain lasts longer. Moreover, each component has its own rhythm of maturation. For example, the maturation of prethalamic fibers is prenatal for audition and perinatal for vision. Also, the maturation of visual thalamic efferent fibers to the primary visual cortex is realized by the end of the fourth month after birth, perhaps sooner, whereas the maturation of auditive thalamic efferent fibers to the primary auditive cortex lasts much longer, approximately four years (Yakovlev & Lecours, 1967). This difference of tempo between the maturation of auditive and visual tracts is very interesting considering the importance of auditive projections in the development of language.

There is not only a difference of tempo in the myelinogenetic maturation of the auditive and visual tracts. Paul Flechsig, one of the most important pioneers to map the myelinogenetic maturation of different areas of the brain (Flechsig, 1901), found that the myelinogenetic maturation of the primary cortices is completed a long time before the one of the association cortices (primordial myelinogenetic fields of Flechsig), and is followed by the myelinogenesis of specific association areas of the supralimbic cortices (intermediate myelinogenetic fields of Flechsig). Finally, the myelinogenesis of the intracortical neuropil, taking place in nonspecific association areas of the supralimbic cortices, begins several years after birth, is very slow to develop, and may last even during adult life (terminal myelinogenetic fields of Flechsig). A good example of this phenomenon is the myelinogenesis of the intracortical neuropil of the angular gyrus, which starts around the age of 5 and continues until the age of 60, according to Theodor Kaes (1907). One should remember that the angular gyrus is very important in comprehension of language presented in visual as well as auditive modalities.

INNATE PROPERTIES AND INVENTIONS

Paul Broca (1865a) was the first to declare that the functional specialization of the left hemisphere for language happens as a result of a genetic programming of the human species. It is now usual to teach in medicine or neurosciences, orthophony or psychology, anthropology or linguistics, that oral language is an immediate expression of an "innate property" of the human brain. It is assumed, however, that written language is an invention of humankind.

This viewpoint has been supported by two observations: on the one hand, when there is no extreme social anomaly, a child is considered biologically abnormal if he or she does not learn to speak in the first years of life; on the other hand, the acquisition of written language, otherwise mandatory, is less perceived as an obligation than the learning of oral language. Someone might be uneducated but biologically normal.

The second argument might be appropriate if the human species were considered as a whole and in the present time. Yet it can be misleading. There are indeed some human subspecies for whom this argument is not suitable. A child would not correspond to the norms of society if he could not learn to write and read in an environment where schooling is obligatory. This phenomenon takes place more frequently in boys than in girls in a ratio of 4 to 1 and there is some reason to believe that it can be caused by a neurogenetic particularism. The child presenting this kind of problem might have a level of intelligence exceeding the "normal" level, and sometimes far beyond the average (Lecours, 1996).

This reconsideration of classical education with an opposition between oral and written language leads to a review of the history and the definition of language. First, the history. Forty thousand years ago, perhaps more, socioenvironmental pressures of unknown origins were exerted on several human communities, already supplied by a genetic potential, to progressively develop speech. Fifty-three centuries later, external pressures of another nature (increasing demography, town development, prosperity, commerce), determined by the impact of the first invention (i.e., speech), forced one or many human communities to extend the genetic potential of the species and to invent writing. In other words, throat and hand languages are equally the expression of human biology. They both actualized themselves along with the evolution of humankind, depending more on the collective pressures than on individual constraints.

The definition of language that we are referring to is different from the one given by several linguists. It seems that language can be defined without the opposition between oral and written aspects as well as without an opposition between the gesture codes of deaf individuals and the digital reading of blind people. In our opinion, the definition of language refers to two parameters. The first relates to human biology and the second is based on social interactions between individuals.

In reference to human biology, language seems to be the behavioral manifestation of at least two different properties. The first property allows the creation of arbitrary links between sent or received conventional signals and external or internal facts; the second authorizes the combination of these signals following other arbitrary conventions.

In reference to social interactions, the genetic program that permits language could not actualize itself without the influence of a human environment. Beyond the invention of language, the control of language implies that someone has to "learn" and the "learning" can only be provided by the outside, by the knowledge of others. This conception of the history and the definition of language supposes that the development of the species follows the same rules as the development of individuals. Genetic potential and environmental pressures both play necessary and complementary roles.

There is a big difference, however, between spoken language and sign language, on the one hand, and between written language and digital reading, on the other. The first of each of these pairs are always learned by reference to the outside world, and the second are learned by reference to the first. In our opinion, this implies that the invention of writing and the learning of written language are dependent on the neuronal systems without any linguistic specialization along with the neuronal systems already committed to oral language. All these considerations raise questions as to the notion of an innate property of the human brain for language. We now believe that the biological substrate of language might be dependent on two innate and interactional, but also distinctive, properties. The first enables an individual to invent or learn and then utilize conventional and abstract signals. These signals are presented to the senses and are arbitrarily matched with the knowledge that individuals have of themselves and their environment and with the knowledge that their peers might have of these signals. If the human species exploits this capacity to create conventional abstractions at a very high level, other animal species share this aptitude in more simple, but always inventive and adaptive, modes.

The second property is unique to humankind. With the adjustment from sounds to words and from words to sentences, and then from sentences to speech, the second property permits combinations of abstractions following conventional rules, also invented and arbitrary. This capacity of the human species increases the possibilities of communication of thoughts. Other kinds of combinations allow several possibilities for the communication of emotions.

DIACHRONY

The individual evolution of language behaviors is, to a certain extent, formed by sociocultural pressures. However, studies conducted in the laboratory of Jacques Mehler have shown that human babies are prepared, in advance, for language. As soon as babies leave the uterus, (a) the voice of their mother is already familiar, (b) they already recognize the noises relevant for language, (c) they are able to make a distinction between their mother tongue and other languages, and (d) the majority process language information with the left hemisphere more than with the right hemisphere (Mehler & Fox, 1985). In other respects, one should not forget that upon leaving the uterus, human babies and babies of other mammals have been listening for a long time, but they have never seen anything. This fact is essential if one wants to have a good appreciation of the impact of environmental information on the actualization of innate properties of the brain.

Undoubtedly, the ontogenesis of language starts very early in the human being, even before birth, and will last for a very long time. First, oral language or the "primary code" will take place, followed, in several societies, by written language or the "secondary code." Beyond the intrauterine mysteries, the "primary code" is first learned by imitation, within the restricted boundaries of family and neighborhood. This functioning allows a normal child to deal with his or her native language, syntax included, around the age of 4 or 5. However, except very rare cases, there is no consciousness, at first, for the rules underlying language. One should remember that the myelinogenetic maturation of thalamic projections to the primary auditive cortex ends around the age of 4 or 5.

It is around the age of 7, that the learning of the "secondary code" begins, only this time the child will follow the formal and explicit conventions of the primary school. This learning will take place in a "transitive" mode, by referring to the sounds of the "primary code." Inherent to the social privilege of schooling, this dichotomy between the "primary code" and the "secondary code" is certainly valid. However, this concept should not hide the fact that after primary school and its formalism, the learning of language will continue without any dependency of written language on oral language. In the case of individuals who integrate written language into their life, it is undeniable that they first learn numerous words and some syntactic arrangements with their eyes, and not with their ears. Later, they will use these elements in writing or in speech.

In summary, the subordination of written language to oral language might decrease with time, when the ontogenesis of language continues. As long as the brain remains healthy, written language can assume a predominant role and can have an important impact on cerebral biology. In democratic countries with a free-market economic system, it is not rare that older people read more and refer more to dictionaries than do younger individuals.

Also part of the ontogenesis of cognition and language is a decline of memory, which is inherent to the normal aging process. It may happen sooner in some individuals than in others. It does not affect all kinds of memory: semantic memory (Smith & Fullerton, 1981) and implicit memory (Light, Singh, & Capps, 1986) are usually spared by the aging process, whereas explicit memory (Flicker, Ferris, Crook, Bartus, & Reisberg, 1986) and the active component of working memory (Salthouse, Mitchell, Skovronek, & Babcock, 1989) capacities seem to decrease. Nothwithstanding these discrepancies in the involution of memory, word access is always altered by age (Belota & Duchek, 1988). In everyday life, this deterioration of memory might affect language: as a compensation mechanism, there may be greater "verbosity" of speech together with periphrases to mask the difficulty of lexical evocation. This is often a "frequency effect" or a more automatic and efficient utilization of syntactic procedures (Obler & Albert, 1984). The deficit of lexical evocation is frequently unconscious. Despite the fact that this phenomenon refers to involution, it can also be considered, in our opinion, as one of the latest stages of the ontogenesis of language.

Sociocultural factors are very important in brain–language interactions. Indeed, they can facilitate or even enforce the development of a certain biological potential of the human brain. We now consider oral language and discuss it in relation to sociocultural factors with two examples.

1. In some parts of the world, in Vietnam, for example, people speak so-called isolating tongues. In these tongues, most of the words only consist of one syllable. The meaning of each word is specified by the tonal pitch given to its vowel. Each word may have as many as six different meanings depending on the pitch of the vowel (which can be increased, decreased, or modulated). The

brain must then adjust its functioning during all its ontogenesis to enable individuals to use that many tonal pitches, which, elsewhere in the species, have no role to play in the comprehension and production of the speech.

2. In other parts of the world, in Poland, for example, people speak so-called flectional tongues. In these tongues, most of the words consist of many syllables that correspond, in several cases, to prefixes and suffixes. The brain must then mold its functioning in a way that enables individuals to juggle an impressive number of syllables all their life. These syllables only play a role in modifying the meaning of the root to which they are tied.

We now consider written language and give three examples about two genetical potentials of the brain sustained by a different biological substrate.

1. "Logographic" writing systems—the one of China, for example—which allow people speaking different languages to read the same newspapers, exploit maximally one of these genetic potentials, which is to memorize written words in their global form.

2. Other writing systems, related to only one spoken tongue, favor the other genetical potential, which permits the processing of written information by the "conversion" of smaller entities of words from their auditive forms to their visual forms or vice versa. Spanish reading and Italian writing are good examples of these systems.

3. Other writing systems—French, English, and Japanese, for example—ask the brain to actualize and utilize the two biological potentials described.

These sociocultural differences may have an important impact on the biology of the normal brain, of course, but also on clinical manifestations of learning disorders in oral and written language. In addition, with comparable cerebral lesions, these sociocultural differences will affect clinical manifestations of acquired diseases of language.

APHASIA

In the absence of wars, aphasia is often the result of cerebral vascular accidents. Normally, Broca's aphasia, characterized first by suppression of speech and followed by agrammatism, is more related to young patients, whereas the "jargonaphasia" of Wernicke is more associated with aging people. This assertion has been supported by a study of Joanette (Joanette et al., 1983), which showed that patients with Broca'a aphasia were 45 to 55 years old and patients with Wernicke's aphasia were 55 to 65 years old. For some people, this discrepancy can be the result of cerebral arterial changes induced by the aging process, thus to the topography of cerebral vascular accidents (Benson, 1978). This interpretation suggests that Broca's aphasia is always the result of a dysfunction of Broca's area and Wernicke's aphasia is always the result of a dysfunction of Wernicke's area.

However, it is now known that a lesion of Wernicke's area can produce a semiology of Broca's type when it happens in a 4-year-old child, for example. It is also known that the same lesion can induce clinical manifestations of a conduction aphasia in a 24-year-old individual, for example. Again, it is a well-known fact that a lesion of Wernicke's area is often the cause of Wernicke's aphasia later in life. Finally, linguistic characteristics of "jargonaphasia" might not be the same if found in an old or in a very old patient. A progressive subcorticalization of biological language procedures, appearing when these procedures become more mechanical, has been suggested as the cause of variations occurring with age in the semiology of aphasia (Benson, 1978). Although this hypothesis is very interesting, it has never been tested. All these clinical changes related to the aging process and observed following similarly localized lesions allow an important deduction: the functional lateralization of language, together with the functional specialization of components of the area of language, occur progressively, over a long period.

Also, the total, or partial, recovery of language capacities following stabilized cerebral lesions in old people as well as young adults and children is possible and can be very spectacular, especially in children. Sometimes, it has been hypothesized, this recovery is assumed by the right hemisphere in right-handed individuals. All these clinical data refer to the notion of a "genetic potential of replacement."

CONCLUSION

It seems beyond doubt now that the ontogenesis of the human brain and language lasts as long as the brain remains healthy. In other words, it ends with death. To a certain extent, sociocultural factors mold the ontogenesis of language and influence the evolution of cerebral biology. This point of view allows the possibility of an ontogenetic recovery, at least for some individuals, and even late in life.

PART I

History of Neurolinguistics

CHAPTER 1

Neurolinguistics from the Middle Ages to the Pre-Modern Era: Historical Vignettes

Harry A. Whitaker

Department of Psychology, Northern Michigan University, Marquette, Michigan 49855

1-1 MEDIEVAL AND RENAISSANCE NEUROLINGUISTICS

Historical observations of disorders of language have been reported in Pharaonic medical papyrus texts circa 3000 B.C., a Hittite cuneiform tablet text circa 1500 B.C., the Hippocratic corpus circa 400 B.C., Latin texts circa the first century A.D., Latin texts in the Middle Ages (10th to the 14th century), Renaissance texts such as the Schenck compendium of 1558, various 17th-century texts such as Schmidt and Rommel and 18th-century ones such as the Wepfer compendium of 1727. The early 19th-century phrenological contributions of Gall, Spurzheim, and Hood, as well as the more orthodox medical work of Lallemand, Bouillaud, Lordat, and Dax, inaugurated a period of intense interest in the brain localization of language functions, culminating in the classic work of Broca, Meynert, Wernicke, Bastian, Jackson, and others during the late 19th century. Although questioned on occasion by holistic views (Head, Goldstein), the localization model persevered until, by the second half of the 20th century, classical neurolinguistics became firmly entrenched in both clinical and experimental neuroscience research. A glance at the cover of the third edition of Kandel, Schwartz, and Jessell's 1991 textbook, *Principles of Neural Science,* will suffice to support this contention.

The most durable model of functional brain localization so far has been the Medieval Cell Doctrine. Derived originally from ideas of Herophilus (c. 270 B.C.) and Erasistratus (c. 260 B.C.), Medieval Cell Doctrine, and its particular variant, ventricular theory, were developed by Galen (130-200), the church fathers Nemesius (c. 400), Posidonius (c. 370), and Saint Augustine (354-430), with additional Arab contributions, particularly from Avicenna (980-1037). Ventricular theory was a model based on fluids and fluid flow for the apparent reason that thoughtful early scientists realized that something in the brain must move around in order to accomplish brain functions—something passes from sense organs to effectors, and the "animal spirits" were as good a candidate as any available. It was sufficiently entrenched to have endured well into the 17th century—for example, in the hydraulic model of brain function advocated by Descartes (1596-1650). Certain principles of Medieval Cell Doctrine and ventricular theory (e.g., modular functions localized in different brain regions, as well as other vestiges of this two-millenia-old idea) remain as features of every model of brain function up to those of contemporary cognitive neuroscience.

From the 11th century through the Renaissance, one finds textual discussions usually in the form of neuropsychological case reports as well as graphic representations of Medieval Cell Doctrine and ventricular theory. Diagrams from the Middle Ages were typically crude and childlike, whereas those from the Renaissance period were often ornate works of art in their own right. These studies typically, although not invariably, accepted the standard model that placed memory in the third ventricle; the authors then "fit" the lesion evidence from brain-damaged subjects to that model. Language, or, more precisely, speech output, was also situated in the third cell or ventricle, an association that may have originated from Galen's belief that motor functions were a property of the cerebellum. The psychological link between memory and language may have been the cause or the result of a conception of memory as verbal memory; whichever, it is clear that they were so closely connected that memory disorders were interpreted as language disorders and vice versa. For example, consider this case from an 11th-century manuscript: Guillaume de Conches (1080-1150) reported that Solinus had spoken of a man who suffered traumatic injury to the last cell of the brain and who fell into an amnesia so profound that he had forgotten his own name. Many of these cases were compiled by Johannes Schenck (1584) and Johannes Wepfer (1727), both of whom are discussed in this chapter. Others are compiled in Jules Soury, *Le système Nerveux Central* (1899).

Eventually, the overt form of Medieval Cell Doctrine and ventricular theory fell by the wayside, the demise beginning in the 17th century, as some of its key components (e.g., animal spirits flowing through hollow nerves or between empty ventricles) were challenged by both experimental and theoretical arguments. However, other aspects of the model—for example, the concepts of a modular functional localization, of information flow, of the close links between memory, language, and motor function, and the idea that memories are stored images—were not only never challenged but seem to have become absorbed into post-Renaissance models of brain function.

From at least the 16th century, a correlation between language disorders and particular cerebral lesions (rather than "ventricular," or simply generally "in the brain") has been recognized. The late-Renaissance physician Johannes Schenck von Grafenberg (1530-1598) published *Observationes medicae de capite humano* in 1584; it is a treasure of neuropsychological observations including many observations on language disorders. It includes contributions from 20 Greek, 5 classic Latin, 11 medieval Arabic and Jewish, 37 medieval Latin, 292 recent Latin, and 139 classic nonmedical authors, as well as 64 unpublished cases reported by Schenck's contemporaries. His encyclopedic collection of clinical information, all with an explicit bibliographic citation, has no counterpart in its time. At least 16 of Schenck's observations mention symptoms of language disorders. Although some case descriptions are superficial, many include quite explicit observations, showing that neurolinguistic knowledge was in fact more advanced than one might have believed. Armand Trousseau, a contemporary of Broca, cited Schenck as an early physician who appreciated the essential nature of aphasia. A sampling of Schenck's cases (see Luzzatti & Whitaker [1996a]) illustrates the level of description in this early neurolinguistic work. Reporting a 14th-century case from Gentile da Foligno and a 16th-century case from Cristobal de Vega, Schenck notes the selective loss of the alphabet and object names, accompanied by loss of reading and writing. Another 16th-century case from Tertius Damianus reports that a patient studied his letters again, as an effort at rehabilitation. Schenck recorded an impressive 16th-century description by Conrad Lycosthenes of his own right hemiplegia and aphasia, described as a loss of voice. The inability to speak lasted 12 days and the severe motor deficits lasted 3 months. Lycosthenes had even lost rote-memorized passages such as the Mass but had retained internal language and thinking; he was able to communicate by pointing, with his left hand, to letters in sequence to form words. Evidently, one of his friends had hit upon the idea of writing the alphabet down on a board as a communication aid. Citing Jakob Oethaeus, Schenck observes that the tongue may retain its motor functions when word memories have been lost, thus establishing quite early the dissociation between articulatory control and language production. Clearly, the Renaissance legacy to the 17th century's Age of Reason and scientific discovery was considerable insofar as neurolinguistics is concerned.

1-2. A PERIOD OF TRANSITION: THE 17TH AND 18TH CENTURIES

Johannes Jakob Wepfer (1620-1695), was an acquaintance of Thomas Willis and, although less well known than Schenck, was in fact one of the most important "neurologists" of the 17th century. He authored several papers and books; the one relevant to our interests is *Observationes medico-practicae de affectionis capitis internis & externis,* which was written around 1690 but only available in a posthumously published version of 1727. The book is notable for its accuracy, colorful clinical descriptions, and systematic organization. In observation 98, "loss of memory," Wepfer describes a transient language disorder with selective impairment of proper names, accompanied by a syntactic disorder; no acalculia nor other neuropsychological deficits were observed. The following is translated and paraphrased from the Latin:

A 53-year-old man returning from a walk vomited several times. He spoke both in Latin and German, but his words did not relate to the topic of conversation, nor were they answers to questions put to him. He continuously repeated the same utterances, stuck on the pronunciations of words, and frequently complained that he could not express his thoughts in words. A month later he suddenly forgot all names, even his own. He could not name any object, neither in Latin nor in German. He gave the impression that he was able to recognise objects and people but the words he uttered were alien and incoherent. After one more month, his memory was almost completely restored; his faculty of hearing and repeating was adequate and he could count, add, subtract and divide without error. Occasionally, he could not find the proper names of a person or a place as quickly as usual and from time to time he could not find some of the little words of the sentence. He is able read again and writes full texts without hesitation. His faculty of judgement is normal and he was able to fulfill his duties as before. When he was talking, I could observe him from time to time violate syntactic rules and, against the structure of German sentences, he would antepose one word to another [word-order errors] and sometimes he could not complete some little word [inflectional morphology errors]. In general, he recognised his errors and corrected them. (Luzzatti & Whitaker, 1996a, pp. 161-162)

In addition to its interest as an early study of agrammatism, Wepfer's case is also exemplary of most, but not all, classical aphasia research on grammatical or syntactic disorders up until the 1970s: a descriptive summary of the omission or misuse of function words or inflections in spontaneous or elicited speech samples of patients with an expressive aphasia. It is exemplary as well as an analysis that made no attempt to explain the disorder but rather to describe it, a tradition in neuropsychology that still has its advocates. Wepfer's observation 102, a loss of memory followed by melancholia, is instructive for the neuropsychological insight:

A most reverend 44-year-old cleric sat in judgement on a controversy proposed by a peasant; however, when he tried to express his judgement, he could not find the words he wanted to communicate his thoughts. From that moment on, he lost his memory so much that he no longer was able to call the onlooker by his name nor he could name any object. He clearly saw the colours and features of objects and drawings presented to him, but he could neither read nor combine letters. All his remaining outward and inner sensations were normal. He recognised his loss of memory and the errors he made while speaking. There was no trace of paralysis; he could pronounce every letter, even the R. The site of suffering is neither in any nerve pertaining to both types of sensation, because he hears, sees, tastes, etc., nor in the nerve subserving speech, because he can pronounce all sounds, even the R, nor in the brain stem or spinal cord, because there is no trace of movement disorder. It seems to be located in that part of the brain where the images of objects are organised. These may be retrieved, as from a store, and, when required, supplied by means of the mnestic faculty. What part of the brain this might in fact be, cannot be defined, but clearly it is not an involvement of the entire brain, because the patient is not drowsy, and has lost neither his ingenuity nor his understanding. (Luzzatti & Whitaker, 1996a, p. 162)

Compared to the similar collection (organization and focus) of clinical cases compiled by Schenck, Wepfer's cases seem to have been more carefully selected; many were followed by an autopsy description, and the etiological-pathogenic commentary is quite advanced in relation to the zeitgeist of the 17th century, particularly in comparison with Wepfer's contemporary Thomas Willis. Wepfer is arguably the first scientist to discuss the nature of cognitive functions using a modern neuropsychological approach. To our knowledge, he was the first to consistently report the side of lesion and the presence or absence of aphasia, although he does not overtly express the relation between language disorder and left hemisphere lesions, despite the fact that he often wrote, *paralisys dextri lateris, cum loquelae impedimentum.* Perhaps the cultural-philosophical milieu in which Wepfer was working did not allow him to understand the relevance of his own observations, or perhaps his silence on the matter of hemispheric asymmetry was out of concern for an unwelcome reception by the church. Galileo had been accused less than 30 years earlier and the squares of Europe still smelled of the stakes of the Inquisition.

Other 17th-century neurolinguistic reports worth mentioning include one by Johann Schmidt (1624-1690), a case of severe fluent paraphasia with right hemiparesis; the patient showed good recovery of oral language and agraphia, but not for his reading disorders.

> A leading citizen among us, Nicholas Cambier, an old man of 65 years, was seized with a very severe attack of apoplexy which all his attendants feared would lead to his death. . . . Upon his return home, it was evident that his right side was paralyzed and that he had difficulty in speaking. He muttered a good deal but was incapable of expressing the feelings of his mind; he substituted one word for another so that his attendants had difficulty in determining what he wanted. . . . He could not read written characters, much less combine them in any way. He did not know a single letter nor could he distinguish one from another. But it is remarkable that, if some name were given to him to be written, he could write it readily, spelling it correctly. However, he could not read what he had written even though it was in his own hand. Nor could he distinguish or identify the characters. For if he were asked what letter this or that was or how the letters had been combined, he could answer only by chance or through his habit of writing. . . . No teaching or guidance was successful in inculcating recognition of letters in him. It was otherwise with a certain stone cutter in our country. Wilhelm Richter came to see me after his apoplexy receded because he was not able to read at all or to recognize letters. However, he learned the alphabetic elements of the language again in a short time. He then combined them and attained perfection in his reading. (Benton & Joynt, 1960, p. 209)

Another 17th-century report is Peter Rommel's (1643-1708) 1682 study of a patient whose spontaneous output was reduced to a few automatisms, but serial speech, prayers, and comprehension were preserved (Benton & Joynt, 1960, pp. 209-210). Considering the cases by Schmidt, Rommel, Schenck, and Wepfer, it is clear that scientists of the 16th and 17th centuries were able to clearly distinguish language disorders from general cognitive deficits, production versus comprehension deficits, lexical from sound or letter errors, and both syntactic and morphological impairments. Although the neuropsychological modeling needed to complete these analyses did not come until the beginning of the 19th century, the 17th century also witnessed a fair degree of progress on the neuroanatomical front. For example:

1. Experimental brain research flourished, as seen in more accurate dissections, more careful observations, and comparisons of brains of different species, while ventricular theory was being replaced by models that localized functions to the brain substance. Nicolaus Steno (a.k.a. Niels Stensen [1638-1686]) could be regarded as a transition figure; he is remembered for his

Discourse on the Brain (1669), which provides some evidence of the state of neuropsychology in the 17th century. Steno was convinced that wherever fibers are found in the brain, they maintain a certain pattern among themselves, of greater or lesser complexity according to the functions for which they are intended. His comments on the brain's white matter should be interpreted as prefiguring David Hartley's ideas of functional specificity and modularity (see Aubert & Whitaker, 1996).

2. More systematic clinico-pathological observations of memory disorders and language disorders, with frequent reports of the co-occurrence of aphasia with right-sided paralysis, accompany the first formal teaching of clinical-pathological correlations for the purpose of determining the causes of disease (Boerhaave), to be later developed by Morgagni in the 18th century.

3. Early ideas of domain specificity of functions in the brain were further developed in the work of David Hartley in the 18th century, incorporating Isaac Newton's vibration model of nerve function; Newton's model depended on the view that nerves were solid, not hollow. The Renaissance view that nerves transmit information to and from the brain was retained, but the idea of encoding of information—different vibrations for different ideas—was added.

1-3. LOCALIZATION COMES OF AGE: FRANZ JOSEPH GALL AND HIS TIMES

Whatever else may be said of the 18th century, it is clear that by its end, medicine had clearly accepted the idea of the discrete localization of functions in the brain (or, in the vernacular of the time, faculties; cf. the philosophy of Thomas Reid). Charles Bonnet (1720-1792) a Swiss naturalist and philosopher approvingly cited by Franz Joseph Gall, authored books on botany as well as zoology; Bonnet's psychological ideas were close to those of Étienne Bonnot de Condillac (1715-1780), a leading exponent of French sensationalism. Bonnet believed that every faculty—sensitive, moral, or intellectual—is connected in the brain to a specific bundle of fibers. According to Bonnet, every faculty has its own laws that determine its mode of action; not only does every faculty have its own fasciculus of fibers in the brain, but every word had its own fiber in the brain. Not only is this directly reminiscent of David Hartley's (1705-1757) neuropsychology (see Aubert & Whitaker, 1996), but it also harks back to René Descartes's model of brain function (see Smith, in press) and prefigures Johannes Mueller's doctrine of specific nerve energies. In light of these ideas from the 17th and 18th centuries, Gall's craniology (organology), or faculty neuropsychology, seems to be a logical if somewhat fanciful development. Much has been written about Franz Joseph Gall's (1758-1828) contribution to neuroscience, but not a lot is known about the roots of his ideas. Christine Grou, in her unpublished doctoral dissertation, demonstrated a close parallel between the faculty psychology of

Thomas Reid (1710-1796) and the faculties of Gall and Spurzheim, and also a commonality between the hundreds of physiognomic characteristics proposed by Johann Kaspar Lavater (1741-1801) and the phrenological faculties (Grou & Whitaker, 1992). Gall's idea that growth patterns of the cortex, hypertrophy or atrophy, would impress themselves on the inner table of the skull and thus be "readable" as bumps on the skull was directly borrowed from Lavater. The great 18th-century naturalist Charles Bonnet proposed a vibration-based theory of memory reminiscent of Hartley; Bonnet also proposed a doctrine of localization of function in the brain that clearly influenced Gall, as the latter cites the former in several of his books. Nonetheless, the details of Gall's indebtedness to these 18th-century scientists remain to be elucidated. Gall had theoretically anticipated a division between two aspects of the language function and had proposed two distinct localizations. On a skull, shelved in the University of Vienna Medical Museum, Gall inscribed in an elegant penmanship the various faculties of the mind; in the "left" bony orbit only (what an extraordinary coincidence! but of course he was likely right-handed) he wrote *Wortsinn,* the memory for words or the lexical component of language, and just above it, *Sprachsinn,* the motoric component or the faculty of spoken language. Although Gall believed that the clinico-pathological correlation of brain lesions to aphasic disturbances could furnish evidence in support of his model, and although he did in fact study two cases of sword injury to the frontal lobe referred by Napoléon's surgeon, Dominique Larrey, Gall himself evinced little direct interest in clincal neurolinguistic cases.

On the other hand, the phrenological model (Spurzheim's terminology) based on Gall's ideas served to preserve ideas of brain–language correlations through the first two-thirds of the 19th century, in the hands of such stalwarts of medicine as Alexander Hood and Jean Baptiste Bouillaud. We have worked out some of the connections between craniology-phrenology and the development of neuropsychology in the period from 1820 to 1860 and we have also analyzed how the early phrenologists helped to found the doctrine of the clinico-pathological correlation of language impairments. Gall's argument that a well-developed language faculty causes a protuberance of the inferior, anterior frontal lobe, which in turn would make the eye sockets shallow— thus, folks with high verbal skills were said to have "cow's eyes"—was not successfully challenged by the scientific community, even though in fact it should be regarded as a scientific mistake, an error in reasoning using his own craniological method. It was well known at the time that the backside of the eye sockets do not abut the frontal lobe—a great deal of sinus cavity lies between the two, and therefore it is quite impossible that a frontal brain bump could impinge upon the eye sockets. Curiously, from the standpoint of "good science," Bouillaud not only accepted this localization but championed it unceasingly right up to 1861 when Broca's publication seemed to vindicate Gall's model. Evidently, it was the accumulating evidence that frontal lesions typically led to speech disturbances, documented by Lallemand, Bouillaud, and others from the 1820s on, that kept the phrenological language model alive until the great paradigm shift of the 1870s. But first, let us consider what the practicing phrenologists had to say.

1-4. PHRENOLOGY STUDIES LANGUAGE SCIENTIFICALLY: ALEXANDER HOOD'S PHENOMENAL CASE STUDIES

The question is, does phrenology, at least in the first half of the 19th century, merit the sort of scathing criticism it is often given? For example, Garrison (1929) said, "exploited by quacks and charlatans, phrenology soon became an object of derision among scientific men" (p. 539). Considering Johann Gaspar Spurzheim's (1776-1832) intellectual legacy, one might usefully evaluate this historical judgment by looking at what practicing phrenologists actually did; one way of doing that is to look at what was published in their own journal. The *Phrenological Journal and Miscellany* (PJM), the house organ of the Edinburgh Phrenological Society, was founded in 1820 by George Combe (1788-1858) following his earlier indoctrination into the ways of phrenology by Spurzheim. In 1824, an obscure phrenologist and medical practitioner from Scotland, Alexander Hood, published in PJM a remarkable case history of one Adam M'Conochie, who, subsequent to a CVA in the left frontal lobe, had become what by the end of the century would be called a "Broca's aphasic." Hood's paper may occasion some historical interest on several counts: its contribution to the technique of clinico-pathological correlation, its indebtedness to the phrenological movement in Scotland launched by Spurzheim and Combe, its use of aphasic data to argue for both the modularity and the fractionation of language, and, of course, the fact that the date in its title—July 24, 1824—was the very one on which Pierre Paul Broca was born (Hood, 1824).

Hood discussed the case of Adam M'Conochie, a 48-year-old gardener from Dankeith, Scotland, who had evidently sustained a mild left hemisphere CVA the morning of May 31, 1824. Hood first saw the patient that same afternoon and took a brief history from his wife. M'Conochie rose from bed later than customary that morning and began looking for something he could not name. For a while his wife could not understand what he was searching for; eventually it turned out to be his socks. Nothing unusual was reported about M'Conochie's ability to dress himself. His wife did notice "something peculiar in his manner and by a striking deficiency that she observed in the use which he made of language, which in fact was limited to a few words" (p. 82). M'Conochie, plainly more concerned about his duty to his master than his recent stroke, proceeded to gravel the walks around the house, while his wife contacted the steward to tell him of her husband's problem. The wife, the steward, and finally the factor were needed to persuade M'Conochie to go home and send for medical assistance. When Hood first entered the room, McConochie was dozing in bed, perspiring a great deal. He quickly got out of bed as though nothing was wrong; his pulse was "natural, but the tongue was foul and the bowels were slow" (p. 82).

Hood questioned him about his complaint and the state of his "feelings" but was unable to get a satisfactory response "on account of his utter inability to give a description in language of the state of his mind together with his internal feelings and sensations." To indicate where the problem lay, M'Conochie obligingly raised his left hand to the left side of his head (recall, of course, that the right hand was weak, if not actually paralyzed) and, pointing to the temple behind the angle of his eye, said there was "something about it." One can imagine Hood's excitement as an aphasic patient signaled the site for the organ of language as having a problem! After repeating the phrase "something about it" several times, M'Conochie said the word "rheumatism," signifying, as it was understood in that epoch, that his complaint was a rheumatic affection of this part. He had no apparent difficulty in comprehending any question that was put to him but his answers were always such as these: "something about it," "plenty about it," "little about it," or "nothing about it" (p. 84), suggesting, in my opinion, not only an anomic impairment but equally the possible preservation of expressions of measure or quantity. When asked whether he had pain or uneasiness in his head, M'Conochie's answer was that there was "something about it." But when similar questions were put to him with respect to the chest and abdomen, he replied by saying, there was "nothing about it." A book was shown to him to see whether he could read; the only word that caught his attention and that he could name was "man." However, he was unable to say aloud the names of the three letters in "man." He was then asked whether he could read fluently prior to the present attack, to which he repled that he knew "plenty about it" and promptly brought Hood a volume of travels that he had lately been reading. When M'Conochie was asked to repeat the Lord's Prayer, his answer was, he "ken'd plenty about it" but this was all the information he could give concerning it or other rote verbal passages (p. 84). On June 14, M'Conochie continued to improve in reading, to which he applied himself assiduously; but Hood noted that there was apparently not quite so much improvement in his powers of conversation. The other symptoms remained nearly the same. By June 21st Hood noted that his patient could now read tolerably well, with the exception of a few words that he could not pronounce accurately. M'Conochie could not, however, repeat a sentence or even a syllable of the Lord's Prayer, or any other portion of the sacred book with which he was formerly familiar. He moved his legs and arms with perfect facility and apparent vigor; he said his right hand was "dumb," though he probably meant "numb." On the whole, he was flat and spiritless, and thought himself little or no better (p. 85). Hood's July 2 observation was that M'Conochie had been rather melancholy; his right arm and leg were affected with weakness and irregular nervous action; he could read pretty well, but could not repeat anything that might have been committed to memory; there were still some words he could not pronounce, though there were much fewer than there had been three weeks earlier. A strong degree of excitement seemed to give him temporary command of a few words that he could not speak in his calmer moments (p. 85). For the entry of July 10, Hood noted that his patient's reading had not materially improved since the last report, but he could read verses much better, having obviously reacquired an extensive addition to his vocabulary. He could use both arms and hands equally well; Hood perceived little difference with regard to strength, but a striking difference with regard to sensation. However, when M'Conochie picked up something with his right hand, he often had to employ his left hand so as to enable him to judge accurately of the temperature of bodies. This day, for the first time by means of language, he was able to make himself intelligible with respect to the state of his feelings since the commencement of his complaint. M'Conochie offered the comment to Hood that, when a book was previously given to him and he was asked to read, he had "numbered" the words, that is, spoken the words "one, two, three, four, five, six," and so on. Apparently, he fully comprehended the import of the actual words given as numbers, although he did not recollect how to pronounce or name them. (pp. 86-87)

"At the end of his case report, given here in much of its detail so that the nature and progression of the symptoms can be appreciated, Hood turned to his observations on brain function. In this section, Hood included parts of the discussion of three other cases, reported earlier in the same journal. To quote him directly:

> An attentive perusal of these cases suggests some important reflections concerning the pathology of the brain. . . . [they] show to what an alarming extent this important organ may be injured without destroying life or the functions of health. It would neither be professional, nor philosophic, however, from a few remarkable cases, to infer that the brain is of little consideration in the animal economy, or totally unconnected with the operations of the mind. Were the patients in all such cases capable of giving a distinct and accurate account of their feelings and sensations, and the medical observer fully adequate to point out every deviation from the healthy condition of body and mind, much of our surprise would cease, and the apparent anomaly would disappear. (P. 88)

"In the case McPherson, an iron penetrated one and a half to two inches into the brain in a transverse direction; there were no effects at first and the patient could walk about, eat, etc., until suppuration (infection) caused a confusion of intellect and a palsy on one side" (p. 88). "The patient showed an unusual degree of Timidity; indeed he was struck on the organ of Caution" (p. 89).

"Case Morton fell from a coach and fractured his skull. The fracture extended over the ear from the temporal bone into the parietal and occipital bones; the arm on the opposite side is still an inch smaller than the other one. Morton is much more irritable than before and not quite so capable of prosecuting any design which requires continued and steady Perseverance" (p. 89) In reference to the sequelae of brain damage, Hood noted that although the parts heal up, they never acquire such a "perfect organization" as that with which they were originally endowed.

> Hence it seems reasonable to suppose, that when a part of the brain, on which some moral or intellectual function depends, has suffered much from inflammation, or external violence, although the part thus weakened or injured may perform the function or functions which depend upon it, in the ordinary states of the system, yet when the whole faculties are roused by intense feeling, or excited by intoxicating liquors, then will the organic lesion be evinced by corresponding derangement in the manifestations of intellect and feeling. (P. 89)

For Case 3 Hood performed an autopsy on the brain of a woman who died and discovered a small spicula of bone in the dura mater on the upper part of the falx where it dips down between "the organs of Self-esteem." In addition, he found a small tumor growing from the dura mater, pressing upon the organ of Hope. He regarded both lesions as continuing sources of irritation of the brain, and explained the patient's impaired behavior prior to death in terms of the effects on these organs:

> Whatever may be thought of this attempt to reconcile the phenomena of the disease with the appearances on dissection, it was almost impossible for any one acquainted with the principles of the new doctrine on this subject, to omit observing the striking coincidence

and apparent corroboration which Phrenology seems to receive from morbid anatomy. We have no faith in the doctrine of chance, and therefore cannot believe that the mental affection and morbid appearances after death, were the result of fortuitous circumstances. (P. 91)

As for Case 4 of M'Conochie, "the last of these cases is not without considerable interest, in as much as it unfolds to us the nature of that affection by which individuals sometimes suddenly lose the verbal recollection of almost every term in the language, without the ideas being lost, or the judgment impaired" (p. 91). In other words, Hood is arguing one may separate or fractionate two features of the organ for language, based on the evidence from this patient. Although the patient had forgotten in general the names of objects with which he was familiar, when the name of a thing was given in his hearing he was capable of recollecting it for a few seconds of time, so as to be able to pronounce it again; Hood notes that this is quite different with case R.W. This latter patient very often could not recollect the word long enough to enable him to pronounce it a second time, thereby demonstrating that the paralytic affection was more severe in this case than the other.

Hood argues that the case of M'Conochie and the case of R.W. considered together make a most pointed analogy with respect to the phenomenon of language. One must bring a degree of credulity to the localization, of course; in Hood's terms, however, it was perfectly credible that a patient could indicate the site of a lesion by pointing to the spot on his head. In one of the two patients the pain "affected a much greater surface"; accordingly, the organ of language was not exclusively affected, for some of the muscles that assist in articulation were in a certain degree paralytic, from the impediment that was conspicuous in the pronunciation of some words. Hood also takes note of the loss of sensation on right side of face, numbness of the arm and particularly of the hand, without any loss of power of motion. Thus, to Hood, R.W. and M'Conochie "mutually illustrate each other." In R.W.'s case the paralytic affection seemed to be confined exclusively to the organ of langauge; in M'Conochie's case, the organ of language was affected, though not quite so severely; however, the paralytic tendency (lesion) extended farther, involving some of the muscles concerned in articulation (p. 92). Hood says: M'Conochie's case, and more particularly R.W.'s, so often mentioned, go far to establish the individuality of the organ of Language, and to confirm the accuracy of the observations of the gentlemen who have discovered its place in the brain to be above and behind the socket of the eye" (p. 92). In sum, by comparing two different aphasic cases, Hood has fractionated the organ of language into one part directly controlling the articulatory muscles, another controlling verbal expression, and a third controlling the memory or ideas for words—a remarkably prescient experimental conclusion, employing a theoretically coherent albeit implausible clinico-pathological correlation. Hood concluded with a homily: one needs to have a good theory before studying the brain so that the evidence of impaired function may be properly related to the elements of the theory. How true! In Hood's case, however, the "theory" is the map of the phrenological organs neatly painted on a plaster bust. In his own words:

When therefore the mental phenomena, in any given case, indicate preternatural and diminished function in a particular part or parts of the brain, and when dissection after death actually exhibits the organ or organs indicated in a diseased state, the nature of the evidence thus obtained is as conclusive, with regard to the accuracy of the mode of the investigation, as any demonstration in Euclid. (P. 94)

1-5. FROM GALL TO BROCA: THE CONTRIBUTION OF JEAN BAPTISTE BOUILLAUD

François Broussais (1772-1838), one of the most prestigious physicians in France, had pronounced in favor of Gall's ideas, eventually offering a full course in phrenology to the medical students in Paris, the lessons of which were published in 1836 (Valentin, 1988). Jean Baptiste Bouillaud (1796-1881), a leading cardiologist in France, also believed in phrenology, while most other recognized scientists had sided with Jean-Pierre Marie Flourens (1794-1867), the doctrine's most famous critic. After Broussais's death, Bouillaud became the major medical figure in France to support the concept of cerebral localization. Between the 1820s and the 1860s, Bouillaud gave several major presentations at the French Academy of Medicine meetings, defending the frontal lobe localization of language, and thus became the most important link between Gall and Broca.

There has been a tendency for historians to jump from Bouillaud's 1825 paper to the Auburtin–Gratiolet–Broca discussion of 1861. One reason, as Clarke and O'Malley (1968) noted, for example, is that Bouillaud's views did not gain wide acceptance during the 3-decade interval in question; however, he was not by any means ignored. In the pages of various French Academy publications, Bouillaud debated the concept of cerebral localization with important clinical and experimental opponents such as Andral, Flourens, Pinel (fils), Cruveilhier, Lallemand, Rochoux, Rostan, and Castel. It seems evident that any medical researcher during this period of some 30 years was frequently confronted with the theory of cerebral localization, particularly that of language. It is also apparent that people were quite aware of the fact that if the proof of the localization of language were to be established, then the entire enterprise of cerebral localization would be viable and legitimate.

Bouillaud's first publication on localization (1825) provided independent clinical evidence supporting Gall's theory of the frontal lobe localization for language. Although Bouillaud has been credited with having made the first observation of 'internal' and 'external' speech, or what he termed the "power of creating words as signs of our ideas and that of articulating these same words" (Harrington, 1987), we saw earlier that Hood deserves that recognition. However, Bouillaud's own words do sound much more like modern neurology: "the tongue and its related organs in the act of speech can be separately paralyzed, that is to say, without the other parts being affected simultaneously"; and, "the loss of speech depends now upon the loss of the memory of words, now upon the loss of the muscular movements by which speech is composed, or, what comes perhaps to the same thing, now upon a lesion of the gray matter and now upon that of the white matter of the anterior lobes" (McHenry, 1969, pp. 490-491). It is not generally noted that Bouillaud did not uncritically accept Gall's theory: for example, he sided with Flourens against Gall on the matter of the role of the cerebellum and further agreed with Flourens that phrenology was a "pseudoscience."

Although Bouillaud had somewhat distanced himself from phrenology per se, he maintained the truth of the theory of cerebral localization and in particular the frontal lobe localization for language. The following is based on a paper by Luzzatti and Whitaker (1996b).

The purpose of Bouillaud's 1825 paper was to verify Gall's theory of the relation between loss of speech and a lesion of the frontal lobes; it summarizes the major issues concerning functional-anatomical correlation particularly for language. Bouillaud discusses the motor control of language and its relation, on the one hand, to the motor control of nonlinguistic movements of the tongue and, on the other, to intelligence. After a brief introduction, Bouillaud vehemently attacks Flourens's skepticism about the hemispheric localization of motor functions. He next argues that even accepting that the hemispheres control motor functions is insufficient because different parts of the brain control different kinds of movements. This is particularly clear for movements of the tongue and eye. Bouillaud claims first to be able to demonstrate that speech movements may be compromised separately from limb movements and second to demonstrate which part of the hemispheres is involved in the control of speech movements. To prove his first claim he states that two actions are functionally separate if they can be disrupted in isolation from each other, which is evident for three cases in which a severe speech disorder was not accompanied by any impairment of limb movement or by any defect of nonspeech tongue movements. Thus, speech has its own motor control center.

Bouillaud then moves directly to the problem of identifying the area of the hemispheres responsible for speech motor control, following Gall's theory, the frontal lobes. Surgical lesions in animals cannot furnish evidence because animals do not have speech, nor can experimental lesions be carried out in humans. He defines the principle of a "necessary and sufficient condition" to test anatomo-functional correlations. If the frontal lobes are crucial for directing speech movements, two conditions must be satisfied: first, when frontal lobes are affected, speech must also be affected; second, when frontal lobes are spared, the same must be the case for speech movements.

Bouillaud next describes a few positive and negative cases where the anatomo-clinical correlation actually corresponded to his expectations, and then he tests his hypothesis of the frontal lobe localization of speech using the neurological and neuropathological case studies reported in Claude-François Lallemand's books (1820, 1823), among others. He analyzed Lallemand's data using the above-described principle of positive and negative observations and claims that he obtained a full confirmation of his theory both for the positive and for the negative observations. He even records his surprise that Lallemand and others did not come by themselves to the same conclusion, faced with such a clear evidence.

Bouillaud differentiates between an "organ of articulated language" and "memory for words," which he considers as a secondary attribute, and argues that this is different from Gall's view (however, recall that Gall had indeed separated word memory from speech; it is not known whether Bouillaud was aware of Gall's distinction). Finally, Bouillaud suggests that the frontal lobes are the site of both these distinct faculties, the grey matter being the organ of verbal (memory) representations and the white matter the locus of speech motor output. Putting aside the question of priority between Gall, Hood, and Bouillaud, it should be emphasized that Bouillaud's description is quite outstanding for a mainstream medical publication of this era. Nevertheless, it is exceedingly curious that he could have based his model on Lallemand's data. Luzzatti and Whitaker (1996b) reanalyzed the Lallemand clinical cases and arrived at a quite different picture from that of Bouillaud.

Claude-François Lallemand described his cases in a series of "letters" (Lallemand, 1820, 1823); it is from the first four letters that Bouillaud (1825) drew his conclusions. Of the 174 case descriptions, some subjects are poorly documented and thus unsuitable for our purpose; some had lesions outside the cerebral hemispheres proper (e.g., in the cerebellum, pons, or medulla). In a few cases, Lallemand provides a good clincal report but the pathology is missing or insufficient, and these cannot be used either. Of the remaining 63 usable cases, 33 have a right hemisphere lesion, 23 have a left hemisphere lesion, and 7 have bilateral lesions.

The almost normal distribution of right and left hemisphere lesions, of course, showed a higher rate of aphasic/dysarthric cases within the LH group against a majority of nonaphasic patients in the RH group (16 out of 23LH versus 13 out of 33RH). The Chi-square is significant (p < .01). The results of the Luzzatti & Whitaker (1996b) analysis are even more impressive if one considers the location of the single lesions in the two hemispheres. Consider first the 33 cases with a right hemisphere lesion only: in agreement with Bouillaud's predictions, there are 2 frontal cases with speech or language disorders and 16 nonfrontal cases without speech disorders. However, there are also 4 frontal cases without speech deficits and 5 cases with speech deficits but a lesion outside the frontal lobe. The data for the left hemisphere lesions are equally problematic for Bouillaud: in agreement with his predictions, there are 7 frontal cases with speech or language disorders, 4 nonfrontal cases without speech disorders, and no frontal cases without a speech disorder. Unfortunately, there are also 7 cases with speech deficits but a lesion outside the frontal lobe. Today, of course, we know the reason why. Summing up the left and right hemisphere lesions, we find that 15 out of 40 patients with a localized unilateral lesion (38%) did not fit Bouillaud's prediction. The question that arises is why did Bouillaud come to such an inaccurate conclusion? It is difficult to believe that this could have been due to differences in how we interpret these cases. Consider, for example, that, in a note added to the 1834 edition of his book, Lallemand made an ironic comment directed to Bouillaud, pointing out that the sites of the lesions in these cases are incompatible with Bouillaud's theory. When we looked at Bouillaud's text and compared it to the case from Lallemand that he purported to be describing, it was clear that he had changed the meaning in some cases or just ignored other cases.

For instance, Bouillaud's observation 7 (Lallemand's Letter 1, Case 11), a 70-year-old male presenting with a language disorder a few days after a traumatic head injury, had multiple lesions involving the frontal lobe and the left posterior part of the brain; Bouillaud refers to this case as only a frontal lesion by reporting only the first part of Lallemand's text. Unfortunately, the only reasonable explanation is that Bouillaud purposefully manipulated Lallemand's case descriptions, that he committed scientific fraud.

Whatever the reason that brought Bouillaud to misrepresent the experimental demonstration of his theory, it is interesting to keep in mind the scientific cultural frame in which Bouillaud developed his hypothesis on localization of language. Besides his general model of a dual aspect of language and its motor control—we might now say, between articulation and language representations—it is quite surprising how little interest Bouillaud showed in the "intellectual" aspect of language. Such a neglect appears to characterize almost all French neuropsychological studies on language until near the end of the 19th century, including Broca himself. We suggest that this approach to brain–language interaction is largely due to the influence of the Gall–Flourens dispute over phrenology and Flourens's overwhelming interest in motor control.

1-6. FROM GALL TO BROCA: DAY-TO-DAY MEDICAL WORK

Pierre Paul Broca (1824-1880) focused on disorders of speech production *(langage articulé)* because, before the 1860s, one did not discuss disorders of language "comprehension." Language was constructed as speech output and the rest, what could have been discussed under the rubric "comprehension," fell under the notion of "mind," which at the time was in the province of philosophy, religion, and nascent psychiatry (the alienists). This, of course, had been tradition since the time of Galen; the primary difference was that the location had shifted from the third cell or ventricle, as in Medieval Cell Doctrine, to the frontal lobe, as in Gall's craniology, later phrenology.

From the 1820s on, both orthodox medical and phrenological medical journals were filled with case reports of expressive language impairments arising from stroke and trauma, with and without autopsy evidence of the involvement or lack of involvement of the frontal lobes. That is to say, Gall's hypothesis that the frontal lobes are the seat of language was hotly debated for more than 40 years prior to Broca's famous articles. The publications of Bouillaud, Lallemand, Marc Dax, and Jacques Lordat are among the more familiar; however, there were dozens of obscure medical practitioners publishing these reports, such as Alexander Hood's case studies, which we have already discussed. Another typical example comes from the journal *The Lancet* (Vol. 1, 1848). A case report on page 260, written by T. E. Tebay, Esq., who was the resident medical officer of the Westminster Hospital, discusses a case of cerebritis, hemiplegia of the right side with a defective memory for words. The patient died on the twenty-fourth day after admission; the autopsy reported a gelatinous softening of the left anterior cerebral lobe. The case details are as follows:

J.W. male, 35 y.o. (admitted March 10) speechless, could sit in his chair and protrude his tongue or move his hand when desired to do so. Sensation on the body surface is unimpaired, the right arm is powerless, the right leg dragged when walking (progression) but the tongue protruded straight. There is no facial paralysis, the mental faculties appeared lost to a considerable extent. By 10:00 A.M. patient appeared to have recovered mental faculties to a considerable extent but cannot give utterance to his wishes for want of words, which he strives in vain to recal [sic]. His articulation of some words is tolerably distinct when he can recal [sic] them. March 12th: patient is much the same; he appears to have almost entirely recovered his intellectual powers with the exception that his memory for words continues very defective. March 14th: the hemiplegia is diminishing. March 28th: motor power in arm is better restored than in leg and recollection of words is returning gradually. March 29th: there were several momentary attacks of tetanic rigidity. March 30th: there were upwards of 20 such attacks. April 3rd: frequent recurrence of attacks of tetanic rigidity (all affecting right side of body as before) were attended by persistent coma. During these attacks the facial muscles are very distorted; the right angle of mouth is drawn downwards and outwards. April 4th: coma and convulsions continue (attended with groaning); patient died at 4 P.M. The autopsy report showed a cavity of the size of a small nutmeg, containing half a teaspoonfull of yellowish sero-purulent fluid in the left posterior lobe, one inch from the posterior surface of the cerebrum and nearly on a level with the corpus callosum; the cerebral substance around was natural in appearance but slightly softened. The membranes covering the anterior lobe, half an inch above the orbit, bore a depressed and puckered appearance over a space of one inch square, owing to a loss of cerebral substance at that part. The cerebral substance (of the anterior lobe) being, to the depth of half an inch or more, greatly softened, partially absorbed and almost converted into a fluid yellowish jelly. The opposite hemisphere (i.e. the right) appeared normal.

From cases like these—and there were literally hundreds of them in the weekly, monthly, and quarterly medical journals from Europe and America—it is clear that the average practicing medical doctor was well aware of the frontal lobe speech center hypothesis. Cases with lesions outside of the frontal lobe, with aphasia, were taken as evidence against the speech localization hypothesis, and not as data for different kinds of aphasia.

What, then, to make of Pierre Paul Broca's series of papers on "aphemia" published in 1861, 1863, and 1865? We already know that Broca's papers are no longer recognized as the first research to provide empirical evidence for the lateralization of language to the left hemisphere. Marc Dax (1770-1837) read a statement to his medical colleagues in Montpellier, France, in July 1836 in which 40 cases with clinico-anatomic correlation showing the role of the left hemisphere in language were summarized. Recent historical research suggests that Broca's papers were also not the first to demonstrate the localization of a language function to a specific site in the frontal lobe, nor were they even the first to do so using an acceptable scientific paradigm, that of clinico-pathological correlation. Perhaps the best that can be said is that Broca was in the right place at the right time. Historical analyses can help us realize that our neurolinguistic models *(a)* have precursors, *(b)* are contextually influenced by the scientific milieu, and *(c)* are relative to the assumptions and constraints of the paradigms we happen to currently accept.

Establishing priority is typically a judgmental enterprise. Consider the following quote from David Caplan (1987, p. 46): "The 1861 paper by Broca is the first truly scientific paper on language–brain relationships." Caplan supports this conclusion (here, as elsewhere, Caplan successfully integrates historical with contemporary research) with three claims: that Broca presented a detailed case history with "excellent gross anatomical findings at autopsy," that Broca had the insight that the gross brain convolutions are constant anatomical features that may be related to particular psychological functions, and that Broca's primary conclusion that expressive speech depends on a small part of the inferior frontal gyrus was a good first approximation, which we generally accept today. Priority in the case of Broca is a matter of scholarly judgment. Alexander Hood, in 1824, did a better job of analyzing expressive language functions and correlating them to a frontal lobe localization; Hood postulated a lexical-phonological level, a phonological-articulatory level, and a motoric level for expressive speech, based on the speech and language impairments he observed in stroke patients. Excellent clinical-pathological studies of aphasic cases are found in the 17th-century studies of Wepfer (1727, p. 14), studies that are so good one may verify the left hemisphere localization of language from them. Lallemand (1820, 1823), Bouillaud (1825), and others long before Broca published detailed autopsy reports of patients with aphasia. In this context, one might also consider the fact that the classical neuroanatomists of the late 19th and early 20th centuries (i.e., after Broca) virtually abandoned the possibility of systematically describing gyral geography because of its evident individual variability, a variability that currently plagues PET researchers who need to coregister sites of PET activation with MRI images in order to compare results across subjects. And, finally, we are obliged to mention the fact that the autopsy of Broca's 1861 patient actually demonstrated an extremely large left hemisphere lesion encompassing frontal, parietal, and temporal cortex. Broca "inferred" that the third inferior frontal part of this large lesion was the one responsible for the patient's aphemia, by estimating the degree of necrosis and trying to back-correlate that with the patient's medical history.

What is important to appreciate here is that it is not a question of disputing the claims of priority, nor even the facts themselves, but of how one chooses to interpret the historical record. The view I prefer is *(a)* that Broca inherited a tradition of clinical-pathological correlation that already presupposed that different brain regions had different functions, *(b)* that Broca was theoretically constrained by a construct of language that placed psychological preeminence on speech production, *(c)* that Broca was immediately challenged and certainly intrigued by the debates involving many famous members of the French scientific community (e.g., Gratiolet, Bouillaud, Auburtin, and Flourens) concerning the role of the frontal lobe in speech and therefore was predisposed to see the age of that lesion as having a significant frontal component, and, most important, *(d)* that Broca had the position, power, and prestige to take advantage of his serendipitous clinical observation.

1-7. DIAGRAMMING AND LOCALIZING IN THE LATE 19TH CENTURY

Although some historical "facts" are subject to interpretation, as we have just seen, some are just plain right or wrong, and, it behooves us to get it right. Consider the following opinion from Morton (1984): "We have a number of lessons to learn from history. If we are lucky we can avoid making the same mistakes as thinkers in the past." A mistake made by the "diagram-makers," according to Morton, was to confuse the goal of representing the elements of language processing in the brain with the goal of determining the localization in the brain of these elements (p. 40). Henry Head (1926) held a similar view; in fact, Head coined the term "diagram-makers" to refer to the classical aphasiologists of the late 19th century.

Let us examine Morton's (1984) claim: did the diagram-makers confuse the psychological (processing elements) with the neurological (localization of elements) goals of their neurolinguistic enterprise, as Morton asserts? Baginsky (1871; see Jarema, 1993), one of the diagram-makers discussed by Morton, believed that he had based his model on the "physiology of speech formation"; Baginsky did not stipulate specific anatomic sites for each of his language "centers," maintaining that "we do not yet have a precise conceptualization" of this relationship (Jarema, 1993). Kussmaul, another diagram-maker discussed by Morton, claimed that his colleagues, particularly Wernicke, were mistaken in trying to localize the various speech centers to specific regions of the brain. Kussmaul was "acutely aware of the limitations of the localizationist approach to linguistic processes" (Jarema, 1993, p. 509); "extraordinarily removed from strictly anatomical and physiological considerations, Kussmaul the physician achieves an understanding of the psychology of language in terms of the concepts which constitute the core of present day models, e.g. the distinction between various levels of representation . . . their respective autonomy yet interconnection, and the notion of linguistic processes" (Jarema, 1993, p. 497). The same may be said of Elder (Whitaker, 1988) and Grasset (Dos Santos, Nespoulous, & Whitaker, in preparation), both mentioned by Morton, as well as of Bastian (Eling, 1994), who, though not discussed by Morton, was one of the best known of the British diagram-makers of the period. Eling (1994) remarked that, in general, characterizing the work of these classical aphasiologists with a few short statements and adjectives does not do justice to their careful analytic description and argumentation. It should be mentioned that Morton's questionable analysis of the model-theoretic assumptions of Baginsky, Kussmaul, Elder, and Grasset may not be his fault; he relied on Moutier's 1908 dissertation as his secondary source material. Moutier was the student of Pierre Marie, who was notorious for his antipathy toward anyone who fractionated language into its component elements and thus anyone who believed that there were several different types of aphasia, obviously the main tenet of the diagram-makers. Ironically—and history sometimes has a penchant for the ironic—it was Pierre Marie who proposed that the insula (Island of Reil) was a functional component of expressive language (Marie's quadrilateral), the evidence for which was finally obtained by electrical stimulation studies (Ojemann & Whitaker, 1978b) of the insula that elicited naming errors. On the matter of psychological versus neurological modeling, Marie's basic objection to those who believed in localization, the diagram-makers, was a psychological one—for example, his slogan "l'aphasie est une." In Marie's view, language was reconstructed again such that it became comprehension (understanding, the lexicon, etc.); what for Broca had been language, speech production, Marie relegated to the status of motoric output. To this day, neurolinguistics has wrestled with the problem of how to incorporate the motor component of language into a model of the expressive aphasias. In the 1960s and 1970s, this was one of the major theoretical disputes between the Mayo School of Darley, Aronson, Brown, et al. (apraxia of speech) and the Boston School of Geschwind, Goodglass, Benson, et al. (Broca's aphasia with dysarthria).

Although it seems quite clear that most of the diagram-makers well understood the different demands of a psychological versus a neurological model, they did occasionally make real mistakes. Laubstein (1993) has elegantly shown that Lichtheim, the "paradigmatic diagram-maker" and one of those discussed in Morton's chapter, had produced a neurolinguistic model that is ambiguous with respect to some predictions of language disorders, that fails to predict some language disorders that had already been described, that is internally inconsistent, and, finally, that cannot be falsified, all in terms of the 19th-century paradigm within which Lichtheim operated. Laubstein's (1993) analysis of Lichtheim's model goes to the heart of the basic model-making assumption of that period and of our own: the correlation between aphasic language data and the components of the processing model of language used to account for such data. What Morton says is the diagram-makers' "mistake" was certainly not characteristic of all of them. There is little to be gained in pointing out how our forebears differed from us; one hopes there is much to be learned in analyzing how they developed and tested hypotheses.

To Broca, language was still thought of as expressive (speech) output, since the time of Galen in point of fact. Things were to change quickly. Beginning at least as early as 1866 with the publication by Theodor Meynert of a case of receptive aphasia with jargon, followed by the dissertation of Arnoldus van Rhijn on aphasia in 1868 in which the notion of "disconnection" was introduced, in turn followed by Henry Charlton Bastian's symmetric model of language input and output in 1869, the stage was set for the reconstruction of language to include comprehension, seen in the work of John Hughlings Jackson and Carl Wernicke starting in 1874.

1-8. INTRODUCING THE CONCEPT OF RECEPTIVE APHASIA: THEODOR MEYNERT

From the historical perspective, the neurolinguistic studies of the 1860s to the 1890s (the classical aphasia period) take on a foreshadowed, albeit interesting place in the story. But there is a twist, as in all good stories. Through the first two-thirds of the 19th century, language was constructed as the power of verbal expression or production and was associated with, or identified as, memory. It should be emphasized that neither Gall nor Broca overlooked or ignored, or failed to grasp, the concept of verbal comprehension: in their epoch, "comprehension"—or, to use more common 19th-century terminology, "understanding"—was not part of language; it was part of the mind, or, as we would now say, cognition. One of the significant contributions of Carl Wernicke in his 1874 paper "The Symptom Complex of Aphasia" (see Eggert, 1977), was the reconstruction of language so that comprehension became a separable component of language and thus a candidate for cerebral localization. Wernicke is traditionally considered the first to describe the features of, and the brain pathology underlying, impaired auditory comprehension and related linguistic dysfunctions. This tradition persists in spite of the fact that in his 1874 paper Wernicke clearly, properly, and repeatedly acknowledged his indebtedness to his mentor, Theodor Meynert. Today, when anyone mentions Meynert and Wernicke, this debt is understood to mean merely that Meynert taught Wernicke neuroanatomy. Wernicke's own words support this interpretation: he said that his 1874 work essentially reflected his application to the study of aphasia, of Meynert's neuroanatomical approach. As usual, the real story is not that simple (Whitaker & Etlinger, 1993).

François Boller (1977) already noted that the concept of receptive aphasia had been discussed in an 1871 publication by Johann Baptist Schmidt. Two years before that, in 1869, Henry Charlton Bastian had proposed that a disturbance in the comprehension of language ought to parallel the disturbance in expression that had been proposed by Broca. Neither Schmidt nor Bastian supported their analyses with autopsy evidence, nor had they suggested what the underlying anatomy of language comprehension might be; thus, neither Schmidt's nor Bastian's work could be said to dislodge Wernicke's claim of priority to the anatomical model of comprehension. However, a virtually unknown work by Theodor von Meynert, "An Anatomical Analysis of a Case of Speech Disturbance," appeared in 1866 in the 22nd Annual Medical Yearbook of the Journal of the Imperial and Royal Society of Physicians in Vienna (pp. 152-187) (see Whitaker & Etlinger, 1993). In this paper, Meynert discussed the case of a 23-year-old servant girl suffering from left coronary valves stenosis. About 2 weeks before her death, she experienced "a sudden inhibition of expressive speech" (Meynert's words, translated from the German). Although she was free of any paretic symptoms and possessed complete mobility of the tongue, she was unable to access the individual words necessary for communication (e.g., *head,* or *hand*), while the missing words' substitutes that emerged from her had no relationship with the intended communication. For example, she produced the word *yellow* in place of the word *hand* (to continue with Meynert's description of the aphasia). At the same time, in many cases when she achieved a congruency between the word and its prevailing image, her word construction was defective. For example, prior to the appearance of her partial speech disturbance, she had always articulated the word *Husten* (= cough) correctly, but ever since she transformed the word into *Hutzen* (has no meaning in German).

It is clear that Meynert described two types of paraphasias that we typically associate with so-called Wernicke's aphasia, the unrelated semantic type and the phonological type. The autopsy showed that an embolus had blocked an artery in the Island of Reil (insula) leading to pathological changes in the most posterior gyrus of the insula and in the posterior portions of temporal lobe above the insula, part of what we now refer to as the planum temporale and part of what we refer to as Wernicke's area. Although it is clear that Meynert was describing sensory aphasia, it is also clear that his main interest in this paper was on anatomy and histology. He used this case, plus five others that had been sent him by colleagues, to argue that since the acoustic nerve terminates in the temporal lobe wall of the Sylvian fossa, this cortex is a "resonance field" or "sound field" and thus is a "central organ for speech." In his summary remarks, he noted that, over a lifetime, memories enter the cortex via facial (visual) and tactile perceptions and these combine with sound patterns to result in such linguistic functions as naming. In the case of this patient, Meynert says: "although her visual memory was intact, she was not able to pull the sound pattern, formerly joined to the visual, over the threshold of consciousness." He argues that the clinical and anatomical facts at present do not permit him to decide between two choices: *(a)* that the sound images were destroyed or *(b)* that they were merely disconnected from their corresponding visual images—a controversy that was to occupy classical neuropsychology for many years to come, even though no one has ever referenced Meynert's paper where one finds the original interpretative dilemma. According to his biographer, Max De Crinis, Theodor von Meynert was resigned to the fact that he would not receive much recognition for his ideas during his lifetime; however, he was convinced that his ideas would be vindicated later. To some extent, this has already happened: today we speak of association and projection fiber systems; many current models rely extensively on proposed white-matter tract connections; the analysis of cell assemblies and comparative neuroanatomy are important approaches in neuroscience; the relationship of blood flow to brain functions is a cornerstone of PET studies; we still make a distinction between reflex and voluntary motor control; and we still think of the frontal lobes as an important substrate of socially correct behavior. It may come as something of a surprise to realize that every one of these ideas was pioneered by Theodor von Meynert. We should add one more to the list, the suggestion that the posterior first temporal gyrus plays a special role in auditory language processing (Whitaker & Etlinger, 1993).

1-9. LOCALIZING AND DIAGRAMMING: HENRY CHARLTON BASTIAN

In a biographical sketch, Marshall (1994, p. 101) noted that during his career H. C. Bastian believed in abiogenesis, the spontaneous formation of life from nonliving substances, and in heterogenesis, the appearance of one life-form as the offspring of another, different life-form. One suspects that these beliefs caused Bastian's colleagues then and later to ignore his contributions to aphasia; except for Marshall's insightful critical Introduction to his ideas (1994, pp. 103-111) and Head's (1926) condemnation of him along with the other "diagram-makers," Bastian is hardly discussed at all in the history of neurolinguistics. This is unfortunate since his contributions can easily be considered among the best of the epoch of the localizers and diagrammers. A sample of the ideas in his work notes that Bastian provided one of the first detailed and persuasive analyses of what was then called "word blindness" and "word deafness" (Marshall, 1994, p. 103). Bastian argued that writing is a more conscious activity than speaking because the muscles of the arm are more voluntary than the orofacial muscles. He distinguished storage from retrieval in verbal memory and proposed three types of verbal memory: auditory, visual and kinesthetic. Bastian understood that "center" was as much a psychological as an anatomical concept and recognized that some language functions did not necessarily have a focal locus and that some language functions were distributed in networks. He was one of the first to call attention to the role of the angular gyrus *(pli courbe)* in visual memory as well as in word recall (note that he did not reference Meynert's earlier work on this subject). In his model, the centers are separable from the places that exert direct "live" control.

In his 1869 paper, Bastian laid out his model of speaking, reading, writing, and their cerebral processes; he began with a discussion of acquisition of language from an associationist-sensationalist perspective but, allowed for hereditary transmission of specialized brain structures for speech (from ancient, historical times). According to his psychological model of language acquisition, reading and writing are added to speaking, again by associationist principles. At this point Bastian brings anatomic centers into the description. Early on he realized that one studies aphasia by looking at the effects of lesions in a center, in a connecting pathway, or in the center's connection to the periphery.

Bastian's model was a dynamic, network model; for example, speaking is explained as follows: (1) excitation in the visual word center (2) arouses, through the visual-auditory commissure, (3) the auditory word center, which in turn incites (4) the glosso-kinesthetic center via a pathway (later, this path was to be called the arcuate fasciculus). His model is a functional, task-oriented one, much more dynamic or interactive than is usually thought of the classical localizers and diagrammers. According to Bastian, words build associations to percepts and concepts, and so on, so that in thought processes the cerebral activity is not limited to narrowly localized centers, but there are widespread processes of activity in varied regions of cortex, in both hemispheres. The data suggested to Bastian that his "centers" actually have subparts, thus paving the way for a more fine-grained linguistic analysis which was to come years later.

Bastian proposed that concepts emerge out of perceptive processes and that this would take place in parts of brain in close relation, structurally and functionally, to the several sensory centers; Bastian called these "Annexes" (a "surround" of brain), and he says that they are the same as the association areas of Flechsig! According to Bastian, these areas (after Flechsig) are parts of prefrontal lobe, a large part of temporal lobe, a large part of posterior parietal lobe, and the island of Reil. These areas seem to have no direct afferent and efferent connections (shades of the Geschwind model), which of course led to the dispute between Bastian, on the one hand, and Broadbent, Lichtheim, Kussmaul, and so on, on the other over a piece of the diagram (the concept, naming, or ideation center), according to:

1. psychological arguments about theory of language
2. clincal data about explanation for types of aphasia

However, Bastian recognized the need to add something for higher-order functions (intellect, concepts, ideas) thus, he accepted the Broadbent/Flechsig proposal of the association areas as a conceptual (psychological) addition to his model, at the same time suggesting the concept of "annexes" around his centers instead of another center. One *implication* is that an impairment of idea/intellect would necessarily be accompanied by an impairment in the proximate sensory center, which of course is an empirical issue! Bastian (1898, pp. 80-113) discussed case histories to show that Broca's aphasia can exist without agraphia. Although, by 1898-1897, Wernicke and Dejerine had evidently rejected the idea of a writing center, for the reason that writing is just the simple copying of visual images stored in visual word centers, Bastian used the case history approach to attack this idea and support the idea of a writing center. But perhaps the most important feature of Bastian's diagrams, one shared by many other diagram-makers, is that he tried to create a dynamic model in which the functions of centers depended on specific kinds of input from other components. And the most important feature of his neurolinguistic research was his recognition that empirical data informed his models; data could support or reject any aspect of his model of language functions.

1-10. EARLY AGRAMMATISM RESEARCH: JACKSON, PICK, HEAD, AND WEISENBURG AND McBRIDE

Impairments of grammatical structures are essentially linguistic aspects of aphasia, relevant to the structure of language at the complex word, phrase, sentence, and discourse levels (Whitaker, 1997). Although most historical discussions of neurolinguistics mention the principal figures in the history of agrammatism, it is more problematic to situate them in the context of classical aphasia studies. The model of sentence structure proposed by John Hughlings Jackson (1835-1911) and employed by Arnold Pick (1851-1924) in particular and Henry Head (1861-1940) to a lesser extent, represents a different line of historical development than the classical models of their contemporaries. One difference is that premodern work on agrammatism (Pick) and syntactical aphasia (Head) employed abstract, mental models of sentence structure that were presumed to underlie observed speech, models that seem to be formally equivalent to contemporary cognitive neuropsychological models of language, albeit vastly less detailed, whereas the aphasia studies of contemporaries such as Broca, Wernicke, Lichtheim, Dejerine, Marie, Bastian, and others were typically limited to taxonomic analyses of observable word- and subword-level phenomena, which were correlated with specific, observable brain sites.

John Hughlings Jackson was interested in how two aspects of speech, voluntary and involuntary expressions, might be localized in the two sides of the brain; in his study of involuntary or automatic speech, he spent much of his time analyzing the recurrent utterances of patients with expressive aphasia. Jackson did not seem to have taken much interest in the different varieties of aphasia with which his contemporaries were engrossed; in his collected papers on affections of speech (Taylor, 1958), aside from Broca, there are no discussions of Wernicke, Dejerine, Lichtheim, and others who contributed to the classical model of aphasia. Yet it was Jackson's notion of the proposition as the essential unit of internal mental structure as well as expressive language that, in the work of Arnold Pick and Henry Head, laid the foundations for the study of agrammatism. In his "Notes on the Physiology and Pathology of the Nervous System" (1868), Jackson first develops the propositional basis of language, as well as his ideas concerning the role of the right hemisphere in automatic (involuntary) expression and perception. That this is a psycholinguistic construct is clear: "the meaning of a proposition does not depend on the mere words which compose it, but on the relations these words have to one another—such a relation that the sentence is a unit" (Taylor, 1958, pp. 234-235).

Arnold Pick wrote the first treatise devoted exclusively to agrammatism, only sections of which have been translated from the German (Spreen, 1973; Friederici, 1994). Although linguistic theory was not well developed in 1913 when Pick completed this monograph, some of his theoretical assumptions are notable. He assumed that a sentence schema, which logically implied a syntactic structure, precedes the selection of lexical items; he suggested that content words were selected prior to function words, after which surface grammatical form is specified (Friederici, 1994, p. 257; see also Spreen, 1973, pp. 147-152, for a more detailed analysis of the hierarchical order in Pick's microgenesis of language). From Pick's monograph on agrammatism:

> that the schematic formulation of the sentence precedes the choice of words, as well as the syntactic formulation and the portion of the grammatical functions that corresponds to it, is shown by the fact that the meaning of a single word, whatever it may be, is determined only by the position it takes or interacts with; therefore the mental framework should in principle be ready in a grammatical sense as well: before the choice of words ensues, the plan has to be determined before the different pieces are put together. (Friederici, 1994, p. 267)

Pick considered the question of agrammatism again in a chapter published posthumously in 1931 (Brown, 1973, pp. 76-86). This represented his final ideas on this subject; we note, of course, that he would have read Henry Head's ideas on the subject by this time. Pick discussed two forms of expressive agrammatism, motor (quantitative) and sensory (qualitative). He attributed both to the loss of and/or impaired control over grammatical devices (Friederici, 1994, pp. 258-259). He associated "motor agrammatism" with motor aphasia, frontal lesions, and a telegraphic style, whereas he linked "sensory agrammatism" with Kleist's "paragrammatism" (thus, with lesions of the temporal lobe), although emphasizing speech output, or what he called "the expressive side" (Brown, 1973, p. 76). From the 1931 chapter:

> This temporally determined form is characterized, in pure cases, by disturbances in the use of auxiliary words, incorrect word inflections, and erroneous prefixes and suffixes. In other words, it concerns all those linguistic devices which serve to express relationships between objects, which differ widely and numerically from one language to another. . . . Regarding telegraphic style of primary origin, discussed in relation to motor agrammatism, in which the word order is not appreciably disturbed, alterations of word order have occasionally been reported as due to a discrepancy between the normal order (most important element first) and the grammatically required order (final position of the verb as in the languages of

children). However, a thorough investigation of this question, based upon linguistics . . . is not yet available. . . . absence of or defect in the auxiliary words which are normally produced in automatic fashion. It is probable that the patient with motor agrammatism at times retains the sentence skeleton since he may not comprehend the prepositions nor be able to write them. . . . The patient attempts to produce the best possible results (that which best makes him understood) with the least expenditure of effort, utilizing the optimal but still automatic application of his linguistic resources. . . . Another likely factor is the attention fixed on the effortful production of speech. If the prepositions are either not automatic or only incompletely so, attention will not suffice for their voluntary production. . . . In paragrammatism, the temporal form of expressive agrammatism, it must be remembered that grammar is by no means a unified process, but contains many factors which may be affected separately or in combination. In this form, the disorder lies one stage deeper than telegrammatism. . . . There is a discrepancy between poor agrammatical speech and better appraisal of ungrammatical sentences presented to the patient. This is explained by the contrast between defective or absent feeling for the language [and] the fact that the patient recognizes what he sees as incorrect, but is incapable of putting it into correct grammatical form. (Brown, 1973, pp. 74-84)

In the following chapter on word and speech deafness, Pick continues his discussion of agrammatic phenomena:

Disorders of the second type concern what is called sensory (impressive) agrammatism (in the narrow sense). Disturbance of stages in the comprehension of sentence meaning which correspond to the grammaticization of the sentence, insofar as both characterize relationships, will disturb to a varying extent the comprehension of those relationships. There may also be a loss of knowledge of sentence form. . . . this latter form is generally not separated from the others, both for clinical considerations (e.g., the coexistence of expressive and sensory agrammatism) as well as psychological considerations (e.g., the same process is at one time centrifugal and at another centripetal). . . . A distinction should be made between lack of comprehension of the grammatical forms of correct speech and inadequate recognition of incorrect forms. (Brown, 1973, pp. 94-95)

Pick (1931 [in Brown, 1973]) placed syntactic organization in a central position between sound and meaning, giving it a primary role in expression and in comprehension. He recognized that there are multiple components to syntactic structure, not all of which are affected at the same time in the varieties of expressive and receptive grammatical impairment. Borrowing from Jackson, he distinguished automatic from voluntary dimensions of syntax, partly to explain the observed differences between agrammatic production and comprehension.

One often remembers Henry Head for his lively, though inaccurate and misdirected, attack on the "diagram-makers," his term for the classical aphasiologists. Head is rarely noted for his criticisms of superficial descriptions of aphasia such as "motor" or "sensory," nor for his effort to avoid uninformative clinical descriptions such as "the patient was said to be able to read and write, although he could not speak" (1926, 1: pp. 197-217). Caplan's (1987) discussion of Head's critique of inclusive diagnostic categories of aphasia points out that Head had observed that actual aphasic performance usually crosses descriptive boundaries, that specific aphasic symptoms are often transient, and that there is wide variation in aphasic performance in patients with quite similar lesions (1987, pp. 83-88). As Caplan thoughtfully observed, these problems remain largely unsolved.

Contemporary researchers correlate Head's syntactical aphasia with the syndrome of Wernicke's aphasia (cf. Lecours, Lhermitte, & Bryans, 1983, pp. 254-255; Goodglass & Kaplan, 1983, p. 77; Goodglass, 1993, p. 24; Benson & Ardila, 1996, pp. 114-115). This may not do justice to Head's theoretical analysis of aphasia. As Weisenburg and McBride (1935) observed, "The syntactical form in Head's conception differs from the sensory aphasia which shows pronounced disturbances of speaking and also from the so-called agrammatism or paragrammatism, but partakes of the nature of these" (1935, 50); Weisenburg and McBride credit Delacroix with the observation of the correspondence between Head's four aphasia categories and the hypothesized stages of thought to spoken expression. This is similar to Pick's (1913) conception, which, according to Spreen (1973), was drawn from the contemporary psychological research of Wilhelm Wundt, Karl Buhler, William James, and others. Following this model, Head characterizes syntactical aphasia as a defect in an internal grammatical arrangement (1935, pp. 50-51). Weisenburg and McBride wrote of Head's system as though it were only a classification of the aphasias; it is likely, however, that Head was trying to develop a psycholinguistic theory with which to describe aphasic disorders. For Head there are four dimensions to symbolic formulation and expression, each or any of which may break down; thus, instead of four aphasic types, there are four aspects to aphasia that would manifest themselves in varying degrees in any one patient. Head was attempting to characterize the nature of language impairments in their psychological context, albeit without benefit of well-constructed linguistic or psycholinguistic theories. Thus, the four categories were exemplars, not categorical types: "No two examples of aphasia exactly resemble one another; each represents the response of a particular individual to the abnormal conditions. But, in many cases, the morbid manifestations can be roughly classed under such descriptive categories as Verbal, Syntactical, Nominal or Semantic defects of symbolic formulation and expression" (Head, 1926, 2: p.x).

In Head's theory, syntactical defects (1: pp. 230-240) represent impaired grammatical phrase structure as well as impaired rhythmic components of symbolic formulation, what we would now label "prosodic." As an example he observed that "the patient talks rapidly, his speech is jargon, and prepositions, conjunctions and articles tend to be omitted" (2: p. xiv). He probably used the term "syntactical" instead of "agrammatic" because he believed that the language impairments are deeper than surface, observable grammatical words; he thought that these deficits affected the basic (and, of course, internal and mental) formation and use of language (1: p. 240). His patients' agrammatic production, both verbal and written, is well described, although Head fails to theoretically link the expressive impairments with the patients' evident comprehension difficulties, a theoretical lapse that Pick did not make. Consider, for example, Head's case of syntactical aphasia number 15:

> Asked what he had done since he came to the London Hospital, he replied, "To here; only washing, cups and plates." Have you played any games? "Played games, yes, played one, daytime, garden." . . . He did not usually employ wrong words and, if the subject under discussion was known, it was not difficult to gather the sense of what he wished to say. Thus, when I was testing his taste and placed some quinine upon his tongue, he said, "Rotten

to drink it. Something medicine or that. Make you drop of water after it, so to take out of your mouth." . . . when asked to say after me short sentences, which he had not heard before, his defective syntax became evident. . . . In this case the disorder of language mainly affected syntax and rhythm. The production of single words and their use as names were not materially disturbed; but groups of words could not be combined into coherent and effective phrases. (1: pp. 174-178.)

When Weisenburg and McBride (1935) completed their clinical psychological study of aphasia, both agrammatism and of paragrammatism had been incorporated into general descriptions of language disorders due in part to the efforts of German researchers such as Bonhoeffer, Goldstein, Salamon, Isserlin, Kleist (who had coined the term "paragrammatism") and Pick. In their analysis of agrammatism, Weisenburg and McBride (1935, pp. 60-61, 71-72) noted the omission of grammatical words (telegraphic style), disrupted word order, and the omission or substitution of prefixes, suffixes, and inflections. They also discussed the "economy of effort" principle as a cause of agrammatic aphasia and, in addition, several of the tests in their aphasia battery explicitly addressed the evaluation of grammatical abilities. It seems evident that Weisenburg and McBride understood the central nature of agrammatical disturbances, as the following comments demonstrate:

> The question of paragrammatism has already been touched on in connection with agrammatism and the cortical form of motor aphasia. It has been said that paragrammatism is characterized by confusions of grammatical forms, of auxiliaries, pronouns, prepositions, and so forth, and by changes in word order. A consideration of the nature and extent of these errors makes it evident that they are not simply the result of word-substitutions, and consequently not like the paraphasic errors. They involve a more extensive change which shows uncertainty, not so much in the choice of words, as in the grammatical and formal aspects of the sentence structure. (1935, p. 72.)

Historically, an interesting aspect of Weisenburg and McBride's research is their recognition that Henry Head had attempted a psycholinguistic analysis of the aphasias, that Arnold Pick had proposed an abstract structural model of agrammatism, and that both Head and Pick had based their ideas on those of Jackson. Perhaps one should regard their commentary as a nascent cognitive tradition in neurolinguistics.

1-11. SOURCES FOR THE HISTORY OF NEUROLINGUISTICS

Although the history of psychology, and, of course, the history of medicine, are well represented in both primary and secondary source material, the history of neuropsychology, and in particular the history of neurolinguistics, are not as yet. In their seminal paper, Benton and Joynt (1960) collected descriptions of language disorders in historical documents from the last 50 centuries. Starting from the famous Case 22 of the Edwin Smith Surgical Papyrus (c. 3000 B.C.), they traced examples of aphasic disorders in the ancient Greek (Hippocratic corpus), Latin (Valerius Maximus, Pliny the Elder), medieval, and Renaissance medical traditions (e.g., Guainerius, 1481), observations from the 17th century (Schmidt, 1673; Rommel, 1682), and the illuministic literature (Linnaeus, 1745; Morgagni, 1762; Gesner, 1770).

O'Neill's (1980) scholarly text argues that, at least through the Renaissance, observations of aphasic symptomatology were hardly part and parcel of any general, coherent theoretical model of brain–language relationships. Her summary of early neurolinguistics ends at the 17th century, leaving us with a number of gaps in the story from the Renaissance to the 20th century, gaps that are only partly filled in by current research on persons who have actually made substantial contributions to the development of neuropsychology and neurolinguistics.

In the readers by Clarke and O'Malley (1968), by Hunter and Macalpine (1982), and by Shipley (1961), one finds selected excerpts from works by Gall, Broca, and Jackson as well as Jean Baptiste Bouillaud (1796-1881), Simon Alexandre Ernest Aubertin (1825-1865), Johann Gaspar Spurzheim (1776-1832), and George Combe (1788-1858), all relevant the the history of neurolinguistics. None of the three are as complete as Eling's (1994) *Reader in the History of Aphasia,* but the overlap is not at all extensive.

Two exemplary models of historical analysis of the trends in the neurosciences as well as insightful expositions of the varying milieu are to be found in Harrington (1987) and Clarke and Jacyna (1987). Caplan (1987) contains two thought-provoking chapters of historical analysis that focus on theoretical models of classical authors from a contemporary point of view. Bouton's monograph, *Neurolinguistics: Historical and Theoretical Perspectives* (1991), begins with ancient Egyptian (Pharaonic) texts and carries the discussion to modern neurolinguistic studies in the 1980s. The 1993 issue of *Brain and Language* (vol. 45, no. 4) is exclusively on historical studies in neurolinguistics; the guest editors for this issue were Gonia Jarema and Roch Lecours. A monumental compendium of significant archival value is Finger's text (1994); it treats the origins of neuroscience from the earliest written records to the latter half of the 20th century. Nearly a third of the chapters bear directly on neurolinguistic issues.

Eling's (1994) *Reader in the History of Aphasia,* excerpts the work of Franz Joseph Gall (1758-1828), Pierre Paul Broca (1824-1880), Carl Wernicke (1848-1905), Henry Charlton Bastian (1837-1915), John Hughlings Jackson (1835-1911), Sigmund Freud (1856-1939), Jules Dejerine (1849-1917), Pierre Marie (1853-1940), Arnold Pick (1851-1924), Henry Head (1861-1940), Kurt Goldstein (1878-1965), and Norman Geschwind (1926-1984). By and large this is a good selection of historical work on brain and language.

PART II

Clinical and Experimental Methods in Neurolinguistics

CHAPTER 2

Methodological and Statistical Considerations in Cognitive Neurolinguistics

Klaus Willmes

Neurologische Klinik—Neuropsychologie, Universitätsklinikum RWTH Aachen, Aachen, Germany

Three broadly characterized phases of (cognitive) neurolinguistic research on the relation of language and the brain are discerned. The modern cognitive neurolinguistic and cognitive neuropsychological research program in general, which dominate the third phase, are outlined with a view on diverging and opposing lines of argumentation. Challenges from the neurosciences and connectionist modeling are mentioned as well. Afterwards, two methodological approaches relevant for conducting technically adequate single-case studies are described in some detail. The concepts of criterion-referenced measurement are shown to be useful for assessing a patient's degree of competence in some task domain. Furthermore, it is argued that randomization tests allow for valid statistical tests in individual subjects. Both approaches are combined to provide operational definitions of three types of (double) dissociations as proposed by Shallice (1988).

The last two decades have witnessed a proliferation of theoretical developments and empirical findings in the fields of neurolinguistics and neuropsychology in general (see, for example, Coltheart, Sartori, & Job, 1987; Denes, Semenza, & Bisiacchi, 1988; Mapou & Spector, 1995; Margolin, 1992; McCarthy & Warrington, 1990a). These have helped to improve an understanding of the complex brain-behavior relationships (or brain-cognitive mechanisms) involved in processing language or performing other higher cognitive functions.

Yet there are ongoing debates, not only about the adequacy of particular models of language processing or the interpretation of particular experimental results—a ubiquitous state of affairs in all (empirical) sciences—but rather about more general methodological, metatheoretical, and epistemological concerns. Central to these debates are two issues. First, this chapter discusses the way patterns of impaired (and spared) language performances in individual patients subsequent to some form of (acquired) brain damage can be employed to reveal the "nature" of language processing per se. Second, many researchers express more or less strong reservations about whether this is possible without recourse to what is known about the structure and functioning of the human central nervous system itself. After characterizing three overlapping phases in neurolinguistics in terms of dominating methodological, statistical, and psychometric approaches, the chapter will focus on an exposition and discussion of the logic of inference and of methodological problems encountered in cognitive neurolinguistics, followed by a discussion of (more specific) psychometric and inferential statistical issues related to the cognitive neurolinguistic and neuropsychological approach.

2-1. RESEARCH PHASES IN NEUROLINGUISTICS

2-1.1. The Classical Clinical Approach

Clinical observations and clinical studies are as old as aphasiology itself. Case reports from the second half of the nineteenth century already were often detailed, but they lacked methodological and psychometric rigor. These reports concentrated on the (clinical) examination of the specific pattern of impairments of the patient under study rather than on properties of the examination process itself, the reason being obvious: there was no elaborate empirical and psychometric methodology at hand. Nevertheless, the Wernicke-Lichtheim model of word processing (Lichtheim, 1885) was an important early accomplishment in theorizing about (impaired) language processing in relation to brain organization and its impairment subsequent to brain lesions. Nevertheless, the "diagram-makers'" approach was subjected to three major criticisms concerning their postulates of precise localization, their inadequate psychological concepts, and their weak empirical accounts (Shallice, 1988, Chapter 1–2).

2-1.2. The Modern Group-Study Approach

In accord with developments of methodology in psychology and advances in theoretical linguistics, psycho- and neurolinguistic experimental studies were carried out with more or less narrowly defined (aphasic) patient groups. These were formed according to type, side, and site of brain lesion and/or more or less broadly defined patterns of language symptomatology (aphasic syndromes). The aim of the studies was to examine specific aspects of differences in level and/or quality of expressive or receptive language impairments under different experimental conditions focusing on specific language modalities, components of language, and stages of language processing. Developments in psychological diagnostics, including the concern for psychometrically controlled assessment procedures, were also taken up in aphasiological studies in that the selection of patients was primarily based on the pattern of performances in comprehensive diagnostic test batteries with well-studied objectivity, reliability, and validity properties (for an overview, see Willmes, 1993 and the references cited there), in what Shallice (1988) termed the "modern aphasia group-study approach." Besides the growing body of psycho- and neurolinguistic assessment procedures, psychological on-line reaction time measurement methodology was introduced. This approach employs the additive factors rationale of Sternberg (1969) to support the existence of sequentially organized stages in language processing. Psychophysical (skin conductance, sonographic blood flow registration), and electrophysiological measurement methods (evoked or event-related brain potentials) are utilized to study brain-behavior relationships when performing language processing tasks. Functional neuroimaging techniques (SPECT, PET, and fMRI) employed in language activation studies are considered to contribute new knowledge about brain-language relationships (Habib & Demonet, 1996).

2-1.3. The Cognitive, Information-Processing Approach

The advent of cognitive psychology in the late 1960s and 1970s brought with it a major change at the theoretical level in trying to get away from the ruling empiricism of experimental psychology. Information-processing models were developed in general psychology and psycholinguistics for various cognitive functions, such as the logogen model of John Morton (1980) for single-word processing in normal subjects. Typically, in this type of model some complex cognitive or language behavior is decomposed into a hypothetical sequence of processing steps, in which information in some representational format flows from one processing component to the next, with specific transformations within the components operating on these representations (Massaro & Cowan, 1993). Although early models were often mute with respect to particular types of representations and processes, this way of theorizing opens up a sensible way to characterize patterns of impaired cognitive performance in general and impaired language processing in particular in terms of damaged (sub)components and/or transmission routes. In early accounts of this approach (for example, Marin, Saffran, & Schwartz, 1976), the expectations concerning the relevance of neuropsychological evidence for the understanding of normal function were rather modest, and hoped to yield a taxonomy of functional or isolable (sub)systems (Posner, 1978; Tulving, 1983) within a broadly modular overall system architecture (Marr, 1982; Fodor, 1983). Marin et al., however, pointed out that the constraints due to the functional and anatomical architecture of the brain have to be taken into consideration. Later on, researchers have become more confident that cognitive neuropsychology can do more than just explain the pattern of impaired and intact cognitive performance seen in brain-damaged patients in terms of damage to one or more of the components of a model of cognitive functioning (Ellis & Young, 1988; Olson & Caramazza, 1991). The approach is deemed powerful enough to allow drawing conclusions about normal, intact cognitive processes; that is, it is assumed that the understanding of normal function can be enhanced by studying impaired function (see Shallice, 1988).

On the methodological side, the study of single cases has regained respect again (Shallice, 1979; Valsiner, 1986 for general psychology). Group studies, in which patients were allocated according to brain lesion or "weak" clinical syndromes (Schwartz, 1984; Marshall & Newcombe, 1984; Caramazza & Badecker, 1989), were not considered informative because of the apparent functional heterogeneity across patients despite frequent co-occurrences of sets of symptoms. It was also argued that the classical aphasia syndromes might even be artifacts of the vascularization of the brain (Poeck, 1983). A similar fate has also happened more recently to more narrowly defined functional syndromes such as surface dylexia (Patterson, Marshall, & Coltheart, 1985), deep dyslexia (Coltheart, Patterson, & Marshall, 1987), or agrammatism (Berndt, 1987). Single-case studies are increasingly considered to provide the most stringent evidence for inferences to normal (language) functioning, often with a special status attributed to (functional) dissociations (Teuber, 1955; Dunn & Kirsner, 1988) in performance and double dissociations, in particular. Shallice (1988) even provides a characterization of (strong) functional syndromes via sets of dissociations as the organizing principle of his book. The most radical ("ultracognitive") position, as held by Caramazza (1986) and McCloskey (1993), is that single-patient studies provide the *only* valid basis for inferring normal cognitive mechanisms from patterns of impaired performances, be they dissociations or associations, a claim that will be discussed in the next section.

2-2. THE COGNITIVE NEUROLINGUISTICS RESEARCH PROGRAM

2-2.1. Basic Assumptions

The heading seems to imply that researchers in the field would fully agree on what constitutes *the* cognitive neurolinguistics (and cognitive neuropsychology) approach. Yet, there is an ongoing debate about the appropriate methodology and logic of inference that has to be followed when taking data from brain-damaged patients both to develop or to test and evaluate cognitive models. There is no need to reiterate the various positions in detail, since the exchange of arguments has been well documented. After an exposition of the "single-case-only approach" (Caramazza 1984, 1986; Badecker & Caramazza, 1985; Caramazza & McCloskey, 1988), a first round of discussion followed, filling a whole issue of *Cognitive Neuropsychology* (Bub & Bub, 1988; Caplan, 1988; Newcombe & Marshall, 1988; Whitaker & Slotnick, 1988). A response to the critics was written by McCloskey and Caramazza (1988), and additional critical contributions were published in other leading journals by Bates, Appelbaum, and Allard (1991), Zurif, Gardner, and Brownell (1989), and Zurif, Swinney, and Fodor (1991). The most extensive and balanced account, in my view, can be found in Chapters 9–11 of Tim Shallice's book (1988).

The next challenge to the position of "strong" cognitive neuropsychology was put forward by authors like Kosslyn and Van Kleek (1990), with a response by Caramazza (1992), as well as Kosslyn and Intriligator (1992), claiming that behavioral data from a single patient cannot provide sufficient evidence for the development of a meaningful cognitive science. Rather, computational models—with the possibility of showing what sorts of new functions can emerge after damage (Kosslyn & Intriligator, 1992, p. 102)—and insights from the neurosciences about the realization of cognitive functions in "neural hardware" (Kosslyn & Koenig, 1992) would be required. It is important to note that Caramazza is not at variance with recognizing that these two sources of information are highly relevant in principle.

Another potentially serious challenge concerns the two fundamental assumptions of cognitive neuropsychology. The *fractionation* assumption states that brain damage can result in selective impairment. More so, the *transparency* assumption claims that brain damage can result in *local* modifications of an otherwise unchanged cognitive system, for which *universality* is assumed, implying that the cognitive systems of normal subjects are basically identical. One line of attack dwells on the property of the brain to operate in a highly nonlinear fashion with multiple areas being involved in processing, often in reciprocal fashion. Farah (1994) therefore goes on to propose a new set of assumptions—in line with the parallel distributed processing (PDP) framework Rumelhart, McClelland, 1986—for drawing inferences about the functional cognitive architecture. Farah states that human information processing is graded, distributed, and interactive. In particular, she discusses a connectionist network model (Farah & McClelland, 1991), which, when lesioned, produces category-specific "naming disorders" similar to the double dissociations between the naming of living and nonliving things reported in the neuropsychological literature.

Yet another line of critique has taken up the dismissal of group studies again (Robertson, Knight, Rafal, & Shimamura, 1993) in the influential *Journal of Experimental Psychology,* followed by a rebuttal from McCloskey (1993) in the same journal. The crucial arguments are not really specific to cognitive neuropsychology. They concern the possibility or impossibility of guaranteeing homogeneity among brain-damaged patients. The whole enterprise of *generalizing* the (significant) effects found in a study of normal subjects rests on the *independent* a priori assumption that the target population is in fact homogeneous with respect to the cognitive mechanisms of interest. McCloskey argues (1993) that in patient group studies there is no way of knowing beforehand or of demonstrating empirically that homogeneity prevails. Consequently, the results from a patient sample cannot be generalized to the target patient population. But this type of generalization is not needed if one's goal is to relate the results obtained in an individual patient to a model of normal processing. Converging evidence must be sought from multiple single-patient studies.

Without going into detail, I want to note that the use of group studies in general cognitive psychological research plays a "technical" role also. Typically, theoretical claims in cognitive psychology, expressed as psychological hypotheses (PH), are about processes and regularities in individual subjects. The hypotheses tested in group studies are, however, often hypotheses about populations of subjects, more precisely about the identity or difference of parameters of the population distribution(s) of some dependent variable under, say, two experimental conditions. This approach makes sense if the population statement can be *deduced* from the PH. (Random) samples of subjects are only employed for the purpose of reducing the influence of nuisance variables that may obscure the relation between independent experimental conditions and the dependent variable of interest. If one were able to control all the relevant nuisance variables adequately when studying just one subject, the results of that single-subject experiment would be fully informative with respect to the psychological hypothesis within the theoretical framework of interest; only, in an inductive framework one is attempting to generalize to a population of subjects. This is particularly troublesome for "open" populations, that is, when lawful statements are assumed to be valid also in the future, rendering true random samples impossible.

2-2.2. Chance Dissociations

To end this section with a more modest topic, McCloskey (1993) also refutes the claim by Robertson et al. (1993) that there is a serious problem in being deceived by chance dissociations, when there is no way of replicating results in another patient, since every brain damage presents a new experiment of nature. McCloskey points out that every measure is taken both through using large numbers of stimuli in a host of experimental tasks (possibly with replication) and through appropriate statistical tests to determine that the results reported are reliable. These two psychometric and statistical topics will be covered in the following two sections.

2-3. PSYCHOMETRIC CONSIDERATIONS

For an in-depth analysis of an individual patient's (language) performance pattern, aimed at tapping the source(s) of some language processing deficit within some (tentative) cognitive model of normal language functioning, assessment procedures are very often specifically tailored to the requirements of a particular investigation. Thus the methodological approach of cognitive neurolinguistics leads to a theory-motivated or theory-driven, yet sometimes ad hoc, collection of items composing the tasks of interest. Obviously, no sound *empirical* evidence concerning the particular reliability (homogeneity and stability) and construct validity properties of the items composing a specific task, as well as the various tasks themselves, is available. Even if some tasks have been employed repeatedly in different single-case studies, the psychometric adequacy of the items and tasks remains a matter of observational impression. This is not meant to generally question the construct validity (Embretson, 1985) of tasks and items. But it is a long way from some language processing model—be it of the boxes-and-arrows or the network type—to the final decision about how to best tap the functions presumably involved in responding to the content of an item, which in addition has to be cast in a specific item format for presentation and response. In general, the functioning of several subcomponents and/or processing routes is implied in any task, let alone the possibility of strategy choices. It is my impression that the arduous task of caring for construct validity, which plays a central role in psychological measurement (for example, Messick, 1980; Wainer & Braun, 1988), is given too little attention in applications of the cognitive neurolinguistic approach. Empirical demonstrations of the psychometric properties of theoretically motivated assessment tools such as PALPA (Kay, Lesser, & Coltheart, 1992; Kay & Franklin, 1995), PAL (Caplan, 1995) or LeMo (Stadie, Cholewa, De Bleser, & Tabatabaie, 1994), which have been used repeatedly in single-case studies, are still widely lacking.

2-3.1. Criterion-Referenced Measurement

In the theoretically—as opposed to psychometrically—oriented cognitive neurolinguistic approach, *content validity* is implicitly taken to be the major criterion for item generation and selection. A test is called *content valid* (Berk, 1980; Klauer, 1987) if it (completely) contains or represents a *universe (domain)* of items, for example, the universe of legal neologisms for a particular language, or the universe of temporal prepositions. Such a domain is defined either by exhaustive enumeration of its elements or, more often, by stating the properties via some generation rule(s) that an item from that particular domain has to fulfill. In both instances, an actual test or task usually consists of some *representative* sample from that domain. For a complete characterization of an item, a stimulus and a response component, that is, an operation the subject has to perform on the stimulus, are combined to yield a specific *item format,* for example, make a lexical decision or choose a temporal preposition from a written multiple-choice set composed of other temporal prepositions in order to complete a sentence.

This general framework is well established in educational psychology and associated with *criterion-referenced measurement* or *mastery testing* (Hambleton & Novick, 1973; Hambleton, Swaminathan, Algina, & Coulson, 1978; Suen, 1990; Crocker & Algina, 1986). It should prove to be useful in cognitive neurolinguistics as well; in fact, it has been employed in some implicit form in the cognitive neurolinguistics-oriented language assessment batteries mentioned earlier as well as in theoretically motivated single-case experimental studies. Two important aspects are the *stratification* of a domain according to one or more aspects in a hierarchical or completely crossed fashion and the formation of *representative samples* from each stratum or combination of strata in order to draw inferences from the item sample about the whole item domain. To give an example, aspects of stratification for reading, writing, and lexical decision tasks are word-frequency intervals, numbers of syllables, concreteness/abstractness ratings, regular/irregular spelling, complexity of initial, middle, final consonant clusters, and so forth. The researcher always has to make an informed choice about which of these parameters he wants to vary systematically and which he is willing to ignore or control for by allocating items with these properties at random to the strata or strata combinations. Subsequently, the proportion of items from each stratum or strata combination has to be specified. Representativeness can often be obtained by drawing a random sample of prespecified size from the subset. This random sampling notion has been debated in earlier psycholinguistic research

(Clark, 1973) from another perspective. Random allocation is only meant to statistically control—not to eliminate—the influence of such unwanted systematical, yet not identified, variation. A similar approach to item generation and selection is provided by facet theory, as proposed by Guttman (Shye & Elizur, 1995; for an application in aphasiology, see Willmes, Poeck, Weniger, & Huber, 1983). Within that approach, a domain is structured according to (content) facets that have the same status as strata. These ways of item generation and selection also provide a convenient way to obtain (data independent) *content valid parallel tests,* by drawing distinct representative samples from the domain of interest, a prerequisite for proper repeated assessment of the same subject when stability of some performance is deemed crucial. The notion of content valid parallel tests is implicit when a set of items is randomly split, for example, for use under different conditions of presentation.

For a given domain of study, the *degree of competence* (achievement, ability) for an individual subject i can now be defined as the probability p_i of processing correctly items from that particular domain. Empirically, the level of competence must be estimated from a representative sample of items from that domain, the properties of the ability estimate being dependent on the particular test model introduced, for example, the binomial model. Besides characterizing a subject by means of some competence estimate, a (dichotomous) classification according to mastery or nonmastery with respect to some domain is useful in cognitive neurolinguistics. *Mastery* can be defined to hold, if the individual p_i is no less than some high, lower-bound criterion probability p_c often fixed at 0.90 or 0.95. Mastery or criterion-referenced testing thus relates individual test performance to a certain (high) level of competence, not to a distribution of scores across some reference population, as does a diagnostic test constructed according to the classical test theory model (Lord & Novick, 1968; Crocker & Algina, 1986). In cognitive neurolinguistics, the concept of mastery can be utilized to empirically demonstrate the intactness of some processing components and/or routes in particular, when there is (almost) no variation in level of performance in samples of control subjects.

2-3.2. The Binomial Test Model

The binomial model is implicitly used in all single-case studies in which the level of competence is estimated for a set of dichotomously scored items as the relative frequency of correct responses. Even for a narrowly defined domain of language processing, items will be of different difficulty rendering the "ability model" interpretation of the binomial model untenable, in which all items in a test must have identical difficulty for any subject tested. Otherwise, the relative frequency of correct responses is not a proper ability level estimate. Luckily, there is also an *"item sampling model"* version of the binomial model (Lord & Novick, 1968, p. 250; Klauer, 1987), allowing for possibly different item difficulties. This version requires that for every subject examined a *new* random sample of items without or with replacement has to be drawn from the finite/infinite item domain. This presents no unsurmountable problem when using a PC for item presentation and test administration.

2-3.3. Operational Definition of Dissociations

Willmes (1990) has shown how the criterion-referenced measurement approach employing the binomial model can be utilized in combination with Fisher's exact (randomization) test for 2-by-2 tables (discussed in section 1.4) to give an operational definition and a worked-out procedure to detect different types/degrees of (double) dissociations in performance as introduced conceptually by Shallice (1988, Chapter 10). For graded scores or metric responses, see Willmes (1995).

For a *classical* dissociation, performance in some task A must be well below the normal range and must be much inferior to performance in task B, which in term must fall within the normal range, possibly close to the premorbid level of competence. For tasks solved very well by almost all normal subjects, therefore, a very high mastery level of $p_c = 0.99$ should be adopted for task B. In case of substantial systematic variability in competence for normal subjects p_c may be fixed at some quantile (25th, 50th) of the score distribution for normal controls, if available. The individual competence level p_i for task A must be much lower (for example, at chance level for multiple-choice items). In any case, it must be demonstrated using inferential statistics with sufficient test power (Cohen, 1988) that there is a significant difference in competence.

A *strong* dissociation does not require performance on task B to be within the normal range. For a task in which non-brain-damaged subjects show ceiling performance, one can set mastery at a competence level no less than, for example, $p_c = 0.80$ for the task assessing the (supposedly) spared function. The competence for the other task A has to be fixed at a considerably lower level, for example, at $p_0 = 1/m$ for an m-choice response format or at a reasonable, yet arbitrarily chosen, value for free-response formats, for example, in production tasks. Again the numbers of items in tasks A and B must be large enough to reveal a real difference $\delta = p_c\text{-}p_0$ with sufficient statistical power.

A *trend* dissociation denotes the weakest form of a reliable difference in individual competence among two tasks. In this case it is particularly mandatory to compute in advance whether some real difference $\delta = p_B\text{-}p_A$ in competence can reliably be demonstrated with the actual number of items chosen for both tasks. In the subsequent statistical evaluation of the data a (trend) dissociation should be taken to hold only if there is a significant difference at the assumed type I error level and if the estimated effect size, that is, estimated difference in competence, is about at least as large as the effect size parameter value δ for which the sample size computation has been carried out.

2-3.4. Double Dissociations

For a valid demonstration of a *double* dissociation, Shallice (1988, p. 234) has argued that in order to safeguard against resource artifacts it is not sufficient to reveal two complementary dissociations on two patients. Rather, it must be demonstrated in addition that both patients exhibit a complementary and significant difference between both tasks. Therefore, the empirical hypothesis of a double dissociation implies a conjunction of four *one*-sided statistical alternative hypotheses. As Westermann and Hager (1983, 1986) have argued, there is no need for a type I error reduction because of multiple testing, in case one is only willing, say, to accept the empirical hypothesis to be corroborated if all four tests yield significant results at the prespecified type I error level of $\alpha = 0.05$. Instead, paying attention to statistical power considerations, the number of items per task must be sufficiently large to warrant enough statistical power for a type II error reduction to $\beta/4$ for each of the four individual tests.

2-3.5. Inherent Item Difficulties

Another methodological point to be discussed in this context is the problem of potential differences in the "inherent" difficulty of two tasks (Shallice, 1988; Blanken, 1988; Sergent, 1988), the term "inherent" indicating that within the particular processing model assumed to hold, one task has to, for example, rely on more and/or more demanding crucial processing steps than the other one, thus requiring more resources or competence, finally leading to higher overt task difficulty. Without a computationally explicit theory of task performance, this notion of inherent difficulty must remain somewhat vague, though. The difficulty argument is particularly important in those instances in which performance is poorer on the supposedly more demanding task and when the reverse pattern of performance has not been observed so far. The problem is made more complicated in those (frequent) instances in which normal subjects exhibit perfect performance on both (language) tasks. We have argued (Huber, Willmes, & Göddenhenrich, 1988) that in those instances it may be informative to look at data obtained from a (clinical) sample of (aphasic) patients—possibly fulfilling more or less detailed inclusion criteria—in order to learn about the "average" task difficulty and the frequency of dissociations in performance (for a limited approach to estimating this frequency see Bates, Appelbaum, & Allard, 1991).

2-4. INFERENTIAL STATISTICAL TESTS FOR SINGLE-CASE STUDIES

2-4.1. Random Sampling versus Random Assignment

The great majority of statistical tests employed in single-case studies in neurolinguistics are conceptually wrong. Tests for contingency tables (2-by-2 tables or others) relating the value of the test statistic to the (central) chi-square distribution with the appropriate number of degrees of freedom as well as t-tests, analysis of variance F-tests, or other parametric tests, all rely on the notion of random sampling from some (infinite) population, mostly requiring the observations to be independent as well. Edgington (1995) has repeatedly argued for more than two decades that these sampling assumptions are not tenable, even for multisubject experiments in most of general psychology, and that they are not required in experimental work. *Random assignment* of subjects from any kind of sample to experimental conditions or random assignment of adminstration times to experimental conditions in repeated measures designs is sufficient for drawing valid inferences about contrasting effects of experimental conditions on some response measures from the results of randomization tests. Inferences about effects of experimental manipulations for other subjects must be nonstatistical in nature, without a basis in probability. Hypotheses tested by randomization tests refer to individual subjects even when many subjects participate in an experiment. Likewise, psychological theories—particularly those in cognitive (neuro)psychology—often explain phenomena in terms of processes within individuals.

2-4.2. Randomization Tests for Single-Subject Studies

The application of statistical tests to single-subject experimental data cannot be justified on the basis of any random-sampling model. But, as Edgington (1995) has argued, any randomization test that can be applied to some multiple-subjects design with random assignment to experimental conditions can be applied as well to data from a single-subject design. In these designs, administration times are randomly assigned to experimental conditions in some fashion. Willmes (1990, 1995) has outlined this approach in the context of single-subject aphasia therapy research for the most basic types of hypotheses concerning the comparison of two types of tasks, each (mostly) composed of a set of homogeneous items, or of two experimental conditions for the same set of items. Randomization tests are available and well known for dichotomous item scorings (Fisher's exact test for 2-by-2 tables; the exact version of the sign test), graded scores (exact versions of the Mann-Whitney U-test for ranks and the Wilcoxon signed-ranks test), and continuous measures such as reaction times (randomization test analogs of the independent and paired observations t-test; see Edgington, 1995).

The most simple situation is one in which the performance of one subject has to be compared for just two different sets of items. The number of administration times for each item set is fixed at the number of items in that set—possibly different for both sets—and within that restriction there is random assignment of administration instances to item sets. In other instances, there may be some additional blocking of application times. The typical null hypothesis tested is as follows: for each of the administration instances, the single subject's response is independent of the impact of the respective item set (experimental treatment) presented at that particular instance. Or, put differently: the association between administration instances and responses is the same as it would have been for any alternative assignment. There is no need for an independence assumption among responses. Only the weaker requirement for *interchangeability* of responses has to be fulfilled. However, rejection of the null hypothesis may be caused not only by differences between the item sets but also by differential carryover effects. Such an effect is present whenever the subject's response to an item from a particular item set is influenced not only by how many items preceded that item (a nonproblematic general carryover effect like a practice effect or fatigue), but also on the particular characteristics of the item sets, from which items were administered at each of the preceding item presentations. Although these differential carryover effects can never be ruled out in principle in any single-subject research design, well-known experimental control measures can be exercised to keep them at a minimum. Edgington (1995) also presents some worked-out examples and applications for several types of restricted randomization designs encountered in behavior modification studies such as intervention and withdrawal designs or multiple baseline designs (see also Onghena & Edgington, 1994).

2-4.3. Computational Aspects

The theoretical basis for randomization tests was already laid out some six decades ago (Pitman, 1937, 1938; see also Fisher, 1935). But these tests require heavy computations, except for very small data sets. The test statistic has to be evaluated for every admissible permutation of the responses observed under the respective null hypothesis in order to determine the observed significance level (p-value). The actual conduct of randomization tests has become feasible only quite recently with the availability of efficient special-purpose PC-program packages (StatXact, SCRT, RAN-DIBM; for references, see Edgington, 1995, pp. 398-399, or PITMAN by Dallal, 1988). Some exact tests have also been incorporated in general statistical packages such as SPSS or SAS. Nevertheless, randomization tests have rarely been employed in cognitive neurolinguistic studies.

2-4.4. Statistical Demonstration of Dissociations

The detection of the different types of dissociations and double dissociations (classical, strong, trend), which is so important for theorizing in cognitive neurolinguistics, can be accomplished entirely with randomization tests. This has been demonstrated in detail for dichotomously scored items and outlined for other types of responses by Willmes (1990, 1995). For dichotomous item scores, a large number of the inferential statistical problems involved can be approached with just the exact version of the sign test (or the exact version of the McNemar test) and Fisher's exact test. However, these tests have to be augmented by power considerations in accord with reflections on a sufficient number of items required to detect true differences in performance or to corroborate the (near) identity of impairments or achievements (for example, in the case of associated disorders).

2-4.5. Example

An exemplary application of this approach to single-case studies in number (word) transcoding is shortly presented for the data of two patients reported in Table 1. From one particular influential model of number (word) processing (McCloskey, 1992), one can deduce that transcoding either spoken verbal numbers or written number words into written Arabic numbers is assumed to rest on the same syntactic processing component. The demonstration of a double dissociation between both types of transcoding tasks would call into question this central feature of the model. Since the same two- and three-digit Arabic numerical stimuli were used for testing, the randomization version of the McNemar test (that is, the exact sign test) was used for comparisons between tasks in one patient and Fisher's exact test for the (one-tailed) comparison between subjects per task. Before carrying out the study, one can examine whether the chosen number of items from the domain ($k = 78$) would be sufficient to detect significant differences with sufficient statistical power. Application of the PC program STAT-POWER (Bavry, 1991) reveals that already for a medium effect size (cf. Cohen, 1988, Chapter 5), which has to be present for a proper strong dissociation, the chosen number of items is large enough to detect an existing difference of that size. (It has to be noted that for a priori power considerations the random-sampling model is more convenient.) As indicated by the same program, the item number is also sufficient for Fisher's exact test for the comparison of two subjects.

Comparing the off-diagonal entries for patient 1 (35 vs 0) using the exact sign test reveals a highly significant difference in performance ($p < .0001$, one-tailed) with a large effect size estimate (according to the criteria put forward by Cohen, 1988) and an empirical power estimate of $1-\beta = 1.00$. In addition, the perfect spoken verbal to Arabic transcoding performance is compatible with a mastery level of $p_c = 0.99$ according to the binomial model, suggesting a classical dissociation in performance. Patient 2 revealed a reverse pattern of performance with a highly significant difference, a large effect size and a power estimate of 1.00 as well. Since the 95%-confidence interval around the ability estimate of $65/78 = 0.833$ for transcoding written number words covers a criterion competence of "only" $p_c = 0.90$, this dissociation is labeled "strong."

Comparison of performances between patients using Fisher's exact test as computed in the StatXact package (Mehta & Patel, 1995) yielded highly significant one-sided differences ($p < .0001$) with power estimates of at least 0.99 and a medium to large effect size for written verbal input and a large effect size for spoken verbal input. Since there are no generally accepted proposals for classifying different degrees of double dissociations, I propose to call this pattern of performances a strong double dissociation according to the weakest single dissociation encountered.

TABLE 1

Crossclassification of Correct (+) and Incorrect (−) Responses of Two Patients in Two Number (Word) Transcoding Tasks Using the Same $k = 78$ Two- or Three-Digit Numerals Stratified According to Seven Complexity Levels

Transcoding		Written verbal to Arabic numbers				
		Patient 1			Patient 2	
		+	−		+	−
Spoken verbal to Arabic numbers	+	43	35	+	38	1
	−	0	0	−	27	12

2-5. FUTURE OUTLOOK

It is impossible to make any precise predictions about the course of development of the whole cognitive neurolinguistic and neuropsychological field. One can, however, be reasonably convinced that as the evidence from cognitive neurolinguistic single-case (and small-group) studies accumulates and language processing models become more elaborate, both areas will continue to be of mutual benefit to each other. The advances in the field will also crucially depend on the refinement of (experimental) language assessment procedures, functional brain-imaging technology with increased spatial and temporal resolution, and a growing concern for psychometric and statistical matters. Neural network modeling and simulation will offer unexpected and productive findings, both when running intact and when "lesioned" systematically or randomly. There are, for example, models of reading aloud by Coltheart, Curtis, Atkins, and Haller (1993) as well as Hinton and Shallice (1991), and of visual object naming by Plaut and Shallice (1993). Cognitive neurolinguistics will also be increasingly helpful for planning and evaluating language therapy, although a theory of cognitively oriented remediation of language impairments and disabilities is still largely lacking and not without basic conceptual problems (cf. Caramazza, 1989; Riddoch & Humphreys, 1994).

CHAPTER 3

Clinical Assessment Strategies

Evaluation of Language Comprehension and Production by Formal Test Batteries

Jean Neils-Strunjaš

Department of Communication Sciences and Disorders, University of Cincinnati, Cincinnati, Ohio 45221

Assessment of language may be conducted to further knowledge about a language disorder and underlying processes, but more commonly, testing is conducted for the patient's benefit. In a patient-centered approach, the goals for testing include determining the normalcy of the patient's language, and, if a language impairment exists, determining the cause of the language impairment. A patient-centered approach will also define the patient's language deficits and ability to communicate in everyday situations. Age, sex, culture, languages spoken, and education have all been shown to affect test performance, and such factors need to be considered in the development of new tests and the collection of normative data.

Assessment of language is perhaps one of the most highly developed components of neuropsychological assessment due to the early development of aphasiology in the late 19th century (Benton, 1994). Despite the well-documented early development of aphasiology, a review of recent literature suggests that there is not a widespread interest in standardized language assessment; much of the recent psychometric research deals with nonlanguage functions (cf. the *Journal of the International Neuropsychological Society,* 1996). However, when language assessment methods are scrutinized, significant problems are revealed. Some of these problems relate to the fact that the purpose of language assessment has evolved over time. Neurodiagnostic testing, for the most part, has replaced the need to localize lesions associated with language deficits. Also, the language-impaired patient population continues to change (Horner et al., 1995). The current aphasic population in North America is older, has more medical problems, and is more culturally diverse than it was 20 years ago. New normative data will need to be collected and some tests may lose their application. This chapter will review rationales for language testing, language tests most frequently described in the literature, limitations of tests, and areas of future research.

3-1. WHY USE FORMAL TESTS OF LANGUAGE?

Clinical assessment requires that the clinician first determine the reasons for the assessment of the patient's speech and language. A rationale for testing will allow the clinician the most judicious selection of formal tests. Without a rationale, the testing will be inefficient and possibly ineffective. Inefficiency can result when the clinician administers more tests than is necessary to meet the goals of the examination. The testing may also be ineffective. The clinician may review the test results and realize that they are meaningless, because the tests that were chosen do not meet the goals of the examination. For example, Caplan (1995) states that one goal of language assessment is to describe the patient's language disorder in relation to the major components of the language processing system; however, according to Caplan, the majority of existing aphasia batteries do not allow for this type of analysis. The following sections will review the most common rationales for conducting language testing. Both limitations of formal language batteries and the need for future research can be elucidated within this framework.

3-2. AN ACADEMIC-EXPERIMENTAL RATIONALE FOR TESTING

In some clinical settings, one purpose for testing a patient's language abilities may be an academic one, that is, for the examiner to understand more about a particular disorder. Much of the information about speech and language disorders was obtained in academic settings where clinicians were oriented toward learning more about their patient's difficulties for the purpose of acquiring new knowledge and sharing this knowledge with their colleagues through grand rounds. Because of developments in neurodiagnostic testing, there is less need to infer lesion location through language testing; however, there is still a need to conduct research to correlate functional neurodiagnostic tests such as SPECT, functional MRI, and PET scan findings with findings from neuropsychological tests, including performance on tests of language (Benton, 1994). Also, neurodiagnostic techniques showing functional change in cerebral function associated with diaschisis, restoration of function over time, and alterations in between-hemispheric communication could be correlated with repeated language measures (Benton, 1994).

3-3. A PATIENT-CENTERED RATIONALE

A second rationale for testing that is common to all settings is focused on improving the patient's quality of life. The primary concern is to answer the question: "what difference will the results obtained from this test make in the quality of the patient's life?" Some specific goals of testing fall under a patient-centered approach. The goals are described in the sections that follow.

3-3.1. To Determine Normalcy

The results of testing may help determine whether the patient's speech and language abilities are within normal limits in comparison to his or her peer group. The patient's peer group may be defined as others of the same age, the same sex, the same educational level, the same socioeconomic status, or any combination of these patient characteristics. Determination of normalcy of language abilities may be important for the patient, and perhaps his or her family and friends. In addition, legal and employment decisions may be influenced by the results of the language evaluation.

There is no reliable method available for determining whether a patient is normal or aphasic using a standardized language battery. Wertz, Deal, and Robinson (1984) found that the Western Aphasia Battery classified five out of 45 patients as having normal language, whereas the Boston Diagnostic Aphasia Examination classified these same patients as having aphasia. Brauer, McNeil, Duffy, Keith, and Collins (1989) present discriminant weights that were found to be accurate for differentiating normal subjects from aphasics; however, the same discriminant weights were not able to differentiate normal performance from that of malingerers. Variability in patient characteristics may account for some of the inconsistent results in determining normalcy.

Ardila (1995) argued that neuropsychological tests, including language tests, have standards that were determined by the performance of middle-class North Americans, and that these standards may not be accurate for other cultures with different values, education, cognitive styles, and possible brain organization. He recommends that tests must be standardized and norms obtained for different age, cultural, and educational groups. As one example of the need for accurate normative data for language tests, a group of researchers from the University of Cincinnati (Neils et al., 1995) evaluated the performance of elderly subjects on the Boston Naming Test and found that the age of the person tested, along with the person's level of education and living environment (e.g., nursing home or independent living), made a significant difference in test performance. Many of the persons tested who were determined normal according to other standards, but who were elderly and had a limited education (6 to 9 years of formal education), obtained "abnormal" scores. Many nursing home residents who were judged to have normal cognitive functioning by other methods also obtained abnormal scores. Another study (Randolph, Lansing, Ivnik, Cullum, & Hermann, 1996) on the Boston Naming Test found that males score higher than females as the result of their performance on specific items.

The majority of the language tests available were not developed or standardized for the culturally diverse, neurologically impaired population that is found in many clinical settings. Often, tests are translated from English to other languages without concern for all but the most superficial aspects of translation. For example, Löwenstein and Rupert (1992) examined the results obtained from the FAS Controlled Association Test given to Cuban Americans. They found that even though the FAS fluency test was administered in Spanish, the performance of many of the Spanish-speaking normal elderly subjects would have placed them in the impaired range according to normative data obtained for English-speaking subjects. Löwenstein and Rupert speculated that the inferior performance of the Cuban Americans may have been due to the frequency of the letters *F, A,* and *S* occurring less often as initial letters in Spanish than in English; however, many other more subtle factors such as the effect of bilingualism could have influenced the speed of word generation.

Based on these studies and others like it, a clinician must compare the age, sex, educational level, living environment, language, and culture of the patient to the population the norms were obtained from. This comparison is especially important when the test has been shown to be gender biased or influenced by other variables as in the case of the Boston Naming Test. Löwenstein, Arguelles, Arguelles, and Linn-Fuentes (1994) recommended methods of decreasing cross-cultural bias in testing by (1) constructing tests that are more culture-fair and salient to diverse cultural groups; (2) modifying or discontinuing tests that are not salient or relevant to particular language/cultural groups; and, (3) developing appropriate age-, education-, and culture-fair normative databases. One test that is consistent with these recommendations is the Bilingual Aphasia Test (BAT; Paradis, 1993). The BAT has 60 language versions that are not direct translations but transpositions that are as linguistically and culturally equivalent as possible. In addition, each of the versions has been normed on native speakers in their respective countries.

3-3.2. To Determine a Differential Diagnosis

In some cases, knowledge of the patient's language may be helpful in the diagnosis of the patient's medical condition. Historically, before the availability of CT (and even more recently, MRI), language testing was more commonly used to identify the location of lesions associated with stroke. However, there still exist some cases where formal language assessment may be helpful to identify the presence of a disease process associated with "silent lesions" or neurochemical dysfunction that causes primary progressive aphasia (Kertesz, Caselli, Graff-Radford, & Miller, 1995), Pick's disease (Caselli, 1995), and early Alzheimer's disease (Grossman et al., 1995). Sass et al. (1992) found that results from the Boston Naming Test could predict whether the epileptogenic focus was in the left or right hemisphere in patients with medically intractable seizures. Still, for the majority of language-impaired patients, the diagnosis will be determined through neuroimaging and other medical tests. For these cases, it is still important to define language abilities and deficits.

3-3.3. To Define Language Abilities and Deficits

A particular patient's language problem may be so apparent that testing is not needed to determine normalcy. Instead, the language tests may be chosen to better define the problem. Definition of the language impairment allows the clinician to describe the language impairment to the patient's family, medical personal caring for the patient, and other clinicians. In addition, the test results often help the clinician determine which aspects of language should be addressed in language therapy conducted by a speech-language pathologist. There are three currently employed methods for defining language impairments related to aphasia: classification of aphasia types or syndromes, determination of overall severity of aphasia, and an information-processing approach.

3-3.3.1. Aphasia Types or Syndromes

Swindell, Holland, and Fromm (1984) found little agreement among aphasiologists in their clinical judgment of aphasia types (e.g., Broca's aphasia) and findings from the Western Aphasia Battery. Moreover, it has been determined that there is little agreement in classification of aphasia according to the two main tests used to classify aphasia: the Boston Diagnostic Aphasia Examination and the Western Aphasia Battery (Wertz, Deal, & Robinson, 1984). A large percentage of patients are unclassifiable according to methods prescribed by the Boston Diagnostic Aphasia Examination (Benton, 1994). A relatively new assessment instrument, the Aphasia Diagnostic Profiles (ADP; Helm-Estabrooks, 1991), is reliable in aphasia type classification, according to its authors; other studies, such as those that have been conducted on the reliability and validity of the Western Aphasia Battery (WAB; Kertesz, 1982) and the Boston Diagnostic Aphasia Examination (BDAE; Goodglass and Kaplan, 1983), are needed before clinicians select this measure over the WAB or the BDAE. Although there is no gold standard for determining aphasia syndromes, the use of syndrome labels remains the most common shorthand method for defining language impairment in both clinical situations and in research reports.

3-3.3.2. Severity of Aphasia

Several tests assess severity of language impairment, among them the Communicative Abilities in Daily Living (CADL; Holland, 1980) and the WAB. Severity of language impairment, in addition to being a useful descriptor, can also help the clinician determine which language tests should be administered to the patient. Both patients with mild language impairments and patients with severe language impairments pose special problems to the examiner. The majority of formal language tests (e.g., Porch Index of Communicative Ability, Revised Token Test) were developed for patients whose language is moderately impaired.

3-3.3.3. Information-Processing Approach

In this approach, language is conceptualized as a flow diagram in which a sequence of operations is activated on one or more levels of language. An information-processing approach begins with a model of the particular aspect of language in question (e.g., naming pictures, written production of single words written to dictation). The clinician then attempts to localize the deficit on the model through administering a series of tests selected to measure the integrity of processes represented on the model. One limitation of this approach is that it has been applied most often in case studies of very specific aspects of language such as naming pictures (Hillis, 1994). However, Caplan (1995) has developed a psycholinguistically oriented language assessment battery that operationalizes an information-processing approach and includes a comprehensive analysis of language, from the lexical level to the discourse level. The Psycholinguistic Assessment of Language (PAL) is "currently being used for research purposes and may become available to clinicians in the next few years" (p. 97). The PAL consists of 27 subtests, including measures of reading and writing. The examiner determines which subtests are performed abnormally, and then makes inferences about the associated language processing components and whether input components are normal. Secondary breakdowns are also determined, that is, components of language that are impaired because they are dependent on primary components.

All three methods of defining language impairment are useful, but for different purposes. The determination of aphasia syndromes is especially useful in medical settings where communication among neuropsychologists, neurologists, and speech-language pathologists is important. Also, in cases where a particular syndrome label neatly describes the patient's language deficits, the label may be useful to include in a research report. However, for any in-depth study of language, for patients who show a very specific language impairment in which a processing component is impaired in isolation, or for unusual profiles of language impairment, an information-processing approach is the preferred method. Research is needed to improve reliability of assigning aphasia syndromes; also, as indicated by Caplan, further research is needed to test "construct validity, sensitivity, and clinical applicability" (p. 97) of the PAL. A measure of severity is an essential supplement to syndrome identification in a patient who is recovering from aphasia or in a patient whose language is progressively declining. A severity rating might also be helpful in assigning patients to language treatment groups. Although several standardized batteries have become available for the severely impaired patient, information on the validity of these measures and their clinical utility is just becoming available (Helm-Estabrooks, Ramsberger, Morgan, & Nicholas, 1989; Cunningham, Farrow, Davies, & Lincoln, 1995). Measures for patients with mild language impairments are needed. Tests used to define the patient's language abilities and deficits may be limited in their ability to describe the patient's use of language for communication purposes.

3-3.4. To Determine the Patient's Ability to Use Language Communicatively

It is logical to question whether the results of testing help to determine the patient's ability to function independently. For example, can the patient communicate on the telephone and negotiate shopping independently? One test that attempts to fulfill this purpose is the test of Communicative Abilities in Daily Living (Holland, 1980). However, Lezak (1995) warns that the severity estimate derived from a standardized battery, even the CADL, may underestimate a patient's ability to communicate in everyday situations because contextual cues, routines, and familiarity with vocabulary are qualities of everyday communication that are difficult to replicate in a standardized examination (Monroe, 1985). For example, Lambrecht and Marshall (1983) report that it is not uncommon for a patient to be unable to point to a cigarette on a test but to produce a package of Camels when asked "Have you got a cigarette?"

Patients with right hemisphere lesions and the majority of patients with traumatic brain injuries have problems with communication more so than with language. One aspect of communication is discourse comprehension, which can be measured with the Discourse Comprehension Test (Brookshire & Nicholas, 1993). In this test, the patient listens to spoken paragraphs and answers yes–no questions about the paragraphs. Tompkins, Baumgaertner, Lehman, and Fossett (1995) suggested that patients with right hemisphere brain damage have poor discourse comprehension because they have difficulty suppressing irrelevant or incompatible information. Further research needs to be conducted to determine if discourse comprehension is more predictive of communication abilities in everyday situations for certain patient groups (e.g., mild aphasics, patients with right hemisphere deficits, and patients with traumatic brain injuries) than results from more broad standardized tests of language.

3-4. REVIEW OF TESTS

Standardized tests of language fall into three general categories: screening tests, comprehensive tests of language functioning, and tests of specific language abilities. Screening tests will not be reviewed here because the usefulness of even the most highly regarded screening tests has been questioned (Benton, 1994; Lezak, 1995). Subtests of standardized batteries, clinical judgment (Lezak, 1995), and consideration for why a language evaluation should be undertaken are effective alternative methods for determining which patients should undergo a complete evaluation of language.

3-4.1. Comprehensive Tests of Language Functioning

Porch Index of Communicative Ability (PICA; Porch, 1971, 1981)   The purpose of this test is to quantify language comprehension, language production, reading, writing, gesture, object awareness, and copying. Spontaneous conversation is not addressed. The PICA is limited as a measure of auditory comprehension (Haber, 1988; Lezak,

1995) and, indirectly, posterior temporal brain dysfunction (Metter, Reige, Hanson, Kuhl, & Phelps, 1984). Because of the PICA's discrete scoring system, the clinician can readminister the PICA over time and measure change as a result of some treatment or therapy. The test is scored using one of the most elaborate scoring systems in neurolinguistic assessment. The scores vary qualitatively and quantitatively according to accuracy, the need for cues, completeness, speed of responding, and whether the patient has motoric limitations. Those who understand the scoring system, and are adept at using it, unanimously agree that it is the most thorough system for describing responses (Haber, 1988). Haber suggests that it is possible to use this scoring system with other more efficient tests of aphasia. This test is not recommended for clinicians who have budgetary concerns because the test itself and the recommended training course are expensive compared to other comparable tests. The PICA is not useful in determining normalcy of language functions because it is not sensitive to mild aphasia (Webb, 1995). Also, the prognostic usefulness of the test for predicting recovery from aphasia has been challenged (Fleming, Hubbard, Schinsky, & Datta, 1982; Webb, 1995; Wertz, Dronkers, & Hume, 1992). The PICA is recommended for clinicians who have a large practice of patients with moderate language impairments as a result of a stroke.

Communicative Abilities in Daily Living (CADL; Holland, 1980)   The purpose of this test is to assess the functional communication skills of individuals with aphasia. Normative data are available for elderly institutionalized and noninstitutionalized populations. In order to administer this test, common objects must be collected or purchased (e.g., a white laboratory coat, a stethoscope). The test offers a single score of functioning that can be compared to a cutoff score. There are also 10 categories for which scores can be derived, but the number of items in each category is either too few to draw meaningful conclusions, or the categories are too broad (e.g., reading, writing, calculating) or too obtuse for interpretation (e.g., performance of sequenced behavior, recognition of cause-effect relationships). The results of this test might be best thought of as a general indicator of normalcy of communication and a screening tool for further exploration of communication problems (Monroe, 1985). It is recommended that the results of the testing be discussed with others with whom the patient interacts to validate the findings.

Boston Diagnostic Aphasia Examination (BDAE; Goodglass & Kaplan, 1983) The BDAE provides diagnostic information about the normalcy of language functions and aphasia syndromes (e.g., Broca's aphasia). There is a revised version (1983) of the original (1972); the actual test revisions, however, are minor. Some items have been reordered, and the organization of the test record booklet has been changed. The new version of the BDAE allows for scaled scores to be represented in percentiles, and normative data is also included. The older version only provides z-scores, which can be compared to z-score profiles of aphasia syndromes illustrated in the test manual. The BDAE is highly recommended for any clinician who sees aphasic patients. It is frequently cited in the literature; therefore, the results obtained from any one patient can often be compared to published case studies and research on groups of patients. In addition, it is relatively inexpensive and easy to store because only a test manual, test booklet, and a set of stimulus cards are required. Normative data for individual subtests allow examiners to give them separately as needed, which may account for the battery's popularity (Abeles, 1995; Lezak, 1995). Van Demark (1982) reported that posttreatment BDAE scores could be predicted with three subtest scores from the BDAE: Confrontation Naming, Complex Ideational Materials, and Body Part Identification. Brookshire and Nicholas (1994) found that by using additional measures of connected speech (in addition to the BDAE's cookie theft picture) higher test-retest correlations could be obtained in measures of words per minute, correct information units (CIUs) per minute, and percent correct information units (calculated by dividing the number of CIUs in a speech sample by the number of words in the sample). Supplementary stimulus items that Brookshire and Nicholas (1994) tested were the picnic scene picture from the Western Aphasia Battery, other action pictures, and open-ended questions, such as "Tell me where you live and describe it to me."

Western Aphasia Battery (WAB: Kertesz, 1982)   The WAB is a comprehensive test of language that allows clinicians to classify aphasic patients according to aphasia syndromes and to determine normalcy of language. An overall score of severity of language impairment (i.e., the Aphasia Quotient) can also be derived. In addition, scores on tests of reading, writing, and cognitive measures can be obtained. Common objects (e.g., screwdriver, matches) must be collected or purchased. Two translations of earlier versions of the battery are available in Spanish and French (Lezak, 1995).

The WAB is similar to the BDAE in that patients can be classified to an aphasia type or syndrome with both measures. The differences are that the WAB is clearer in its directions and easier to score than the BDAE; also, the stimulus items of the WAB will be more familiar to most patients than the stimulus items of the BDAE. (Compare the picnic scene picture to the cookie theft picture.) On the other hand, the BDAE contains more challenging items to test auditory comprehension than the WAB. Therefore, when the examiner aims to capture subtle auditory comprehension problems, the BDAE is recommended over the WAB. Both the WAB and the BDAE incorporate ratings of fluency based on spontaneous speech samples. Trupe (1984) studied the reliability of the spontaneous speech ratings on the WAB and describes an improved scoring system for the content in spontaneous speech. Trupe (1984) reported, however, poor reliability in judging fluency of speech, and recommends the BDAE rating of fluency which, unlike the unitary scale of fluency on the WAB, is rated with multiple scales.

Minnesota Test for Differential Diagnosis of Aphasia (MTDDA; Schuell, 1965)
The MTDDA is a comprehensive test of aphasia and was developed by one of the most prominent aphasiologists of the 20th century, the late Hildred Schuell. The tests are grouped into five sections: tests for auditory disturbances, tests for visual and reading disturbances, tests for speech and language disturbances, tests for visuomotor and writing disturbances, and tests for disturbances of numerical relations and

arithmetic processes. The sections have limited value for describing areas of deficit. For example, "repeating digits" and "repeating sentences" are in the "auditory disturbances section," although these tests could also be considered measures of repetition or immediate memory. Individual subtests may provide clinically useful material; for example, sections on using money may be helpful in planning remediation; and indeed, clinicians frequently select individual subtests for administration (Webb, 1995). The test does provide a means for classifying patients; however, the categories, which are based on severity and accompanying symptoms, are idiosyncratic and generally unacceptable (Lezak, 1995).

Multilingual Aphasia Examination (Benton & Hamsher, 1989)   The Multilingual Aphasia Examination is a collection of short tests that evaluates the presence, severity, and qualitative aspects of aphasia disorders. It was developed from its parent battery, the Neurosensory Center Comprehensive Examination of Aphasia (Lezak, 1995). There are nine separate procedures, and along with two ratings of the patient's speech production and written production, 11 scores are obtained. Translations of this test are available in French, Italian, German, and Spanish as well as English, but norms are only widely available in English and Spanish. The battery includes useful measures of single-word spelling in three output modalities: spelling with block letters, written spelling, and oral spelling. Two of the tests, the controlled oral word association test and the Token test, are types of measures commonly used as supplementary tests of language. Individual subtests are useful snapshots, but the overall battery does not include a range of items to assess language comprehension and production along a continuum of complexity—a quality considered important for comprehensive aphasia batteries (Webb, 1995).

3-4.2. Tests of Specific Language Abilities

Jackson and Tompkins (1991) surveyed 58 experienced aphasiologists and found that 98% of the respondents used the Boston Naming Test, 74% used the Word Fluency Test, and 62% used the Revised Token Test. These were the three most commonly used tests of specific aspects of auditory-verbal language and will be reviewed in the following sections.

The Boston Naming Test (BNT; Kaplan, Goodglass, & Weintraub, 1976, 1978, 1983)   The BNT is one of the most frequently used tests for determining confrontational picture-naming abilities in patients with suspected focal or diffuse brain damage. The most recent version contains 60 line-drawn pictures that the patient attempts to name spontaneously; if the response is incorrect, the patient may receive cues. Patterns of errors may be examined to further investigate impaired processes that might underlie the anomia (Kremin, 1990; Strub & Black, 1988). The test can potentially yield a significant amount of information about a patient's language abilities in a short period of time with little effort on the part of the examiner. At the same time, the test is one of the most heavily criticized for the directions the manual provides

(Jackson & Tompkins, 1991; Nicholas, Brookshire, MacLennan, Schumacher, & Por-razzo, 1989), its validity (Jackson & Tompkins, 1991), score stability or test-retest reliability (Jackson & Tompkins, 1991), and the normative data that it provides (Neils et al., 1995). Examiners should refer to normative data in published research articles (e.g., Neils et al., 1995).

3-4.3. Word-Fluency Measures

Word-fluency measures are often used in diagnostic evaluations of neurologically im-paired patients because they are one of the most sensitive to neurological impairment and can differentiate a number of patient groups (aphasic, right-brain-injured, trau-matic brain injury, and Alzheimer's disease) from normal subjects (Adamovich & Henderson, 1984; Wertz, Dronkers, & Shubitowski, 1986). These measures, however, are less useful for discriminating among brain-damaged patient groups (Wertz, Dron-kers, & Shubitowski, 1986). The most frequently used controlled association word-fluency tasks include the animal naming subtest of the Boston Diagnostic Aphasia Examination (BDAE) (Goodglass & Kaplan, 1972); the initial letter (*F, A,* and *S*) word-fluency subtest of the Neurosensory Center Comprehensive Examination for Aphasia (NCCEA; Spreen & Benton, 1977) and the initial letter *(S, T, P,* and *C)* word-fluency measures by Wertz, 1979 (Adamovich & Henderson, 1984). The animal naming subtest of the Western Aphasia Battery is also frequently used and is identical to the BDAE subtest. In relatively mild patients who do not have oral motor impair-ment, fluency tests are efficient measures of documenting one component of language production.

Revised Token Test (RTT; McNeil & Prescott, 1978)   This is a test of auditory processing and comprehension. Some of the subtests also assess auditory memory because they involve lengthy commands (Caplan, 1995; Curtiss, Jackson, Kempler, Hanson, & Metter, 1986). The patient is required to follow commands using small plastic "tokens" that are of different colors and shapes. This test also employs essen-tially the same scoring system as the PICA; thus, it shares this strength. The standard Revised Token Test was designed to improve on older versions of the Token Test by increasing the number of items and by standardizing the procedures. This is a long test, especially in light of the fact that it is a specific measure of auditory processing. In an attempt to make the testing procedures of the RTT more efficient, Arvedson, McNeil, and West (1985) examined shortened forms of the RTT and found that "the five-item mean overall scores are so close to the standard RTT that a direct substitution can be made" (p. 62). With the five-item procedure, the examiner only administers the first five items of each of the 10 subtests. It would be useful to compare the reliability and validity of the shortened version of the RTT proposed by Arvedson et al. (1985) with older versions of the Token Test. Both of these shorter versions are recommended over the administration of the complete Revised Token Test, except in cases where auditory processing is the primary or only suspected impairment.

3-5. SUMMARY AND FUTURE OUTLOOK

Clinical assessment of language in an adult patient may be conducted to learn more about language disorders, to provide a service to the patient, or both. Standardized batteries and specific tests of language should be selected according to the rationale(s) for testing. The patient-centered rationales for testing include determination of the normalcy of language, the cause of the language impairment, definition of the language deficits and abilities, and the patient's ability to communicate in everyday situations. Definition of the patient's language deficits and abilities may be viewed according to type of aphasia syndrome, severity of aphasia, and an information-processing approach. A review of the most commonly used standardized tests of language suggests that word-fluency measures, the BNT, the Revised Token Test, and the BDAE are best at determining normalcy of language functions. These same tests, along with tests of memory, would be most useful for differential diagnosis of primary progressive aphasia and Pick's disease from normal language functioning and early Alzheimer's disease (Kertesz et al., 1995). The majority of the standardized language batteries are useful for defining aspects of language production and comprehension. The WAB and the BDAE are most useful for defining aphasia syndromes although future modifications of these batteries are required to resolve conflicts in the disparity in syndrome labels assigned. Along with these batteries, the PICA and the CADL provide severity ratings, but they fail to capture mild language impairments. Other batteries, such as the Boston Assessment of Severe Aphasia and the Assessment of Communicative Effectiveness in Severe Aphasia, may prove to be more effective in defining the language abilities of severely impaired patients than other tests of aphasia, although comparison studies are needed. Beyond a description of language deficits and abilities, the clinician should obtain a measure from which the patient's ability to function in everyday situations can be inferred. Although the CADL was developed for this purpose, clinical impression (Monroe, 1985; Lezak, 1995) suggests that it falls short. Discourse comprehension may be a better tool for inferring the patient's ability to function in everyday situations; however, research is needed to test this hypothesis. Finally, in all future research, the patient population that a test or normative data will be used for needs to be considered. Such factors as age, sex, culture, languages spoken, and education have all been shown to affect test performance in a profound way.

CHAPTER 4

Research Strategies
Psychological and Psycholinguistic Methods in Neurolinguistics

Chris Westbury
Center for Cognitive Studies, Tufts University, Medford, Massachusetts 02155

Neuropsychological research in the past century has increasingly relied upon a functional definition of language processing whose broad outlines are now widely accepted. Neurolinguistic research within this functionalist framework has three main goals: to enumerate and classify the fundamental components of the human language system (mainly by studying damaged language systems); to study how those units normally function; and to model our understanding of the complex process of language use in a compact and comprehensible way. In this chapter, examples of various common methods for achieving these goals are presented and discussed.

What is language? What are its natural components? How are those components structured, interconnected, and processed? How is language structure instantiated and processed in the brain? How does that structure develop? How does it support the different input and output modalities of language? What rules or principles determine how language processing operates on the structure? How is that processing affected by extralinguistic resource demands? How is it affected by neurological damage?

These are some of the questions addressed by neurolinguistics. In order to answer such a broad spectrum of questions, neurolinguists have developed an equally broad spectrum of research methods. In this chapter we will consider those methods that have been most commonly or most fruitfully employed in the recent history of neurolinguistic inquiry. Before enumerating these methods, it will be useful to consider the neurolinguist's answer to the first question: what is language?

4-1. WHAT IS LANGUAGE?

This misleadingly simple question has been a topic of debate for centuries. Neurolinguists today content themselves with a functional working definition of language, that is, with a definition given in terms of how language is processed by normal human beings. Such a definition completely ignores some important aspects of language, including its cultural and social aspects. By concentrating mainly on the study of sublexical and lexical access, current functionalist definitions of language also tend to greatly downplay (though they certainly do not totally ignore) the importance of even such defining aspects of language use as grammatical syntax and semantic reference. Although the fine details and precise organization of a complete functional linguistic system are still unknown, there is a broad consensus within the field of neurolinguistics regarding its general outlines. Almost all researchers who use a functionalist definition of language would agree that the outline will look something like the hierarchical outline in Table 1 (for similar classification schemes, see Caplan, 1992; Howard & Franklin, 1988; Patterson, 1986; Roeltgen & Rapcsak, 1993; Shallice, 1988).

At the top level of the classification, there is a distinction between the two main language modalities: spoken and written language. Within each of the modalities, language breaks down roughly into input and output functions, although this distinction is more complex than it might appear at first pass. The input and output classes may each be broken down into two subhierarchies. Both classes include one subhierarchy linking that modality to semantics. The second subhierarchy consists of a set of components that function to access the elements of language defined by a hierarchical deconstruction of the input and output streams relevant to that modality, into sentences, words, and letters in the written modality, and into sentences, words, and phonemes in the auditory modality. Both of these subhierarchies may themselves be further decomposed, as outlined in sections A and B of Table 1. Note that some of the components (marked with a black dot) serve a special role, which is translating between input and output, or between the written and oral modalities.

This hierarchically structured definition of language does not capture all aspects of language that are of interest to the neurolinguist. Special consideration of some particular word types is required to deal with categories of words for which there is some evidence of bimodal special treatment, including (but not limited to) affixed words (Miceli & Caramazza, 1988), abstract words (Breedin, Saffran, & Coslett, 1994), low-frequency words (Monsell, Doyle, & Haggard, 1989; Seidenberg, Waters, Barnes, & Tanenhaus, 1984), and, more controversially, words from particular semantic categories (Warrington & Shallice, 1984). Neither the functional status nor the correct placement within the functional hierarchy of the resources devoted to processing these

TABLE 1
A Functional Definition of Language

A) Auditory-oral modality
    a) Input/comprehension
        i) Semantic access
           1) Words
              Word-picture matching; Synonym judgment
           2) Sentences
              Pragmatic comprehension of simple sentences
        ii) Access to language elements
           1) Phonemes
              Phoneme discrimination
           2) Words
              • Repetition of word and nonwords; Auditory lexical decision
           3) Sentences
              • Sentence repetition; Syntactic comprehension of sentences
    b) Output/production
        i) Semantic access (as above, in the written modality)
        ii) Access to language elements
           1) Words
           2) Sentences
B) Written modality
    a) Input/comprehension
        i) Semantic access (as above, in the written modality)
        ii) Access to language elements
           1) Words
              • Reading aloud; Written lexical decision
           2) Sentences/grammatical structure
              Written sentence comprehension
    b) Output/production
        i) Semantic access
           1) Words
              Naming; Definition/synonym production
           2) Sentences
              Syntactic Production
        ii) Access to language elements
           1) Words
              • Written copying of single words; • Writing words to dictation (Spelling)
           2) Sentences
              • Written copying of sentences; • Writing sentences to dictation; Written sentence
                production
    c) Word types
        i) Affixed versus root words
        ii) Abstract versus concrete words
        iii) Low frequency versus common words
        iv) Category-specificity (e.g., organic/inorganic)
        v) Function versus content words
        And probably others not yet identified

Note: Components marked with a black dot serve a special role, translating between input and output, or between the written and oral modalities.

word types is yet clear. It may be speculated that the observed dissociations in processing these word types are due to quantitative rather than qualititative disparities of the representation. This speculation finds some support from Warrington and Shallice's (1984) hypothesis that the distinction between biological and nonbiological kinds is dependent on the degree to which each class relies on functional versus perceptual representations.

4-2. ORIGINS OF THE NEUROLINGUISTIC MODEL

The functional definition of language, as outlined in Table 1, has evolved from a research program that arose in response to the failure and subsequent rejection of the aphasic taxonomic systems developed in the nineteenth century. Although the details behind the hotly debated rejection of those classification systems need not concern us here, that failure had three related results that are of direct relevance in understanding modern neurolinguistic research methodology. The first result was that neurolingistics was forced to return to "first principles," redefining itself (beginning in the late 1970s) with a research program that focused on developing an abstract and detailed description of the way language breaks down as a result of neurological injury or degeneration. The second result was that the study and classification of language deficits in and of themselves fell out of favor. Research efforts began to focus increasingly on relating the language behaviors observed following neurological injury to the normally functioning language system (see Marshall, 1986). This shift of focus was extremely fruitful methodologically, as researchers studying language began to borrow and adapt methods from other subfields of experimental psychology. They were aided in this task by the third result of the rejection of nineteenth-century classification systems: the rise of computational theories of psychological function, influenced by the widespread adoption in psychology of the information-processing paradigm as a metatheory for psychological theorizing (see Gardner, 1985). The rise of computational theories has a great many implications that have impacted on the evolution of the field's methodology (Ellis & Young, 1988; Morton & Jusczyk, 1984; Parisi, 1985; Shallice, 1988). By far the most important of these is the assumption of modularity, which states that large computations should be implemented in small subparts that are as independent as possible within the constraints imposed by the function being computed. The bulk of neurolinguistic research may be aptly characterized as an attempt to confirm and clarify the assumption of modularity, most importantly by cataloging in precise terms what the smallest nondecomposable functional modules of the language system are.

These three results define three intertwined goals of current neurolinguistic research:

1. to map out the space of all possible linguistic deficits;
2. to buttress and extend the explanatory implications of that research with evidence from normal human subjects;

3. to unite the findings obtained from studying normal and brain-damaged subjects with a neurologically plausible model of how language is processed in the brain.

In the remainder of this chapter, we will consider the methods employed in the pursuit of each of these three goals.

4-3. METHODS

4-3.1. Mapping Out the Neurolinguistic Deficit Space

The mapping of neurolinguistic deficit space does not have any necessary priority over the other two goals of neurolinguistics. A model of neurolinguistic functioning or an experiment on normal subjects may reveal something important about the way that language must be instantiated, making a prediction about the way the actual language system must fail as a result. For example, the discovery of separate paths for processing real words and nonsense words was due largely to results from experiments with normal subjects. However, in practice the documentation of linguistic deficits in a clinical population has almost always served to generate the basic scientific facts of neurolinguistics. In this section we will consider four important classes of methods that are in widespread use: direct measures, reaction time measures, measures of stimulus discrimination/confusability, and measures of priming and interference. Before we consider these four types of clinical research methods, we should note in passing that there exists a major methodological debate among neurolinguistic researchers who study patient populations: the debate over the status of group studies (in which data from patients deemed to have the same syndrome is averaged together) versus single-case studies (in which individual patients are described in detail, often without bothering to classify their "syndrome"). Although this complex and subtle issue is of considerable importance in understanding current neurolinguistic methodology, we will not address it here (see Willmes, this volume).

4-3.1.1. Direct Measures

It might seem that the documentation of neurolinguistic deficits should be as straightforward as simply asking a patient to perform a linguistic task and then recording whether or not he or she can do it. There would appear to be little room for debate about whether or not such a measure might be a good one, since (unlike many other measures in psychology) the relation of the measure to the function being measured is not open to dispute. However, in practice, testing a language function is never as simple as merely asking a patient to perform a language task and scoring his or her performance. There may be any number of explanations for a patient's low score on a particular task, since it is impossible to find words that have exactly one feature or tests that test precisely one function. As a result, no single language test can be definitive. A great deal of research effort in the field has been (and is still being) devoted to defining the optimal battery of mutually controlling tests of language function. A neurolinguist working in English has a choice of a multitude of aphasia batteries. (For information on language tests and batteries in current usage, see Spreen & Strauss, 1991; Lezak, 1995; and Neils-Strunjaš, this volume).

Because no test battery can ever be absolutely conclusive or exhaustive, directly documenting the linguistic deficits of a single patient is a necessarily complex and time-consuming task. Although the precise methodological details of the documentation process will vary from case to case, the process will inevitably involve duplication of measurement, and carefully controlled elimination of confounding variables, such as the effect of word type or modality. Often, this time-consuming but preliminary analysis will be followed by the creation, normalization, and study of patient performance on tasks specific to the individual patient or set of patients under examination. As well as comparing performance to matched controls or to the subject's own performance on other tasks, many studies are devoted to the comparison of a subject's performance on language tasks to performance by other subjects who are of interest either because they come from a known diagnostic group, or a different language group, or (more controversially) because they have a known lesion site. Excellent examples of how to undertake a systematic detailed study of a linguistic deficit are given by Howard and Franklin (1988) and Caramazza and Hillis (1989b). It is mainly these kinds of painstakingly detailed descriptions of patient deficits which have contributed the raw data which has been used to construct models of language processing.

When used carefully, with appropriate controls and with comprehensive coverage of the language system, direct measures allow for excellent description of deficits at a high level that is ecologically valid. In other words, they are useful for describing *what* aspects of the system may be damaged. However, they are of minimal use in explaining exactly *how* the language system is failing at a mechanistic level. They provide no insight at all into the actual process of language production or comprehension, since by their very nature they can only report on the language system retrospectively, after processing is completed. The remaining methods we will consider in this section each have the advantage of being aimed at a lower level of analysis.

4-3.1.2. Reaction Time Measures

One lower-level measure that is used very often with patient populations in contemporary neurolinguistics is reaction time, a precise measure of the time it takes for the patients to produce a response on a task. Reaction time is commonly used as an indirect measure of task difficulty, on the assumption that a task that requires more processing will take a longer time to complete. Differences in reaction time between normal and patient groups may shed light on both the role of specific brain areas and on the functional deficits associated with particular identified groups. Reaction time measures have been used to deduce deficits in such diverse functions and subject groups as lexical decision in closed head injury patients (Münte & Heinze, 1994), syntactic processing in an aged population (Obler, Fein, Nicholas, & Albert, 1991), and interhemispheric transfer time in reading-disabled children (Davidson, Leslie, & Saron, 1990).

Reaction-time measures are more sensitive than direct measures, in the sense that reaction-time differences in processing different types of linguistic stimuli may be documented even when there are no differences in score. Documentation of such differences suggests that the different stimuli types must be represented and/or processed in different ways. In this way, reaction-time data have the great advantage of reflecting information about the true underlying representation of a stimulus, though they would of course be misleading in the unlikely case that two different processes coincidentally took the same time, at the level of measurable difference, to run their course. Technically, chronometric data can be difficult to gather (see Sergent, 1986), since the computers that are typically now used to gather such data are subject to technical limitations that can muddy the data, especially if those data depend on very subtle differences. Another complication is that many disease processes (as well as normal human aging) can impact on reaction time, confounding its use as a measure of language processing.

4-3.1.3. Discrimination Measures

One way around some of the limitations of reaction-time data is to gather chronometric information relevant to the timing of language functions without requiring a reaction from the subject, by asking subjects to make distinctions between stimuli that differ in subtle temporal characteristics. This technique has been used to show that some aphasic patients have disturbances in very low-level perceptual categories (Blumstein, Baker, & Goodglass, 1977) and to show that phonemic processing deficits may be related to lower-level auditory processing deficits (Aurbach, Allard, Naeser, Alexander, & Albert, 1982).

Measures that rely on stimuli discrimination not only have the advantage over reaction times of allowing for the gathering of chronometric data that is independent of motor-reaction time, but also allow researchers to test for ability to make very fine, very low-level distinctions that might be difficult to test using other methods. Subtle differences in phonemes or click rates might not be apparent if one did not hear the stimuli in rapid succession.

Asking a subject to differentiate between ambiguous or confusable stimuli can be informative for purposes other than testing fine chronometric distinctions. Since confusability rates can serve as a measure of similarity, confusion data have one major advantage over other types of data: if one can gather enough repeated measures, it is possible to process such data with multidimensional scaling packages in order to derive information about the abstract organization of the stimuli within the language system. Arguin, Bub, and Dudek's (1996) documentation of the relation between stimuli complexity and semantic organization in a patient with category-specific anomia for fruits and vegetables was inspired by their analysis of which fruits and vegetables that patient confused with other fruits and vegetables. Methodologically analogous work has been done in studying rates of confusion of phonemes in aphasic patients (see Caplan, 1992, p. 53, for a review of this literature). One major weakness of such statistical representations of confusability data is that it is not possible to gain much information about the actual neuronal representations (as opposed to the abstract organization) of the confused stimuli. Representing data in a non-Euclidean multidimensional space does not tell us how or why the brain has instantiated that particular map, any more than our ability to describe planetary motion using calculus tells us how gravity works.

4-3.1.4. Measures of Priming and Interference

Another technique in widespread use is priming, in which exposure to one stimulus is used to facilitate or (in the case of negative priming) interfere with access to a second stimulus, which is related to the first one either lexically, syntactically, or semantically. The amount of facilitation or interference (often assessed by measuring differences in reaction time to the second stimulus) is used to glean information about the way the target words and the primed words are organized. Facilitative priming tasks have been used, for example, to document disruption in lexico-semantic access in quite different ways in patients with Huntington's disease (Smith, Butters, White, Lyon, & Granholm, 1988) and Alzheimer's disease (Chertkow, Bub, & Seidenberg, 1989; Chertkow, et al., 1994). The paradigmatic example of a negative priming (interference) effect is that seen in the Stroop task. In that task, color names were shown to interfere with access to the color in which those names were printed, suggesting that the lexical and perceptual semantic systems do not function independently of each other. By comparing the extent of the interference between analogous tasks involving systematic interference, it is possible to deduce how tightly coupled two functions are, and how that coupling is affected by different neurological conditions. For example, Revonsuo (1995) studied an aphasic patient who was unable to match written color names to the corresponding color, and demonstrated that he nevertheless showed a normal Stroop effect, suggesting that there can be implicit processing of unrecognized words. Lackner and Shattuck-Hufnagel (1982) used another kind of interference task (immediate repetition of words presented at varying rates, called "shadowing") to demonstrate that penetrating wounds of the left cerebral hemisphere could leave long-term language deficits even if none were apparent in normal discourse. Many other variations on manipulating the difficulty of an interfering task are possible. An original use of the general method is exemplified by Arguin, Bub, and Dudek's (1996) experiment, in which they presented evidence showing that the increasing number of dimensions of variation of a class of stimuli interfered with the ability of an agnosic patient to access semantic information about that class.

Experiments that rely on priming or interference measurement can only give us information about the time course and the extent of interference between competing cognitive processes or representations of different stimuli. They cannot, in themselves, tell us much about what those processes and representations are, how they function, or exactly why they prime or interfere with each other. There is currently no clear understanding of exactly what it means, at a neuronal level, for one psychological process to facilitate or interfere with another. Interpretation of the functional significance of priming and interference effects must therefore be made with caution.

4-3.2. Obtaining Evidence from Normal Human Subjects

Testing performance on language tests is of minimal use in studying language organization in normals, since normal adults tend to show a ceiling effect on most of the subtests that are commonly used with patient populations. Research on normals thus depends largely on studying the effect of indirect measures that perturb the language system in some systematic way. Many of the measures discussed in the previous section are also commonly used in studying normal language organization. In many cases, information may be gleaned from using such measures to compare two different normal populations, such as unilingual speakers of languages that differ in theoretically interesting ways (for example, Katz, Rexer, & Lukatela, 1991) or subjects from different age groups (for example, Gropen, Pinker, Hollander, & Goldberg, 1991). The number of experiments that have been conducted on normal language processing is far too vast to allow them to be described in any detail here. Reviews may be found in Foss (1988) and Gernsbacher (1994).

4-3.2.1. Natural Errors

One rich source of information about language processing in a normal population is the study of naturally produced errors and hesitations. As Meringer and Mayer put it over a century ago, by such study "the cover is lifted from the clockwork and we can look in on the cogs" (cited in Aitchison, 1987, p. 18). Much research on children's malapropisms suggests that such errors are produced in a rule-based fashion (for example, Gropen et al., 1991; see Aitchison & Straff, 1982 for an analysis of how adult errors differ).

The study of natural language is important because it serves to constrain the set of possible models of language organization. Consistency in the kinds of errors shown by children and adults suggests that such errors may be revealing something of the fundamental architecture of language. Inconsistencies across development reveal something about how the language system evolves over a human life. The study of natural language errors is difficult and expensive, however, as the necessary data are rare and time-consuming to collect.

4-3.2.2. Other Methods

The difficulties inherent in studying language processing in normals have led to the recent development of a number of innovative measures for studying language processing. For example, advances in eye tracking have been used (e.g., Just & Carpenter, 1993) to measure a multitude of new variables (including the likelihood that a subject has fixated on a word, the fixation duration, the number and kind of regressive eye movements, and the total time devoted to examining a word) and recent advances in electrical and magnetic recording devices allow for the measure of language event-related potentials (ERPs) (e.g., Nobre & McCarthy, 1994, 1995; Segalowitz & Chevalier, this volume) and evoked magnetic fields (EMFs) (see Papanicolaou, Simos,

& Basile, this volume). With continuing advancements in technology, we can expect that the number of unusual dependent measures being evaluated for their relation to language processing will increase in the coming years. Much work will be necessary to determine what they may reveal about the principles underlying language organization.

4-3.3. Building Neurologically Plausible Models

The last of the three goals of neurolinguistics under discussion is model generation. As they do in any science, models in neurolinguistics serve two main roles, both as mnemonic devices to represent complex data sets in ways that make those data sets humanly comprehensible, and as explanatory devices that relate data from one domain to a different domain that is either more general, better-understood, or more easily amenable to empirical study. The theoretical level at which neurolinguistic models are defined ranges widely, from very abstract, structured "top down" text-based models, to detailed dynamic "bottom up" computational simulations.

The majority of models in neurolinguistics serve the first role, not so much adding new data to our understanding as helping to organize the data that underlie that understanding. "Box and arrow" models have played a fundamental role in the recent history of neurolinguistics, mainly as mnemonic devices. A well-known example is Patterson's (1986) model of single-word access, which instantiates a great deal of information about the role that various abstract components have been inferred to play in word access, and how those components must be connected given the observed pattern of breakdowns in single-word access. However, the model makes no claims about how the individual functional components might work, or about how they might be instantiated in the brain, and makes no explicit predictions about new ways in which the language system might fail.

At the opposite end of the modeling spectrum—and increasingly important as a methodological tool in neurolinguistics—are the connectionist models, which use artificial neural networks to simulate language behavior (see Besner, Twilley, McCann, & Seergobin, 1990; Harris & Small, this volume; Harley, 1993; Reilly & Sharkey, 1992; Seidenberg, 1993, 1994 for overviews). Connectionist models have been used to model a diverse set of language phenomena, including the Stroop effect (Cohen, Dunbar, & McClelland, 1990), category-specific agnosia (Farah & McClelland, 1992), and naming errors in optic aphasia (Plaut & Shallice, 1993).

Many published models fall somewhere in the range between fine-grained dynamic connectionist models and high-level static models. For example, Friederci (1995) proposes a static theoretical model of syntactic activation that is constrained by experimentally gathered neurophysiological timing evidence. There is no current consensus in the field about which kind of models are most appropriate given the current state of knowledge (see Gupta & Touretzky, 1994; Besner et al., 1990 for dissenting viewpoints on this issue).

Models may be a mixed blessing in neurolinguistics. It is possible to build computational and conceptual models that can accurately reproduce phenomena without being sure that the models have captured anything essential about the way those phenomena are produced naturally: Compare Farah and McClelland's (1988) model of category-specific agnosia to the very different explanation of that phenomenon offered on the basis of experimental evidence by Arguin, Bub, and Dudek (1996). So litttle is known about the underlying representation of language in the brain that it is often very difficult to assess the validity of assumptions and parameter settings in models of neurolinguistic phenomena. Nevertheless, modeling serves an important role by producing "existence proofs" of how language might be represented. Such models serve as hypotheses that can guide experimental research directed at testing the implicit or explicit assumptions that went into a successful model.

4-4. CONCLUSION

The literature on psychological approaches to neurolinguistics is now so large that a thorough review is no longer possible. The cursory overview provided in this chapter may help to convince the reader that the fundamental methods serve one of three main purposes. The most important purpose at the current stage of knowledge—the sine qua non of all neurolinguistic research—is to exhaustively list and classify the fundamental components of the human language system. Progress is being made toward this end mainly by a carefully controlled documentation of ways in which the language system can fail. As neurolinguists gain increasingly fine-grained knowledge of the fundamental units of the language system, we can achieve our second purpose, which is to study how those units act and interact in the normally functioning human being. This is achieved mainly by perturbing the normal system in various ways, while measuring the effect of such perturbation on the fundamental linguistic functions that have been previously identified. Finally, as we understand more about the language system, we attempt to consolidate, extend, and demonstrate our understanding by constructing models that summarize our knowledge in a compact way, or in a way that reproduces observed real-world linguistic phenomena. Although the methods of neurolinguistics will certainly be extended and refined in the coming years, these three fundamental goals will surely continue to guide the field.

CHAPTER 5

Event-Related Potential (ERP) Research in Neurolinguistics: Part I

Techniques and Applications to Lexical Access

Sidney J. Segalowitz and Hélène Chevalier

Department of Psychology, Brock University, St. Catharines, Ontario, Canada L2S 3A1

The past decade or so has seen a very large increase in the use of event-related potential (ERP) techniques to study the brain's responses to the processing of linguistic information. In this chapter, we discuss some of the strengths and weaknesses of ERP techniques in neurolinguistic research and the ways they have been applied to brain theory. We also review specific ERP paradigms addressing questions of lexical access, word-class distinctions (open versus closed class and noun versus verb), and phonological access.

During the 1970s and 1980s, event-related potential (ERP) techniques were greatly expanded to include investigations beyond primary sensory coding and simple attention. Coupling this with developments in the microcomputer and electronics fields that dramatically reduced the costs of an ERP laboratory, many new investigators from a variety of fields joined with electrophysiologists to broaden the field. Over the past decade or so, there has been an enormous growth in ERP studies of psycholinguistic

95

issues alone. A summary from the PsychLit abstracts of all papers published from 1986 to 1995 for the key words "language" and "ERPs" (and their synonyms) produces more than 350 journal articles and 45 book chapters. Many of these make use of language-related ERPs to study other phenomena, such as memory or psychopathology, or the characteristics of ERPs themselves. Still, there are far too many to provide an exhaustive review of the literature. Our purpose here is to outline the ERP paradigm and the kinds of neurolinguistic and psycholinguistic questions being addressed in scalp-recorded ERP paradigms with a focus on research since 1990, and some of their strengths, weaknesses, and possible future directions. In this chapter, we will also describe work on lexical and phonological access. In the next chapter, we will give a review of ERP studies of semantics, syntax, reading, developmental, and second language issues. More detailed reviews of specific research questions and methodological issues can be found elsewhere (Garnsey, 1993; Hagoort & Kutas, 1995; Halgren, 1990; Osterhout & Holcomb, 1995; Kutas & Van Petten, 1988, 1994; Metz-Lutz, 1995; Pulvermüller, 1996).

5-1. OVERVIEW OF THE ERP PARADIGM

5-1.1. The EEG Signal and the Theory of Averaging

The EEG records the fluctuating voltage changes at various scalp sites with respect to some common reference location, often the earlobes or an average of all the scalp sites. The scalp voltage reflects primarily synaptic activity on the dendrites of cortical pyramidal cells because pyramidal cells align their dendrites vertically and this allows the voltage potential to summate (Martin, 1991). Because of this parallel arrangement of pyramidal dendrites, the scalp potential may reflect the activity of these connections in their cortical columns (Picton, Lins, & Sherg, 1995). Thus, EEG does not reflect presynaptic activity (that is, axon potentials), nor much of the activity of nonpyramidal neurons whose dendrites are not aligned. Therefore, we should not think of the EEG as a full reflection of the brain's processing of information. In this way, the EEG does not reflect the same activity as do metabolically related brain-imaging systems such as PET, SPECT, or fMRI.

When EEG is recorded to a series of events, we refer to these EEG segments as event-related potentials. Normally, we group such ERPs to many similar events into an overall average for each subject. In this chapter, we will use the term "ERP" to refer to these averaged waveforms. Averaging the EEG across many events reduces the effect of the nonevent-related EEG, that is, it averages out other ongoing activation, so that the waveform that remains is more likely to reflect the effects of the particular information processing induced by the stimulus event. An example of a standard ERP waveform is given in Figure 1, where the stimulus was a series of easily identified high-pitch tones. The ERP is characterized by a series of positive and negative components (or peaks) that we have learned, after many years of study, reflect somewhat specific aspects of information processing, the earlier ones more likely to reflect the automatic aspects of processing the stimulus and the later ones more likely to reflect processing contingent on the subject's cognitive strategy (for overviews, see Polich 1993, and Picton & Hillyard, 1988). Each component often actually reflects a number of subcomponents that can only be teased apart by careful experimental designs and by examination of the pattern of ERPs across the scalp. This pattern across the scalp is referred to as the scalp topography, and it is sometimes crucial for determining the interpretation of the ERP components. Thus, it is very important to know the testing paradigm that the ERP comes from to fully evaluate the result. Just as it is important to know the physical challenge on the body to evaluate cardiovascular data, we must know the cognitive challenge to evaluate the neuroelectric result. (A note on terminology: Negative and positive peaks are designated N and P followed by a number. Single digits refer to the sequence of first, second, etc., peaks and do not reflect the timing, for example, P1, N1, P2, N2, P3, while others indicate the timing in milliseconds from the onset of the stimulus, for example, N280, P600. Sometimes the initial specification of timing has been recently dropped in favor of a sequence number because a component has become well accepted, and so a peak can have two labels, for example, P3 and P300, N4 and N400.)

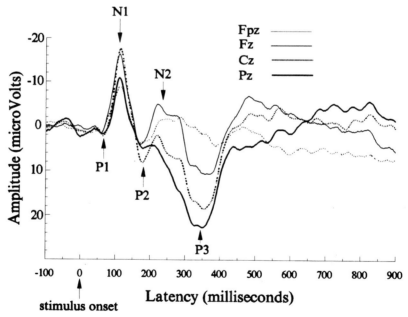

FIGURE 1   Auditory ERPs averaged over 49 trials taken to target tone stimuli in a discrimination task from a subject who pressed a key when the target was heard. The major peaks P1, N1, P2, N2, and P3 are labeled, as are the four electrode sites Pz, Cz, Fz, Fpz, which indicate standard parietal, central, frontal, and frontal pole midline locations. Note that the amplitude of the P3 peak is greater the more posterior the site, whereas this is not the case with earlier peaks. Negative is up. (Unpublished data from the authors' laboratory.)

5-1.2. Benefits and Limits of Using ERPs in Linguistic and Neurolinguistic Research

There are a variety of benefits of ERP technology over standard behavioral methods in investigating psycholinguistic phenomena with respect to brain structures and functions. First of all, it is one of the few noninvasive ways, and certainly the least expensive, of getting fairly detailed measures of brain activity as they relate to information processing, whether in normals or with those with some brain disturbance (Gevins, Leong, Smith, Le, & Du, 1995). In addition, ERPs allow examination of online processing with respect to a mental chronometry, so that with ERPs we are in a position to make some (often cautious) statements about the timing of various cognitive processes (Jennings & Coles, 1991). ERPs also permit measurement of online brain activity in contexts where behavioral output is problematic, for example, with respect to speech or language parsing with aphasic patients and with infants. Whereas other brain-imaging techniques (such as PET and fMRI) are more suitable for localizing activity in the brain, ERP technology provides more flexibility in terms of stimulus presentation and more resolution in the time domain. In addition, there have been dramatic improvements in methods for spatial localization using ERPs (Gevins et al., 1995; Tucker, Liotti, Potts, Russell, & Posner, 1994), that is, for determining the physiological sources of ERP components.

ERPs also have inherent limitations, which are circumvented only with a good deal of ingenuity and organization. First of all, we must remember that ERPs are responses to time-locked stimulus presentations, with millisecond accuracy, and they are not appropriate for circumstances where a longer-term mental state is of interest, such as over minutes. Several methods have been developed to bypass this limitation, however, by using electrophysiological methods that are indirect with respect to the stimuli. One such technique involves ERP probes, which are averaged waveforms that are time-locked to some unobtrusive set of events, such as tones or light flashes, which do not require a response from the subject, who is engaged in some cognitive task of interest such as identifying phoneme sequences in one condition and emotional tone of voice in another condition (for example, Papanicolaou & Johnstone, 1984). By examining the pattern of how the ongoing probe ERPs alter as a function of the experimental condition, for example, listening for phonemes or emotions, one can draw conclusions about which scalp sites seem sensitive to the distinction (and therefore presumably which brain areas are involved). Another such technique is spectral analysis, which breaks down the ongoing EEG signal into frequencies from very slow to very fast (delta, theta, alpha, beta, and gamma waves). Again, by examining the pattern of changes associated with the experimental conditions, one can draw conclusions about the relative sensitivity of various sites to the experimental manipulation (Pulvermüller, 1996). These alternative methods, however, lose much of the richness of the standard ERP paradigm because they do not provide a mental chronometry of the stimulus-processing event.

Another practical limitation in the collection of ERP data is the requirement of having the subject suppress motor actions that mask the cognitive ERP, especially eye movements. Various techniques have been used (Picton et al., 1995), the most common of which is for the subject to withhold blinks and responses until after the stimulus is presented. The only limitations here are that normal behavior is restricted, and of course this may not be possible with all subjects, for example, those with neurological disorders that produce uncontrollable movements, and sometimes with children. Although these limitations in the ERP techniques are surmountable with effort, comparisons across laboratories are sometimes difficult because of different choices of how to deal with the problems.

There are other limitations that we should be aware of. The ERP is not an absolute reflection of the brain's electrical activity but is a reflection as seen through the particular data-collection equipment. The particular montage of electrode sites influences the pattern of ERP components seen in the signal. Similarly, the choice of reference site influences the ERP morphology. Also, the characteristics of the amplifiers determine the limitations of the signals recorded: The frequency cutoff, determined at the low end by the time constant of the amplifiers, affects the degree to which the system is sensitive to the slow wave components of the ERP. For example, slow-wave activity is attenuated unless the amplifiers allow very low frequencies to pass through. Cutoffs at the upper end smooth the signal, making the ERP easier to score and less "noisy," but also delay somewhat the peak latencies of the major slow components, such as the P300, and can completely mask components of short duration. Although these amplifier parameters are usually held constant within laboratories, they often differ across laboratories and thus we acknowledge that there will be differences in research results between research groups for reasons that are not related to experimental factors.

In addition, different researchers have preferences for different analysis techniques, the two prominent ones being a search for particular well-studied components, such as the P300 or N400, versus the search for whatever segments of the data differentiate the independent variables of the study. The latter can be done by careful visual comparison of the critical experimental with appropriate control conditions (with subsequent statistical follow-up), or by the Principal Components Analysis-Analysis of Variance (PCA-ANOVA) or similar technique that searches for components that account for major portions of the variance in the ERP and then relates these to the independent variables. There are pros and cons to each approach, and we must at least be aware of data analysis as a potential source of variation across studies (Segalowitz & Berge, 1995).

5-2. RELATIONS BETWEEN ERPS AND LANGUAGE BEHAVIOR

An overview of this literature leads us to recognize two uses of ERPs in neurolinguistic research. First, ERPs can be used as a dependent variable to test hypotheses about language processing, language distinctions, and language development or acquisition, just as behavioral psycholinguists might use word knowledge, lexical access response times, and sentence production. In such a case, electrophysiological *theory* is less important than it might otherwise be since no serious conclusions of a physi-ological sort are being made. This is not the case with the second use, where ERPs can be used as a tool to examine brain localization of function. In this situation, we must assume that the researcher has a phenomenon well established and only wishes to ask the question, "Can we find dissociable brain sites for this distinction?" In such a case, we must have at least an implicit model of language in the brain.

5-2.1. Organization of *Language* in the Brain

A large number of psycholinguistic questions that have occupied behavioral research-ers have now become amenable to ERP testing. Such questions include basic issues of linguistic analyses: *(a)* the psychological reality of the noun versus verb distinction; *(b)* the content versus function word distinction; *(c)* divisions between syntax and semantics; and *(d)* specific questions of grammar theory. Others involve psycholin-guistic issues of language acquisition: *(e)* comparability of the linguistic competence of first and second languages (L1 and L2) in bilinguals; *(f)* how to assess linguistic competence without behavioral production (for example, in infants and nonspeaking children); and *(g)* longitudinal relationships between very early speech processing and later language acquisition. There are also a large number of specific questions, such as the relationship between word reading and phonological access, and acoustic versus phonological coding. We will address the first two issues in this chapter as they serve to introduce the paradigm. The others we will cover in the next chapter.

5-2.2. Organization of Language in the *Brain*

Neurolinguistic theory has traditionally seen language as represented in the brain in two main centers: an anterior one that deals with word output, phonological planning, and basic syntactic constructions, and a posterior one that deals with lexical access, word meaning, and the linking of lexical semantics to world knowledge. These divi-sions come, of course, from clinical aphasia research, with assumed extrapolation to subjects with normal language and intact brains. ERP studies have allowed us to bridge this gap to some extent by testing brain localization hypotheses in normals. Many of the issues outlined in the previous section have also been addressed with respect to brain localization (when adequate electrode placement has permitted). In addition, there are specific research questions that are amenable to ERP testing, such as the locus of language representation in aphasics who have recovered some linguistic abil-ity (for example Papanicolaou, DiScenna, Gillespie, & Aram, 1990; Papanicolaou, Moore, Deutsch, Levin, & Eisenberg, 1988), but we will not cover them here.

5-2.3. A General Model of Brain-Language Relationships

Pulvermüller (1996) outlines a general approach that is flexible enough to deal with the various models currently available. He takes as his starting point the now well-accepted notion proposed by Hebb in 1949, namely, that psychological constructs have as their counterpart in the brain neural networks that form assemblies of cells that can be quite far-ranging in terms of tissue distance. To the extent that psycholinguistic distinctions are manifested in cell assembly distinctions, then we should be able to find distinctive firing patterns. It need not be the case, of course, that the firing patterns will be distinctive enough for ERP techniques to distinguish them, but it is always possible that they might be so. Given this possibility, we can ask empirically whether such distinctive patterns can be found for specific questions.

5-3. SOME LEXICAL SEMANTIC ISSUES

Lexical semantics involves the coding of word meanings (Caplan, 1987). We take this to include semantic features (for example, + animate, + object, − action) that also have implications for grammatical use. Thus, lexical semantics includes the specifications of the word that promote the use of the word as a noun versus a verb (with crossing over being more likely in some cases than in others, and being subject to further morphological rules). Because of this, it may be difficult to separate neurolinguistically the grammatical from semantic specifications.

The noun/verb distinction is primary to much of linguistic theory and clearly is a suitable distinction in the context of sentence processing. Given this, one might expect that such a distinction would be respected in terms of how the brain codes lexical items or processes them, that is, whether they depend on different tissue in the brain. Daniele, Giustolisi, Silveri, Colosimo, and Gainotti (1994) present cases of left frontal damage leading to impairments of verb processing and left temporal damage leading to impairment of noun processing. Hillis and Caramazza (1995) argue that noun and verb dissociation is further embedded within input and output systems. Such a noun/verb dissociation has also been demonstrated with ERPs in normals performing lexical decisions (deciding whether a stimulus string of letters is a legitimate word in the language), with verbs showing significantly greater positivity around 200 to 230 msec after word presentation at anterior electrode sites (Preissl, Pulvermüller, Lutzenberger, & Birbaumer, 1995) and nouns showing a significantly greater effect at occipital sites (in a reanalysis of the data presented in Pulvermüller, 1996). This contrast is attributed to the semantic associations of the words used in the study—the nouns were chosen to be more concrete and imagistic, verbs more motor-associated—but, as mentioned earlier, it may be impossible to really separate semantic and grammatical features. Interestingly, there were no hemispheric effects, suggesting that areas outside the classical language areas contribute to the lexical semantics in the wider "world-knowledge" sense. (It is not yet clear what implications this has for the concept of lexical semantics in the narrow linguistic sense.) Similarly, Samar and Berent (1986) found that an ERP component centering on 220 msec poststimulus distinguishes nouns from verbs, again with verbs obtaining greater positivity. In this case, there were no interactions with site, and analyses at separate sites were not presented.

Preissl et al.'s (1995) use of nouns and verbs confounds semantic space with word class, since only verbs denoting motion and nouns denoting concrete objects were used. From a linguistic viewpoint, nouns and verbs are not defined this way; "democracy" is just as much a noun as is "table," and "to think" is just as much a verb as is "to push." Defined in this way, the presumed cell assemblies reflect the denotative subsets as much as grammatical class. Perhaps a further way to examine grammatical class per se would be to look at ERP patterns resulting from noun/verb homographs, easy in English (*rebel, love, hate, hope, house,* etc.), controlling of course for word frequency and imagery value, or with a systematic inclusion of abstract nouns and verbs.

Lexical distinctions can be made within word class too. For example, Wetzel and Molfese (1992) recorded ERPs to two factive verbs *(noticed, revealed)* and two nonfactive verbs *(maintained, supposed)* embedded within appropriate sentence frames; for example, "Dick noticed Bob was gone," about which the subject answered a question. Bilateral frontal and left temporal sites distinguished verb type late around 1100 msec, but there was also a distinction at central-parietal sites bilaterally at 140 msec. Although the latter seems early enough as a marker of syntactic processing, the former is perhaps more a reflection of memory and processing requirements resulting from the verb being factive versus nonfactive.

Lexical semantics also involves semantic features, such as word concreteness. This has been explored with a view to testing the notion that abstract words make use primarily of neural assemblies in the left hemisphere while concrete words make use of bilateral representation. Kounios and Holcomb (1994) showed that concrete items are associated with more negativity at 300-500 and 500-800 msec than abstract items (with the two time frames providing separate components). In both time ranges, ERPs to concrete and abstract words were different over the right hemisphere sites but were not different at left hemisphere sites, consistent with dual-coding theory, with the effect being more anterior at the 300-500 msec period and more evenly distributed at 500-800 msec. Pulvermüller (1996) has applied EEG spectral techniques to these same issues with much success.

5-4. THE CONTENT VERSUS FUNCTION WORD DISTINCTION

Most of the studies on the noun/verb distinction described so far present stimulus words singly for lexical decision, except for Samar and Berent (1986), who presented words in short phrases. Researchers have not suggested that the ERP results would change dramatically if the stimulus words were presented in sentence frames, although this is conceivable for reasons discussed later, namely, that online lexical-frame processing need not tap the same neurological processes as lexical lookup.

The same difficulty may hold for the major word class distinction of open-class (OC) versus closed-class (CC) words. Open-class (or content) words are composed of nouns, verbs, and adjectives, and technically they also contain adverbs. Closed-class items (or function words) are the limited set representing pronouns, relative pronouns, articles, conjunctions, prepositions, and, technically speaking, auxiliary verbs, but in isolation it is sometimes hard to distinguish some auxiliaries from verbs and so they are usually omitted in experiments. Similarly, one may want to exclude prepositions because they have more "content," although they also have lower frequency counts and therefore are better matches to open-class words on this dimension.

Despite the dissociation of CC from OC words in aphasia subtypes (Caplan, 1987) and on reading speed measures (Chiarello & Nuding, 1987), it has been suggested that CC are simply a special subset of OC words that have high frequency, low imageability, short length, and low stress (see Neville, Mills, & Lawson, 1992 for a discussion and references). On the other hand, finding qualitatively different patterns of ERPs to the different classes, if we could control for frequency, word length, semantics, and so forth, would justify the division on the basis of the grammar. Doing so would be extremely difficult, of course.

Neville et al. (1992) presented sentences for sensibility judgments, that is, determining whether or not the final word was semantically appropriate. ERPs were taken to every word except the first and last in each sentence and averaged to form an OC average (nouns, verbs, adjectives, n = 446) and a CC average (articles, prepositions, conjunctions, etc., n = 391). What they found accords well with expectations from the aphasic literature. OC and CC words were differentiated significantly by 150 msec in the ERPs at the frontal sites and slightly later at posterior sites at 170 msec. CC words produced a significantly greater negative peak over frontal cortex, while the OC words produced the larger negative peak posteriorly. This early distinction was greater over the left hemisphere. In addition, there is a clear distinction between the latency of OC and CC negative peaks during the 235-400 msec window: The negativity associated with CC words comes earlier (N280) and with OC words later (N350), clearest over temporal and anterior temporal sites (see Figure 2). The amplitude of the CC N280 was greatest over frontal and anterior temporal sites of the left hemisphere, while the OC N350 amplitude was largest over posterior sites of both hemispheres, and in fact tended to be larger at the right hemisphere sites. The late positive component (LPC) between 300 and 600 msec peaked 50-100 msec earlier for CC than OC words, and this effect was significant only at the left hemisphere sites. LPC amplitude and later negativities (N400-700) also differentiated CC and OC words with electrode site interactions.

Because it is important to consider the confounding of word class with frequency of occurrence in the language (CC being more frequent), Neville et al. (1992) also checked high- and low-frequency words within word class, and found that the N170 and N350 were both sensitive to frequency (less frequent words elicited greater negativities over posterior regions). Over anterior regions, word frequency did not affect the ERP parameters, but of course CC words are on the whole more frequent than OC words so it may be impossible to rule out a frequency range factor affecting peak latency, amplitude, and topography. Word length also affected ERP results but not in a way that would suggest that the word-class effect is really a word-length effect.

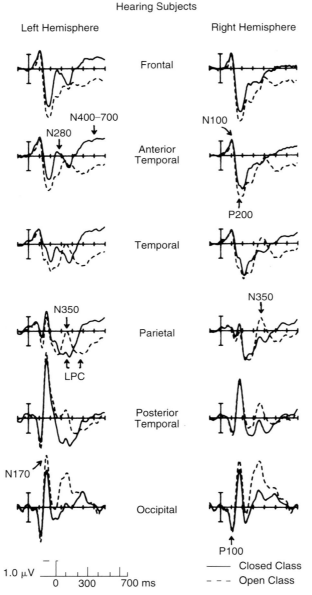

FIGURE 2 ERPs averaged across 17 normal-hearing adults, elicited by closed- and open-class words in the middle of visually presented English sentences. Closed-class words elicited an N280 response over left anterior regions; open-class words elicited a bilateral, posterior N350 response. Negative is up. From Neville, Mills, & Lawson (1992). Fractionating language: different neural subsystems with different sensitive periods. *Cerebral Cortex,* 2, 244-258. (Reprinted by permission of Oxford University Press.)

5-4.1. Is the Anterior–Posterior Distinction for OC and CC Words Due to Lexical Access or Sentence Context?

Neville et al.'s (1992) results from normals are consistent with the notion stemming from the aphasia literature that grammatical dysfunction is associated with left hemisphere anterior damage, and that OC and CC words serve different functions and are dependent on different storage and linguistic factors. They are also compatible with the faster, more automatic access people have to CC words. However, by averaging across all trials, some individual words are more represented than others in the averaged ERP. This would be especially so among CC items where we would expect *the* and *a* to appear more often than any other item. In addition, the subjects presumably could anticipate the word type that is about to occur. Therefore, what was being averaged was OC and CC word frames as well as words themselves. If the results really reflect such word frames, then they support the grammar/semantics distinction, since such grammatical frames seem to be associated with frontal and lateralized negativities, whereas semantic access is related to a later bilateral posterior negativity, the N400.

This "word-frame" interpretation of the results is mitigated by other studies demonstrating somewhat similar effects in lexical decision tasks, where the words can be balanced for presentation frequency and are randomized so as to preclude word-class anticipation. Pulvermüller, Lutzenberger, and Birbaumer (1995) found that the ERP differentiated OC and CC words by 160 msec poststimulus onset and that CC words produced a hemisphere interaction but OC words did not. There are some differences, however, from the Neville et al. (1992) study. The timing of the components and their morphology were different, with Pulvermüller et al. (1995; see their Figure 3) not finding N280 and N350 components distinguishing CC and OC words in the same way. In contrast to Pulvermüller's results, Nobre and McCarthy (1994) also examined ERPs to CC and OC words in isolation in a task where subjects were reading for meaning (instead of lexical decision), and they do report an anterior negativity (N288) with the CC items and an N410 posteriorly with OC items. Similarly, the ERPs to function words reported by Kluender and Kutas (1993b) also show a negative peak at the more anterior sites at around 300 msec that disappears posteriorly, and is stronger at the left sites. Their anterior negativity (which was not the subject of their investigation or statistical analyses) was different from subsequent N400 responses, which were primarily posterior and stronger at right hemisphere sites. This earlier anterior negativity was absent for sentence-final content words. Thus there is considerable circumstantial evidence to support Neville et al.'s (1992) findings of ERP markers for OC and CC, but not necessarily when people are making lexical decisions. Although it is true that the processing demands of word frames and frequency of occurrence may account for these ERP markers, this does not mitigate their existence in normal language processing, but may rather explain the basis for finding them.

Further studies will be needed to determine whether this ERP distribution is the same in single-word presentations as in sentence presentations. In a way, we should not expect identical results, since words presented in sentence context have anticipation working where lexical decision tasks do not. But the anterior/posterior differences in the studies that require reading for meaning do suggest that the anterior/posterior effect is attributable to the full lexical lookup in contrast to lexical decisions, which may not fully require full lexical access (that is, to both full phonological and semantic representations). Another logical alternative, although one for which we have no empirical support, is that lexical decision requires much more than lexical lookup and this additional processing masks important ERP components.

This issue of the comparability of processing words in isolation versus in sentence contexts is a major one (Kutas, 1993). Van Petten (1995; Van Petten & Kutas, 1990, 1991) showed that ERP differences related to the frequency effects (larger N400 amplitudes to low-frequency items) are lost when meaningful sentences provide contextual constraints. The same data set shows no indication of a frontal N280 for CC words, although there does appear to be a posterior N350 for OC words and also some late frontal (400-800 msec) sensitivity to context for CC words but not for OC words. Again, this is suggestive of a CC-frontal connection and an OC-posterior connection, although they seem highly sensitive to test conditions. In real sentence contexts, semantic features, grammatical expectations, and word-frequency effects are closely linked. These studies demonstrate the rich dynamics of word processing.

5-5. STUDIES ON PHONOLOGICAL PROCESSING USING ERPS

There are fewer studies of phonological processing using ERPs than of semantic or syntactic processing, which is probably due partly to the particular linguistic interests of the ERP researchers (psycholinguistics has a longer history in semantic and syntactic issues) and partly to technical requirements of the presentation of speech stimuli compared to visually presented words. However, there are a variety of questions concerning phonological processing that are amenable to ERP methodology. Because of space limitations, we will present some of them only briefly, with the implications being, we hope, self-evident. Praamstra and Stegeman (1993) showed that the N400, a component usually associated with semantics, can reflect the phonological level of processing: subjects produced an N400 on rhyming mismatch trials, whether the spoken stimuli were real words or nonwords (eliminating lexical access as a necessary component) and when the items to be matched were spoken by different voices (eliminating the physical-acoustic factor), and in lexical decisions with real words but not in lexical decisions with nonwords (suggesting that lexical access prompts phonological access). Further, Praamstra, Meyer, and Levelt (1994) found that ERP data were *more* sensitive to phonological priming than were behavioral measures. Using visual presentations, Valdés-Sosa, Gonzalez, Xiang, Yi, and Bobes (1993) reported N400 effects for rhyming versus nonrhyming primes in Chinese, where the logographs do not directly code the sounds at all. This would support the position that word reading would certainly activate phonological representation in languages (like English) whose writing systems place more emphasis on phonological properties of the words.

Whereas these studies all focused on the phonological level of processing, acoustic (not categorical phonological) factors have also been examined with ERPs, and are reflected in a MisMatch Negativity (MMN) component, which is a negativity spreading over the time period for the P2 and N2 (Kraus et al., 1993; Sharma, Kraus, McGee, Carrell, & Nicol, 1993; Sams, Aulanko, Aaltonen, & Näätänen, 1990). Similarly, Connolly and Phillips (1994) found a negativity between 270 and 300 msec that reflected phonological expectation independent of semantic expectations. Aaltonen, Tuomainen, Laine, and Niemi (1993) used the MMN to show that although all their aphasics show MMN discrimination of differing pure tones, those with left posterior damage do not show acoustic discrimination at this level of information processing, while their anterior aphasics did.

Molfese (1978, 1980b) found that it was the right hemisphere and not the expected left hemisphere sites that were sensitive to voice onset time (VOT) distinctions in consonant discrimination, whereas place of articulation (POA) discriminations were found at left hemisphere sites. This was replicated in preschool children (Molfese & Molfese, 1988), and Segalowitz and Cohen (1989) replicated this in adults with three consonant (C)-vowel (V) contexts: CV, VC, and VCV. Molfese (1980a) argues that this right hemisphere effect is due to a general acoustic parser that is sensitive to onset time, whether produced in speech or pure tones.

Intensive study of the speech discrimination abilities of neonates has also been successfully explored using ERPs, even though clearly the experimenters have no control over subjects' cognitive sets. For example, Molfese, Burger-Judisch, and Hans (1991) found that neonates' ERPs discriminate POA. Molfese and Molfese (1985, 1994) also found that using neonatal ERPs to test for discrimination of stop consonants predicts later language development, which fits with Tallal's (Tallal et al., 1996) finding that children who are at risk for language development are aided by procedures that help discriminate precisely those stop consonants. ERP analysis has also been used to examine the location of generators in the auditory cortex of infants that produce ERPs sensitive to speech signals (Novak, Kurtzberg, Kreuzer, & Vaughan, 1989).

5-5.1. Summary

Thus, ERPs have been shown to be sensitive to various levels of speech analysis: phonological and acoustic, and both explicit and implicit processing. ERP data can be more sensitive than behavioral responses to phonological effects, presumably because they reflect such levels that are inaccessible to conscious responding. In addition, some studies have had success in providing some information on physiological generators of speech sounds.

5-6. CONCLUSIONS

5-6.1. Do ERPs Reflect Local Brain Processes?

We would like to be able to confidently draw conclusions from ERP data about the contributions of specific areas of the brain to the underlying linguistic processes. There is currently a serious debate as to the extent to which we can make such conclusions. Although many researchers are satisfied with 8 or 12 electrode sites to differentiate the contributions of different brain areas, others argue that at least 64 or 128 sites are needed for reasons of the physics of the electrical activity about the scalp, not to mention for capturing the electrical dynamics inherent in the signal (Gevins et al., 1995; Tucker et al., 1994). Although the field has yet to form a consensus on this issue, it is clear that ERPs are useful for differentiating linguistically related processes, and can provide at least a gross sense of localization.

5-6.2. Where Are Words in the Brain?

ERPs provide a way to reflect the activity of brain circuits that deal with language processing. This method appears to complement or actually be richer than the traditional behavioral methods. Of course, ERPs do not reflect all neuronal activity of the cortex, nor do they provide a set of absolute markers because the details of the ERP components are sensitive to the technical parameters of the recordings, and to the cognitive context of the stimulus presentation. Much the same can be said, however, for behavioral responses that are sensitive to response modality and cognitive task requirements. Using these electrical brain responses as dependent measures, we can ask psycholinguistic questions about cognitive architecture and we can ask neurolinguistic questions about brain localization and representation.

Applying ERP technology to studies of subclasses of words can be important for several reasons. First of all, although it is obvious that words have to be represented in the brain in some manner, it is not at all obvious how they should be subdivided. Traditional grammars differentiate open and closed classes of words and differentiate nouns and verbs, but finding ERP components sensitive to these differentiations is not automatically guaranteed. The aphasia literature suggests a posterior–anterior difference on sensitivity to open- and closed-class words, respectively, and the ERP evidence presented seems to agree, especially with respect to the anterior N280 and posterior N350 components. These differences are more likely to arise when the grammatical context provides the subject with some expectation, and thus these ERP components may not reflect purely the lexical access of open- and closed-class words. However, the ERP results reinforce the conclusion that there is a clear distinction involved in open and closed classes of words. Further work will have to address the extent to which the ERP distinction relies on word frequency and word semantics rather than grammatical class only.

Differentiations of nouns from verbs on the basis of semantic factors also shows up in the ERP, although it is not clear yet whether this is due to the semantic or the grammatical class difference. Testing for ERPs sensitive to grammatical class in a way that is independent of semantic features is not simple and may turn out not to be possible.

Another conclusion from the ERP studies that relates to the testing paradigm concerns whether lexical decision promotes lexical access in the same way as reading for meaning in a phrase or sentence context. The resultant ERPs are not identical. It may be that ERP methods will help settle the debate within the field over the extent to which the lexical decision paradigm requires phonological and semantic access (Balota & Chumbley, 1984; den Heyer, Goring, Gorgichuk, Richards, & Landry, 1988).

Acknowledgments

We would like to thank Friedemann Pulvermüller for helpful comments on an earlier draft of this paper. This chapter was completed while the first author was a visiting scholar at the MRC Applied Psychology Unit in Cambridge, UK, and a visiting fellow at Churchill College. We would like to thank the MRC-APU, Cambridge University, and Churchill College for their cooperation and support.

CHAPTER 6

Event-Related Potential (ERP) Research in Neurolinguistics: Part II
Language Processing and Acquisition

Sidney J. Segalowitz and Hélène Chevalier
Department of Psychology, Brock University, St. Catharines, Ontario, Canada L2S 3A1

In this chapter, we present a review of how event-related potentials (ERPs) have been used to reflect semantic and syntactic processing in normal adults and children, both to provide evidence of a neural distinction between semantics and syntax and to evaluate specific constructs in linguistic theory. ERPs have also been used to test a variety of neurolinguistic hypotheses concerning language acquisition (both first and second), reading development, and vocabulary knowledge in subjects who cannot give responses. We also consider whether ERPs reflect specifically linguistic processes or only general responses to cognitive information processing.

6-1. OVERVIEW OF THE ISSUES

Event-related potentials (ERPs) have been successfully applied to questions of lexical access (see Segalowitz & Chevalier, this volume). They have also been used to test specific hypotheses concerning semantic and syntactic processing. The first task has been to define which electrophysiological components seem sensitive to semantic processing and which to syntactic processing, and whether these are different from each other in interesting ways. For nearly two decades, Kutas and her colleagues have been exploring how a late negativity in the ERP, apparent at posterior electrode sites, reflects the subject's detection of a semantic anomaly. This has become a standard technique to examine semantic sensitivity, even in subjects who cannot respond in the usual fashion (e.g., infants). Detection of syntactic anomalies, on the other hand, produces somewhat different patterns of ERP components, allowing one to argue for a neural separation of semantic and syntactic processes. Linguistic theory further suggests many types of syntactic operations, and an examination of the ERPs resulting from violation of each type suggests different neural activity for each, although some have questioned whether we are seeing a reflection of the linguistic process or its cognitive ramifications.

These techniques have been applied to paradigms examining first and second language acquisition, for example, concerning critical period hypotheses about certain aspects of language acquisition. We also discuss ERP studies testing neurolinguistic hypotheses about reading and its acquisition using ERPs.

6-2. DIVISIONS BETWEEN SYNTAX AND SEMANTICS

6-2.1. The N400 and Semantic Processing

Maintaining that there is a division between semantics and syntax is not too controversial, although identifying the dividing point is more difficult. Finding ERP sensitivity to the division would confirm its suitability in neurolinguistics. One ERP component that clearly responds to semantic processing is the N400, a large negativity that covers a broad bilateral posterior area when the subject perceives a semantic anomaly but not a syntactic one (Kutas & Hillyard, 1983). For example, the final word in "He went to the post office to mail the house" would elicit a large negativity posteriorly and bilaterally around 400 msec after the presentation of the final word, whereas an expected word does not (see Figure 1). This N400 to semantic anomalies has even been reported in an aphasic patient who showed no explicit comprehension, raising the issue of unconscious understanding (Revonsuo & Laine, 1996). The semantic anomaly producing the N400 can be incidental as well as the focus of attention (Kutas & Hillyard, 1989; but see Bentin, Kutas, & Hillyard, 1995). It also arises in other contexts (e.g., in mismatches in phonology: Kraus et al., 1993; in faces: Barret & Rugg, 1989; Bobes, Valdés-Sosa, & Olivares, 1994; Olivares, Bobes, Aubert, & Valdés-Sosa, 1994; in color patches: Katayama & Yagi, 1992; in pictures: Nigam, Hoffman, & Simons, 1992), but that does not diminish its sensitivity to linguistic semantics. This negativity is somewhat earlier in natural speech than in reading (Holcomb & Neville, 1990, 1991). The N400 component suggests that an expectation

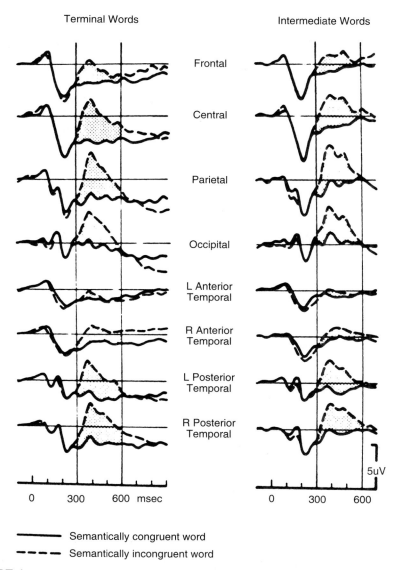

FIGURE 1 Typical N400 response to anomalous versus nonanomalous words in a sentence context. Negative is up. From Kutas & Hillyard (1983). Event-related brain potentials to grammatical errors and semantic anomalies. *Memory and Cognition, 11,* 539-550. (Reprinted by permission. Copyright by The Psychonomic Society.)

for the word that does appear lessens the negativity (or promotes a positivity) (see Rugg, 1995; cf. Van Petten, Kutas, Kluender, Mitchener, & McIsaac, 1991). It has been suggested that the semantics-related N400 is linked to activation, at some point in the process, in the left angular gyrus, as seen through cerebral glucose metabolism rates (Nenov et al., 1991).

Full sentences are not needed to elicit an N400. Münte, Heinze, and Mangun (1993) presented word pairs for synonym judgments and found N400s for both verb and noun trials (e.g., spend–squander, gangster–robber versus excavate–migrate, parliament–cube). When they asked the subjects to rate grammatical acceptability of two-word phrases (e.g., "you spend" versus "*your write"), they did find a late negativity, but this was 50-80 msec later than in the semantic task and had a greater frontal distribution, a pattern that has arisen in other studies too. The ERP topographical differences between the semantic and syntactic tasks illustrate that there are differences in the processing.

An N400 is also produced by random word lists. Morphologically rectifying the list so as to produce a syntactically correct but meaningless sentence does not reduce the N400 at all, as in "He ran the half-white car even though he couldn't name the raise" (Van Petten & Kutas, 1991). Thus, we can use ERPs to reflect semantic access, something that is very popular in the ERP literature on semantic disturbances in Alzheimer's dementia (e.g., Hamberger, Friedman, Ritter, & Rosen, 1995; Friedman, Hamberger, Stern, & Marder, 1992) and schizophrenia (e.g., Adams et al., 1993).

The N400 has been used to answer questions of how ambiguous words are processed in context. Van Petten (1995) found that sentence context constrained the sense of the word from its multiple meanings enough to reduce the N400 effect from the priming of a word following the sentence. Similarly, Garnsey, Tanenhaus, and Chapman (1989) used the N400 reaction when confronted with implausibility in sentences to show that readers try to interpret a syntactic ambiguity rather than waiting for later information to disambiguate it.

6-2.2. Semantic versus Syntactic ERP Components

Osterhout and Holcomb (1992) contrasted semantic anomaly with syntactic anomalies by presenting sentences some of which had a final word that was semantically anomalous (e.g., *The car rolled down the hill and complained), some of which had inappropriate causal complements (*The woman persuaded to answer the door), some of which were normal acceptable nonanomalous simple active sentences (e.g., The woman struggled to prepare the meal), and some of which had a variety of other grammatical errors. Subjects had to indicate by a button push, after a visual prompt appeared, whether the sentence was acceptable or not. The word at which the error in the causal complement became apparent elicited a positivity in the 500-800 msec range, which they termed the P600. They replicated this in a second study that expanded the type of grammatical violation, where again, the point at which the subject can realize the "garden-path" aspect of the sentence elicited a P600 (replicated and extended in Osterhout, Holcomb, & Swinney, 1994). They noted that in their sentences, the P600s were elicited to closed-class (CC) or "function" words (which marked the divergence from grammatical expectation), whereas the classic N400 is elicited to open-class (OC) or "content" words. As they argued, however, one can get N400s to CC words given appropriate presentation, suggesting that the P600 is really an indication that syntactic violation has cognitive repercussions different enough from semantic anomaly to affect the ERP differentially. Osterhout and Holcomb (1993) extended their work to auditory presentation and again found a positivity in reaction to being "garden-pathed." This time, however, the positivity often began quite a bit earlier (as do N400s to auditory presentation) and appeared over much of the 1000 msec ERP duration.

Similar to Osterhout and Holcomb's (1992) demonstration of a positivity in response to syntactic anomalies, Hagoort, Brown, and Groothusen (1993) report positivities in response to difficulties in parsing sentences. Concerned that the overt making of a grammatical judgment influenced the ERP, they had subjects simply read (without overt response) sentences half of which had either a subject-verb nonagreement (*On a rainy day the old man *buy* a life insurance), verb subcategorization error (e.g., *The tired young man *elapsed* the book on the floor), or noun phrase structure order violation (e.g., *Most of the visitors like the colourful *very* tulips in Holland). Note that in this study the subject was alerted to the anomaly by OC items, in contrast to Osterhout and Holcomb (1992, 1993). About 500 msec after the subject-verb nonagreement, a posterior positivity arose that was sustained through the following word and then became a sustained negativity. The other grammatical violations (verb subcategorization and noun phrase structure violation) produced some results difficult to interpret, such as sustained shifts *prior* to the grammar violation. For the interpretable results, Hagoort et al. (1993) suggest that their Syntactic Positivity Shift (akin to Osterhout's P600) reflects the difficulty the grammatical parser was having dealing with the sentences. A similar interpretation was placed on an earlier-occurring positivity (P345) that arose in the context of making sense of subject- versus object-relative clauses (Mecklinger, Schriefers, Steinhauer, & Friederici, 1995).

This tidy picture of N400s for semantic anomalies and P600s for syntactic anomalies is clouded somewhat by the findings of Friederici, Pfeifer, and Hahne (1993, much of which was replicated by Rösler, Pütz, Friederici, & Hahne, 1993), who examined ERPs to spoken sentences with or without a semantic error (incorrect selectional restriction), morphological error (incorrect verb inflection), or syntactic error (phrase structure violation). They found large N400s to both semantic and morphological errors at all electrode sites (left and right "Broca's" sites, left and right "Wernicke's," and Fz, Cz, and Pz), but the morphological errors also produced a primarily posterior positivity after about 700 msec. The syntactic violations elicited a small anterior negativity at about 180 msec, primarily at the Broca's-left site, followed by a frontal negativity in the N400 time frame. Although the late positivity is reminiscent of the P600 notion, why it did not emerge for the syntactic violations as well is not clear. The semantic-syntactic ERP distinction is further complicated by the results of Osterhout's (in press) study showing that, in response to a sentence-final syntactic anomaly (on an OC word), some subjects are prone to producing a P600 and others an N400. This raises the issue of whether subjects have the option of treating the anomaly in one of two ways (as a syntactic reparsing difficulty versus a semantic misfit) or whether subjects have different organization for language.

6-2.3. Specific Questions of Grammar Theory

Although the N400 was originally taken as a reflection of linguistic semantic processing (Kutas & Van Petten, 1988), this position has changed to include expectation in more contexts of meaning (Kutas & Van Petten, 1994), that is, the N400 may not be a reflection of lexical semantics per se but of the mental processing requirements that arise when a revision of expectation is required, in a general sense. In time the same attribution of processing requirements may befall the P600. Using the N400 and the like to index processing effort opens up an entirely new way to investigate psycholinguistic phenomena. This use of ERP components has been applied to online sentence processing in studies designed to investigate dissociations predicted by syntactic theory. The assumption is that cognitive events that are mediated by different computational substructures should elicit differing ERPs.

One of the complexities of grammatical decoding is the pairing of relative pronouns with the place of their missing referent. This place is called the "gap" and can be illustrated by the sentence "What did you say _____ to John?" where the initial *wh*-word is an element that has been moved from where the line appears, leaving a "trace" behind. Kluender and Kutas (1993b) refer to this as a "filler" for the object that is missing later on in the sentence. Various *wh*-questions can be distinguished from each other according to Government and Binding (GB) linguistic theory (Chomsky, 1981) on the basis of the rules used to construct them, which is the focus of Neville, Nicol, Barss, Forster, and Garrett (1991).

Neville et al. (1991) examined whether three types of syntactic deviance predicted by GB theory are reflected in different ERP patterns (taken to the italicized comparison words), in a way that contrasts with semantic anomaly (e.g., *The boys heard Joe's *orange* about Africa). The three syntactic deviation types were

1. Phrase structure violations: The prepositions *of* and *about* appear left instead of right of the head noun, producing the form
[NP's *of* Noun NP]
e.g., *The man admired Don's *of* sketch the landscape.
or
[NP's *about* Noun NP]
e.g., *The student sang Lisa's *about* songs freedom.
2. Specificity violations: *Wh*-phrases cannot be extracted from an NP with a specific reference, so,
e.g., What did the scientist criticize a *proof* of?
is acceptable but,
e.g., *What did the scientist criticize Max's *proof* of?

is not. Since specificity is a semantic notion, this sentence's violation is in both semantic and syntactic aspects.

3. Subjacency violations: It is ungrammatical to extract a *wh*-phrase from inside the subject NP, a point that is subsumed under the *Subjacency Condition*. So,
e.g., Was a sketch of the landscape *admired* by the man?
is acceptable but,
e.g., *What was a sketch of *admired* by the man?
is not. Subjects were asked to make acceptability judgments, and ERPs were taken from the onset of the comparison words.

Semantic anomalies elicited an N400 (300-500 msec) that was bilateral and larger posteriorly, as expected (see Figure 2). Phrase structure violations elicited an increased N125, limited to left anterior regions, followed by a left temporal and parietal N400, followed by a large bilateral positivity starting at around 500 msec. Specificity violations elicited an increased N125 over left anterior regions and the N400 over the left frontal and temporal regions. Subjacency violations increased the P2 (200-300 msec) at all but occipital sites and the 500+ msec positivity at all sites. Thus, each type of syntactic deviation elicited a specific pattern of ERP alterations, supporting the hypotheses. Neville et al. (1991) also concluded that the classic N400 (bilateral, posterior) only appeared to semantic anomalies, although there were other topographical patterns of negativities in this time frame. It may be that distinctive patterns of ERP changes are associated with different types of syntactic violation because of the different location of linguistic difficulty, both cognitively and physiologically. Alternatively, as Neville et al. (1991) point out, such differences may also be due to differences in the ease of recovery from the violations. Note that there was no ERP change that was general to all syntactic disturbances, so the different ERP "markers" of the deviations may be a result of the standard ERP being affected by the various time frames in the recovery function. However, we can note that the two types of violations (phrase structure and subjacency) that produced the late positivity (perhaps a P600) were the ones where the word immediately before the deviation forms an unacceptable sequence with the deviating word (e.g., "Don's of" and "of admired"), while the word pair at the specificity violation forms an acceptable phrase (e.g., "Max's proof").

Another approach is to use *wh*-words because they force the reader to keep the filler in working memory until the gap appears. Varying the length of this gap and seeing the effect on ERPs was the subject of Kluender and Kutas' (1993a, 1993b) studies. Like Neville et al. (1991), Kluender and Kutas (1993b) took ERPs to online sentence processing, visually presenting one word at a time. This study was done in the context of whether some of the principles associated with subjacency should be attributed to specific innate linguistic mechanisms or to a more general processing limitation related to the nature of the grammatical requirements. They hypothesized that N400 amplitude would index processing effort, a notion that contrasts with the interpretation outlined earlier of the ERP reflecting lexical access rather than the ongoing working memory requirements (see also Gunter, Jackson, & Mulder, 1995 for

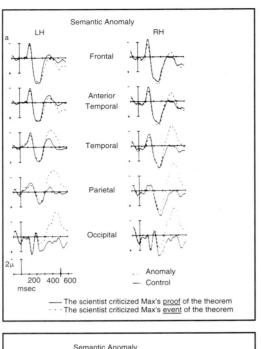

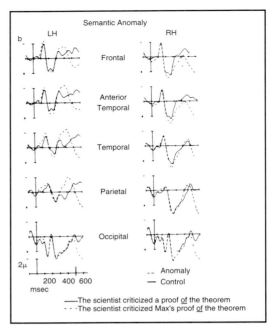

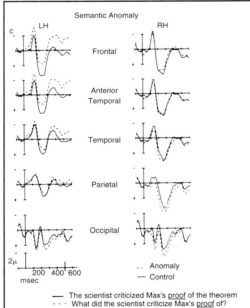

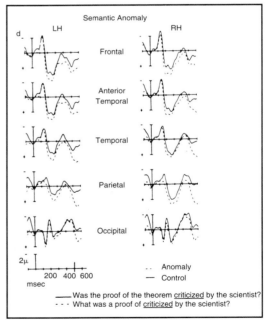

FIGURE 2 ERPs averaged across 40 subjects. Responses in dashed lines were elicited by words (underlined) that rendered the sentence either semantically *(a)* or grammatically *(b, c, d)* deviant. Solid lines represent ERPs that were elicited by comparison words in semantically and grammatically acceptable sentences. Note that at 500 msec, another word was presented. Negative is up. From Neville, Nicol, Barss, Forster, & Garrett (1991). Syntactically based sentence processing classes: evidence from event-related brain potentials. *Journal of Cognitive Neuroscience, 3,* 152–165. (Reprinted with permission. Copyright by the Massachusetts Institute of Technology.)

a treatment of N400 as a reflection of working memory requirements of sentence processing). Kluender and Kutas (1993b) found different N400 amplitudes for the different function words *that, if,* and *what/who,* concluding that within the CC word category, different items have different processing requirements. *If* and *what/who* differed in N400 amplitude despite similar grammatical frames and word frequencies. They also found that the processing cost of holding a filler in working memory was reflected in a left anterior negativity (LAN) associated with the hypothesized ordering of difficulty, namely, *who/what > if > that.* Similarly, Kluender and Kutas (1993b) cite other work from their laboratory indicating that object relative clauses show greater LAN compared with the relatively easy subject relative clauses. Object relatives require a filler to be kept in working memory, for example,

> The reporter *who* the senator harshly attacked _____ admitted the error.

whereas subject relatives are immediately assigned to the adjacent gap, for example,

> The reporter *who* _____ harshly attacked the senator admitted the error.

with the negativity starting at precisely the time when the filler has to be maintained. As would be expected, individuals with poorer verbal working memory produce larger LANs in this context (King & Kutas, 1995). Other LAN effects are reviewed by Kluender and Kutas (1993b, pp. 623–624) and more supporting data are presented by Kluender and Kutas (1993a).

6-3. DEVELOPMENTAL ISSUES AND SECOND LANGUAGE ACQUISITION

The notion of developmentally sensitive periods is well established in neurobiology, and there is no reason to think the neurobiology of language should prove an exception. What has been generally found is that what we think of as standard neural organization (for example, of the visual system) is in fact dependent on standard experiences within a standard time frame. Alteration of the earliest visual experiences (Freeman & Thibos, 1973) or the removal of early auditory experiences (Neville, Schmidt, & Kutas, 1983) results in changed neural organization for visual experiences. In a series of studies, Neville and her colleagues also examined the OC/CC distinction, semantic processing, and sensitivity to syntactic anomalies in American Sign Language (ASL) and English in deaf and hearing adults, and in English in Chinese–English bilinguals who acquired English at different ages.

The deaf adults had acquired ASL as a first language (L1) and English as a second language (L2), and some of the hearing controls also learned ASL as L1, having been born to deaf parents. Semantic anomalies in English elicited N400 ERP components in deaf subjects essentially identical to those from hearing subjects, although the peak negativity was about 25 msec later (Neville, Mills, & Lawson, 1992). Similarly, ERPs to OC words were also identical in deaf and hearing groups in terms of P100, N170,

N350, and LPC components. However, CC words did not elicit the asymmetrical N180 among deaf subjects as they did in hearing subjects, but rather a symmetrical non-highly focused scalp distribution (Neville et al., 1992).

Neville et al. (1997) have also examined OC/CC distinctions in ERPs elicited to ASL signs among profoundly deaf adults, hearing native users of ASL, and fluent hearing late learners of ASL. When tested in ASL, deaf subjects reacted to OC and CC words in much the same way as hearing subjects react to English OC and CC words, except that the early negative component to CC words (N280) in English that is absent to OC words (a) is also present but reduced for ASL OC words in deaf subjects, (b) shows less asymmetry in deaf subjects, and (c) is slightly earlier (N250).

Hearing adults who are native signers produced ERPs to OC and CC signs that were much the same as the deaf adults. The differences were independent of word-class membership and therefore probably due to the hearing loss in the deaf adults.

Hearing late learners of ASL (after 17 years of age) produced the N250 anteriorly and the N400 posteriorly, as did hearing native signers, but showed no word-class effect for the N250! Thus the anterior ERP marker for the OC/CC distinction was sensitive either to age of language acquisition or to the L1/L2 distinction.

To further examine the effects of age of language acquisition on ERP components sensitive to semantic and syntactic structures, Weber-Fox and Neville (1996) presented the sentences with three types of syntactic violations and with semantic anomalies described earlier from Neville et al. (1991) to 61 Chinese–English bilingual adults. They classified subjects with respect to the age at which they were exposed to English (1-3, 4-6, 7-10, 11-13, and after 16 years of age). Age had a major effect on grammaticality judgments and associated ERP components, even in those whose exposure to English began in the earliest period. In contrast, the N400 and semantic anomaly judgments were altered only in subjects exposed to English after age 11, and in these groups there was simply a delay in N400 peak amplitude. For the syntactic judgments, however, those exposed to English later in development had ERPs that were different in both the morphology and the distribution of components. Thus age of acquisition, rather than the L1/L2 distinction per se, appears to be the essential factor modulating the representation of grammatical processes in the brain.

6-4. ASSESSING RECEPTIVE LINGUISTIC COMPETENCE WITHOUT BEHAVIORAL PRODUCTION

Since ERPs reflect ongoing brain activity in response to stimulation without the need for overt behavioral response, the technique should be applicable to subjects who for some reason cannot respond, for example, studies of speech perception in infancy (see Segalowitz & Berge, 1995 for a review). Molfese and his associates have also shown that ERPs are sensitive to whether children know the stimulus words or not (Molfese, 1989, 1990; Molfese, Wetzel, & Gill, 1993); they also taught new words to babies as young as 14 months and found ERPs sensitive to this new learning (Molfese, 1989; Molfese, Morse, & Peters, 1990; Molfese, & Wetzel, 1992). Similarly, Mills, Coffey-Corina, and Neville (1993) found that ERPs discriminated known from unknown words among 20-month-olds, and that a pattern of hemispheric specialization increased with increased language ability.

Applying this approach to children with language production difficulties, Byrne, Dywan, and Connolly (1995a) showed that a nonverbal child with cerebral palsy produced N400s to nonmatching picture–word pairs compared to matching ones, showing that he understood much more vocabulary than he could demonstrate behaviorally. Similarly, children show N400s in this paradigm at a level commensurate with their expected vocabulary level (Byrne, Dywan, & Connolly, 1995b).

6-5. SOME QUESTIONS ABOUT READING

There are a large number of questions in the study of reading that can be addressed with ERPs. Halgren (1990) presents an integrative model of neural and cognitive bases for reading at the many levels of processing involved, showing how ERPs have been used at each level. Sentence context versus the reading of isolated words is explored with ERPs by Kutas (1993) and Van Petten (1993). Rugg, Doyle, and Melan (1993) give ERP evidence for the hypothesis that heard words generate a phonological representation only, whereas read words generate both orthographic and phonological representation. There is clearly a great deal that can be addressed in the study of reading using ERPs. We will briefly discuss only a few hypotheses.

6-5.1. Phonological versus Semantic Access in Reading

Peres-Abalo, Rodriguez, Bobes, Gutierrez, and Valdés-Sosa (1994) suggest that semantic access is faster to printed words than is phonological access, contrary to traditional reading theory. They showed this by comparing ERPs to match and mismatch trials in a semantic versus rhyming condition and found that the difference between the ERPs to semantic match and mismatch trials was significantly earlier than the difference for the phonological trials. Thus, the subjects' ERPs were sensitive more quickly to semantic information than to phonological information from printed words.

6-5.2. Hemispheric Processes in Reading and Dyslexia

Bakker (1990) outlines his research with colleagues involving ERPs from children at various stages of learning to read their native language (Dutch). They reported an initially larger ERP response to words over right hemisphere sites that switches during their first year of schooling to a large left hemisphere response. They argue that children shift from a slower right hemisphere dominance in processing letter information to a faster left hemisphere one as they become more fluent. They also report that some children have difficulty learning to read because they read extremely slowly and appear to not shift from this right hemisphere dominance. Other children, whose reading is hindered by excessive rapidity and error proneness, switch too quickly. Bakker (1990, 1994) has developed a remedial strategy based on this hypothesis.

Another hemispheric-reading hypothesis concerns the issue of two routes to reading, that is, the notion that reading English involves a phonologically based route and a visually based route (Coltheart, 1985). Reasoning that these may be based on a left-versus right-hemisphere mechanism, respectively, Segalowitz, Menna, and MacGregor (1987; Segalowitz, 1989) presented letter-string pairs and subjects were to indicate whether they were identical or not. Pronounceable letter strings elicited ERPs at left hemisphere sites that were greater in amplitude than at right hemisphere sites, whereas nonpronounceable letter strings elicited the opposite. They reasoned that pronounceable letter strings would be more likely to induce phonological access, whereas nonpronounceable letter strings would more likely induce visual memory processes. Thus, the results supported the two routes–two hemisphere model. Segalowitz, Wagner, and Menna (1992) then replicated these results in 15-year-olds, where good readers showed this pattern but poor readers did not. However, neither was their degree of reading difficulty predicted by their degree of "inappropriate" hemispheric asymmetry in the ERP. Rather, it was the size of the interstimulus negativity in the matching task, presumed to be generated in the prefrontal cortex, that was predictive of their reading skill.

6-6. CONCLUSIONS

It is now evident that ERPs have been a highly useful tool to test linguistic or psycholinguistic hypotheses concerning syntactic theory and first or second language acquisition. In addition, ERPs have provided some support for working hypotheses about the location of cortical sensitivity to semantic and syntactic processes, and again how these may differ in first versus second language acquisition. We must also be sensitive to the controversies about how to get the most from ERP data and about the interpretation of ERPs as reflections of brain processes as opposed to cognitive processes.

6-6.1. Getting the Most from Linguistically Driven ERPs

It would be nice if ERPs reflected ongoing processing of linguistic phenomena qua linguistic phenomena and not as general cognitive phenomena. Buried in the ERP signal may be some pattern-capturing variance specific to individual phonemic (perhaps even phonetic) features of words, specific to the features activated through lexical access, specific to the mental computations associated with grammatical parsing. These patterns may be discernible with the precision available with truly multivariate analysis of the ERP patterns across the scalp, that is, taking into account all scalp sites in a single multivariate data space. This is the goal of the PCA-ANOVA technique, where the ERPs from all sites and all conditions are entered into a Principal Components Analysis in order to find these sources of variances. This has been done with some success in language studies (e.g., Molfese, 1978; Segalowitz & Cohen, 1989), but also is open to difficulties in statistical analysis and interpretation that may limit its use to only quickly processed phonetic stimuli (see Segalowitz & Berge, 1995 for a discussion of some concerns in using this technique). New techniques along these lines seem to be focusing on finding physiological generator sources (Picton, Lins, & Sherg, 1995), which is of course not exactly the same problem but may be amenable to some aspects of neurolinguistic issues (e.g., Pulvermüller, Preissl, Lutzenberger, & Birbaumer, 1996).

6-6.2. Do ERPs Reflect Linguistic or Cognitive Phenomena?

Whether or not an adequately sophisticated statistical procedure can be found, we still have to face another serious concern. The assumption in ERP research is that there is sufficient variance in the ERP to capture the psycholinguistic features of interest. However, we know that the largest variance in the ERP is common to many paradigms—the N1, P2, N2, P3, N400, late positivity—and these are usually interpreted as reflecting primarily attentional processing demands. Thus any overlay of *purely* psycholinguistic variance will be highly obscured in the general ERP paradigm by factors related to working memory demands, attentional allocation, general stimulus processing, response and task demands, degrees of stimulus familiarity, and so on. We have seen some shift along these lines already, with left frontal negativity being attributed to working memory demands and N400 and P600 amplitudes being attributed to the processing efforts required from the subject. Although they do not provide the "magic bullet" that will cleanly characterize the brain activities specific to linguistic computation, ERPs nonetheless provide a rich window on the cognitive repercussions of language processing.

Acknowledgments

This chapter was completed while the first author was a visiting scholar at the MRC Applied Psychology Unit in Cambridge, UK, and a visiting fellow at Churchill College. We would like to thank the MRC-APU, Cambridge University, and Churchill College for their cooperation and support.

CHAPTER 7

Electrical Stimulation Mapping of Language Cortex

Harry A. Whitaker
Department of Psychology, Northern Michigan University, Marquette, Michigan 49855

Three techniques for electrical stimulation mapping of language cortex in the human brain are discussed: how they are done, the advantages and disadvantages of each and both the general and specific results of using these techniques in connection with epilepsy surgery.

Focal electrical stimulation of the human brain (fESB) for the purpose of mapping functions such as the location of motor or sensory cortex and the identification of language or memory cortex is an adventitious research technique usually incidental to the clinical requirements of surgery for epilepsy. During certain other neurosurgical procedures, such as removing an arteriovenus malformation (AVM) or occasionally during tumor surgery, there is also a need for and opportunity to conduct fESB. The purpose of fESB is simple: before removing a piece of damaged or malfunctioning brain, it is important to know whether or not that part plays a critical role in motor, sensory, language, or memory functions. Experience has shown that such functions may be found in epileptic brain, in an AVM, or in a tumor; information about these functions may modify a neurosurgeon's operative plans. Experience has also shown that, on occasion, a region of brain that, based on textbook maps, one supposes to be critical for motor, sensory, or language functions is in fact silent and thus could be removed.

Thus, in assessing fESB's contribution to neurolinguistics, one should keep in mind that there are structural brain abnormalities with accompanying neurological impairments in all subjects studied using this technique. On the other hand, it is equally important to remember that fESB has been applied to both presumed normal and abnormal brain regions with quite comparable results. Furthermore, electrical stimulation results have been comparable no matter what the lesion type—tumor, AVM, gliosis, and so on.

7-1. TECHNIQUES

Over the past quarter-century, certain features of fESB have become standard: current is delivered across bipolar electrode contacts, using pulsed square waves with equal positive and negative excursion so that the net coulombs entering brain tissue is zero; stimulus duration is typically on the order of seconds, and thus there is no demonstrable tissue damage associated with the fESB. Because the distance between electrodes is minimal, usually on the order of 5 mm, one may regard fESB as creating a temporary, immediately reversible, focal area of functional loss, one discrete location at a time. These facts, coupled with the experimental rigor employed by most researchers using fESB, argue for a fair degree of confidence in the general results.

There are three different fESB techniques, one acute and two chronic: *(a)* a hand-held electrode, usually bipolar, that briefly contacts the brain's surface during a neurosurgical operation; *(b)* a flexible strip or grid of electrodes that is placed subdurally on the surface of the brain for varying periods from hours, intraoperatively, to weeks, interoperatively; and *(c)* a small-diameter rod (depth electrode) with several contacts along its length that is inserted into the brain for varying periods, from days to weeks, interoperatively. Hand-held electrodes can only be used intraoperatively and only on the brain surface that has been exposed when a section of the cranium has been removed. Both grid and strip electrodes may be slipped under the dura mater to make contact with, for example, mesial frontal lobe, inferior temporal lobe, or posterior occipital lobe, areas of brain that are adjacent to the area exposed by the skull opening; strip electrodes may also be inserted through burr holes, thus not requiring a full skull flap procedure. Depth electrodes are usually placed through small burr holes, which may be made in just about any part of the cranium lying over the brain; unlike the other two techniques, depth electrodes are designed to access deep brain structures, and, unlike the other two types, depth electrodes cause a small lesion along the track of the electrode.

7-2. ADVANTAGES AND DISADVANTAGES OF EACH TECHNIQUE

Although in the past different neurosurgical centers tended to use one or another of these techniques predominantly, today most neurosurgeons will adapt their technique to the exigencies of each particular case. Strip electrodes are commonly used by everyone because they can be placed in areas and in ways that simply cannot be duplicated by hand-held or depth electrodes. Depth electrodes, as well as strip electrodes, have the advantage of being able to access both hemispheres at the same time, whereas the hand-held electrode is restricted to the exposed craniotomy field during an operation, thus to a single hemisphere. Strip electrodes have varying numbers of contacts, from a few to 5 dozen; because the electrode contacts are embedded in a flexible plastic sheet, they may be trimmed to different shapes—from a long strip with a single line of contacts to a large square with a field of contacts. Thus, in terms of area of coverage, strip electrodes are quite useful; the contacts are fixed in the plastic sheet, but the whole strip can be repositioned. Thus a varying number of contacts will actually not be active, fortuitously being positioned over a sulcus, an artery, or a vein. For excellent control over the precise site to be stimulated, the hand-held electrode is superior. Depth electrodes, of course, are able to access brain structures below the surface; since they must be inserted directly through brain substance and brain substance is highly vascularized, there is greater risk of medical complications with depth electrodes than with the other two types. Except in certain cases of chronic, thalamic pain, depth electrodes are typically not left in place for long periods of time, although they could be; strip electrodes are often left in place for a week, which, of course, greatly increases the opportunity for clinical and experimental observation of fESB. The hand-held electrode is limited to use during neurosurgical procedures.

7-3. BACKGROUND: EARLY FINDINGS

Focal electrical stimulation was first used in the late 19th century, in an attempt to answer the question originally posed by Albrecht Haller as to whether the cerebral cortex was "excitable" or passive (Neuberger, 1897/1981). The first report of the effects of faradic current applied directly to the human brain was by Bartholow (1874). In reviewing the status of localization studies at that time, Bartholow stated:

> The most important results as regards localization of functions, have been obtained by far-adization of limited parts of the brain. The demonstrations recently made in this way by Fritsch and Hitzig (*Archiv f. Anat. Physiol. u. Wissenschaftliche Medizin, 3,* 1870) and by Ferrier (*West Riding Lunatic Asylum Reports,* vol. 3) are entirely opposed to the well-known experiments of Magendie, Longet, Flourens, Vulpian and others, which had apparently shown the inexcitability of the cerebral hemispheres. (1874, p. 306)

Stimulating both the left and right anterior parietal lobes through a tumor-caused skull defect, Bartholow observed contralateral muscular contractions in hand, arm, leg, and neck muscles in his patient. It was quickly realized that the technique could be used to functionally map the motor and sensory regions of the cortex, and soon fESB was considered to have furnished the proof of functional cortical localization (Bolton, 1911). The first report of stimulation mapping in a conscious human patient undergoing neurosurgery was made by Cushing (1909); he stimulated the postcentral gyrus and produced sensations in the contralateral limbs.

7-4. MODERN STUDIES

The most extensive and best-known premodern series on the effects of electrical stimulation, including speech and language effects, are those of Penfield and his colleagues, collected in Penfield and Erickson (1941), Penfield and Jasper (1954), Penfield (1967), Penfield and Roberts (1959), and Penfield and Perot (1963). The modern era of fESB using the hand-held electrode was ushered in by the work of John van Buren and colleagues (Ojemann, Fedio, & Van Buren, 1968), continued in the work of George Ojemann and colleagues (Ojemann, 1975; Ojemann & Whitaker, 1978a; Ojemann & Whitaker, 1978b; Ojemann & Mateer, 1979), surveyed in Ojemann (1983), and summarized in Ojemann, Ojemann, Lettich, and Berger (1989). Modern studies using the depth electrode technique (pioneered by J. Talairach) were developed by Jean-Marc Saint-Hillaire and colleagues (Richer, Martinez, Robert, Bouvier, & Saint-Hillaire, 1993) and the modern studies using chronically implanted strip electrodes were pioneered by Ron Lesser and his colleagues (Gordon, 1997; Gordon, Hart, Lesser, & Arroyo, 1996; Gordon et al., 1994; Crone, Hart, Boatman, Lesser, & Gordon, 1994; Hart & Gordon, 1992; Boatman et al., 1992; Gordon et al., 1991; Lueders et al., 1991; Gordon, Lesser, & Hart, 1989; Lueders et al., 1987; Lesser et al., 1987.

7-5. COMPARISON WITH OTHER MAPPING TECHNIQUES

Optical image mapping takes advantage of the fact that a decrease in the light reflectance properties of cortex correlates with increased neuronal activity. Optical image studies are similar to fESB in that they also exhibit discrete sites for motor control of the vocal tract and for naming responses. Studies that have compared the location of optical changes with naming with the fESB location of naming sites have shown a close correspondence of results (Haglund, Ojemann, & Hochman, 1992; Hochman, Whitaker, Haglund, & Ojemann, 1995); and fESB identification of motor cortex corresponds to magnetic resonance images of gyral topography (Berger, Cohen, & Ojemann, 1990). Overall, fESB localization data is consistent with data obtained from other sources such as PET, fMRI, regional electrical activity recorded directly from the cortex, magnetoencephalography, and single-unit data (Gordon, 1997; Ojemann, Creutzfeldt, Lettich, & Haglund, 1988).

7-6. GENERAL RESULTS

The most systematic and interesting set of critical commentaries of fESB—technique and results—is to be found in the discussions following the target article by Ojemann (1983). The comments by Cooper (1983), Marshall (1983), and Studdert-Kennedy (1983) were particularly relevant to the interpretation of fESB data. Many of the questions raised in these commentaries have been addressed in electrical stimulation research since the early 1980s; combining the studies of Ojemann's and Lesser's teams, more than 200 subjects have been studied using fESB. Broadly speaking, fESB tends to support the classical model in neurolinguistics; as stated by Gordon et al. (1994), "language functions need to be understood in terms of their underlying componential processes. For the most part, the gross outline of these componential processes (e.g., for reading: a visual orthographic lexicon, semantics, a semantic-to-phonologic conversion process, and output phonology) resembles the general outlines currently popular" in neurolinguistics (p. 5). They continue: "evidence suggests the existence of very fine subdivisions in the representation of language (perhaps analogous to the multiple maps seen in the visual system" (ibid.) and conclude that the fESB results indicate that the more finely grained a linguistic function is, the more narrowly localized is its cortical representation.

7-7. SPECIFIC RESULTS

Franz Joseph Gall's organology of the early 19th century was, above all, a neuropsychology of individual differences. The question of individual differences also occupied the minds of the late-19th- and early 20th-century anatomists who, by all accounts, had given up hope of describing a consistent gyral geography of the human brain (Whitaker & Selnes, 1976). Individual difference has returned with a vengeance in fESB research; if anything has been clearly demonstrated by electrical stimulation studies, it is that no two individual brains have the same geographic/functional location for language. In addition, fESB has provided evidence of linguistic functioning in regions of the brain heretofore suspected but not known to be language cortex, namely, the insula, anterior temporal lobe, inferior/basal temporal lobe, and ventrolateral thalamus. Studies of patients who speak more than one language (Ojemann & Whitaker, 1978a, 1978b; Ojemann, 1983; Rapport, Tan, & Whitaker, 1983) have clearly demonstrated that multiple languages are differentially localized in the cortex. Since these reports were published, a dozen more bilingual patients have been studied, with the same results: there are sites where one or the other language is represented exclusively, sites where both languages are equally represented, and sites where both are represented, but to an unequal degree (Ojemann & Whitaker, unpublished data).

Dissociations of components of language processing are the meat and potatoes of neurolinguistic research; in fESB research, dissociations are commonplace, due probably to the precision with which fESB can functionally disrupt a relatively small piece of brain, as noted by Gordon et al. (1994). Some of the observed dissociations are a pure impairment of input phonology, a pure impairment of auditory comprehension, a pure impairment of spontaneous speech, a pure impairment of visual confrontation naming, and a pure impairment of single-word reading, where "pure" means that fESB did not impair any other language functions at that time and at that site. It is, of course, also common to find two or more linguistic functions interrupted by fESB at the same site (Ojemann & Whitaker, 1978b; Ojemann, 1983; and the various papers by Gordon and colleagues). The patterns of single and multiple functional disruption appear to be consistent with a network of interrelated nodes for different functional systems. Some of the mixed functional localizations are rather interesting in that they have not been reported in the lesion literature, for example, an auditory comprehension impairment and impaired repetition with impaired visual confrontation naming and naming to definition but with completely preserved spontaneous speech and reading

(Gordon et al., 1994). Occasionally an unusual localization is found, one that seems quite different from the standard lesion literature, for example, the finding of impaired word reading only when anterior inferior temporal lobe was stimulated (Gordon et al., 1995). Functional electrical stimulation has also been employed in the nonlanguage-dominant (right) hemisphere; although some studies report no language functions, others report that simpler, overlearned lexical tasks may employ nondominant hemisphere systems (Bhatnagar & Andy, 1981; Andy & Bhatnagar, 1984).

Ever since Penfield and Jasper (1941) reported unusual psychic phenomena associated with fESB, there have been unique, clinical observations of linguistic phenomena elicited by electrical stimulation. Until these are experimentally supported, they remain tantalizing windows on future research, but their intrinsic interest warrants some mention. In collaboration with George Ojemann and colleagues, the present author has observed effects of fESB such as the following: jargon aphasic responses and semantic paraphasic responses in visual confrontation naming tasks, function word and derivational affix errors in reading tasks and semantic category errors (i.e., fruits and vegetables versus musical instruments being differentially affected by focal electrical stimulation). Coupled with the increased availability of optical and metabolic imaging techniques that can be correlated with the results of fESB, it would seem that the future of this research is bright and promising.

CHAPTER 8

Tomographic Brain Imaging of Language Functions

Prospects for a New Brain/Language Model

Jean-François Démonet
INSERM U455, Hôpital Purpan, Toulouse Cedex, France

Advanced brain mapping techniques such as Positron Emission Tomography and functional Magnetic Resonance Imaging represent a major step toward better understanding the neural correlates of language functions, as these studies, conducted in normal subjects while performing language activation tasks, produce results that are independent of the classical clinico-anatomic method in aphasic patients. Both functional mapping and lesion-based methods may be combined to study the functional basis of recovery from aphasia. However, the changes in brain regions that are recorded by these techniques actually reflect very complex phenomena related to functions (and dysfunctions) of widely distributed and adapting neural ensembles that subserve language processes. As a new scientific field, functional imaging studies require the combination of many different expertises, from cognitive psychology to physics, in order to get accurate and comprehensive interpretations of their results. Some key methodological issues and examples of results are presented in this chapter.

Since the 19th century, many attempts have been made to establish clear-cut relationships between aphasia and brain lesions, using aphasia as a paradigm to build up a model of brain/language relationships. Although some parts of this lesion-based model are quite well established (e.g., the critical role of Wernicke's area for decoding and encoding spoken words), many other aspects still remain poorly specified due to several factors of complexity. First, the relationships between lesion anatomy and language disorders appear more stable when aphasic *symptoms*—rather than syndromes—are considered (Poeck, 1983; Willmes & Poeck, 1993). Second, in accord with current theories on the neural correlates of cognitive functions (Damasio, 1989; Mesulam, 1990), many aphasic symptoms, for example, anomia, correspond to various lesion sites and are, in fact, linked to a "distributed anatomy" of lesions (Alexander, Hiltbrunner, & Fischer, 1989), the same language disorder being induced by lesions of different nodes of a network distributed over the left hemisphere, if not the entire brain. Third, a variety of "additional" factors have been invoked to account for "exceptional aphasias" (Basso, Lecours, Moraschini, & Vanier, 1985), that is, cases that are at variance with the classical teaching on lesion-symptom relationships. Subject-specific factors (e.g., handedness, age, gender, cultural factors) together with recovery and compensation phenomena are very likely to influence the aphasiological model. Because of highly complex interactions between these factors and the consequences of brain lesions, the heuristic value of the aphasiological model by itself appears limited; and modern functional imaging techniques, for example, PET or fMRI, have the potential to renew this research because they generate data that are *independent* of the aphasiological framework. These techniques, which have dramatically improved recently in terms of sensitivity, resolution, and reliability, can be combined with a *new paradigm,* namely, *cognitive brain activation,* to provide information on the location of functional changes in the brain while normal subjects perform language tasks. Language activation studies in normal subjects should therefore fruitfully complement previous aphasiological views on the brain correlates of language functions. Most important, these techniques should, sooner or later, help us understand more about recovery from acquired cognitive dysfunctions such as aphasia.

8-1. ISOTOPIC AND NONISOTOPIC TOMOGRAPHIC IMAGING: PHYSIOLOGICAL BACKGROUND, TECHNICAL AND METHODOLOGICAL ISSUES, AND THEIR IMPLICATIONS FOR COGNITIVE STUDIES

8-1.1. Using Blood Flow as a (Relatively) "Local" Index of Neural Activity

Isotopic techniques—for example, PET—have been used for about 30 years to measure local changes in functional indices of human brain metabolism and blood flow (Lassen & Ingvar, 1961). Computed from measurements of the dynamics of radioactivity counts emitted from the brain by short-lived isotopes such as oxygen 15, regional Cerebral Blood Flow (rCBF) is the most frequently used functional index of metabolic activities in underlying neural populations. Local changes in brain arterioles and capillaries are related, although not proportionally (cf. section 8-1.3. on functional Magnetic Resonance Imaging), to variations in energy metabolism (involving, among others, glucose and oxygen) of the corresponding local populations of neurons ($n > 10^5$), and above all, their synapses, within submillimetric functional domains of the cerebral cortex (Grinvald, Lieke, Frostig, Gilbert, & Wiesel, 1986). Adaptive changes—for example, dilation and blood flow increase within local vasculature after energy demands in synapses have begun—are observed over a time period of few seconds: a similar phase shift is observed for rCBF after synaptic activities have returned to their initial level.

8-1.2. Positron Emission Tomography: Progress and Limitations

Not only is the intimate regulation of such adaptive vascular changes not well understood, but also techniques such as PET have general limitations in terms of spatial and temporal resolution that should not be overlooked. By comparison to SPECT or first generation PET techniques, the sensitivity of PET cameras have recently improved together with their spatial resolution (about 5 mm in 3 axes) (Townsend et al., 1991; Bailey, Jones, Friston, Colebatch, & Frackowiak, 1991) so that it is currently possible *(a)* to coregister PET and morphological, millimetric MR images onto the same stereotaxic space and *(b)* to conduct analyses in individual subjects. This eliminates the problems associated with intersubject anatomical and functional variability and the confounding effects of across-subjects averaging (as an example of a subject-based language activation study, see Herholz et al., 1996). These improvements also allow one to reduce the radiation dose administered for each PET measurement; the number of PET measurements that can be made in an individual is usually about 12 in one pass.

The temporal resolution of this technique, even considering recent improvements (Silbersweig et al., 1993, 1995), remains very poor (greater than 10s) compared to the typical time scale of cognitive processes (less than 1s). Prolonged periods of data acquisition with PET imply major constraints on the design of language activation paradigms. Activation tasks must be designed so that the psycholinguistic processes induced by the task do not vary over the rCBF recording period (typically about 60s), and that habituation or automatization does not progressively appear, unless these phenomena are purposefully under consideration in the study. For instance, habituation may originate from tasks that are too easy (for a given subject) or by repetition of the same stimuli in the same scanning session. In general, tasks that are likely to keep subjects firmly attending, therefore, are preferable to so-called passive tasks in which subjects are asked to "only perceive" stimuli. Such a vague instruction actually means that the nature of the cognitive components of the task is just ignored and therefore makes the interpretation of any between-task differences in brain activity difficult a priori. Instead, the cognitive components that are involved in different experimental tasks should be defined as precisely as possible. Behavioral indices (such as response accuracy or reaction time) recorded during PET scanning also help to demonstrate that stimuli were actually processed by the subjects in a manner compatible with the theoretical assumptions underlying the task design (Démonet, Chollet et al., 1992). As a drawback of this methodological option, it follows that such activation language tasks require a strong attentional control, and are very artificial when compared with "natural" language processing. The interest of studies specifically devoted to more automatic modes of language processing should certainly not be overlooked (Price, Wise, & Frackowiak, 1996), and we shall see that automatic access to language representations may have an important role even in quite controlled experimental conditions. Moreover, investigations on "real" language—that is, continuous, discursive language samples—is obviously the final goal of functional brain-imaging studies of human language (as examples of tentative approaches to the supralexical and discourse level of language processing, see Démonet, Celsis et al., 1992; Kawashima et al., 1993; Mazoyer et al., 1993; Bottini et al., 1994). However, even at the discourse level, precisely controlled activation tasks seem still needed before moving on to more "ecological" experimental conditions.

8-1.3. Functional Magnetic Resonance: An Imaging Technique Faster Than PET but with Some Drawbacks as Well

Nonisotopic tomographic functional imaging, that is, fMRI, does not use radioactive compounds and has a better time resolution than PET (although fMRI time scale still is poorer than that of electrophysiological techniques). For the time being, it seems that fMRI complements rather than substitutes for PET in brain-mapping studies. Indeed, even though increasingly sophisticated fMRI studies (using high-field magnets) have appeared in the literature, this method also has its drawbacks, such as spurious signal changes related to various factors, namely, head motion (even millimeters in amplitude), heart and respiratory rates, and so on. The sensitivity of the method vis-à-vis cognition-related neural activities, together with its ability to explore, in a one-pass fashion, the entire brain volume, still appear inferior to those of the most recent PET equipment. The key advantage of fMRI compared to PET is rather the number of separate measurements of functionally significant signals that can be achieved in a short period of time. The most widely used method, the Blood Oxygenation Level-Dependent (BOLD) effect (Ogawa, Lee, Kay, & Tank, 1990), is based on the paramagnetic properties of DeOxyHemoglobine and takes advantage of the fact that the local vascular response to "energy demands" of neurons typically exceeds cellular needs (Fox & Raichle, 1986), and consequently induces local decreases in the blood concentration of DeOxyHemoglobine (even though its metabolic rate actually increases due to energy consumption in the nearby synapses). It follows that the observed fMRI BOLD signals actually increase in an activated region (i.e., a region with increased synaptic firing).

Using the fast echo-planar imaging (EPI) acquisition technique (Mansfield, 1977) and the BOLD method, the main part of the cortical "language area" (about 10 axial slices 5mm thick) can typically be scanned in 1.4s (less than 0.14s per slice); this is repeated many times in a scanning session for one subject (e.g., 80 repetitions per session). Multiple acquisitions therefore may be used to analyze several behavioral changes in one subject over a single session. In general, binary alternations of "resting" and "active" conditions are used that last about 10s or more and involve several acquisitions (and stimuli) (e.g., 10 acquisitions per condition).

These alternations are in fact considered as the entry function (main frequency of about 0.05 Hz or less) that can be related to the observed signal changes; fMRI signal processing thus relies on correlational methods showing voxels (elementary volume units of the image matrix) in the explored brain volume whose value "oscillates" with similar frequencies (Bandettini, Wong, Hinks, Tikofsky, & Hyde, 1992). These voxels are thought to reflect local neural responses to experimental behavioral changes. The correlational approach (see section 8-2.2. on the correlational method) permits one to compensate for the strong background noise of fMRI signals (not to mention the strong *acoustic* noise produced by the machine) and their relatively small signal-to-noise ratio (especially using routine, 1.5 Tesla MRI machines) that considerably lower the statistical power of the classical, subtractive across-task comparisons (e.g., t tests between signal amplitudes, averaged over blocked conditions, namely, statistics on stimulation minus rest comparisons).

However, acquisition speed using the EPI acquisition mode also has its drawback, as it tends to reduce the spatial resolution of functional MR images compared to that of morphological MR images. Finally, it should be noted that the most sensitive fMRI methods are those using high-field permanent magnets (e.g., 3 or 4 Tesla, which is about 3 or 4 times 10^4 greater than the earth's magnetic field) that *(a)* are expensive machines only available in few research centers, and *(b)* may cause biological side effects that have not yet been fully investigated.

The reader may find additional information on techniques in two recent contributions (Bandettini & Wong, in press; Toga & Mazziotta, 1996).

8-2. THE ACTIVATION PARADIGM IN NORMAL SUBJECTS

There are several ways to use functional brain imaging to expand our knowledge beyond the limits of the classical lesion-based model. The principal one is to address whether the regions related to impairments of specific language functions when damaged might be *activated* in *normal* subjects exhibiting the *same functions* while they undergo a particular language task.

For instance, several authors, such as Geschwind (1965), Heilman, Rothi, McFarling, & Rottmann (1981), and Cappa, Cavalotti, & Vignolo (1981), claimed that phonological disorders are associated with lesions close to the left sylvian fissure whereas lexical semantic disorders are linked to lesions of regions that are more distant from the fissure, such as the inferior parts of the parietal or temporal lobes. In several PET activation studies, Démonet and his colleagues (Démonet, Chollet, et al., 1992; Démonet, Price, Wise, & Frackowiak, 1994a; Démonet, Price, Wise, & Frackowiak

1994b) addressed the question whether such a topographical segregation between phonological and lexical semantic processes might be observed in normal subjects when comparing brain activation induced by two language tasks. Each task crucially depended on one of these two processes, the other process being less important. The results were based on "subtractive" comparisons between the two activation tasks (see Figure 1). In a given comparison, significant local increases of blood flow were computed that corresponded to one task against the other, which was considered as reference. The results proved to be in good accord with previous observations on aphasic symptoms and the topography of the corresponding lesions. However, a satisfactory convergence was not observed in all activation studies (nor between them) over the past few years, thus suggesting difficulties in using such simple across-task subtractions.

8-2.1. Limitations of the Subtraction Method

Many recent PET language-related activation studies now provide a large amount of data on the functional anatomy of language, mainly based on the "subtractive" method (Petersen, Fox, Posner, Mintun, & Raichle, 1988; Posner, Petersen, Fox, & Raichle, 1988; Petersen, Fox, Posner, Mintun, & Raichle, 1989; Petersen, Fox, Snyder, & Raichle, 1990; Wise et al., 1991; Démonet, Chollet, et al., 1992; Howard et al., 1992; Sakurai et al., 1992; Sergent, Ohta, & Mac Donald, 1992; Sergent, Zuck, Levesque, & Mac Donald, 1992; Zatorre, Evans, Meyer, & Gjedde, 1992; Mazoyer et al., 1993; Paulesu, Frith, & Frackowiak, 1993; Sakurai et al., 1993; Bottini et al., 1994; Démonet et al., 1994a, 1994b; Price et al., 1994; Price, Moore, & Frackowiak, 1996; Price, Wise, & Frackowiak, 1996; Raichle et al., 1994; Buckner, Raichle, & Petersen, 1995; Damasio, Grabowski, Tranel, Hichwa, & Damasio, 1996; Fiez et al., 1995; Fiez, Raichle, Balota, Tallal, & Petersen, 1996; Herholz et al., 1996; Martin, Wiggs, Ungerleider, & Haxby, 1996; Price, Moore, & Frackowiak, 1996; Price, Wise, & Frackowiak, 1996; Warburton et al., 1996; Zatorre, Meyer, Gjedde, & Evans, 1996). Despite their interest and richness, these data have proved to be increasingly difficult to interpret; and sometimes they are contradictory. Such discrepancies have different origins. They partly relate to various, and often subtle, alterations of some experi-

---

FIGURE 1 Statistical maps showing, on transparent views of a standardized brain, areas of significant blood flow increases (p < .001, after corrections for multiple comparisons) in a lexical semantic task when a phonological task is taken as baseline (top) and vice versa (bottom). These tasks both consisted in monitoring targets among heard stimuli that were adjective-noun pairs in the lexical semantic task and polysyllabic (consonant-vowel) non-words in the phonological task. Lexical semantic processes induced predominant activation in the left middle temporal and inferior parietal regions (probably organized in a complex network involving also left prefrontal, posterior cingulate and right temporo-parietal regions). Phonological processes resulted in enhancement of activities in a more restricted left-sided network localized in the banks of the left sylvian fissure (see Démonet et al., 1994b for details). These features are in good accord with predictions drawn from studies of lesional localization in aphasia.

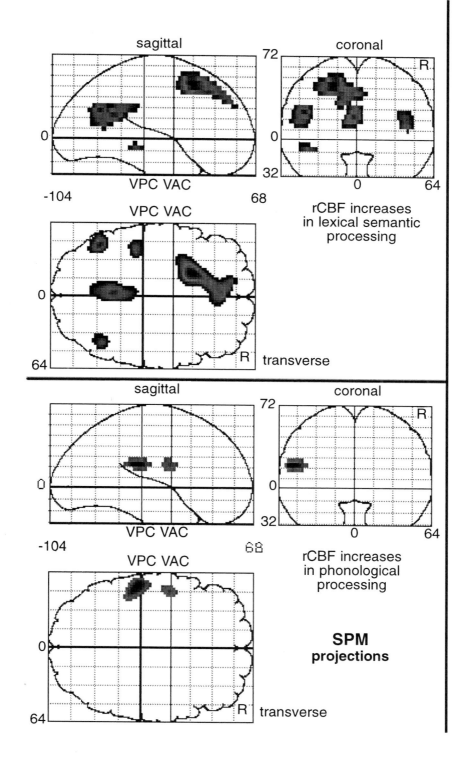

sagittal

coronal

VPC VAC

-104   68

VPC VAC

rCBF increases
in lexical semantic
processing

transverse

sagittal

coronal

VPC VAC

-104   68

VPC VAC

rCBF increases
in phonological
processing

**SPM
projections**

transverse

mental aspects of the activation procedure that may have a major impact on the results. Two examples of these seemingly "secondary" factors that are actually crucial determinants of brain activation (for a review, see Démonet, 1995) will be discussed here as their influences have recently been investigated in detail.

The first one relates to conditions of stimulation such as rate of presentation and exposure duration of stimuli, systematically explored by Price et al. in a series of papers (1992, 1994, 1996). In their first study, these authors described a linear relationship between rate of auditory stimulation (listening to words) and amount of activation in the primary auditory cortices, whereas the rCBF response was unaffected by the rate in the left association auditory cortex (Wernicke's area). Further, complex interactions were described in reading tasks between presentation rate, exposure duration, explicit instructions, and implicit processes involved in the tasks.

Another very important factor is related to the degree of familiarity with the task. Raichle et al. (1994) described dramatic changes in rCBF activation when comparing verb generation in naive subjects and after the task had been overlearned and automatized in these subjects.

Apart from these complex, and largely unexpected, experimental factors influencing the results of language activation studies, the major source of difficulty in summarizing the available activation data and accounting for discrepancies originates from the subtractive method itself when applied to complex cognitive processes such as language functions. In its most simplistic version, across-task subtraction implies that its net result will isolate *the* brain correlate of a particular language component as this is thought to be "present" in the "experimental" task and "absent" in the reference task. Although seemingly straightforward, such simple comparisons cannot account for language-related functional changes for two main reasons related to the relevant neural and cognitive architectures, respectively.

First, the neural correlates of language functions seem to be widely distributed over large portions of the brain, and therefore are not likely to be anatomically restricted to a single (or even small sets of) activation peak(s). However, to reduce the number of false positives, statistical analysis most frequently consists in thresholding the peaks of rCBF changes at their very top, corresponding to very low p values (Bailey et al., 1991). Conservative statistics applied to reduce the alpha risk may therefore overlook functionally significant activation signals corresponding to lower and more widespread rCBF changes (Poline & Mazoyer, 1993), especially when dealing with cognition-related activations in association cortices.

Second, because of the fundamental parallelism of language processes and because any language task will automatically engage many of these processes (perhaps to various degrees of intensity), a single subtractive comparison can hardly isolate a given language component or function. Rather, only the combination of several comparisons can allow for a *relative* isolation of such a component, taking into account not only blood flow increases but also *decreases* (for a discussion of the functional significance of blood flow decreases, see Démonet, 1995; Wenzel et al., 1996) between, for instance, an "experimental" active condition (supposed to especially involve the cognitive component under study), "reference" active condition(s) (in which this component is *less active,* rather than totally absent), and a "resting" condition (whatever "rest" might mean in a living brain) (for further discussion of such technical issues, see Démonet, 1995; Démonet, Fiez, Paulesu, Petersen, & Zatorre, 1996).

8-2.2. The Correlational Approach to Brain Activation and Its Implication in fMRI Studies

Alternative approaches exist that can bypass some of the shortcomings of the "subtractive" method. The correlational method for exploring functional anatomy in the brain has been proposed for several years (e.g., Horwitz, Duara, & Rappoport, 1984; Metter, Riege, Kuhl, & Phelps, 1984; Moeller, Struther, Sidtis, & Rottenberg, 1987) but has been applied only in a few cognitive activation experiments using PET (Bartlett, Brown, Wolf, & Brodie, 1987; Frith, Friston, Liddle, & Frackowiak, 1991; Horwitz et al., 1993) and non-PET techniques (Goldenberg, Podreka, Steiner, & Willmes, 1987; Démonet, Celsis, et al., 1992; Lagrèze, Hartmann, Anzinger, Schaub, & Deister, 1993).

Linear correlations can be studied in an across-subject within-task design or vice versa. In an across-task correlational design, Frith et al. (1991) demonstrated that this method provides interesting findings that complement the results obtained by subtraction. In a verbal fluency task compared to a lexical decision task, these authors observed significant decreases in the superior temporal regions, whereas increases were found in the left dorso-lateral prefrontal cortex. The correlational analysis showed that the focus of rCBF increase in the dorso-lateral prefrontal cortex was negatively correlated with the superior temporal region, thus suggesting the existence of a functional link between these two areas, with a possible inhibitory influence of the dorso-lateral prefrontal cortex on the superior temporal cortex.

Friston, Frith, Liddle, & Frackowiak (1993) stressed, from a theoretical point of view, the interest of correlational methods to study brain functional connectivity, that is, functional networks such as those supporting language functions. Most important, they mentioned that these methods might be a useful formal tool to describe functional data acquired by different techniques (e.g., PET, functional MRI, and electrophysiology) using different time scales. They described a nonlinear, nondirected (i.e., not referring to any particular focus of "activation") correlational method using principal component analysis that was applied, as an illustration, to an activation experiment on an orthographic verbal fluency task, compared to a phoneme repetition task. This study identified two factors accounting for a large portion of the observed variance of rCBF, corresponding, respectively, to two functionally independent sets of distributed brain components. Whereas the second factor probably related to practice effects over the series of 12 PET scans, the first one was identified as an "intentional" system that corresponded to the verbal fluency task and involved the left dorso-lateral prefrontal cortex, Broca's area, the thalami, and the cerebellum. Interestingly, the two subcortical regions were not found to be activated in previous similar studies conducted in the same laboratory but analyzed in a classical, subtractive way. This method seems a powerful method to assess the brain activation as it concerns signal changes in a *series* of measurements over which the influence of not only categorical but also graded variables (e.g., psychophysical variables such as rate of stimulation, exposure duration of stimuli, or even psycholinguistic variables such as lexicality, imageability of words, and so forth) can be explored. One limitation of this method using PET, however, is that the number of repeated measurements can hardly be more than 12 because of the radiation dose that can be safely administered to subjects.

This drawback does not exist with fMRI in which the number of repeated measurements only depends on the length of time that subjects agree to stay in the machine.

Some fMRI studies were oriented to clinical applications (Desmond et al., 1995; Cuenod et al., 1995); language-related fMRI experiments have mainly consisted so far of replications of previous PET studies using the same experimental designs such as fluency tasks and across-task subtractions. For instance, McCarthy, Blamire, Rothman, Gruetter, & Shulman (1993) focused on activation in the left inferior prefrontal cortex during a verb-generation task and supported Petersen et al.'s (1988) results regarding this region. Binder et al. (1995) and Shaywitz, et al. (1995), exploring lexical semantic and phonological tasks, respectively, confirmed a series of previous PET findings (e.g., Démonet, Chollet, et al., 1992; Démonet et al., 1994a). We have also partly confirmed our PET results, using the same activation paradigm with fMRI equipment (Berry et al., 1995) and the same applies to a tentative replication of previous PET results on brain correlates of verbal working memory by Paulesu et al. (1993) and Paulesu et al. (1995).

As already mentioned, the correlational method applied to fMRI takes advantage of the great number of repeated measures provided by the technique. Correlations are therefore studied between the time series of recorded signals and various external, behavioral variables that can be used to modelize an input function, for example, an alternation between stimulated and unstimulated states that is thought to be reflected in observed fluctuations of fMRI signals. Interest in this method has been stressed by Binder et al. (1994), who described activation signals in both superior temporal regions significantly correlated to various auditory language stimulation. However, the validity or usefulness of this type of analysis still remains to be explored in more sophisticated paradigms, involving three (or more) cognitive activation conditions; indeed, these conditions might interact in a complex way, and the modelization of such interactions as input function is likely to be far more complex than that of the binary paradigm used by Binder et al. (1994).

8-3. IMAGING CORRELATES OF LANGUAGE FUNCTIONS IN DAMAGED BRAINS

8-3.1. Resting State Studies

Compared to activation experiments, these studies represent a more classic approach to the brain/language problem as they investigate metabolic abnormalities that are induced by lesions and are seen in functional images during prolonged measurements of regional brain metabolism (mostly using the FluoroDeoxyGlucose PET technique), during which patients do not receive any particular stimulation and are therefore considered to be in a "resting" condition.

A fair number of glucose steady-state PET studies were done in the eighties, especially in North America, with relatively small groups of patients (Metter et al., 1981; Metter et al., 1983; Metter et al., 1984; Metter, Hanson, et al., 1986; Metter, Jackson, et al., 1986).

One of the major contributions of these studies was to demonstrate the existence of massive remote effects of lesions with metabolic depression occurring far away from the anatomical site of the actual lesion. The most striking example of these remote effects relates to the so-called subcortical aphasias in which hypometabolism in the ipsilateral cortex is very frequently observed (Baron et al., 1986).

Some of these studies also reinforced previous findings that direct or indirect damage to a specific region such as the left posterior temporal cortex is a crucial factor for both aphasia type and prognosis (Karbe et al., 1989).

Finally, some follow-up studies have been done (Cappa, et al., 1991) and some others are currently reported to be ongoing. However, these longitudinal data are still unclear, if not contradictory. In general, the functional significance of changes in brain metabolism observed at rest, remains to be clarified. For instance, remote hypometabolic effects may represent at least two different phenomena. On the one hand, the affected regions may be only de-afferented but still can participate in functional activation via other connections or networks. On the other hand, these hypometabolic regions, particularly when they lie not too far away from the morphological lesion, or within the same vascular territory, may be actually affected by neuronal loss, due to ischemic "penumbra," resulting in an irreversible loss of function.

8-3.2. Activating Damaged Brains

The shortcomings of resting state PET studies obviously lead one to explore the functionality of the undamaged parts of the brain by using activation tasks to test whether some spared territories could be involved when patients performed these tasks. In fact very little has been done so far using up-to-date methodological standards of PET activation that is high-resolution rCBF recordings using the Oxygen 15 technique.

Weiller et al. (1995) conducted one of these studies. They studied 6 Wernicke-type aphasic patients with retro-rolandic lesions and good recovery. By comparison to the activation observed in covert nonword-repetition and verb-generation tasks in normal subjects, aphasic subjects of course demonstrated no activation in the damaged region and increased, supranormal activations in the right hemisphere, both in the superior temporal and the inferior frontal regions on both tasks.

Although appealing at first glance, this type of study soon appeared particularly complex because it combines two main sources of variance: one is related to brain lesions and aphasia, and their factors of complexity were briefly mentioned earlier in this chapter; the other comes from the many "secondary" experimental factors that may distort the results of cognitive activation even in normal subjects, as already addressed earlier.

In general, such a complexity suggests that activation should only be explored using single-subject studies; this is possible today thanks to technical improvements in imaging methods (see, for instance, Engelien et al., 1995).

The major factor likely to affect brain activation in aphasic patients is obviously the lesion. In cases of massive tissue loss or impairment, a pervasive hypometabolic effect is frequently seen over the entire hemisphere, and a (poorly specific) functional activation restricted to the right hemisphere is in fact the only finding that one could possibly predict as a result of any stimulation.

However, in cases of less massive lesions, the question can be raised whether such right-sided activations, together with left-sided ones among the spared territories, might represent neural activities specifically related to language functions and furthermore illustrate in some way the mechanisms of recovery and compensation following brain lesion.

Involvement of right-sided structures during language activation tasks has frequently been reported in the literature on activation in normal right-handed subjects. We described right temporal activations in our own studies using auditory comprehension tasks (Démonet, Chollet, et al., 1992; Démonet et al., 1994a) and interpreted this finding as an automatic engagement of early acoustic-phonetic aspects of stimuli processing, as these signals tended to vanish in the most difficult, highly specific variants of our tasks. It could be speculated, however, that these right-sided foci of activation may represent potential substrates for compensatory mechanisms after left-sided lesions.

In aphasic patients, available functional imaging data relevant to the question of mechanisms for compensation and recovery are too scarce at present to draw even a general sketch of such phenomena.

To assess the specificity of activations observed in aphasic patients during language tasks and their relevance to functional compensation, further studies should obviously seek correlations between such imaging findings and behavioral changes. Moreover, it should be addressed whether, in follow-up studies, functional changes observed in undamaged brain regions (e.g., right temporal regions) may parallel specific language improvements (e.g., efficiency of lexical semantic decoding processes) over time.

8-4. CONCLUSION

Whatever the immense complexity of cognitive imaging studies in normal subjects or brain-damaged patients, functional imaging studies based on the single-subject approach now offer the unique opportunity to design experiments specific to the observed cognitive disorders in a given aphasic patient (see, for instance, Cardebat et al., 1994). Although clearly very stringent, the methodological requirements that have been underscored in this chapter are thought to maximize the heuristic potential of advanced functional imaging studies and, last but not least, reconcile cognitive neuropsychology and functional neuroanatomy.

CHAPTER 9

Applications
of Magnetoencephalography
to Neurolinguistic Research

Andrew C. Papanicolaou, Panagiotis G. Simos, and Luis F. H. Basile
Department of Neurosurgery, University of Texas-Houston Health Science Center, Houston, Texas 77030

This chapter describes the basic principles and procedures used for recording minute changes in the magnetic field that are generated by neuronal aggregates in the context of an experimental task. This technique, magnetoencephalography (MEG), is entirely noninvasive and combines high temporal resolution with great accuracy in the localization of intracerebral events. The advantages and shortcomings of the MEG technique as a tool for source localization are described. Moreover, the utility of MEG measurements in localizing the sources of activity associated with the subjects' engagement in a variety of linguistic tasks is critically reviewed. These tasks included phonetic detection, sentence reading, semantic categorization of word pairs, and picture naming. An attempt is made throughout this chapter to identify the kinds of questions that MEG procedures are capable of addressing and those that MEG techniques are not in a position to address in a satisfactory way.

In recent years we have witnessed the emergence of several new and impressive methods for studying neural function in the normal, living brain. Each of these methods measures a particular aspect of brain physiology, such as blood flow, consumption of oxygen or glucose, and intracellular currents within neurons. These techniques have demonstrated an unprecedented precision in localizing cerebral events within relatively small anatomical regions. The explosion of new data that followed the incorporation of imaging methods into neuropsychology led to the belief that functional images were capable of uncovering the brain *mechanisms* of language and thought. The term "mechanism" refers not only to a set of anatomical units that display a characteristic pattern of activity, but also to the algorithm or code that governs the distributed activity that produces a particular result. When this outcome entails recognizing a written word or a face, understanding the meaning of a sentence, solving a problem, and so on, then the algorithm is the neural analogue of what we call, by common agreement, a *cognitive operation*.

This being the case, the expectation that any functional image, which under the best circumstances is simply an index of the relative degree of activation of large populations of neurons, will reveal operations or mechanisms of language and thought is rather unwarranted. On the other hand, it is entirely reasonable to expect that the functional imaging methods will be making unique contributions toward the achievement of the distant goal of understanding such mechanisms. First of all, by virtue of their noninvasiveness, these methods allow us to appreciate the pattern of brain activity in experimental situations that afford a greater degree of ecological validity than those using invasive procedures. The functional images that we can now record, though they do not reveal mechanisms, do give us an outline, however fuzzy and incomplete, of functional systems that mediate, somehow, certain cognitive operations that, we hypothesize, are necessary for performing particular types of tasks. Finally, one could expect that, with the addition of information from other methodologies (for example, single-cell recordings, in vitro neurochemistry) regarding the mode of operation of smaller functional units and of their interconnections, it may become possible to achieve realistic descriptions of the mechanisms themselves.

9-1. TECHNICAL ASPECTS OF MAGNETOENCEPHALOGRAPHY

Traditionally, moment-to-moment variations in cerebral activity associated with a particular experimental event were recorded noninvasively in the form of time-varying voltages sampled from the scalp surface, known as event-related potentials (ERPs). Although useful in certain respects, ERPs do not provide sufficient information for the precise localization of the sources of the observed voltages. To a great extent this problem is related to the fact that measurements of scalp-recorded electrical activity reflect both primary and secondary cerebral currents. The former are produced by postsynaptic neurochemical events in active neurons and correspond to intracellular currents. The latter correspond to extracellular currents that are passively conducted through the medium that surrounds their source (that is, the primary current) and they are consequently affected by the inhomogeneities in the conductivity of the surrounding tissues. A consequence of this is an irregular voltage distribution on the scalp that creates serious problems in the correct estimation of underlying sources.

9-1.1. The MEG Recording System

To some extent, these problem can be addressed with the use of magnetoencephalography (MEG), which entails recording minute changes in the magnetic fields generated within the brain, time-locked to an external stimulus or event. These measurements are made entirely noninvasively with the use of the neuromagnetometer (Cohen, 1972) an instrument sensitive to magnetic fields produced by activity in small cortical patches (Chapman, Ilmoniemi, Barbanera, & Romani, 1984). Neuromagnetic recordings were made possible with the development of the superconducting device known as SQUID (Superconducting Quantum Interference Device). The SQUID is connected to a magnetic detection coil or magnetometer, both immersed in a liquid helium bath at temperatures near absolute zero. Magnetic fields that reach the surface of the scalp and pass through the detection coil produce small currents along the superconducting wire by induction. To reduce the relative contribution of environmental magnetic sources, various configurations of magnetic coils have been used. The most widely used device, known as gradiometer, consists of a detection coil placed near the scalp and one or more compensation coils wound in the direction opposite to the detection coil and placed further away from the scalp. The magnetic flux that passes through both sets of coils induces opposite currents that cancel each other out (see inset in Figure 1). This is true for magnetic fields generated at locations far from the sources of the signals of interest. In contrast, magnetic fields generated near the lower coil are more likely to induce detectable currents. The spatial separation of the two coils also determines the relative sensitivity of the magnetometer to near and far cerebral sources. Most commercially available systems are primarily sensitive to superficial sources, but they can also detect strong subcortical sources. For reasons that are beyond the scope of this chapter, deeper sources are associated with greater error in localization.

The superconducting coils are oriented parallel to the surface of the scalp. Therefore, they are almost exclusively sensitive to the component of the magnetic field that is normal to the scalp. This component reflects primary cerebral currents generated by sources that are tangential to the scalp surface (see Figure 1). It appears that the primary current measured by MEG is generated in the apical segments of pyramidal cells, which are typically oriented perpendicular to the cortical layers and are arranged in vertical columns (Lewine, 1990). Pyramidal cells located on the walls of cortical sulci are tangentially oriented relative to the scalp surface and in a position to make a strong contribution to the magnetic field normal to the scalp recordable by the neuromagnetometer. In contrast, the magnetic flux produced by radial sources is tangential to the scalp surface and therefore is not detectable by the magnetometer. To further improve the signal-to-noise ratio of the magnetic recordings, several MEG records are obtained in response to repeated presentations of the experimental event

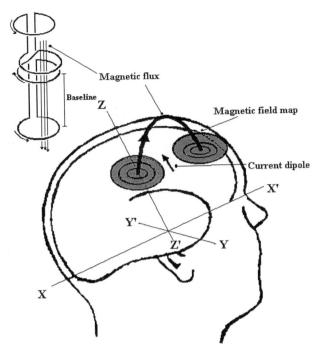

FIGURE 1 A head diagram demonstrating the reference points and planes that define the Cartesian coordinate system used for source localization. Also shown is a schematic representation of a current dipole and of the magnetic field that it produces on the head. In this example, a current dipole located in the right frontal lobe is associated with a region of magnetic outflow (maximum extremum) at fronto-central scalp locations, and a region of magnetic inflow (minimum extremum) at anterior frontal locations. The inset shows one of the most commonly used configurations of the magnetic coils known as a second-order gradiometer. The distance between the lower and the higher coils (baseline) determines the relative sensitivity of the instrument to superficial brain sources as opposed to deep sources and ambient magnetic noise. Curved arrows indicate the direction of the current induced by a distant magnetic source.

and averaged together. These records are known as evoked magnetic fields (EMFs). Averaging has been a standard procedure in ERP and MEG research designed to yield more reliable estimates of the underlying activity. The first MEG machines were capable of recording EMFs from a small number of scalp locations simultaneously (usually one to seven locations). In order to cover the entire surface of the scalp, several successive placements of the recording device were necessary. This procedure relied on the assumption that the parameters of the sources of the magnetic activity of interest remain constant over time. However, variations in the subject's state, combined with habituation and practice effects, were likely to introduce considerable extraneous variability in the data. In more recent years, neuromagnetometers equipped with large detector arrays were developed that have the capacity to detect signals from 24 to 72 scalp locations simultaneously (for example, Gallen et al., 1994). Also, systems that have the capacity to record EMFs from the entire surface of the head simultaneously have been developed (Ahonen et al., 1993; Nakasato et al., 1996).

9-1.2. Source Localization

A key issue in the use of MEG to determine the sources of cerebral activation associated with a particular stimulus or task is constructing an adequate model of the underlying source(s). Thus far, most researchers have modeled the sources as single current dipoles in order to account for the distribution of magnetic field measured at the scalp surface. The problem of calculating the parameters of intracranial sources (that is, their spatial coordinates, strength, and orientation) from the distribution of magnetic flux measured from the scalp is known as the *inverse problem.* Unfortunately, the inverse problem does not have a unique solution because, in principle, there is an infinite number of source combinations that can produce the same magnetic flux pattern. If the characteristics of the source(s) were known, then one could calculate the magnetic flux generated by the source currents at a given distance from the source using the Biot-Savart law, which is expressed by the following equation: $B_o = (\mu_o Q \sin\theta)\backslash(4 \pi r^2)$, where B is the magnetic field strength in units of Tesla, r is the distance between the source and the point of measurement in meters, Q the strength of the source in nanoAmpere·meters, and μ is the magnetic permeability constant of free space that is the same for biological tissues. In practice, to address the problem of source localization, researchers start with a hypothetical set of dipole parameters and calculate the magnetic flux pattern that the dipole in question would produce. In other words, the first step in this procedure involves solving a forward problem. Then they use a least-squares algorithm to compare the actually recorded pattern with the one calculated from the hypothetical dipole parameters. If a good match is not found, the dipole parameters are iteratively modified and the forward problem is repeatedly solved until a satisfactory correspondence between the predicted and the observed magnetic flux pattern is achieved. A set of computed dipole parameters is accepted only when the correlation between predicted and observed magnetic fields exceeds a preset criterion value, which ranges between .85 and .95 across studies.

To further ensure that a given solution to the inverse problem is valid, source localization data are compared, whenever possible, with information obtained from other research paradigms, such as lesion studies, intracranial recordings in epileptic patients, and other functional neuroimaging techniques. When these precautions are taken, the technical limitations of the MEG method do not pose serious restrictions on its source localization capacity. The validity of source localization techniques based on MEG data has been affirmed in several investigations of the intracranial sources of activity associated with sensory stimulation (for example, Pantev et al., 1988; Papanicolaou, Rogers, Baumann, Saydjari, & Eisenberg, 1990; Romani, Williamson, & Kaufman, 1982). In these studies the estimated sources of the magnetic equivalent of early ERP components such as the N100 response (known as the N100m component) elicited by simple sounds were localized in the vicinity of the primary auditory cortex in accordance with knowledge of the anatomy and physiology of the auditory system.

Subsequent studies examined the magnetic correlates of components that are presumably associated with engagement in specific tasks (such as the P300 response; Rogers et al., 1991; Rogers, Basile, Papanicolaou, & Eisenberg, 1993). In general, there was close agreement between the anatomical locations of estimated dipolar sources of these components and data from human lesion studies (Johnson, 1989; Knight, Scabini, Woods, & Clayworth, 1989), and from intracranial recordings (McCarthy, Wood, Williamson, & Spencer, 1989).

9-1.3. MEG versus Other Imaging Techniques

The MEG has a number of advantages compared to other widely used imaging techniques such as PET, SPECT, rCBF, and fMRI. First, its temporal resolution is several orders of magnitude higher than any of these methods. Second, MEG is entirely noninvasive, whereas PET, SPECT, and rCBF require inhalation or injection of radioactive tracers. Third, the MEG data do not depend on the assumption of linear additivity of brain activity due to different cognitive operations. This assumption provides the rationale for implementing the subtraction method, which is a necessary step in all PET and SPECT studies and in many fMRI applications. Finally, MEG is believed to provide a direct measure of primary electrical currents that result from local changes in neuronal transmission. Therefore, MEG findings can be more directly related to changes in local brain activity associated with a particular experimental event or tasks. Cerebral metabolism or cerebral blood flow data, on the other hand, can only serve as indirect measures of primary neurophysiological events.

Another advantage of MEG is that it can yield data that are meaningful at the single-subject level, that is, without requiring group averaging, which is often necessary in similar PET and fMRI applications. In this way, MEG is more readily applicable in the detection of individual differences than other imaging methods. Such differences have been reported in the precise location of dipoles (for example, Simos, Basile, & Papanicolaou, in press), in the morphology of EMF waveforms (for example, Basile, Simos, Tarkka, Brunder, & Papanicolou, 1996), and in the temporal course and the laterality profiles of estimated sources of activity (Salmelin, Hari, Lounasmaa, & Sams, 1994). Among the potential sources of variation that could account for differences in source localization are the following: *(a)* variability in the functional organization of cortical areas (for example, Ojemann, Ojemann, Lettich, & Berger, 1989); *(b)* anatomical variability leading to differences in the geometry of the cortical regions that contribute to the scalp-recorded magnetic fields during a given task (Steinmetz & Seitz, 1991); and *(c)* individual differences in the prominent cognitive strategy (-ies) employed by each subject for performing the experimental task (for example, Cohen & Freeman, 1978). Differences in waveform morphology might be better understood in light of the possibility that the externally recorded magnetic fields receive contributions from subpopulations of cells that show distinct temporal firing patterns (for instance, "tonic" versus "phasic" cells: Fuster, 1989; Watanabe, 1990). It is possible that under conditions that are not currently understood, the relative contribution from each subpopulation to the EMFs varies across subjects producing distinct temporal patterns in their averaged waveforms.

9-1.4. Summary

To summarize, evoked magnetic fields reflect primarily changes in intracellular electrical currents time-locked to external events. A standard practice has been to model the intracranial sources of EMFs as equivalent current dipoles that can be defined in terms of spatial coordinates, strength, and orientation. Dipole parameters are estimated from the best solution to the inverse problem. Although the solutions to the inverse problem are not unique, the capacity of the MEG method to localize sources within anatomically plausible cerebral regions has been validated in several MEG studies.

9-2. APPLICATIONS OF MAGNETOENCEPHALOGRAPHY TO THE STUDY OF LANGUAGE

The search for electrophysiological correlates of language has received much attention over the last three decades. Guiding that search was the notion that the ability to perceive speech depended on certain "mechanisms" that were language-specific (that is, specific to speech sounds and specially evolved for language; for example, Eimas, 1975; Liberman, Cooper, Shankweiler, & Studdert-Kennedy, 1967). In addition, it was assumed that the special status of language was somehow related to the asymmetric capacity of the two cerebral hemispheres to perform linguistic tasks (for example, Lenneberg, 1967). It seemed natural, then, to ERP researchers to search for parameters in the ERP signal that would be specific to language stimuli and, at the same time, be asymmetrically distributed across the two hemispheres. These early attempts had focused on comparisons between speech (for example, consonant–vowel syllables, words) and nonspeech sounds (such as tones) (for example, Molfese, Freeman, & Palermo, 1975; Morrell & Salamy, 1971). However, apart from failing to produce replicable results (for example, Grabow, Aronson, Rose, & Greene, 1980), such studies failed to take into account important acoustic parameters that distinguished the two classes of sounds. Consequently, the occasional positive findings could not be directly related to the hypothetical operations used by the auditory system to encode speech.

9-2.1. EMFs in Response to Speech and Nonspeech Stimuli

With the advent of the MEG method, researchers sought to explore its superior performance in source localization, in searching for stimulus-specific changes in activity generated within the human auditory cortex. It was anticipated that the new method would not only permit the detection of hemispheric asymmetries, but it would also allow for the within-hemisphere differentiation of sources that might display stimulus-specific activity. Unfortunately, most of the published MEG reports to date have not escaped from the same pitfalls that characterized the early ERP research. Most of these studies attempted comparisons between the magnetic responses elicited by speech and nonspeech sounds (Kaukoranta, Hari, & Lounasmaa, 1987; Eulitz, Diesch, Pantev, Hampson, & Elbert, 1995). In the most recent of these investigations, Eulitz et al. (1995) presented series of synthesized vowel sounds randomly intermixed with pure tones. The subjects' task was to count the occurrences of the vowel /œ/. The EMFs that were recorded from 11 right-handed adults were characterized by an early deflection that peaked at the same latency as the electrical N100 response (that is, the N100m peaking at approximately 100 ms poststimulus onset) and was followed by a sustained field (SF) that lasted for a few hundred milliseconds. The sources of both components were localized in the vicinity of the auditory cortex. A notable finding was that the amplitude of the SF, as well as the strength of its estimated dipole source, was larger for vowels than for tones, and that this difference was significant only in the measurements made over the left hemisphere (LH). Although the possibility exists that the observed interaction was due to hemispheric asymmetries in the distribution of speech-related sources of activity, as the authors claimed, the data are open to a number of alternative interpretations because the two types of stimuli differed in many important aspects besides the verbal/nonverbal dimension. These included acoustic properties (such as the number of spectral components and their bandwidth) and the amount of attention that the subjects were likely to devote to each class: tones were always task-irrelevant stimuli.

Magnetoencephalographic correlates of consonant sound discrimination were investigated by Aulanko and associates, who presented series of consonant–vowel (CV) syllables that were perceived as either a /bæ/ or a /gæ/ in a phonetic discrimination task (Aulanko, Hari, Lounasmaa, Näätänen, & Sams, 1993). The relative frequency of occurrence of the two CVs was 20 and 80% in one block of trials and was reversed in the other block. In order to introduce an additional acoustic dimension, the fundamental frequency (f_0) of each CV was varied randomly within blocks. The strength of the N100m deflection of the EMF was larger for infrequent than for frequent (that is, "standard") tokens, an effect that was identified with the magnetic equivalent of the Mismatch Negativity (MMN) known as the Mismatch Field (MMF).[1] The authors concluded that this effect reflected the detection of a phonetically relevant acoustic difference between the two CVs (that is, the direction of the second formant transition), while they remained unaffected by the task-irrelevant changes in fundamental frequency. The MMF effect was preserved when subjects were instructed to ignore the stimuli and read a text instead. However, based on these data, one cannot conclude that changes in f_0 had no effect on the amplitude of the MMF, because this variable was not included in the analyses. Therefore, the reported findings simply indicate the sensitivity of the MMF to a particular acoustic dimension (that is, the direction of the second formant transition) without permitting any conclusions regarding the relation between the activity in the auditory cortex indexed by the MMF and the ability to

---

[1]The MMN effect refers to a change in the amplitude of the early portion of the ERP elicited by a physically deviant auditory stimulus that is embedded in a series of similar or standard stimuli (Näätänen, Simson, & Loveless, 1982). The MMN does not require the subject to be engaged in an active discrimination task.

perform phonetic distinctions based on this acoustic cue. Indeed, the MMF appears to be sensitive to the direction of frequency modulation in nonspeech contexts as well (Pardo & Sams, 1993). Once again it will be emphasized that the selection of stimuli in these investigations did not permit a separation between the contribution of acoustic and linguistic variables to the observed EMF patterns.

9-2.2. EMFs in Response to Complex Linguistic Tasks

Another study attempted to examine asymmetries in brain activity associated with engagement in a task that presumably involved higher levels of linguistic analysis: a semantic-category detection task with auditorily presented words as stimuli (Hari et al., 1989). The EMF waveforms on both sides of the scalp were characterized by the N100m deflection that was followed by a sustained field that reached maximum amplitude between 400 and 500 ms after stimulus onset. Dipole sources estimated at the peak of the sustained field were located more posterior in the LH than corresponding sources found in the right hemisphere. Since individual MRI scans were not available, these comparisons were based on the absolute coordinates of the respective dipole sources. Once again, the relevance of these differences to language-specific neural processes could not be ascertained on the basis of the available data.

9-2.2.1. Source Identification Guided by Waveform Morphology: The N400m

One way to address this problem is to examine components of event-related brain activity whose sensitivity to linguistic parameters has been empirically established in previous research. One such component is the N400, which has been repeatedly shown to respond to semantic anomalies in visual (Kutas & Hillyard, 1980a, 1980b) and auditory sentence-reading paradigms (McCallum, Farmer, & Pockock, 1984). The amplitude of the N400 response is believed to be an index of the degree that a potentially meaningful stimulus is perceived as being "appropriate" within a particular linguistic context (Kutas & Van Petten, 1988). Therefore, identifying the intracranial sources of the scalp-recorded N400 deflection might reveal some of the brain areas that participate in the evaluation of contextual congruity. A preliminary attempt to record the magnetic equivalent of the N400 (N400m) in the standard sentence-reading paradigm was successful (Schmidt, Arthur, Kutas, & Flynn, 1989). However, because of a number of methodological limitations (such as the lack of individual MRI scans, and the small number of subjects and MEG recording locations), the study did not succeed in providing conclusive source localization information. In order to provide a clearer picture of the N400m sources, we undertook a more systematic investigation using the same sentence-reading task (Simos et al., in press). The subjects were instructed to read sentences that had highly predictable endings and were presented one word at a time (Kutas & Hillyard, 1980a, 1980b, 1984; Polich, 1985). They were asked to focus on sentence meaning, without making any explicit decision on the presented stimuli. Half of the sentences ended with a semantically inappropriate—but grammatically and syntactically appropriate—word that typically elicits a reliable electrical N400 response.

Neuromagnetic measurements were obtained from 49 locations covering the left side[2] of the scalp with a 7-channel neuromagnetometer, in eight right-handed adults. The ERP data showed clear N400 responses in every subject (Figure 2A), that were significantly larger in amplitude in the Incongruous condition compared to the Congruous condition, in agreement with the existing ERP literature. Satisfactory solutions to the inverse problem of dipole fitting were obtained from seven subjects in the Incongruous condition at several contiguous time points around the latency of the N400 peak. As expected, the N400m response elicited by incongruous endings was larger in field strength as well as in the strength of its estimated dipole source compared to the response elicited by congruous endings (see Figure 2B & 2C). In every case N400m sources were localized in temporal lobe structures, including neocortical areas on the lateral surface of the left hemisphere (that is, near the temporo-parietal junction, and in the middle temporal gyrus), and medial temporal paleo- and archeocortical regions, that is, in the vicinity of the hippocampus and the parahippocampal gyrus.

In one subject (S#5), dipole regions extended more posteriorly into the lateral temporo-occipital association cortex (Brodmann's area 37).

The anatomical locus of estimated sources could be accounted for on the basis of knowledge regarding the possible role of these cortical areas in relation to the characteristics of our task. Specifically, several lesion studies have established a link between the left temporo-parietal cortex and word recognition (Damasio & Damasio, 1983; Greenblatt, 1973). Moreover, data from lesion and PET studies suggest that certain middle temporal regions are involved in reading and semantic comprehension of written material (Hart & Gordon, 1990; Howard et al., 1992; Price et al., 1994). Finally, activity in more medial aspects of the temporal lobe synchronous with the scalp-recorded N400 response has been verified using intracranial recordings in epileptic patients (McCarthy, Nobre, Bentin, & Spencer, 1995; Smith, Stapleton, & Halgren, 1986). Activity in these structures may be related to the presumed role of the hippocampus in recent memory (Squire, 1986; Damasio & Damasio, 1989, p. 39). The latter is an important component of processes that contribute to word recognition (such as contextual integration and subjective expectancy).

9-2.2.2. Source Identification Guided by Variables Other Than Waveform Morphology

Although the approach to source identification that focuses on particular ERP/EMF components may prove useful in addressing certain issues in neurolinguistics, it relies heavily on the assumption that the components under investigation reflect operations

---

[2]Data from earlier reports (Schmidt et al., 1989; Smith Stapleton, & Halgren, 1986) and our own preliminary data did not warrant systematic examination of magnetic fields on the right side of the head. This issue, however, requires further investigation.

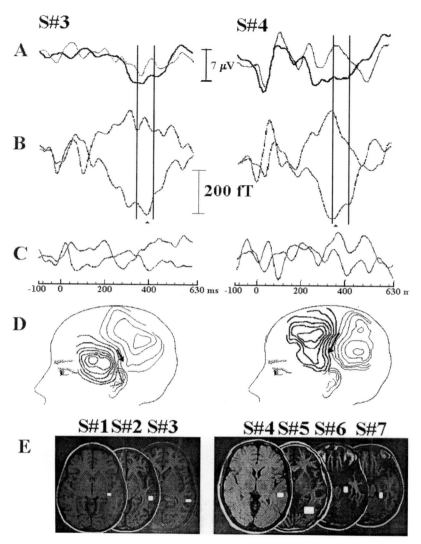

FIGURE 2 A: Averaged ERP waveforms from two subjects in response to final words in the sentence-reading task. Thin lines: Congruous condition; thick lines: Incongruous condition. The vertical lines mark the latency window used to identify the N400m response. B-C: Examples of averaged EMF waveforms recorded at the two extrema of the respective magnetic field distribution in response to anomalous (B) and congruous endings (C). Upward deflections indicate the maximum extremum, whereas downward deflections indicate the minimum extremum. D: Isofield contour maps of the magnetic field distribution from the same subjects displayed in 2B–C. The latency at which these maps were constructed is indicated by the arrowheads in 2B. The presumed direction of current flow is indicated by a solid arrow. Thin lines: region of magnetic outflow; thick lines: region of magnetic inflow. E: Locations of estimated dipole sources for the N400m response in the Incongruous condition superimposed on the subjects' own MRI scans (left part of figure), or on MRIs that best approximated each subject's head size (right part of figure). The squares mark the areas in which reliable dipole solutions were localized for each subject across the range of time points enclosed by the vertical lines in 2B.

that are specific to language—an assumption that may not be entirely justifiable. More-over, it is possible that by using this strategy important information regarding the spatiotemporal dynamics of brain activity may be lost. One way to address these problems is to follow a source identification approach that is not guided by waveform morphology. Initially, the studies that adopted this strategy examined whether partic-ular tasks could activate reliably anatomical regions that are known—from other sources—to be involved in the execution of these tasks.

For instance, Salmelin and associates sought to identify the cortical areas that be-come active as subjects attempt to attach verbal labels to a series of common objects (Salmelin et al., 1994). Magnetic fields were recorded from six right-handed adults in a picture-naming task using a whole-head (122-channel) magnetometer. A prominent finding was that the estimated dipole sources were almost exclusively located in temporo-parietal and occipital cortices during the first 400 ms after stimulus onset. In some cases, during the second half of this period sources were also localized more anteriorly, in superior temporal regions, possibly in the vicinity of Wernicke's area. During the next 200 ms period, dipoles were found exclusively in peri-Sylvian areas including temporal and frontal regions. The observed spatiotemporal pattern of activity was in general agreement with the Wernicke-Geschwind model (Geschwind, 1965). This model predicts that during object naming cortical engagement proceeds from posterior "visual" to anterior "speech" areas. A clear laterality profile was not ap-parent in the data, a finding that was in sharp contrast with reports from studies using invasive methods (such as the Wada test), which often indicate that the performance of naming tasks depends largely on left hemisphere structures (for example, Loring et al., 1990). Despite this discrepancy, which could be partially accounted for by the large amount of intersubject variability that was present in the data, this study was important because it demonstrated the capacity of MEG to reveal the time course of cerebral activation associated with a particular task, a quality unmatched by any other imaging technique to date.

A slightly different approach to functional imaging using MEG is to search for indices of regional specialization within a particular area of the brain. Previous find-ings had suggested that distinct regions of the frontal cortex are involved in the prep-aration and execution of different nonverbal tasks (Basile, Rogers, Bourbon, & Pa-panicolaou, 1994). In a more recent study, we extended these findings with the use of a semantic categorization task (Basile et al., 1996). This study was designed to test the prediction that consistent sources of magnetic activity can be found in the left, but not in the right, prefrontal cortex as right-handed subjects become engaged even in a simple verbal task. A semantic categorization task was used, in which the subjects were asked to decide whether a pair of printed words, which were presented in suc-cession on each trial, belonged in the same category, and to keep a running count of the number of category matches. All six subjects showed clearly dipolar isofield maps (see Figure 3B) on the left side of the head, and in all cases the estimated sources were located in the left posterior frontal cortex in the vicinity of Brodmann's areas 6 and 44 (Figure 3C).

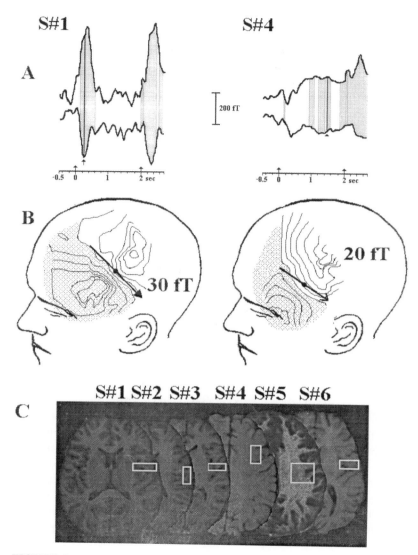

FIGURE 3   A: Examples of averaged EMF signals recorded at the extrema of the magnetic field distribution from the left side of the head in two subjects. The shaded areas mark the time intervals during which satisfactory solutions to the inverse problem were obtained and estimated dipole sources were localized in the left frontal cortex. The onset of the first word was at zero and of the second word at 2 s. B: Isofield contour maps from the same subjects that correspond to the highest magnetic field amplitude (indicated by arrowheads in A). The increment between adjacent isofield lines is given in femtoTesla. The orientation and strength of the equivalent current dipoles that best accounted for each subject's fields are indicated by arrows. C: Horizontal MRI sections from each subject. The range of estimated dipole loci that were obtained during the trial duration is represented by the rectangle areas.

These results were in agreement with accumulating evidence showing increased activation in postero-lateral frontal areas during a variety of language tasks (for example, Demb et al., 1995; Price et al., 1994). Dipolar magnetic field distributions were also observed on the right side in most subjects. However, the estimated sources of these fields were never localized in frontal regions. Rather, they corresponded to the magnetic equivalents of ERP responses (such as the N400/P300 complex) with sources located typically in temporal lobe regions (Rogers et al., 1991; Rogers et al., 1992; Rogers et al., 1993; Simos et al., 1996). A small-scale replication of the experiment with two of the subjects produced essentially identical results.

Despite their consistency, the conclusions that can be drawn from these findings remain limited. Although one can argue with certainty that some aspect of the verbal task was associated with frontal lobe activity that was clearly lateralized, there is greater uncertainty regarding the implications of these results for current notions of the organization of language in the brain. Experiments that are now under way attempt to identify the specific components of the verbal task that were more closely associated with the frontal sources observed in this study (for instance, orthographic, phonological, semantic encoding). Other experiments will examine the contribution of variables such as memory involvement (by manipulating stimulus set size and length of the interstimulus delay interval), complexity of the comparison decision (by manipulating the number of task-relevant stimulus dimensions), and the type of task (for example, by comparing decisions on phonological similarity with decisions on the similarity between spatial patterns made of the same stimuli).

9-2.3. Summary

To summarize, the MEG studies reviewed in this section fall into two broad categories. Studies in the first category attempted to identify EMF parameters that could distinguish speech from nonspeech stimuli. Their findings showed that the source parameters of the magnetic fields did not differ dramatically as a function of stimulus type. Even when differences were observed, it was not clear, due to the choice of stimuli, if they were related to activity in neuronal aggregates specialized for the analysis of language stimuli. Studies in the second category examined the characteristics of magnetic fields elicited when subjects were engaged in complex linguistic tasks such as sentence reading and picture naming. The estimated sources of these fields became active several hundred milliseconds after stimulus onset and were generally located in cortical regions that, on the basis of evidence from other sources, presumably play an important role in language-related operations. A second conclusion that emerges from these studies is that clear-cut functional hemisphere asymmetries are not invariably seen in linguistic tasks. When such asymmetries are present, they involve activity elicited during a limited portion of the EMF waveform and are restricted within certain brain regions.

9-3. CONCLUSIONS

Magnetoencephalography, as a functional imaging technique, has certain unique features. It combines the very high temporal resolution offered by ERPs with the capacity to localize neurophysiological events generated within relatively small cerebral regions. In addition, it is suitable for studying individual differences in functional neuronal activity. Its spatial sensitivity is superb when mapping well-localized events, such as the activation of primary sensory cortices within the first 100-200 ms following the onset of a stimulus. A different situation emerges when researchers try to analyze the magnetic fields generated as subjects become engaged in complex linguistic tasks. When interpreting these MEG findings, it is important to remember that any observed dipolar field distribution reflects activity of extended cortical patches that become involved in the execution of distinct operations required for the execution of the task. The estimated location of the current dipole derived from the data could well represent the unevenly weighted center of the distribution of local currents that are oriented tangentially with respect to the surface of the head. In other words, the estimated dipole coordinates are bound to depend on the geometry of the cortical surface, which determines, in part, the distribution of local currents that can be recorded from the scalp. To overcome these difficulties, new computational techniques are being developed that permit the modeling of intracranial EMF generators either as extended sources or as multiple simultaneously active current dipoles (for example, Eulitz et al., 1994). It is expected that in the future these approaches will provide more realistic descriptions of cortical activation during complex tasks.

Despite its limitations, the MEG technique has proven its utility for identifying the sources of brain activity associated with the engagement in a variety of linguistic tasks with reasonable anatomical precision. Further, with their superb temporal resolution, MEG recordings can provide unique information on the time course of regional cortical involvement in a particular task. Finally, because MEG data are meaningful at the level of the individual subject, they may offer valuable information regarding the neurophysiological correlates of individual differences in the performance of various tasks. However, thus far the contribution of MEG findings to our understanding of the cerebral mechanisms of language has not entirely fulfilled the expectations triggered when this technique was first introduced. One reason for this apparent shortfall is that these expectations were not altogether reasonable: as has been stressed repeatedly, functional images do not reveal the mechanisms of language. Unless researchers have a clear picture of the entire circuitry of the brain, the algorithm that governs the brain activity, which in turn mediates linguistic performance, such mechanisms will remain elusive. To this date, MEG, much like the other functional imaging methods, has been successful in corroborating a number of traditional conceptions regarding the anatomical substrates of language. This is an important achievement in that it establishes the worth of the method and enhances our confidence that its future application may result in the emergence of new concepts and more realistic models regarding the brain mechanisms of language.

Acknowledgments

This research was supported in part by grant NS 29540-005A1 to the first author from the National Institutes of Health, Washington, DC.

CHAPTER 10

Vertical Integration of Neurolinguistic Mechanisms

Phan Luu[1] and Don M. Tucker[2]

[1]Department of Psychology, University of Oregon, Eugene, Oregon 97403; [2]Department of Psychology, University of Oregon, and Electrical Geodesics, Inc., Eugene, Oregon 97403

Contributions from subcortical circuits to language functions are acknowledged but are often considered separately from the contributions of the cortex. Brain organization and the evidence from studies of primate vocalization, as well as evidence from aphasic studies, argue for a framework in which contributions from the entire extent of the neuraxis to language is considered. Lateralization of language functions may be a result of asymmetries in brainstem neuromodulator projection systems and corticolimbic organization. These asymmetries appear to reflect the different functioning of two elementary, regulatory systems. These regulatory systems and their underlying subcortical–limbic–cortical circuits provide mechanisms by which emotion, memory, and motor control, as well as language functions, are regulated and integrated across the neural hierarchy.

Fundamentally, mental representation is analogical. A memory is formed as an analogue of perceptual experience or behavioral action. Language, however, is a remarkably arbitrary code. The representational element, the word, has an iconic, analogical relationship to its referent only in exceptional cases. Perhaps because of this arbitrary, symbolic nature, language is often approached in neuropsychological studies as if it were somehow removed from the more elementary neurophysiological mechanisms regulating alertness, sensation, and action. The neurophysiology of language is almost invariably assumed to be a neurophysiology of the neocortex—specifically, the "higher" or "association" areas of the neocortex—with only the most basic support functions (such as arousal and sensorimotor relays) provided by subcortical neurophysiological circuits. The unique features of the human cortex, such as left hemisphere specialization for language, are often seen as the major clues to understanding neurolinguistic mechanisms.

Animal communication often takes an arbitrary form. An arbitrary action, such as avian head bobbing, becomes routinized through evolutionary selection for its communication value (Tinbergen, 1951). However, the control of even arbitrary communication in other animals, including higher mammals, is not restricted to the neocortex but engages multiple levels of the neural hierarchy, including primary adaptive control by the subcortical structures of the telencephalon (limbic structures and basal ganglia) and both regulatory and pattern-generation contributions from the diencephalic and mesencephalic levels of neural organization.

In this chapter, we consider general themes of neural functioning that link together various cognitive operations, such as memory, attention, motor control, and language. These aspects of cognition are typically studied separately. However, because the human brain evolved, it may be possible to describe a set of primitive control processes and how more complex cognitive functions differentiated from them. We begin with a brief overview of cortical organization and its implication for functional integration. We then discuss our theoretical efforts to understand the mechanisms underlying hemispheric specialization for emotion and motivation. These mechanisms include what appear to be asymmetries of the brain-stem neuromodulator projection systems and asymmetries in the organization of corticolimbic networks. Early on, it became clear that these basic mechanisms may explain important aspects of hemispheric specialization for cognitive functions, including language. The elementary systems that evolved for activating and motivating vertebrate behavior remain important to controlling human behavior, including cognition as well as action. By tracing these themes of neural organization across brain-stem, limbic, and cortical levels, we sketch a theoretical outline of neurolinguistics that is speculative, and that lacks detail at several levels of analysis. However, this outline frames a neurolinguistics that includes not just the superficial articulation of language but the motivational basis that shapes its meaning.

10-1. MOTILITY AND CORTICAL ORGANIZATION

From his embryological and comparative studies, Herrick (1948) concluded that the major influence on the differentiation of the brain in evolution has been the acquisition of increased motor articulation. In the nervous system of amblystoma (the tiger salamander), Herrick observed that the regions of the brain not directly involved in sensory or motor functions, which he called the intermediate zones and include the cerebral hemispheres, increased in size and articulation as the system for action became more complicated. For mass movements, actions are competently instantiated within the primitive neuropil (synaptic network) of the sensory and motor zones, namely, the tectum and tegmentum of the brain stem, respectively. In contrast, increased sophistication in action, articulated movements removed from reflex arcs, necessitated an equivalent increase in the size and differentiation of the intermediate structures responsible for further development and planning of these movements.

Yakovlev (1948) also recognized the importance of motility and behavior to cerebral development. He noted that the vector of behavior, taken as an integrated whole, is formed as each sphere of motility gives rise to the next, operating from internal states to produce action on the environment. In this context, actions arise through a progression from axial, proximal movements to discrete, distal articulation, tracing the anatomical organization and control of motility. The first sphere of motility, expressed as internal homeostatic and visceral functions, is a basic property of all living organisms. For Yakovlev, social communication of internal states, including human language, is considered part of behavioral motility and thus belongs to the second sphere. By placing social communication within this sphere, Yakovlev clearly emphasizes the dependency of language on more primitive functions and recognizes that communication is an extension of motility. The third sphere of action consists of behaviors that are directed toward changing the environment.

Within this framework of the organization of behavior, the vertebrate brain shows a concentric organization, from the inside out, reflecting the central importance of the adaptive control of motility in evolution. Yakovlev's model of concentric organization (see also Pribram, 1960) parallels closely the approach we have taken to vertical integration (Tucker, 1993). The inner structures or lower centers are driven by homeostatic functions and the outer structures or higher centers orchestrate action sequences. The core of the brain, organized around homeostatic centers of the hypothalamus and periventricular system, regulates visceral states and directs the action of the second ring. The second ring is organized around the brain-stem reticular network and the forebrain limbic and striatal structures. This level coordinates the activity of the homeostatic core, and it sets up biases that will determine how internal states are expressed in behavior, (Pribram, 1960). The third, most exterior ring is the cortex of the cerebral hemispheres mediating the "higher-order" negotiation of internal states with the environmental context.

10-2. ANATOMY OF PRIMATE COMMUNICATION

The general framework of concentric/vertical organization of the brain and behavior emphasizes the continuity and unity of action. Research on primate vocalization has mapped each level of neural organization and its contribution to the vocalization process. Ploog and his associates examined the neural control of vocal communication in the rhesus monkey with a systematic combination of electrical stimulation and lesion methods (Ploog, 1981, 1992). Elementary motor patterns of vocalization are organized at the brain-stem (mesencephalic) level. The more complex patterns of vocalization that characterize species-specific communications are organized at the midbrain (diencephalic) level, providing control and inhibition of the brain-stem centers. However, the emotional coloration and modulation of these communication patterns is missing unless the diencephalic circuits receive input from the limbic (telencephalic) structures. The paralimbic (archicortical) cingulate cortex appears to provide an executive regulation of the limbic and midbrain influences, affording the monkey with the capacity for what Ploog terms "voluntary call initiation." At a still higher level of the neural hierarchy, the neocortex enables a more discrete articulation of the vocalization, a "voluntary call formation," in which segments of the vocal pattern may be differentiated from the holistic midbrain expressive pattern.

10-2.1. The Vertical Hierarchy and Functional Anatomy of Language

The challenge for the brain is to achieve and maintain what may be called a vertical integration of the communication process, in which the direction and dominance of control is complex and distributed across the multiple levels of evolutionary organization (Tucker, 1993). A similar neurophysiological analysis may be required to understand the control of language in humans. The adaptive use of language may involve substantial influences mediated through limbic and midbrain pathways. Ploog (1992) points out that patients with lesions of the anterior cingulate cortex initially show profound deficits of initiative (akinetic mutism), and later show aprosodic and impoverished speech, implying that the voluntary call initiation and emotional modulation provided by the rhesus cingulate may be paralleled by cingulate contributions to human language as well. Ploog notes that certain areas of the limbic system have grown disproportionately large in humans, including the anterior thalamic limbic nuclei. These serve the supplementary motor area and premotor and orbital frontal cortices, suggesting that increased limbic input to the cortex, as well as increased input from the cortex to the limbic structures, has been integral to the evolution of the human cortex.

In the clinical analysis of human language disorders, the requirement for support from intact subcortical circuits (brain-stem, thalamic, striatal, limbic) is certainly acknowledged, but it is often considered as a separate issue from the language operations themselves. However, Brown (1988) has argued that an adequate neuropsychology of language must include not only the final articulation of the expressive and receptive structure of language in the neocortex, but the full developmental process through which subcortical and limbic structures initiate and motivate an action pattern that then achieves articulation for expression or comprehension within the neocortex.

Brown describes language generation as *microgenetic,* following Werner's assertion that there is a developmental progression, from syncretic and global to differentiated and articulated, in each cognitive event (Werner, 1957). Whereas Werner kept the evolutionary-developmental metaphor at an abstract, psychological level, Brown advances a concrete anatomical interpretation. He proposes that the microgenetic organization of each linguistic production recapitulates the phylogenetic organization of the vertebrate brain. The initiative for the vocalization (or idea) is recruited in the rostral brain-stem. At the midbrain and striatal levels, the action pattern is formed and modulated by limbic motivational influences. At the cortical level, there is a progression from global organization in paralimbic networks toward increasing articulation of linguistic content as the representation takes form in neocortical networks.

Brown's theoretical approach frames language within the anatomical framework of primate vocal communications. Within this framework, Brown suggests that the left hemisphere's cortical networks, the classical neurolinguistic centers, become important only in the final articulation of the microgenetic process, as the neocortex is engaged. However, in our theoretical work on the vertical integration of motivational and emotional functions, we have come to the conclusion that the lateral asymmetry of the human brain is not restricted to the neocortex, but may be an essential feature of multiple levels of the human neuraxis (Liotti & Tucker, 1994; Tucker, Luu, & Pribram, 1995).

10-3. MOTIVATIONAL BASIS OF HEMISPHERIC SPECIALIZATION

In the evidence on hemispheric asymmetries for human emotion and motivation, several findings suggest that the cognitive specialization of the left and right hemispheres is closely linked to their differing motivational biases. For example, the left hemisphere's analytic cognition and capacity for sequential control of cognition appear closely linked to a person's anxiety level (Tucker, 1981). Evidence from normal emotion shows that anxious subjects tend to be highly focal and detail-oriented in their perception (Tyler & Tucker, 1982). Evidence from psychopathology shows that patients with high anxiety often show deficits of language function (speech and test anxiety), but also show an analytic and intellectualized cognitive style that may degrade to obsessions and compulsions (Shapiro, 1965) in extreme cases. Evidence from neuropathology shows that patients with a left temporal lobe epileptic focus may show a disorganized motivational control of verbal cognition that in some cases appears schizophreniform (Flor-Henry, 1969) but more typically presents as an intellectualized cognitive style in which verbal cognition is infused with emotional significance, leading to a preoccupation with religious or philosophical themes (Bear & Fedio, 1977).

If there is an inherent link between the motivational bias of anxiety and the linguistic cognition of the left hemisphere, then persons with pathologically low anxiety, such as psychopaths and impulsives, would be expected to show poorly developed verbal cognition. The lower verbal than performance IQ of these persons is a well-known observation from the traditional psychometric literature (Tucker, 1981).

10-3.1. Self-Regulation through Activation and Arousal

These various findings of individual differences raise the question whether the relation between the motivational bias of anxiety and left hemisphere cognition is a strictly trait phenomenon, or whether a certain state or level of anxiety is required for effective neurolinguistic function (Tucker, 1981). In this sense, "anxiety" would not refer to a discrete emotional state experienced rarely, but would be a quality of waking consciousness that may be integral, at some level of activation, to a motivated focus of attention. In attempting to examine the qualitative influences on attention and cognition that may arise from states of motivational arousal, Tucker and Williamson (1984) examined the catecholamine (dopamine, DA, and norepinephrine, NE) projection systems that mediate important aspects of the brain-stem control of alertness.

In the Tucker and Williamson model, the focus on DA and NE to the exclusion of other neuromodulators, particularly acetylcholine, is a major limitation. Furthermore, the literature on the functional roles of neuromodulator systems has been both confusing and controversial, and no clear resolution has emerged (see Le Moal & Simon, 1991). However, in the behavioral observations on the effects of manipulating DA or NE function there are indications of qualitative effects on behavior; that is, these major regulatory controls do not simply "activate" brain activity and behavior in a quantitative sense, but alter its character qualitatively. Drawing on the Pribram and McGuinness (1975) concepts of unique attentional effects of elementary activating systems, Tucker and Williamson (1984) formulated distinct cybernetic biases associated with each catecholamine projection system. The NE system appears to be sensitive to novelty and habituates rapidly. Tucker and Williamson theorized that this may form the basis for a phasic arousal system, underlying the perceptual orienting response, and operating under a habituation bias. The inherent affective quality to phasic arousal is elation. In contrast, the DA system is integral to initiating and maintaining action, and at high levels of function it leads to behavioral stereotypy. This system was characterized as a tonic activation system, supporting motor readiness, and operating under a redundancy bias. Its inherent affective quality is anxiety.

These primitive neural control systems produce different modes of allocating attention and working memory. These modes are theorized to be differentially important to the left and right hemispheres. The DA tonic activation system, by biasing working memory toward redundancy, maintains a focused attention that may form a basis for the left hemisphere's analytic cognition. The NE phasic arousal system, by its habituation to constancy, selects for novel input and thereby produces an expansive attentional mode that may be integral to the right hemisphere's holistic cognition (Tucker & Williamson, 1984).

The right hemisphere's contributions to language, as in humor and metaphorical comprehension (Brownell, Michel, Powelson, & Gardner, 1983), may draw on this more holistic semantic representation. Consistent with the notion that an elated emotional state engages the right-lateralized NE phasic arousal system, a positive emotional state has been shown to be associated with a more expansive, broadly associated mode of semantic representation (Isen, Niedenthal, & Cantor, 1992).

The left hemisphere's specialization for complex motor sequencing has been seen as an important foundation for its specialization for language (Kimura, 1979). The redundancy bias of the tonic motor activation system provides a primitive cybernetic control that may be essential for the sequencing, routinization, and focusing of perception and action that results in a stereotyped, arbitrary communication pattern such as language. In verbal cognition, ideas and experiences must be analyzed into informational units that can be verbally represented, and sequenced in the culturally shared, and thus predictable, pattern of grammar.

In the normal brain, an adequate level of tonic activation—and anxiety—may provide the neurolinguistic mechanism for the crisp delineation of ideas that is essential for analytic cognition and critical thinking. More excessive levels of tonic activation may produce the overly focused and restricted attention of the anxiety disorder patient, or the cognitive routinization and behavioral stereotypy of the obsessive-compulsive (Tucker & Williamson, 1984). The psychopathologies may be instructive for neurolinguistics because they manifest exaggerated control mechanisms that are hidden within the balanced control structure of the normal brain.

10-4. ASYMMETRIES IN CORTICOLIMBIC EVOLUTION

That hemispheric specialization involves an asymmetry in corticolimbic networks was first proposed by Bear (1983). He recognized that the right hemisphere's spatial skills must draw on the spatial processing abilities of the dorsal cortical pathway. Liotti and Tucker (1994) agreed with this speculation, and argued in addition that the left hemisphere's analytic cognition, and its skill in codes such as language and mathematics, must draw heavily on the ventral cortical pathway that supports foveal vision and object recognition. The dorsal and ventral pathways of the cortex represent distinct patterns of connectivity between sensory and motor systems and the memory operations of limbic structures. This is because they reflect the evolutionary roots of the neocortex at the dual points of origin in the archicortex (dorsal pathway) and paleocortex (ventral pathway; see Pandya, Seltzer, & Barbas, 1988; Sanides, 1970).

These theoretical efforts may suggest a new perspective on the interesting questions of hemispheric differences in corticolimbic interaction in the literature on lateralization and psychopathology (Bear & Fedio, 1977; Flor-Henry, 1969). They may also suggest additional ways of understanding the adaptive basis of hemispheric neurolinguistic mechanisms. The semantic base of neurolinguistic organization within the left hemisphere appears to be formed by the paralimbic networks (Brown, 1988). If the left hemisphere draws preferentially on the ventral corticolimbic pathway, important aspects of the lexical organization of neurolinguistics may be understood to emerge from the more elementary capacity of this pathway.

10-5. ADAPTIVE CONTROL OF THE DUAL PATHWAYS

10-5.1. Limbic Circuits

Papez (1937) proposed a cortical-subcortical circuit involved in the experience of emotion. This circuit is centered on the medial aspect of the cerebral hemisphere and involves the cingulate gyrus and hippocampus (both are cortical structures of the dorsal pathway), the anterior nucleus of the thalamus, and the hypothalamus. It is interesting to note that the posterior cingulate, hippocampus, and hypothalamus are preferentially innervated by NE. Brown, Crane, and Goldman (1979) found that the hypothalamus received the densest NE innervation out of all the subcortical structures that they studied. A second major limbic circuit centered on the basolateral surface was proposed by Livingston and Escobar (1973). We refer to this circuit as the ventrolateral circuit to be consistent with its anatomical position in relation to the dual trends of cortical evolution. This ventrolateral circuit involves the orbitofrontal cortex and temporal pole (both belong to the ventral cortical pathway), the amygdala, and the mediodorsal nucleus of the thalamus. The orbitofrontal cortex, amygdala, and temporal poles are structures that are preferentially innervated by DA (Levitt, Rakič, & Goldman-Rakič, 1984).

The two limbic circuits appear to have different roles in motivational and emotional experience. Kleist in 1934 (cited in Valenstein, 1990) noted that lesions to the mediodorsal areas of the frontal lobe result in apathy and indifference. Indeed, it has been reported that large lesions to the anterior cingulate in man result in inattention, apathy, and ultimately a lack of self-initiated action or speech (Barris & Schuman, 1953). Lesions to orbitofrontal regions, on the other hand, result in symptoms of disinhibition and difficulty in maintaining social relations due to lack of social constraint and planning (Damasio, Tranel, & Damasio, 1990; Eslinger & Damasio, 1985). Livingston and Escobar (1973) suggested that the ventrolateral circuit, with its connections to the temporal lobe, is involved in emotional interpretations of sensory information. In contrast, these authors believed that the mediodorsal circuit is involved in the regulation of activity. Disorders to this circuit can result in hypo- or hyperkinesia.

The similarities between the proposed functioning of the mediodorsal and ventrolateral limbic circuits and the right-lateralized NE arousal system and the left-lateralized DA activation system, respectively, are striking. Thus, the activation and arousal systems appear to be made up of structures spanning the extent of the neural hierarchy. The ventrolateral limbic circuit involves the DA cell groups of the midbrain, the mediodorsal nucleus of diencephalon, the amygdala of the subcortical telencephalon, and the temporal pole and orbitofrontal cortex. The locus coeruleus of the midbrain, the anterior nucleus of the thalamus, and the hippocampus and cingulate of the telencephalon make up the circuit of the NE arousal system.

10-5.2. Motivation of Dual Memory Systems

The role of the hippocampus and the medial temporal lobe in memory formation is well known (Squire & Zola-Morgan, 1991). However, the hippocampus may be more important to the formation of a particular aspect of memory than others (Nadel, 1992). Mishkin and colleagues (Bachevalier & Mishkin, 1986; Mishkin, 1982; Mishkin & Murray, 1994; Mishkin & Phillips, 1990) have outlined two memory circuits that involve different cortical and subcortical structures. One circuit consists of the mediodorsal nucleus of the thalamus, the amygdala, and the ventromedial aspects of the prefrontal lobe. We refer to this circuit as the ventrolateral circuit to emphasize the paleocortical components within this system. The second circuit involves the anterior nucleus of the thalamus, the hippocampus, and the mediodorsal prefrontal lobe (cingulate). We refer to this circuit as the mediodorsal circuit to emphasize the cingulate cortex and the hippocampus, both belonging to the archicortical trend.

Although the differences in memory functions subserved by each of these circuits are not fully understood, there is reason to believe that the mediodorsal system is involved in spatial memory, which may be generalized to memory for context, and that the ventromedial system is involved in categorical memory, that is, memory based on identity of objects (Nadel, 1992). Based on findings that the hippocampus is important to the processing and encoding of nonreinforced stimuli, Pribram (1991) suggested that the hippocampus supports the contextual representation in which behaviors occur. Similarly, Nadel (1992) believes that the hippocampal memory system emphasizes the unique aspects of an event to be remembered. Thus, memory supported by the hippocampal system would resemble a key aspect of episodic memory. The asymmetric involvement of the right hemisphere in retrieval of episodic memory, as revealed by positron emission tomography, has been documented (Tulving, Kapur, Craik, Moscovitch, & Houle, 1994). In contrast to the contextual aspects of memory supported by the dorsomedial system, the ventrolateral system supports memories based on object classifications and thus emphasizes generalizations (Nadel, 1992).

Because these memory circuits are the same as the limbic circuits described earlier, the acquisition of memory within each system must be regulated by the emotional and motivational control biases of each trend. A characteristic of the mediodorsal memory system is that learning within this system is quickly acquired and subject to rapid extinction (Nadel, 1992). Nadel suggests that the rapidity in acquisition and extinction must be dependent on and regulated by exploratory motivations. This aspect of motivation is emphasized by the phasic arousal system outlined by Tucker and Williamson (1984) in which habituation occurs quickly and novel events are attended.

In contrast, the ventrolateral memory system must be motivated by self-regulatory processes that are involved in evaluating information about objects in the environment. The ventrolateral circuit evaluates information received from the inferior temporal lobe and the amygdala appears to be critical to the experience of object familiarization (Pribram, 1991). We propose that the tonic activation system of Tucker and Williamson provides the redundancy bias necessary to support the slow learning that occurs within the ventrolateral memory system (Nadel, 1992).

10-5.3. Motivation of Dual Motor Systems

The evidence from cytoarchitectonic studies shows that there are two premotor areas in the frontal lobe, each differentiating from its respective cortical origin (Goldberg, 1985; Pandya et al., 1988; Pandya & Yeterian, 1990). The anterior cingulate gives rise to the supplementary motor area (SMA), and the insular cortex, which is derived from the paleocortex, gives rise to the arcuate premotor area (APA). The SMA and the APA can be argued to represent cortical areas that have differentiated from the limbic circuits described earlier. Each new addition allowed the systems of self-regulation to be extended into the sphere of articulated action.

Goldberg (1985) argued that the SMA and APA represent key components of a mediodorsal motor system and a ventrolateral motor system, respectively; each system has a different mode of motor control. The mediodorsal system has a mode of control that is characterized as feedforward. The action plans are internally generated and organized around a contextual framework. The guidance of action within this system is internally generated, based on probabilistic models of the future. The ventrolateral system, on the other hand, controls actions in a feedback mode in which external input, including information about an object, is evaluated and acted upon. This system is dependent on information about external stimuli to guide action. In contrast to the anticipatory nature of motor control of the mediodorsal system, the ventrolateral system is more interactive (responsive) in its mode of action. Elsewhere, we have suggested that the differences between these motor systems may reflect the differences in the functioning of the dual limbic circuits (Tucker et al., 1995); each limbic circuit would preferentially provide specific motivational biases for motor control.

10-6. MOTILITY AND LANGUAGE

Language may be seen as the human extension of the development toward increasingly articulated expressions (Yakovlev, 1948). In lower vertebrates, such as fishes, salamanders, and some reptiles, motility is mostly axial. The development of limbs allowed for motility to be transformed from purely axial movements to appendicular (involving the appendages) locomotion. Furthermore, the progression of motility from the horizontal axis to the vertical axis freed the forelimbs from locomotion and allowed them to manipulate objects for the purpose of effect (to use objects as tools). In an evolutionary process known as preadaptation, the forelimbs became available for use as a means of communication (Kimura, 1979). Without the availability of forelimbs, communication must be largely limited to gestural postures.

The emergence of pulmonic respiration provided the necessary foundation for the evolution of strategies for vocal communication (Lieberman, 1984; Yakovlev, 1948). The sounds that humans produce during speech reflect the additional development of the supralaryngeal tract (Lieberman, 1984). The ability to communicate vocally afforded a major adaptive advantage over that of communication by gestural means: communication can occur in the absence of visual contact, and communication and tool use can occur simultaneously.

The development of the supralaryngeal tract in humans can be regarded as a step in a system of communication that is uniquely human. Yet, the control of phonemic combinations is still yoked to mechanisms controlling the more fundamental ability of the motor system to organize and manipulate objects. From her work and that of others, Greenfield (1991) noted that children's ability to hierarchically manipulate objects, phonemes, and sentences develops in stages and that there is a coincidence in the developmental progression of these stages. Greenfield argued that the developmental coupling of the stages is due to the fact that a common neural system, centered on prefrontal connections with Broca's area, underlies the control of object, phoneme, and sentence manipulation. Similarly, according to Kimura (1979), the control system involved in speech disorders is not specific to speech control. This system also controls the performance of motor imitation and meaningless gestures, both manually and orally. Kimura calls this system the "praxis" system to emphasize its relation to the apraxias observed after lesions to certain cortical regions, including regions associated with Broca's area.

The idea of language control being rooted in systems that were initially adapted for motor control has been extended to the cerebellum (Ito, 1993; Leiner, Leiner, & Dow, 1993), a structure that is traditionally believed to be predominantly related to motor functions. Leiner et al. (1993) traced the lateral expansion of the cerebellum and the development of the neodentate, a cerebellar nucleus, to their corresponding connections with frontal and prefrontal regions, including Broca's area. Furthermore, there are massive, but indirect, connections from these cortical areas back to the cerebellum. One pathway that Leiner et al. (1993) noted is the cortical projections to the red nucleus. The red nucleus in turn projects to the inferior olive. The inferior olive then relays to the cerebellum. The red nucleus, in most mammals, projects to the spinal cord and is thus involved in motor functions. However, in human brains this spinal cord projection is diminished and the projections to the inferior olive reflect a major output route instead. Leiner et al. argued that the reciprocal connections between the cerebral cortex and the cerebellum allowed the cerebellum to be involved in language functions. The involvement may occur by adapting cerebellar mechanism of motor control for the purpose of regulating language functions. Ito (1993) outlined a model of cerebellar function in which the cerebellum models and automates motor sequencing and control. This cerebellar model is general enough that it can be adapted for language and higher-order processes. Thus, Ito argues that ideas and concepts are manipulated in a manner similar to the ways limbs are manipulated.

10-7. A DUAL ROUTE INVOLVED IN LANGUAGE REGULATION

Aphasia is a disorder of language often observed after cerebral lesions to the language areas of the dominant hemisphere. The deficits vary from comprehension to production, depending on the site of the lesion. Of the subtypes of aphasia, symptoms related to conduction aphasia and transcortical motor aphasia (TCMA) show marked differences. Conduction aphasia is characterized as having spontaneous speech intact. The ability to repeat words, however, is severely impaired (Benson & Geschwind, 1985). In contrast, TCMA is characterized by impairments in spontaneous speech. The ability to repeat words and sentences are left intact. McCarthy and Warrington (1984) noted that in tasks that emphasized semantic processing the conduction aphasics excelled. In contrast, tasks that minimized semantic processing improved the speech production in the TCMA patients. McCarthy and Warrington postulated that the difference between conduction aphasia and TCMA is due to damage to different routes of speech production. They argued that patients with conduction aphasia must have damage to a route that handles speech perception and articulation. On the other hand, they argued that the TCMA patients have damage to a route that handles semantic analysis and speech production.

In keeping with the notion of a fundamental relation between language and motor functions, Goldberg (1985) suggests that the semantic analysis route is related to the mediodorsal system of motor control. This system is responsible for the initiation of speech from an internal source. The ventrolateral motor system is related to the phonological perception loop. This system is responsible for language functions requiring feedback loops, such as repetition. Indeed, Freedman Alexander, and Naeser (1984) observed that transcortical aphasia usually involves the separation of the SMA from the rest of the language areas of the frontal lobe. This observation led the authors to emphasize the important relation between the cingulate and SMA in speech function. Freedman et al. believed that this pathway represents a means by which limbic functions initiate speech. These authors contrast this pathway to another pathway that determines how language is produced. Clearly, Freedman et al. recognized the importance of limbic contributions to language processes of the mediodorsal system. However, we believe that both pathways are influenced by limbic inputs and that these inputs constitute biases that determine how language is produced. In other words, the way language is organized and produced within the ventrolateral system reflects the biases of the underlying motivational system. We propose that motivational bias of the tonic activation system provides the requisite influence to the ventrolateral speech system. Unlike the internal, contextual nature of the dorsal system's bias, the tonic activation system bias language functions in such a manner that feedback is the dominant form of control.

The dual system of speech production may extend to language representation as well. Landau and Jackendoff (1993) analyzed the linguistic representations of objects and spatial relations and found that the number of words that represent objects overwhelmingly outnumber those that represent spatial information. Furthermore, words describing objects capture detailed object features, whereas words describing spatial relations lack detail specificity. Landau and Jackendoff argue that the nature of linguistic representation of space and objects is directly influenced by the nature of representations within the ventral and dorsal visuomotor systems, respectively. Indeed, as argued by Jeannerod (1994), only sparse information concerning object features is represented in the dorsal, "pragmatic," visuomotor system.

10-8. LANGUAGE EMERGENT ACROSS THE NEURAXIS

In previous papers from our group (Liotti & Tucker, 1994; Tucker, 1992), we have theorized that hemispheric specialization has evolved through differential development of the dorsal and ventral trends within the two hemispheres. Specifically, the left hemisphere seems to have organized its analytic cognition in line with the focused attention and object recognition skills of the ventral (paleocortical) cortical pathway. There is initial evidence suggesting that DA is preferentially distributed within ventral cortical structures (Levitt et al., 1984). This would be consistent with our speculation that the DA mode of tonic activation and motor readiness is integral to the left hemisphere (Tucker & Williamson, 1984). Understanding the preferential emphasis on the cognitive skills of the ventral cortical trend within the left hemisphere may help to explain the integral role of object memory in the left hemisphere's contributions to language. In turn, understanding the preferential role of DA tonic activation in the ventral cortical pathway may help explain how the focused attention caused by the tonic activation system is integral to parsing objects from the perceptual stream.

A similar continuity may be found across the memory, attention, and arousal processes of the right hemisphere. The spatial memory skills of the dorsal pathway may be integral to the right hemisphere's cognition, including its contributions to language. The broad attentional scope produced by the NE phasic arousal system may be a primitive control mode that sets up a foundation for spatial memory, at one level of organization, and for broad semantics, at a higher level. The right hemisphere's role in such language functions as humor and story comprehension (Brownell et al., 1983) may require this form of broad semantics.

In everyday language, the organization of the communication process requires the coordination of both left- and right-lateralized attention/memory systems. Within each system there must be an effective vertical integration across the neuraxis to achieve elementary arousal control, the appropriate scope of working memory, and the recruitment of cortical systems in the representational process. This vertical integration requires that neurolinguistic mechanisms are not restricted to neocortex, but must operate across the phylogenetic hierarchy of the brain.

The microgenetic approach (Brown, 1988; Werner, 1957) suggests that vertical integration is achieved in a developmental fashion in each cognitive act. A foundation of initiative and arousal in brain-stem and midbrain regions becomes elaborated with emotional significance through engaging limbic circuits, and then a progression of paralimbic and neocortical networks is recruited, tracing the evolutionary hierarchy of the cortex. The developmental progression results in a kind of recapitulation of the phylogenetic order. This recapitulation traces the development of cognitive process from global organization, centered on core-brain motivational mechanisms, to articulated selections, drawing from the discrete control afforded by pyramidal motor pathways. When this approach is applied to language, it becomes clear that the product of the overt communicative behavior only reflects the final contribution of neocortical areas. As it occurs in the context of daily life, language is a motivated process, organizing the effective contributions of all levels of the neural control hierarchy.

10-9. CONCLUSION

We have outlined how motivational systems may provide fundamental biases to guide motor and cognitive processes. These biases may be crucial to understanding the motivational basis of hemispheric specialization for language. We have considered the possibility that the lateralization of these motivational systems may reflect a more fundamental asymmetry in corticolimbic organization of paleocortical and archicortical pathways. The organization of the catecholamines and limbic circuits according to the dual cytoarchitectonic trends may prove to be an important clue to the primitive basis of lateral asymmetry. For communication to be coherent, the final product must reflect successful elaboration across the neural hierarchy. Each self-regulatory system applies an inherent bias, such as the redundancy bias of the tonic activation system, that may shape not only the motivational direction, but the cybernetic mode of the communication, such as the tight parsing of the articulatory process. Language is an evolved process, and its phylogenetic roots are embodied in the neurophysiology of each psychological manifestation.

Acknowledgments

This work was supported by NIMH grants MH42129 and MH42669 and by a grant from the Pew Memorial Trusts and the James S. McDonnell Foundation to support the Center for the Cognitive Neuroscience of Attention.

CHAPTER 11

Brain Lesion Analysis in Clinical Research

Nina F. Dronkers[1] and Carl A. Ludy[2]

[1]Veterans Affairs Northern California Health Care System, Martinez, California 94553 and University of California, Davis, California 95616; [2]Veterans Affairs Northern California Health Care System, Martinez, California 94553

Brain lesion analysis has been used in clinical research for more than a century. From the discoveries of Broca and Wernicke to the offerings of modern neuroimaging, the method of lesion analysis has been the source of current theories regarding the processing of language in the brain. This chapter discusses current methodological issues in lesion analysis and reviews work in this area since 1990. Moving beyond attempts to localize aphasia syndromes, lesion analysis is now exploring the specific deficits within such areas as semantics, syntax, and speech production and is beginning to guide in the diagnosis and prognosis of aphasic patients.

The analysis of brain lesions in relation to behavioral deficits has been popular for centuries. In fact, most of what we know about how language is processed in the brain is due to lesion analysis. The theories of Broca (1861a; 1865), Wernicke (1874), and later, Geschwind (1965) were all based on correlations between the speech and language disorders of brain-injured patients and the locations of these brain lesions. Although these models are now undergoing some revision, the foundations of neurolinguistics and their modifications are still largely dependent on lesion analysis.

"Lesion analysis" simply means comparing the brain structures lesioned in a patient to the behavioral deficits exhibited by that same patient. These days, this involves obtaining a CT or MRI scan on the patient (historically, autopsy data were used), defining the location and extent of the lesion, and then describing the structures involved either verbally or by depicting the lesion on a template. At the same time, careful behavioral testing is done with the patient to assess the extent of the deficit: which functions are spared and which are not. Conclusions can then be drawn as to the involvement of that brain area in the specific behaviors now seen to be affected by the injury. Some lesion analyses are on a case-by-case basis and are important because they represent interesting or unusual deficits that can offer rare insights into brain/behavior relationships. Other analyses are of groups of patients who exhibit similar deficits, such as aphasia syndromes or specific speech or language deficits. In this type of analysis, computer reconstructions of patients' lesions are sometimes overlapped to determine whether they might reveal a common lesion shared by all patients with the disorder.

The growing sophistication of both neuroimaging techniques and behavioral measurement has advanced lesion analysis to a new level. Not only has lesion analysis given us theories to work with, it now also allows us to test and refine these theories. *In vivo* neuroimaging shows the patient's lesion in great detail and permits us to expand our areas of consideration beyond just Broca's and Wernicke's areas. Improvements in measuring the language deficits are freeing us from such elementary contrasts as "expressive" versus "receptive" deficits. We are learning that aphasia syndromes are not the best guide for describing localizable deficits, because syndromes have not correlated well with isolated brain regions. We are also seeing that subcortical structures may play an important role in language processing. Finally, lesion site information is becoming useful in predicting response to treatment and recovery of language functions.

The remainder of this chapter will discuss some of the methodological considerations in lesion analysis and the contributions of new work using this technique from 1990 to the present, including work from our own Aphasia Research Laboratory. Work prior to that time can be found in other volumes such as Ardila and Ostrosky-Solis (1989), Damasio and Damasio (1989), and Kertesz (1983). Our focus will be on humans who have lost the ability to process some aspect of speech or language, not on lesion studies of animals, since the comparison to complex human language is still difficult to draw. In addition, we will refer to lesion analysis here as the evaluation of the effects of structural damage, not functional or metabolic changes. For a summary of PET studies with aphasic patients, see the review by Metter and Hanson (1994).

11-1. METHODOLOGICAL ISSUES IN LESION ANALYSIS

As with any technique, the success of lesion analysis depends entirely on how well it is carried out. Damasio and Damasio (1989) point out that the method is only as good as the sophistication of the experimental hypotheses and design, the behavioral tests used, the underlying neurological models, and the anatomical resolution used to define the lesions. All of these have improved dramatically over the past two decades. Our methods of investigating language disorders, for example, have been refined by the involvement of linguists and psycholinguists and their knowledge of language and linguistic systems. Improvements in the clinical assessment of language disorders have helped us to understand more of the range of deficits in these patients. Progress in neuroscience has changed the manner in which we think about the brain and brain function, as the other chapters in this handbook can attest. Finally, neuroimaging techniques now allow researchers and clinicians to evaluate the location of the lesion *in vivo,* while continuing to investigate the behavioral deficits of the patient.

Still, researchers must be aware that the correlation of a behavioral deficit to an area of injury is not proof of localization of function to that area. The effects of diaschisis (the temporary effects of damaged areas on healthy ones), as well as plasticity and reorganization of function (the flexibility of healthy brain areas to assume the functions of damaged ones) must also be taken into account (see Kertesz, 1994a, for a review). However, the consistent and reliable association of behavioral deficits to lesion site cannot be overlooked and must, at the very least, imply a strong relation between a brain area and a behavioral deficit.

11-1.1. The Use of CT versus MRI Scans

The advent of neuroimaging techniques has taken the lesion method a big step forward. From the radioisotope studies of the 1960s and 1970s (e.g., Benson & Patten, 1967; Kertesz, Lesk, & McCabe, 1977), computerized tomography (CT) of the 1970s and 1980s (e.g., Kertesz, Harlock, & Coates, 1979; Naeser & Hayward, 1978; Poeck, de Bleser, & von Keyserlingk, 1984), magnetic resonance imaging (MRI) of the 1980s (e.g., Damasio & Damasio, 1989), to the three-dimensional MRIs of the 1990s (e.g., Damasio, 1995), the techniques for localizing brain lesions for comparison with speech or language disorders have improved tremendously. They have, in many ways, surpassed the advantages of autopsy data for lesion localization, as the resolution of new scanners is very high while the rate of consent for autopsy is very low. In addition, *in vivo* imaging can be done closer to the time of testing without the risk of additional neurologic changes that may occur later in life (Damasio & Damasio, 1989).

The decision of whether to use CTs or MRIs in lesion analysis depends largely on the purpose of the study and the availability of the scans. The technology of the two differs significantly. CT scanning of the brain involves rotating an X-ray source and image receptor systematically around the head. A computer uses information about the attenuation of the X-ray beam to reconstruct a two-dimensional image representing a horizontal slice through the brain. Chronic strokes appear as dark regions, since they are filled with cerebrospinal fluid, which is less dense than healthy tissue (see Figure 1).

MR imaging, on the other hand, involves no ionizing radiation, but instead uses nuclear magnetic resonance to produce high-resolution images. The powerful magnetic field of the scanner causes the spinning nuclei of atoms to align. A specific target is

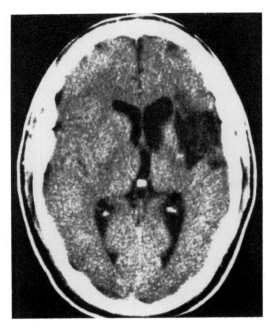

FIGURE 1   A sample of a horizontal slice rendered by a CT scanner is shown in this figure. The lesion of this patient through the insula and Broca's area is seen as a dark area within the left side of the brain. By radiologic convention, the brain is imaged from the patient's perspective, with the left hemisphere of the brain on the right side of the figure, and the right hemisphere on the left.

then stimulated by external radio-frequency waves, causing the atoms to wobble in a process known as "resonance." When the external source is turned off, the atoms return to their alignment (known as "relaxation") and release energy that can be detected and computer-reconstructed into two-dimensional images. Two different measurements of relaxation times can be made and are known as T1- or T2-weighted images. One of the advantages of MR imaging is that images can be rendered in three planes: horizontal (similar to the CT scanner), coronal, and sagittal. Thus, if a lesion cannot be seen clearly from one angle, it can be examined from one of the other two. In addition, MR images can be rendered in three dimensions with the appropriate protocols and software (see Figure 2). The resolution of MR imaging also far surpasses that of CT.

Several studies discuss the general benefits of CT versus MRI scanning. For example, Mohr et al. (1995) found that CT and MRI were both good at the early detection of stroke, though they did not compare the two techniques at later times post-stroke, nor did they focus on their use in defining lesion boundaries. For case studies investigating lesion-deficit correlations, the use of either technique is generally accepted, provided that the lesion can be clearly seen.

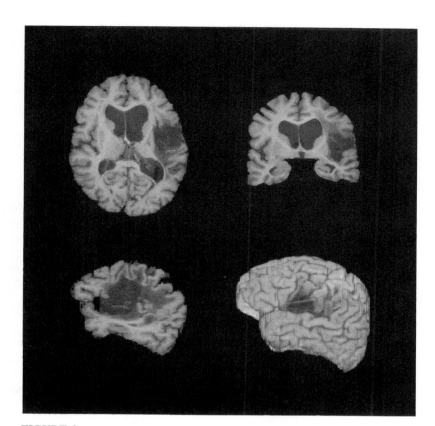

FIGURE 2   These images represent samples of three-dimensional MRI lesion reconstructions of a patient with anomic aphasia and apraxia of speech. The upper left image is a horizontal slice similar to those seen in CT images. The upper right image depicts a coronal section, and the lower left, a sagittal section. The lower right image shows a 3-D representation of the lesion on the lateral surface of the left hemisphere.

For accurate definition of lesion boundaries, Naeser and Palumbo (1994) argue that the horizontal slices of CT scans show the depth of the lesion better than MRIs. They also point out that the differently weighted images (T1 and T2) of MRIs often disagree as to the extent of the lesion, with T1-weighted images underestimating the border and T2-weighted images exaggerating the border. However, the superior resolution of the images and the addition of coronal and sagittal views from MRI scans are of great assistance in determining the involvement of specific gyri or structures, particularly when these can be computer-rendered in three dimensions.

For group studies, it is preferable to stay with one type of scan, either CT or MRI, without mixing the two types. This is especially true when comparisons between patients will be made. However, this is not always possible, particularly when one has to depend on existing scans from radiology files, as in retrospective studies, or when the expense or inconvenience to the patient is not medically warranted. For example, some patients find it difficult to lie still in the scanner for extended periods of time and cannot be re-scanned. Other patients with metal in their bodies, such as from pacemakers or surgical clips, should not undergo MRI scanning, because of the risk of the magnet dislodging the metal. The advantages and disadvantages of each technique must be considered for each study, and must be determined by the purpose of that study.

11-1.2. Time Post-Onset to Scan

The time at which the scan is performed can make a significant difference in how well the lesion can be visualized and in judging its boundaries. Scans obtained on the first day after the injury are rarely useful in defining the specific lesion location, though they may differentiate between etiologies of the illness (e.g., stroke versus tumor). We have found that 3 weeks post-onset of the illness is the minimum amount of time to wait before obtaining a scan that can be used in determining lesion boundaries. This time frame is also advocated by Kertesz (1989). Kinkel and Jacobs (1976) advise waiting a full month, while Naeser and Palumbo (1994) recommend 3 months before the lesion has stabilized.

In many cases, the timing of the scan depends simply on what scans are already available in the patient's files or for what purpose the scan will be used. If lesion boundaries are important because the lesion will be rendered onto templates or compared to other patients, then it is best to use chronic scans, at least 3 weeks post-onset, but preferably closest to the time of behavioral testing. However, as Damasio and Damasio (1989) point out, "data obtained during the acute or periacute period are not necessarily useless. Provided early anatomical and neuropsychological observations coincide, their relations are still interpretable, and may be quite revealing. It is the cross-correlation of data from different epochs that is methodologically unacceptable" (p. 121).

11-1.3. Time Post-Onset to Language Testing

As with scanning, the amount of time post-injury at which behavioral testing is performed can make a significant difference in the findings. It is well known that patients can make significant recovery in the few weeks and months after injury, with language profiles changing considerably. For example, the language deficits of patients initially diagnosed with global aphasia will often evolve into a Broca's aphasia, and Broca's aphasic patients will often evolve into anomic aphasics. Patients with a Wernicke's aphasia immediately after their injuries almost never retain the Wernicke's aphasia, but evolve into conduction or anomic aphasics. Thus, studies that intend to associate areas of brain injury to behavioral deficits must first be certain that the lesions *and* the disorders have stabilized and will not evolve into other, milder deficits.

Those studies that have succeeded in localizing speech or language functions to specific areas of the brain have correlated chronic deficits with chronic lesions. Studies that use acute patients usually yield results with low correlations between lesion location and the deficit under study. For example, Basso and colleagues (Basso, Lecours, Moraschini, & Vanier, 1985) found that as many as 17% of their patients with aphasia did not support traditional models of language localization. Their study included patients who ranged from 1 to 78 months post-onset and may have included those who were still evolving. Willmes and Poeck (1993) also found a poor correlation between aphasia type and lesion location in patients who were 1 to 340 months post-onset. They acknowledge that they mixed acute and chronic patients in their overall analysis, and point out that "unfortunately, localization studies are not based on predetermined and universally accepted dates of examination post-onset" (p. 1538).

Kertesz et al. (1979) made a distinction between acute and chronic patients and found that the more stable relationships between deficits and lesions were seen after a year post-onset. We also recommend waiting at least 1 year post-onset, when the disorders have become fairly stable. This does not mean that patients do not show any change after 1 year, since gradual recovery continues even after many years. However, it is well accepted that the most significant gains are made in the first few weeks and months post-onset, with few major changes thereafter in most cases.

11-1.4. Regions of Interest

Predefining regions of interest involves stipulating, in advance, the areas to be investigated, excluding all others. This approach is appropriate when certain regions are being compared to behavioral correlates and when specific hypotheses about areas are being tested. For example, Willmes and Poeck (1993) used broadly defined regions of interest in their analysis of lesion localization and aphasia syndromes. They were interested in whether lesions in specified brain areas would produce the aphasia syndromes predicted by traditional theories. Naeser and colleagues (Naeser, Gaddie, Palumbo, & Stiassny-Eder, 1990) investigated the involvement of Wernicke's area versus the subcortical temporal isthmus in their study of recovery of auditory language comprehension skills in globally aphasic patients. Both studies were interested in specific areas and thus focused only on those regions.

Sometimes an investigator will not restrict the analysis to certain regions of interest and will let the data determine which brain areas are related to a certain deficit. This is the approach we tend to take in our group, particularly in exploring relationships that have not previously been examined. For example, Dronkers (1996) compared the overlapped lesions of 25 patients with an articulatory planning deficit known as "apraxia of speech" with those of 19 patients without the disorder. The result led to the correlation between this motor speech disorder and a brain area previously unrecognized for this function. By not confining the analysis to predetermined regions, a new functional location for one aspect of speech processing was identified. Thus, the region-of-interest approach tests given areas of theoretical importance, while the lesion overlapping method is less theory-dependent and allows the examination of areas not previously identified.

11-1.5. Lesion Reconstruction Templates

Some of the oldest templates are those of the renowned anatomist Dejerine (Dejerine, 1901). These are carefully detailed templates with the structures clearly labeled and were used by Vanier and Caplan in their study of lesions in agrammatism (Vanier & Caplan, 1990). These classic templates are easy to use, but are not as readily adaptable to computer applications as are some of the other templates available.

Naeser and colleagues use templates for CT scans (see Naeser & Palumbo, 1995), which are done at 15–20° to the canthomeatal line, the imaginary line drawn from the canthus of the eye to the auditory meatus. They specify certain regions of interest on these templates and calculate the extent of the lesion within each area on a scale of 0 to 5. With this method, they have analyzed numerous features of aphasia as well as the relation of certain lesions to recovery of these deficits.

Willmes and Poeck (1993) used the templates and software described in Willmes and Ratajczak (1987) derived from the Matsui and Hirano (1978) atlas. Their method stores lesion information in a grid system of 16 templates that records the presence or absence of a lesion within each grid point. Neighboring grid points can be combined into regions of interest or specific anatomic areas. Lesions are overlapped by computing a "lesion similarity coefficient." This method can be helpful if predefined regions of interest are to be tested.

Damasio and Damasio (1989) describe their system whereby different sets of templates are used to accommodate the various angles that might be obtained for different patients during scanning. The advantage is that reconstruction is easier when the angle of the template matches the angle of the scan. The disadvantage is that lesions from different patients are difficult to compare when the templates themselves differ between patients. However, this method is an excellent way of portraying a lesion and is used in a number of published case studies.

The method used in our lab is that developed by Frey, Woods, Knight, Scabini, and Clayworth (1987). A behavioral neurologist trained in neuroimaging and a coauthor of the reconstruction program reconstructs the lesions onto standardized templates derived from the atlas of DeArmond, Fusco, and Dewey (1976). These reconstructions are then entered into a mini- or microcomputer with software designed to depict the lesions in horizontal and lateral views and calculate lesion volume (see Figure 3). Because the templates are all at the same angle (0° to the orbital-meatal line), lesions can also be overlapped when common areas of injury among patients exhibiting similar deficits are sought.

All of these template systems are perfectly acceptable. Which one to use is entirely a matter of the type of experiment being conducted. The important thing to remember is to use the same template throughout the study so that lesions can be compared easily across patients.

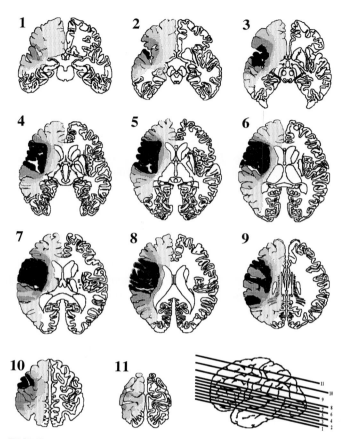

FIGURE 3   This computer-generated lesion overlay depicts the combined lesions of 12 patients with Broca's aphasia. Darker areas indicate greater numbers of patients with lesions to that region, with solid black indicating involvement of 100% of the patients. These templates are examples of some of the ones used in lesion analysis. Computerized templates such as these also allow for the projection of the lesion on the lateral surface of the brain and the calculation of lesion volume.

11-2. RECENT FINDINGS IN LESION ANALYSIS

11-2.1. Review Articles

Several articles since the early 1990s have surveyed progress in the lesion analysis of language disorders. Tranel (1992) covers work from 1990 to 1991, and Damasio (1992) provides an excellent review of aphasia and discusses the brain areas involved as determined by lesion analysis. Dronkers and Pinker (in press) present the aphasias in a description that is based largely on the results of lesion analysis. Books on aphasia, such as those by Goodglass (1993) and Benson and Ardila (1996), offer reviews of traditional interpretations of lesion studies of language deficits, while edited volumes such as that by Kertesz (1994b) offer reviews of specific aspects of language functioning.

11-2.2. Lesion Analysis of Aphasia Syndromes

Lesion analysis research on language in the 1970s and 1980s was largely devoted to testing whether traditional autopsy-based models of language localization could be supported by new structural imaging techniques. Specifically, researchers were looking to see if the major types of aphasia were indeed associated with the lesion sites predicted by these theories. Early studies, such as Naeser and Hayward (1978), found traditional theories to be upheld when comparing aphasia classifications on the Boston Diagnostic Aphasia Examination to the patients' lesions on CT scans. However, the reverse correlation of predicting aphasia type from lesion site was not possible. Others, such as Basso et al. (1985), also found a poor correlation between lesion site and aphasia type. They found classical correlations less absolute than predicted and estimated a 12 to 17% exception rate, with all types of dissociations found.

Willmes and Poeck (1993) took another look at the localization of aphasia syndromes and found even more disagreement than did Basso and colleagues regarding the probability of certain aphasia syndromes occurring in conjunction with defined brain regions. In a retrospective study of 221 patients, they compared the results of the Aachen Aphasia Test (Huber, Poeck, & Willmes, 1984) with lesions reconstructed onto their standardized grids previously described. Aphasia categories evaluated were global, Wernicke's, Broca's, and amnesic aphasia with two additional categories for those patients who were "nonclassifiable" and those who belonged to smaller "other" categories such as conduction, mixed transcortical, and transcortical sensory aphasia. Lesions were reconstructed onto grids derived from the Matsui and Hirano (1978) atlas and tested against traditional hypotheses of lesion localization. Some of their findings included a probability of only .48 that a posterior superior temporal lesion (including Wernicke's area) would result in a stable Wernicke's aphasia. Similarly, only a .35 probability was found for the hypothesis that anterior lesions would result in a Broca's aphasia.

Several studies looked at single aphasia syndromes. In a paper describing CT scan correlates of a severe form of global aphasia, de Renzi, Colombo, and Scarpa (1991) found that only 35% of their 17 cases had the large lesions including Broca's and Wernicke's areas that are predicted with global aphasia. Rapcsak and colleagues (Rapcsak, Krupp, Rubens, & Reim, 1990) presented two cases of acute mixed transcortical aphasia with lesions restricted to the frontal lobe, without the additional parietal involvement usually expected. In a study from our lab (Dronkers, Shapiro, Redfern, & Knight, 1992), we found that lesions to Broca's area did not necessarily result in a Broca's aphasia, nor did the diagnosis of a Broca's aphasia reliably predict involvement of Broca's area. These studies demonstrate the difficulty of correlating aphasia syndromes to lesion sites.

In fact, this lack of an association between aphasia syndromes and lesion site is not surprising. Aphasia syndromes are constellations of many different problems. For example, Broca's aphasic patients have many deficits: fluency, comprehension, repetition, naming, and motor speech. It is unlikely that all of these functions would be processed in only one area of the brain. Although syndromes are a convenient shorthand for describing groups of deficits, they are not particularly useful as a means of investigating language processing mechanisms. As the next section will show, framing the investigation by using more specific criteria for defining deficits is a more productive approach.

11-2.3. Lesion Analysis of Specific Deficits

Several studies since 1990 addressed the localization of specific deficits in aphasic patients. These concerned various aspects of semantics, syntax, and speech production. The studies described in this section exemplify the direction this work is taking.

11-2.3.1. Semantics

In an effort to better characterize the semantic deficits that affect many aphasic patients, Hart and Gordon (1990) sought patients with low scores on the Boston Naming Test and further probed semantic processing. They evaluated three levels: (1) superordinate/categorization (do *hammer* and *wrench* belong to the same category?); (2) equivalence/single-word meaning (which is synonymous with *engine: gas* or *motor?*); (3) subordinate/property judgment (what has the same color as a *skunk:* a *pig* or a *penguin?*). Hart and Gordon found three patients who showed isolated impairment of single-word semantic comprehension, without deficits in speech perception or production. They then used the Naeser and Hayward templates to reconstruct the available CT and MRI scans and overlap the lesions. In doing so, they found an area in the posterior temporal lobe and a small portion of the inferior parietal region that were always lesioned in the three patients with this deficit, but never in those patients without the disorder. The authors speculate that this area may be concerned with multimodal processing and integration of language.

Rapcsak and Rubens (1990) report an interesting case with semantic jargon following a frontal lesion. The patient showed impairment on language output tasks that required semantic mediation, whether in spoken or written modalities. Language tasks that did not tax the semantic output system were unaffected. Speech and writing were fluent but incoherent, and writing showed "semantic agraphia" with good spelling to dictation but severe impairment of homophone spelling, which requires semantic disambiguation of phonologically identical words (e.g., *see/sea*). Comprehension of speech and reading seemed intact. CT scanning revealed a large, left prefrontal hematoma, extending from the orbito-frontal region up to the level of the lateral ventricle. Three months later, language retesting showed that all functions had returned to normal, including the homophone spelling test. The authors repropose Luria's (1976) idea that the prefrontal cortex acts as a semantic "monitoring device" with an important role in programming and monitoring ongoing cognitive activity, including language production.

Damasio and colleagues (Damasio et al., 1996) tested 127 patients with single brain lesions on a naming task using three different types of items: famous faces, animals, and tools. They found that deficits in naming each stimulus type were correlated with lesions in specific areas of the temporal lobe. Deficits in naming faces were associated with lesions in the left temporal pole, difficulty in retrieving animal names were related to lesions in the anterior inferior temporal gyrus, and errors in naming tools correlated with lesions in posterior inferior/middle temporal gyri and the temporo-parieto-occipital junction. The authors take this finding to support their theory that categories of words are represented in separate regions contained within a larger network. They also provide data from a study in normals using positron emission tomography that support approximately the same brain regions in the processing of these three types of information.

11-2.3.2. Syntax

The syntactic processing deficits of aphasic patients have been of great interest to neurolinguists. In previous years, investigations sought to locate a single syntactic processor in the brain. Broca's area was considered the most likely area, though later work failed to confirm a unique role for this or any other single brain region. In a study from our lab, Dronkers, Wilkins, Van Valin, Redfern, and Jaeger (1994) found that a new, unexpected area emerged as important to sentence comprehension. The anterior portion of Brodmann's area 22 (just anterior to auditory cortex on the superior temporal gyrus) was lesioned in all patients with severe deficits in sentence comprehension, but spared in those with only a mild disorder. However, it was not the only area lesioned in these severe patients. Broca's or Wernicke's areas were involved in some, but not all, patients, as were other frontal regions. This suggested that the processing of syntax involves contributions from several brain areas, presumably because of the involvement of many aspects of language and cognition, including memory and attention.

Similar conclusions were reached by Caplan, Hildebrandt, and Makris (1996). In their large-scale study, they tested 60 patients with left or right hemisphere strokes on their ability to comprehend various syntactic structures. Using an object manipulation task, patients acted out the thematic roles in a set of 25 sentence types varying in complexity. The first interesting finding was that both left and right hemisphere-injured patients performed worse than normal controls on the more complex sentence types. This implies that the right hemisphere also contributes to normal sentence processing. Localization of the lesions in 18 of the 60 patients was also examined. Five regions of interest were evaluated, including both parts of Broca's area (Brodmann's areas 44 and 45), the superior temporal gyrus (area 22), the angular gyrus (area 39), and the supramarginal gyrus (area 40). Lesions to all of these areas were found to affect syntactic processing abilities, leaving the authors with the conclusion that these areas all contribute to a larger neural system underlying sentence comprehension.

11-2.3.3. Production

Baum, Blumstein, Naeser, and Palumbo (1990) compared the temporal parameters of speech production in patients with anterior lesions (Broca's area, anterior limb of internal capsule, lower motor cortex) versus those with posterior lesions (Wernicke's area, subcortical temporal isthmus). They used the Naeser and Hayward templates with these regions predefined. In general, they noted that speech production deficits could be found in both groups, with some differences. Patients with anterior lesions had more motoric difficulty with, for example, the interaction of the laryngeal system and supralaryngeal vocal tract. Posterior patients showed subtle phonetic impairments related to more global, durational patterns of speech, and showed more variability of production. Patients with lesions in both areas showed a combination of these deficits. The study demonstrated how even broad anatomical resolution inherent in a large-scale region-of-interest approach can combine with careful behavioral analysis to yield results that extend our understanding of speech production processes.

Our group was interested in whether a common lesion could be found in patients who exhibit apraxia of speech, a disorder believed to affect the planning of articulatory movements (Dronkers, 1996). The lesions of 25 patients with this deficit were over-lapped, all showing involvement of a particular area of the insula. Nineteen patients without this disorder also had widespread lesions, but they spared the same small region of the insula that was involved in the speech apraxic patients. This is another example of how modern lesion analysis can identify new functional areas.

11-2.4. Recovery from Aphasia

Naeser and colleagues (Naeser et al., 1990) examined recovery in patients with global aphasia as tested with the Boston Diagnostic Aphasia Exam. They looked specifically at lesions in Wernicke's area and the temporal isthmus, two areas previously found to relate to poor recovery in Wernicke's aphasic patients. They found that most patients with lesions that involved at least half of Wernicke's area had a poorer recovery than most of the patients with lesions sparing Wernicke's area but including the temporal isthmus. Their regions of interest had not included other temporal areas that are now being investigated with regard to language comprehension and lexical retrieval (Damasio et al., 1996; Dronkers, Redfern, & Ludy, 1995). However, this study was one of the first to apply the results of lesion analysis to issues of prognosis and recovery in aphasic patients.

Kertesz, Lau, and Polk (1993) also examined recovery in 22 Wernicke's aphasic patients. Using templates derived from Matsui and Hirano (1978), they defined 31 regions of interest within the left cerebral hemisphere. They found that persisting Wernicke's aphasia involved the supramarginal and angular gyri in the parietal lobe as well as the superior temporal area. The group with the best recovery tended to spare the superior temporal and middle temporal gyri.

We have also found temporal lobe cortex to be important in recovery (Dronkers et al., 1995). Five patients with severe and persisting Wernicke's aphasia were all found to have extensive lesions in the middle temporal gyrus and underlying white matter. A sixth patient had only deep white matter involvement of the temporal lobe. Although the patients with the cortical involvement have shown little or no change in their language deficits, the patient with the lesion restricted to deep temporal fibers is beginning to change after 4 years. How much recovery will be made remains to be seen.

11-2.5. Response to Treatment

One of the rewarding outcomes of lesion analysis is the possibility of applying its results to the treatment of the patients who contributed to the findings. One example of this is a study by Naeser and Palumbo (1994), who evaluated the success of using a computer-assisted visual communication treatment program known as C-ViC. Seven patients fell into groups of good versus poor responders. No one lesion site could separate the groups, but a combination of sites did differentiate between them. The patients with lesions in both the supraventricular area and the temporal isthmus responded poorly to this method of treatment, while those patients whose lesions spared either of these areas showed a good response. Therefore, this treatment technique will not yield optimal results for patients with this combination of lesions. The authors suggest that outcome goals should be lowered in these cases.

11-2.6. Size versus Site of Lesion

Large lesions were often seen to result in severe and persistent language deficits. This finding led to the belief that lesion size is a critical factor in predicting outcome. However, Willmes and Poeck (1993), in their study of 221 patients, did not find any relation between lesion size and aphasia severity. Naeser and Palumbo (1994) summarize their work and that of others and conclude that lesion size is generally not helpful in making predictions about recovery. Recent research has focused less on lesion size and more on precise location. This is not to say that lesion size is irrelevant. The larger the lesion, the more chance that it will encompass critical sites. It is our increasing understanding of these sites that gives us the most information about language mechanisms in the brain and about how to use this information in treating patients with these disorders.

11-3. CONCLUSION

The method of lesion analysis has done much for the study of language and the brain. A century ago, lesion analysis taught us that certain brain regions could play specialized roles in language processing. The identification of Broca's and Wernicke's areas was due entirely to lesion analysis. Modern approaches, taking advantage of better behavioral measurements and high-resolution *in vivo* neuroimaging, continue this exploration into the neural mechanisms of language.

The last few years have taught us much about language and the brain. First, modern techniques have allowed us to discover that, although aphasia syndromes may be a convenient way of summarizing a group of deficits, they are not particularly helpful in localizing language functions. The failure to find unitary brain areas underlying syndromes led to a better definition of the behaviors that contribute to these constellations of deficits. By specifying the components of the disorder, correlations between brain areas and behavioral deficits are becoming more reliable. In turn, theories about language and the brain are gradually evolving to incorporate these new findings.

Lesion analysis has also turned the focus away from traditional areas such as Broca's and Wernicke's and has widened our view of the brain as a system of interacting parts that involves many more brain regions than those originally considered. These must all work in concert to produce a behavior as complex as language. The extent of the lesion is significant, not because of its sheer size, but because of the parts affected and their roles in this integrated network.

Our interpretations of the consequences of specific lesions also enhance our ability to diagnose individual patients' deficits and to prescribe the most effective means of remediation. Lesion information can also assist patients and their families in understanding the long-term effects of their injuries and how best to manage them. As information provided by lesion analysis becomes more accessible, it can only be used to greater and greater advantage.

Lesion analysis requires a multidisciplinary approach. It has developed beyond its previous limitations because of the sophistication of the behavioral and imaging methods contributed by numerous fields. This methodology epitomizes the interaction of the different specialties that have led to the success of neurolinguistics as a new scientific enterprise. In this age of exciting new developments in brain research, lesion analysis remains an important benchmark against which new methods must be tested, and will certainly continue to play an important role in the study of language and the brain.

CHAPTER 12

The Sodium Amytal (Wada) Test
Procedural and
Interpretative Considerations

Julie A. Fields and Alexander I. Tröster
Department of Neurology, University of Kansas Medical Center, Kansas City, Kansas 66160

This chapter familiarizes the reader with the sodium amytal (amobarbital) test, both as a technique to establish hemispheric language capabilities in epilepsy surgery candidates, and as a procedure to gain insights into the neural correlates of language. After a brief historical overview, the chapter reviews the rationale underlying the use of the amobarbital procedure. Because the exact nature of the amobarbital procedure varies from center to center, we next provide a description of how the procedure is performed at the University of Kansas Medical Center's Comprehensive Epilepsy Center. Some of the problems and considerations involved in the administration and interpretation of the amobarbital test are identified, and recent developments aimed at overcoming some of these problems are highlighted. A discussion of the sodium amytal test's contribution to our understanding of inter- and intra-hemispheric differences in the organization of language concludes the chapter.

12-1. A BRIEF HISTORICAL NOTE

The intracarotid sodium amytal (amobarbital) test (ISA) was first described by Juhn Wada (1949) and thus is often referred to as the "Wada test." Wada originally developed this technique to study the interhemispheric spread of epileptiform discharges in patients undergoing unilateral electroconvulsive therapy. Based on his observation that aphasia resulted when the language dominant hemisphere was injected with amobarbital, he reasoned that this technique might be useful in determining hemispheric language dominance in neurosurgical candidates (and consequently avoid speech and language dysfunction in patients undergoing dominant hemisphere surgery). Favorable results were reported early (Branch, Milner, & Rasmussen, 1964; Wada & Rasmussen, 1960) and the ISA is widely used (Rausch et al., 1993).

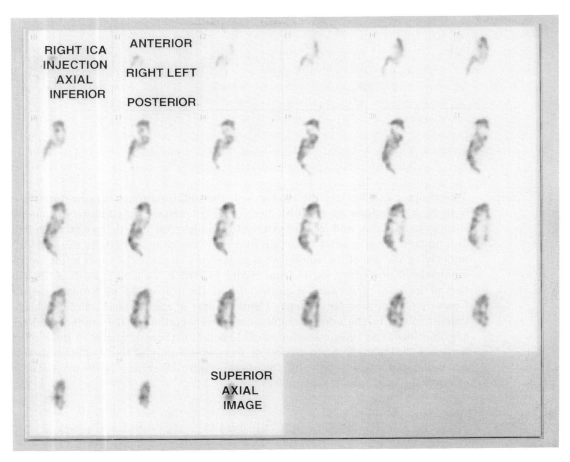

FIGURE 1a Tc99mECD SPECT Scans (axial) illustrating the cerebral territories perfused by amobarbital after injection of the right internal carotid artery. Because the radioactive tracer is injected along with the amobarbital, darker regions took up more amobarbital than lighter regions. The left side of the brain is on the right side of the scan.

12-2. RATIONALE UNDERLYING THE ISA PROCEDURE

The sodium salt of amobarbital, a di-alkyl substituted oxybarbiturate, is very lipid-soluble. Consequently it crosses the blood–brain barrier easily, and its anesthetic effect is rapid. Injection with amobarbital of the internal carotid artery (ICA) leads to pharmacologic inactivation of brain areas in the distribution of the ipsilateral anterior and middle cerebral arteries, and the anterior choroidal artery. The territories supplied by the posterior cerebral artery (PCA) can be anesthetized by separate PCA amobarbital injection when supplementary information is desired about memory function. Figure 1 presents SPECT scans showing the territories inactivated by right ICA and PCA amobarbital injections.

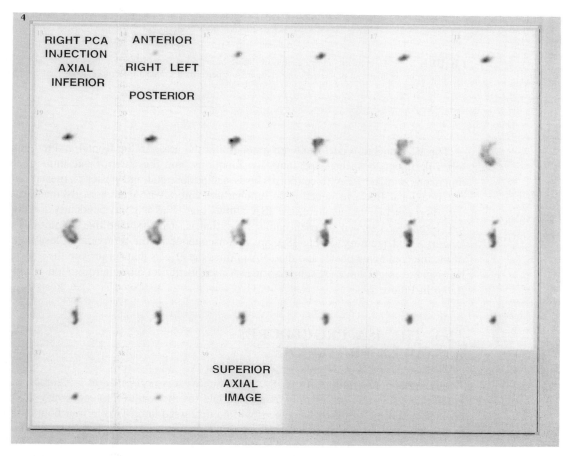

FIGURE 1b   Tc99mECD SPECT Scans (axial) illustrating the cerebral territories perfused by amobarbital after injection of the right posterior cerebral artery. Because the radioactive tracer is injected along with the amobarbital, darker regions took up more amobarbital than lighter regions. The left side of the brain is on the right side of the scan.

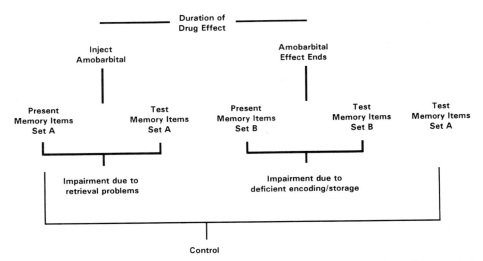

FIGURE 2 Diagrammatic representation of timing of memory stimulus presentation and memory testing during the amobarbital procedure, and comparisons performed to make inferences about memory processes affected.

The ICA injection perfuses with amobarbital the areas in the frontal and temporal lobes pertinent to speech and language functions, and the anterior one-third of the hippocampus. The PCA injection perfuses occipitoparietal and posterior mesial temporal regions. Because injection of amobarbital into an ICA temporarily inactivates only one cerebral hemisphere, the ISA allows one to assess independently the cognitive functions served by each hemisphere; that is, it is assumed that disruptions of language and memory during ISA are a consequence of the temporary "lesioning" of the injected hemisphere, and that ISA mimics the effects that surgery on the injected hemisphere might have. A schema for ISA administration and interpretation is illustrated in Figure 2.

12-3. THE ISA PROCEDURE: A SAMPLE PROTOCOL

Following is a description of the ISA procedure as it is performed at our center. It is emphasized that this serves only as an example for the reader. Protocols vary across centers, and some of these differences will be discussed later. Two to four hours prior to the first ISA injection, a baseline cognitive evaluation is completed using one of five alternate protocols. This not only serves to establish a baseline score against which to compare ISA results, but also to familiarize the patient with the protocol and to obtain informed consent. Baseline evaluation also allows the clinician to observe whether any conditions exist that might contraindicate ISA testing (e.g., if the patient has had a seizure in the last 24 hours, it is possible that post-ictal confusion is present; the patient might not have complied with fasting requirements; he/she might demonstrate signs of anticonvulsant toxicity; or he/she might have a mood disturbance making it difficult to comply with the procedure). Importantly, baseline evaluation affords the opportunity to provide the patient with memoranda, the recall of which is assessed during the ISA procedure.

At our center we evaluate serial speech, word reading, color naming, object naming to visual and tactile confrontation, praxis, sentence repetition, aural sentence comprehension, preposition comprehension, syntactic comprehension, identification of a facial expression, and memory for visually and aurally presented stimuli.

For ISA testing, the patient is transported to the radiology suite. Following local anesthesia, the right femoral artery is punctured. A guide wire and catheter are advanced under fluoroscopic guidance through the femoral artery and aorta to a point just distal to the ICA's origin. An angiogram is performed to visualize the vasculature and identify any potential abnormalities, and to determine if the contrast material perfuses vessels other than those intended (e.g., whether cross-flow to the other hemisphere occurs, or if the posterior cerebral artery fills via the posterior communicating artery).

Typically, the left ICA is injected first (other centers often first inject the hemisphere ipsilateral to surgery), using a dose of 100 mg amobarbital. The amobarbital is mixed with a small amount of radioactive material (Tc99mHMPAO or Tc99mECD), so that following ISA a SPECT scan can be performed to visualize distribution of the amobarbital. At time of injection, the patient, who is supine, is instructed to hold his/her arms straight up and to begin counting. Upon injection, the patient almost immediately becomes hemiplegic or hemiparetic contralateral to side of injection, and the patient's arm opposite to the side of injection becomes limp. Following language-dominant hemisphere injection, the patient typically evidences speech arrest, and subsequently dysphasic speech. Testing continues using a protocol similar to the one outlined in Table 1.

EEG is monitored continuously for delta-wave activity, and arm strength is checked periodically throughout testing to ascertain drug effect, which can last anywhere from 90 to 300 seconds (Rausch & Risinger, 1990). After hemiplegia has resolved, and the EEG has returned to baseline, the patient's recall and recognition of stimuli presented prior to and during drug effect are assessed.

The second hemisphere is injected 60 minutes after first injection to ensure recovery from first injection. Scores obtained on the memory protocol are expressed as a percentage relative to baseline score. We consider a memory score below 80% of baseline a failure. In addition, performance on memory and language items is compared after left and right ISA injection, as is occurrence of speech arrest, so as to make a judgment about hemispheric language dominance and possible asymmetries in memory performance.

TABLE 1

The University of Kansas Medical Center's Comprehensive Epilepsy Center Cognitive Evaluation Protocol for Use in Sodium Amytal Testing

Part I	Part II (amytal effects cleared)[a]
Patient raises arms and begins counting	Recall of last number (not scored)
Injection	
Serial speech (e.g., months of the year)	
Recall of single word (established during baseline)	Recall of single word
Tactile object naming	Recall/recognition of object
Visual object naming	
Color naming	Recall of object's color
Assessment of arm strength or visual fields	
Reading of compound word	Recall/recognition of read word and color of paper
Visual picture naming	Recall/recognition of picture
Auditory comprehension of sentence	Recall/recognition of sentence
Execution of auditorially presented praxis command	
Recall of picture	
Assessment of arm strength or visual fields	
Sentence repetition	Recall/recognition of sentence
Auditory comprehension of sentence	Recall/recognition of sentence
Recall of home telephone number	
Token test item	
Assessment of arm strength or visual fields	
Execution of praxis command	
Design recognition	Design recognition
Identification of appropriately-colored object	Recall/recognition of colored object
Identification of facial expression	Facial recognition
Assessment of arm strength or visual fields	
Syntactic comprehension	
Prosody (sing *Happy Birthday*)	Recall of song (not scored)
Semantic verbal fluency	
Digit span forward (to 4) and backward (to 3)	
Assessment of arm strength or visual fields	
Word association (4 items)	
Assessment of arm strength or visual fields	Recall of hemiparesis/hemianopia (not scored)

[a]Part II items are not listed in order of presentation, but indicate the referent item in part I.
From Tröster et al. (1996). Reprinted with permission. Copyright Lippincott-Raven Publishers.

12-4. PROBLEMS AND CONSIDERATIONS IN ISA PROTOCOL ADMINISTRATION AND INTERPRETATION

Although the rationale underlying the ISA seems straightforward, the performance and interpretation of the test are not. Both the determination of speech dominance and that of memory representation are highly subjective. Protocols for the ISA vary from center to center (Loring, Lee, & Meador, 1994; Rausch et al., 1993; Snyder, Novelly,

& Harris, 1990), making meaningful data comparisons difficult. Some of the more important between-center and intersubject differences, which affect ISA language test result interpretation, are discussed in the following sections. Readers interested in the controversy about the validity and reliability of ISA memory assessment are referred to reviews by Jones-Gotman et al. (1993), Loring, Lee, and Meador (1994), and Rausch et al. (1993). Some of the considerations detailed here, however, apply equally to language and memory assessment.

12-4.1. Patient Selection

When only certain patients are subjected to ISA, the generalizability of the findings remains in question, thus complicating evaluation of the ISA's predictive power. Some centers select for ISA only those patients showing evidence of bilateral brain disease or left-handedness (Snyder et al., 1990), and others exclude right-handed patients being considered for right temporal lobectomies (Jones-Gotman, 1987). Some centers perform the ISA on only one hemisphere (Rausch et al., 1993), which becomes a significant issue in cases of bilateral speech representation. Without assessing both hemispheres, the conclusion that language is solely in one hemisphere is invalid (Dodrill, 1993; Loring, Lee, & Meador, 1994), and candidates who undergo surgery based on results obtained from a single injection may be at greater risk for postoperative language deficits.

12-4.2. Drug Administration

Most frequently, sodium amobarbital is the anesthetic agent of choice. However, Coubes et al. (1995) report that in France, where amobarbital has not been approved for use, methohexital is used. This drug's shorter effect allows both hemispheres to be evaluated within 2 hours. Unfortunately, the quantity of test stimuli must be decreased, possibly limiting confidence in the test results. Furthermore, given methohexital's epileptogenic potential, the drug is usually injected along with an anticonvulsant.

Drug parameters—for example, amobarbital dosage and concentration, volume of amobarbital and saline mixture, rate of delivery (steady or incremental), and method of delivery (hand or automated injection)—also vary widely (Loring, Meador, & Lee, 1992a; Rausch et al., 1993) and can lead to discrepant findings. Drug parameters affect extent and duration of anesthesia: A faster rate or larger volume of injection will typically perfuse a more extensive vasculature, thus compromising more domains of function; smaller volumes, or slower rates of injection will lead to a greater concentration of drug in a smaller area, possibly leading to more intense or prolonged drug effects. Anesthetization should not be so sedating or persistent that the patient cannot respond, yet should be sufficient to create a condition modeling as closely as possible the effects of surgery, and long enough to permit presentation of an adequate number of test items and thus valid inferences from test results.

Occasionally, there will be cross-filling of the contralateral hemisphere or the posterior cerebral artery, leaving uncertainty about the neural bases of elicited responses

(Hart et al., 1991; 1993). Interpretation of the ISA is predicated on the assumption that brain regions supplied by the anterior and middle cerebral arteries are inactivated. A neuroimaging study by Hart et al. (1993) suggests that this assumption should not be made in all cases. Using SPECT, they reported that there is great interindividual variability in the regions actually perfused by amobarbital after ICA injection. Although some centers determine likelihood of cross-flow of amobarbital into the contralateral hemisphere or perfusion of other territories by amobarbital via angiography, the correlation between contrast medium and amobarbital distribution is limited by differences in methods of injection of contrast media and amobarbital (e.g., Rausch et al., 1993).

12-4.3. Determination of Adequacy of Anesthesia and Timing of Stimulus Presentation

Among many features of ISA protocols that differ across centers (Benbadis, Dinner, Chelune, Piedmonte, & Lüders, 1995; Dodrill, 1993; Jones-Gotman, 1987; Loring, Lee, & Meador, 1994; Rausch et al., 1993; Snyder et al., 1990) are the determination of when an adequate drug effect is evident, timing of stimulus presentation, types of stimuli and response formats, and criteria used to infer adequacy of language and memory. It is agreed across centers that presentation of stimuli is contingent on adequate hemispheric anesthetization. In order to achieve a model of how the brain will function if tissue were removed, testing should occur during maximum drug effect. It has been shown, for example, that memory items presented earlier, during maximal drug effect, are more sensitive to lateralizing seizure onset and predicting lateralized temporal lobe impairment than are items presented later, when drug effects have worn off to some extent (Loring, Lee, & Meador, 1994; Loring, Meador, et al., 1994; Rausch et al., 1993).

Unfortunately, means of determining onset and duration of adequate anesthesia differ across centers. One or more of the following might be used to infer adequate anesthesia: contralateral hemiparesis (although different centers consider anywhere from 0/5 to 4/5 as adequate), grip strength, loss of antigravity tone, and marked EEG slowing. Yet other centers simply present stimuli during a predetermined, standard interval (Rausch et al., 1993). Even if anesthetic effect were similarly defined across centers, there would still be disagreement about timing of stimulus presentation. When the speech-dominant hemisphere is injected, and speech arrest ensues, some clinicians wait for speech return before proceeding with testing, whereas others continue stimulus presentation regardless of speech difficulty. Each approach has advantages and disadvantages. Waiting too long for speech impairments to resolve means that other cognitive functions will also be recovering, thus increasing the probability of a false negative result on the memory test (i.e., incorrectly concluding that the hemisphere opposite to the one inactivated can adequately sustain memory). On the other hand, if items are administered during acute sedation, speech arrest, or aphasia, it becomes difficult to determine why memory items are failed. Consequently, the potential for a false positive result with respect to memory increases (i.e., one might incorrectly conclude that the hemisphere opposite to the one injected cannot adequately sustain memory). Although Lesser, Dinner, Lüders, and Morris (1986) reported that the presence of aphasia does not necessarily imply that stimuli will not be remembered, this does not imply that memory failure is not attributable to aphasia.

12-4.4. Criteria to Establish Hemispheric Language Dominance

What exactly constitutes evidence for the presence of language in a cerebral hemisphere? Speech arrest might not be a sufficient criterion for determining the presence of language because it may reflect motor or motivational factors rather than language disruption. From a clinical standpoint, speech arrest might, however, be sufficient to suggest that hemispheric language mapping by electrical stimulation be undertaken. Most frequently, disruption of object naming is employed as a measure of speech representation, followed by response to verbal commands, ability to count, unspecified dysphasic signs, and word/phrase repetition (Dodrill, 1993; Rausch et al., 1993; Snyder et al., 1990). Although these criteria enable establishment of left or right speech dominance with relative ease (Rausch et al., 1993), less clear are the criteria to establish bilateral language representation.

Benbadis et al. (1995) attempted to provide objective criteria for determination of hemispheric language dominance. Three sets of criteria based on speech arrest were employed: absolute duration, side-to-side difference, and a laterality index computed as $(L-R/L+R)$. The three methods yielded the same classification of hemispheric language dominance in 86% of 142 patients. In the remaining patients, the three methods differed with respect to unilateral or bilateral language classifications. Although the establishment of objective criteria represents an advance, these criteria were derived from ISA data without validation against an external standard. A more serious concern is that the cutoff scores used to determine hemispheric language dominance were derived from a subset (123) of the 142 subjects in the sample. Consequently, the accuracy of these criteria awaits cross-validation in an independent sample.

12-5. RECENT DEVELOPMENTS AIMED AT IMPROVING ISA VALIDITY AND RELIABILITY

12-5.1. Drug Administration

In an attempt to overcome the difficulties posed by possible oversedation in patients receiving bolus injections of amobarbital, Levin, Cantrell, Soukup, Crow, and Bartha (1994) evaluated an incremental drug administration procedure. In this procedure, a concentration of 10 mg amobarbital per ml saline was injected at the rate of 2-3 ml/sec until a flaccid hemiplegia developed. A comparison of seven patients undergoing the incremental procedure, and seven undergoing the bolus procedure, revealed that the incremental procedure produced less sedation, required less total drug, and yet adequately demonstrated hemispheric differences in language and memory.

12-5.2. Supplementary Procedures

Several procedures have been employed to enhance reliability and validity of the ISA (Coubes et al., 1995; Hart et al., 1993; Jack et al., 1989; Jeffery et al., 1991; Jones-Gotman et al., 1993; Lee, Loring, Smith, & Flanigin, 1995; Loring, Meador, & Lee, 1992b). Because the predictive validity of these procedures remains to be determined, they are all best considered supplements rather than alternatives to the ISA (Kurthen, Solymosi, & Linke, 1993).

12-5.2.1. EEG and Electrocorticography

Some centers consider surface EEG during ISA helpful in locating and monitoring of slowing after amobarbital injection, whereas others do not (e.g., Dodrill, 1993). Some centers perform the ISA after intracranial electrode implantation so as to permit EEG recording directly from subdural grid and/or strip electrodes, as well as depth electrodes implanted in the mesial temporal lobes. Gotman, Bouwer, and Jones-Gotman (1992), for example, reported that the standard ISA injection produced adequate EEG slowing in the anterior and middle hippocampus. In addition, although ISA also was associated with slowing in the hemisphere contralateral to the injected one, this slowing was less pronounced and briefer than in the injected hemisphere.

12-5.2.2. Neuroimaging

Concern has been expressed about particular brain regions inactivated during ISA: several studies (Coubes et al., 1995; Hart et al., 1993; Jeffery et al., 1991) have shown that the cerebral distribution of amobarbital and methohexital varies among individuals. Perhaps it is for this reason that not all patients who develop expressive aphasia during dominant hemisphere ISA also develop language comprehension deficits (Hart et al., 1991), and that only some patients demonstrate memory failures after ISA. Hart et al. (1993), for example, found that SPECT indicated mesial temporal inactivation during ISA in only 28% of cases. SPECT scanning accompanying ISA might assist in providing verification of brain regions perfused, and thus assist interpretation of ambiguous cognitive evaluation results (Hart et al., 1993; Kurthen et al., 1988). Functional magnetic resonance imaging (fMRI), a noninvasive technique, may be useful in corroborating ISA test results, at least in the domain of language (Binder et al., 1996; Desmond et al., 1995). However, the evidence to date is best considered preliminary until a larger number of individuals, especially with atypical language representation, are studied.

12-5.2.3. Selective Amobarbital Injections

Sometimes it is difficult to interpret the ISA because of excessive sedation or language impairment. One option is to readminister the ISA using an equal or lower dose of amobarbital (Loring, Meador, Lee, & King, 1992; Loring, Lee, & Meador, 1994).

Another option is to evaluate cognition after injection of a vessel supplying a more limited cerebral territory, a technique also helpful in identifying extratemporal seizure origin (e.g., Brundert, Elger, Solymosi, Kurthen, & Linke, 1993). Selective injections are used typically to evaluate memory rather than language, although Tröster et al. (1996) report on letter-by-letter reading following PCA injection of the language-dominant hemisphere.

12-5.3. Validation Studies

The validity of the ISA to establish hemispheric language dominance is well accepted (see Dodrill, 1993; Kurthen, 1992). The interesting question raised by recent studies is whether cases undergoing ISA, and consequent cortical mapping, have better outcomes than do cases without language mapping. Among left-hemisphere language-dominant patients who underwent ISA and then conservative left temporal lobe resection without language mapping, no significant post-operative language decrements were observed by Davies, Maxwell, Beniak, Destafney, and Fiol (1995). Hermann and Wyler (1988), comparing outcome in two groups of 13 patients who underwent temporal lobectomy on the language-dominant hemisphere either with or without functional mapping, found visual naming decrements to be more pronounced in the group that did not undergo mapping, despite the fact that the extent of resection was more conservative in this group. In a larger follow-up study of 162 patients who underwent temporal lobectomy without mapping, Hermann, Wyler, Somes, and Clement (1994) observed a postoperative dysnomia in 7% of left lobectomy patients. Furthermore, an association between later age at onset of epilepsy and postoperative dysnomia was observed in the left lobectomy group. However, it is not clear whether outcome would have been different in a group undergoing language mapping. Thus, although as a group, patients undergoing lobectomy without mapping might fare well postoperatively, the fact that a subgroup of individuals with postoperative dysnomia can also be identified suggests that mapping after ISA might be the prudent course of action to take until more definitive data are available.

12-6. A CAUTIONARY NOTE CONCERNING THE ISA IN PEDIATRIC POPULATIONS

Research about the ISA in pediatric populations is extremely limited. Discussion about the ISA in children and adolescents is frequently couched in terms of adult research findings (e.g., Bernstein, Prather, & Rey-Casserly, 1995). However, a few empirical studies suggest that the ISA protocol might need to be modified for children and adolescents. Szabo and Wyllie (1993) noted that language dominance was established in all children who had bilateral injections and at least borderline intelligence, but in only about half of the children with mental retardation. These authors and Williams and Rausch (1992) also caution that children under 13 years of age might demonstrate an inappropriately high memory failure rate, suggesting the possible need for a lower cutoff score. Westerveld et al. (1994) reported more encouraging data. Using amo-barbital doses of 100 or 130 mg, they considered the ISA to yield unambiguous data concerning language dominance in children as young as 7 years.

12-7. HAS THE ISA TAUGHT US ANYTHING OF THEORETICAL SIGNIFICANCE ABOUT LANGUAGE?

The ISA bears several advantages as a research tool relative to traditional lesion studies (Tröster et al., 1996). Specifically, it is possible to obtain baseline data in the individual to be studied, to compare lesion effects on both hemispheres in the same individual, to gather group data where the "lesion"-to-assessment interval is relatively homogeneous, and to reverse the lesion. Disadvantages of the ISA entail the brevity of the drug effect and the possibly limited generalizability of findings. When gathering clinically relevant information is of paramount importance, there is, within the time constraints of the ISA, little opportunity to manipulate the nature of stimuli. Further-more, because the ISA is carried out in individuals with typically chronic cerebral dysfunction, it is unclear if findings obtained apply to normal populations, or even to groups with other forms of cerebral dysfunction.

12-7.1. Handedness and Hemispheric Language Dominance

One domain of inquiry richly informed by ISA studies is the relationship between handedness and hemispheric language dominance. A study commonly cited to refer-ence the distribution of handedness and speech lateralization in the general population (Rasmussen & Milner, 1977) reported that 96% of right-handed individuals were left-hemisphere speech-dominant. Among left-handed individuals, 70% were left-hemisphere dominant, 15% right-hemisphere dominant, and 15% were deemed to have bilateral speech representation. However, this sample may not be representative of a normal population, or even of all individuals with epilepsy, given the criteria employed to select patients for ISA (Woods, Dodrill, & Ojemann, 1988).

Although right-handedness is highly correlated with left-hemisphere speech dom-inance in a normal population, ISA studies reveal that brain injury has a tendency to alter this relationship. Specifically, early brain injury is often associated with a shift in language dominance (Rey, Dellatolas, Bancaud, & Talairach, 1988; Satz, Strauss, Wada, & Orsini, 1988; Woods et al., 1988), but a shift in language dominance does not necessarily entail a shift of handedness, or vice versa. Satz et al. (1988) reported that only 68% of patients who shifted language or hand preference shifted both. When only handedness or language shifted, language was more likely to shift.

Several studies have identified factors associated with shifts in language and hand-edness: injury extent and location, and age at injury. Woods et al. (1988) found that left-hemisphere injuries sufficient to cause hemiparesis were associated with both left-handedness and right or bilateral speech representation. Among patients without hemiparesis, sinistrality and right or bilateral language representation were associated with extratemporal pathology. Left-handers with left-hemisphere speech dominance were less likely to have had hemiparesis or extratemporal injuries. Kurthen et al. (1994) too reported that extratemporal lesions, and multiple or bilateral seizure foci (but not handedness or laterality of seizure focus), were related to atypical speech representation.

Another factor underlying atypical language representation is the age at which injury occurs. Because resiliency of the brain is maximal during the early years of life (Satz et al., 1988), injury at a very early age might induce either the intact or damaged hemisphere to reorganize (i.e., accommodate speech and language zones) so as to compensate for functional losses (Liederman, 1988; Satz et al., 1988). Satz et al. (1988) found that intrahemispheric language reorganization was more likely than interhemispheric reorganization after age 6 (speech was ipsilateral to the lesion in 78% of these cases). Interhemispheric organization is more likely to occur at earlier ages, but unfortunately, not without possible cost to nonverbal functions (i.e., crowding).

Strauss, Satz, and Wada (1990) evaluated verbal and nonverbal functions during ISA in epileptic patients with early onset of left-hemisphere dysfunction. They found that patients with atypical speech representation performed as well as patients with left-hemisphere speech on most measures of language function. However, the patients with atypical speech patterns performed more poorly than the "left-hemisphere" patients on nonverbal tasks. It appeared that transfer of language "crowded out" functions normally under mandate of the right hemisphere. Thus, a right–left maturation gradient, characterized by a slower development of the right than the left hemisphere, might permit the still more plastic right hemisphere to assume language functions, but not without disturbing the typical organization of right-hemisphere functions.

Amobarbital studies of patients with epilepsy have also challenged the notion that hemispheric language dominance falls into discrete categories. Rather, it is more likely that language representation falls on a continuum from strongly left-hemisphere dominant to strongly right-hemisphere dominant, just as handedness seems to fall on a continuum from strongly dextral to strongly sinistral (Loring, Meador, Lee, Murro, et al., 1990; Snyder et al., 1990). Kurthen et al. (1994), however, offer data suggesting that cases might not fall along all points of such a continuum. Specifically, these authors proposed that there are four major patterns of hemispheric speech representation. Toward the left-dominant end of the spectrum, there are indeed several clusters representing degrees of left speech dominance. However, individuals with bilateral or right speech dominance tended to center at these respective points, without the gradation observed for left-hemisphere representation.

Amobarbital studies have also indicated that crossed aphasia (the occurrence of aphasia in right-handed individuals after insult to the right hemisphere) should not necessarily be taken to imply right-hemispheric language dominance in these right-handers. Both Lanoe, et al. (1992) and Loring, Meador, Lee, Flanigin, et al. (1990) determined via ISA that their patients with crossed aphasia actually had bilateral language representation.

12-7.2. Bilingualism and Sign Language

Several studies have shown that the interhemispheric organization of both languages in bilinguals is complementary, that is, hemispheric dominance for the two languages is similar (Berthier, Starkstein, Lylyk, & Leiguarda, 1990; Gomez-Tortosa, Martin, Gaviria, Charbel, & Ausman, 1995). However, the intrahemispheric organization of the native and second language is likely different. Berthier et al. (1990), on the basis of the observation that the patient's second language recovered before the native language after left middle cerebral artery amobarbital injection, speculated that the second language might be organized within the central sylvian core, whereas the first language might be represented in more distant perisylvian areas. One might, however, find a diametrically opposing interpretation more plausible. Electrical stimulation studies (for review, see Ojemann, 1983) confirm that at least object naming in first and second languages is differentially represented within the language-dominant hemisphere, with the second language being represented in a larger area, more peripheral from the sylvian fissure. Thus, one might conclude, assuming that amobarbital effects dissipate earlier in more distant areas, that Berthier et al.'s (1990) findings indicate the first language to be more centrally represented.

Sign language studies in nondeaf individuals also indicate interhemispheric organization of signed and spoken language to be similar, at least in that the same hemisphere is dominant for both forms of communication. The existence of subtle interhemispheric differences in the organization of signed and spoken language, however, is not settled by amobarbital studies. One amobarbital study lends support to Poizner and Battison's (1980) contention that sign language is characterized by greater bilateral representation than spoken language. Homan, Criswell, Wada, and Ross (1982), in a right-handed male who had learned speech and signing at the same time, found that hemispheric dominance for speech and signing were the same (left). However, lateralization for signing was not as strong as for speech, in that the right hemisphere also contributed to the propositional and emotional components of signing. In contrast, other studies have found the interhemispheric organization of signed and spoken language to be highly similar. Mateer, Rapport, and Kettrick (1984), in a left-hander, found similar mixed dominance for both spoken and signed language. Whereas right-hemisphere ISA disrupted motoric aspects of both languages, left ISA disrupted grammar and semantic usage in both languages. Damasio, Bellugi, Damasio, Poizner, and VanGilder (1986) found that left ISA in a right-handed individual led to marked aphasia for both American Sign Language (ASL) and spoken English, and that right temporal lobectomy left ASL and English unperturbed.

Because a substantial proportion of deaf individuals apparently have experienced some cortical reorganization, Wolff, Sass, and Keidan's (1994) report of ISA test results in a deaf individual are of particular significance. Their study provides tentative data about whether findings from normal-hearing individuals concerning the organization of sign language apply to deaf individuals. Wolff et al. (1994) reported complete left-hemispheric dominance in a right-handed individual for ASL, signed English, and finger spelling. Thus, these authors did not find evidence of bilateral representation for sign language.

12-8. SUMMARY

The ISA has been shown to be a reliable and valid method of determining hemispheric language dominance in the hands of skilled clinicians. Results from electrical mapping of cortical language areas, as well as from functional MRI, support the concurrent validity of the ISA. Although the ISA's predictive validity is established since early studies, the question has been raised whether postoperative language outcome is different in individuals undergoing language-dominant temporal lobectomy with and without cortical mapping. The answer appears to be that a small subgroup of patients not undergoing language mapping might develop a dysnomia, the persistence of which has not yet been documented.

Lack of uniformity of test protocols across centers makes it difficult to compare research findings from different centers. However, drug administration variables, the criteria used to define adequate anesthesia and hemispheric language dominance, and individual differences in cerebral drug distribution are important considerations in interpreting ISA results.

The ISA has also advanced our understanding of the relationship between hemispheric language dominance and handedness: age at injury, as well as the injury's extent and location influence whether language dominance and handedness will be reorganized inter- or intrahemispherically. Interhemispheric reorganization of language is more likely to occur before age 6, and such reorganization might exact a cost on nonverbal functions (i.e., "crowd" these functions). Amobarbital studies have also been helpful in demonstrating the generally complementary cerebral organization patterns of first and second languages in bilinguals, and of signed and spoken language.

PART III

Experimental Neurolinguistics

CHAPTER 13

Phonetics and Phonology

Jackson T. Gandour
Department of Audiology and Speech Sciences, Purdue University, West Lafayette, Indiana 47907

Phonological and phonetic studies of aphasia are evaluated with respect to issues of neural representation of speech and language. The traditional dichotomy between anterior and posterior syndromes has been challenged in recent years. In speech production, deficits at the phonological level are discussed in relation to psycholinguistic models of speech production, phonological features, underspecification, markedness, syllable structure, and sonority. Segmental deficits at the phonetic level are discussed primarily in relation to temporal parameters of consonants and vowels, segmental coarticulation, and speaking rate effects. The data challenge the traditional view that articulatory implementation deficits are circumscribed to anterior lesions, planning deficits to posterior lesions. Hemispheric specialization is assessed for linguistic aspects of prosody and brought to bear on a number of competing hypotheses. Production deficits are examined for phonemic stress, contrastive stress, lexical tones, tonal coarticulation, intonation, and foreign accent syndrome. The bulk of the evidence suggests minimal involvement of the right hemisphere in mediating linguistic prosody. In speech perception, the role of phonology in lexical access is highlighted as are breakdowns in stress, tone, and intonation.

Current psycholinguistic models of language production distinguish three major stages in the production of words and connected speech: access, planning, implementation (e.g., Caplan, 1992; Kohn, 1993; Levelt, 1989, and references therein). Much of the evidence in support of this functional architecture comes from in-depth case studies of aphasic patients. However, there is some disagreement over details of mechanisms (Buckingham, 1992), and some even question the validity of the research paradigm itself (Butterworth, 1992).

The traditional dichotomy between anterior and posterior aphasia syndromes has been challenged in recent years. Much of the data to support alternative models of language specialization in the brain has come from the area of speech production. The neural representation of speech motor control is perhaps more diffusely represented than heretofore believed.

The majority of the experimental studies covered in this review have used the terms "Broca's aphasia" and "apraxia of speech" more or less synonymously, so we will not be able to determine whether any differences in articulatory deficits are to be ascribed to these two diagnostic categories (Blumstein, 1994; Kohn, 1993).

This chapter will focus on topics of theoretical interest since 1990 in the realm of phonetics and phonology, especially various aspects of prosody. Excellent reviews of earlier work in phonetics and phonology are readily available (Blumstein, 1990, 1991, 1994; Ryalls, 1987a; Whitaker, 1988).

13-1. SPEECH PRODUCTION

13-1.1. Deficits at the Phonological Level

Nearly all aphasic patients produce phonological errors in their speech output. All types of phonological errors can be found across diagnostic categories of aphasia (see Blumstein, 1990, 1991, 1994, for reviews and references therein). Kohn (1993) has argued that when differences in experimental tasks, stimuli, and linguistic analyses are taken into account, it is possible to distinguish the phonological output behavior of different aphasia syndromes by means of analysis of their phonological errors (see Butterworth, 1992, for detailed criticism).

Segmental error patterns reflect disruption at different stages of speech production that may be associated with different aphasic syndromes (see Kohn, 1993, for review). Wernicke's aphasics' difficulties stem primarily from impaired access to underlying phonological representations (stage 1); conduction aphasics, on the other hand, have problems primarily in constructing the phonemic representations (stage 2; cf. Béland, Caplan, & Nespoulous, 1990); whereas the error patterns of Broca's aphasics reflect primarily a phonetic disturbance (stage 3). Various psycholinguistic mechanisms have been hypothesized to account for phonemic paraphasias at stage 2 (see Buckingham, 1992, for extended discussion).

A few studies have focused on phonological errors that occur in the context of multiple repeated attempts at a single target word (conduite d'approche). The successive approximations of conduction aphasics show a tendency of steady progression toward the target, which supports the notion that they suffer from a stage 2 deficit

(Gandour, Akamanon, Dechongkit, Khunadorn, & Boonklam, 1994; Kohn, 1989; Valdois, Joanette, & Nespoulous, 1989). In contrast, the error sequences of Wernicke's aphasics actually deteriorate from the intial to the final attempt, linking their segmental errors to a lexical problem or stage 1 deficit.

13-1.1.1. Phonological Features and Underspecification Theory

Recent theories of generative phonology posit that features are organized in a hierarchical tree structure (see Halle, 1992, and references therein). Blumstein (1990) found that the majority of phoneme substitution errors reflect feature changes within a single tier rather than across tiers. Evidence for an independent laryngeal tier has been shown in aphasia (Dogil, 1989; Dogil, Hildebrandt, & Schürmeier, 1990; Gandour, Holasuit Petty, & Dardarananda, 1989; Gandour, Akamanon, et al., 1994).

Theories of underspecification have been proposed that aim to explain when features may be absent in underlying or derived representations (see Steriade, 1995, for review). Béland et al. (1990) showed how phonemic errors made by a conduction aphasic are related to syllabification processes that interact with underspecified segmental representations at different stages in phonological encoding. An analysis of phonemic paraphasias in French-speaking aphasics supported the underspecified nature of coronal consonants (Béland & Favreau, 1991). Underspecification theory has also been applied to predict the location and quality of epenthetic vowels in phonemic paraphasias produced by French-speaking aphasics (Béland, 1990).

13-1.1.2. Markedness

It has been argued that markedness theory can explain the pattern of phonemic paraphasias in aphasia. Among other criteria, "marked" segments are considered to be more complex articulatorily and universally rarer than "unmarked" segments. Substitution errors of aphasic patients have been observed to involve replacement of marked segments with unmarked ones (see Blumstein, 1990, 1991, for review and references therein). The pattern of phonemic paraphasias was indistinguishable across clinical syndromes. More recent work, however, has challenged the unitary treatment of different aphasic syndromes as well as the claim that markedness considerations play a critical role in accounting for the error patterns (Béland et al., 1990).

13-1.1.3. Syllable Structure

Recent theories of phonology recognize the syllable as a phonological constituent (see Blevins, 1995, for review). Phonemic substitution errors appear to be sensitive to syllable-internal hierarchical branching structure. Few substitutions occur when the consonant is part of a consonant cluster (see Blumstein, 1990, for review). The majority of consonant substitution errors occur when the consonant is preceded or followed by a vowel. These findings are inexplicable in a theoretical framework that treats the syllable as an unanalyzable whole. Analyses of phonemic paraphasias across word boundaries show that the affected phonemes occur in like syllable positions (e.g., onset, nucleus, coda). Not all constituents of the syllable are equally prone to disruption. An analysis of neologistic jargon produced by two Wernicke's aphasics revealed that the coda is more susceptible to impairment than the onset; the nucleus is the most stable of the syllable-internal constituents (Stark & Stark, 1990).

Kaye, Lowenstamm, and Vergnaud's (1985) theory of phonological government distinguishes two types of consonant clusters that have different syllabic representations and are characterized by different governing domains. Valdois (1990) found that aphasic errors support the claim that obstruent-liquid clusters and other cluster types have different syllabic representations and, moreover, suggest that segments in the governed position are more likely to be involved in the destruction or creation of clusters than segments in the governing position.

13-1.1.4. Sonority

Sonority refers to the perceptual prominence of one phoneme relative to another. Along a sonority scale, obstruents are considered to be the least sonorous, vowels the most sonorous. The sonority sequencing principle holds that segments are optimally ordered so as to achieve a rise–fall sonority profile: an increase in sonority from the beginning of the syllable to the vowel peak followed by decrease in sonority from the vowel peak out to the end of the syllable. Sonority sequencing principles have been invoked to explain phonological error patterns frequently seen in fluent aphasia, especially in neologistic jargonaphasia (see Christman, 1994, and references therein).

13-1.2. Deficits at the Phonetic Level

The traditional view is that anterior aphasics have a tendency to produce phonetic errors; posterior aphasics have difficulty primarily at the phonological level. The underlying cause of the phonetic errors has usually been attributed to a breakdown in articulatory implementation, whereas phonemic errors have been ascribed to a breakdown in access or planning the appropriate phonological output (Blumstein, 1990, 1991, 1994; see Ryalls, 1987a, for reviews of the earlier literature on phonetic patterns of dissolution in aphasia).

13-1.2.1. Temporal Parameters of Consonant and Vowel Production

Acoustic, physiological, and perceptual investigations of consonant production deficits in aphasia support a dichotomy between phonological planning and phonetic implementation (see Blumstein, 1990, 1994; see Blumstein & Baum, 1987, for reviews and references therein). Anterior aphasics exhibit primarily deficiencies in articulatory implementation, whereas the consonant production deficits of posterior aphasics generally reflect problems of a phonological nature. Anterior aphasics' difficulties do not lie with particular phonetic features. The "voicing" feature provides a case in point. In English, this feature is associated with word-initial stop and fricative consonants as well as vowel duration as a cue to postvocalic voicing in stop consonants. Results indicated that voicing in initial stops and fricatives was impaired, whereas vowel duration was relatively spared (Baum, Blumstein, Naeser, & Palumbo, 1990; Baum, 1996).

Crosslinguistic comparisons have reinforced the notion that the phonological status of the durational attribute has little effect on speech production impairments of anterior aphasics. For example, Blumstein, Cooper, Goodglass, Statlender, and Gottlieb's (1980) seminal findings on voice onset time (VOT) in word-initial stop consonants have been replicated in many other languages. There is a marked compression of the VOT continuum regardless of the number of stop voicing categories that are present in the phonological system (Gandour et al., 1992b). Conversely, vowel duration at the syllable level appears to be relatively spared following brain damage regardless of whether it is used contrastively at the lexical level (Baum et al., 1990; Gandour et al., 1992c).

Besides an impairment in the timing of independent articulators, anterior aphasics may have difficulties with the implementation of laryngeal gestures that closely interact with the supralaryngeal vocal tract system (Blumstein, 1990, 1991, 1994). In contrast, laryngeal gestures that may be implemented independently of the supralaryngeal tract appear to be relatively spared. Indeed, it is argued that disruption of laryngeal gestures is a secondary manifestation of a more general timing disorder (Danly & Shapiro, 1982; Gandour, 1987). Nevertheless, a patient who exhibited a dissociation between oral and laryngeal gestures has been described by Marshall, Gandour, and Windsor (1988).

Other work, however, has challenged the notion that the timing disorder is restricted to instances of temporal integration or coordination between two independent articulators. Ryalls (1987b) concluded that the degree of disturbance of vowel duration in aphasia varies depending on the size of the linguistic unit within which vowels are produced. Baum (1992) similarly reported that anterior aphasics fail to reduce the root syllable duration in longer words at the same magnitude as normals. An inability to increase speaking rate in longer, multisyllabic utterances may underlie these timing abnormalities (Baum, 1993; Baum & Ryan, 1993). Gandour et al. (1992c; Gandour, Dechongkit, Ponglorpisit, Khunadorn, & Boongird, 1993; Gandour, Dechongkit, Ponglorpisit, & Khundorn, 1994) found that anterior aphasics, in contrast to posterior aphasics, exhibit a deterioration in speech timing regardless of the size of the linguistic unit. The fact that temporal patterns of dissolution are not the same across multiple tasks of varying length and linguistic complexity suggests that their timing deficit emanates from different underlying mechanisms.

13-1.2.2. Segmental Coarticulation

The study of coarticulation provides insights into the size of the planning units that can be programmed in the production of syllables and words as well as the dynamic aspects of speech production. Acoustic, physiological, and perceptual investigations of coarticulatory phenomena involving consonants and vowels present mixed findings on the issue of mechanisms underlying any coarticulatory deficits in brain-damaged patients (see Katz, Machetanz, Orth, & Schönle, 1990a, for review of earlier studies). Katz (1988) found no evidence of a deficit in anticipatory coarticulation in anterior and posterior aphasic English-speaking patients. Anticipatory labial coarticulation was also found to be essentially intact in two German-speaking anterior aphasics (Katz, Machetanz, Orth, & Schönle, 1990a, 1990b). In contrast, anticipatory coarticulation was found to be disturbed in two English-speaking apraxic patients as well as one of two conduction aphasics (McNeil, Hashi, & Southwood, 1994). It is possible that these conflicting findings might be due to individual subject differences, to problems of taxonomic classification, or to differences in measurement methods—acoustic, kinematic, or perceptual (see Katz, 1988; Katz et al., 1990a, for detailed discussion).

13-1.2.3. Speaking Rate Effects

Another interesting aspect of temporal control is the manipulation of speaking rate and its effects on segmental and suprasegmental production. Although their changes in rate were of smaller magnitude than those for normal subjects, both anterior and posterior patients appear to be relatively unimpaired in their ability to manipulate speaking rate (Baum, 1993, 1996; Baum & Ryan, 1993; McNeil, Liss, Tseng, & Kent, 1990). Yet temporal deficit patterns between speaking rate conditions are not the same for anterior and posterior patients. For example, vowel durations of nonfluent aphasics were not significantly different between slow and fast conditions (Baum, 1993). In contrast, fluent aphasics, like normal subjects, produced shorter vowel durations at a faster speaking rate. The emergence of these differences in performance at the segmental level suppports the view that distinct impairments underlie the anterior and posterior patients' temporal deficits.

13-1.2.4. Subtle Phonetic Deficit in Posterior Aphasics

Acoustic, physiological, and perceptual studies over the past decade and a half or so have reported instances of a "subtle phonetic deficit" in left hemisphere-damaged (LHD) patients with posterior lesions (see Vijayan & Gandour, 1995, for review and references therein). These data challenge the traditional view that articulatory implementation deficits are circumscribed to anterior lesions, planning deficits to posterior lesions. Although the speech-production difficulties of posterior patients seem to be primarily phonological in nature, it is claimed that they also display "phonetic impairments [that] are not clinically perceptible but emerge only upon acoustic analysis" (Blumstein, 1994, p. 32). Acoustic analysis of phoneme errors made by a conduction aphasic, however, has revealed that most perceived phoneme substitutions exhibited acoustic characteristics appropriate to the substituted sound, and thus most likely reflect true phoneme selection errors (Baum & Slatkovsky, 1993).

Timing deficits in LHD posterior patients have been attributed to a subtle phonetic impairment (Baum, 1996; Baum et al., 1990; Blumstein et al., 1980; McNeill et al., 1990). These deficits, however, are not necessarily the same in anterior and posterior patients. Whereas anterior patients experienced timing difficulties at the syllable, word, and sentence levels, posterior aphasics exhibited normal-like timing in isolated monosyllables, but had deviant timing patterns for longer words and sentences (Gandour et al., 1992b, 1992c; Gandour, Dechongkit, et al., 1993; Gandour, Dechongkit, Ponglorpisit, et al., 1994). Yet acoustic analysis has revealed that anticipatory and perseverative tonal coarticulation were disturbed in LHD posterior patients as well as anterior patients (Gandour, Potisuk, Ponglorpisit, Dechongkit, et al., 1996).

13-1.3. Deficits in Speech Prosody

It is well known that speech prosody can can be used to signal linguistic and emotional information (see Van Lancker & Pachana, this volume, for discussion of emotional or affective prosody; see Weniger, 1993, for earlier review of disorders of prosody in aphasia). A number of competing hypotheses have been proposed to account for the lateralization of linguistic prosody. The "functional load" hypothesis states that prosodic cues are processed differently in the left hemisphere depending on whether they serve a linguistic or a nonlinguistic function (Van Lancker, 1980). A more restrictive version of this hypothesis stipulates that the left hemisphere is specifically engaged for lexical contrasts (Packard, 1986). The "parallel processing" hypothesis posits that both hemispheres may simultaneously participate in processing various components of the speech signal for which the hemisphere is specialized. The left hemisphere has been hypothesized to mediate temporal information, the right hemisphere spectral information (Robin, Tranel, & Damasio, 1990; Robin, Klouda, & Hug, 1991; Van Lancker & Sidtis, 1992). The "attraction hypothesis" (Shipley-Brown, Dingwall, Berlin, Yeni-Komshian, & Gordon-Salant, 1988) states that differential lateralization occurs as a result of an interaction between the acoustic property and its function. F_0 may be mediated by the right hemisphere in nonlinguistic functions, but may be "drawn" to the left hemisphere when it is used to signal linguistic prosody. As a corollary to the attraction hypothesis, Behrens (1988, 1989) has emphasized that size of the utterance or task demands influences the lateralization of linguistic prosody.

13-1.3.1. Stress

Stress is especially interesting for testing hypotheses about hemispheric specialization of prosody because of its multiple acoustic correlates: duration, amplitude, and F_0. An across-the-board deficit in all three acoustic correlates would imply a functional level disturbance at high level of linguistic representation. Selective deficits in any one or more of these multiple correlates, on the other hand, would suggest that there may be differential lateralization for the processing of different prosodic parameters. Acoustic investigations of LHD and right hemisphere-damaged (RHD) patients' ability to produce lexical and phrasal stress contrasts have generally found a stress deficit in LHD patients only (Emmorey, 1987; Behrens, 1988; Ouellette & Baum, 1994; Bryan, 1989). Instead of characterizing the deficit as a stress deficit, LHD patients appear to have difficulty with the manipulation of durational cues, whereas the other acoustic correlates of stress remain relatively intact (Ouellette & Baum, 1994; Emmorey, 1987). Ouellette and Baum's findings lend support to the hypothesis of differential lateralization for the processing of different acoustic parameters. It is possible that deficits in the processing of linguistic prosody attributed to LHD patients may be a secondary consequence of a more basic impairment in speech timing. This finding is consistent with previous acoustic investigations of other aspects of speech prosody (Danly & Shapiro, 1982; Gandour et al., 1989). The isolated impairment of duration as a cue to prosody in LHD patients further suggests that the primary deficit may not necessarily be restricted to the prosodic domain.

Acoustic investigations of contrastive stress have similarly reported impaired processing mechanisms for LHD patients only (Behrens, 1988; Ouellette & Baum, 1994). Using the same subjects as in the phonemic stress experiment, Ouellette and Baum's finding of a selective deficit in durational cues associated with contrastive stress reinforces the view that the timing deficit cuts across various linguistic representations irrespective of the size of the linguistic domain. For lexical or phrasal stress, planning involved only one lexical unit or a short phrase; for contrastive stress, planning involved a sentence-length unit. These findings are in apparent conflict with earlier claims that the stability of acoustic patterns in the speech of brain-damaged patients depend primarily on the size of the temporal domain over which the linguistic unit is programmed (Behrens, 1989; Gandour et al., 1989; Gandour, Dechongkit, et al., 1993; Gandour et al., 1996).

13-1.3.2. Tone

Tones, like consonants and vowels, may be disrupted secondary to left hemisphere damage (see Gandour, 1987, in press, for earlier review). Tonal paraphasias, like consonant and vowel paraphasias, are in evidence across tone languages (see Gandour, in press, for review). It has been argued that tonal paraphasias are phonologically equivalent to consonant and vowel paraphasias (Packard, 1986). This hypothesis is difficult to reconcile with findings that tones were corrected earlier than either consonants or vowels in sequences of phonemic approximations (Gandour, Akamanon, et al., 1994), and were preserved in the running speech of a Broca's aphasic despite disruption of consonants and vowels (Gandour et al., 1989).

Although tonal breakdown has been observed in Mandarin Chinese, Cantonese, and Thai (see Gandour, in press, and references therein), variability in the magnitude of the deficit appears to be related primarily to aphasia severity and time since onset of the stroke (Gandour et al., 1992a; Yiu & Fok, 1995). Nonfluent aphasics generally experience more difficulty in tonal production than fluent aphasics.

Tonal production, however, appears to be spared following unilateral lesions to the right hemisphere (Gandour et al., 1992a; Hughes, Chan, & Ming, 1983). Tones may also be disrupted subsequent to a left-hemisphere subcortical lesion (Gandour & Ponglorpisit, 1990). In Norwegian, a *pitch accent* language, findings similarly support the view that linguistically significant F_0 contrasts are mediated by the left hemisphere (Ryalls & Reinvang, 1986; Moen & Sundet, 1996).

There is some evidence for a selective deficit of types of tones based on articulatory complexity. In Mandarin Chinese, Tone 3, a tone with a complex rising F_0 contour, was especially vulnerable to disruption (Packard, 1986). In Thai, almost all tonal substitutions resulted in a mid, low, or falling tone (Gandour et al., 1992a). These findings are compatible with the notion that rising tones are more complex articulatorily than falling tones. Yet errors appear to be evenly distributed across the six lexical tones of Cantonese (Yiu & Fok, 1995).

13-1.3.3. Tonal Coarticulation

Anticipatory tonal coarticulation in Thai bisyllabic noun compounds was found to be essentially intact in both LHD and RHD patients (Gandour, Ponglorpisit, et al., 1993). Using the same subjects, Gandour et al. (1996) found that anticipatory and perseverative tonal coarticulation was disturbed in longer sentence contexts in both anterior and posterior aphasics. Tonal coarticulation was absent in the former, markedly reduced in magnitude and temporal extent in the latter. Coarticulatory effects in RHD patients, on the other hand, were indistinguishable from those of normal controls. Together, these findings suggest that the planning of tonal coarticulation may deteriorate in longer sentence contexts in the speech of LHD patients. Like consonants and vowels, some deficiencies are observed in the coarticulation of tones.

13-1.3.4. Intonation

Contrary to the view that all aspects of prosody are mediated by the right hemisphere, much of the acoustic phonetic evidence seems to indicate that intonation is more vulnerable to disruption in the speech of LHD than RHD patients (see Cooper & Klouda, 1987, and Ryalls & Behrens, 1988, for earlier reviews). Although the deficit is more severe in Broca's aphasics (Danly & Shapiro, 1982), abnormalities in intonational patterns have also been observed in Wernicke's aphasics (Danly, Cooper, & Shapiro, 1983). Not all aspects of intonation are vulnerable to impairment; for example, the sentence-final fall in F_0 (Danly & Shapiro, 1982).

The extent to which sentence declination patterns deviate from those of normals appears to depend largely on the size of the utterance (Cooper, Soares, Nicol, Michelow, & Goloskie, 1984; Danly & Shapiro, 1982). Declination patterns were more likely to deteriorate in longer sentences. A narrowing of the F_0 range that has been observed in anterior LHD patients of nontone languages (Cooper et al., 1984) does not generalize to tone languages. Not only is there no compression of the F_0 range in Thai-speaking aphasics (Gandour et al., 1992a), but their tone space actually has a wider than normal range. An intonation deficit in the speech of a nonfluent Chinese-speaking aphasic appears to be a secondary manifestation of a more pervasive timing disorder (Packard, 1993).

These crosslinguistic findings suggest that disturbances in F_0 patterns in aphasic patients may be a secondary consequence of an underlying timing impairment (Danly & Shapiro, 1982; Cooper & Klouda, 1987; Gandour, 1987). Consistent with this hypothesis, a Thai-speaking Broca's aphasic exhibited aberrant timing at the sentence level in tandem with preserved F_0 contours of the five tones (Gandour et al., 1989; cf. Moen, 1991).

The right hemisphere may be implicated in the control of global aspects of sentence intonation (see Ryalls & Behrens, 1988, for review of earlier studies). Baum and Pell (1997), however, found intonational characteristics to be spared in RHD patients as well as LHD patients, leading them to conclude that some aspects of intonation may be under the control of subcortical structures. Gandour (1995) and Gandour et al. (in press) similarly found certain aspects of intonation to be spared in LHD and RHD Thai-speaking patients.

Yet, the evidence seems to support the notion that the right hemisphere modulates dominantly the graded, affective components of language. All patients manifested a dense, nonfluent aphasia during the left-sided Wada test (Ross, Edmondson, Seibert, & Homan, 1988); during the right-sided Wada test, they exhibited a severe to moderately severe flattening of spontaneous speech. In a pre- and poststroke comparison, Blonder, Pickering, Heath, Smith, and Butler (1995) similarly found their RHD patient to exhibit a more restricted variability in F_0 and pause duration during spontaneous speech 6 weeks after the stroke, leading them to suggest that gradient properties of intonation are disrupted in RHD patients.

13-1.3.5. Foreign Accent Syndrome

The foreign accent syndrome (FAS) is a rare speech disorder that is characterized by the emergence of a perceived foreign accent in speech following left hemisphere brain damage (Blumstein, Alexander, Ryalls, Katz, & Dworetzky, 1987; Kurowski, Blumstein, & Alexander, 1996, and references therein). Blumstein et al. hypothesized that acoustically anomalous features are linked to a common underlying deficit relating to speech prosody. Prosodic impairments, however, were minimal in Kurowski et al.'s (1996) patient, leading them to challenge the notion that a general prosodic disturbance is the sole underlying mechanism in FAS. It has proved to be very difficult to determine what constellation of phonetic features typically characterizes FAS.

13-2. SPEECH PERCEPTION

Much of the earlier literature on speech perception attempted to determine whether segmental or phonemic perception is impaired in aphasia and whether such perceptual deficits underlie auditory language comprehension deficits (see Blumstein, 1991, 1994, for reviews). These findings suggest that aphasic patients do suffer from deficits in processing segmental contrasts, yet surprisingly, neither speech perception deficits nor auditory deficits appear to underlie language comprehension impairments. More recent work has attempted to show how the interaction between the phonetic features and the lexicon could ultimately contribute to deficits in language comprehension.

13-2.1. Role of Phonology in Lexical Access

Milberg, Blumstein, and Dworetzky (1988) investigated the extent to which phonological distortions affect semantic facilitation in a lexical decision task. Anterior aphasics showed semantic facilitation only in the undistorted real word condition, but not in the phonologically distorted, nonword conditions; in contrast, posterior aphasics showed facilitation in all phonologically distorted, nonword conditions at a level equivalent to that in the undistorted real word condition. These findings suggest that anterior aphasics have a higher threshold for lexical access, whereas posterior aphasics have a lower threshold (Gordon & Baum, 1994).

Similarly, the lexical status of a word has been shown to differentially affect how aphasic patients perform phonetic categorization (Blumstein, Burton, Baum, Waldstein, & Katz, 1994). Blumstein et al. found that anterior aphasics showed a larger-than-normal lexical effect, whereas posterior aphasics showed no lexical effect. Their findings suggest that anterior aphasics place a greater reliance on the lexical status of the stimulus in making their phonetic decisions than on the perceptual information in the stimulus. In contrast, posterior aphasics appear to be unable to use lexically based heuristic strategies in phonetic categorization.

13-2.2. Deficits in Perception of Speech Prosody

13-2.2.1. Stress

The perception of lexical, phrasal, emphatic, as well as sentence-level stress contrasts appears to be more disturbed following left-hemisphere than right-hemisphere damage (Emmorey, 1987, and references therein; Bryan, 1989). Across studies, no differences have been reported between the performance of anterior and posterior aphasics.

13-2.2.2. Intonation

The role of the left and right hemispheres in the processing of intonation contours is much less clear. Several perceptual studies have argued in favor of the notion that the right hemisphere is dominant for processing the acoustic correlates of sentence intonation (Shipley-Brown et al., 1988, and references therein). These findings have generally been interpreted to support the parallel processing hypothesis. Perkins, Baran, and Gandour's (1996) findings support the functional load or attraction hypothesis. When unfiltered pitch contours were presented for identification in a linguistic context, RHD patients performed better than LHD patients. Conversely, when filtered pitch contours were presented for discrimination in a nonlinguistic context, LHD patients performed better than RHD patients. And finally, when unfiltered pitch contours were presented for discrimination, a task that allowed for either linguistic or nonlinguistic processing, no differences in performance emerged between LHD and RHD patients.

13-2.2.3. Tone

Crosslinguistic findings from perceptual investigations of LHD and RHD patients who are speakers of tone languages consistently point to the left hemisphere as dominant for the perception of lexical tones: Mandarin Chinese (Hughes et al., 1983), Cantonese (Yiu & Fok, 1995), Thai (Gandour & Dardarananda, 1983), Toisanese Chinese (Eng, Obler, Harris, & Abramson, in press). The left hemisphere also emerges as the dominant hemisphere for the perception of pitch accent distinctions (Moen & Sundet, 1996). Those tones that are most susceptible to perceptual confusion are precisely those that are phonetically similar in height and shape to others in the tone space. Dichotic listening studies have similarly indicated left-hemisphere specialization for the perception of lexical tones and pitch accents (Van Lancker & Fromkin, 1973; Moen, 1993).

13-3. CONCLUSION

Phonological analyses of aphasic production data generally confirm recent proposals for features, syllable structures, and underspecified representations. Similarly, these analyses support psycholinguistic models of speech production that allow for three stages of phonological encoding: access, planning, and implementation. Such analyses, however, are scarce, and limited to only a few patients. It remains to be determined whether or not these findings can be generalized to clinical populations at large. If not, it is possible that "idiosyncratic preferences reflect the functioning of backup systems rather than the normal systems themselves" (Butterworth, 1992, p. 283).

Aphasic data also suggests that the phonological/phonetic components of the grammar may be more diffusely represented in the language-dominant hemisphere than heretofore believed. Phonetic evidence clearly points to a temporal deficit in speech production subsequent to left-hemisphere lesions, though the exact nature of the timing disorder remains an empirical question. New experimental designs that incorporate multiple tasks of varying type and complexity with the same patients should help to clarify whether the subclinical disturbances observed in posterior patients are due to a breakdown in phonetics or something else.

Not all aspects of linguistic prosody are subject to disruption following unilateral lesions to the left hemisphere. Therefore, the term "dysprosody" is an invalid label for characterizing the phenomenon. The deficit does not appear to reside at a linguistic level. Instead, evidence appears to weigh in favor of hypotheses that allow for differential lateralization of prosodic cues. One hypothesis that is deserving of further investigation is that timing impairments underlie suprasegmental as well as segmental deficits. In this way, it may be possible to achieve a unitary account of speech production disorders in LHD patients. Although the issue has not been resolved conclusively, the bulk of the evidence suggests minimal involvement of the right hemisphere in mediating linguistic prosody.

As is the case with segmental contrasts, anterior and posterior LHD patients suffer a perceptual deficit in prosodic contrasts. RHD patients, on the other hand, are generally spared in the perception of linguistic prosody. The attraction hypothesis seems to best account for the disparate array of findings in perceptual studies of stress, tone, and intonation. Impairments in the use of phonological information to access the lexicon appear to underlie at least some of the deficits in auditory language comprehension. The relationship between perception and production in aphasia remains unclear.

CHAPTER 14

The Breakdown of Morphology in Aphasia
A Cross-Language Perspective

Gonia Jarema

Department of Linguistics, Université de Montréal, Montréal, Québec, Canada and Centre de Recherche de l'Institute Universitaire de Gériatrie de Montréal, Montréal, Québec, Canada H3W 1W5

The purpose of this chapter is to characterize the breakdown of morphology in the performance of aphasic subjects in sentence and in single-word processing. Evidence from an array of languages investigated within a unilingual or cross-linguistic approach is presented and several recent interpretations are reviewed. It will be argued that the introduction of theoretical linguistic and psycholinguistic considerations into the study of aphasia contributes to a better understanding of the way in which pathology affects the morphological component of language.

In this chapter, we will consider disturbances affecting the morphological structure of words in aphasia. Disorders of morphologically complex, or multimorphemic, words have been reported to exist in the production and comprehension of aphasic subjects, at both the sentence and the single-word levels. We will discuss morphological impairments during sentence generation, as well as in single-word processing. Our effort will be to characterize the nature of these impairments, drawing on research that relies on a variety of tasks and languages and that shows a growing interest in theoretical linguistic models of word formation and psycholinguistic models of lexical processing. We will cite only a limited number of the many studies that have reported on the various forms of morphological disturbances in aphasia over the last two decades. Overall, they have provided converging evidence for the following assumptions:

1. Single-word processing disturbances can dissociate from morphological breakdown in sentences.
2. Morphological deficits can occur in all modalities of linguistic performance.
3. Inflectional and derivational morphology, and compounding, can be selectively impaired.
4. At least some words are decomposed in comprehension and composed in production.
5. In stage-by-stage models of sentence production (Garrett, 1980, 1982), on-line morphological breakdown is related to the positional level of sentence processing—the level at which phrasal frames are specified, together with bound morphemes.
6. Language-specific features contribute to the error patterns observed in aphasia.
7. In reading, morphological impairments can mask underlying semantic or visual impairments since morphologically similar items are also semantically and visually similar.
8. Frequency, imageability, abstractness, word class, and phonological or orthographic complexity influence morphological processing.

In the pages that follow, we will present the empirical evidence and the arguments that underlie these assumptions and that have contributed to the advancement of our understanding of the breakdown of morphology in aphasia.

14-1. MORPHOLOGICAL IMPAIRMENTS IN SENTENCE PROCESSING

14-1.1. The Case of Agrammatism

The most extensively investigated aphasic patients presenting with morphological disturbances are classified clinically as "agrammatic." Agrammatism is mainly defined as a disorder at the level of grammatical formatives—free (auxiliaries, determiners, pronouns, and some prepositions) or bound (inflectional affixes); agrammatic speech is generally effortful and reduced, that is, sentences are short and characterized by an absence of embeddings. Comprehension deficits tend to parallel production deficits, and cross-patient as well as within-patient performances can vary significantly. Several diverging accounts in terms of either a representational or a processing impairment have been proposed in the literature (for a review and critique of the current debate, see Kean, 1995).

The now widely reported behavioral variability in the performances of so-called agrammatic patients has led some authors (Badecker & Caramazza, 1985, 1986; Miceli, Silveri, Romani, & Caramazza, 1989) to challenge the very existence of agrammatism as a distinct aphasic syndrome. However, our discussion does not hinge on this issue because our goal is not to characterize a clinical entity or to contrast aphasic syndromes, but rather to specify the nature of morphological deficits in patients typically demonstrating difficulties with bound morphology. Furthermore, agrammatism has been defended as a theoretically coherent aphasic category (Caplan, 1986, 1991) since agrammatic patients share difficulties at the level of closed-class vocabulary (function words and bound morphology) while performing well on open-class vocabulary (major lexical categories), irrespective of intersubject differences, and since these difficulties arise at a specific stage of processing, that is, the stage at which function words and bound morphemes are accessed in sentence production. (See Garrett, 1980, 1982, for a model of speech production in which the generation of phrasal frames is closely tied to that of inflectional morphology.)

14-1.2. Morphological Deficits in Spontaneous Speech: The CLAS Study

The main feature of agrammatic speech is the omission and/or substitution of affixes. This feature has received increasing attention since Goodglass and Hunt (1958) first reported on the production of the English suffix -s in agrammatism (see Goodglass, 1976, for a review of the early modern literature on morphological disturbances in aphasia), and has been studied extensively in the spontaneous speech of patients (e.g., Tissot, Mounin, & Lhermitte, 1973; Menn & Obler, 1990). Menn and Obler (1990) coordinated a large-scale crosslinguistic investigation (the Cross-Language Agrammatism Study, or CLAS) based on comparable corpora of conversational speech, storytelling, and picture description in 14 languages. The study's detailed analysis of agrammatic speech production clearly indicates that breakdown patterns vary greatly across subjects and languages. For example, in Finnish (Niemi, Laine, Hänninen, & Koivuselkä-Sallinen, 1990), patients appear to be syntactically rather than morphologically agrammatic. A small number of morphological errors are manifested in the form of substitutions of verb inflections, the only category affected. Niemi et al. argue that bound grammatical morphemes are not lost in a synthetic language such as Finnish, in contrast with analytic languages such as English. In a nonconcatenative language such as Hebrew, on the other hand, the reverse pattern is demonstrated (Baharav, 1990), with syntactic abilities being more spared than morphological ones. Bound morphemes give rise to substitution only, mainly in verbs. Because roots are consonantal in Hebrew, vocalic morphemes must be introduced for words to become pronounceable; thus omissions are barred from the language. Interestingly, in Japanese (Sasanuma, Kamio, & Kubota, 1990), where bare verb stems also cannot stand alone and where, therefore, only substitutions can be predicted to occur, neither omissions nor substitutions are reported. Sasanuma and her colleagues argue that, in Japanese, inflections are determined word-internally and are thus more robust, unlike in Indo-European languages, where agreement rules extend across words and phrases. Another finding relevant to the issue of cross-language variablility is that in highly inflected languages such as Polish (Jarema & Kadzielawa, 1990) and Serbo-Croatian (Zei & Sikic, 1990) in which word order is relatively free and syntactic relations are signaled by case, the analysis of subjects' performances on bound morphemes clearly indicates that case-marking inflectional markers are not affected. The conclusion drawn from the variable patterns of morphological breakdown observed in Menn and Obler (1990) is that disturbances are determined by language-specific features and that our understanding of aphasic speech benefits greatly from investigations that take into account theoretically motivated descriptions of the language(s) under study.

Menn and Obler's (1990) corpora also include samples of narrative production in the written modality. Overall, the results obtained for the languages studied confirm the clinical observation that the breakdown patterns observed in the writing of agrammatic aphasic patients generally parallel those found in their oral productions. However, performance on the written narrative was strikingly spared in one of the Japanese-speaking agrammatic patients described in the study (Sasanuma et al., 1990). By contrast, a dissociation between impaired writing and spared oral reading, which were tested on the same narrative, was reported for several languages, for example, for both Japanese-speaking patients and for one of the two Finnish-speaking patients (Niemi et al., 1990). Although the supplementary data on writing and reading discussed in Menn and Obler (1990) represent only small samples, they nevertheless indicate that oral and written impairments tend to cluster in agrammatic patients, whereas reading aloud is generally more resilient.

14-1.3. The Role of Parametric Variation

The problem of variability in aphasia has been linked to the role of language-specific features in language processing, among other factors. In the domain of morphology, Miceli, Mazzucchi, Menn, and Goodglass (1983) demonstrate the importance of parametric variation. In this study, agrammatic subjects systematically substituted rather than omitted inflections in Italian, a language in which Ø-inflected stems generally result in nonwords. Similar results are reported by Kehayia (1990) and Kehayia, Jarema, and Kadzielawa (1990) for Greek, a language in which bare stems are never legal. This pattern contrasts sharply with the one reported for the more typically studied English-speaking agrammatic aphasics, whose frequent omission of inflections results in the production of lexically well-formed bare stems (e.g., *flower* instead of *flowers,* or *walk* instead of *walked*). Grodzinsky (1984) cites Miceli et al. (1983) and discusses Russian data from Tsvjetkova and Glozman (1978) to support the view that patients revert to Ø-forms only when the stem is a word in the language, otherwise misselecting the required inflection. However, it has been shown that in Icelandic (Magnúsdóttir & Thráinsson, 1990), bare stems are almost never produced, even when a Ø-affixed form exists in the language, as is the case with the imperative and the typical first person singular present form of verbs. This observation speaks against the thesis that agrammatic aphasics' omissions of bound affixes reflect a default mechanism, that is, that in reality they represent substitutions of inflected forms by a Ø-affixed form (Grodzinsky, 1984). The cross-linguistically robust finding, reported in

Menn and Obler (1990), that morphological substitutions are generally off-target by only one feature further indicates that the systematic selection of Ø-forms is an unlikely procedure since it would wrongly predict that in some instances a very distant form would be selected. For example, in the Polish data (Jarema & Kadzielawa, 1990), we do not see any cases where the Ø-inflected singular imperative form replaces a tensed form, e.g., the past tense, which is inflected for mood, aspect, tense, person, number, and gender (six features).

Grodzinsky (1984) further proposes that, in contrast with many Indo-European languages, only inflectional substitution can be resorted to in Semitic languages, in which words are formed by inserting discontinuous vocalic morphemes into discontinuous consonantal roots.[1] To illustrate this structural property, the author discusses an example from Hebrew in which the three consonants *s, m,* and *l* are shared by the related words *simla* (a dress) and *smalot* (dresses). Omitting the vocalic base would result in unpronounceable nonwords, that is, forms that are both phonologically and morphologically illegal.

Investigating another Semitic language, Algerian Arabic, Mimouni and Jarema (1995) obtained results similar to those put forth in Grodzinsky (1984) in the performance of three agrammatic subjects. In their spontaneous speech, as well as in sentence repetition and reading aloud, subjects resorted to substitutions and never produced nonwords. However, in this study, it was observed that omissions of affixes can and do occur in Semitic languages and that in cases where omitting the affix would result in a morphologically ill-formed bare stem, subjects substituted the nonword stem with a real-word stem. Consider examples 1a and 1b below.

	TARGET	APHASIC PRODUCTION
1a	yə-drab →	*drab*
	(He hits) →	(He hit)
1b	xədm-ət →	*xdam*
	(She worked)	(He worked)

In 1a, *drab* is a word, and the omission of the prefix *yə* results in a substitution of tense. In 1b, however, **xədm* is a nonword and, as a result, is not produced in isolation. Instead, subjects who have difficulties with the prefix revert to *xdam,* a related, and legal, unaffixed form yielding a gender substitution.

By contrast, in French, uninflected verbs surface to the level of the word, as is the case in English. Jarema and Kehayia (1992) show that, in a constrained sentence production task, French-speaking agrammatic aphasics do omit verb inflections resulting in Ø-inflected, but grammatical, tensed forms. However, when the omission of inflections would result in a different root,[2] subjects revert to substitution of the affix, while retaining the target root (cf. 2a and 2b below).

[1]Inflectional morphology in Semitic languages also includes prefixation and suffixation (see McCarthy, 1981).

[2]In French, verbs from the first conjugation have a single root and are fully transparent throughout the verbal paradigm, while verbs from the second or third conjugations are characterized by the presence of two or more roots in the verbal paradigm.

TARGET   APHASIC PRODUCTION
2a *mangera* →   *mange*
Root:/mãz/   /mãz/
(will eat, 3p.sg.)   (eats, 3p.sg.)

2b *vendra* →   *vendre*
Root:/vãd/   /vãd/
(will sell, 3p.sg.)   (to sell)

In 2b, subjects did not substitute the future form of the verb with the present form (/vã/, a Ø-inflected form) as in 2a. This would have required access to a different root, that is, the root /vã/, which is used in the formation of the singular forms of the present tense of the verb. In line with a strong lexicalist approach to morphological representation,[3] this observation has implications for the internal organization of the mental lexicon in terms of morphological "families." More specifically, it implies the listing of morphologically related lexical items around a "nucleus" ("Satellite Hypothesis," Lukatela, Gligorijevic, Kostic, & Turvey, 1980) or under a "head," possibly the root, as proposed by Jarema and Kehayia (1992). In the case of the French verb system, verbs featuring a single root are listed under this common root, while verbs featuring more than one root have separate, but linked, listings under each root. These separate roots act as the head of a morphological subfamily (Figure 1).

To summarize, patients never produce items that would result in nonwords in their language, that is, they omit inflections only if the language permits. In cases where stripping a stem does not yield a real word, they resort to substitution when unable to access the correct form. Furthermore, the substituting inflection generally is selected within the target's paradigm; for example, a verb inflection is substituted by another

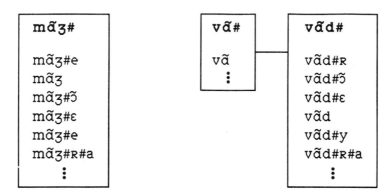

FIGURE 1   Lexical representation and organization of a French verb with a single root (*manger,* "to eat"), and of a French verb with two roots (*vendre,* "to sell").

[3]The Strong Lexical Hypothesis (Jackendoff, 1975; Lieber, 1980; Scalise, 1984, Lapointe, 1985; Di Sciullio & Williams, 1987) posits the existence of an autonomous morphological component consisting of a set of listed items and a set of word-formation rules and principles; both derivational and inflectional operations take place in this component.

verb inflection, indicating that subcategorization features of lexical items are not violated. Thus, importantly, not only phonotactic and phonological well-formedness criteria, but also constraints on word-formation processes, appear to be respected in the erroneous productions of agrammatic subjects.

14-1.4. Predictions

Overall, the crosslinguistic data discussed above indicate that aphasic subjects with morphological disorders do not produce nonwords. As a consequence, one can propose the following generalization, in accordance with, and expanding on, Grodzinsky's (1984) proposal. In languages that feature roots that surface to the level of the word (e.g., English), subjects resort to either omission or substitution of inflections, but omission tends to predominate. In languages in which roots are generally nonwords (e.g., Icelandic), subjects exceptionally choose an existing Ø-inflected form if and only if that form is a paradigm-internal form; otherwise, they substitute inflections. In languages in which the distributions of Ø-inflected word and nonword stems are relatively balanced (e.g., Polish), both omissions and substitutions occur with a high degree of variability in individual trends. In languages in which roots are never words (e.g., Greek), only substitutions are observed. These predictions were borne out in a crosslinguistic study (Kehayia et al., 1990) that investigated morphological errors in the repetition, comprehension, and production of sentences in English-, Greek-, and Polish-speaking agrammatic aphasics. Morphological markers were omitted in English, substituted in Greek, and either omitted or substituted in Polish. More specifically, in Polish, omissions and substitutions were found in words that allow roots to surface to the level of the word. However, only substitutions were found in words derived from Ø-affixed roots that are illegal. Note that when the Ø-affixed form features phonological complexity (e.g., vowel epenthesis), it was not produced. This points to the interaction between morphology and phonology in morphological disturbances.

The claim, then, is that morphological omissions are *variably* present *only* when the inflection *can* be stripped, that is, when the stem produced is phonologically and morphologically well formed. Crucially, this claim extends to cases such as that described in 1b, where a legal bare stem replaces the illegal stem of the stripped target item. Thus, the omission of bound morphemes attaching to nonword stems *is* possible at some early stage of processing difficulty; however, it is masked by the substitution of the stem. This observation further confirms the resilience of morphological well-formedness criteria observed in language pathology.

The data reported here thus support the view that parametric variation plays a role in the error patterns observed in aphasia and provides evidence against Grodzinsky's (1990) claim that patterns of omission and substitution are likely to be item-bound rather than language-bound. In his revised account of inflectional deficits in agrammatic speech, the author does away with the "default mechanism" he had proposed earlier (Grodzinsky, 1984). Instead, he stipulates the existence of an "agrammatic grammar"—a grammar restricted by the underspecification of syntax-level values of inflectional features (e.g., agreement). In combination with intact word-formation rules, this reduced grammar automatically yields omissions when Ø-inflected items are well formed (i.e., there is no need to generate an inflection for a feature that is underspecified if the uninflected word is well formed), and in all other cases it yields substitutions compatible with the items' representation. However, the data presented above clearly demonstrate that in some languages substitutions occur in spite of a Ø-inflection option; thus the language rather than the item appears to be relevant to the behavioral patterns manifested in aphasia.

14-1.5. Implications for Theories of Lexical Representation and Access

Importantly, all of the studies reviewed above not only underscore the role of language-specific lexical features in the patterns of morphological breakdown, but also bear upon theories of lexical representation and access. If one were to adopt a full-listing model of lexical representation (Butterworth, 1983) according to which all lexical items, simple and complex, are listed in the mental lexicon, several claims could be made.[4] When some lexical families exhibit several roots, as is the case in French, links between members of the family will not be equal and substitutions will reflect this inequality. Furthermore, the status of the "basic" lexical element of a family, if its role is validated, may vary from language to language. Lukatela et al. (1980) have, for example, proposed the nominative singular Ø-inflected form for Serbo-Croatian masculine nouns. This form is the nucleus around which all related forms are organized in a satellite-like structure. As shown above, Jarema and Kehayia (1992) suggest that, in French, verb forms may be headed by their root(s). For Semitic languages, the positionally loose consonantal root, the only link between cross-categorial members of a lexical family, has been proposed as the mother node of all the nodes it heads; each daughter node in turn heads a morphological paradigm in a treelike structure. (See McCarthy, 1981, for a theoretical discussion of Semitic morphology.) For Algerian Arabic, Mimouni and Jarema (1995) further propose that for verbs, the Ø-affixed perfective third person singular form, the citation form and the form patients produce most frequently when making errors (e.g., *drab* or *xdam* in the first example above), is the head of a verb paradigm in the language. Thus, the organization of morphologically related lexical items in the mental lexicon and the nature of a putative base form may vary crosslinguistically and, consequently, may influence access patterns in normal and pathological performance. The relationship between morphological disturbances in aphasia and issues of lexical representation and access has received much attention in the study of single-word processing, which we will address in the following section of this chapter.

---

[4]Dual route models such as the Augmented Addresses Morphology (AAM) model of Caramazza, Laudanna, and Romani (1988) or the Morphological Race (MR) model of Frauenfelder and Schreuder (1992), which assume the existence of a direct access (whole-word) route *and* of a morphological decomposition route (to deal with the problem of novel words and nonwords for which there are no lexical entries), are also compatible with our view of the psycholinguistic implications of cross-linguistic evidence, since lexical items are fully listed in these models.

14-2. MORPHOLOGICAL IMPAIRMENTS IN SINGLE-WORD PROCESSING

An important observation with respect to morphological impairments in aphasia is that some patients appear to have difficulties with bound morphemes only when processing sentences, but not when processing words in isolation.

14-2.1. The Dissociation between Single-Word and Sentence Production

A dissociation between impaired sentence and intact single-word production in French is described dramatically in Nespoulous et al. (1988). However, the subject's sentence production difficulties at the lexical level were shown to be mild and restricted to the class of freestanding grammatical morphemes (mainly pronouns), and bound morphemes were not explicitly tested in the single-word production tasks used.

A prototypical asymmetry along a strictly morphological dimension is reported by Caramazza and Hillis (1989a). The authors describe a patient who, in her oral and written narratives, as well as in three constrained sentence production tasks (oral reading, repetition, and writing to dictation), presented with inflectional errors, in contrast with her performance on single-word production tasks (reading, repetition, and spelling to dictation). The authors interpret this observation, together with evidence from a number of other tasks, to reflect a disorder at the level of representation where grammatical morphemes are specified in sentence constructing, that is, the "positional level" in Garrett's (1980, 1982) sentence production model.

The role of syntactic structuring in morphological disturbances is further investigated by De Bleser, Bayer, and Luzzatti (1996), who report on the performance of two German- and two Italian-speaking agrammatic aphasics in single-word and sentence production. In both languages, subjects showed relatively well-preserved lexical morphology, but impaired inflectional morphology in phrasal structures when nonlocal operations were involved. De Bleser et al. (1996) suggest that even in severely impaired patients the lexicon can be intact and that regular inflectional forms are available in local domains. Furthermore, the presence of overt cues, such as verbal number cues, can "extend the threshold of structural analysis" in agrammatic performances. They thus propose that lexical information is available and that bound morphology becomes vulnerable only when patients fail in constructing phrasal structures.

14-2.2. The Dissociation between Single-Word Recognition and Sentence Comprehension

The coexistence of intact processing of morphologically complex words when tested in isolation and of impaired sensitivity to inflectional affixes when tested in sentences has also been shown to exist in the auditory modality. Tyler and Cobb (1987) and Tyler, Behrens, Cobb, and Marslen-Wilson (1990) present agrammatic patients with normal recognition of polymorphemic words who were unable to detect morphological errors in context in a word-monitoring task. Unlike the group of control subjects, the patients were not faster in monitoring for words immediately preceded by a correctly inflected test item rather than by an incorrectly inflected one. The authors concluded that the patients were not impaired in their ability to recognize complex words. Rather, they were unable to process inflectional morphology on-line; that is, they could not use syntactic and semantic information encoded in the affixes in sentence computing. Friederici, Wessels, Emmorey, and Bellugi (1992) report on the ability of four German-speaking agrammatic aphasics to process inflectional morphology. They present two experiments also using the word-monitoring paradigm and claim, in line with Tyler and her colleagues, that agrammatic patients are unable to retrieve the full syntactic information encoded in inflectional affixes on-line. Friederici et al. (1992) further argue that this does not imply that syntactic knowledge cannot be derived at all. Indeed, their first experiment crisply demonstrates that patients are sensitive to inflectional morphemes as markers of grammatical category. In the experiment, violations of grammatical category were introduced by using tensed verbs derived from stems that are homophonous with simple nouns (e.g., *tanz-*, dance) in a context requiring a verbal form (e.g., *ER TANZ "he dance"; control: ER TANZTE "he danced") or by attaching an illegal verbal affix to a noun in a position where the presence of a verbal form yields an ungrammatical sentence (e.g., *EIN TANZTE "a danced"; control: EIN TANZ "a dance").[5] However, the study's second experiment in which within-category inflectional violations (*ER TANZTEST "he danced": 2nd pers. sing.) were presented did not yield a grammaticality effect. Friederici et al. (1992) contend that agrammatic aphasics suffer from a "selective disruption of the temporal organization of language processing." The authors' thesis is that there is a dissociation between structural knowledge, which is preserved as evidenced by off-line tasks (Shankweiler, Crain, Correl, & Tuller, 1989; Lukatela, Crain, & Shankweiler, 1988; Wulfeck, 1987), and time-dependent computational knowledge in these patients.

14-2.3. Parallelism between Single-Word and Sentence Production

Alongside cases of asymmetrical performances, patients presenting with morphological disturbances in the production of both sentences and single words are reported in the literature. This symmetry points toward a breakdown in lexical processes rather than in accessing bound morphemes during sentence planning. Miceli and Caramazza (1988), for instance, describe a patient who could be classified clinically as "agrammatic" and who showed a 50% error rate in repeating affixed words presented in isolation. Stems, prefixes, and derivational suffixes were intact, while inflectional suffixes were severely affected. The authors claim that these findings support the view that lexical items are decomposed (Taft & Forster, 1975; Taft, 1979),[6] that morpho-

---

[5]Upper-case fonts were used in the study since in German all common nouns are capitalized.

[6]In sharp contrast with the Full Listing Hypothesis (Butterworth, 1983), this theory postulates that multimorphemic words are recognized through an affix-stripping decomposition process that isolates the root (e.g., un-*think*-able) and initiates its lookup in a lexicon of roots.

logical operations are located in the lexicon, and that inflectional and derivational processes are functionally distinct, that is, that morphological subcomponents can be selectively impaired in aphasia.

14-2.4. The Asymmetry between Derivational/Inflectional Affixation and Compounding

Luzzatti and De Bleser (1996) used a set of carefully constructed off-line experiments aimed at further probing the issue of morphological dissociations in two previously studied Italian-speaking agrammatic aphasics (see discussion of De Bleser et al., 1996, above). The subjects were asked to provide the inflectional markers of gender and number for simple, derived, and compound nouns, as well as to produce prepositions in prepositional compounds *(pasta [al] forno)* and to supply adjectival derivational suffixes *(una eruzione vulcan[ica]).* Results revealed a number of theoretically relevant dissociations. The first patient, MG, was unimpaired at the inflectional and derivational levels, while his performance on compounds was very poor. The second patient, DR, also showed problems with compounds and, like MG, performed flawlessly on derivation. However, although DR's regular inflections seemed to be spared, his capacity to produce irregular affixes was highly affected and manifested itself in the form of overgeneralizations of regular affixational rules. Both patients could thus construct derived words but not compounds, and both could make use of lexical rules for inflecting and deriving words, but again not for compounds. In view of these asymmetries, and in view of the fact that the only impaired morphological subcomponent is the one where syntactic principles are involved at the word level, Luzzatti and De Bleser interpret their findings as evidence against a strictly morphological account of agrammatism and as support for the thesis that the underlying deficit is syntactic in nature. This assumption is further supported by De Bleser et al.'s (1996) study reviewed earlier in which morphological breakdown is syntactically determined.

14-2.5. Single-Word Disturbances in Dyslexia: The Issue of Confounding Factors

A further observation is that aphasic patients who present with morphological deficits in oral production sometimes also cannot read bound morphemes correctly. Badecker and Caramazza (1987) report on a dyslexic patient whose speech was nonfluent and reduced and who produced morphological errors in the form of affix deletions and substitutions. In reading, his performance included visual, semantic, inflectional, and derivational errors. Given that morphologically related words are often visually similar, and because they are also semantically related, the authors were concerned that reading errors that are identified as morphological errors may in fact be visual or semantic errors. They constructed a number of highly controlled reading tasks that included lists of morphologically simple and complex (affixed) words, monomorphemic pseudosuffixed words *(wicked),* and words containing pseudostems, that is, words embedded in words *(pierce).* Stimuli were matched in frequency and letter length. The subject made more errors reading polymorphemic affixed stimuli than monomorphemic pseudoaffixed or pseudostem stimuli. Badecker and Caramazza (1987) interpret these results as evidence in favor of a morphological rather than a visual account of the subject's errors. The authors caution readers, however, that high-stem or pseudostem frequencies, as opposed to whole-word frequencies, may have influenced the differences obtained. Further testing with other stimuli yielding a significant dissociation along a purely linguistic dimension led Badecker and Caramazza (1987) to suggest that at least some of the reading errors in acquired dyslexia are morphological in nature. The importance of this study lies in underscoring that an array of factors, such as visual similarity, semantic relatedness, frequency, concreteness, phonological complexity, or grammatical category, can contribute to reading words erroneously. Each of these factors must be considered in the light of linguistic and psycholinguistic theories of the lexicon before the claim that reading errors reflect a disturbance of morphological processing can be made. Badecker and Caramazza reinforce their cautionary stand by surveying two early studies that reported on a morphological processing deficit in the oral reading of dyslexic patients (Patterson, 1982; Job & Sartori, 1984) and by arguing that the authors fail to provide valid support for their claim that their subject's errors can be attributed to a breakdown at the level of morphology.

14-2.6. Single-Word Disturbances in the Written Modality

Cases of patients producing morphological errors in single-word tasks in the written modality have also been observed. Badecker, Hillis, and Caramazza (1990) describe a patient with acquired agraphia who deleted, substituted, transposed, and inserted letters asymmetrically toward the end of monomorphemic words, but whose error rate declined in the region of the suffix in multimorphemic words. Similarly, his error rate was low in the initial portion of prefixed words and only started rising at the beginning of the stem. In compounds, his errors were equally distributed; that is, error rates were higher at the end of both conjuncts. This compositionality effect was not observed in length-matched monomorphemic words containing pseudostems (e.g., *yearn, dogma).* Badecker et al.'s (1990) findings thus provide strong evidence for morpheme-sized rather than whole-word processing units in writing.

14-2.7. Limited Processing Resources: Evidence from On-Line Tasks

The patterns of performance reported in the literature seem to be consistent with the view that, at least in some aphasic patients, knowledge of the structure of multimorphemic words, that is, knowledge at the level of lexical representation, is not entirely "lost." Rather, limited processing resources appear to be responsible for the deficits observed. This is not only revealed by within-patient variation and by modality-specific disturbances, but also by the error patterns observed in the processing of multimorphemic words. For example, Libben (1990) describes an agrammatic Broca's aphasic who produced repetition errors only in nontransparent multimorphemic words, such as *illegible* and *regularity,* as opposed to transparent forms such as *unhappiness* and *materialism.* His erroneous productions—for example, *inlegible* (failure to assimilate segments) and *régularity* (failure to shift stress)—are most revealing. In these cases, the patient decomposed the word and produced the form of the underlying stem—which must thus be represented—without making the required phonological adjustments in the affix or in the stem. Libben interprets these findings as evidence for a decrease in processing resources resulting from the formal complexity of multimorphemic words, and for "doing morphology" on-line. Reporting on the same patient, Libben (1994) shows that when tested in a naming latency experiment using legally (e.g., *birmity*) and illegally affixed nonsense stems (*rebirmity,* where the suffix *-ity* attaches to the root *birm* forming the noun stem *birmity,* and where morphological well-formedness is violated since the prefix *re-* can only attach to verbs), the patient performs more slowly and produces more errors in naming the illegally affixed items than the legally affixed ones. Libben argues that this behavioral pattern is an indication that selectional restrictions of affixes are available during word production.

Further evidence for the availability of lexical information is found in the parallel results obtained for normal and aphasic subjects in lexical decision tasks (Kehayia & Jarema, 1994; Mimouni, Kehayia, & Jarema, 1992). In both studies, although patients showed considerably slower reaction times, response patterns were very similar to those obtained from normal subjects; that is, patients were remarkably sensitive to the language-specific features of Greek and Polish, and of Arabic, respectively.

14-2.8. Summary

Overall, patients seem to be sensitive to word-internal structure, as evidenced through their performance on off- and on-line tasks conducted in a variety of languages. Furthermore, the observation that morphemes tend to be recognized or retrieved together with their accompanying features and that morphological breakdown generally does not lead to word-formation violations[7] points toward the availability of linguistic rule knowledge. Finally, although multimorphemic words can be impaired in isolation, they seem to be particularly vulnerable when brought to interact with the domain of syntactic processing.

14-3. CONCLUDING REMARKS

In this chapter, we have demonstrated that descriptions of morphological disturbances in aphasia are increasingly enhanced by linguistic and psycholinguistic theory. We have cited studies of impaired morphological processing that incorporate theoretically motivated linguistic distinctions and that reveal properties of the mental representation and organization of multimorphemic items and specify the cognitive processes involved in word recognition and production. With respect to the structural aspects of the mental lexicon, one is led to infer a representational system not only from the

[7]See Laine, et al. for a counterexample.

patterns of errors found in language pathology, but also from the lexical access or retrieval procedures that these patterns suggest. Thus, a whole-word mechanism implies the listing of all lexical items, whereas a decomposition–composition mechanism compels one to assume a listing of either *(a)* all atomic forms, roots and affixes alike, *(b)* all lexical simple and complex forms, with some form of representation of word-internal structure for multimorphemic entries, or *(c)* a listing of roots, each specified for all the possible affixes that can attach to it. Furthermore, the full-listing assumption, whether entries are morphologically opaque or transparent, has been further refined to include evidence of modes of lexical storage of morphologically related items. Formal aspects of lexical organization, such as the construct of "morphological family," together with the concepts of "link" and of "nucleus" (or "basic unit of representation"), have been developed. Finally, hybrid systems, such as the dual-route models in which both whole-word and morpheme-sized units are represented, feature full listing of chunk-type multimorphemic words with pointers toward listed component morphemes, however, without representation of morphological relatedness.[8]

In conclusion, it is heartening to witness a convergence of efforts to capture the nature of morphological disturbances in aphasia across languages, tasks, and theoretical frameworks, particularly in view of the considerable challenge researchers are faced with in designing highly controlled experiments that will shed light on the interaction between morphological errors and various processing factors and that will also reliably dissociate morphological deficits from other word-level deficits. Undoubtedly, the growing body of experimental work on word recognition and production will further the emergence of increasingly satisfactory proposals on the organization of the mental lexicon and on the computational mechanisms involved in normal and impaired morphological processing. This in turn will ensure a more informed description of morphological breakdown in aphasia.

[8]In this chapter, we have not discussed network models (e.g., Fowler, Napps, & Feldman, 1985; Dell, 1986), which are representational systems that do not rely on parsing. In such models, all morphologically related lexical items are represented as interconnected nodes, with connections between nodes varying in weight and governing the patterns of activation across nodes.

CHAPTER 15

The Shadows of Lexical Meaning in Patients with Semantic Impairments

Peter Hagoort

Max Planck Institute for Psycholinguistics, Nijmegen, The Netherlands

Central in the processing of words is the retrieval of their meaning. Neurological diseases can affect the processing of lexical meaning in a number of ways. First, various examples of semantic impairments are discussed, including semantic paraphasias, deep dyslexic reading errors, and category-general and category-specific impairments in the semantics of concrete nouns. Next a number of theoretically relevant issues in studies on semantic impairments are discussed: *(a)* the concreteness effect, *(b)* impairments of perceptual versus functional information, *(c)* amodal versus multiple semantic systems, and *(d)* access impairments versus storage deficits. On the basis of a review of the neuropsychological evidence, it is proposed that the representation of lexical meaning consists of conceptual structures tied to models that are tailored to the requirements of the different sensory systems and the motor system.

A central notion in all models of language production and language comprehension is the mental lexicon. The mental lexicon refers to the knowledge of the language user about the words of his/her language(s). This knowledge specifies not only the sound pattern and the orthography of words, but also their grammatical properties (e.g., word class, gender), their morphological structure, and their meaning. Both neuropsychological evidence and recent brain-imaging studies have shown that these different types of word knowledge are represented in widely distributed areas of the brain.

The role of the mental lexicon in language processing is to mediate between different representational domains, including form (sound, orthography) and meaning. To do this effectively, the mental lexicon has to contain information about a large number of words. For instance, an adult speaker of English has an estimated passive vocabulary of at least 40,000 words (Nagy & Herman, 1987). On the basis of the spoken or written input, the best matching word form in the mental lexicon is selected. As a result, the adult speaker of English gains access to all the information associated with this particular word form. For instance, he or she knows that it is a noun or a verb, that it refers to an animal that flies, or to the act of flying, and so forth.

This chapter focuses on the processing of a central aspect of words, namely, lexical meaning. Although there is some disagreement about whether the meaning of words should be treated as lexical knowledge per se, there is no disagreement about the centrality of meaning in word processing. The centrality of meaning is further supported by the neurolinguistic literature, which provides ample evidence that brain damage can affect the ability to select (in speaking and writing) and recognize (in listening and reading) the appropriate lexical-semantic representations.

15-1. SEMANTIC IMPAIRMENTS

Ideally, a neurolinguistic theory of semantic impairments is based on an explicit account of how lexical meaning is represented in the unaffected language system. However, surprisingly enough, the nature of meaning representations has not played a major role in studies of semantic impairments. In these studies, the two central themes of the last decade are *(a)* category and modality specificity of semantic impairments and semantic representations, and *(b)* degraded representations versus access impairments. Often these issues are addressed without clear statements about the notion of meaning representation that underlies the interpretation of the results. However, it is not at all clear that satisfying answers to the questions involved can be given in the absence of an explicit account of meaning representations.

To set the stage for discussing a number of theoretically relevant issues in the semantic impairment literature, I will first describe a few symptoms and syndromes associated with impairments at the level of lexical-semantic processing.

15-1.1. Semantic Paraphasias

A common symptom in patients with a Wernicke-type aphasia are so-called semantic paraphasias. In this type of paraphasia, a word is produced that deviates in meaning from the intended word (Poeck, 1982). The actually produced word very often has some semantic similarity to the intended word. For instance, an aphasic patient whom

I asked to describe a picture of a man reading a newspaper replied, "The man reads the radio." Another aphasic patient said in an interview about his disease that he "became deaf in his eye." When tested in a word–picture-matching task, such patients regularly correctly match words like *newspaper* and *radio* to the relevant picture, indicating that the semantic specifications of the intended words are still largely intact. The observation that semantic paraphasias are often in the same semantic field as the intended word (e.g., *radio–newspaper, deaf–blind*) suggest that they are not unlike a particular type of speech error in normal speech, namely, word substitutions (cf. Levelt, 1989).

15-1.2. Deep Dyslexia

Word substitutions on the basis of sensory instead of conceptual input are found in patients with deep dyslexia. Deep dyslexia is an acquired reading disorder with semantic errors as its most striking symptom. A patient with deep dyslexia might, for instance, read the orthographic string RIVER as "ocean." However, the actual error pattern is complicated by the presence of a number of co-occurring symptoms. These include visual errors (SCANDAL read as "sandals"), morphological errors (SELL read as "sold"), and a better performance for concrete than for abstract words. Different explanations have been given for this clustering of symptoms. One possibility is that in fact we are faced with a number of independent deficits that co-occur for anatomical reasons. The brain damage happened to affect anatomically proximal areas subserving independent language functions. However, no evidence has been obtained in direct support of this hypothesis.

An interesting attempt to explain the co-occurrence of semantic and visual errors as well as the concreteness effect is found in a connectionist model of acquired dyslexia (Hinton & Shallice, 1991; see Plaut & Shallice, 1993a, for a detailed account of the model). In this model, patterns of activation in the layer of orthographic units are converted into associated patterns of activity in the layers representing the semantic space. If the initial pattern of activation within the semantic space falls within the region of a particular meaning, the pattern of activation converges on the pattern of activation representing that meaning (see Figure 1). These regions of convergence are called the basins of attraction.

In the model, the basins of attraction for words with similar meanings tend to be in close proximity. In addition, the attractor basins for similarly spelled words tend to be close to each other. Similarly spelled words are thus mapped onto nearby points in semantic space (see Figure 1). Damage to this network (e.g., removal of subsets of the nodes or connections, or addition of noise to the weights on the connections) can change the boundaries of the basins of attraction. As a result, the initial pattern of activation in semantic space might fall into an incorrect basin of attraction. Because both semantically and orthographically related words are likely to have nearby basins of attraction, when the initial pattern of semantic activation falls into the incorrect basin, either a semantic (CAT read as "dog"), a visual (CAT read as "can") or a mixed error (CAT read as "rat") might result.

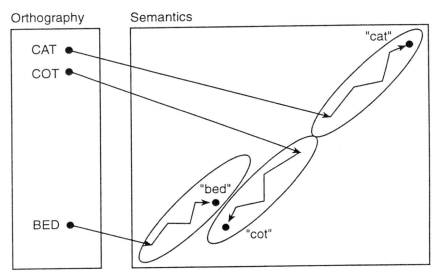

FIGURE 1   A representation of the basins of attraction at the semantic level of the attractor network for reading (after Plaut & Shallice, 1993). The solid ovals depict the basins of attraction. Information is converted from the orthographic layer (CAT) to a position somewhere in semantic space. As a result of activity in an additional layer of the network (i.e., the cleanup layer), this position is drawn to the point corresponding to the closest meaning ("cat"). Whenever the network's initial semantic output appears within a basin of attraction, the network's state will inexorably be drawn to one position within the region. This position is the point in semantic space that represents the meaning (e.g., "cat") that is associated with the orthographic input (e.g., CAT). However, a lesion in the network can change the boundaries of the basins of attraction. As a result, CAT might fall into the basin of attraction of "cot."

Although this connectionist model nicely accounts for the clustering of errors in deep dyslexia, there is very little independent evidence to date for a nonarbitrary mapping between orthographic and semantic representations (Rueckl & Dror, 1994).

15-1.3. Semantic Dementia

In recent years, an increasing number of patients have been reported with a progressive, degenerative brain disease that initially shows up as a selective semantic impairment. This disorder is known as semantic dementia (Hodges, Graham, & Patterson, 1995; Hodges, Patterson, Oxbury, & Funnell, 1992; Snowden, Goulding, & Neary, 1989). Brain-imaging data indicate that temporal lobe damage, predominantly in the left hemisphere, is crucially involved in the disease process. Semantic dementia should be distinguished from a syndrome called primary progressive aphasia (PPA; Mesulam, 1982). PPA is also a progressive, degenerative brain disease. However, in this case the impairment of the phonological and syntactic aspects of language is most prominent, whereas comprehension is relatively preserved.

Patients with semantic dementia can often only produce the category name when confronted with a picture or a name (e.g., *geese*—"an animal, but I've forgotten precisely"). On the whole, these patients show a fairly consistent pattern of breakdown, with loss of knowledge about semantic attributes, but relatively long-term preservation of superordinate category information. Hodges et al. (1995) tested patients with semantic dementia longitudinally during their progressive disease on a series of semantic tests, including naming of the Snodgrass pictures. A characteristic longitudinal performance pattern is illustrated by patient J.L., who responded as follows to the picture of an *elephant* in four different sessions over a 1.5-year period: (1) "elephant," (2) "horse," (3) "horse," (4) "animal." There is a clear progression from the specific, correct response to a prototypical instance of large animals ("horse") to the generic "animal" response. Semantic dementia patients usually show a parallel decline in tests of comprehension and production. This parallel decline suggests that a central semantic deficit underlies the semantic impairments observed in different language modalities.

I have given a few examples of neurological syndromes with semantic impairment as one of their most salient characteristics. In the remainder of this chapter, I will focus on a few theoretically interesting aspects of the reported semantic impairment symptomatology (for an excellent review, see Saffran & Schwartz, 1994). These are *(a)* the concreteness effect, *(b)* impairments of perceptual versus functional information, *(c)* amodal versus multiple semantic systems, and *(d)* access impairments versus degraded representations.

15-2. THE CONCRETENESS EFFECT

Brain-damaged subjects often show a better performance in the production and comprehension of concrete words than abstract words. Normal subjects also show a processing advantage for concrete words, but this effect is strongly amplified in patients. It has been known for a long time that, in general, aphasic patients are more impaired in retrieving abstract words than concrete words. The same holds for patients with reading disorders such as deep dyslexia.

One can find at least three different accounts of this concreteness effect in the literature (Breedin, Saffran, & Coslett, 1994). According to Paivio (1991), the advantage for concrete words is due to the existence of a dual code for this class of words, one verbal and the other nonverbal (imaginable). Abstract words, in contrast, only activate a verbal code.

The context availability account of Schwanenflugel (1991) explains the effect by assuming that concrete words are embedded in a larger body of associated contextual-perceptual information in memory than are abstract words.

Finally, a third possibility is that concrete words are represented by a larger set of semantic features than abstract words (Plaut & Shallice, 1993a). When there is an impairment of the mapping of orthography onto semantics in the Hinton and Shallice model of acquired dyslexia (Hinton & Shallice, 1991), the reading of abstract words is more affected than the reading of concrete words; that is, lesioning the network has the strongest consequences for words with a relatively sparse semantic representation.

All three accounts thus assume that concrete words activate a richer representational structure in memory than do abstract words. In general, this makes concrete words easier to access and remember, and less vulnerable to brain damage. However, this shared feature of the accounts of the concreteness effect cannot easily explain the cases of patients who are better at retrieving abstract than concrete word information. At least five such cases have been reported in the neuropsychological literature (Breedin et al., 1994). This reversal of the concreteness effect has potential implications for our understanding of the nature of semantic or conceptual representations.

Breedin et al. (1994) argue that a crucial difference between concrete and abstract words resides in the sensorimotor (including perceptual) attributes of the former. On this account, attributes of perception and action are part of the representations of concrete but not of abstract words. Selective impairment of these sensorimotor attributes might be a core aspect of the semantic disorder of patients with a reversed concreteness effect. The observed double dissociation between impairments of concrete and abstract words implies that there is more than just a quantitative difference in their representational structure. To account for the full set of patient data, it seems necessary to assume and specify qualitative rather than only quantitative representational differences between concrete and abstract words.

15-3. IMPAIRMENTS OF PERCEPTUAL VERSUS FUNCTIONAL INFORMATION

Further evidence for the importance of sensorimotor attributes in the representation of concrete nouns comes from patients with category-specific deficits. Elizabeth Warrington and her colleagues especially (for an overview, see McKenna & Warrington, 1993) have reported a number of single cases in which semantic knowledge of either objects of nature (e.g., animals, fruits, vegetables) or man-made artifacts (e.g., tools, furniture, kitchen utensils) was selectively impaired. So far, more cases have been reported with a loss of knowledge of living things than of artifacts. However, generally the number of cases is too limited to determine whether this distributional difference is meaningful.

Patients with category-specific deficits show, among other things, striking differences in the adequacy and specificity of their definitions for words from the two classes. One patient (J.B.R.), for instance, described a *briefcase* as a "a small case held by students to carry papers," but when asked to define *daffodil,* J.B.R. could only come up with "plant." Although it has been suggested that these category-specific deficits can be attributed to a lack of control over differences in the familiarity and visual aspects of the stimuli, it is unlikely that one can attribute all the reported cases to imperfect testing procedures (Saffran & Schwartz, 1994).

A theoretically important issue raised by the reports of these category-specific deficits is to determine the precise nature of the semantic dimension that cuts the semantic impairment pie into living and nonliving things. The best candidate seems to be the distinction between perceptual and functional attributes. Warrington and Shallice (1984) suggested that sensory attributes are very salient features for the identification of living things such as animals or fruits. In contrast, functional attributes are probably more important than perceptual characteristics for the identification of artifacts such as tools. Recent empirical evidence supports the claim that visual features are more salient in definitions of living things than of artifacts (Farah & McClelland, 1991).

In a further refinement of this account, Warrington and McCarthy (1987; McCarthy & Warrington, 1990b) have proposed that the contributions of sensory (perceptual) and motor (functional) channels are differentially weighted not only between but also within categories. For instance, within the category of artifacts, small manipulable tools are associated with a repertoire of skilled movements, and hence rely more heavily on motor channels than do large man-made objects such as airplanes. Airplanes are probably not too different from birds in their reliance on sensory channels for identification and categorization.

15-3.1. Evidence from Brain-Imaging Studies

Supportive evidence that there is a relation between conceptual knowledge and brain systems for perception and action comes from a PET study by Martin, Wiggs, Ungerleider, and Haxby (1996). In this study, subjects were asked to name pictures of animals and of tools. For pictures of both kinds, bilateral activation was obtained in ventral regions of the temporal lobes. The naming of animals resulted in additional activation in the left medial occipital lobe, an area involved in visual processing. In contrast, the naming of tools led to additional activation in the left premotor area and an area in the left middle temporal gyrus. These areas are close to cortical tissue that is active when using objects and perceiving motion. The authors conclude that the brain circuitry underlying the conceptual representation of objects includes regions that are particularly well suited for the processing of their most salient meaning aspects (perceptual, functional).

This evidence is largely compatible with an analysis of the lesion data of patients with disorders in the identification of living things versus man-made artifacts. On the basis of a review of the lesion data of the known cases, Gainotti, Silveri, Daniele, and Giustolisi (1995) conclude that the lesion distribution of these two patient types suggests a dominance of areas for visual object processing (living things) versus areas that are especially important for somatosensory and motor functions (man-made artifacts).

In conclusion, the neurological evidence seems to be compatible with the distinction between perceptual and functional attributes as an important metric for semantic categorization.

15-3.2. Implications for Lexical Semantics

Although the category-specific deficit data support the importance of a distinction between perceptual and functional attributes for concept retrieval, their full interpretation with respect to the processing of lexical meaning is dependent on one's theory of semantics. For instance, one influential theory of semantics makes a distinction between the core of a concept and nondefining features that are usually used for a quick identification of the concept (Armstrong, Gleitman, & Gleitman, 1983; Osherson & Smith, 1981). Neither the patient data nor the PET results allow one to determine whether it is the core aspects of meaning that are represented in the area of lesion or activation. The alternative option is that the identification procedures are localized in the visual and sensorimotor areas, whereas the core meaning aspects are represented in other brain areas. The lack of an explicit account of meaning in many neuropsychological studies of semantic impairments certainly adds to the confusion about the interpretation of the results. The same holds for recent brain-imaging studies in the area of semantics (Caramazza, 1996; Damasio, Grabowski, Tranel, Hichwa, & Damasio, 1996; Martin, Haxby, Lalonde, Wiggs, & Ungerleider, 1995; Martin et al., 1996; Ungerleider, 1995).

Jackendoff's account of conceptual-semantic structure might be of help here. Jackendoff (1987) argues that the representation of a word in long-term memory includes not only a description in a propositional format, but also an abstract visual-geometric description (a 3-D model, in Marr's terminology). This claim "reflects the intuition that knowing the meaning of a word that denotes a physical object involves in part knowing what such an object looks like" (p. 104). A similar idea is present in some instances of prototype theory in which word-meaning representations contain an image of a stereotypical instance. But the proposed 3-D model is a more abstract representation in which objects are spatially decomposed into parts and subparts in a viewer-independent orientation.

The neuropsychological data and the results from brain-imaging studies, however, indicate that Jackendoff's theory of lexical semantics needs further extension. Not only is a 3-D model part of word knowledge in semantic memory, but, for certain classes of words, the functionally relevant motor aspects might be represented as well. According to Jackendoff (1987), these functional aspects can be dealt with in the 3-D model too. However, the empirical data indicate that functional aspects might actually need to be specified in a structural description of another kind, more tailored to the properties of the motor system. Crucially, Jackendoff's account points toward ways of enriching the lexical-semantic representations of concrete nouns with specifications of their perception and action attributes in terms of their respective formats.

15-4. MULTIPLE SEMANTIC SYSTEMS?

An issue that has led to considerable controversy in the last decade is whether there is one amodal semantic system, or different, independent, and modality-specific semantic representations. The empirical evidence for this latter option consists largely of semantic impairments that only occur in one input modality (for relevant ERP data, see Ganis, Kutas, & Sereno, 1996). McCarthy and Warrington (1988) reported a patient (T.O.B.) with a selective impairment for animals. However, this impairment was only observed when animal names were presented verbally, but not when presented as pictures. For example, T.O.B. gave as definition for the spoken word *rhinoceros:* "Animal, can't give you any further function." However, when shown a picture of a rhinoceros, he gave a much more specific description: "Enormous, weighs over one ton, lives in Africa." Warrington and McCarthy (1994) have reported a patient with a category-specific deficit for common objects such as knives, cups, and so forth, but only when visually presented. On the basis of these case reports, Warrington and McCarthy (1994; see also Shallice, 1993) hypothesize that there are modality-specific meaning systems (e.g., visual semantics and verbal semantics). On the basis of the occurrence of similar category-specific deficits in the verbal and the visual semantic systems they claim that the organizational principles (e.g., sensory versus functional core aspects) are the same in these two modality-specific semantic systems.

The notion of multiple modality-specific semantic systems has been strongly criticized (e.g., Caplan, 1992; Caramazza, Hillis, Rapp, & Romani, 1990; Rapp, Hillis, & Caramazza, 1993; Saffran & Schwartz, 1994). A major problem is that no specification of visual semantics is given. Is it something like Marr's 3-D model? Jackendoff (1987) has convincingly shown that a 3-D model alone is insufficient as a conceptual representation of objects. In the absence of an articulated account of visual semantics, it is unclear what the notion of visual semantics will do for us. In addition, the data clearly do not exclude the possibility that it is modality-specific access (or identification) procedures that are impaired rather than modality-specific conceptual knowledge itself.

Since this chapter is about lexical meaning, the focus is on semantic representations that are tailored to the verbal system. Therefore, it is sufficient to conclude that, intriguing as these modality-specific semantic impairments are, the implications for representational and processing accounts of lexical meaning are far from clear.

15-5. IMPAIRED ACCESS VERSUS LOSS OF SEMANTIC KNOWLEDGE

So far in my discussion of the different semantic impairments there has often been the implicit assumption that these impairments impact the lexical-semantic representations, whatever their characterization. However, this is not the only possibility. It might well be that the brain damage affects the access or retrieval operations rather than the representational structures themselves. Therefore, it is important to determine whether an observed semantic deficit is due to the loss or degradation of word-meaning representations (a storage deficit), or to a failure in the routine procedures called upon to access and exploit these representations in real time (an access deficit).

It is not easy to formulate criteria that allow an unambiguous distinction between these two general options (for an in-depth treatment, see Shallice, 1988). One reason for this is that the nature of neither lexical-semantic representations nor access operations is sufficiently specified (cf. Rapp & Caramazza, 1993). Nevertheless, a number of criteria have been defined to distinguish between impaired access and loss of semantic knowledge (e.g., Shallice, 1988; Warrington & Shallice, 1979; for criticism of these criteria, see Caplan, 1992; Rapp & Caramazza, 1993).

15-5.1. Consistency in Performance

One important criterion for distinguishing between impaired access and loss of semantic knowledge is item consistency in performance. In the case of degraded semantic representations, it is predicted that the inability to identify or name an item should be consistent over time and across tests. In contrast, an access impairment is associated with performance variability across time and tests for the same item. For patients with dementia of the Alzheimer type (DAT), consistency in performance has been reported in a number of cases. The same names they failed to produce in a picture-naming task were the ones they failed to recognize in a word-to-picture-matching test (Chertkow & Bub, 1990; Huff, Corkin, & Growdon, 1986). In addition, Chertkow and Bub (1990) reported that on a retest, a group of 10 DAT patients failed to name 92.5% of the items that they were unable to name during an earlier testing session. This consistency in performance across tests and over time is in marked contrast with the absence of consistency across a range of semantic tests in a group of aphasic patients with impaired language comprehension (Butterworth, Howard, & McLoughlin, 1984).

Consistency measures suggest that in Alzheimer patients we see a real breakdown of semantic representations, whereas in aphasic patients an access impairment is more likely than a semantic breakdown. However, since only a limited number of studies have explicitly tested item consistency, these effects need replication. In addition, more stringent tests are required to unambiguously determine the locus of the knowledge degradation. For instance, the inability to name a picture might be due to a loss of the word concept, but it could also arise from an impairment at the level of word-form representations.

15-5.2. Semantic Priming

In the last decade or so, most studies that have addressed the issue of impaired access versus degraded representation have used the semantic priming technique. Since the early seventies (Meyer & Schvaneveldt, 1971), it has been a well-known and robust phenomenon in psycholinguistics that the processing of a word benefits from the presence of a preceding word to which it is related in meaning. For instance, subjects name the word *dog* faster when they have just seen (or heard) the word *cat* than when they have seen (or heard) the word *bar*. This processing advantage of the primed word *(dog)* is often attributed to the increase in activation that its concept node inherits from the related prime *(cat)*.

In the influential model of Collins and Loftus (1975; see Figure 2), lexical meanings are thought to be organized as a network of concept nodes, with the network's wiring diagram determined by semantic similarity: there are stronger and more direct connections for semantically related nodes than for unrelated nodes. Upon reading or

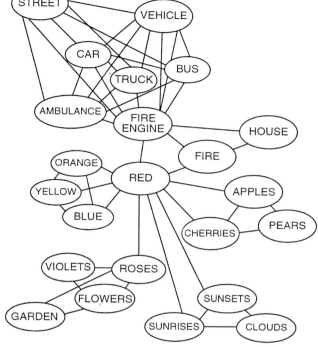

FIGURE 2   A nondecompositional model of word-meaning representations (after Collins & Loftus, 1975). Each node represents one concept. The wiring of the network is determined by semantic similarity. The links between the nodes represent the meaning relations between the concepts. The strongest semantic relations exist between concepts with short and direct links. Activation can spread from one concept to another via the links. The more direct a link between nodes, the more activation is received from a node related in meaning.

hearing the prime, its concept node gets activated and spreads parts of its activation to nearby nodes in the network. In this way, related words increase their levels of activation and require less processing time when these words are subsequently read or heard. This effect is very short-lasting, probably not longer than about 400 msec (Neely, 1977).

In recent years, an increasing number of neurolinguistic priming studies have appeared in the literature. In these studies, either aphasic patients with impaired comprehension or DAT patients were tested. These patients often show marked impairments when they are required to perform semantic judgments or other forms of explicit semantic evaluation. In marked contrast to their performance in explicit semantic tasks, a substantial number of studies report priming effects in such patients. In general, these priming effects have been taken as evidence that in many aphasic patients with semantic impairments, lexical-semantic representations are largely unaffected, but that they have a problem in accessing these representations (see Hagoort, 1993, for an overview). For DAT patients the results of priming studies are less clear-cut, but the overall conclusion has been the same as for the aphasic patients (see Chenery, 1996, for an overview). However, for DAT patients the interpretation of the priming results is in contrast to conclusions based on their consistency scores (see Section 15-5.3).

Overall, the results of priming studies with aphasic and DAT patients suggest that an access impairment underlies their semantic deficits. However, some caution is warranted with respect to the conclusions of these priming studies, since significant priming effects as such are not necessarily clear evidence for an access deficit (cf. Moss & Tyler, 1995). I will discuss two concerns about the interpretation of priming effects.

The first concern is related to the locus of the observed priming effects. Most patient studies used words that were related either associatively (e.g., *bread–butter*) or both associatively and semantically (e.g., *dog–cat*). These associations might be due to frequencies of co-occurrence in the language input, so it cannot be excluded that links between associatively related words exist at the level of word forms. In the absence of firm evidence that the priming effects actually arise at the concept level, significant priming effects in patients are insufficient evidence for the integrity of lexical-semantic (concept) representations. In a few studies, priming in purely semantically related word pairs (e.g., *church–villa*) was compared to priming in associatively related pairs (Hagoort, Brown, & Swaab, 1996; Ostrin & Tyler, 1993). The comforting outcome of these studies is that the aphasic patients showed the same pattern of results for both types of relations.

A second concern is related to the priming mechanisms involved. Priming effects can be caused by quite different underlying processes (see for an overview, Neely, 1991). The automatic spreading of activation between concept nodes in semantic memory is one of the contributing mechanisms. But there are others, such as a postlexical matching of the prime and target for semantic overlap. This latter mechanism can easily explain priming under conditions of degraded representations. As has been observed by Warrington and others, in the absence of detailed semantic knowledge about a particular concept, patients still are often able to produce the category name (e.g., *animal*). It is exactly this kind of generic information that might be left intact in the case of a degraded semantic representation. Clearly, for pairs such as *dog–cat* a postlexical matching would still detect the category overlap, even in the absence of a more specific lexical-semantic structure. This illustrates the importance of testing priming under conditions that mainly or exclusively tap the spreading of activation through the lexical-semantic network itself (Hagoort, 1989, 1993). These conditions are realized by having very short time intervals between primes and targets, by presenting only relatively few related word pairs, and by avoiding as much as possible task components over and above the natural ones of listening or reading (Hagoort et al., 1996; Hagoort & Kutas, 1995). Priming that is obtained under these circumstances can be relatively safely interpreted as indicating intact lexical-semantic knowledge.

Priming studies can substantially contribute to our understanding of the nature of the semantic impairment. One major advantage of the priming technique is the implicit nature of the task. What subjects are required to do (e.g., lexical decision, or, in the case of recording event-related brain potentials [ERP], just listening or reading) is in no way related to the actual question at stake, for example, whether or not patients are sensitive to the semantic relations between primes and targets. Sensitivity to or lack of sensitivity to these relations cannot easily be explained by reference to specific aspects of the task. However, priming studies will only contribute further to our understanding of the nature of semantic deficits if an attempt is made to tap the full richness of lexical-semantic structure by testing a whole range of semantic relations (e.g., antonyms, category-member, co-hyponyms, and so forth) for concrete as well as abstract words.

15-5.3. Consistency versus Priming

The complexities of determining whether the impairment is due to an access problem or to degraded representations become especially clear in the case of Alzheimer patients. On the basis of item consistency in their performance, it has been claimed that these patients suffer from a loss of semantic knowledge. However, the significant priming effects in most studies with DAT patients are taken as evidence for an access impairment (Chenery, 1996). Possible reasons for this inconsistency of results are that most priming studies with DAT patients have used only associatively related word pairs, that the presentation conditions of some studies have been insufficient to guarantee the contribution of automatic spreading of activation to the priming effects, and that still other studies have found excessively large priming effects (hyperpriming). Hyperpriming is indicative of a representational deficit under the assumption that degraded representations benefit more from spreading activation provided by the prime than unaffected semantic representations (Chertkow, Bub, & Seidenberg, 1989). However, Chenery (1996) has argued in her review that even when these factors are taken into consideration, the weight of evidence still favors an account in terms of an access impairment. Additional, finer-grained studies are needed to resolve the remaining inconsistencies and to determine the functional locus of semantic impairments in DAT patients.

15-6. CONCLUDING REMARKS

The literature on semantic impairments indicates that theories of lexical-semantic representations of nouns should be able to account for the qualitative differences in the semantics of abstract and concrete words, and for the perceptual versus functional attributes of different categories of concrete nouns.

To date, most accounts of semantic impairments suffer from vagueness about the presupposed nature of lexical-semantic representations and lexical-semantic processing. Warrington and Cipolotti (1996) define semantic memory as "a system which processes, stores and retrieves information about the meaning of words, objects, facts and concepts" (p. 611). However, nothing is said about the nature of the semantic representations for words, objects, concepts, and facts. What are the differences and commonalties between the semantic representations of these seemingly different memory items? These are issues that need to be addressed as well. It is, however, an unfortunate aspect of the semantic memory tradition that the fractionation of semantic memory into different components has received more attention than the representational structure of its content. Unless more explicit accounts of (lexical) semantics are given, it will remain difficult to decide whether patient data support explanations of semantic impairments in terms of multiple versus central semantic systems, in terms of access versus storage deficits, and so forth.

With respect to impairments in the processing of lexical meaning, we also need a better understanding of what exactly our task configurations tap. One thing that has become clear in recent years from brain-imaging studies on language is that seemingly subtle task differences have overriding consequences on the patterns of brain activation obtained (e.g., Price et al., 1994). This should make us realize that for answering questions about the underlying nature of semantic impairments, we need to be more explicit about the requirements for performing the task at hand over and above accessing lexical meaning. In essence, we often ask patients to solve a problem imposed by the experimenter, such as to match a word to a picture, to give a verbal definition on the basis of a picture or word input, or to pantomime a concept. Too often the peculiarities of these task requirements are not taken into consideration enough in the interpretation of the results.

Finally, I have suggested that a revised version of Jackendoff's theory of lexical semantics (Jackendoff, 1983, 1987, 1992) might be able to account for most of the empirical data. According to Jackendoff, word meanings are decomposed into a restricted set of primitive conceptual features, paired with an abstract visual description (a 3-D model). I have proposed extending this account with other matched pairs of conceptual structure and nonvisual sensory models, and models of action specified in a format that is tailored to the requirements of the motor system.

In my view, the major challenge of the coming years will be to close the gap between the intriguing findings in studies of different types of semantic impairment, the more recent brain-imaging data on semantic processing, psycholinguistic data on real-time processing of word meaning, and theories of word meaning. Only then will we begin to see the contours of a neurosemantic theory.

Acknowledgments

I thank the following friends and colleagues for their comments on an earlier version of this chapter: Melissa Bowerman, Colin Brown, Bart Geurts, Stef van Halen, Pim Levelt, Theo Meyering, Ardi Roelofs, Brigitte Stemmer, Harry Whitaker.

CHAPTER 16

Disorders of Syntax in Aphasia
Linguistic-Descriptive and Processing Approaches

Herman Kolk

Nijmegen Institute for Cognition and Information, University of Nijmegen. Nijmegen, The Netherlands

Current research in the field of syntactic disorders takes one of the following *two theoretical approaches*. The first is to look for a theoretically based linguistic description that captures the empirical distinction between "affected" and "non-affected" linguistic behaviors. The group of agrammatic patients is typically the target of this approach. The second approach is to analyze the disorder of agrammatic and/or nonagrammatic patients in terms of processing limitations. Evidence obtained within each approach is reviewed separately for comprehension and production disorders. Discussion centers around linguistic concepts such as traces, case, and functional categories, as well as processing notions such as capacity, syntactic work space, and processing rate. In addition, the relevance of a number of dissociations is discussed. In conclusion, it is pointed out that the two approaches could merge into an integrated account of syntactic disorders in aphasia.

Disorders of syntax in aphasia have most often been studied in Broca's aphasics and the term "agrammatism" has been used to refer to the difficulties in sentence processing that these patients exhibit. It is therefore natural to take this syndrome as a starting point for the discussion. However, grammatical problems are apparent in fluent aphasics as well, and these will also enter our discussion. Usage of the term "agrammatic" does not imply that all patients referred to as such have a common deficit. In particular, as the result of Badecker and Caramazza's (1985) criticism of the concept of agrammatism, researchers have become more aware of the possibility of multiple deficits. This has resulted in an increasing number of individual case studies that suggested dissociations within the sentence processing system. A number of these dissociations will be discussed in this chapter. Perhaps the most convincing one is that between comprehension and production: several patients have been described who are agrammatic in production but unimpaired in sentence comprehension (Miceli, Mazzucchi, Menn, & Goodglass, 1983; Kolk, van Grunsven, & Keyser, 1985; Nespoulous et al., 1988; Caramazza & Hillis, 1989b; Druks & Marshall, 1991). In view of this dissociation, I will treat disorders in each domain separately.

The chapter is organized as follows: I begin with disorders of sentence comprehension. Various linguistic approaches are reviewed, in particular the trace-deletion hypothesis and its reformulations, as well as a proposal that locates the deficit in the Case module. Then, a section follows on processing approaches to comprehension deficits, such as the mapping hypothesis and various proposals that assume a limited capacity for syntactic processing. The next section is on production. It begins with a brief summary of the main production parameters: morphology, sentence construction, and speech rate. Next, a number of linguistic and processing approaches are discussed, focusing on how they deal with the three production parameters.

16-1. SENTENCE COMPREHENSION: THE LINGUISTIC-DESCRIPTIVE APPROACH

16-1.1. The Trace-Deletion Hypothesis and Its Reformulations

It has long been established that agrammatic patients have difficulties with sentence comprehension. These difficulties are apparent primarily if sentences are "reversible" (e.g., Caramazza & Zurif, 1976). A sentence is reversible if, after major noun phrases have exchanged position, the sentence still makes sense (e.g., "the dog is chewing the bone" is not reversible, but "the dog is chasing the cat" is). Not all reversible sentences are equally difficult, however. It is well known that active sentences are much easier than their passive counterparts (e.g., Schwartz, Saffran & Matin, 1980). To account for this within-category variation, Grodzinsky (1986, 1989, 1990) proposed that not the whole syntactic component is implicated (e.g., Caramazza & Zurif, 1976), but only particular representational elements, the so-called *traces*. These traces result from movement transformations, according to the theory of Government and Binding—GB theory for short (Chomsky, 1981). For instance, in a passive sentence like "John was kissed by his mother," the D-structure has the NP "John" represented after the verb "kissed." In S-structure, "John" (also called an "argument" of the verb) is moved from the postverbal to the subject position. However, a trace of the moved argument is left behind and this trace carries an index. The presence of traces becomes critical at a later stage, when "roles" are assigned ("who does what to whom"). In the case of the passive, the assignment algorithm expects the NP playing the "theme" role (the entity affected by the action) to be at its D-structure position, that is, right after the verb, and does not find it. However, it does find a trace with an index. The subject NP (John) is found to carry the same index, and therefore the theme role can be assigned to the proper NP.

Grodzinsky postulates that these traces are absent from the agrammatic representations (hence, "trace-deletion hypothesis," or TDH). As an immediate consequence, normal assignment of moved arguments is no longer possible. However, assignment of *nonmoved* arguments is unimpaired. Thus, in the case of the passive, the agent role can be assigned by parsing the by-phrase. One would expect that with verbs like "kiss," which take two arguments, the theme role would be assigned to the remaining NP, "the boy." Such an assignment is overruled, however, by the obligatory application of a word-order heuristic. According to the "default principle," whenever the first NP of a clause has not received a thematic role as a consequence of trace deletion, it is automatically assigned the agent—or some other appropriate—role. But the agent role is already assigned to the NP in the by-phrase. Hence a conflict arises, which the patient is assumed to resolve by guessing. This guessing results in chance-level performance, which Grodzinsky claims to be present with all sentences having moved arguments and noncanonical word order. Thus chance-level performance would be observed with passives, object relatives, and object clefts (e.g., "it was the boy that the mother kissed"). Above-chance-level performance would be present with sentences having moved arguments but canonical word order, such as subject relatives (e.g., "the boy that kissed the girl was tall") and subject clefts (e.g., "it was the girl that kissed the boy").

Grodzinsky's work has stimulated others to look at agrammatic comprehension from the point of view of theoretical linguistics (cf. Hickok & Avrutin, 1996; Frazier & Macnamara, 1995; Saddy, 1995). Here we will concentrate on the research directly related to the TDH. One relevant criticism that has been leveled against this hypothesis concerns the default principle. Since the heuristic is not based on grammatical principles, why would the agrammatics stick to the agent-first heuristic, even if it is in conflict with information obtained from role assignment to unmoved arguments? This was one important reason for the development of two reformulations of the TDH, one by Hickok (Hickok, 1992; Hickok, Zurif, & Canseco-Gonzales, 1993) and the other by Mauner, Fromkin, and Cornell (1993). There are a number of differences between these two proposals with respect to linguistic representation, but both replace the default principle with strategy, by which the patients choose between the possibilities that remain after direct assignment has been made.

Another major criticism concerns the empirical evidence for the TDH. First of all, comprehension is not as predicted in the case of *(a)* passives without a by-phrase (Martin, Frederick-Wetzel, Blossom-Stack, & Feher, 1989), *(b)* Hebrew passives with postverbal subjects (Druks & Marshall, 1991), and *(c)* the matrix clause of sentences with a center-embedded relative clause (Hickok et al., 1993; Kolk, & Weijts, 1996).

Second, passives elicit chance-level performance only in a subset (roughly one-third) of agrammatic patients, as has been shown by Berndt, Mitchum, and Haendiges (1996) in a meta-analysis of 64 patients, described in the literature.

16-1.2. The Case-Deficit Hypothesis

Of particular interest is a subset of patients in the Berndt et al. study (1996), the ones that performed better on passives than on actives. Druks and Marshall (1995) reported on one other case of this kind, B.M., who performed at chance on actives and above chance on passives, on three different kinds of comprehension tests. Another patient, M.H., showed the reverse (TDH) pattern. To account for both performances within one linguistic framework, Druks and Marshall propose an alternative to the TDH. It holds that the Case module, as defined in GB theory, is impaired in agrammatic patients. GB theory distinguishes between two types of Case. First, there is *structural* Case, which is assigned at the level of S-structure. Examples are nominative and accusative Case. In English, these Cases are morphologically expressed only occasionally (e.g. "he" and "him"). Second, there is *inherent* Case, a lexical phenomenon, assigned at D-structure by prepositions. Inherent Case is not expressed in English, but it is in other languages (e.g., German: "in *dem* Glass"—in the [dative] glass). In the passive by-phrase, "by" assigns inherent case to the embedded NP. Without appropriate Case assignment, thematic role assigment is disrupted and comprehension is impaired. Druks and Marshall propose that in patients such as M.H., with chance performance on passives and above chance on actives, inherent Case is impaired; when the reverse pattern is observed, as in B.M., the impairment would be in structural case assignment. A difficulty with this account is that B.M. is significantly *below* chance on passives in two out of three tests. Therefore, Druks and Marshall are forced to invoke an agent-first heuristic. They claim that B.M. follows this strategy because he would be impaired in both types of case. In M.H., on the other hand, who obviously does not employ the strategy (he is at chance in actives), the use of the strategy is blocked because M.H. has only one type of impairment. It is important to realize, however, that, just as with Grodzinsky, the use or nonuse of the strategy is not linguistically motivated. Furthermore, it is not psychologically motivated either. In particular, why would a patient with "only" a structural Case impairment not follow the strategy if this is the only way he can make sense out of an active sentence? The notion of a word-order heuristic appears to be a stumbling block for this approach, as it is for Grodzinsky's (cf. Zurif, 1996; Druks & Marshall, 1996).

16-2. SENTENCE COMPREHENSION: THE PROCESSING APPROACH

16-2.1. Mapping Deficit or Limited Capacity?

Grammatical difficulties exhibited by agrammatic patients not only vary with sentence type but also depend on the type of task. One important type of task variation was discovered by Linebarger, Schwartz, and Saffran (1983). These authors demonstrated that agrammatics who perform at chance on sentence–picture-matching tasks with reversible passives can obtain very high scores on a task in which they are asked to indicate whether a sentence is grammatical or not, not only with passive sentences but with other complex sentences as well (Linebarger, Schwartz, & Saffran, 1983; see also Linebarger, 1989, 1990, 1995). This result is difficult to explain for Grodzinsky's and Hickok's hypotheses, although not for the one proposed by Mauner et al., discussed in the previous section. According to Linebarger et al., (1983), there are two ways to account for this dissociation between comprehension and judgment. The first is that the agrammatic problem is not in parsing but in mapping the syntactic representation onto a semantic one. The second is a limited-capacity hypothesis, which holds that comprehension, requiring both parsing and mapping, overloads the processing capacity of the agrammatic patient; for judgment, requiring only parsing, there is enough capacity. Schwartz, Linebarger, Saffran, and Pate (1987) rejected the limited-capacity hypothesis because, in their group of agrammatic patients, they only found an effect of canonicity and not an effect of phrase-structure complexity. Kolk and Weijts (1996), however, maintained that this rejection was premature, because with another operationalization of complexity (embedding instead of conjoining of sentences) and a control of strategy effects, they did observe a complexity effect, which was as large as the canonicity effect.

One advantage of the limited-capacity notion is that, since the limitation can vary in degree, variation in severity can be accounted for (Caplan, Baker, & Dehaut, 1985; Kolk & van Grunsven, 1985). This feature contrasts sharply with the linguistic approach, which applies only to patients with chance-level performance on critical sentence types.

Quite a number of recent proposals can be considered as variants of the capacity approach. The one that corresponds most closely to the Linebarger et al. (1983) proposal is the Just and Carpenter capacity theory of language comprehension (cf. Just & Carpenter, 1992; Miyake, Carpenter, & Just, 1994; see Haarmann, Just, & Carpenter, in press, for a computer model). Miyake et al. (1994) presented normal subjects with sentences under heavy time constraint (120 or 180 ms per word). They found that the rank order of difficulty over sentence types, varying from simple to complex, was very similar to the one found by Caplan et al. (1985) for an unselected group of aphasics. Furthermore, there was a large effect of "severity": subjects who scored relatively low on a language capacity test (the so-called reading span test) made many more errors than subjects who scored relatively high. Miyake et al. (1994) conclude that there is a single limitation in working memory capacity that can vary in severity and that underlies not only variation in comprehension ability within the normal population but also aphasic sentence comprehension; perhaps the comprehension difficulties of older people can be accounted for in this way as well (Carpenter, Miyake, & Just, 1994).

Of special interest is Miyake et al.'s (1994) finding that different subgroups showed different performance profiles: a particular sentence type could be relatively easy for one subgroup but relatively hard for another, thus creating a "double dissociation" within unimpaired normal adults. Some of these dissociations dissappeared after repeated testing, whereas others remained stable over time. These findings present a challenge to the way in which behavioral dissociations in neurolinguistic patients are routinely ascribed to specific deficits at the level of particular components of the language system. As Miyake et al. suggest, a particular neurolinguistic dissociation could also be due to stochastic noise (Haarmann & Kolk, 1991a; McClelland, 1993)—unless, of course, it is demonstrated to be stable over time, which is not always done—or to consistent idiosyncratic comprehension strategies.

16-2.2. Capacity Limitation: General or Specific?

The Just and Carpenter capacity theory has been called into question by Caplan and his colleagues. They tested sentence comprehension in Alzheimer patients, who presumably also suffer from such a language capacity reduction. It was found that performance in these patients was poorer only for sentences that had two propositions (described two events) as compared to one. There was no effect of canonicity; that is, in contrast with what is generally found with aphasics, sentences with noncanonical word order (e.g., passives or cleft objects such as "it was the dog that the horse passed") were not significantly harder than sentences with canonical word order but that included the same number of words, propositions, verbs, and thematic roles (Rochon, Waters, & Caplan, 1994). Even when a dual task (digit recall) is added—which, according to Just & Carpenter (1992), should reduce the amount of capacity available for parsing—a canonicity effect does not arise (Waters, Caplan, & Rochon, 1995). This dissociation between canonicity and number of propositions suggests that we must distinguish between (at least) two kinds of language-related capacity: a purely syntactic one, and one that deals with verbal reasoning. A similar picture arises from a study by Martin and Romani (1994). They looked at the effect of adding extra material to sentences on performance in two tasks: a sentence–picture matching and a grammaticality judgment task. A.B., an aphasic patient, showed an effect only on matching; another aphasic patient, M.W., did so only on judgment. This dissociation again supports the existence of separate capacities for syntactic and semantic analysis. (For more discussion on the Just and Carpenter theory, see Caplan & Waters, 1995; Martin, 1995; Miyake et al., 1995; Waters & Caplan, 1996.)

In line with these latter findings, a number of researchers have claimed that the capacity limitation in aphasic patients is restricted to parsing (cf. Kolk & van Grunsven, 1985; Caplan & Hildebrandt, 1988; Shankweiler, Crain, Gorell, & Tuller, 1989; Friederici & Frazier, 1992). The parsing limitation is thought to apply to agrammatics only (e.g., Friederici & Frazier, 1992; Zurif, Swinney, Prather, Solomon, & Bushel, 1993), to agrammatic as well as paragrammatic (Wernicke) patients (Kolk & Friederici, 1985), or to all aphasics with comprehension disorders (Caplan & Hildebrandt, 1988). Parsing, of course, is still a multifaceted phenomenon, and it is conceivable that capacity limitation is even more selective. Zurif et al. (1993), for instance, assume that the capacity limitation underlying the parsing difficulties of agrammatics is basically a lexical one. Or, to take another example, normal parsing requires rapid access to the thematic properties of verbs, but the capacity to do so does not seem to be reduced in agrammatics (Shapiro & Levine, 1990; Shapiro, Gordon, Hack, & Killackey, 1993).

Specific-capacity approaches vary with respect to the nature of the limitation: spatial or temporal. Caplan & Hildebrandt (1988) put forward an account of the first type—"the parsing work space" hypothesis: aphasics would be limited in the size of their syntactic buffer. Temporal hypotheses generally come in two kinds. It is assumed either that activation of grammatical information is slowed down (Gigley, 1983; Kolk & van Grunsven, 1985; Friederici, 1988; Friederici & Kilborn, 1989; Haarmann & Kolk, 1991a, 1991b; Hagoort, 1997; Cornell, 1995; Swaab, Brown, & Hagoort, 1997; Swinney, & Zurif, 1995) or that the parsing results are subject to pathologically fast decay (Gigley, 1983; Kolk & van Grunsven, 1985; Haarmann & Kolk, 1994). A similar distinction between slow activation and fast decay has been made by Schwartz, Dell, Martin, and Saffran (1994) in a computer simulation study of aphasic word production.

What the proper characterization of the nature and domain of the processing limitation is is still a matter of debate (see Kolk & Weijts, 1996, for an overview). For one thing, not all hypotheses account equally well for the judgment data. Second, the effect of canonicity may be due to extra computational load, along with the effect of phrase-structure complexity, or to the effect of word-order strategy, since such a strategy facilitates processing of sentences with canonical word order at the expense of sentences with noncanonical word order.

16-3. SENTENCE PRODUCTION: THREE PERFORMANCE PARAMETERS

Discussions about agrammatic disorders of sentence production center around three manifestations of this deficit. The best known is the morphological aspect: agrammatic patients tend to omit or substitute grammatical morphology. Grammatical morphology refers to both free (function words) and bound morphemes (inflections). It is also referred to as "closed-class" morphology, because its number is relatively fixed in a given language. The closed-class vocabulary is not impaired as a whole, however: there is considerable within-category variation (DeVilliers, 1974; Menn & Obler, 1990; Haarmann & Kolk, 1992). (I will come back to this within-category variation later.) The production of derivational morphology (-ion in destruction) need not be impaired (Micelli & Caramazza, 1988). The "open-class" vocabulary (content words) is relatively spared, but it has often been noted that Broca's aphasics have more difficulty with verbs than with nouns in naming tasks; several anomics show the reverse problem (McCarthy & Warrington, 1985; Miceli, Silveri, Villa, & Caramazza, 1984; Zingeser & Berndt, 1990). It is important to realize, however, that fluent aphasics are also reported to have more difficulties with verbs than with nouns (Berndt 1991; Williams, & Canter, 1987). For more detailed analysis of verb impairment in aphasia, the reader is referred to Kegl (1995) and Breedin and Martin (1996).

The other aspect is *constructional;* it corresponds with what Goodglass and Kaplan (1983) call "reduced variety of grammatical form." In Broca's aphasics, this variety is very limited and we typically observe declarative sentences with canonical word order, no embedding or phrasal elaboration, and little tendency to add adjuncts in the form of prepositional phrases, adverbs, or nonfinite verb complements (cf. Goodglass, Christiansen, & Gallagher, 1994).

There is a third aspect that is commonly associated with agrammatism in production: slow rate of speech (nonfluency). Saffran, Berndt, and Schwartz (1989), for instance, observed a speech rate of about 30 words per minute in their agrammatic group, versus 130 in their controls. Although agrammatics typically have articulatory impairment, speech rate can also be reduced in patients without such impairment. This suggests a relation between speech rate and sentence production impairment.

Although these three aspects typically go together in agrammatic speakers, dissociations have been reported. Most critical are the two cases that are sometimes claimed to demonstrate normal syntax and impaired morphology (Miceli, Mazzuchi, Menn, & Goodglass, 1983; Kolk et al., 1985). The speech of both patients, however, was characterized by a high number of finiteness omissions: clauses that either lacked a verb (e.g., "because a big pain in the chest") or had only a nonfinite verb (infinitive or past participle; e.g., "But not feel [inf.] well, always have [inf.] these pains here"). For such clauses, it is hard to argue that they are syntactically "normal": in the absence of finiteness, there is no evidence that a sentence has been produced. A convincing case of normal syntax and seriously impaired morphology has not yet been observed.

16-4. AGRAMMATIC PRODUCTION: THE LINGUISTIC-DESCRIPTIVE APPROACH

I will start with the morphological aspect. Until recently, linguistic approaches to agrammatic production treated the closed-class vocabulary as a whole (e.g., Kean, 1977; Grodzinsky, 1984, 1986). Hagiwara (1994, 1995), however, has made a proposal that does accommodate to within-category variation. Her account follows recent developments in generative linguistic theory (Pollock, 1989; Chomsky, 1993). It extends to agrammatic comprehension, but I will focus on production. The basic claim is that agrammatic production is caused by an impairment of *functional categories.* Functional categories refer to those nodes in the phrase-structure hierarchy that dominate grammatical morphology. The AGR-node, for example, dominates agreement morphology of the verb; the C-node dominates complementizers such as *that, if, whether* and *for,* which introduce an embedded clause. Each node has its position at a particular level in the hierarchy. C-nodes, for instance, are located higher in the tree than AGR-nodes. Every patient can be characterized as having a deficit with respect to a particular level in the hierarchy. Grammatical morphemes dependent on nodes lower than this level are spared, whereas the ones dependent on higher nodes are affected. This explains the existence of the within-category variation referred to earlier (though not necessarily its specific form). Hagiwara, for instance, showed that her Japanese agrammatics did not omit negation (dependent on a relatively low NEG-node) but did omit complementizers (dependent on the much higher C-node).

Of particular interest is the connection Hagiwara assumes between this linguistic description and processing capacity. Structures dominated by lower nodes would require a fundamental structure-building operation called "Merge" (Chomsky, 1994) to be carried out fewer times. These structures are therefore more economical and more accessible to the agrammatic. Agrammatics suffer from a reduced linguistic capacity to carry out this operation, and the amount of reduction determines at what level in the hierarchy functional categories become inaccessible. Hagiwara thus has an account of severity variation, which, as we saw earlier, was absent from linguistic-descriptive approaches to agrammatic comprehension.

A related proposal has been put forward by Friedmann and Grodzinsky (1994, 1997). They report on a Hebrew-speaking patient who, in sentence repetition and sentence completion tasks, makes almost no errors with subject–verb agreement but makes many errors with tense (a mismatch between a temporal adverb and the verbal inflection). Friedmann and Grodzinsky account for this dissociation by assuming that in the syntactic tree, AGR-nodes are located lower than T(ense)-nodes, which dominate tense-related morphology, and that their patient has a deficit at the level of T. Like Hagiwara, they assume that different patients can have impairments at different levels in the hierarchy, and that the higher the level, the greater the number of grammatical elements that are affected. They do not conceive of this severity variation as being a variation in the amount of available capacity. Instead, they assume that if a patient has an impairment with respect to a particular node, higher levels are also affected because, in the process of verb movement, the verb cannot cross over the impaired node.

Besides the morphological aspect, we have the constructional and the fluency aspect to deal with. With respect to the latter, I can be brief: nonfluency has received little or no attention from theoretical linguists. Regarding the constructional difficulties, the functional-category hypothesis suggests that complementizers, depending on (high-level) C-nodes, will be impaired, with the consequence that sentence embedding is prohibited, at least in languages that require such complementizers (Friedmann & Grodzinsky, 1997). Finally, Hagiwara's hypothesis predicts that phrasal elaboration will also be limited. It is known, for instance, that agrammatics have difficulties with double-adjective constructions (e.g., "a large white house"; cf. Gleason, Goodglass, Ackerman, Green & Hyde, 1975), with noun phrases that include a prepositional phrase, and with compound nouns (Luzatti and De Bleser, 1996). Complex noun phrases like this would require more capacity than is available to the agrammatic. In this way, at least some aspects of syntactic simplification could be accounted for by the same principles invoked to explain the morphological difficulties.

16-5. AGRAMMATIC PRODUCTION:
THE PROCESSING APPROACH

My exposition of the processing approach begins with the morphological difficulties. We saw an example of such an approach earlier: Hagiwara's hypothesis that particular syntactic nodes are less accessible to the agrammatic. In the same vein, both Stemberger (1984) and Lapointe (1985) have proposed a reduced availability of syntactic structures to underlie agrammatic production. However, neither presents an explicit account for within-category variation. In the processing theory worked out by Kolk and his colleagues, on the other hand, an attempt is made to explain this variation as a side effect of the reduced access/availability of syntactic structures.

In this theory, changes in internal syntactic processing rate are supposed to underlie production difficulties, in the same way as this was postulated for comprehension (see earlier). These temporal changes are thought to have two effects. First, slow activation and/or fast decay of phrase structure information would cause desynchronization of parts of the syntactic tree, just as in the Haarmann and Kolk (1991a) model of agrammatic comprehension (Kolk, 1987). Synchrony would be more difficult to achieve, the greater the complexity of the phrase structure to be produced. A second effect occurs in, for instance, simple structures, in which synchrony is obtained in spite of the slow activation. In such situations, phrase-structure delivery is still slower than in the normal speaker. Now, correct production of grammatical morphology requires another kind of synchrony: that between a syntactic slot (e.g., the DET slot) and the lexical item (e.g., "the"). If slot delivery is delayed, the lexical element is already decaying and is suffering from response competition with other lexical items. If a morpheme is selected at this stage, there is a high chance that it will be an erroneous one.

Within this model, variation within the closed-class vocabulary is explained as follows. Slots that are dependent on a relatively complex part of the syntactic tree will be delivered later. For the production of a plural inflection, for instance, the minimally required structure is an NP, whereas for a verb inflection, an S-node is required: accordingly, plural noun inflection is easier than verb inflection (Goodglass & Berko, 1960; Haarmann and Kolk, 1992). Kolk (1995) indicates how other within-category differences, observed by Haarmann & Kolk (1992), can be explained as effects of syntactic complexity.

Besides morphological difficulties, there is the symptom of syntactic simplification capacity approaches have to deal with. According to Kolk (1995), the higher the degree of complexity, the higher the chance of desynchronization within the syntactic tree. As a consequence, the chance that an agrammatic patient will successfully produce a complex phrase or sentence is relatively low. Thus, in this theory, the distinction between "morphological" and "syntactic" agrammatism (cf. Saffran, Berndt, & Schwartz, 1989) is one of degree. The bias agrammatic patients have toward canonical word order can be explained by assuming that, for the production of noncanonical word order, extra processing time is required, connected, for instance, to a movement operation or to the production of a less frequent structure. Consistent with this hypothesis is Kolk and van Grunsven's (1985) demonstration that the production of noncanonical word order is dependent on syntactic complexity.

16-5.1. The Role of Adaptive Strategies

One empirical phenomenon in agrammatic speech is not yet accounted for: the effect of task variation. It has been demonstrated that, at least in Dutch- and German-speaking agrammatics, the speech output observed in free conversations radically differs from the output during constrained tasks such as the cloze task and various forms described earlier and various forms of picture description tasks (Kolk & Heeschen, 1992; Hofstede & Kolk, 1994; see also Nespoulous & Dordain, 1991 for a study in French; for a partial failure to replicate this result in English, see Hesketh & Bishop, 1996; for a criticism of this study, see Kolk & Heeschen, 1996). In particular, in free speech, one almost exclusively observes omissions of function words, whereas in constrained tasks, the number of omissions goes down and that of substitutions goes up: the speech pattern changes from agrammatic to paragrammatic. Similar shifts have been reported to occur spontaneously (Bastiaanse, 1995) and upon request (Kolk & Hofstede, 1994). To explain this task effect, an economy hypothesis is proposed: in free speech, the patients adapt the syntactic complexity of their utterances to their limited capacity. They would do this by employing a subset of the normal inventory of syntactic forms, the production of which requires less capacity: the repertoire of ellipsis. This is the repertoire normal speakers use when they want to be very brief: in informal conversations (e.g., "everything paid") or in addressing children ("naughty girl!"), or among foreigners ("passport not good"). In accordance with this hypothesis, Kolk and Heeschen (1992) have shown that the morpho-syntactic properties of normal ellipsis and agrammatic speech largely overlap. In the output observed in constrained tasks, when patients are adapting to a lesser extent, the elliptical features are less prominent; in fact, their output—at least at the morphological level—is not very different from the output of fluent Wernicke patients (Kolk & Heeschen, 1992). This suggests that Broca's and Wernicke's have a similar grammatical impairment, but that their spontaneous speech is so different (agrammatic in the first case and "paragrammatic" in the second) because Broca's adapt to their performance and Wernicke's—for some reason—do not (Heeschen, 1985; Kolk & Heeschen, 1990). In accordance with this prediction, Haarmann and Kolk (1992) observed that the difficulty profile for the various types of morphology that they studied was the same for their group of Broca and Wernicke patients.

The adaptation approach also offers a processing account for an aspect of sentence production not yet discussed: nonfluency. Kolk and van Grunsven (1985) have suggested that nonfluency, just as telegraphic speech, is a symptom of adaptation. According to Kolk (1995), when the generation of a phrase structure is disrupted, this process is simply restarted. Since the process can profit from rest activation that remains from previous attempts, restarting leads to a higher chance of successful encoding (for some experimental evidence, see Kolk & Hofstede 1994).

16-6. CONCLUSION

We have discussed two approaches to syntactic disorders in aphasia. Linguistic proposals with respect to comprehension deficits involve the representation of traces and Case. Apart from the question of what type of representation is affected in what patient, these proposals may suffer from two shortcomings. First, there is no room for severity variation. Second, the status of word-order heuristics is not clear, neither empirically, nor theoretically. Processing approaches to comprehension disorders vary in (a) the domain of the presumed capacity limitation (e.g., language comprehension versus syntactic analysis) and (b) the nature of the limitation (spatial or temporal). Severity variation is explained by assuming variation in available capacity. Here too, however, there is uncertainty about the role of word-order heuristics. In particular, the effect of canonicity may be due to extra computational load, along with the effect of phrase-structure complexity, or to the effect of strategy. Both linguistic and processing approaches to production disorders have involved attempts to explain variation within the category of grammatical morphemes. In both proposals, the notion of phrase-structure hierarchy plays a crucial role: either a particular node is selectively damaged, or processing of structure up to that node overloads the processing system. The two accounts appear closely related. This similarity suggests that linguistic and processing approaches could well merge into an integrated theory of aphasic disorders of syntax not only in production but also in comprehension. Such a theory should specify: (a) the linguistic representations involved, (b) the computational load connected to the processing of these representations, (c) the "capacity bottlenecks" in aphasia: in what aspects of sentence processing capacity demand is higher than what is available to a particular patient or group of patients, and (d) the strategies employed by particular patients under particular conditions to circumevent the consequences of capacity limitations."

Acknowledgments

The author wishes to thank Rob Hartsuiker and Marco Havekort for commenting on an earlier version of this manuscript.

CHAPTER 17

Impairments of Discourse-Level Representations and Processes

Yves Chantraine,[1] **Yves Joanette,**[1] **and Dominique Cardebat**[2]

[1]Centre de Recherche de l'Institut Universitaire de Gériatrie de Montréal, and École d'orthophonie et d'Audiologie, Faculté de Médecine, Université de Montréal, Montréal, Quebec, Canada H3W 1W5; [2]INSERM U455, Hôpital Purpan, Toulouse, France

Language is used to communicate. In this chapter, some discourse-level components and processes of verbal communication are examined, with a focus on narrative and conversational discourses. The theoretical frameworks of discourse are introduced in the first section, where the importance of Kintsch and van Dijk's model of discourse comprehension is critically acknowledged and the importance of context in conversation is stressed. In the second section, the main findings of recent neurolinguistic studies devoted to discourse troubles in aphasia, right-hemisphere damage, and dementia of the Alzheimer type are reviewed. The conclusion emphasizes the need for clinical research to be inspired by theoretical discourse models, as complex as they are, since they integrate all cognitive abilities required for interpersonal communication.

The capacity to communicate with others implies more than the mere apposition of sounds or the ability to construct correct sentences. Specifically, one needs to cogitate the message to be expressed or understand the message that has been received, one has to organize the elements of information into connected speech, and one must take into consideration the general and the specific contexts in which this exchange takes place. Discourse-level components and verbal communication processes refer to these fundamental aspects of language as a means of communication. It is essential that these aspects be considered since acquired language impairments are increasingly examined in the context of the handicap situations they generate for the individuals in their everyday life (Lacroix, Joanette, & Bois, 1994).

The incorporation of discourse-level aspects of cognition into neurolinguistics is relatively recent. In fact, conceptual frameworks and tools addressing these components and processes have only been systematized with the joint evolution of thought in the philosophy of language, text linguistics, sociolinguistics, and cognitive psychology. Because incorporating a discourse perspective in neurolinguistic research is still under construction, the second section of this chapter will provide the reader with the most frequently referred to theoretical frameworks.[1] The third section is devoted to the contribution of these discourse-level frameworks within the current status of neurolinguistics. Such a contribution will be examplified through a series of studies that have examined narrative or conversational impairments in individuals with acquired brain lesions.

17-1. THEORETICAL FRAMEWORKS OF NARRATIVE AND CONVERSATIONAL DISCOURSE

Discourse can be defined as a sequence of natural language expressions representing a piece of conceptual knowledge that a speaker/writer wants to communicate to a listener/reader. There are several types of discourse, such as narrative, conversational, procedural, descriptive, argumentative or expository, the first two having been the most looked at among patients with acquired brain damage. Narrative discourse is, by far, the type of discourse that has generated the largest number of models and theories. On the other hand, until recently, conversational discourse had not been widely studied in neurolinguistics nor had it benefited from a comprehensive and global theoretical framework. It is being increasingly examined as it is believed to be more representative of natural language.

17-1.1. Aspects of Narrative Discourse

Historically, at least in aphasiology and in neurolinguistics, narrative discourse was primarily the object of formal measures. Presently, however, it is obvious that the analysis of surface parameters such as lexical and morphosyntactical measures (e.g., verb or adjective/noun ratio, Type–Token ratio, mean length of utterance, mean number of clauses, or analyses of syntactical complexity) corresponds not only to a very

---

[1]For more information on this issue, the reader is referred to the writings of Levinson (1983), Atkinson and Heritage (1984), and van Dijk (1985).

partial approach of discourse but also one that misses its essence, namely, its semantic organization. This is not the case for measures of cohesion, such as the correct use of pronominal reference (anaphora and cataphora), ellipsis, lexical repetition, substitution, synonymy, and conjunctions. In fact, the semantic organization of discourse is expressed through the appropriate use of these cohesive forms, which are tightly linked with a coherent representation of the discourse in the speaker's mind and a coherent interpretation by the addressee.

17-1.1.1. Structure versus Content: Story Grammar, Scripts, Schemata, and Causal Networks

In the mid-seventies, scholars such as Mandler and Johnson (1977) attempted to transpose the generative grammar approach to discourse by formalizing the regularities they found in narratives. They proposed a set of rules involving a finite set of supraphrasic units (episodes: setting, goal, attempt, outcome, ending, and so forth) related by a finite set of relations, such as causality and temporal sequence. This underlying structure of stories was assumed to be invariable and independent of the content of specific stories. A mental counterpart, the story schema, was invoked to account for comprehension, memorization, and production of stories. Although the notion of a narrative structure was heuristic, this approach rapidly appeared to be too simplistic and not in accordance with actual story processing.

Other researchers argued that the mental representation of a text first and foremost depended on its content, and that the text superstructure proposed by story grammarians was only a by-product of a content-based organization. With the concept of script, Schank and Abelson (1977) asserted that knowledge about common situations (e.g., going to a restaurant and ordering a meal) did not need a superstructure but rather was a sequence of concrete scenes in which typical roles perform specific actions. The importance of causal relationships (a global plan subdivided into a sequence of goals and subgoals) and the chronological progress between events was also emphasized. Schank and Abelson proposed the concept of causal network in which the most important events are those more closely connected to other events in the story. Although appealing, causal networks are restricted to situations where strong causal relations exist between events; moreover, none of the approaches mentioned until now have dealt with the cognitive processing of stories.

17-1.1.2. The Cognitive Processing of Narrative Discourse

The distinction between microstructure and macrostructure proposed by Kintsch and van Dijk has been of great heuristic value (Kintsch & van Dijk, 1978). Using narrative recall as a basic task, text comprehension is viewed as a real-time process constrained by the limited capacity of short-term memory (STM) that proceeds in cycles, with only one clause or proposition in STM at any one time. Reading or listening to a narrative starts with microprocessing, that is, parsing the text into *micropropositions* formed by one predicate, such as a verb, and one or more of its arguments, such as an agent, goal, or object. These micropropositions are linked together if they share one argument and if they co-occur in STM. The thematically more important propositions stay in STM for several cycles, where they are hierarchically organized into a network, with a topical proposition as the superordinate node. The reader attempts to represent a text with a given network; if there is a lack of argument overlap, s/he searches for the missing propositions in long-term memory or makes inferences[2] to complete this single network. Otherwise, s/he creates a new network. The next step in this top-down process consists of building the macrostructure of the text.

In macroprocessing, the most relevant micropropositions are retained, generalized, or constructed in the form of *macropropositions*. The resulting macrostructure of the text is composed of an ordered set of macropropositions, which constitutes the unified underlying semantic representation of the text. In addition to microstructure and macrostructure, the text base also includes superstructure information retrieved from long-term memory. This abstract cognitive superstructure consists of the conventional elements from the different types of discourse (e.g., for narratives: setting, complication, and resolution) as well as scripts.

Kintsch and van Dijk's model was partially validated by several experiments. It also attracted criticisms, two of which should be mentioned. First, it has been argued that the argument overlap in STM is insufficient to account for textual coherence. In fact, coherence can be local (intrasentential and actualized by cohesive ties) or global, and referred to as contextual plausibility; in this second case, the story as a whole is linked with the reader's general knowledge about the world. Actually, it has been shown that a text could be locally coherent through the correct use of cohesive ties while absolutely implausible on a global level (Garnham, Oakhill, & Johnson-Laird, 1982).

The main criticism, however, concerns the propositional representations of both micro- and macrostructure and their overdependence on the text. Kintsch and van Dijk were considered to have underestimated the role of world knowledge, not only in making inferences or in maintaining coherence, but more generally, in the construction of a mental model of the story, a level of representation that is independent of the linguistic structure of the text (Ehrlich & Tardieu, 1993). Although van Dijk and Kintsch (1983) did propose a "situation model" that was presumed to be separate from the macrostructure, it was considered too largely propositional and still too text-based. In 1988, Kintsch eventually advanced a connectionist "Construction–Integration" model of discourse comprehension. In this approach, there is a loose and context-free Construction step whose nodes and links are possibly weak or ambiguous. Then, through spreading activation, an Integration step taking discourse context into

---

[2]As noted by Kintsch (1993), inference is not a unitary phenomenon. Some inferences are uncontrolled (e.g., automatic bridging inferences) while others are controlled. In these two categories, some inferences generate new information while others expand on knowledge from long-term memory. Moreover, whereas some inferences add information, others reduce it (e.g., the generation of main points in macroprocessing).

account strengthens the contextually connected nodes of the network and eliminates the irrelevant ones. Such theoretical frameworks, however, have never been used to account for narrative discourse impairments in neurolinguistics.

17-1.1.3. The Conceptual Level

Although more complex, and less used until recently, the multilayered model of discourse processing advanced by Frederiksen, Bracewell, Breuleux, and Renaud (1990) appears much more suited for neurolinguistics and its general goals. This model views discourse processing as a complex cognitive process operating simultaneously on four different levels (linguistic, propositional, semantic, and conceptual), while stressing the importance of the latter level. This conceptual level (which corresponds more or less to the "situation model" in van Dijk and Kintsch 1983) is a mental representation, or model, of the reality referred to by the discourse that maps only partially onto the textual representation itself. Taking the linguistic, propositional, and semantic levels as a starting point, the conceptual level includes information from long-term memory (inferences using general and specific knowledge of the reader) and from discourse context. Contrary to scripts, it is *constructed* incrementally and revised as the story unfolds; as such it constitutes the interpretative context for the rest of the discourse.

17-1.2. Aspects of Conversational Discourse

Conversation is the most widely used form of interindividual communication and represents the main source of communicative handicap situations for brain-damaged patients. Conversation can be defined as a goal-directed alternating cooperative interaction between two, or more, interlocutors sharing a spatiotemporal context. In examining conversational discourse, it clearly appears that one cannot express or understand discourse without relying on its context. However, despite many research studies (e.g., Levinson, 1983), the lack of availability of a conversational frame per se as well as the absence of a detailed theoretical model have so far limited neurolinguistic studies to individual aspects of conversation. In this section, three pragmatic aspects of conversational discourse are briefly reminded—Grice's cooperative principle, the dialogic importance of the addressee, and the role of coherence in topic management—before presenting more applied aspects, namely, turn taking and an example of a request sequence.

17-1.2.1. Some Pragmatic Aspects

Conversation is, by nature, cooperative. According to Grice (1975, 1978), individuals observe a general cooperative principle, based on the four maxims of Quantity, Quality, Relation, and Manner. These maxims request the speaker to provide a contribution that contains the right amount of information, that is true, that is relevant to the conversation, and that is perspicuous.[3] Actually, people communicate much more than they literally say, since part of the meaning of their utterances is implied and has to be inferred by the listener through conversational *implicature* requiring inference. In fact, the speaker is not even forced to observe Grice's maxims. Relying on the addressee's cooperativeness and inferential abilities, the speaker can decide to deliberately flout any one of them, and yet convey a meaningful statement. Irony, for instance, relies greatly on the violation of the Quality maxim ("Try to make a true contribution") such as in "The meal with Edgar was a complete success" when the speaker actually intends to convey that this meal was a total disaster.

The addressee is another key piece of conversational context; while in narrative discourse the listener could be considered a passive partner, in conversation, the interlocutor is de facto active. The speaker has to consider what s/he believes to be the addressee's knowledge, just as the addressee has to rely on the context and on his/her beliefs about the speaker's knowledge in order to interpret discourse. This is made easier because, besides general world knowledge, interlocutors mutually share some degree of common knowledge, beliefs, and suppositions (Clark, 1985). The assessment of this common knowledge relies on three kinds of evidence: linguistic (what the interlocutors have said previously), perceptual (what happened in the conversation) and community membership (what is known, believed, supposed in the various communities to which the interlocutors both belong). Relevant utterances are then integrated by the interlocutors in their mental model of the ongoing conversation and increase their common ground.

The local coherence and global plausibility of these relevant utterances are of prime importance for topic management (e.g., opening, developing, closing). According to Charolles (1986), coherence is based on four main rules: overlap of elements (to avoid absence of link), thematic progression (to avoid redundancy), logical noncontradiction, and pragmatic relevance to the discourse context. The speaker is forced to manage the topic and this avoids, for instance, abrupt topic changes. If the speaker wants to shift to a new topic, s/he is expected to explicitly inform the addressee of his/her will and to connect the new topic with the current topic (Crow, 1983).

17-1.2.2. Elements of Conversational Analysis

The most salient structural feature of conversations is probably its turn-taking organization, which was described as a locally managed system by Sacks, Schegloff, and Jefferson (1974). Usually, one person talks at a given time, and overlaps between interlocutors, though common, are brief. A speaker can either "select" the next speaker (e.g., by asking a question), or the conversational partner "selects" himself or herself by picking up a turn at an appropriate conversational signal. Turn taking often occurs at the end of the current speaker's turn (at "relevant transition place"),

---

[3]These conversational maxims have been discussed at length, notably with the principle of politeness advanced by Leech (1983) or the more radical reworking proposed by Sperber and Wilson (1986).

which can be difficult to detect, thus causing overlap. Generally, however, interlocutors' cooperativeness ensures smooth turn transitions.

Another conversational feature is that turns often come in pairs where one utterance depends on the form and the content of the other; examples of such adjacency pairs are greetings–greetings, question–answer, offer–acceptance (or refusal), and so forth. The structure of a conversational encounter such as, for example, opening and closing sequences, or the main ("business") part of a conversation as well as repair sequences, are other aspects that have been investigated (see Levinson, 1983 for a review). The type of utterances produced by a speaker (for example, speech acts, such as requests or apologies) and the influence particular contextual variables have on the way such speech acts are interpreted, has been another area of investigation. Requests are considered speech acts that initiate a turn. Requests have commonly been characterized as being "direct" or "indirect" depending on how "direct" the intention of the speaker is expressed. Some authors have ranked requests on a continuum of directness with direct requests at one end ("Open the door"), followed by conventionally indirect ones ("Can you open the door?") and nonconventionally indirect requests at the other end ("It's cold in here"). Whether an utterance is formulated as a "direct," "conventionally indirect," or "nonconventionally indirect" request depends on several contextual factors such as the "cost" of the request for the addressee, the likelihood and the right of the addressee to comply with the request, the social power of the addressee, and the familiarity between the interlocutors (for a review, see Stemmer, 1994). When the situational context requires a request to be formulated in an indirect form, the request is often framed by opening and/or supporting moves to form a whole request sequence.

17-2. DISCOURSE IMPAIRMENTS IN BRAIN-DAMAGED INDIVIDUALS

Different aspects of discourse-level components can be impaired in patients with acquired brain damage.[4] Even though deficits of discourse and/or conversational abilities are somewhat typical of the verbal communication impairments that can be seen in patients with focal right hemisphere damage or in those suffering from dementia (e.g., dementia of the Alzheimer type), they are also frequent among aphasic individuals having suffered a left-hemisphere insult. However, discourse impairments are not usually recognized as *aphasic* mostly because these abilities were not systematically examined when the classical description of aphasias was introduced more than one hundred years ago. The result of this historical bias is that discourse impairments are assimilated to aphasic signs inasmuch as the impaired individual has been otherwise

---

[4]Because of space limitations, only aphasic, right hemisphere-impaired and demented individuals with discourse disabilities shall be discussed here. For other degenerative dementias (Pick, Parkinson, Huntington), or the effect of closed head injuries on discourse abilities, see the other chapters in this volume.

diagnosed as aphasic. On the other hand, they are considered as "*nonaphasic*" when they occur in isolation, that is, in the absence of properly linguistic (e.g., phonological, syntactical) impairments. Other authors would argue that discourse impairments are nonaphasic in nature because they can be the expression of disturbances in nonlinguistic components of cognition (e.g., working memory, executive function). This sectorial approach to cognition cannot be defended seriously since, for instance, most of what is classically considered core linguistic impairments in aphasia (e.g., agrammatism, anomia) also coexpress the malfunctioning of components of cognition nonexclusive to language (e.g., working memory impairments in agrammatism, conceptual semantic problems in anomia). Be that as it may, for purely conventional reasons, discourse-level impairments among nonaphasic brain-damaged individuals have been referred to as *verbal communication* disorders.

17-2.1. Discourse Impairments in Aphasia

Most of the descriptions of discourse impairments in aphasia were formulated with reference to narrative discourse, many of them focusing on formal aspects of verbal production. Thus, the early description of discourse-level impairments in aphasia focused more on the form rather than the semantics of discourse, whereas the current trend is in the reverse direction.

Although presented earlier as obviously limited, narrative abilities among aphasics have been frequently described formally. The literature thus contains the following observations:

1. Word ratios among distinct grammatical classes (e.g., verb/noun or adjective/noun ratios) have been shown to be different among various aphasic types. For example, the verb/noun ratio is increased among Wernicke's aphasics, whereas it is low among Broca's aphasics (Berko-Gleason et al., 1980).
2. Sentences are syntactically less rich in terms of length, complexity, and correctness (Ulatowska, North, & Macaluso-Haynes, 1981).
3. Deictics (pronouns and adjectives) are present in larger number, thus expressing a more descriptive strategy when asked to produce narratives on the basis of pictures (Dressler & Pléh, 1988).
4. Deficits in the referential system are particularly expressed through impairments in the processing of anaphors. Thus, anaphoric pronouns are frequently introduced without a clear reference (e.g., Cardebat, 1987). In some cases, aphasics tend to use general word knowledge in order to compensate for the lack of reference (Chapman & Ulatowska, 1989).
5. A reduction occurs in the number of lexical items used by aphasics in order to express the narrative message that would principally affect the story episodes that are more marginal within the story structure (Berko-Gleason et al., 1980).

Because of their complexity, theoretical discourse models are underexplored. The purpose of most clinical descriptions of narrative abilities remains descriptive and prosaic and, actually, far from being a test ground of theoretical constructs. For instance, recently published articles include research on the informativeness of aphasics' connected speech (Nicholas & Brookshire, 1992), on the optimal speech sample size of connected speech (Brookshire & Nicholas, 1994), or on the informativeness and grammaticality of aphasics' narratives of the "Cookie theft" picture (Menn, Ramsberger, & Helm-Estabrooks, 1994).

However, many contributions have looked at the organization of the semantic content of narrative, relying partly or totally on a theory-driven approach. These studies have shown that aphasic individuals can exhibit impairments in their narrative abilities that do not require linguistic processing per se, but rather the contribution of other components of cognition to the processing of the content of narratives. Thus, aphasics have been shown to present impairments at a purely semantic level based on their difficulties in rearranging a set of pictures that could form a coherent story, or to choose the appropriate picture in order to end a story coherently (Huber, 1990). However, it is impossible to describe in a unitary fashion the possible discourse impairments in aphasia. In fact, while looking at the coherence violations in the narratives of three groups of aphasics (conduction, anomic, and Wernicke's aphasics), Christiansen (1995) has shown that the patterns of violations were qualitatively different for the three groups. The first two groups were characterized by discourse impairments hypothesized as compensations for their surface-level impairments, while Wernicke's aphasics uttered significantly more nonpertinent information, which could signal troubles at a deeper level. Among other conclusions, this research stresses the importance of carefully distinguishing the subjects' profiles when including them in a group study.

Many of the descriptions of narrative impairments in aphasia have been made by reference to the micro- versus macrostructural aspects of narrative processing (see the section on aspects of narrative discourse). For instance, Ulatowska, Freedman-Stern, Doyel, Macaluso-Haynes, and North (1983) have demonstrated that the macrostructure of aphasics' narratives was preserved even when the microstructure was impaired. It remains to be known, however, if indeed aphasic patients have no macrostructure impairments since it is possible that the results obtained by Ulatowska and colleagues could indicate that the subjects simply relied on their world knowledge in order to come up with the correct macrostructure. In fact, more recently, Ulatowska and Chapman did report disturbances of macrostructure in aphasic individuals (Ulatowska & Chapman, 1994). Related to this is the fact that other studies have shown aphasics to exhibit preserved script abilities either in the production of procedural discourse (Ulatowska, Doyel, Freedman-Stern, Macaluso-Haynes, & North, 1983) or in the processing of the semantic content of scripts (Armus, Brookshire, & Nicholas, 1989; Lojek-Osiejuk, 1996). Such results have been confirmed in studies investigating conversational discourse abilities (Ulatowska, Allard, Rayes, Ford, & Chapman, 1992). In still other studies, it has been shown that aphasics' discourse abilities were unaffected by pragmatic determinants such as the existence of shared knowledge with the listener (Brenneise-Sarshad, Nicholas, & Brookshire, 1991; but see Perkins, 1995 for contradictory results) or the familiarity of the listener (Fergusson, 1994).

In summary, the analysis of discourse impairments in aphasia tends to show that although there can be definite and otherwise recognized impairments at the surface level, some aspects of the semantic content appear to be preserved, such as the organization of the macrostructure. However, in other cases, evidence suggests that there can be some degree of disturbance in the semantics of narratives. The latter raises the question of the source of the narrative impairments in aphasia, whether they are properly linguistic by nature or the expression of more general cognitive deficits. Regardless of the answer to this question, narrative impairments can genuinely interfere with the planning and the implementation of the communicative exchange. The multiplicity of cognitive impairments that can result in a semantic impairment of narrative discourse is an indication that some of these impairments are properly neurolinguistic while others are not. Again, semantics of narrative discourse is one of many examples of the fact that there is no definite border in cognition between properly linguistic components and other components of cognition.

17-2.2. Discourse Impairments among Right Hemisphere-Damaged Nonaphasic Individuals

The analysis of discourse abilities of right hemisphere-damaged (RHD) patients has certainly contributed a great deal to the current knowledge about the possible discourse deficits that can occur following a brain lesion.[5] The reason for this is that RHD patients have a focal, well-delineated lesion but without gross linguistic impairment. Thus, sophisticated tasks can be used with these patients in order to truly understand the nature of their discourse impairment. Whether these impairments are exclusive to RHD individuals still has to be demonstrated. In fact, the same tasks used in RHD patients are frequently inapplicable to aphasics. Moreover, even when they are, it is possible that the subgroup of RHD patients who do actually show discourse impairments should be compared to the subgroup of left hemisphere-damaged aphasics who simply cannot be studied in the same way due to the importance of the surface linguistic impairments.

17-2.2.1. Surface Linguistic Impairment

Despite the absence of surface linguistic impairment, RHD patients have been shown to be impaired in their ability to adequately produce the information contained in a narrative (Joanette, Goulet, Ska, & Nespoulous, 1986). In fact, the narratives of RHD patients are poorer in informative content despite the absence of lexical or syntactic deficiencies. Another characteristic of RHD individuals' narratives is the abundance of confabulations, embellishments, and unnecessary and repetitive details (Hough, 1990; Sherrat & Penn, 1990). According to Hough (1990), the latter characteristics are more frequent among RHD individuals with anterior lesions. In one of the few

---

[5]For a thorough description, the reader is referred to Joanette, Goulet, and Hannequin (1990), Tompkins (1995), and Brownell, Gardner, Prather, and Martino (1995).

studies investigating discourse comprehension in brain-damaged individuals and relying on the use of a multilayered model of discourse processing, Stemmer and her colleagues (Fredericksen & Stemmer, 1993; Stemmer & Joanette, in press) showed that the three RHD patients investigated were incapable of making "bridging" inferences in a text recall procedure. In contrast to aphasic patients and healthy individuals, the RHD patients were not able to reconcile two episodes in a text to construct a new mental model. However, despite ongoing efforts, there is currently no satisfactory comprehensive account of the narrative discourse impairments in RHD individuals.

17-2.2.2. Individual Differences

It has to be emphasized, however, that not all RHD individuals demonstrate narrative deficits. For example, Joanette et al. (1986) observed that only half of their 42 RHD patients did indeed exhibit a narrative deficit. This frequently overlooked aspect of the problem is a reminder that communication deficits are not always present among RHD patients, just as aphasia is not present in all left hemisphere-damaged individuals. Moreover, it is not always the case that the specific narrative cues examined can identify those RHD individuals with discourse impairments. For example, Tompkins and her colleagues (1992) were unable to capture the objective differences between RHD and normal individuals' discourse abilities where there appeared to be definite qualitative differences. Furthermore, even when present, the specific cognitive impairment that could affect discourse-level processing is not the same in all individuals (Kennedy, Strand, Burton, & Peterson, 1994), just like different left-hemisphere lesions result in different types of aphasia (Joanette, Goulet, & Daoust, 1991).

17-2.2.3. Nonliteral Language

A number of studies have documented discourse-level impairments among RHD patients related to the nonliteral interpretation of connected speech such as metaphors, sarcasm, humor, or indirect speech acts. These are examples of discourse segments that require some reference to the immediate or general context in order to be adequately understood. For instance, Stemmer, Giroux, and Joanette (1994) refined the analysis of RHDs' impairments with the processing of indirect requests. They showed that RHD individuals essentially had the most problems with nonconventional indirect requests. According to the authors, this impairment could express the presence of a conceptual problem when the patient is required to integrate surface and textual mental models with a conflicting model based on the situation depicted in the scenario. In this context, the observation that RHD individuals tend to accept as plausible some events or elements of information that are otherwise constrained by the communicative context (e.g., Kaplan, Brownell, Jacobs, & Gardner, 1990; Rehak, Kaplan, & Gardner, 1992) seems to support the hypothesis that RHD individuals are likely to have problems with processing at the level of constructing or modifying mental models.

17-2.2.4. Presence of an Inferential Deficit?

One possible interpretation proposed for the observed discourse-level impairments among RHD individuals refers to an inferential deficit. The possibility of such a deficit among RHD individuals has been discussed by Brownell, Potter, Bihrle, and Gardner (1986). However, there does not seem to be strong direct evidence in favor of such a deficit. In fact, many, if not most, of the studies that looked at nonliteral language processing relied on multiple-choice task procedures that could be interpreted as expressing the presence of a plausibility metric efficiency (Gardner, Brownell, Wapner, & Michelow, 1983). According to this explanation, RHD patients would rather exhibit difficulty rejecting implausible events, that is, events whose probability of occurrence according to the context is low. Direct evidence of inferencing deficits among RHD patients is still under debate. In fact, a number of studies have clearly shown that RHD patients are not impaired either in logical inferencing (i.e., syllogism) or in partly semantic knowledge-based inferencing (Joanette et al., 1990).

In summary, the occurrence of a right-hemisphere lesion, though not the cause of an aphasia proper, can lead to discourse ability impairments, either when processing incoming information or when producing connected speech. Although they express impaired or limited functioning in one or many aspects of cognition, or a reaction to cognitive impairments, the exact source of such discourse impairments is largely unclear. Overall, verbal communication among RHD patients is usually normal with regard to properly linguistic aspects, but can be impaired with regard to the supraword semantics and the ability to process the content of discourse according to the context in which it belongs.

17-2.3. Discourse Impairments among Individuals with Dementia

Discourse processing in demented patients has mostly been studied by assessing their narrative abilities (Ulatowska et al., 1988; Bloom, Obler, De Santi, & Ehrlich, 1994, Chapters 9-13). Unlike RHD patients, individuals with dementia usually show impairments in the formal dimensions of language, such as more or less discrete phonological or syntactic impairments. However, these impairments are much less pronounced than those found in aphasia, at least in the earliest stages of the disease.

At the interface between narrative form and content, demented patients have been shown to exhibit impairments with the processing of anaphors. Furthermore, narratives have been shown to lack cohesion, whether they express the intrusion of paragnosias (when narratives are presented with a visual support), the presence of semantic paraphasias, or the expression of a genuine cohesion impairment (Chenery & Murdoch, 1994). In general, impairments at the level of the semantic structure of narratives in dementia tend to parallel impairments of other components of language (e.g., naming, writing to dictation, reading; Tomoeda & Bayles, 1993).

However, the most important difference between narratives and conversations of demented patients compared to that of other brain-damaged individuals is the fact that the macrostructure can also be impaired. Specifically, crucial components of a narrative are simply omitted, or components from other narratives emerge and become a source of intrusion in a given narrative. In many cases, these intrusions can also correspond to tangential speech motivated by the patient's personal story. These impairments at the macrostructural levels are not only observed in narratives. They have been shown to be present in tasks requiring script production (Grafman et al., 1991) and also in the spontaneous discourse of demented patients (Blanken, Dittmann, Haas, & Wallesch, 1987).

Macrostructure impairments at a conversational level are evidenced by topic shifts and topic maintenance, which are reported as difficult for patients with dementia (Garcia & Joanette, 1994; Mentis, Briggs-Whittaker, & Gramigna, 1995). It has been shown that a deficit in the ability to maintain a conversational topic had to be compensated for by the interlocutor, which undoubtedly contributed to the burden of exchange with demented patients (Garcia & Joanette, 1994).

Other aspects of communication abilities have also been found to be impaired. Analyzing the evolution of the use of clarification requests in her conversations with an Alzheimer patient, Hamilton (1994) noted that, with the passing years, the patient used fewer clarification requests, which paralleled a decrease in her explicit references to memory problems. At the same time, the patient uttered more potential requests for confirmation of information, that is, she seemed to lack implicit information that the speaker considered to be common (given) to them; this last aspect was considered as an early marker of pragmatic troubles in understanding indirect discourse, where much has to be inferred by the addressee (Hamilton, 1994).

Despite a number of studies addressing the discourse abilities in demented patients, the question still remains whether the impairments express a decay within the semantic memory content or a progressive impairment in the different cognitive components necessary in order to comprehend or produce well-formed and relevant discourse. Many more theory-driven efforts will be needed in order to provide satisfying answers to such questions.

17-3. CONCLUSION

Understanding discourse-level impairments is among the most recent chapters in the history of neurolinguistics. Indeed, the interest in the ability to comprehend or produce connected speech that is relevant by reference to the immediate as well as the general context is less than two decades old. Current knowledge appears to be limited by the fact that explicit and implementable theoretical frameworks are only starting to be applied. However, when applied, the relevancy of such systematic analyses of discourse-level abilities has been demonstrated both for clinical and theoretical purposes. In the latter case, discourse processing certainly represents an aspect of verbal communication that will allow a bridge between more classically language-related aspects of cognition and other aspects of cognition that are not exclusive to language but nonetheless allow for optimal interpersonal communication abilities.

Future studies in discourse impairments in brain-damaged individuals will have to systematically incorporate an appreciation of all cognitive abilities that are necessary in order to achieve normal interpersonal communication (e.g., working and semantic memory, attention, inferential abilities, visual skills). It will also become increasingly important to propose approaches that will enable capturing inter- and intrasubject variability. The application of theoretical models will have to be conducted in a more rigourous manner (Kahn, Joanette, Ska, & Goulet, 1990). Finally, and most important, future discourse studies in neurolinguistics will definitely have to be explicitly inspired by available theoretical models. As with other aspects of neurolinguistics, and despite the particularly difficult nature of discourse models and their intricate links with cognition, it is hoped that with reference to a sound theoretical framework, an increasingly greater number of bridges will produce a deeper understanding of discourse-level impairments.

Acknowledgments

This work was made possible by an MRCC grant to Yves Joanette (no. MA-13135). Yves Chantraine is a Postdoctoral Fellow of the Medical Research Council of Canada. We also thank Colette Cerny and Elizabeth Ohashi for editing the manuscript.

CHAPTER 18

Attention as a Psychological Entity and Its Effects on Language and Communication

Zohar Eviatar

Department of Psychology, University of Haifa, Haifa, Israel

Recent views of the structure and functions of attention and their relationship to our conceptions of language disorders are presented. First, a brief review is presented of the methodologies used in the study of attention. In the next section, three aspects of recent findings and thoughts about attention are reviewed: the structure of the attentional faculty, allocation and movement of attention, and what attention *is*. Findings from both language-impaired and normal populations are presented. The final section discusses specific attentional disorders that are believed to be involved in both developmental and acquired language disabilities.

Our view of attention as a psychological entity has undergone quite radical changes over the past decade. A large portion of attention research focuses on selective attention (Posner, 1995), conceiving of attention as a mechanism that enhances processing of certain stimuli while inhibiting processing of others. A somewhat different conception of attention divides cognitive processes into those that are automatic, and therefore do not require attention, and those that are controlled, and have a "cost" in terms of mental resources. Allport (1993) has called into question two of the basic questions that have guided the investigation of attention: the search for the point at which attention impinges on perceptual and higher processes (i.e., early versus late selection), and the distinction between automatic and controlled processes. The most interesting recent conceptualizations of attention have made these issues moot, proposing connectionist models in which early and late processes interact, and redefining the controlled/automatic dichotomy as a continuum.

There are three main issues around which much of the newer research can be organized: (a) conceptualizations of the functions and structure of the attention faculty; (b) research into the mechanisms involved in the allocation and movement of attention; and (c) attempts to define attentional processes, or, what attention *is*. Following a brief survey of the basic methods used to study attention, recent findings and thoughts around each of these foci will be reviewed. In the final section several attentional disorders and their implications for language functioning will be described.

18-1. METHODOLOGY

There is a wide variety of tools available for the study of attention. Researchers are continuing to use classic experimental tasks such as the Stroop task, letter cancellation tasks, and visual detection tasks. In addition, dual-task paradigms, which usually pair visual attentional detection tasks with auditory language tasks such as phoneme monitoring or text shadowing (e.g., repeating a taped text as the participant is hearing it), are also used with both normal and pathological populations. Performance on the detection task alone is compared to performance when it is paired with the language task. Changes in performance between these two conditions are interpreted as reflecting the costs of dividing and allocating attentional resources.

Covert visual attention is examined with an experimental paradigm developed by Posner and his colleagues (the COVAT paradigm; Posner & Raichle, 1994). Participants are required to respond manually to targets presented in the right or left visual field. The display usually consists of a central fixation point and two squares or circles presented 4 to 8 degrees of visual angle laterally. Targets are stars or dots that appear in one of the squares. On approximately 80% of the trials, a spatial cue (the brightening of the contours of one of the squares) indicates where the target will appear (a valid cue). On the remaining trials, the spatial cue appears on the other side (an invalid cue). In some experiments there are no-cue trials as well. The important dependent variable is the response time to valid and to invalid cues. This paradigm has been used extensively in the study of visual spatial attention and the relationship between executive and modality-specific functions of attention.

Investigations of auditory attention use directed attention procedures with the dichotic listening task (Mondor, 1994). Two competing auditory stimuli are presented simultaneously, one to each ear, and the participant reports what she or he hears in both ears, or in one. Dichotic verbal stimuli result in more accurate reports of the stimuli presented to the right ear (a right ear advantage [REA]), while nonverbal stimuli often result in a left ear advantage (LEA). The general interpretation of these ear asymmetries is that they reflect hemispheric specialization for these types of stimuli (see Bryden, 1988 for a review). In general, for verbal materials, there seems to be an attentional bias toward the right ear, or right hemispace stimuli, even with monaural stimulation (see Bradshaw & Nettleton, 1988 for a review). When normal subjects are asked to pay attention to left ear verbal stimuli, however, the dichotic perceptual asymmetry switches, and a LEA is found. Interestingly, Zaidel (1983a) has reported that attentional manipulations affected ipsilateral pathway suppression in a split-brain subject, suggesting that the effects of attention are in the modulation of the interaction of structural elements that result in the perceptual asymmetry.

Analyses of the performance of brain-damaged patients in the neuropsychological tradition of generalization from the damaged to the normal brain are an important source of both data and models. Functional brain imaging (e.g., PET and fMRI; see Posner & Raichle, 1994, for an excellent explanation of these techniques) during the performance of attentional and language tasks has been especially useful in localizing attentional functions in the brain. Analyses of both event-related potentials (ERPs) and running EEG data have also been used to measure neural activity associated with both sensory and cognitive functions in populations with language and attentional impairments.

Finally, many important models of attention utilize computer simulations of attentional and language functions. These artificial neural–network models are based on networks of interconnected processing units. The units are connected to each other in specific structures, and are activated or inhibited as a response to input from other units. Representations are manifested by patterns of unit activation. These types of models have been described extensively (e.g., Rumelhart, McClelland, & the PDP Research Group, 1986), and, as will be demonstrated, have provided some of the most interesting new conceptualizations of attention as a psychological phenomenon.

18-2. THE STRUCTURE AND FUNCTIONS OF THE ATTENTION FACULTY

18-2.1. Attentional Functions

Two influential conceptions of the structure of attention and its relationship to brain functioning have arisen in somewhat different contexts. Researchers who rely mainly on behavioral assessment batteries of pathological and brain-damaged populations have proposed a heuristic factorial structure of attention (e.g., Mirsky, Fantie, & Tatman, 1995). Researchers who work in the context of neuroimaging and other experimental paradigms have proposed a network of attentional systems (e.g., Posner, 1995; Posner & Peterson, 1990). There are both similarities and differences between these theoretical structures, based largely on the somewhat different methods and definitions used to construct them.

18-2.1.1. The Factorial Structure of Attention

Mirsky and his colleagues (Mirsky et al., 1995; Mirsky, Anthony, Duncan, Ahearn, & Kellam, 1991) have noted that impaired attention is one of the most common behavioral disturbances in pathological populations (e.g., schizophrenia, affective and anxiety disorders, lead intoxication, petit mal epilepsy, traumatic brain injury, AIDS-related dementia). They constructed a battery of tasks whose execution seems to depend on attentional abilities and applied them to several populations of subjects. The scores on these tests were used to generate a heuristic factorial model of attentional functions with four factors: The *focus–execute* function refers to the ability to attend and respond selectively; the *shift* function refers to the ability of patients to shift from attending to one aspect of a stimulus to another; the *sustain* function refers to the ability to maintain vigilance; and the *encode* function is interpreted as the ability to register, manipulate, and recall numerical information. Mirsky and his colleagues have proposed a brain localization scheme for these factors based on their use of the attention battery with various patient populations, on previous case reports, and work with monkeys. The *focus–execute* function is believed to rely on the integrity of the inferior parietal, superior temporal, and striate cortexes; the *shift* function on prefrontal cortex; the *sustain* function is subserved by midbrain structures; and the *encode* function by the hippocampus and amygdala.

18-2.1.2. Attentional Systems

Posner and his colleagues (e.g., Posner, 1995; Posner & Peterson, 1990) have proposed that there is an attentional system in the brain that is at least somewhat anatomically distinct from other specific functional systems, and that this system is composed of a network of anatomical areas. This model is based on research with nonhuman primates and the use of imaging techniques of human brain functioning within experimental paradigms. The model proposes three major attentional functions: orienting, detecting, and maintaining an alert state. *Orienting* to sensory stimuli has been studied largely in the visual modality because of its relation to eye movements (see Posner, 1995 for review), and is proposed to have three dissociable elements: disengagement from the present target (this function is believed to be subserved by areas in the posterior parietal lobe), movement of attention (subserved by the area of and around the superior colliculus), and engagement of the new target (subserved by the pulvinar nucleus in the thalamus). Several studies have shown that these elements are also useful in describing orienting to auditory stimuli (Bedard, Massioui, Pillon, & Nandrino, 1993; Hugdahl & Nordby, 1994; Mondor, 1994). The *detecting* function is thought to be an executive network that involves awareness, intentional semantic processing, and the allocation of attentional resources (Posner, Sandson, Dhawan, & Shulman, 1989; Fuentes, Carmona, Agis, & Catena, 1994). This system has been localized in the anterior cingulate gyrus on the basis of metabolic neuroimaging findings (Posner & Raichle, 1994). *Maintaining an alert state* increases the speed of action taken toward a target that has been detected. The frontal regions of the right hemisphere have been proposed to contain the neural substrate for this function.

As already noted, there are both similarities and differences between these models. The *focus–execute* function in Mirsky et al.'s (1995) model is equivalent to the *orienting* function in Posner's model, while the *encode* function may be seen as including both elements of the *orienting* function and the *detecting* or executive function in

Posner's model. Mirsky et al.'s *shift* function is analogous to the functions of the *detecting* or executive attentional network. Both models posit a separate vigilance factor.

In summary, the present consensus in the field is that attention is composed of several dissociable factors, or elements, which are subserved by different areas of the brain. These areas are anatomically separate from those subserving processing mechanisms specific to the task that the person is performing (reading, talking, etc.). These factors include an orienting/focusing element, a detecting/encoding element, a system for allocating attention, and a vigilance factor.

18-2.2. Allocation and Movement of Attention

18-2.2.1. Movement of Attention and the Eyes

A number of studies have specifically explored the relationship between attention and eye movements, and have used reading as the experimental task (e.g., Inhoff, Pollatsek, 1995; Posner, & Rayner, 1989; Rayner, Sereno, Lesch, & Pollatsek; 1995; Vitu, O'Regan, Inhoff, & Topolski, 1995). The general finding is that attention (in terms of the ability to extract information) is asymmetrical in the direction of scanning (that is, to the right in Latin scripts and to the left in Hebrew; Pollatsek, Bolozky, Well, & Rayner, 1981) and is limited to the line of text (i.e., little information is extracted from below the line of text; Pollatsek, Raney, Lagasse, & Rayner, 1993). Although it has been shown that this asymmetry of attention does not compromise measures of left hemisphere specialization for visually lateralized verbal materials (a right visual field advantage has been found in Hebrew as well; see Faust, Kravetz, & Babkoff, 1993, for a review and discussion), several studies (e.g., Eviatar, 1997; Vaid & Singh, 1989) have suggested that reading scanning habits may affect performance asymmetries for nonlanguage tasks. Eviatar (1995) has suggested that these effects reflect an underlying learned attentional bias that skews covert attention to the side from which reading usually begins and then proceeds. This interpretation is based in part on the occulomotor readiness hypothesis proposed by Rizzolatti and his colleagues (Rizzolatti, Riggio, Dascola, & Umiltà, 1987) that posits that both movement of attention and saccades are mediated by the same neural circuits. Interestingly, Hoffman and Subramaniam (1995) have reported that it is not possible for subjects to move their eyes to one location and covertly attend to another location, emphasizing the point that visuospatial attention and the mechanism of generation of voluntary saccades are intimately connected.

The hypothesis that reading scanning habits and saccade programming are related to attentional mechanisms is supported by several types of data. Many individuals with deficits in one area also have deficits in the other. Fischer and his colleagues (Fischer, Biscaldi, & Otto, 1993; Fischer & Weber, 1990) have reported that developmental dyslexics have erratic eye movements not only when they read, but also when they are performing a noncognitive task. In addition, a growing number of studies are showing that children with developmental dyslexia and dysphasia differ from normal controls in their ability to process and produce sequences of both acoustic and visual rapidly changing stimuli (Tallal, Galaburda, Llinas, & von Euler, 1993) and have suggested that this is related to a general deficit in the controlled, attentional aspects of such tasks. Thus, such children may have a deficit in the attentional aspects of occulomotor programming that affects their ability to read. On the other hand, scanning habits may result in the relative sparing of attentional functions. Speedie et al., (1995) compared left and right hemisphere-damaged patients who had read only Semitic (e.g., written from right to left) or non-Semitic (left-to-right) languages on a line bisection task. Right hemisphere (RH) damage often results in the syndrome of unilateral spatial neglect, where patients do not attend to the left side of objects or space (Rafal & Robertson, 1995). One of the defining symptoms of unilateral spatial neglect is the failure to bisect lines; patients with RH damage tend to put the midpoint to the right of the true midpoint. Speedie et al. (1995) report that patients with RH damage who had read only right-to-left languages before the age of 15 bisected lines closer to the true midline than patients who had read left-to-right languages. There were no significant language-related differences in their LH-damaged patients. Thus, again, it appears that reading scanning habits may affect the mechanisms (specifically RH mechanisms) that subserve the allocation of spatial attention.

18-2.2.2. Movement of Attention without Eye Movements

Using the COVAT paradigm, Posner and his colleagues (see Posner, 1995 for a review) have identified the elements of visual attentional orienting that are involved in the control of covert attention. They have also identified the probable neuroanatomical substrates of these elements by showing that patients with different lesions reveal deficits selectively in these components of the task. What is of relevance to us is that they have shown that the usual invalidity effect (the slowing of responses to targets when the invalid cue was in the opposite visual field) disappears when both normal individuals and patients with parietal lobe damage (who usually show exaggerated invalidity effects in the contralesional field) are performing either a phoneme-monitoring task or shadowing a text at the same time. They interpret these results to suggest that there is an executive attentional system that controls resource allocation for both auditory language functions and visuospatial attention in the left hemisphere, and that it is located in the anterior cingulate gyrus.

This hypothesis is supported by the behavior of several types of populations. Research with aphasics (mostly with anterior damage—Broca's aphasics) using both the COVAT procedure (Petry, Crosson, Gonzalez-Rothi, Bauer, & Schauer, 1994) and dual-task paradigms (Tseng, McNeil, & Milenkovic, 1993), together with dual-language task procedures with normal subjects (Fuentes et al., 1994), supports the hypothesis that there is an attentional allocation system for language in the frontal portion of the LH. The most interesting and controversial support comes however, from studies testing covert orienting in schizophrenics. Originally, Posner, Early, Reiman, Pardo, and Dhawan (1988) found that with short cue-target intervals (100 ms

SOA) schizophrenics show the same asymmetry in responses to invalid cues as patients with left parietal damage (and the aphasics described earlier). However, with long SOAs (800 ms) there was no such asymmetry. They concluded that schizophrenics have damage to the LH anterior attentional system. Unfortunately, these findings have had a somewhat uneven replication record, with some authors (e.g., Gold et al., 1992) interpreting the results to support RH deficits in schizophrenia. However, Maruff, Hay, Malone, and Currie (1995) have replicated Posner et al.'s (1988) findings with a very tightly controlled study. They found this short SOA asymmetry only in unmedicated or recently medicated patients, not in chronic, long-term medicated patients, which suggests that subcortical dopaminergic mechanisms altered by long-term neuroleptic medication are involved in the functioning of the anterior attentional system. Converging evidence that attention allocation mechanisms in the LH of schizophrenics are deficient is presented by Green, Hugdahl, and Mitchell (1994), who show that both hallucinating and nonhallucinating schizophrenics are unable to shift attention from one ear to the other in directed attention conditions of a dichotic listening task.

In summary, many studies have explored the relationship between eye movements and attention movements, and have supported hypotheses that the neural mechanisms involved in eye movements are also involved in the movement of attention. In addition, several studies have suggested that lifetime eye movement habits (such as reading scanning direction) may affect attentional mechanisms in the right hemisphere (Eviatar, 1995; Speedie et al., 1995). It must be noted that Maruff et al. (1995) report that although their schizophrenic subjects showed deficits in the allocation of attention, and deficits in eye movement measures, there were no correlations between these measures in their subject samples. The incompatibility of these two pieces of information may be reduced by the finding that, for visual scanning at least, there are independent attentional mechanisms in the two cerebral hemispheres of split-brain patients (Luck, Hillyard, Mangun, & Gazzaniga, 1994). Reading scanning habits may affect right and left hemisphere mechanisms in different ways. Several routes of investigation converge to support the hypothesis that in the anterior portion of the LH there is an executive system that is involved in the allocation of attentional resources to language functions and to visuospatial functions, and that language functions take precedence over visual detection tasks in dual-task conditions.

18-2.3. What *Is* Attention?

It is clear that stimuli that are attended to are processed differently from stimuli that are not. Attention modulates signal detectability so that stimuli that are attended to are detected more quickly and accurately than those that are not attended to. The question is, what is being modulated? Several possibilities have been suggested: Attention can affect the order by which information is read out of the early sensory processing stage, such that attended stimuli are processed first and result in higher-quality representations for subsequent processing (Hawkinset et al., 1990; Yantis & Johnson, 1990); attention can enhance sensory gain from inputs at the attended location relative to inputs from unattended locations (Hawkins et al., 1990; LaBerge, Brown, Carter, Bash, & Hartley, 1991; Giard, Collet, Bouchet, & Pernier, 1994); attention is an exclusionary mechanism that filters out noise and inhibits the processing of inputs from unattended locations (LaBerge et al., 1991; Shui & Pashler, 1995) or representations of features (Treisman & Sato, 1990). These mechanisms are not conceived of as exclusive, and it is probable that all of them contribute to enhanced responses to attended stimuli.

Previous attempts to define attention have utilized the distinction between controlled and automatic processes. Automatic processes are done quickly, in parallel, and do not require attention, while controlled processes are intentional, slower, more serial, and therefore require resources (see Allport, 1993 for a review). The automatic/controlled distinction has been especially useful in psycholinguistic models. In models of auditory language processing, attention is defined as a controlled process that mediates the inhibition of competing candidates so that the correct word is recognized. The major theoretical framework posits that the incoming stream of speech is segmented in some way (e.g., by syllables; Segui, Dupoux, & Mehler, 1990) and that these segments activate a cohort of candidates (e.g., Vroomen & de Gelder, 1995). This process of activation is believed to be automatic (i.e., it does not require attentional resources). The correct candidate is chosen through listener expectations that are based on both the semantic (Chiarello, Maxfield, Richards, & Kahan, 1995) and the syntactic (Deutsch & Bentin, 1994) context of the utterance, and serve to inhibit the "wrong" candidates that have been activated. However, some psycholinguistic models of reading have questioned the distinction between automatic and controlled processes (e.g., Tabossi & Zardon, 1993). Cohen, Dunbar, and McClelland (1990) have suggested that automaticity in word recognition can be seen as a continuous variable, with attention as a modulation of processing in specific pathways, both facilitating the attended pathway and/or inhibiting processing in unattended pathways. Chiarello et al. (1995) have reported findings compatible with this model, supporting the idea that word recognition involves numerous parallel interactive processes, and that attentional modulation of these processes is possible at all stages. These issues are directly addressed by connectionist models of attentional and of language processing (Cohen et al., 1990; Mozer & Behrmann, 1990; Phaf, Van der Heijden, & Hudson, 1990). These authors conceive of attention as a modulation of pathway strength and an additional source of input to the interconnected processing units that affects their activation thresholds. Attention enhances the activation of input features within the attended area relative to those outside the attended area. Whether this occurs via active inhibition or not is a specific attribute of each model, however; most do not specify whether active inhibition is a necessary prerequisite for selectional processes.

18-2.4. Summary

Attention is a faculty that has different components: orienting, detecting, shifting, and maintenance of vigilance. Movement of attention (at least of visual attention) seems to be related to the mechanisms that subserve eye movements, and specific brain areas have been implicated in the allocation of attentional resources between very different tasks. Attention enhances processing of the stimuli that are being attended, and may be necessary to inhibit the processing of representations that have been activated but are not relevant to the task that the individual is performing. The next section concentrates on attentional and language disorders and summarizes our conceptions of their interactions.

18-3. LANGUAGE AND ATTENTIONAL DISORDERS

18-3.1. The Frontal Lobes: Attention, the Executive System, and Language

The frontal lobes compose between a quarter and a third of the human cerebral cortex and are the most readily distinguishable difference between the primate brain and that of other mammals. Damage to the frontal lobes results in a variety of symptoms that have generally been classified as deficits in the executive control functions or supervisory system. As mentioned earlier, the anterior attentional system proposed by Posner and his colleagues is believed to be subserved by the frontal portion of the cingulate gyrus. As many frontal lesions result in aphasic symptoms, researchers have looked at the relationship between executive and language functions. In general, the finding has been that executive dysfunctions are separate from language dysfunctions. Glosser and Goodglass (1990) found that aphasics with or without lesions extending to the dorsolateral regions of the left frontal lobe do not differ in language tasks, but the former are more impaired on tests of executive control. Awh, Smith, and Jonides (1995) found deficits in verbal working memory with left frontal damage and in spatial working memory with right frontal damage. Patients with right frontal damage are usually not aphasic, but can display aprosodia and deficits in discourse and pragmatic aspects of language (Alexander, Benson, & Stuss, 1989). Degeneration of either frontal lobe results in reduced verbal output and finally in mutism (dynamic aphasia), without specific deficits in the structural aspects of language (Neary, 1995). Attention is an important factor in executive or supervisory functions, and thus seems to affect language behavior indirectly, via these executive functions.

18-3.2. Neglect Dyslexia

Neglect dyslexia is an acquired reading disorder that occurs in the context of unilateral spatial neglect. It is a syndrome that most often follows damage to the right parietal lobe, and can be characterized as a deficit in the allocation of attention to the left side of space (Rafal & Robertson, 1995). The major symptom of neglect dyslexia is the omission of text on the left side of the page, the line of text, or the omission or misreading of the initial letters of single words. Of special relevance is the phenomenon of extinction—a patient can detect or read a single word in the contralesional visual field (usually the left visual field) when it is presented alone, but will ignore it if another word is presented in the ipsilateral field. A large number of studies have revealed that this left-side extinction is subject to a number of very interesting effects: it is sensitive to cueing, so that forcing patients to pay attention to the left side (by asking them to report the left word first, for example) reduces left-side extinction (Behrmann, Moscovitch, Black, & Mozer, 1990); to lexical status, such that nonwords are misread more often than words (Sieroff, Pollatsek, & Posner, 1988), and even nonorthographic features of letters (e.g., the color of the ink in which they are printed) are reported more reliably from words than from nonwords (Brunn & Farah, 1991); and to semantic context (Tegner & Levander, 1993; Ladavas, Paladini, & Cubelli, 1993). Misreadings of words often preserve word length and position, such that the word BEACH may be read as PEACH or REACH rather than as EACH (Caramazza & Hillis, 1990a).

Behrmann and Mozer and their colleagues (Behrmann, Moscovitch, & Mozer, 1991; Mozer & Behrmann, 1990) have proposed a connectionist model of neglect dyslexia that shows effects of deficits in the allocation of attention on both early and late stages of processing. In the model, damage is to bottom-up connections between an attentional mechanism and input feature maps, and is graded monotonically from left to right, such that it is most severe at the extreme left and least in the right. Neglect of the left is not all or none, such that top-down processes are able to use poorly activated information from nonselected items. This accounts well for the finding that words are less vulnerable to extinction than nonwords, and for other effects of semantic variables in the neglected visual field. The results of Speedie et al. (1995), which show a mitigation of neglect in right-to-left readers, may suggest that lifetime reading habits can affect the slope of this gradient.

18-3.3. Attentional Dyslexia

The four patients with attentional dyslexia who have been described in the literature—F.M. and P.T. (Shallice & Warrington, 1977), B.A.L. (Warrington, Cipolotti, & McNeil, 1993), and P.R. (Price & Humphreys, 1993)—all present an acquired dyslexia without dysgraphia. The defining symptom of attentional dyslexia is the ability of patients to read whole words better than to name the letters in a word. Patients can name single letters, but not a row of letters; they can read single words, but not a row of words. In the two patients described by Shallice and Warrington (1977) and in P.R., described by Price and Humphreys (1993), this impairment was not restricted to verbal stimuli. For example, they were better at naming a single shape or a picture than a row of shapes or three pictures presented simultaneously. These deficits cannot be attributed to a primary spatial deficit, as all of these patients could point to the letter they were attempting to identify and had no deficits in scanning attention in a cancellation task. Both Shallice (1988) and Price and Humphreys (1993) suggest that the deficit in these cases resides in the setting of the appropriate focus or size of the selective attentional window, which allows output from perceptual analysis of stimuli other than the target to activate higher stages of processing, which results in response interference.

Warrington et al. (1993) report that B.A.L. has trouble only with verbal stimuli and not with pictures. They suggest that, in this patient, the locus of the attentional deficit lies after a stage at which letters or words have been accessed as units, at the stage of transmission of information from a visual word-form system to a semantic or phonological stage of processing. This interpretation converges with psycholinguistic models of visual word recognition in which attention modulates processing at all stages (e.g., Monsell, Patterson, Graham, Hughs, & Milroy, 1992; Tabossi & Zardon, 1993). It may be relevant that Chiarello and her colleagues (e.g., Chiarello, 1991) have shown that semantic priming occurs on a broader scale in the RH than in the LH, suggesting that the LH may have some sort of mechanism to halt spreading activation. It may be the functioning of this mechanism on various levels that is damaged in attentional dyslexics.

18-3.4. Alzheimer's Disease

One of the first nonmemory neuropsychological consequences of Alzheimer's disease is a decline in attentional capacities. This decline has been documented in all of the attentional functions defined earlier (see Jorm, 1986 for a review). Several studies have suggested that certain language deficits seen in patients with dementia of the Alzheimer's type (DAT) are the result of specific attentional deficits, not of a deterioration of linguistic abilities. These suggestions make sense within an interactive/connectionist model of language processes, in which attention is a modulating factor on patterns of activation within the language system. Balota and Ferraro (1993, 1996) have suggested that mildly demented Alzheimer's patients are deficient in their ability to inhibit partially activated information, which causes the mispronunciation of exception words during reading and large priming and inhibitory effects in lexical decision tasks. Neils, Roeltgen, and Greer (1995) have shown that deficits in measures of selective attention and vigilance predict the degree of characteristic spelling errors made by DAT patients to a greater degree than a measure of language ability. Waters, Caplan, and Rochon (1995) have shown that lowered performance on a sentence comprehension task in DAT patients is the result of inefficient executive control and allocation of attention to the propositional structure of sentences, and not of a deficit in syntactic processing. It will be interesting to follow the theoretical utility of connectionist models of language processing that interact with an attentional faculty in the elucidation of language breakdown in these patients.

18-3.5. Attention Deficit Disorders

A large area of research focuses on attention deficit disorders (ADD), which are diagnosed in (usually) school-aged children. Such children exhibit lowered, or age-inappropriate, levels of attention, heightened impulsivity, and overactivity, together with poor modulation and self-regulatory behavior. Unfortunately, the majority of the research in this area is not typically done within the supporting structures of an attentional theory or model, and the final results of such projects often end up as orphaned lists of empirical findings from which it is hard to generalize.

Recent research (e.g., Elbert, 1993; Goodyear & Hynd, 1992) has focused on differentiating clinically distinct subclasses among children with ADD, specifically, those who have a co-occurring motor hyperactivity (ADD+H) and those who do not (ADD-H). Many authors have noted that there is a high co-morbidity of ADD with learning disabilities, specifically with reading and writing abilities (e.g., Ackerman, Dykman, Oglesby, & Newton, 1994; Elbert, 1993; Whyte, 1994). However, the relationship and directionality between these disorders are not clear. Studies looking at the reading and writing ability of children with ADD reveal that these skills are below normal (Elbert, 1993), while studies looking at children with dyslexia reveal that many of them have attentional disorders (Ackerman et al., 1994; Whyte, 1994). On the other hand, many studies using the directed attention manipulations in the dichotic listening paradigm find, in fact, that reading-disabled children are able to raise the performance of the left ear when attending to it even more than normally reading controls (see Morton, 1994 for a review). Morton (1994) has shown that the ability to direct attention in the dichotic listening task can interact with a subtype of dyslexia and specific characteristics of the stimuli. In these studies, the authors either specifically chose non-ADD subjects, or did not categorize their subjects by behavioral attentional criteria. Thus, further research in both areas, developmental dyslexia and ADD, is needed to clarify the relationship between them.

Measurement and interpretation of event-related potentials (ERPs) have provided a useful tool to measure neural activity associated with both sensory and cognitive processes in these subject populations. The component known as P300 has been interpreted as an index of amount and allocation of attentional resources. Duncan et al., (1994) measured both auditory and visual ERPs from a group of adult developmental dyslexics and normally reading controls. They report that abnormally small P300 components appeared only in a visual task, and only in the subgroup of dyslexics that had also suffered from ADD+H in childhood. The dyslexics who had not, revealed brain potentials that were indistinguishable from those of the control subjects. Interestingly, the two subgroups of dyslexics were not distinguishable on the behavioral measures. These authors also found a hemispheric asymmetry in the P300 component, where both normal subjects and non-ADD dyslexics revealed higher amplitudes above the right hemisphere (RH), whereas the ADD+H dyslexics revealed an opposite pattern.

The ERP data, together with the hypothesis that RH parietal areas are involved in reading, implicate the RH in disorders of reading and attention. A large amount of evidence points to the centrality of the RH in the control of visual attention (e.g., Hellige, 1993a). RH-damaged patients also have difficulties with the pragmatic aspects of language: understanding metaphors and humor, and in the use of contextual cues to interpret conversations (Brownell, Carroll, Rehak, & Wingfield, 1992). Several groups of researchers have investigated cognitive and social behavior of children with Developmental Right Hemisphere Syndrome (Gross-Tsur, Shalev, Manor, & Amir,

1995; Voeller & Heilman, 1988). These are children who present with neurological signs that suggest a deficit in the functioning of the RH. In these studies, together with paralinguistic deficits, almost all of the children were diagnosed with ADD. However, Branch, Cohen, and Hynd (1995) report that behavior rating scale scores and the frequency of diagnosis of ADD+H are not different among children believed to have left or right hemisphere damage. As with the epidemiological data already described, these findings are suggestive of possible relationships between reading as a visual attentional task and the specialization of the right hemisphere for spatial attention. However, there is a need for a stronger empirical and theoretical basis for such hypotheses.

18-4. CONCLUSION

The goal of this chapter has been to summarize developments in the investigation of attention, and to show how these are influencing our thinking about the relationship between the attention and language faculties of the mind. We have seen that there is reason to believe that an executive attentional system that is particularly involved in language processes, and also in allocating attentional resources in general, is subserved by neural mechanisms in the anterior left hemisphere. There is also a visuospatial attentional system that is subserved by the posterior parietal area in the right hemisphere, which may be involved in the visual aspects of reading. In addition, the mechanisms involved in attentional movement seem to be also involved in eye movement. In general, recent conceptualizations of attention represent it as a modulating factor within connectionist models of cognitive processes, and of language processes in particular. Developments that use these types of theoretical frameworks are now beginning to yield new and interesting ways of looking at language disorders.

CHAPTER 19

The Role of Working Memory in Language and Communication Disorders

Martial Van der Linden and Martine Poncelet
Neuropsychology Unit, University of Liège, Liège, Belgium

This chapter examines the neuropsychological evidence relating working memory, and more specifically the phonological loop and the central executive, to some aspects of language processing: sentence comprehension, speech production, and vocabulary acquisition. It is argued that some aspects of Baddeley's (1986) working memory model need to be elaborated in order to accommodate some empirical findings, in particular the effect of semantic information on short-term memory performance. In addition, a more precise specification of the central executive system is clearly required.

Working memory refers to a limited capacity system that is responsible for the temporary storage and processing of information while cognitive tasks are performed. Baddeley's model represents the most extensively investigated and the best-articulated theoretical account of working memory (Baddeley, 1986). This model is composed of a modality-free controlling central executive that is aided by a number of subsidiary slave systems ensuring temporary maintenance of information. Two such systems have been more deeply explored: the phonological loop and the visuospatial sketchpad. The phonological loop system is specialized in storing verbal material and is composed of two subsystems: a phonological store and an articulatory rehearsal process. The phonological store receives directly and unvoidably any information auditorily presented and stores it in terms of a sound-based code. Although material in this store is subject to decay and interference, it can be maintained and reinforced through the articulatory rehearsal mechanism. The phonological store is also able to receive visually presented items but these must first be converted into an articulatory form before gaining access to the store. These items are conveyed to the store by the articulatory rehearsal process. The visuospatial sketchpad system is assumed to be involved in setting up and maintaining visuospatial images. According to Logie (1995), the visuospatial system consists of a visual temporary store that is subject to decay and to interference from new incoming information and a spatial temporary store that can be used to plan movement and also to rehearse the contents of the visual store.

The core of the working memory model is the central executive. The central executive is assumed to be an attentional control system responsible for strategy selection and for control and coordination of the various processes involved in short-term storage and more general processing tasks. An important characteristic of this system is that its resources are limited and divided into different processing and storage functions. Baddeley (1986) has suggested that the Supervisory Attentional System (SAS) component of the attentional control of action model proposed by Norman and Shallice (1986) might be an adequate approximation of the central executive system.

Convincing evidence for the existence of different components included in the working memory model comes from the study of brain-damaged patients with specific short-term memory impairments (for a review, see Van der Linden, 1994). Some patients had a deficit of auditory short-term memory that was attributed to a selective impairment of the phonological loop (e.g., Vallar & Baddeley, 1984). Other patients had a specific impairment of the visuospatial sketchpad (Hanley, Young, & Pearson, 1991) or of the central executive (Van der Linden, Coyette, & Seron, 1992; Baddeley, Logie, Bressi, Della Sala, & Spinnler, 1986).

Numerous findings from experimental, neuropsychological, and developmental sources suggest that working memory makes significant contributions to different domains of cognition, such as reading, auditory comprehension, vocabulary acquisition, counting and mental arithmetic, reasoning, problem solving, and planning (see Logie, 1993). The aim of this chapter is to examine the neuropsychological evidence relating working memory, and more specifically the phonological loop and the central executive, to some aspects of language processing, namely, sentence comprehension, speech production, and vocabulary acquisition.

19-1. WORKING MEMORY AND SENTENCE COMPREHENSION

A great number of studies have investigated whether a deficit selectively affecting the working memory system might be the source of language comprehension difficulties. In particular, there has been a great deal of interest in the consequences of a phonological loop dysfunction for sentence processing.

19-1.1. Phonological Loop and Sentence Comprehension Deficits

Patients with a defective phonological loop typically show very poor performance in immediate recall of all strings of unconnected auditory–verbal materials (digits, letters, words). Some patients have been reported to show short-term memory (STM) deficits consistent with a disruption of the phonological store (e.g., Vallar & Baddeley, 1984), while other patients have been identified to appear to show a deficit in rehearsal (e.g., Belleville, Peretz, & Arguin, 1992). In addition, several findings (e.g., Waters & Caplan, 1995) show that the rehearsal deficit observed in some patients occurs at the level of phonological output planning processes that operate on accessed lexical phonological forms.

There exists a lot of evidence suggesting connections between severe phonological loop impairments and language comprehension deficits. The existence of relationships between both types of impairment is particularly supported by a follow-up study of a patient, T.B., who showed a parallel recovery on digit span tasks and sentence comprehension tasks (Wilson & Baddeley, 1993). However, the precise role of the phonological loop in language comprehension remains controversial and a subject of debate. In fact, two main classes of interpretations have been proposed.

19-1.1.1. The Phonological Short-Term Memory as a Support to First-Pass Language Processing

One view suggests that phonological storage is necessary to support the first-pass language processing, in which the subject carries out the initial syntactic analysis of the sentence. However, a phonological representation of an incoming message should be consulted only when the correct syntactic interpretation of a sentence is particularly difficult to achieve (e.g., Saffran & Marin, 1975). For example, using a sentence repetition task, Saffran and Marin (1975) found that I.L., a patient with impaired memory spans, could frequently paraphrase (and thus understand) sentences that she could not correctly repeat verbatim. However, her productions were often semantically incorrect when the sentences to be repeated were either passive constructions such as "The boy was kissed by the girl" or center-embedded sentences such as "The man the boy hit carried the box." I.L.'s errors frequently revealed an incorrect syntactic interpretation that seemed to be dependent on the use of a simple pragmatic strategy in which syntactic relations between elements were assigned in a simple Subject-Verb-Object sequence. Other studies also reported that patients with severe STM deficits showed comprehension difficulties when the sentences are lengthy (Baddeley, Vallar, & Wilson, 1987), are semantically reversible (Caramazza, Berndt, Basili, & Koller, 1981), and contain many content words (Martin & Feher, 1990). On the other hand, numerous findings indicated that patients whith STM problems showed normal performance for short sentences and sentences containing simple syntactic constructions (see Gathercole & Baddeley, 1993).

In a similar theoretical context, Vallar and Baddeley (1987) suggest that the phonological store functions as a "mnemonic window," which preserves word order by a verbatim phonological record. Accordingly, patients with phonological STM deficit should experience sentence processing difficulties restricted to sentences where word order is critical. The authors showed that P.V., a patient with a severe phonological store deficit, performed well on short sentences with a wide range of syntactic constructions and had no problem with semantically reversible constructions. However, in a sentence verification task, she failed to reject false sentences when they were constructed by reversing two relevant items from true sentences (e.g., "The world divides the equator into two hemispheres, the northern and the southern"). On the other hand, P.V.'s scores were better when shortened versions of the complex sentences containing word reversals were presented.

19-1.1.2. The Phonological Short-Term Memory as a Support to Second-Pass Language Processing

The other view concerning the role of the phonological STM in sentence comprehension suggests that phonological storage is used in second-pass analysis after syntactic analysis has been conducted, but before the sentence structure has been fully interpreted. For example, according to Caplan and Waters (1990), misanalyzing sentences such as "The boat floated down the river sank" reflects the operation of first-pass processing. Recovery from such a "garden path" (a sentence that produces temporary syntactic ambiguity in the course of parsing) is a second-pass processing that involves reconsideration of the structure already assigned and this requires the consultation of the phonological representation of the sentence. In the same vein, Waters, Caplan, and Hildebrandt (1991) suggest that the phonological storage is necessary for a postsyntactic check of the assignment of nouns to their thematic role. They described a patient, B.O., who showed a deficit of the articulatory rehearsal process and possibly of the phonological store. She showed excellent comprehension of a wide variety of syntactic structures, indicating that articulatory rehearsal is not needed for assignment of syntactic structure and thematic roles. By contrast, she had a particular difficulty in processing sentences containing three proper nouns such as "Patrick said that Joe hit Eddy," while she had no problem with sentences that used animal names in the place of proper nouns. It was argued that this pattern of performance indicates a role for phonological STM at a postsyntactic stage of sentence comprehension, in which the assignment of nouns to their thematic roles is checked.

This "second-pass" interpretation is also supported by the performance of two patients, R.A.N. and N.H.A., reported by McCarthy and Warrington (1990a). These patients with poor STM performance showed difficulties in comprehension when they had to deal with large amounts of condensed (and relatively arbitrary) information in a relatively brief message (such as Token Test instructions, for example, "Before picking up the green circle, touch the red square"; De Renzi & Vignolo, 1962), when sentences broke conventional conversational rules, and when an understanding of the sentence required supplementary cognitive operations to be performed on the spoken input (for example, simple comparative judgments such as "What is green, a poppy, or a lettuce?").

On the other hand, Caplan and Waters (1990) reinterpreted the performance of P.V., the patient reported by Vallar and Baddeley (1987), in terms of a deficit in second-pass language processing. According to Caplan and Waters, P.V.'s difficulty in processing sentences such as "The world divides the equator into two hemispheres, the northern and the southern" may lie at the stage of processing in which the patient must choose between the syntactically and lexicopragmatically derived meanings of the sentence.

However, some data do not support Caplan and Waters's (1990) hypothesis. For example, Martin (1993) observed that patients with phonological short-term memory deficits showed good comprehension of garden-path sentences.

In conclusion, it appears that patients with impaired phonological STM do not necessarily share the same comprehension difficulties with the same types of structures. This might suggest that the phonological loop does not actually play a role in sentence comprehension. Indeed, the existence of an association between poor immediate retention of verbal material and sentence comprehension deficits is not sufficient to state that phonological storage is involved in language comprehension. Such an association might arise simply from the anatomical proximity of the neural structures involved in both types of activity. In that perspective, Butterworth, Campbell, and Howard (1986) reported the case of R.E., who showed a developmental deficit of the phonological STM. She was tested on a wide range of tasks requiring syntactic analysis, memory, and comprehension of long and complex material. Despite her phonological STM deficit, she showed no deficit in sentence comprehension. Butterworth et al. (1986) argued that, as a phonological STM impairment did not automatically produce deficits in comprehension, it was not critically involved in sentence comprehension. However, some authors (Vallar & Baddeley, 1989) questioned the sensitivity of the comprehension tests administered to R.E. It was also suggested that R.E. might have developed highly atypical strategies for language comprehension as a consequence of a lack of STM capacity.

It could also be argued that the heterogeneity of STM patients' performance in sentence comprehension is, at least partly, the result of differences in the severity or nature of the phonological loop disorders. In fact, most STM patients who were explored had a deficit affecting the phonological store. However, the specific nature and severity of their phonological store deficit might differ. Finally, this heterogeneity might also indicate that factors other than phonological memory capacity intervene in the relationships between STM and comprehension deficits.

19-1.1.3. Involvement of Semantic and Syntactic Short-Term Memory Components in Sentence Comprehension

Most researchers who attempted to explore the performance of patients with reduced verbal memory span in sentence comprehension assume that memory span is dependent on the phonological loop component. However, some evidence has suggested that lexical and semantic storage also plays a role in span performance (e.g., Howard, 1995). Thus, it appears that memory span is a complex phenomenon drawing on different levels of representations. In that perspective, Martin, Shelton, and Yaffee (1994) explored the short-term abilities and sentence comprehension in two brain-damaged patients (A.B. and E.A.). Although both patients showed similar span reduction, one patient (A.B.) showed better performance on span tasks for phonological than for semantic information, while the other patient (E.A.) showed the reverse. For example, Martin et al. compared both patients on immediate recall for words and nonwords. Typically, normal subjects perform better on words than nonwords (Hulme, Maugham, & Brown, 1991), presumably because word retention depends on both phonological and semantic information, whereas nonword retention mainly depends on phonological information. Moreover, the authors used a task in which word lists were presented auditorily and patients were asked to judge either whether a probe word rhymed with one of the words in the memory list or whether the probe word was in the same category as one of the memory list items. The probe rhyming task required retention of phonological information while the probe category task required retention of lexical-semantic information. In lists recall, E.A. showed a superiority for words over nonwords whereas A.B. showed little advantage for words. In addition, A.B. performed better than E.A. on the probe rhyming task, that is, when retention of phonological information was required, whereas E.A. performed better on the category probe task, that is, when retention of semantic information was required. These results suggest that E.A.'s STM deficit affected mainly the retention of phonological information whereas A.B.'s affected mainly the retention of lexical–semantic information. On the other hand, the performance of both patients on the sentence-processing tasks was consistent with these contrasting STM deficits. On sentence repetition, the patient with the phonological retention deficit (E.A.) was more impaired than the patient with the semantic retention deficit (A.B.), whereas on sentence comprehension, the patient with the semantic retention deficit (A.B.) was more impaired. In another study, Martin and Romani (1994) described a patient, M.W., who showed a difficulty in sentence comprehension that was attributed to a deficit affecting specifically the retention of syntactic structures.

According to Martin and Romani (1994), these results suggest that memory span should not be seen as dependent on the operation of a specialized STM store but rather as drawing on the operation and storage capacities of a subset of components involved in language processing. In other words, the different levels of representation involved in memory span and language processing draw on specific resources, which may be conceptualized either as buffers specialized for particular types of representations, or in terms of rates of decay that may differ for different levels of representation. The phonological, semantic, and syntactic components interact and support each other and their respective contribution depends on the nature of the task. For example, syntactic and lexical–semantic information could be linked to phonological information to support verbatim recall, resulting in a larger span for words than for nonwords and a larger span for sentences than for word lists.

This view clearly differs from Baddeley's (1986) working memory conception in that it considers that verbal STM is an integral part of the language system, whereas for Baddeley, the working memory components are not strictly tied to any particular cognitive system. Furthermore, Martin et al.'s (1994) view postulates the existence of specialized storage resources for semantic and syntactic information, in addition to the phonological component. Within the framework of Baddeley's model, the temporary storage of semantic (or even syntactic) information might be plausibly viewed as residing in the central executive system or as resulting from a temporary activation of long-term memory information (a long-term working memory; Ericsson & Kintsch, 1995) that is coordinated by the central executive. Such a conception clearly raises the question of the central executive system's role in sentence comprehension deficits.

19-1.2. Central Executive and Language Comprehension

The work of Daneman and Carpenter (1980) strongly contributed to the view that comprehension of language places demands on working memory considered as a limited pool of general-purpose resources that can be used to serve both processing and storage of information. This conception of working memory appears to correspond rather closely to the central executive component in Baddeley's (1986) working memory model.

According to Daneman and Carpenter (1980), limitations on cognitive resources can account for many types of individual differences and processing strategies in language comprehension (see Carpenter, Miyake, & Just, 1995). Similarly, it could be argued that a deficit affecting the central executive system should impair language comprehension. However, a major difficulty in the exploration of the relations between language deficits and the central executive system is that a wide range of cognitive abilities have been ascribed to the executive system: control, processing, and even storage activities. Another problem is to find a way to specifically explore the central executive system without confounding its function with that of the slave systems.

Research conducted by Morris (see Morris, 1994) and Baddeley et al. (1986) provided some criteria indicating the existence of a specific central executive deficit (see also Van der Linden et al., 1992). Patients with an impaired central executive should show a mild reduction of span, normal effects of phonological similarity and normal length effects on span (indicating the integrity of the phonological store and the articulatory rehearsal process), and impaired performance on dual tasks.

19-1.2.1. Alzheimer's Disease and Language Comprehension

On the basis of these criteria, several studies suggest that patients with Alzheimer's disease (AD patients) have impairments of the central executive component of working memory (see Morris, 1994). More specifically, it appeared that AD patients showed a deficit affecting one important component of the central executive, that is, the capacity to coordinate two or more subprocesses (Baddeley et al., 1986). Waters, Caplan, and Rochon (1995) explored the relationships between processing capacity and sentence comprehension in AD patients. Patients were administered a sentence–picture-matching task. The to-be-interpreted sentences differed in syntactic complexity and number of propositions. Subjects were tested on this task, on the one hand, with no concurrent task and, on the other hand, while concurrently remembering a digit load that was one less than their span or equivalent to their span. The AD patients met Morris's (1994) and Baddeley et al.'s (1986) criteria indicating the existence of a central executive impairment. The results show an interaction between subject group and size of the digit load in the digit recall measurement, which indicates that the performance of AD patients was more affected by dual-task conditions than that of the controls. This is consistent with the view that AD patients have a reduced capacity of the central executive system. However, the patients were not disproportionately impaired on the sentence types that were syntactically more complex. The authors interpret this absence of dual-task effect in AD patients by postulating the existence of different processing resource pools for different types of verbal operations. More specifically, they argue that AD patients showed a reduction in central processing resources but that this reduction did not affect the availability of resources involved in syntactic processing. In addition, the results show that the patients performed more poorly on sentences with two propositions than did the controls, and more poorly on sentences with two propositions than on sentences with one proposition. The authors interpret this disproportionate "proposition" effect by suggesting that the patients have a deficit affecting postinterpretative processing (as opposed to assignment of sentence meaning itself), and more specifically the cognitive processes involved in matching propositional content to a picture. Finally, although the AD patients showed a decrement in dual-task performance, they did not show a disproportionate decrease of performance on two-proposition sentences compared to one-proposition sentences, under digit load conditions, compared to controls. This pattern of results cannot be explained by a reduction of processing resources since such a reduction should have led to a three-way interaction between size of digit load, sentence type, and group, which was not observed. According to Waters et al. (1995), the results could be interpreted by postulating that AD patients have impairments in a control mechanism that shifts attention between tasks and this control deficit could lead to a decline in performance under dual-task conditions, without this decline being greater for more demanding sentences. This study clearly suggests that the investigation of the central executive's role in sentence comprehension requires a precise specification of the resource pools and of the resource allocation. Waters et al.'s (1995) study also indicates that different aspects of the central executive system may be affected, for example, the processing capacity or the control component (flexibility), with different consequences on performance.

19-1.2.2. Language Comprehension in Aphasic Patients

Miyake, Carpenter, and Just (1994) argued that comprehension breakdown in aphasic patients may be attributed to a severe reduction of a general-purpose verbal working memory system. This proposition was based on results issued from two sentence comprehension experiments using rapid serial visual presentation in normal subjects, which they consider simulate important features of aphasic patients' comprehension of syntactic structures. However, this view has been challenged by Caplan and Waters (1995) and Martin (1995). These authors point out that a theory hypothesizing damage to separable components more easily accommodates findings of double dissociations in aphasic patients' performance on different sentence types, and that Miyake et al.'s (1994) results may also be accounted for by a theory that assumes separable processing resource systems involved in language comprehension and other verbal tasks.

19-1.2.3. Aging and Language Comprehension

Finally, a large body of research also suggests that language performance differences (for example, in the comprehension of and memory for discourse) between young and old adults are due to age-associated differences in working memory capacity (see Hupet & Nef, 1994). However, other studies indicate that speed of processing and inhibitory efficiency might be more fundamental factors than working memory (e.g., Kwong See & Ryan, 1995). In this perspective, Hasher, Zacks, and their colleagues have conducted a series of studies showing that age differences in the inhibition of irrelevant internal thoughts and external stimuli underlie age difference in language performance (see Zacks & Hasher, 1994). In fact, they argue that people with inefficient inhibitory mechanisms will allow information that is off the goal path to enter into working memory. For example, older adults have been found to be less likely than young subjects to inhibit initial inferences subsequently made untenable by text context. More generally, differential ability to inhibit information has been proposed as an alternative to resource theories in interpretations of developmental and individual differences (see Dempster & Brainerd, 1995). Most of these inhibition theories seem to consider inhibition as a passive, automatic event. However, recent findings suggest that inhibition is (at least partly) a product of controlled resources and that group differences in inhibition may result from differences in controlled attentional resources, and not from inefficient inhibitory processes (Engle, Conway, Tuholski, & Shisler, 1995). This view is consistent with the idea of a central executive as being a general attentional system, whose function would be the inhibition of irrelevant information as well as activation and maintenance of information relevant to the task.

In summary, there exists some evidence relating central executive dysfunctioning to language comprehension deficits. However, the central executive is certainly not a unitary system and, as a consequence, it could be damaged in different ways. In that perspective, future research should be conducted to explore the comprehension abilities of patients with different types of central executive disorders.

Language comprehension certainly is the domain of language activities in which the role of working memory has been the most extensively explored from a neuropsychological point of view. In the last part of this chapter, we will examine, more briefly, whether neuropsychological data also suggest a contribution of working memory to other types of language deficits.

19-2. WORKING MEMORY AND SPEECH PRODUCTION DEFICITS

The neuropsychological findings favor the view that the phonological loop is not involved in the planning and production of spontaneous speech. Indeed, normal speech production has been reported for several patients with acquired STM deficits. For example, the patient J.B., described by Shallice and Butterworth (1977) had a severe deficit of verbal STM but her speech production abilities were normal: her utterances were syntactically well formed and of normal phrase length, her pause duration was within the normal range, and the content of her speech was reasonably informative. On the other hand, there exists some evidence issued from studies conducted in normal subjects suggesting that the central executive could be involved in constructing the conceptual content of speech (see Gathercole & Baddeley, 1993). At this time, no neuropsychological data directly corroborate this hypothesis.

However, Martin and Romani (1995) reported the case of a patient, Alan, who had a very specific deficit in the short-term retention of semantic representations and who developed problems in producing (and also in understanding) speech. The patient's STM deficit affected only the retention of the meanings of individual words (or lexical–semantic representations). In other words, he did have difficulty in keeping simultaneously in mind the meanings of several words, whereas he was able to hold verbal information in a phonological form. His comprehension abilities were normal as long as he was able to integrate word meanings with each other as each word was perceived. However, he had particular difficulty in comprehension when he was forced to retain individual meanings in STM. On the other hand, his speech production was very poor whenever he tried to produce more than single-word utterances. For example, he was unable to produce an adjective–noun phrase correctly, whereas he was normal at naming individual objects. It was argued that Alan's production deficit derived from an inability to retain several individual word meanings simultaneously while the phonological representations for the words are accessed. Whether the retention of word meanings depends on a specific storage resource or rather on the general-purpose central executive system, it nevertheless remains that these data suggest the existence of a relationship between STM and speech production deficits.

19-3. WORKING MEMORY AND VOCABULARY ACQUISITION

In contrast to speech production, neuropsychological studies have provided strong evidence suggesting that the phonological loop plays a critical role in vocabulary acquisition. Typically, patients with defective phonological STM may have preserved performance on standard episodic memory tasks, requiring the acquisition of verbal information. However, a study conducted on patient P.V., who had a selective phonological store deficit, suggests that learning and retention of new words (that is, pronounceable letter strings without any preexisting lexical–semantic representations) is severely defective (Baddeley, Papagno, & Vallar, 1988). Her verbal long-term learning and retention of individual words and connected discourse were normal. Baddeley et al. (1988) explored P.V.'s abilities to learn familiar words and new words by means of an auditory paired-associate learning procedure. P.V. was completely unable to learn word–nonword pairings (an Italian word and its Russian translation, for example, *rosa–svieti*), whereas her learning of pairs of Italian words was normal. Baddeley et al. (1988) concluded that long-term memory of new phonological forms requires the integrity of the phonological store. Similar difficulties in acquiring new words have also been reported by Baddeley (1993a) in a 23-year-old graduate student, S.R., who showed a developmental deficit of phonological working memory.

The hypothesis that phonological working memory plays a role in vocabulary acquisition was also supported by a study of Vallar and Papagno (1993) conducted in a 25-year-old Italian woman affected by Down's syndrome. F.F. was able to learn three languages (Italian, English, and French) in spite of poor general intelligence and defective visuospatial skills and episodic memory. In addition, F.F. has an entirely normal phonological short-term memory. Finally, contrary to P.V., F.F. was able to learn word–nonword pairs at a normal rate, whereas her learning of word–word pairs was defective. Barisnikov, Van der Linden, and Poncelet (1996) obtained similar results with a 19-year-old woman suffering from Williams syndrome and showing relatively preserved language abilities in spite of a severe visuospatial deficit. These data corroborate the existence of a link between the phonological loop and vocabulary acquisition and also indicate that learning of new vocabulary may occur even in the presence of a defective verbal episodic memory.

Finally, numerous developmental and experimental studies have also contributed to verify the neuropsychological indications that the phonological loop is critically involved in the acquisition of new vocabulary. In particular, it appears that poor vocabulary acquisition associated with developmental language disorders may be due to deficits of phonological working memory (see Gathercole & Baddeley, 1993).

19-4. CONCLUSION

A number of neuropsychological studies indicate that a deficit in working memory may result in some language impairments, such as sentence comprehension, acquisition of vocabulary, or even in speech production. These relationships between working memory and language have mainly concerned the phonological loop system and the central executive. In fact, the visuospatial sketchpad might also play a role in some aspects of language processing, for example, in text comprehension, inasmuch as mental imagery might be involved. Other data issued from developmental or experimental studies have confirmed these neuropsychological indications but have also shown that working memory might be implicated in other domains of language processing, such as reading (see Gathercole & Baddeley, 1993).

Baddeley's (1986) working memory model has played an important part in neu-ropsychological studies of patients with language impairments. However, it appears that some aspects of Baddeley's model need to be elaborated in order to accommodate some empirical findings, for example, the effect of semantic information on STM performance. In addition, a more precise specification of the central executive system is clearly required.

CHAPTER 20

The Influence of Emotions on Language and Communication Disorders

Diana Van Lancker[1] and Nancy A. Pachana[2]

[1]Department of Neurology, University of Southern California Medical Center, Los Angeles, California 90033, and Veterans Affairs Outpatient Clinic, Los Angeles, California 90012; [2]Department of Psychology, Massey University, Palmerston North, New Zealand

Language, cognition, and emotion have traditionally been viewed as separate elements of mental processing. Although language and emotion differ essentially, evidence from normal and disordered linguistic expression suggests that attitudinal and emotional information is naturally present in speech. Verbal communication expresses attitudes and emotion via prosody, syntax, lexicon, and the pragmatics of discourse. Emotional and psychiatric disturbances resulting from dysfunction of the cerebral hemispheres, limbic system, and the basal ganglia are manifest in speech/language function. Disorders of affective communication, prosodic, syntactic, lexical, and pragmatic disturbances, are associated with brain damage at various sites, especially subcortical, bilateral prefrontal, and right hemisphere.

In one view, language, a structural system, has only a tangential relationship to emotions. Strictly speaking, phonology, syntax, and the semantic lexicon provide tools, equally capable of verbalizing about the world, a mathematical theorem, or an emotion. From this perspective, language and emotions differ intrinsically. In another view, language has firm biological roots in emotional expression. Jespersen (1969) speculated that human language originated in love song. Bolinger (1964) believed that the "primitive dance" of human emotional life resides in intonation, the foundation of speech. The recent field of "pragmatics," the orderly use of language in everyday situations, now investigates the role of nonliteral meanings, humor, themes, and inferences, including inference about attitudes, in language processing.

There is much evidence to support the second view—that emotions, moods, affect, and attitudes naturally underlie and inform normal spoken expression. Further, neurolinguistic studies indicate that disorders of emotional processing result in communicative deficits. This chapter provides an overview of the relationship between emotions and language in normal and disordered language and communication.

20-1. SOME DEFINITIONS

Language and cognition (or thought) require only minimal characterization for the purpose of this chapter, as the terms are used fairly consistently by professionals and laypersons. The term "language" refers to the speakers' internal knowledge of phonological, morphological, and syntactic rules with an accompanying lexicon; "speech" refers to the external expression of language. "Cognition" refers to mental experience of all kinds—processing, memory, attention, concentration, images, planning. For many years, cognitive psychology proceeded without mention of emotion or affect (see Zajonc, 1960), but recently the importance of emotional matters in thinking has been recognized broadly (Damasio, 1994; Goleman, 1995).

The terms for emotional phenomena are the most resistant to satisfying systematization. In psychology and psychiatry, emotional states are characterized in terms of the patient's "mood" or "affect." Mood generally refers to the patient's internal subjective feeling state, either in the moment or over time; mood may not be visible to the examiner, as outward clues to internal feeling states are often misleading in the presence of neurologic or affective disorders. "Affect" may be used to refer to an emotional state before it is endowed with awareness; "affect" has been used by others to refer to manifestation of emotion, especially facial expression, gesture, and tone of voice (Benson, 1984). Some behaviors, such as screaming and tearfulness, may occur without a veridical tie to internal mood or emotional state, as seen in pseudobulbar emotional states (Ivan & Franco, 1994). In contrast, "unconscious" affect may account for speech errors, unintentional puns, choice of words, or tone of voice that does not match the words. The "Freudian slip" theory of speech errors (Freud, 1960) has been partially overruled by scientific analysis of speech error data, which indicates that linguistic–structural factors, rather than unconscious motivation, generally account for the errors (Cutler, 1982). We use the terms "emotional behavior" and "affective behavior" for outward manifestations, and "emotional experience" and "affect" to refer to inner states, which may or may not be conscious, and may or may not have outward signs. In fact, frank emotions emerge from speech much less than the broad range of speaker attitude, such as approval, dislike, concern, impatience, and so forth. Attitude is more commonly manifested in ordinary talk, but is often neglected in research on language and expression.

20-2. HUMAN EMOTIONAL EXPERIENCE

Emotional states may be viewed in terms of their bodily manifestations (e.g., a smile) or the subjective experience of the emotion (e.g., feeling happy). Of course, emotions correspond to patterns of brain activity, and some knowledge of cerebral structures and circuits important in emotional experience has been derived from animal lesion experiments and from studies of humans with localized brain damage. Physiological manifestations of emotional states include changes in the autonomic nervous system and the release of neurotransmitters, neuroendocrine secretions, and hormones. Gesturing and changes in posture, facial expression, and voice modulation represent another set of bodily changes in response to emotional states. Other verbal signs may occur—distinctive word or nonverbal vocal choices, such as swearing, sighing, or emitting exclamations that may have begun as a reflexive vocalization but have evolved to onomatopoeic words: *ow!, Jeez!*

Attempts to categorize various emotional states resulted in a plethora of systems, in part because of the difficulty in coming to agreement on definitions, and in part because of the number and range of human emotional experiences (Ekman, 1992). In 1890, William James articulated a theory of emotions that later became known as the James–Lange theory of emotions. This theory states that the subjective experience of emotion is produced by the awareness of physiological changes in the face of arousing stimuli. Thus, when confronted by a snarling dog, we are aware of sweating palms, ragged breathing, and a pounding heart; this leads to a subjective feeling of the emotion of fear. The important relation of body states to feeling states has been revisited (Damasio, 1994; Ekman & Davidson, 1993). But since neuroscientists have noted that visceral responses to arousing stimuli are often similar, how does one explain the great variety and subtle range of human emotional experiences?

A different theory of emotions, which emphasizes cognitive factors, was offered by Schacter and Singer (1962). In their theory, autonomic arousal is shaped into specific emotional experience via an attribution process, utilizing circumstances and context to a great extent. Both physiological arousal and cognitive mediation play important roles in the process of the subjective feeling of emotions and in the subsequent verbal and nonverbal displays.

20-3. LANGUAGE, THOUGHT, AND EMOTION— THE TRIAD

Language, emotion, and thought may form autonomous modules in the mind of the philosopher, but in reality they are intertwined. Thoughts are ordinarily imbued with emotional tone, and language interfaces intimately with cognition and emotion. The actual intimacy of cognition and emotion, with emotion often preceding an apparently autonomous cognitive process, has been eloquently discussed by Zajonc (1960). Current neurobehavioral opinion has argued persuasively for the interdependence of thought, emotion, and body in that all may be disrupted by damage to one cortical area.

Attempts at attributing structure to any one component of the "triad" have been the most successful in the case of language, which is made up of phonemes, morphemes, words, sentences, discourse, and rules for organizing the smaller units (grammar or syntax). Structural models in cognition (e.g., information processing, schemata, models of memory and attention) have clarity and coherence. Of the triad, emotional phenomena have been the least amenable to codification. Linguistic structure has been said to access thought through an intermediate level called "mentalese" by Fodor, Bever, and Garrett (1974). Some cognitive scientists have avidly debated whether "thought" can occur without language, and others have insisted that language feeds back onto thought, determining its conceptual categories (Whorf, 1961). In sharp contrast, the question does not arise regarding language and emotion: no one doubts that emotion can occur without language. It might be argued that "language" as a structural system can occur without the influence of emotion, but actual verbal communication seldom does. Whereas language has been recognized as a major partner of cognition, by contrast, language and emotion are best described as having an uneasy coexistence, being disparate entities as well as coworkers in the business of communication. However disparate language and emotion may be, *emotional and attitudinal nuances tinge all but the most neutrally constructed linguistic expression.*

20-4. DOES LANGUAGE EXPRESS EMOTIONS?

Language does indeed express emotion both indirectly and directly, implicitly or explicitly, using different "levels" of linguistic form: phonetic, lexical, syntactic, semantic, pragmatic. Sounds and words carry valences, expressing attitudes or emotions implicitly, even in the earliest stages of communicative development. For example, a range of attitudes and emotions is identifiable in the melodies of mothers' speech to their infants (Fernald, 1985). A prime lexical example of implicit emotional information lies in connotative meanings; connotations convey attitudes, adding to the referential meaning. The lexical items "slim," "slender," "skinny," and "scrawny" all denote physical thinness, but the first two connote approval, whereas the others are derogatory. Osgood (1980) contributed the "semantic differential" to our understanding of how word meanings are mentally stored with emotional significance. A salient example of specific choice of verbiage to convey attitudinal and emotional meaning is cursing (Jay, 1992).

The speaker's intentions and attitude are very important components of speech, and may be the least studied in neurolinguistics. The term "attitude" in social psychology refers to a stable mental position consistently held by a person toward some idea, or object, or another person, involving both affect and cognition. In language, word choice and intonation convey the attitude of the speaker: irony, incredulity, contempt, disdain, approval, and sarcasm. Attitude can be directed toward the speaker himself/

herself, toward the listener, or toward the utterance. Further, the speaker can express an attitude toward either the form or the content of the utterance. For example, the speaker can ridicule a word or phrase by using prosody and gestures that appear to place it in quotes, as in: "She calls herself a 'sanitary engineer'!"; or he/she may communicate an attitude—approval, contempt, disbelief—about the idea expressed by the statement, as in: "He studies the 'emotional life' of 'rats'!" Attitude can be communicated and perceived unintentionally in ways discovered by sociolinguistic studies (e.g., Erickson, Lind, Johnson, & O'Barr, 1978). For example, facts stated in female voices were rated less important than the same factual statements said by males (Geiselman & Crawley, 1983). Speaking voice is a rich source of impressions about attitude, mood, and personality (Banse & Scherer, 1996; Hecker, 1971; Kreiman, 1997; Sapir, 1926-1927; Scherer, 1979).

20-4.1. How Does Language Express Emotion Actively?

Prosody, or the melody of speech, takes first place in the list of media for actively expressing emotion in language. Its authority will usually override contradictory verbal content. For example, if the statement "I'm *not* angry!" is made with raised amplitude and pitch, increased duration, vocal tension, and hyperarticulated phonetic elements (all indicating anger), the medium—the prosodic underlay—is the message, and the words are unbelievable. Both face and voice join forces to contribute affective information (Massaro & Egan, 1996), but the presence of a smile can be heard in the acoustics of the speech signal with no visual cues (Tartter, 1980). The word "right" can imply a broad range of meanings, from casual agreement to condemning rejection: "right" said with rapidly rising/falling, higher pitch means agreement and reinforcement; spoken with low, falling pitch, longer vowel and creaky voice, the same word indicates sarcastic repudiation. Lexicon, of course, provides a rich source of terms to express attitudinal information about oneself, one's reaction to the world, one's judgments, one's feelings, and so forth. Word order can be used to convey attitude and nuance; importance is given an element by placing it at the beginning or end of the sentence. Nonliteral language (e.g., idioms and proverbs) and other nonpropositional expressions—slang, clichés, exclamations, and expletives—also express and evoke varying degrees of emotion.

Written language, being restricted to words and syntactic structures, is relatively impoverished in its ability to communicate emotion, with some aid from punctuation (exclamation marks, italics, underscoring, and hyphens) in the formal style. Informally, iconic distortions of letters (e.g., large and/or boldface type to indicate loudness, wavy to indicate fear), reduplicated question or other punctuation marks (e.g., "You're leaving???" to indicate surprise or shock), and other creative variations may be used to communicate emotional and attitudinal meanings (see Figure 1). Discourse units, which can use words and grammar to build scenes and ideas, provide the greatest emotional force in the written medium. In a demonstration of physiological emotional correlates of written expression, Hughes, Uhlmann, and Pennebaker (1994) recorded skin conductance changes while subjects wrote about traumatic experiences.

Figure 1 Creative use of punctuation and type styles to represent attitudinal and emotional meanings. (Reprinted with special permission of King Features Syndicate.)

20-5. ROLE OF BRAIN DAMAGE IN EMOTIONAL ASPECTS OF LANGUAGE AND COMMUNICATION

Emotional experience and behavior involve many brain structures in ways yet to be clearly characterized. The neuroanatomical substrate for elemental emotional behaviors includes both cortical and limbic structures. Connections to the prefrontal cortex, cingulate gyrus, and amygdala are believed to be especially important in emotional expression and experience (Adolphs, Tranel, Damasio, & Damasio, 1994; Clark, 1995; Damasio & Van Hoesen, 1983; Devinsky, Morrell, & Vogt, 1995; LeDoux, 1995).

Hemispheric differentiation is a distinctive feature of the human brain and has been of major importance in the study of higher cortical functions such that the location of language and practic functions in the left hemisphere of right-handers is well established. The right hemisphere in the hemidecorticate child may well learn language, but subtle differences in linguistic skill remain (Dennis & Whitaker, 1976). More recent work has located such disparate phenomena as visuospatial skills (Bogen, 1997) and recognition of familiar faces (Benton & Van Allen, 1968) and voices (Van Lancker & Canter, 1982) in the right hemisphere.

Hemispheric specialization for emotional experience and its outward expression and comprehension is less clear (Bear, 1983). Emotional experiencing is regulated to a great extent by bilateral subcortical structures with multiple cortical connections. Both the left and right hemispheres play important roles with respect to the comprehension and expression of emotion (Davidson & Tomarken, 1989; Gainotti, Caltagirone, & Zoccolotti, 1993). When considering "subcomponents" of approach- and withdrawal-related emotion, Davidson (1994) proposes that approach- and goal-related behaviors, both involving positive affect, are associated with left prefrontal cortical function. Verbal reports after viewing positive emotional film clips were more intense in persons with stable frontal brain electrical asymmetry (Wheeler, Davidson, & Tomarken, 1993). Some neurobehavioral observations suggest that damage to the left hemisphere leaves a patient prone to syndromes involving negative affect (catastrophic reactions, anxiety, depressive mood), while damage to the right hemisphere is associated with indifference, denial, and sometimes inappropriate jocularity *(Witzelsucht)* (Kolb & Taylor, 1981; Robinson, Kubos, Starr, Rao, & Price, 1984; Sackeim et al., 1982).

Besides through facial expression as described earlier, emotional information is conveyed in prosody (Van Lancker, 1980); words (Borod, 1993); "nonpropositional" (Bogen, 1969; Code, 1987; Jackson, 1874; Van Lancker, 1988) and nonliteral expressions (Myers & Linebaugh, 1981; Van Lancker & Kempler, 1987; Winner & Gardner, 1977), and in larger "discourse" units, such as conversation and stories (Brownell & Martin, in press). Communicative deficits involving emotional information occur primarily in spoken production or comprehension modes. Although less common, selective deficits in processing emotional content in reading (House, Dermot, & Standen, 1987) or writing can occur, and repetition (Speedie, Coslett, & Heilman, 1984) of affective–prosodic information may be impaired.

Deficits in prosodic production (dysprosody) and comprehension following right-hemisphere damage have been actively studied in recent decades (Heilman, Scholes, & Watson, 1975; Ross, 1980; Van Lancker, 1980). Although impaired "melody of speech" was first associated with left-hemisphere damage (Goodglass & Kaplan, 1972; Luria, 1966; Monrad-Krohn, 1947), there is evidence for a role of the right hemisphere in processing stimuli containing affective information (Bryden, & Ley, 1983). So far, no overall model accounts for the many disparate reports in studies of prosodic function: questions arise about validity of assessments, the meaning or pertinence of prosodic measures chosen, categorical versus dimensional appreciation of affective intonations (Peper & Irle, 1997), other task and modality variations (Blonder, Bowers, & Heilman, 1992; Bowers, Coslett, Bauer, Speedie, & Heilman, 1987; Tompkins, 1991a, 1991b), the transitory nature of the prosodic change following damage (Darby, 1993), or the relative role of linguistic prosody (Behrens, 1989; Shapiro & Danly, 1985; see Ackerman, Hertrich, & Ziegler, 1993 and Baum & Pell, in press, for reviews). A special difficulty lies in the lack of a model of normal prosodic performance, encompassing acoustic and psychological attributes, although considerable progress toward a descriptive framework has been made by Scherer and his colleagues (see Banse & Scherer, 1996; Pittam & Scherer, 1993). Subcortical dysfunction may account for many cases of observed affective dysprosody (Cancelliere & Kertesz, 1990; Cohen, Riccio, & Flannery, 1994; Lalande, Braun, Charlebois, & Whitaker, 1992; Starkstein, Federoff, Price, Leiguarda, & Robinson, 1994; Van Lancker, Pachana, Cummings, Sidtis, & Erickson, 1996). The speech-timing deficits resulting from ataxic disturbances in cerebellar disease may also convey an impression of affective prosodic disturbance.

With acknowledgment of a significant role of the basal ganglia in affective–prosodic behaviors, a new look at the phenomena of "flat" speech or impaired prosody becomes possible. The importance of subcortical structures to intact prosodic production and comprehension is seen in Parkinson's disease (Blonder, Gur, & Gur, 1989; Darkins, Fromkin, & Benson, 1988; Scott, Caird, & Williams, 1984), Huntington's chorea (Speedie, Brake, Folstein, Bowers, & Heilman, 1990), and other dysarthrias arising from basal ganglia deficits (Darley, Aronson, & Brown, 1975). The importance of the basal ganglia in regulating the facial and prosodic expression of motivation and mood (Ali Cherif et al., 1984; Bhatia & Marsden, 1994; Mayeux, 1986; Mendez, Adams, & Lewandowski, 1989; Poncet & Habib, 1994), especially in association with connections to the frontal lobes (Cummings, 1993), gives new impetus to prosodic analysis. In describing the relationship of phylogenetically older limbic sources of vocalization, Robinson (1976) concluded that the limbic system remains "a significant factor in the functioning of human speech" (p. 769).

20-5.1. Varieties of Linguistic Phenomena as Sequelae of Brain Insult

"Flat" or monotonous speech occurs in various neurological disorders, and has been associated with right- and left-hemisphere damage as well as subcortical damage; as yet no clear picture has emerged (Kent & Rosenbek, 1982; Meerson & Tarkhan, 1988; Ryalls, 1988; Ryalls & Behrens, 1988). Changes from normal values in mean fundamental frequency (F_0) as well as F0 variation have been seen in both left (Danly & Shapiro, 1982) and right (Shapiro & Danly, 1985) hemisphere-damaged subjects, as well as those with damage to basal ganglia structures. Patients with left-sided and right-sided damage did not differ in reading aloud material with emotional content (House et al., 1987). Postmorbid dysprosody in production may be due to mood changes, cognitive programming failure (apraxia), motor dysfunction (dysarthria), or motivational defect (abulia), while deficits in affective–prosodic comprehension may be attributable to deficient pitch analysis or to deficits in comprehension of verbal–emotional content. How the known superiority of the right hemisphere in complex pitch (Robin, Tranel, & Damasio, 1990; Sidtis, 1980; Sidtis, & Volpe, 1988; Twist, Squires, Spielholz, & Silverglide, 1991; Zatorre, 1988; Zatorre, Evans, Meyer, & Gjedde, 1992) and musical functions (Gordon, 1970; Gordon & Bogen, 1981; Sidtis, 1984; Tramo & Bharucha, 1991) pertains to right-hemisphere performance on prosodic tasks is also unclear. A better picture of prosodic functioning will emerge when more is known of (a) normal human affective states; (b) prosodic–affective expression types; and (c) acoustic differentiation of vocal affective expression (Johnson, Emde, Scherer, & Klinnert, 1986). Studies of dysprosody are hampered by the fact that "the manner in which (acoustic and phonetic parameters) combine to signal linguistic and affective meaning remains elusive" (Bowers, Bauer, & Heilman, 1993, p. 436).

Deficits in processing emotional lexical content in production and comprehension following focal brain damage have been described in various ways (Cicone, Wapner, & Gardner, 1980; Cimino, Verfaellie, Bowers, & Heilman, 1991; Rapcsak, Comer, & Rubens, 1993). A lack of emotional content in the spoken lexicon, associated with difficulty identifying and describing feeling states, is referred to as alexithymia. Alexithymia (Nemiah, Freyberger, & Sifneos, 1976) refers to a dearth of emotional–expressive words in speech production. A lower percentage of affect-laden words and adjectives in written and spoken language was observed in cerebral commisurotomy patients, probably because of the lack of right-hemisphere information available to the speaking-disconnected left hemisphere (TenHouten, Hoppe, Bogen, & Walter, 1986). There is much evidence that the right hemisphere organizes its lexicon according to contextual, affective, and idiosyncratic or personalized principles (Chiarello, 1988; Drews, 1987; Rodel, Cook, Regard, & Landis, 1992; Sidtis, Volpe, Holtzman, Wilson, & Gazzaniga, 1981; Zaidel, 1977). Wechsler (1973) was the first to report deficient output of emotional words in right-hemisphere damage. More recently, Cimino et al. (1991) found that autobiographical reports generated by right hemisphere-damaged subjects were rated as less emotional than reports by control subjects. These and like observations (Cicone et al., 1980) have led to the proposal of an "affect lexicon" as resident in the right hemisphere (Bowers et al., 1993).

Communicative functioning called "pragmatics"—the use of language in everyday situations—may have significant representation in the right hemisphere (Brownell & Joanette, 1993; Joanette & Brownell, 1990; Joanette, Goulet, & Hannequin, 1990; Perecman, 1983). Inability to perceive or to produce figurative meanings and verbal humor contributes to an impoverished emotional communicative function, and is often associated with right-hemisphere dysfunction (Bihrle, Brownell, Powelson, & Gardner, 1986; Brownell, Michel, Powelson, & Gardner, 1983; Myers & Linebaugh, 1981; Van Lancker, 1990; Van Lancker & Kempler, 1987). Difficulty with appreciating polite and indirect requests—statements that ask for an action implicitly but are structured as a question or factual statement—has also been identified with right-hemisphere damage (Weylman, Brownell, Roman, & Gardner, 1989). Persons with severe nonfluent aphasia frequently produce well-articulated speech formulas and expletives, which may or may not appear to express affect (Code, 1997; Van Lancker, 1988).

Failure to process familiar content, or personal relevance, which involves affect, forms another communicative deficiency, and has also been associated with damage to the right hemisphere; examples are in processing of proper nouns denoting familiar–famous and familiar–intimate persons (Ellis & Young, 1990; Van Lancker & Nicklay, 1992; Wallace & Canter, 1985). The right hemispheres of split-brain patients were observed to recognize familiar–famous and familiar–intimate names and faces and to do so with considerable affective response (Sperry, Zaidel, & Zaidel, 1979). Brain damage to the right parietal lobe has been found to impair familiar voice recognition but not the ability to discriminate between unfamiliar voices; an autonomic correlate of this dissociation has been reported for face recognition (Van Lancker, Kreiman, & Cummings, 1989).

Psychiatric conditions have been associated with prosodic deficits in production (Alpert, Rosen, Welkowitz, Sobin, & Borod, 1989) and comprehension (Murphy & Cutting, 1990), as well as florid content in grammar and lexicon, producing bizarre words; this is especially true of schizophrenia and mania (Critchley, 1964; Cummings, 1985). Alexithymia as a "disturbance in emotional awareness" was found to be measurably present in patients with bulimia nervosa (de Groot, Rodin, & Olmstead, 1995). Hypophonic and slowed speech is characteristic of the depressed patient; some researchers (e.g., Greden, Albala, Smokler, Gardner, & Carroll, 1981) have noted increased speech pause time and decreases in volume at the end of sentences in depressed patients. Diagnostic subtypes of depressed patients perform differently on acoustic-perceptual tests (Bruder et al., 1989). As depression is a prominent psychiatric symptom in many diseases of the extrapyramidal system (e.g., Huntington's chorea, Parkinson's disease), changes in speech similar to those seen in depression are observed in these patients as well. Some aphasic disturbances, having multiple etiologies (mutism, echolalia, perseveration), may resemble language disorders associated with psychiatric disease (Duffy, 1995); frustration, depression, and disordered self-awareness may also occur as sequelae of aphasia, sometimes evolving to paranoia (Benson, 1973).

Vocal and speech automatisms as well as affective automatisms (epileptic laughter, crying) are sometimes seen in those affected with temporal lobe epilepsy (Serafetinides & Falconer, 1963; Sethi & Rao, 1976). Hypergraphia with broad affective content and excessive verbal religiosity may form part of the profile of the temporal lobe epileptic patient. Lesions of the right temporal lobe were associated with excessive talk (Kolb & Taylor, 1981). A strong case for the role of the right hemisphere in psychiatric disorders, as manifested by some of the defective communicative functions described here, has been made by Cutting (1990).

Dementing disorders produce a range of disturbances in cognitive and affective functioning that may severely limit the comprehension and production of meaningful linguistic and emotional communication. Language disturbances form part of the diagnostic criteria for probable Alzheimer's disease, the most common of the cortical dementias. In the cortical dementias, speech, while well articulated, becomes relatively empty of affective and informational content (Cummings & Benson, 1992) while grammatic function is relatively preserved. Ability to interpret idioms and proverbs, containing complex affective and ideational material, is impaired very early in the course of the disease (Kempler, Van Lancker, & Read, 1988), while anecdotal evidence suggests that familiar speech formulas such as greetings and leave-taking are used fluently, if not always context-appropriately. Impaired theme and topic maintenance in pragmatic communication are an early symptom (Tomoeda & Bayles, 1993). Interestingly, spouses adopt a specialized speech register when communicating with adults with probable Alzheimer's disease, reducing syntactic and semantic complexity, and increasing references to more obvious elements in their descriptions (Kemper, Anagnopoulos, Lyons, & Heberlein, 1994). In the subcortical dementias, speech articulation and prosodic range may be affected, and grammatical forms may become restricted to simple sentence types; writing is also affected (Cummings, 1986). Deficits in spontaneity may also affect use of speech formulas in conversational interaction.

20-6. CONCLUSIONS

Emotion, affect, and attitudes have an immense influence on normal speech/language ability and in communication disorders. These qualities consistently inform and underlie speech, provide much of the expressive and informative lexical content of language, and constitute much of the fabric of communicative pragmatics.

Specific communicative disorders selectively involve the affective content of speech, language, or pragmatics. Dysprosody refers to failed signaling or identification of affective and attitudinal cues in the physical speech signal. For language, alexithymia involves defective recognition or retrieval of lexical items for emotion. Impaired processing of personally familiar entities impacts processing of proper nouns. In communication, pragmatic deficits involve weakness in the affective glue that holds together discourse, including nonliteral meanings, inference, interactional terms, theme, and humor.

Although emotion can proceed without language, verbal communication is ordinarily and normally imbued with affective and attitudinal nuances. In the communicative traditions of science, business, politics, and the like, management of affect in linguistic expression becomes part of every seasoned practitioner's goal. If the linguistic units that make up the structure of discourse—phonemes, words, syntactic phrases—could be considered akin to cloth, then attitudes and emotions are like the patterns printed on the cloth. The choice of particular words or a particular grammatical construction can change sackcloth to silk. But if one is not sensitive to the emotional nuances of the speaker, one can miss the way those emotions color the words, and thereby miss some, if not all, of the full communicative meaning. Similarly, when the speaker has lost the ability to project the pattern of his/her attitudes and emotions onto the linguistic expression, the listener will be able to understand the words without being able to discern their emotional pattern. One of the greatest challenges in neurolinguistic research lies in understanding how brain structures integrate to permit verbal comprehension and production of attitude and emotion in speech.

CHAPTER 21

The Processing of Sign Language
Evidence from Aphasia

David Corina

Department of Psychology, University of Washington, Seattle, Washington 98195

Each year in the United States roughly 80,000 persons suffer an acquired aphasia (Brody, 1992). Some small percentage of these cases involves deaf individuals whose primary means of communication is a visual–manual language; in the United States, this is likely to be American Sign Language (ASL). Studies of sign language aphasia have advanced our understanding of the contributions of symbolic, motoric, and linguistic processes in human language. Importantly, these studies have helped alert the medical profession to the potential effects of this malady in the deaf population. This chapter provides an overview of the effects of left- and right-hemisphere damage in users of signed languages, and explores the theoretical implications of these findings. One aim of this chapter is to document the current status of the field and to foreshadow issues that are likely to be of interest in future studies.

No doubt by disease of some part of his brain the deaf-mute might lose his natural system of signs.

H. Jackson, 1878/1932

The possibility of a "sign language aphasia" has long fascinated neurolinguistic re-searchers. Cross-modality comparisons of human language provide some of the strong-est evidence to date for biological determination of human language. These studies provide keen insight into the determination of hemispheric specialization, neural plas-ticity, and the contribution of symbolic, motoric, and linguistic processes in human language. This chapter discusses current issues related to sign language aphasia. Pre-vious reviews can also be found in Kimura (1981), Poizner and Kegl (1992), and Poizner, Klima, and Bellugi (1987).

21-1. ASL AND THE DEAF COMMUNITY

American Sign Language is the language used by most of the Deaf[1] community in North America. ASL is a natural language, autonomous from English. ASL is only one of the many sign languages of the world, but it is the one that has been the most extensively studied. Under ideal circumstances, ASL is acquired as a native language by deaf children from deaf signing parents. However, the majority of deaf individuals have hearing parents and siblings. These deaf children are traditionally socialized into the signing Deaf community either when they enter residential school (often not until age 5 or later) or at some later point as adults (Meadows, 1980). It is often assumed that all deaf people can sign and that all deaf people can read lips fluently. In reality, deaf people's language skills vary considerably. Biological hearing loss alone does not ensure competence in signing, nor does education ensure competence in English. Some factors that contribute to language skills in the deaf include parental input, age of language acquisition, schooling, and affiliation with other deaf signing individuals (for discussion of these issues, see Grosjean, 1982; Kettrick & Hatfield, 1986; Markowicz & Woodward, 1978; Newport & Meier, 1985; Meier, 1991; Padden & Humphries, 1988; Ramsey, 1989). Clinicians assessing deaf patients should be cog-nizant of the factors that influence language competence in this population.

21-2. LINGUISTIC STRUCTURE OF AMERICAN SIGN LANGUAGE

Within the last decade there has been a monumental increase in our knowledge of the formal structure of signed languages. Advances in sign language linguistics come at an opportune time; there is growing awareness that profiles of neurolinguistic impair-ment must be characterized against a backdrop of the language-specific variation (see for example Bates & Wulfeck, 1989). In this section, we review characteristics of the linguistic structure of ASL.

---

[1]The capitalization of the word *deaf* in this context serves to signify recognition of deaf indi-viduals as a distinct cultural group, who function within a minority community and whose primary language is signed (see Lane, 1984; Padden, & Humphries, 1988 for discussions).

21-2.1. Sign Phonology

Phonological organization refers to the patterning of the formational units of the expression system of natural language (Coulter & Anderson, 1993). For signed languages, these formational units refer to abstract articulatory and visual primitives. ASL signs are constructed from a limited set of formational elements drawn from four main articulatory parameters: movement, location, handshape, and orientation (Stokoe, Casterline, & Corneberg, 1965). Recognition of a syllabic unit of organization in ASL phonology has given rise to the identification of "consonantal" and "vocalic" components of sign forms (see Corina & Sandler, 1993, for an overview).

21-2.2. Sign Morphology

ASL has complex grammatical and lexical morphology. ASL lexical morphology permits creation of new sign forms, while grammatical morphology serves to modulate aspects of meaning and permutes the shape of signs in accordance with syntactic requirements. Morphological markings in ASL are unusual as they are expressed as dynamic movement patterns superimposed on a more basic sign form. The prevalence of nested morphological forms, in contrast to the linear suffixation common in spoken languages, is considered a reflection of the modality of expression on the realization of linguistic structure (Emmorey, Corina, & Bellugi, 1995; Klima & Bellugi, 1979). Two forms of ASL lexical morphology have received increased attention in the sign aphasia literature: the ASL classifier system and the ASL grammatical facial expression system.

21-2.3. ASL Classifiers

In many languages (spoken and signed), object and action descriptions require use of obligatory grammatical morphemes that specify salient semantic or visual properties of noun and verb referents. In ASL, classifiers mark semantic categories, such as human, animal, and vehicle, and visual properties, like flat, thin, and round. Classifiers in ASL function as verbs of motion and location, specifying the path and the direction of movement of their noun referent (Newport & Supalla, 1980; Supalla, 1986). The system of classifiers is unusual in its conflation of linguistic and visual object properties. The classifier system is one of the most difficult aspects of ASL grammar, and poses a particular problem for nonnative users of ASL. Recent investigations suggest that right-hemisphere lesions may in some cases selectively impair use of this unusual morphological system.

21-2.4. ASL Facial Expression

Another unusual aspect of ASL morphology concerns the grammatical function of facial expressions. Certain syntactic constructions (e.g., topics, conditionals, relative clauses, questions) are marked by specific and obligatory facial expressions. ASL facial expressions also signal adverbial markings, co-occurring with and modifying verb phrases. Linguistic facial expressions are distinct markers that differ significantly in appearance and execution from those universal affective facial expressions identified by Ekman and Friesen (1978). Recognition studies of facial expression by deaf signers indicate bilateral mediation of facial expressions, which contrasts with the strong right-hemisphere advantage shown by hearing persons (Corina, 1989).

21-2.5. ASL Syntax

ASL makes use of two strategies for distinguishing grammatical relations: word order (which is SVO in the majority of cases) and inflectional morphology. A prominent feature of ASL syntax is the use of signing space for the depiction of grammatical relations. At the syntactic level, nominals introduced into the discourse are assigned arbitrary reference points in a horizontal plane of signing space. Signs with pronominal function are directed toward these points and many verb signs obligatorily move between these points in specifying grammatical relations (subject of, object of) (see Lillo-Martin, 1991; Lillo-Martin & Klima, 1990 for discussions). Thus grammatical functions served in many spoken languages by case marking or by linear ordering of words are fulfilled in ASL by spatial mechanisms (Klima & Bellugi, 1979).

21-2.6. ASL Discourse

The system of spatial reference is also used in the service of discourse relations. Consistency in cross-sentential spatial indexing serves as a means of discourse cohesion. The status of spatial locations used in the service of syntactic and discourse function in ASL is not well understood. An important question is whether spatial locations, which serve coreferential and anaphoric functions, should be treated as part of the linguistic representation or whether these physical spatial locations are merely deictic in nature (see Emmorey et al., 1995; Liddell, 1990, 1994). As will be discussed later in this chapter, this seemingly esoteric distinction has implications for the interpretation of syntactic deficits observed in signers who have incurred brain damage.

21-2.7. Fingerspelling

In addition to American Sign Language, there exists the American manual alphabet. In this system, the letters of the English alphabet are represented by 26 distinct configurations of the hand, and meaningful units (English words as represented by their letters) are conveyed by sequences of these configurations. Fingerspelling provides one avenue for borrowing new English vocabulary into ASL (Klima & Bellugi, 1979). The execution and comprehension of fingerspelling incorporates processes involved in spelling and reading (Wilcox, 1992).

In summary, ASL exhibits formal organization at the same levels found in spoken language. In signed languages, we find sublexical structure analogous to a phonemic level, and a level that specifies the precise ways that meaningful units are bound together to form complex words and sentences analogous to morphological and syntactic levels. Recent studies have described discourse conventions in ASL, as well as unusual morphological constructions including ASL classifiers and linguistic facial expressions. These studies provide a basis for evaluating aphasic symptoms in deaf signers.

21-3. A SUMMARY OF SIGN LANGUAGE APHASIA

Sixteen cases of deaf or signing individuals who have incurred left-hemisphere brain damage and five cases of right hemisphere-damaged signers are summarized in Table 1. These case studies vary greatly in their contribution to understanding the brain processes involved in signing. Many of the early case studies were hampered by the lack of understanding of the relationships among systems of communication utilized by deaf individuals. For example, several of the early studies compare disruptions of fingerspelling and only briefly mention or assess sign language use. Studies often fail to establish premorbid language histories. These case studies include reports on the congenitally deaf, adventitiously deaf, native and nonnative signers, right-handed and left-handed signers, making generalizations from these cases difficult. Anatomical localization of lesions is often lacking or confounded by the existence of multiple infarcts, and rarely are audiological reports presented. Nevertheless, with careful reading, general patterns emerge and contribute to our understanding of sign language representation in the brain. More recently, well-documented studies are starting to provide a clearer picture of the neural systems involved in language processing in users of signed languages. Our discussions will be largely based on this later series of subjects.

21-4. HEMISPHERIC SPECIALIZATION IN DEAF SIGNERS

Two prominent issues concerning brain organization in signers are whether left-hemisphere structures mediate sign languages of the deaf, and whether deaf individuals show complementary hemispheric specialization for language and nonlanguage visuospatial skills. One conclusion that can be drawn from the sign aphasia literature is that right-handed deaf signers, like hearing persons, exhibit language disturbances when critical left-hemisphere areas are damaged. Of the 16 left-hemisphere cases listed in Table 1, 12 cases provide sufficient detail to implicate left-hemisphere structures in sign language disturbances (cases 2, 4, 5, 6, 7, 10, 11, 12, 13, 14, 15, 16) (see Table 1). Five cases provide neuroradiological or autopsy reports to confirm left-hemisphere involvement, and provide compelling language assessment to implicate aphasic language disturbance (cases 2, 5, 10, 12, 16). Five cases of signers with right-hemisphere pathology have been reported (cases 17, 18, 19, 20, 21). All five of these signers showed moderate to severe degrees of nonlanguage visuospatial impairment. In contrast, none of the left hemisphere-damaged signers tested on nonlanguage

TABLE 1
Case Histories of Signers with Left- and Right-Hemisphere Damage

Case number	Reference	Handedness	Age, gender, initials	Onset of deafness	Neurological signs	Etiology
1	Grasse (1896)	n/a	n/a male	n/a	right arm weakness	stroke
2	Burr (1905)	n/a	56 female	early childhood	right paralysis, right homonymous hemianopsia	tumor
3	Critchley (1938)	right	42 male	14 years	right paralysis	stroke
4	Leischner (1943)	right	64 male	congenital	right arm and right leg weakness	stroke
5	Tureen, Smolik, & Tritt (1951)	right	43 male	congenital	partial right paralysis	tumor & hemorrhage
6	Douglass & Richardson (1959)	right	21 female	congenital	right paralysis, right-sided facial weakness	stroke
7	Sarno, Swisher, & Sarno (1969)	right	69 male	congenital	right central facial weakness, moderate right hemiplegia	stroke
8	Meckler, Mack, & Bennet (1979)	left	19 male	hearing signer	right hemiparesis	closed head injury
9	Underwood & Paulson (1981)	left	57 male	congenital	right hemiplegia	stroke
10	Chiarello, Knight, & Mandel (1982)	right	65 female, "L.K."	6 months	moderate right-sided weakness of arm, leg, and face	stroke
11	Poizner, Klima, & Bellugi (1987)	right	81 male, "P.D."	5 years	right hemiparesis	stroke
12	Poizner, Klima, & Bellugi (1987)	right	38 female, "G.D."	congenital	right hemiparesis	stroke
13	Poizner & Kegl (1992)	right	48 male, "N.S."	congenital	partial right hemiparesis	compound skull fracture
14	Brentari, Poizner, & Kegl (1995)	right	81 female, "E.N."	congenital	n/a	stroke
15	Hickok, Klima, Kritchevsky, & Bellugi (1995)	right	62 female	18 months	right homonymous hemianopia, spastic right hemiparesis	stroke
16	Corina, Poizner, Feinberg, Dowd, & O'Grady (1992)	right	76 male, "W.L."	congenital	right-sided weakness	stroke
17	Poizner, Klima, & Bellugi (1987)	right	71 female, "S.M."	congenital	paralysis of left arm and leg	stroke
18	Poizner, Klima, & Bellugi (1987)	right	75 female, "B.I."	congenital	dense left arm paralysis	stroke
19	Poizner, Klima, & Bellugi (1987)	right	81 male, "G.G."	5 years	lower left facial weakness, abnormal left-sided reflex	stroke
20	Corina, Kritchevsky, & Bellugi (1996)	right	61 male, "J.H."	congenital	dense left hemiparesis	stroke
21	Corina, Bellugi, Kritchevsky, O'Grady-Batch, & Norman (1990); Kegl & Poizner (1991); Poizner & Kegl (1992)	right	35 female, "D.N."	hearing signer	lower left quadrant visual field cut	AVM & stroke

Anatomy	Premorbid language/ language environment	Clinical language assessment	Comments
Possible corticospinal motor system (see Kimura, 1981 for some discussion)	n/a	Impaired right-handed fingerspelling. Loss of right-handed motor power and coordination	Case reported in Critchley (1938)
Posterior frontal lobe with extension into basal ganglia	Signing, reading, & writing	Global sign language aphasia	Repeated strokes, briefly documented
n/a	Signing, fingerspelling, & lip reading	Impaired fingerspelling production and comprehension, intact sign language, speech dysarthria	British two-handed fingerspelling
Left frontal and parietal region, (supramarginal gyrus and angular gyrus)	Native signer, Czech and German bilingual, signing and reading and writing	Fluent aphasia, neologistic output	Reported in Kimura (1981). Autopsy reveals both left- and right-hemisphere damage
Second and third frontal convolutions (posterior), anterior tip of the internal capsule	Signing, fingerspelling, & lip reading	Impaired fingerspelling comprehension and production Preserved reading ability Intermittent sign comprehension problems	Praxic function retained in left hand except for communication
Probable frontal and rolandic lesion with possible parietal involvement	Signing, fingerspelling, and writing, limited reading	Impaired production and comprehension of signing and fingerspelling. Sign and fingerspelling paraphasias	Detailed study for its time, no apraxia
Probable left fronto-parietal lesion	Learned sign at age 7, fingerspelling, lip reading, writing, & reading	Good sign comprehension with mild impairment in production, impaired fingerspelling production and comprehension	Detailed study, no apraxia
Traumatic cerebral contusion	Speech, sign, & fingerspelling	Global aphasia for speech, initially sign language less impaired	English-ASL bilingual, left-handed
n/a	Learned sign at age 7, fingerspelling, lip reading, & functional writing abilities	Production problems in both sign and fingerspelling, jargon and paraphasias are noted also in writing	left-handed
Left inferior parietal lobule, with left frontal subcortical extent undercutting Broadman areas 4, 3, 1, and 2	Learned sign at age 5 fingerspelling and some functional writing ability	Severe anomia and comprehension loss for sign, fingerspelling and writing. Sign paraphasia and gesturing	See also Poizner et al. (1987) case "KL."
Subcortical lesion with anterior focus deep to Broca's area, including basal ganglia, posteriorly, white matter of the parietal lobe	Spoken and sign language abilities	Aphasic-paragrammatic, fluent production, syntactic and morphological errors	See also Kimura, Battison, & Lubert (1976)
Left frontal lobe, including Broca's area and anterior portions of the superior and middle temporal gyri	Signing, fingerspelling, reading, & writing	Broca-like signing, impaired production with preserved sign comprehension	
Left parietal lesion	Deaf sibling, residental school	Sign comprehension deficit, mild production problems, pyramidal deficit	No apraxia
Distribution of left posterior cerebral artery, posterior limb of the internal capsule, posterior thalamus and left mesial occipital cortex	Native signer	Motorically fluent sign production with sublexical and semantic paraphasias	Possible case of subcortical sign aphasia
Left posterior cerebral artery stroke. Medial temporal and occipital lobes, left occipital pole, and splenium of the corpus callosum	Signing, fingerspelling, reading, & writing	Severe sign language comprehension deficit (esp. sentence-level). Inability to read words. Preserved sign production	Disconnection syndrome
Fronto-tempo-parietal lesion, including Broadman area 44 and 45 and white matter tracts, white matter deep to the inferior parietal lobule	Signing, fingerspelling, reading, & writing	Severe sign language production and comprehension problems. Semantic and phonological paraphasia	Sparing of nonlinguistic pantomime ability
Distribution of the middle cerebral artery	Signing, fingerspelling, reading, & writing	Severe visuospatial disruption, no aphasia	
Distribution of right middle cerebral artery	Signing, fingerspelling, reading, & writing	Severe visuospatial disruption, no aphasia	
Temporal-parietal lesion, cortex and underlying white matter in the superior temporal gyrus, lower inferior parietal lobule	Signing, fingerspelling, reading, & writing	Severe visuospatial disruption, no aphasia	
Central portions of the frontal, parietal, and temporal lobes, associated deep white matter and basal ganglia structure	Signing, fingerspelling, reading, & writing	Severe visuospatial disruption, no aphasia, profound left neglect	Extinction for objects but not for sign language
Upper portion of the occipital lobe (medial) and superior lobule, white matter involvement	Early sign language exposure, interpreter for the deaf	Deficits in sign language discourse and use of ASL classifier morphology. Subtle visuospatial disruptions	Unusual lesion site

visuospatial tests were shown to have significant impairment. A recent group comparison reaches similar conclusions (Hickok, Bellugi, & Klima, 1996). Taken together, these findings suggest that deaf signers show complementary specialization for language and nonlanguage skills. These studies demonstrate that development of hemispheric specialization is not dependent on exposure to oral/aural language.

21-5. NEUROLINGUISTICS OF SIGN LANGUAGE APHASIA

21-5.1. Sign Language Production

A common question concerning the assessment of aphasia in deaf signers is the apparent effect of hemiplegia or motoric weakness on signing behavior. Sign languages use both hands for sign formation but anatomical handedness is not contrastive in ASL. There is an asymmetrical relationship between the two hands in ASL signing. Only one hand takes the role of the "dominant" articulator, the other hand acting as a mirror copy of the dominant hand or in some cases as a passive place of articulation, with a reduced inventory of possible handshapes (i.e., the dominant moves to and contacts the passive, nondominant hand). In everyday activities, it is often the case that both hands are not available for signing. Sign languages have subsequently developed adaptations to compensate for this natural state of affairs (see Battison, 1978; Vaid, Bellugi, & Poizner, 1989 for a discussion). In informal signing, there is a general tendency to reduce many two-handed signs to more colloquial one-handed forms. Thus, signers who have become hemiplegic are not overly challenged by the lack of an articulator. Moreover, we are confident that sign aphasia is not due to motoric factors when, for example, paraphasic sign formation errors appear on the nonhemiplegic hand. Finally, linguistic analysis of errors is able to differentiate lack of articulatory or motoric agility from linguistic substitutions (see Sections 5.2 and 5.3).

21-5.2. Phonemic Paraphasias in ASL

Spoken language phonemic paraphasias arise from substitution or omission of the sublexical phonological components (Blumstein, 1973). In ASL, sublexical structure refers to the formational elements that compose a sign form: handshape, location, movement, and orientation. Sign language phonological paraphasias result in substitutions within these parameters. Phonological paraphasias in sign aphasia do not appear to affect the syllabic integrity of the sign forms (Brentari, Poizner, & Kegl, 1995).

In principle, selectional errors could occur among any of the four sublexical parameters. For example, subject L.K. (case 10) produced paraphasic signing errors in which substitutions were found in all four major parameters. The sign ENJOY, for example, which requires a circular movement of the hand, was articulated with an incorrect up-and-down movement. Substitutions in the parameters of orientation, handshape, and location are also reported (see Poizner, Klima, & Bellugi, 1987). However, the most frequently reported errors are those affecting the handshape parameter.

Corina, Poizner, Feinberg, Dowd, and O'Grady (1992) describe in some detail the phonemic errors produced by W.L. (case 16), which almost entirely affect handshape specifications. For example, W.L. produced the sign TOOTHBRUSH with the Y handshape rather than the required G handshape, and produced the sign SISTER with an F handshape rather than the required L handshape (see Figure 1). Based upon a linguistic analysis of these types of errors, Corina, Poizner, et al. (1992) argue that these handshape substitutions are phonemic in nature rather than simply phonetic misarticulations.

The preponderance of handshape substitution errors is interesting in that recent linguistic analyses of ASL have suggested that static location and handshape specifications may be more consonantal in nature while movemental components of ASL may be analogous to vowels. One of the most striking asymmetries in spoken language phonemic paraphasia is that errors overwhelmingly favor consonant distortions. Whether the vulnerability of consonants relative to vowels reflects a neural difference in representation or results from the statistical differences in inventory or articulatory/acoustic variables (e.g., degree of freedom before misarticulating a consonant or a vowel is detected as an error) is not well understood. The fact that across spoken and signed languages phonemic errors involve "consonantal" attributes of language is intriguing.

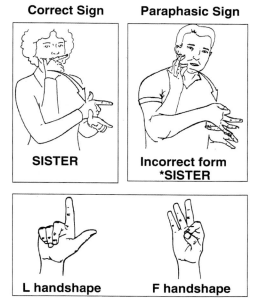

Linguistically Motivated Errors

FIGURE 1   ASL handshape paraphasia. The "L" and "F" handshapes are structurally similar in underlying phonological representation.

21-5.3. Morphology and Syntax Errors

A common error pattern in spoken language aphasia is the substitution and/or omission of bound and free morphemes. Languages differ in the degree to which they use morphology to mark obligatory grammatical distinctions (case and gender, subject and object agreement, etc.), and patterns of impairment may be more striking in some languages than others (Bates, Wulfeck, & MacWhinney, 1991; Menn & Obler, 1990). ASL is a highly inflected language; morphosyntactic agreement distinguishing grammatical subject and object requires directional movement trajectories in certain classes of ASL verbs. In the absence of grammatical movement trajectories, a verb sign will be produced in an uninflected "citation" form. Poizner et al. (1987) have investigated their patients particularly thoroughly in this morphosyntactic realm. Poizner's patient G.D. (case 12), consistently omitted required inflectional morphemes in her spontaneous signing, and instead produced uninflected "citation" verb forms. Poizner's patient P.D. (case 11) produced both omissions in inflectional morphology and inconsistent verb agreement; that is, P.D. failed to maintain consistent verb movement trajectories to spatial locations as is required by syntactic and discourse conventions. G.D. had a large left-hemisphere lesion that involved most of the convexity of the left frontal lobe, including Broca's area. P.D. had a subcortical lesion in the left hemisphere with anterior focus deep to Broca's area, and the lesion extended posteriorly into the white matter in the left parietal lobe. The general pattern of omissions versus substitutions in signers G.D. and P.D., respectively, is consistent with profiles of agrammatic and paragrammatic impairment reported for users of spoken language.

In addition to the morphosyntactic processes, ASL makes rich use of morphological devices to express aspectual distinctions. Subject G.D. omitted morphologically complex forms across the board, omitting not only morphosyntactic forms but temporal and distributional inflections as well. Poizner et al. (1987) report in detail some unusual aspectual morphological errors produced by signer P.D. These errors involved semantically inappropriate inflections on a variety of lexical forms that do not subcategorize for these inflections. For example, they report the sign BRILLIANT as undergoing inflection for predispositional aspect, which only applies to adjectives with transitory quality (but here has been applied to an adjective with inherent quality). To my knowledge, constructions of morphological neologistic forms that violate subcategorization has not been reported for spoken language, even in languages such as Hungarian (MacWhinney & Osman-Sagi, 1991) or Turkish (Slobin, 1991) with highly complex morphology. Whether these ASL errors reflect differences in the patterning of morphological impairment in signed versus spoken languages or represent non-aphasic symptomology in this older subject is not well understood.

21-5.4. Sign Language Comprehension

In hearing individuals, severe language comprehension deficits are associated with left-hemisphere posterior lesions, especially posterior temporal lesions. Similar patterns have been observed in users of signed languages. For example, W.L. (case 16) and L.K. (case 10), who have damage to posterior temporal structures, evidenced marked comprehension deficits. W.L. showed a gradation of impairment, with some difficulty in single sign recognition, moderate impairment in following commands, and severe problems with complex ideational material. Subject L.K. showed severe comprehension problems, especially of multipart commands. In each of these cases, comprehension was assessed against a backdrop of aphasic production deficits. In contrast, Hickok, Klima, Kritchevsky, and Bellugi, (1995) have reported a case of a signer with severely impaired sign comprehension, but well preserved nonaphasic language production. This subject had a left-hemisphere lesion involving medial temporal and occipital lobes, left occipital pole, and white matter tracks that give rise to the splenium of the corpus callosum. Sign comprehension at the sentence level was severely impaired, while single sign comprehension was reasonably intact, with occasional semantic field errors in confrontation naming. Hickcok et al., (1995) argue for a comprehension deficit based on a disconnection syndrome in which visual information from the intact right visual field was unable to project to the language region of the left hemisphere (like alexia without agraphia).

In summary, evidence from lesion studies of deaf signers indicates left-hemisphere dominance for sign language processing, and the familiar patterns of comprehension and production deficits following posterior and anterior damage, respectively. These findings provide evidence that language impairments following stroke in deaf signers are indeed aphasic in nature, and do not reflect general problems in symbolic conceptualization or motor behavior. In addition, the dissociation syndrome described by Hickok et al. (1995), illustrates the intimate relationships between neural systems involved in visual processing and sign language processing, not unlike those posited for auditory and spoken language processing (Geschwind, 1965). Finally, there is good evidence for hemispheric specialization for both language and nonlanguage functions in deaf signers. We now turn to a more thorough description of sign language aphasia and return to a detailed discussion of neuroanatomical specialization for sign language.

21-6. WITHIN-HEMISPHERE LOCALIZATION

Hemispheric specialization in deaf signers broadly mirrors the specialization found in hearing individuals. A more challenging question concerns the extent to which the neuroanatomical systems responsible for mediating auditory languages participate in visual sign languages. This question has important theoretical implications. Common anatomical substrates would provide support for a modality-neutral neural system of human language processing. The existence of such a system has been taken by some to lend validity to the constructs of languages as biologically determined and cognitively encapsulated (Bellugi, Bihrle, & Corina, 1991; Fromkin, 1991). On the other hand, differences in the anatomical substrates of spoken and signed languages suggests some degree of plasticity in the neural systems underlying human languages.

Significant anatomical differences in language representation may indicate that there are many ways to instantiate functional human language in the brain. Behavioral evidence to date suggests a role for both biologically determined mechanisms and influences of environmental constraints in the development of signed and spoken languages (Neville, 1990).

21-6.1. Nonfluent Aphasia

It is generally accepted that lasting Broca's aphasia in a hearing person requires a lesion encompassing the cortical Broca's area (pars opercularis and pars triangularis) of the left frontal lobe, often extending posteriorly to include the lower portion of the motor strip. The lesion must extend in depth to the periventricular white matter, because a purely cortical or shallow lesion produces only a transient disorder (Mohr et al., 1978; Goodglass, 1993). One question that arises in the study of Broca-like sign aphasia is the degree to which there is overlap in the premotor regions. Levine and Sweet (1982) provide evidence that for users of spoken language, pure cases of Broca's aphasia without associated agraphia may be limited to inferior regions of the precentral gyrus, while more extensive lesions extending dorsally may result in a Broca's aphasia with associated agraphia. Thus, agraphia may result from involvement of arm and hand representations of the sensorimotor cortex. It may be possible to use agraphia as a metric for determining degree of overlap between signed and spoken language aphasia; that is, is it the case that all Broca-like sign aphasics also have associated agraphia or are the linguistic representations underlying writing and signing functionally distinct and dissociable? Two cases in the literature provide sufficient detail to make some tentative observations concerning anatomical areas involved in nonfluent Broca-like aphasia in sign language.

Tureen, Smolik, and Tritt (1951) (case 5) describe production problems following hemorrhage of a tumor in the frontal lobe affecting the second and third frontal convolution and the tip of the internal capsule. Importantly, this subject was unable to use his nonaffected left hand for signing, fingerspelling, or writing. Poizner et al.'s (1987) subject G.D. presents with Broca-like signing and has damage to areas 44 and 45 of the frontal lobe. G.D. evidences agrammatic signing and written language output following her stroke. In addition, G.D. shows considerable problems with fingerspelling. Finally, and more speculatively, is Douglass and Richardson's (1959) subject (case 6). This patient showed impairment of sign production and fingerspelling, and is described as dysgraphic. Perseverative errors and substitutions were found in all modes of communication. Unfortunately, neuroanatomical localization is only inferred from behavioral deficits in this patient. Thus, in three cases where there are well-described productive sign language impairments, we find associated disturbances in both writing and fingerspelling. The similarities of the disturbances in the nonaffected hand across modalities of manual expression are suggestive of a common disturbance. These few studies may imply that nonfluent signing aphasias require involvement of classic Broca's area and encroachment upon cortical and subcortical motor areas of the precentral gyrus involved in hand and arm representations.

21-6.2. Fluent Aphasia

Fluent aphasias are associated with lesions to posterior temporal-parietal regions. Wernicke's aphasia, for example, is typically mapped to the posterior region of the left superior temporal gyrus. However, it is not uncommon for lesions associated with Wernicke's aphasia to extend onto the lower second temporal gyrus and into the nearby parietal region (Damasio, 1991). Dronkers, Redern, and Ludy (1995) have reported that a lesion of the posterior half of the middle temporal gyrus is associated with chronic Wernicke's symptoms. Two prominent features of Wernicke's aphasia are impaired comprehension with fluent, but often paraphasic, output (both semantic and phonemic). The persistent neologistic output sometimes associated with severe Wernicke's aphasia is associated with the supramarginal gyrus (Kertesz, 1993).

Three cases of sign language aphasia resulting from posterior lesions are sufficiently detailed to permit some tentative comparisons. Leischner's (1943) case study (case 4) presents anatomical data based on autopsy, and reports damage to left posterior regions involving cortex and white matter associated with angular and supramarginal gyri. This patient produced the closest description of jargon aphasia currently in the literature. The patient produced a great deal of signing, much of it wrong or nonsensical, with frequent perseverations, and meaningless signs. Unfortunately, comprehension measures are not reported. Subjects L.K. (Chiarello, Knight, & Mandel, 1982; Poizner et al., 1987) and W.L. (Corina, Kritchevsky, & Bellugi, 1992) each present with severe comprehension difficulties and relatively fluent paraphasic output. Interestingly, neither have lesions in the classic Wernicke's area, but rather involve more frontal and inferior parietal areas. In both cases, lesions extend posteriorly to supramarginal gyrus. Lesions associated with supramarginal gyrus alone do not typically result in severe comprehension loss in users of spoken language. These two cases suggest that sign language comprehension may be more dependent on inferior parietal areas in the left hemisphere. This difference may reflect within-hemisphere reorganization for cortical areas involved in sign comprehension (Leischner, 1943; Poizner et al., 1987; Chiarello et al., 1982).

In summary, these observations show that the frontal–nonfluent/posterior–fluent dichotomy holds for users of sign language. However, there is some suggestion that within-hemisphere reorganization may be present in the deaf. Broca-like sign production impairments may encroach upon cortical association areas involved in hand and arm representations. Sign comprehension deficits may be more common following damage to left-hemisphere inferior parietal areas not classically associated with speech comprehension deficits. These data suggest subtle differences in cortical organization of sign and speech that must be further validated. Unfortunately, the mapping of critical areas important in sign language mediation is hampered by the bias of positive cases in the literature. Equally interesting would be cases of left-hemisphere damage with lesions in classic language areas that do not result in sign impairment.

21-7. RIGHT HEMISPHERE AND LANGUAGE

In discussing anatomical differences between users of signed and spoken language it is worthwhile to reconsider the evidence for right-hemisphere contributions to sign language processing. As stated, five right hemisphere-damaged signers show frank disruptions in nonlinguistic visuospatial processing. Typically, these subjects are reported to have well-preserved language. Poizner et al. (1987) stated:

> There are several lines of evidence that sign language is intact in right-lesioned signers. The first (and most powerful) line of evidence lies in the fact that their signing is flawless and without aphasic symptoms and is in contrast to the signing of deaf patients after left-hemisphere damage where clear and marked disruption is found. (P. 153)

However, more recent studies suggest that this picture may be changing. We focus on three types of language-related problems in right hemisphere-damaged signers. First, as is the case with hearing persons, right-hemisphere damage in the deaf may disrupt the metacontrol of signed language use as evidenced by disruptions of discourse abilities (Kaplan, Brownell, Jacobs, & Gardner, 1990; Rehak et al., 1992). Second, there is new evidence that right-hemisphere structures may be important in use of the ASL classifier system. Third, we examine the contribution of the right hemisphere in comprehension of spatialized syntax.

21-7.1. Impaired Discourse Abilities

There is growing evidence that the right hemisphere plays a crucial role in the discourse abilities of deaf signers. Analysis of language use in right hemisphere-lesioned subject J.H. (case 20) (Corina et al., 1996) reveals occasional nonsequiturs and abnormal attention to details, which are characteristics of the discourse of hearing patients with right-hemisphere lesions (Delis, Wapner, Gardner, & Moses, 1983). Subject D.N. (case 21) reported in Poizner and Kegl (1992) shows another pattern of discourse disruption. Although D.N. is successful at spatial indexing within a given sentence, several researchers have noted that she is inconsistent across sentences (Kegl & Poizner, 1991; Poizner & Kegl, 1992; Emmorey et al., 1995); that is, she does not consistently use the same index point from sentence to sentence. In order to salvage intelligibility, D.N. uses a compensatory strategy in which she restates the noun phrase in each sentence, resulting in overly redundant content within her discourse strategy. The cases of J.H. and D.N. suggest that right-hemisphere lesions in signers can differentially disrupt discourse content (as in the case of J.H.) and discourse cohesion (as in the case of D.N.).

21-7.2. Impaired Classifier Production

Recent investigation of use of the ASL classifier system by right hemisphere-damaged signer D.N. reveals systematic impairment (Corina, unpublished observation). At jeopardy were D.N.'s depictions of direction and movement. For example, when asked to portray a small barrel hopping down a platform, D.N. signed the equivalent of a barrel hopping up an incorrectly inclined plank. In this example, depiction of the action requires coordination of the right and left hands (e.g., one to represent the barrel, the other, the plank). In many instances when classifier descriptions required two hands, D.N. showed hesitancy and repeated self-corrections. Importantly, D.N. showed no motor weakness that would explain these errors. Another common error is the simplification of classifier handshapes. In simple descriptions, with only one object present, D.N. correctly used the complex classifier handshape forms. However, in descriptions involving a precise orientation of an object, especially those including multiple object relations, D.N. often substituted an unmarked "generic" object classifier. This indicates that she is not unaware of the correct classifier handshape (as is often seen with nonproficient users of ASL), but instead adopts a strategy that reduces the spatial demands imposed by the classifier system.

In cases of impaired discourse production and classifier use, it is reasonable to suppose that right-hemisphere cognitive deficits happen to manifest themselves in aspects of language use. Under this interpretation, right-hemisphere damage does not disrupt linguistic function per se, but may cause trouble in the execution and processing of linguistic information in sign language, in which spatial information plays a particularly salient role. The issues become more complicated, however, when we consider syntactic aspects of ASL.

21-8. ASL SYNTAX IN RIGHT- AND LEFT-HEMISPHERE-DAMAGED SIGNERS

Disturbances in the syntactic processing of ASL have been attested after *both* left- and right-hemisphere damage. In the case of left-hemisphere damage, we find problems in the production and comprehension of spatialized syntax. Left-hemisphere damaged subjects P.D., W.L., and K.L. all demonstrated problems in the comprehension of syntactic relationships expressed via the spatial–syntactic system. Even more surprising, however, is the finding that some signers with right-hemisphere damage also exhibit problems. Two of the right-hemisphere damaged subjects, S.M. (case 17) and G.G. (case 19), tested by Poizner et al. (1987), show performance well below controls on two tests of spatial syntax. Indeed, as pointed out by Poizner et al. (1987), "Right lesioned signers do not show comprehension deficits in any linguistic test, other than that of spatialized syntax." A fundamental question, then, is whether these comprehension deficits are similar across left and right hemisphere-damaged signers. Unfortunately, from the current literature we cannot be certain at this time. One crucial question, as yet unanswered, is whether the left- and right-hemisphere subjects show deficits in the grammatical processing of ASL syntax when it is *not* expressed spatially. Poizner et al. (1987) speculate that perceptual processing involved in the comprehension of spatialized syntax involves both hemispheres; certain critical areas of each hemisphere must be relatively intact for accurate performance.

The deficits in syntactic comprehension in the right- and left-hemisphere subjects raise an interesting theoretical question, namely, should these deficits in comprehension be considered aphasic in nature, or should they be considered secondary impairments arising from a general cognitive deficit in spatial processing? These questions invoke the theoretical status of linguistic structure in ASL. Are the spatial locations of the verb system part of the syntactic representations, or are these physical locations merely deictic? (See Emmorey, 1996 for discussion.) If one argues that spatial locations are part of the linguistic representation, then one is forced to conclude that right-hemisphere damage in deaf adults leads to morphosyntactic impairment in language comprehension. Further work is required to tease apart these complicated theoretical questions.

21-9. APRAXIA AND SIGN LANGUAGE

The relationship between impairments of language production and motor behavior is a complicated and, at times, contentious issue. When one considers impairments in sign language production, the issues become even more difficult to isolate, as the surface appearance of the disorder frequently suggests impaired motor control. There is great interest in studying deaf signers in reference to the distinction between aphasia and apraxia. One hope is that as the manual articulators are directly observable, quantification of the deficits involved may be more objective (see, for example, Brentari, Poizner, & Kegl, 1995).

Convincing dissociations between sign language impairment and well-preserved ideomotor and ideational movements, notably conventionalized gesture and pantomimed object use, are reported in Kimura, Battison, and Lubert (1976), Sarno, Swisher, and Sarno (1969), Chiarello et al. (1982), Poizner et al. (1987) (cases P.D. and K.L.), Poizner and Kegl (1992), and Corina, Poizner et al. (1992). Subject W.L., for example, had a marked sign language aphasia affecting both production and comprehension of signs. However, W.L. produced unencumbered pantomime, often involving stretches of multisequenced pantomime to communicate ideas for which a single lexical sign would have otherwise sufficed. Moreover, both comprehension and production of pantomime were found to be better preserved than comprehension and production of sign language. These cases reemphasize the fact that language impairments following left-hemisphere damage are not attributable to undifferentiated symbolic impairments. More importantly, these cases demonstrate convincingly that linguistic gesture (e.g., ASL) is not simply an elaborate pantomimic system.

A more controversial issue concerns differential impairment in meaningless manual movements and sign language ability. According to Kimura (1993), the left hemisphere appears to be essential for selecting most types of movement postures. Thus, Kimura treats language production impairments following left-hemisphere damage as secondary to an impairment in sequential movement programming. Two left hemisphere-damaged language-impaired signers showed impairment in copying complex nonlinguistic movements (Kimura et al., 1976; Chiarello et al., 1982). These findings are taken as support for Kimura's position that signing disorders are a manifestation of movement disorders. However, dissociations have been reported. Poizner et al. (1987) retested P.D. (who was the subject of the original Kimura et al., 1976 study) and reported that copying nonlinguistic movements was intact at the time of testing; however, P.D. remained severely aphasic. More recently, severely aphasic patient W.L. also performed within normal limits on an abbreviated version of the Kimura copying task (the same test used by Poizner et al., 1987), suggesting dissociation between linguistic impairment and sequential movement disorders (Corina, Poizner, et al., 1992).

In summary, there is good evidence for dissociation between ideomotor and ideational movements and signing. There is some evidence for dissociation between sequential movement disorders and sign language use; however, more rigorous testing is required to conclusively evaluate these claims.

21-10. CONCLUSION

The studies to date provide ample evidence for left-hemisphere mediation of sign language in the deaf. Sign language disturbances following left-hemisphere damage show linguistically significant breakdown that is not attributable to more general problems in motor or symbolic processing. There is also growing evidence for the role of the right hemisphere in sign language processing, especially in the realm of comprehension. Whether these language deficits reflect general spatial–cognitive deficits or whether right-hemisphere neural systems function as part of a linguistic system in the deaf requires further investigation. Future advances in our understanding of sign language structure and processing will provide new answers. Recent *in vivo* imaging of blood flow in normal congenitally deaf and hearing signers also points to an increased role of the right hemisphere—not entirely expected, given previous lesion studies (Soderfeldt, Ronnberg, & Risberg, 1994; Neville et al., 1995; Hickok et al., 1995). Case studies of neurologically impaired signers may illuminate the necessary and sufficient brain areas required for sign processing, while functional imaging may provide a picture of sign processing in the normal, nonchallenged brain. Together, these studies provide important insight into the biological foundation of human languages, and reveal degrees of neural plasticity in systems underlying functional human language.

CHAPTER 22

Levels of Representation
in Number Processing

Stanislas Dehaene[1] and Laurent Cohen[2]

[1]INSERM, CNRS, EHESS, and Laboratoire de Sciences Cognitives et Psycholinguistique, Paris, France;
[2]Service de Neurologie, Hôpital de la Salpêtrière, Paris, France

Numbers can be manipulated mentally in several formats such as Arabic notation, spelled-out numerals, and an abstract quantity representation. Neuropsychological models of number processing have attempted to specify the architecture in which mental representations of numbers are interconnected. McCloskey
and his colleagues initially proposed a simple model with a single central abstract
quantity representation interfaced by notation-specific input and output modules.
Although this model sucessfully accounts for several of the peripheral deficits of
number processing, difficulties arise with the hypothesis of an obligatory step of
semantic interpretation of numbers. Several problematic cases suggest that there
must be direct asemantic processing routes parallel to the semantic processing of
numerical quantities. An alternative model, Dehaene and Cohen's triple-code
model, describes both the functional architecture and the neural substrates of
number processing. It successfully accounts for many types of numerical deficits,
including the peculiar dissociations found in pure alexia, in callosotomy cases, in
Gerstmann's syndrome, and in subcortical acalculias.

There is a domain of language in which we are all, in some sense, bilinguals: the
domain of numbers. Like any other category of words, numbers can be spoken
(/four/, /forty/) or spelled out (FOUR, FORTY). However, we are all familiar with a

third symbolic notation, Arabic numerals (4, 40). Furthermore, numbers can also be conveyed in nonsymbolic ways such as sets of dots (::).

The human brain must contain mental representations and processes for recognizing, understanding, and producing these various notations of numbers and for translating between them. Hence, the number domain provides a manageably restricted area within which to study the representation of symbolic information in the human brain and the interplay between verbal and nonverbal formats of representation. It is hoped that the lessons that numbers might teach us about the organization of symbol systems in the human brain will turn out to be generalizable to other linguistic domains.

In the past decade, much progress has been made toward understanding the internal organization of the mental representations of numbers, their interconnections, and their neural substrates. In this chapter, we first describe some of the evidence for a quantity representation in humans, and the central role that is attributed to this quantity representation in McCloskey's modular model of number processing. We then discuss several single-case studies of patients with number-processing deficits, some that support McCloskey's assumptions and others that seem to invalidate it. We close by briefly presenting the triple-code model and how it accounts for the majority of these cases.

22-1. THE QUANTITY REPRESENTATION

Moyer and Landauer's (1967) seminal study of number comparison provided the first strong evidence for a quantitative representation of numbers in human adults. Measuring the time that it takes to select the larger of two digits, they found a *distance effect:* number comparison becomes systematically slower as the distance between the two numbers decreases. It is easier to decide, say, that 9 is larger than 5 than to decide that 9 is larger than 8. The effect was later extended to two-digit numerals (Dehaene, Dupoux, & Mehler, 1990). The reaction time curve for deciding whether a two-digit number is larger or smaller than 65 is remarkably smooth and shows no significant discontinuities at decade boundaries (Figure 1). Reaction times are influenced by the ones digits even though the tens digit is sufficient to respond: subjects respond "smaller" more slowly to 59 than to 51, although the 5 in the decades position readily indicates that both of these numbers are smaller than 65.

These results suggest that subjects do not compare numbers digit by digit. Rather, it was hypothesized that subjects mentally convert the target Arabic numeral into a continuous quantity, which they then compare to the reference quantity using a psychophysical procedure similar to the one used for comparing line lengths, weights, or other physical quantities. Several models of number comparison suppose that the human brain incorporates an *analogical representation of numerical quantities* that may be likened to a number line. In this representation, numbers are not represented by discrete symbols such as digits or words, but by distributions of activation whose overlap indicates how similar the quantities are. Several experiments with normal

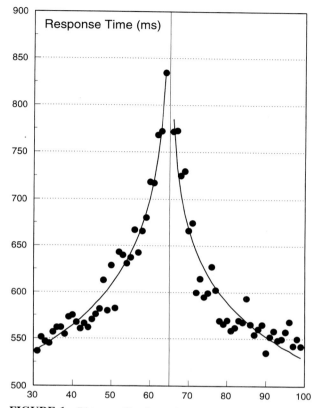

FIGURE 1 Distance effect in number comparison (data from Dehaene, Dupoux, & Mehler, 1990).

subjects confirm that the conversion from digits or number words to the corresponding quantities is fast and automatic and even occurs unbeknownst to the subject (review in Dehaene & Akhavein, 1995).

22-2. McCLOSKEY'S MODEL

What is the neuropsychological architecture in which Arabic, verbal, and quantity representations of numbers are embedded? McCloskey and his colleagues (McCloskey, Caramazza, & Basili, 1985; McCloskey, 1992) proposed a simple modular model that was to become, for a decade, the reference model for studies of number processing. The model places the quantity representation at a central point in processing and assumes that it is interfaced by specialized input and output modules. On the input side, a panoply of modules serves to convert Arabic digits and written or spoken numerals into internal quantities. On the output side, conversely, other modules are called for to write down Arabic digits and number words or to say numerals aloud. The most important assumption of the model is that each mental operation on numbers involves a conversion to the abstract semantic quantity representation. Arithmetic fact retrieval and calculation procedures, in particular, are thought to operate exclusively on the quantity format (Figure 2).

McCloskey's modular stance was remarkably successful in classifying and in interpreting neuropsychological impairments of number processing as well as in discovering new dissociations. One of the first cases to be explored was patient H.Y. (McCloskey, Sokol, & Goodman, 1986). H.Y. made errors in reading Arabic numerals aloud—for instance, reading 5 as *seven* or 29 as *forty-nine*. A careful analysis indicated that the deficit was highly specific and modular. First, the patient still understood the Arabic numerals that he failed to read, as attested by his ability to compare them, to select a number of chips corresponding to an Arabic digit, to match an Arabic digit to a written word, or to verify written calculations. According to the logic of the model, this proved that the conversion of numerals into quantities had to be intact. Hence, the number reading deficit had to originate from the output side, in the process of converting quantities into spoken numerals.

Second, McCloskey and his colleagues showed that writing down Arabic digits in response to various tasks was fairly intact (2-4% errors), while speaking aloud the numerical answers was significantly more impaired (8-14% errors). For instance, when asked the number of eggs in a dozen, the patient wrote 12 while saying *sixteen*. This suggested a rather selective deficit of the spoken production module.

Third, a careful analysis of H.Y.'s reading errors indicated that they mostly constituted substitutions of one word for another, while the grammatical structure of hundreds, decades, and units was well preserved (e.g., stimulus 902, response *nine hundred six*). Furthermore, ones words were almost always replaced by other ones words,

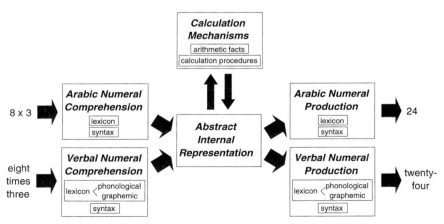

FIGURE 2   Schematic functional architecture of McCloskey's model.

teens words by other teens words, and tens words by other tens words. In the final analysis, the deficit was pinpointed to an impairment in using quantity information to select the phonological form of the word in lexical stacks for ones, teens, and tens words. When reading 15, the patient prepared to read aloud the fifth element of the teens category (eleven, twelve, thirteen, fourteen, *fifteen*), but he mistakenly selected, say, the eighth element *eighteen* instead. Although some details of the word substitutions remained unexplained (see Campbell & Clark, 1988; Sokol, Goodman-Schulman, & McCloskey, 1989), the bulk of the deficit could be reduced to this highly restricted lesion of the modular model.

McCloskey's model successfully predicted several other types of number processing deficits (reviews in McCloskey, 1992; McCloskey, & Caramazza, 1987; Caramazza & McCloskey, 1987). In number reading, patients similar to H.Y. were found who selected words in the wrong numerical category but at the correct ordinal location in the category (e.g., reading 2 as *twelve* or 15 as *fifty*). Others had preserved selection procedures for single words, but generated an incorrect syntactic sequence of words when producing complex numerals (e.g., reading 218 as *two thousand one hundred and eight*).

Similar modular deficits and dissociations were found in calculation. Some patients' only deficit was in identifying the operation signs +, ×, and so on. For others, the deficit was limited to arithmetic fact retrieval—they failed to retrieve single facts such as 2×3 or $5 + 5$ in memory, but could otherwise execute the procedures for multidigit addition and multiplication to near perfection. And conversely, other patients suffered from a selective impairment of calculation procedures, while their memory for individual facts was intact. Hence there was a double dissociation between facts and procedures in calculation, compatible with McCloskey's hypothesis that these correspond to different number-processing modules.

Elegant support for the model was also accrued when Macaruso, McCloskey, and Aliminosa (1993) showed that they could account not only for the qualitative dissociations of patients with highly selective deficits, but also for the quantitative error rates of a patient with multiple deficits. Patient R.H. exhibited mild to severe deficits in several number-transcoding tasks. For instance, he made 25% errors in reading Arabic numerals aloud, 21% errors in writing Arabic numerals to dictation, and 88% errors in converting Arabic numerals to their spelled-out form. These deficits could clearly not be accounted for by a single lesion site in the model. However, Macaruso et al. (1993) found that the patient's quantitative error rates in transcoding back and forth from Arabic, spelled-out, spoken, and dot notations of numbers could be explained by assigning probabilities of errors to each component of the modular model. For instance, the 25% error rate in Arabic numeral reading resulted from a 3% error rate in the Arabic identification module and a 22% error rate in the spoken production module. The error types also conformed to this "multiple-deficits" analysis. For instance, the patient made word substitution errors whenever the task required spoken production of numbers, whether the input was in Arabic, in word, or in dot format—a finding compatible with the existence of a lesioned "number production module" shared between all these tasks.

22-3. SOME PROBLEMATIC CASES

In spite of these successes, McCloskey's model still generates much controversy. In particular, the hypothesis of a central semantic representation of numbers, accessed in all tasks involving numbers, is hotly debated. To take an extreme example, the model predicts that the mere repetition of a spoken number word necessarily involves going through the semantic representation—an implausible prediction, as acknowledged by McCloskey himself (McCloskey, 1992). The absence of a direct surface transcoding route, for instance, going directly from the orthographic form of a number word to its pronunciation, conflicts with neuropsychological evidence from language impairments outside the numerical domain.

A first specific case that was found difficult to explain within McCloskey's framework was Cohen and Dehaene's (1991) patient Y.M. This patient showed a reading impairment characterized by digit substitutions (3 was read *eight*), an effect of visual similarity on errors (3 was rarely substituted with a visually dissimilar digit such as 1), and a spatial gradient of errors (errors mostly affected the leftmost digits of multidigit numerals). In McCloskey's framework, the fact that visual variables affected reading suggested a deficit within the visual Arabic identification module. Yet several observations conflicted with this hypothesis. Most notably, the patient was perfect in selecting the larger of two Arabic digits, and also surprisingly good in verifying additions such as $2 + 2 = 9$. The data, while not fully conclusive, suggested that the semantic route from Arabic digits to quantities was intact, while the route from Arabic digits to spoken words was impaired—a pattern forbidden in McCloskey's model.

Multiple routes also seemed necessary to explain a case reported by Cipolotti and Butterworth (1995). Their patient S.A.M. made frequent errors both when reading aloud Arabic and spelled-out numerals and when writing the same numbers down to dictation. Yet he was remarkably accurate in calculation tasks involving multidigit addition, subtraction, and multiplication problems, including tasks that required him to say the very same numbers that he had failed to read aloud! For instance, S.A.M. could say the result of $396 + 837$ without error, although he made errors in reading these numbers aloud. This striking dissociation suggests that the ability to produce Arabic or verbal numerals can be specific to particular task demands. Cipolotti and Butterworth (1995) suggest that reading aloud and writing down numbers to dictation makes use of direct asemantic routes different from those used in a calculation context.

Perhaps the clearest incompatibility with McCloskey's model came from a study of two patients with pure alexia (Cohen & Dehaene, 1995). Patients G.O.D. and S.M.A. suffered from highly similar infarcts of the left ventral occipito-temporal area, a region involved in high-level visual identification. Both were totally unable to read words. Arabic numerals were slightly better preserved: multidigit numerals were misread on 60-80% of trials and single digits on 8-18% of trials. Calculation on written operands was also impaired. For instance, when presented with $2 + 3$, the patients might say *seven*. That these calculation errors were due to a misidentification of the digits was shown by *(a)* the patients' perfect ability to perform the same calculation with spoken operands, and *(b)* the patients' reading errors in calculation: for instance,

2 + 3 was read *two plus five* and then solved as *seven,* indicating that the patients correctly computed the sum of the operands that they had misread.

So far, this deficit might be understood as a selective impairment of the Arabic number identification module. Contradicting this hypothesis, however, was the fact that both patients compared Arabic numerals with remarkable accuracy. When presented with 44 pairs of two-digit numerals, for instance, neither of them made any errors in pointing to the larger number, whereas they made, respectively, 89% and 91% errors in reading the very same pairs aloud. Some of the reading errors inverted the order of the numbers. For instance, 78 76 was read as *seventy eight, seventy nine*— yet the patient correctly pointed to 78 as the larger number.

This pattern of dissociation is clearly inconsistent with McCloskey's model. In this model, the preserved comparison of Arabic numerals should imply that the Arabic number identification module is intact. Similarly, the perfect performance with spoken inputs and outputs should mean that the verbal number comprehension, production, and calculation modules are intact. With all these intact components, the modular model should predict the reading aloud of Arabic numerals to be perfectly preserved— yet it was severely impaired. Cohen and Dehaene (1995) argued that these dissociations could only be explained by postulating two Arabic numeral identification systems, one involved in identification for the purpose of naming (which is impaired in pure alexia), and the other involved in identification for the purpose of accessing the quantitative semantic representation (which is preserved; more on this later).

A functionally similar case was presented by McNeil and Warrington (1994). Their patient H.A.R. also suffered from alexia without agraphia (although he was also aphasic and suffered from naming difficulties). Just like patients G.O.D. and S.M.A., he experienced severe difficulties in reading Arabic numerals aloud or in calculating with them. Yet he could calculate with near perfection with spoken operands and he could also compare Arabic numerals. An additional twist was that when calculating with written Arabic digits, only additions and multiplications were severely impaired: the patient was excellent in subtracting two digits that he failed to read aloud. It seems that the same procedure that enabled him to convert Arabic numerals into quantities during number comparison could also be used for subtraction—but not for addition or multiplication. According to McCloskey's model, such an operation-specific *and* number-specific deficit should not exist.

22-4. THE TRIPLE-CODE MODEL

The triple-code model was introduced by Dehaene (1992; Dehaene & Cohen, 1995) in an effort to account for these peculiar cases. The model postulates three main representations of numbers (Figure 3):

1. A visual Arabic code, localized to the left and right inferior ventral occipito-temporal areas, and in which numbers are represented as identified strings of digits. This representation, which we call the visual Arabic number form by

analogy with Shallice's visual word form, subserves multidigit operations and parity judgments (e.g., knowing that 12 is even because the ones digit is a 2).

2. An analogical quantity or magnitude code, subserved by the left and right inferior parietal areas, and in which numbers are represented as points on an oriented number line. This representation subserves semantic knowledge about numerical quantities, including proximity (e.g., 9 close to 10) and larger–smaller relations (e.g., 9 smaller than 10).

3. A verbal code, subserved by left-hemispheric perisylvian language areas, in which numbers are represented as a parsed sequence of words. This representation is the primary code for accessing a rote verbal memory of arithmetic facts (e.g., "nine times nine, eighty-one").

Figure 3 shows the patterns of interconnections that are postulated in the model. In the left hemisphere, all three cardinal representations (Arabic, verbal, and quantity) are interconnected by bidirectional translation routes, including a direct asemantic route for transcoding between the Arabic and verbal representations. In the right hemisphere, there are similar routes for translating back and forth between Arabic and quantity representations, but there is no verbal representation of numbers. Finally, it is assumed that the homologous Arabic and quantity representations in the two hemispheres are interconnected by direct transcallosal pathways.

How does the model account for pure alexia patients, who were found so problematic for McCloskey's model? Their lesion affected the left ventral occipito-temporal area and therefore predictably impaired the left visual word and number identification system. Hence, left-hemispheric verbal areas could not be directly informed about the identity of the visual stimulus, resulting in a severe impairment of reading aloud. Since the memory retrieval of rote arithmetic facts is assumed to depend on the verbal format, this reading deficit entailed a calculation deficit—the patients could not turn 3 × 3 into the verbal format *three times three,* which was necessary for them to retrieve the result *nine.* Yet the verbal circuit itself was intact, only deprived of visual inputs. Hence the patients could still calculate when an arithmetic problem was read

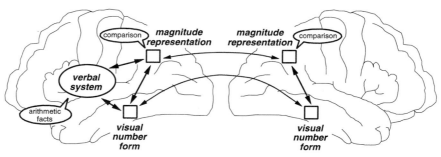

FIGURE 3   Schematic anatomical and functional architecture of Dehaene and Cohen's (1995) triple-code model.

aloud to them. And finally, their right hemisphere was fully intact, including its visual identification and magnitude representation areas. According to the triple-code model, these preserved right-hemispheric circuits underlie the patients' ability to compare the magnitudes of Arabic numerals that they fail to read (Cohen & Dehaene, 1995).

Very similar interpretations account for the above-described patients Y.M. (Cohen & Dehaene, 1991) and H.A.R. (McNeil & Warrington, 1994). Although we know very little about how mental subtraction is performed, clearly we do not learn subtraction tables by rote as we do for addition and multiplication. Dehaene and Cohen (1995) speculate that subtraction is more similar to number comparison and involves quantitative manipulations rather than rote verbal retrieval. This may explain why patient H.A.R. could still subtract Arabic numerals, while his inability to read them aloud severely affected his ability to add or multiply them.

A strength of the triple-code model is that it specifies cerebral localizations, however coarse, for the various representations of numbers. As a result, the model can predict the effect of specific anatomical lesions on number processing. A case in point is the split-brain syndrome. When the corpus callosum is severed, the model predicts that the left hemisphere should remain able to perform all sorts of calculations, because it contains the three cardinal representations of numbers (Arabic, verbal, and quantity). The right hemisphere, however, should be able to recognize Arabic numerals, to retrieve the quantity that they represent, but not to read aloud nor to perform calculations dependent on the verbal code. This predicted pattern of results is exactly what is found, both in the classical literature on split brains (Gazzaniga & Hillyard, 1971; Gazzaniga & Smylie, 1984; Seymour, Reuter-Lorenz, & Gazzaniga, 1994) and in a single-case study of number processing following an infarct of the posterior half of the corpus callosum (Cohen & Dehaene, 1996). When digits 5 and 6 are flashed in the left hemifield, the patients' right hemisphere can decide that 5 is smaller than 6, but it cannot read the digits aloud nor calculate $5 + 6$. Similar results suggesting a right-hemispheric ability restricted to quantitative processing have been obtained in patients with extended left-hemispheric lesions (Dehaene & Cohen, 1991; Grafman, Kampen, Rosenberg, Salazar, & Boller, 1989). Patient N.A.U., for instance, could not tell whether $2 + 2$ was 3, 4, or 5—but he knew that it had to be smaller than 9.

The triple-code model also accounts for some of the various types of acalculias and their anatomical correlates. According to the model, there are two basic routes through which a simple single-digit arithmetic problem such as $4 + 2$ can be solved. In the first, direct route, which works only for overlearned addition and multiplication problems, the operands 4 and 2 are transcoded into a verbal representation of the problem ("four plus two"), which is then used to trigger completion of this word sequence using rote verbal memory ("four plus two, six"). This process is assumed to involve a left cortico-subcortical loop through the basal ganglia and thalamus. In the second, indirect semantic route, the operands are encoded into quantity representations held in the left and right inferior parietal areas. Semantically meaningful manipulations are then performed on these internal quantities, and the resulting quantity is then transmitted from the left inferior parietal cortex to the left-hemispheric perisylvian language network for naming. The model assumes that this indirect semantic route is used whenever rote verbal knowledge of the operation result is lacking, most typically for subtraction problems.

The two types of acalculic patients predicted by these two routes for calculation have now been identified (Dehaene & Cohen, in press). There are several published cases of acalculia following a left subcortical infarct (Whitaker, Habiger, & Ivers, 1985; Corbett, McCusker, Davidson, 1988; Hittmair-Delazer, Semenza, & Denes, 1994). In a recent case of a left lenticular infarct, Dehaene and Cohen (in press) showed that, as predicted by the model, only rote verbal arithmetic facts such as 3×3 were impaired (together with rote verbal knowledge of the alphabet, nursery rhymes, and prayers). Quantitative knowledge of numbers was fully preserved, as shown by the patient's intact number comparison, proximity judgment, and simple addition and subtraction abilities. Hittmair-Delazer et al. (1994) likewise observed a preservation of conceptual, quantitative, and algebraic manipulations of numbers in their severely acalculic patient with a left subcortical lesion.

Conversely, the model predicts that lesions of the left inferior parietal area, typically resulting in Gerstmann's syndrome, affect calculation because they destroy quantitative knowledge while preserving rote verbal abilities. Indeed, Dehaene and Cohen (in press) observed a Gerstmann-type patient who could still read aloud numbers and write them down to dictation, still knew rote addition and multiplication facts such as $2 + 2$ and 3×3, and yet failed even in the simplest of quantitative tasks. He made 16% errors in larger–smaller comparison, 75% errors in simple subtractions (including gross errors such as $3 - 1 = 3$ or $6 - 3 = 7$), and 78% errors in number bisection (stating, for instance, that 3 falls in the middle of 4 and 8). The deficit was highly specific to the category of numbers, since the patient performed quite well in finding the middle of two letters, two days of the week, two months of the year, or two notes of the musical scale. We have now observed several such cases of severe number bisection deficits following a lesion of the inferior parietal area. These cases support the triple-code model's hypothesis that this area is critical for representing and manipulating number as quantities.

22-5. SOME OPEN ISSUES IN NUMBER PROCESSING

In spite of the progress made in understanding the architecture for number processing in the human brain, many gaps remain to be filled.

First, while we begin to understand the large-scale networks for number processing, further studies are needed to better understand the internal structure of representations, particularly the verbal representations used for accessing rote arithmetic facts.

Second, the issue of hemispheric specialization is still largely unsolved. Both hemispheres are clearly able to identify Arabic numerals and to represent the associated quantities. Yet, are the left and right representations equivalent, or are there differences between the magnitude representations available to the left and right hemispheres? Would the left hemisphere be categorical while the right would hold a more continuous, analogical representation of numbers? This important question remains to be addressed.

Finally, an issue that has emerged is the extent to which other types of mathematical knowledge involve yet other cerebral circuits than the ones we have described in this chapter. In addition to quantitative and rote verbal knowledge about numbers, most people also have encyclopedic knowledge (e.g., knowing that 1789 is the date of the French Revolution) and algebraic knowledge (e.g., knowing that $(a + b)^2 = a^2 + 2ab + b^2$). There are good indications that encyclopedic numerical knowledge can remain partially preserved in Gestmann's syndrome (Dehaene & Cohen, in press) and in patients with extensive left-hemispheric lesions (Cohen, Dehaene, & Verstichel, 1994). Hittmair-Delazer and her colleagues (1994, 1995) have even shown that conceptual algebraic knowledge was largely preserved in two severely acalculic patients. This observation suggests, against all intuition, that the neuronal circuits that hold algebraic knowledge must be largely independent from the networks involved in mental calculation. Their cerebral substrate, however, remains almost totally unknown.

CHAPTER 23

Computational Models of Normal and Impaired Language in the Brain

Anthony E. Harris and Steven L. Small
Intelligent Systems Program, Center for the Neural Basis of Cognition, and the Department of
Neurology, University of Pittsburgh, Pittsburgh, Pennsylvania 15261

Computational models have been used to foster insight into mechanisms of normal and abnormal language processing in the brain. Although most of the classical computational models have used symbolic, discrete structures to model linguistic representations and sequential, functional processing to model linguistic processing, the new style of connectionist modeling may assist in bridging the gap between descriptive models of linguistic behavior and mechanistic models of language processing in the brain. This chapter reviews some of the types of connectionist models and examines representative models accounting for language processing, as well as the deficits in such processing that can occur with brain injury. We will also discuss the strengths and weaknesses of the connectionist approach concerning its representational power and its adequacy in providing accounts for linguistic behavior in terms of neural functioning.

Almost two decades ago, Arbib & Caplan (1979) argued that "neurolinguistics must be computational." They believed that computation could bring together under one framework the opinions of several historically important neurolinguistic communities, including the localizationalists (Lichtheim, 1885; Geschwind, 1971), the holists (Jackson, 1878), and the functionalists (Head, 1926; Luria, 1973a). At that time, research in artificial intelligence was providing new insights into the formal representation of human knowledge (including linguistic knowledge), and except for some early low-level (synaptic) models by Arbib himself (Arbib, 1970), there was little interest in or information about parallel distributed modeling. Subsequent work on connectionist modeling (Ackley, Hinton, & Sejnowski, 1985; Dijkstra & de Smedt, 1996; Feldman & Ballard, 1982; Hopfield, 1982; McClelland & Rumelhart, 1981) has brought together parallel processing aspects of the earlier frameworks with the representational knowledge gained through artificial intelligence work. These results may make a computational neurolinguistics possible.

While the majority of work in computational neurolinguistics considers disembodied, abstract models, connectionist (neural network) models take a more grounded approach to the problem. Many computational models regard cognition and language as being deep and difficult to study, with other functions of the brain (such as sensation, motion, and sensorimotor interfacing) considered more straightforward. At the very least, traditional models consider "higher" aspects of cognition wholly separate from low-level somatomotor and perceptual questions, so that questions of cognition are studied completely independently of lower brain functions. While most computer models attempt to explain linguistic capabilities in terms of cognitive structures, connectionist style modeling can hypothesize about neural structures underlying language functioning. Certain researchers have even explained neurolinguistic syndromes by explicitly modeling brain structures in which injury is thought to be causally related to the deficits found (Pulvermüller & Preibl, 1991). By providing a modeling framework for cognition that also has been used profitably to model primary sensory and motor functions, connectionism has suggested some hope for bridging the gap between neurons and cognition.

The ideal approach thus focuses on the interplay between computational modeling and the appropriate neurological and neuropsychological data, the whole couched in connectionist mechanisms that map naturally to what is known of the neurophysiological structure of the brain. Ideally, computational processes, behavioral mechanisms, and anatomical structures are all specified explicitly, enabling studies of performance, damage, and rehabilitation.

In the remainder of this chapter, we will focus on the application of connectionist modeling techniques to the study of neurolinguistics. In particular, we will describe some representative network architectures and learning algorithms, and show how each example network has been used to explain both normal cognitive linguistic phenomena and the abnormal phenomena associated with acquired language deficits. The models thus provide theoretical sketches of aspects of the underlying neural substrates for various components of the language processing system.

23-1. NEURAL NETWORK MODELS

Neural (connectionist) networks are increasingly applied to studies in cognitive neuroscience (Sejnowski, Koch, & Churchland, 1988). A connectionist model consists of simple units of activation, which communicate by sharing their activation with other such units to which they are connected. Each such connection has an associated strength, called a weight. Units thus compute a new level of activation by combining their previous level of activation with the information shared with them through these weighted connections. Input and output of a network are provided by input units, with externally imposed activation levels, and output units, which contain the results of the network computation. All other units are referred to as hidden units of activation. The overall behavior of a model is determined by the pattern of connections, the weights on these connections, and the ways in which units compute their activation levels. Figure 1 shows the main features of an artificial neural network.

The important computational features of the network are these computational units and the weighted connections between them. An example unit in Figure 2 shows most of the important features of the computing elements. The activation, or output value, of a unit is a numerical value associated with that unit. For each hidden or output unit, the new activation value is computed as some function of the activations of the

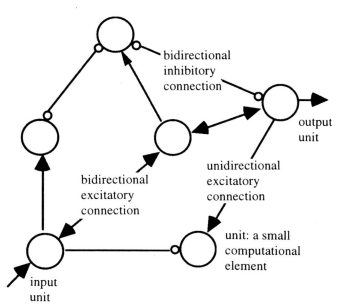

FIGURE 1   A connectionist network that demonstrates the salient features of such a system. All units not marked input or output units are hidden units.

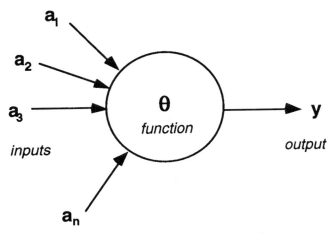

FIGURE 2   An example unit with inputs a_1 to a_n, and output y.

units feeding into it. First the net input is computed, which is the weighted sum of the activations of those units that feed into it. Second, this net input is passed through an activation function to compute the new activation value. While arbitrary functions may be used, the most common is the logistic function of Figure 3.

There are two main aspects of a network that determine its behavior. The first concerns how new unit activation values are computed given some input. For example, the network architecture, or the pattern of connectivity between units, in part deter-

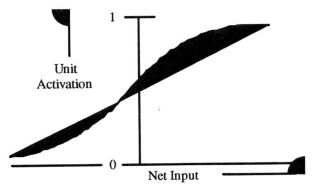

FIGURE 3   The behavior of the typical unit activation function, the sigmoidal "squashing" function. As the amount of unit input rises, the output also rises, but has the sigmoidal shape, with values always within the range of 0 and 1.

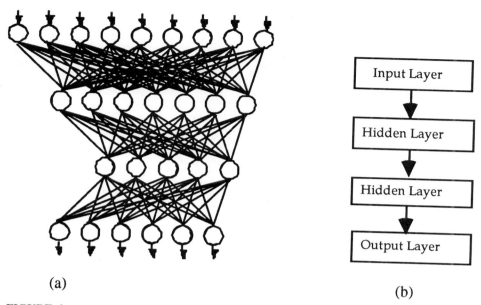

(a)

(b)

FIGURE 4   A feedforward network with two hidden layers. Figure 4(a) shows the full network with every unit in one layer connected to every unit in the next layer, a pattern of complete connectivity. Figure 4(b) shows the abbreviated notation for the network of Figure 4(a).

mines its computations. The units may be arranged in a sequence of layers, with previous layers feeding exclusively forward to subsequent layers (a feedforward architecture, see Figure 4), or units may be allowed bidirectional connections or other loops (a recurrent architecture; see Figure 5 for an example). The mechanisms determining the specifics of network computations can be used to model particular aspects of language representation and processing.

The second aspect of a neural network that determines its behavior is whether or not the connection weights adapt in response to environmental experience. If a network's weights do adapt, the mechanism determining how they do so is called the learning algorithm. Two broad classes of learning algorithms exist. In *supervised learning,* the network is presented with a training set of input/output pairs to be associated. The algorithm then compares the output computed by the network with the desired output (or the teaching signal) and calculates the difference, called the error of the network. This formal error measure is then used to adjust the weights so that subsequent presentation of the input will result in the network computing the proper output. In *unsupervised learning,* no external teaching signal is employed. Instead, the network is exposed to inputs, and the goal of the network is to build internal representations that are in some sense optimal given the input ensemble statistics. Adaptive networks may be used to model aspects of language acquisition and development.

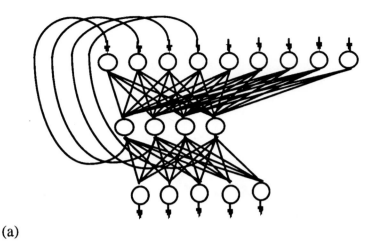

(a)

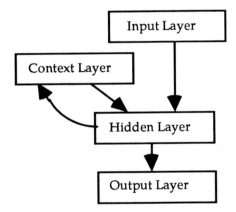

(b)

FIGURE 5 One popular and useful type of simple recurrent network. This network has a single hidden layer. Additional hidden layers could be added after the first if desired. Figure 5(a) shows the complete network with full connectivity between layers. Figure 5(b) shows the abbreviated way to write the network of Figure 5(a).

23-2. HARD-WIRED NETWORKS

The first type of network we will examine is the hard-wired network in which the network structure and connection strengths are preset by the experimenter and no network learning occurs. One example is the interactive activation model (IAM) (McClelland & Rumelhart, 1981). A second type of hard-wired approach is the structured connectionist model (Feldman, Fanty, Goddard, & Lynne, 1988), an early example of which is a model of sentence disambiguation (Cottrell, 1985).

The interactive activation model was first presented in McClelland and Rumelhart (1981). In this model, units are grouped into levels, and units within a level act as representational elements for various aspects of the cognitive phenomena. Connections may be within a level or between adjacent levels. Connections may be feedforward or bidirectional. The connections are hard-wired so that consistent elements of the network representation mutually reinforce each other (via positive, or excitatory, connections), while inconsistent elements mutually suppress one another (via negative, or inhibitory, connections). Thus the units cooperate or compete with each other in groups depending on their consistency as representations of the external environment. Individual units continually build up activation in time by adding their current net input to some decayed fraction of their previous activation value. Thus units with a very rapid decay rate essentially sum their net inputs from zero at each time step, while units with a slow decay build up their activation over time.

The first neurolinguistic data modeled by the interactive activation model concerned the effects of context on letter perception (McClelland & Rumelhart, 1981). The basic experimental finding was that letters are identified more accurately when embedded in the context of a word than when presented in the context of a nonword or a random sequence (Reicher, 1969). In their original formulation, units were grouped into levels corresponding to letter-feature detectors, letter detectors, and word detectors. Units that were consistent would reinforce one another via excitatory connections; those that were inconsistent would inhibit one another.

The context effects of words on letter perception were then explained as follows. If a degraded letter was presented to the network in the context of a word, activation from all the nondegraded letters in the word drove the activation of a set of possible word units to a high value, which in turn suppressed all other word units not consistent with the current input. These activated word units then sent feedback to the letter units, so that only letter units consistent with the candidate word-units received positive feedback, while inconsistent letter units received negative feedback. The combination of positive feedback from activated word units and positive activation from the nondegraded features of the letter drove the correct letter unit's activation above the incorrect letter unit's activation, thus suppressing the incorrect candidates' activations further by the intralevel negative feedback. By this mechanism, contextual effects enhanced letter perception.

Another style of modeling using hard-wired networks is the structured connectionist model (Feldman et al., 1988). One example used a constraint satisfaction network to model disambiguation phenomena (Cottrell, 1985) (Figure 6). Individual units were used to represent lexical items, case levels, and word senses. Compatible units were linked with a positive connection, incompatible ones by a negative connection. As lexical items entered the system and activated the corresponding lexical units, these sent positive activation to all senses and cases with which they were compatible. Due

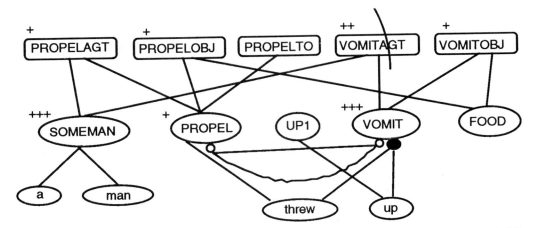

FIGURE 6 A structured connectionist network for parsing the sentence "The man threw up a ball." In this diagram, the number of "+" signs adjacent to a unit indicates its relative level of activation. Connections can be excitatory (—) or inhibitory (—°), with some excitatory conjunctive connections (—·—) requiring two connections to be active at the same time.

to negative interactions, a coalition of units representing the proper senses and cases overwhelmed the incorrect units. Thus, by modeling compatible interpretations with positive links and incompatible ones with negative links, the system was able to perform the linguistic task of disambiguation in exactly the same manner as models of disambiguation of the two Necker cube interpretations (Feldman et al., 1988).

A model of deep dysphasia was presented in Martin, Saffran, Dell, and Schwartz (1994), based on an interactive activation model of normal lexical retrieval (Dell, 1986). Dell's model described lexical selection in production with a three-level model: At the top was a semantic level, which fed into a lexical level, which in turn had bidirectional connections with a phonemic level. The target lexical node was primed by the semantic representation, and in turn sent activation to its associated phonemic representation. Feedback reinforced the appropriate target lexical node, but also activated phonologically related lexical competitors. Correct lexical retrieval usually occurred due to the feedback to the correct lexical item; normal speech errors occurred when inherent variability in the system raised phonologically related lexical competitors' activations.

Subsequent work using this model (Martin et al., 1994) accounted for the symptomatology of the deep dysphasic patient N.C. Deep dysphasia is characterized by a production of semantic errors in repetition, combined with an inability to repeat non-words. N.C.'s behavior also included a high rate of formal paraphasias (phonologically related lexical substitutions) (Martin & Saffran, 1992). The authors used the same model as above, with the only change being an increased decay rate. Initial priming of target lexical nodes by the semantic level led to activation of the phonological level, which fed back to target lexical items and competitors. However, due to the quicker decay rate, the initially primed target lexical node would decay back to normal. Thus the advantage of the primed lexical node was lost, making phonologically related errors more likely. The semantic errors in repetition were similarly explained by having activation originate in the phonological nodes. Finally, the rapid decay of phonological nodes accounted for the inability to repeat nonwords. By positing a pathologically high decay rate, the model accounted for deep dysphasic impairments.

23-3. SUPERVISED LEARNING NETWORKS

In contrast to hard-wired networks, several network paradigms exist in which connection strengths adapt in response to environmental experience. Such models may be used to investigate developmental aspects of language acquisition. In addition, the emergent representations that develop due to certain network learning rules can be nontrivial, and add interesting insights to putative linguistic coding mechanisms in the brain.

As described earlier, supervised learning utilizes a teaching signal to generate an error function with which to train the network. One example of this type of algorithm, the Boltzmann Machine (Ackley et al., 1985), uses stochastic units, and the network learns to reproduce the input data's probability distribution. The most widely used algorithm, however, is backpropagation of error, or backprop (Rumelhart, Hinton, & Williams, 1986). In the algorithm, the network starts with random initial weights, and compares its output from a given input with the teaching signal. Backprop uses gradient descent to alter the weights so as to traverse this error surface. It changes the weights so as to move in the direction of steepest descent (or "downhill"), and stops once the weights have settled to a local minimum of the error. At this point, the algorithm is said to have converged. The main problem with backprop, as with any hill-climbing algorithm, involves these local minima. Since the algorithm always moves downhill on the error surface, the weights are driven to whatever local minimum happens to be in a direct path from the random initial starting conditions. At this local minimum, the error may still be unacceptably high, while there might exist some other, lower-lying minima. The only way to reach these is to reinitialize the weights and run the algorithm again. Much current research involves modifying "vanilla" backprop or devising completely new algorithms to overcome the problem of local minima.

This section will examine examples of supervised learning. The first is a model of word recognition and naming implemented as a backprop model (Seidenberg & McClelland, 1989). In the second, a connectionist account of category-specific semantic deficits is presented (Small, Hart, Nguyen, & Gordon, 1995). The next examples will examine recurrent networks, in which arbitrary connectivity is permitted. We will first look at a recurrent network that encodes temporal contingencies concerning sentence structure (Elman, 1991). Finally, a fully recurrent network is used to explain deep dyslexia (Hinton & Shallice, 1991).

23-3.1. Feedforward Architecture

To model word recognition and naming, Seidenberg and McClelland (1989) constructed a parallel distributed processing model of the mechanisms by which orthographic representations are mapped to phonological ones. The model provides putative mechanisms for the two major aspects of acquisition of word-recognition skills. The first aspect that must be learned is spelling/sound correspondences, or the systematic correspondences between the written and spoken forms of language. The second aspect concerns the distribution of letter patterns in the lexicon, or the distribution of permissible letter combinations making up the written words in the language. One key feature of each of these aspects of word recognition is their inconsistency. The spelling/sound correspondences are not always followed, and the cues as to phonological, syllabic, and semantic information contained in the orthographic representations are not always reliable. Such a system is termed quasi-regular, where the relations among entities are statistical rather than categorical. Hence, connectionist models, which can encode statistical relations between representational units in its connection strengths, are particularly appropriate for investigating the word-recognition domain.

Several supervised learning models have examined issues of abnormal development and effects of brain damage on normal adults (Brown, in press). In one example, Small et al. (1995) presented a model accounting for category-specific naming deficits without assuming explicit categorical knowledge in the brain. Category-specific losses of semantic memory generally arise after temporal lobe damage (Hart & Gordon, 1992), and have involved selective loss of categories such as concrete objects, inanimate objects, animate objects, animals, and fruits and vegetables (Hart, Berndt, & Caramazza, 1985; Hart & Gordon, 1992; Sartori & Job, 1988; Warrington, 1981a; Warrington & McCarthy, 1983; Warrington & Shallice, 1984a). In normals, processing of superordinate categorical information can be slower than for object-level information (Collins & Quillian, 1969). A common account of these findings is that categorical information is explicitly encoded in the brain, with the agnosias corresponding to loss of this information. However, cortical imaging studies and direct cortical stimulation studies have met with difficulty in finding explicit, localized categorical information (Gordon et al., 1990; Ojemann, Ojemann, Lettich, & Berger, 1989; Small, Noll, Perfetti, Xu, & Schneider, 1994). The authors presented an alternative formulation, in which distributed categorical information was coded implicitly in featural information. A feedforward backprop net was trained to match input feature vectors with objects on its outputs. After training, the input feature space and the emergent hidden unit representations were analyzed. They found that the input feature vectors implicitly contained categorical information, due to the statistical structure found in natural clustering of features. In addition, using principal components analysis on the hidden unit representations, it was found that many of the higher principal components encoded for intuitive categories. The result of the study was to demonstrate that categorical information could exist in a distributed fashion without explicit encoding, in contrast to standard information processing accounts.

23-3.2. Recurrent Architecture

While the previous models were essentially feedforward, with a unidirectional flow of information, we now turn to models in which the direction of information is unconstrained. We will examine two particular models: the simple recurrent network (SRN) (Elman, 1991) and a more fully recurrent attractor network (Hinton & Shallice, 1991).

The SRN consists of a three-layer, feedforward backprop network, with the hidden units copied back to a set of context units at each time step (see Figure 5). These context units are combined with the current input for the net input to the hidden units at each time step, and the net is trained with backprop. But due to the context units, which encode information at each time step about the previous time step, temporal contingencies may be represented, in order to address issues of sequential processing of stimuli. In Elman (1991), the author used this architecture to map each sentence in a training corpus onto a time-shifted copy of itself. Representations reflecting lexical categories were found in the analysis of the emergent hidden unit representation, due to the information implicit in the sequential ordering of the inputs. The network could also predict the next word in spite of a training corpus of sentences in which subjects and verbs could be separated by arbitrary numbers of embedded clauses, predicting proper subject–verb agreement and argument structure for verbs. Thus the network developed representations reflecting constituent structure, argument structure, grammatical category, grammatical relations, and number. The same SRN architecture was used to perform lexical ambiguity resolution by encoding and then using contextual information (Small, 1990).

A second model using a recurrent network was presented in Hinton and Shallice (1991). In this model, bidirectional connections are permitted, and activation is passed between units until the network settles into a stable state, or attractor. Each attractor has a basin of attraction around it, so that if the network is begun in any pattern of activation within that basin, the network state will move to the attractor. Thus the network can retrieve content-addressable memories and deal with degraded input, that is, if the net is begun in a noise-corrupted pattern, the network will "clean up" its input to reach the attractor stored in its memory. In this study, the net was used to model aspects of deep dyslexia, a syndrome in which patients seem to obtain the phonological representation of a written word via its semantics. This results in frequent semantic paraphasias (semantically related lexical substitutions). These patients also exhibit many visual errors, and in fact show more mixed errors (both semantic and visual) than would be expected by chance. The authors used a network in which grapheme units were mapped to recurrently connected sememe units (representing features). Thus each featural semantic encoding had a basin of attraction around it. If the network was begun near a semantic memory, the net converged to the stored attractor. Errors in deep dyslexics were explained as follows: Due to damage, input orthographic representations were mapped to semantic representations near the target. Thus damage can cause visual input to be mapped to semantically related output. To account for the visual and mixed errors, the idea is that connectionist networks tend to map similar input encodings to similar output encodings. The damaged net was likely to map nearby orthographic patterns to nearby semantic patterns, resulting in a mixed error. Thus damage that moves the basins of attraction in semantic space can cause visual or mixed errors.

23-4. UNSUPERVISED LEARNING NETWORKS

In the second broad class of learning algorithms, unsupervised learning, no teacher is present for the network. An example of this approach was presented earlier in the discussion of competitive learning of a space of semantic features (Small et al., 1995). In unsupervised learning, the network learns to compute representations that are in some sense optimal based purely on the statistical structure of the inputs. Several unsupervised learning algorithms exist, one of which, the Kohonen network (Kohonen, 1982), will be considered here.

In the Kohonen net, the network learns a topological mapping such that similar inputs in the high-dimensional input space are mapped onto nearby units in the physically two-dimensional layout of the output space. Inputs are presented one at a time, and the output unit with the largest net input, or the "winner," is computed. Then the weights to the winner are shifted toward the current input by a small amount, so that on subsequent presentation of a similar pattern, this unit will respond even more strongly. In addition to the winner shifting its weight, all the units in a local spatial neighborhood around the winner also shift their weights closer. Thus, units that are spatially contiguous come to represent similar inputs. The network becomes a topographic map, where similar inputs are mapped to nearby outputs.

A recent model of the lexical system, called DYSLEX, was presented using modular, topographic maps (Miikulainen, in press). In this model, orthographic, phonologic, and semantic input and output lexicons were modeled with separate, modular topographic maps. Each unit stood for a lexical entry. The maps were trained with distributed feature vectors that encode similarities between entries, and through the learning process these feature vectors were mapped onto nearby units in the appropriate lexicon. Each of these modules was then associated with each other with weights that were learned by a Hebbian learning rule. In such a rule, units with strong correlations (i.e., that tend to co-occur) form strong connections, while units with anti-correlations form small or even strong inhibitory connections with each other. Through this means, orthographic and phonologic input lexical items were mapped to their appropriate semantic lexical items. Because nearby units encode for related items, noisy inputs or connection strengths made it likely for related items to be accessed. Combinations of errors are also possible. Thus an orthographic input could propagate to a visually related word, which could in turn propagate to a semantically related word. By combining modular connectionist substructures in a biologically plausible system, the model suggests an account for several observed linguistic deficits.

23-5. SUMMARY

Neurolinguistics has had a substantial history of computational modeling. Most traditional models have used static, symbolic data structures on which functions act to produce linguistic output. An alternative paradigm, the connectionist approach, uses graded, dynamic representations that are processed by simple mechanisms. It has provided sufficient accounts of high-level behaviors using distributed networks of simple processing units.

While providing demonstrations of higher-level cognitive capacities, connectionist models have also been used to model aspects of primary motor and sensory systems. Thus, by providing a single framework, they have given some insight into how functional cognitive capacities may arise from neural hardware. Although connectionist models have had much success, there are still several problems in bridging the gap between cognition and brain. Although connectionist units are "neuron-like" elements, they are still abstract entities that only approximate real neurons. For example, although the activation levels are often taken to represent the average firing rate of a small group of real neurons, it may be that the temporal characteristics of neuronal spike trains encode information, which would make a single real activation value insufficient to model real neurons' representations. In addition, connectionist models still show some deficiency in moving "up a level" to the computational capacities of symbolic models. In particular, issues such as generativity, systematicity, and productivity are all rather awkwardly handled by connectionist models (Fodor & Pylyshyn, 1988). However, recent work in addressing the binding problem (Mozer, Zemel, Behrmann, & Williams, 1992; Shastri & Ajjanagadde, 1993) may help connectionist models overcome these shortcomings.

The main features of connectionist networks are multiple, simple processing units computing in parallel. Different types of networks include hard-wired networks and adaptive networks, in which connection strengths automatically adjust in response to the environment. Another breakdown can be made on the direction of information flow, with either feedforward or recurrent architectures. Finally, adaptive networks can be characterized according to the type of learning they perform. One class is supervised networks, in which an explicit teaching signal is used to map inputs to outputs. Another is unsupervised learning, in which no teacher is present, and the network learns interesting representations based purely on the statistics of the inputs.

In addition to providing plausible models of normal linguistic functions, these models have provided accounts for linguistic deficits. Models of deep dysphasia, deep dyslexia, developmental dyslexia, and category-specific aphasias have been presented, as well as others. In all of these, accounts of complicated syndromes have been presented without invoking simultaneous damage to modular subsystems. The ability of connectionist models to suggest parsimonious accounts provides support for these models.

Although connectionist models cannot currently account for all aspects of higher-level cognition, they have accounted for many features of language processing. They have done so while maintaining a link to how real structures in the brain behave. Although these models are not strictly biologically plausible, they have suggested means by which neural structures may compute cognitive functions.

Acknowledgments

This work was supported in part by the National Institute of Deafness and other Communication Disorders (NIDCD) of the National Institutes of Health (NIH) under award K08-DC-0054.

CHAPTER 24

Brain Lateralization across the Life Span

Merrill Hiscock

Department of Psychology, University of Houston, Houston, Texas 77204

Functional specialization of the cerebral hemispheres in humans may develop gradually in childhood, reach a zenith sometime in adulthood, and undergo further change in senescence. Or brain lateralization may be established prior to birth and remain invariant throughout life, despite the effects of maturation, experience, and disease. This chapter summarizes evidence from various sources regarding the developmental course of lateralization in infancy and childhood as well as in normal and pathological aging. Major issues are identified, and guidelines for interpreting findings are suggested. Although the question of life-span change versus invariance cannot be answered definitively without attaining a much better understanding of the functions and processes that are thought to be lateralized, and how they change over time, the existing evidence favors the hypothesis of invariant lateralization.

The ontogeny of hemispheric specialization has been a prominent topic in neurology, neuropsychology, and neurolinguistics for many years. The idea of a gradually developing lateralization of function (Lenneberg, 1967) is appealing, perhaps because it lends credibility to the assumption that hemispheric specialization is associated with behavioral complexity and behavioral refinement (Hiscock & Kinsbourne, 1994). Accordingly, full lateralization of brain functions, along with full realization of behavioral potential, would be attained only in adulthood (Luria, 1973b) and perhaps even late in adulthood (Brown

& Jaffe, 1975). The same assumption may motivate attempts to show that humans are more completely lateralized than nonhumans (cf. Glick, 1985), and that socially or academically advantaged humans are more completely lateralized than disadvantaged humans (e.g., Boliek & Obrzut, 1994; Geffner & Hochberg, 1971; McGlone, 1980).

Another implication of the putative connection between lateralization and behavioral excellence concerns aging in humans, namely, the possibility that diminution of cognitive abilities in the aged is accompanied by hemisphere-specific cerebral degeneration, or at least by more rapid deterioration of one hemisphere than of the other (Klisz, 1978). Thus, the optimal specialization of the hemispheres, which is realized at the beginning of adulthood or sometime thereafter, would be altered by unequal deterioration of the hemispheres during senescence.

Empirical data thus far have provided little support for the hypothesis of gradually developing lateralization (nor for a systematic progression of lateralization across species or differences between advantaged and disadvantaged groups). There is even less evidence of a general tendency for one cerebral hemisphere to deteriorate more rapidly than the other in dementia or normal aging.

24-1. EVIDENCE REGARDING DEVELOPMENTAL CHANGES

24-1.1. Brain Morphology

24-1.1.1. Asymmetries in the Immature Brain

The relevance of neuroanatomical asymmetries to functional asymmetries in either the immature brain or the mature brain rests on the assumption that the size of a cortical region reflects its adaptive significance, the complexity of its information processing, or some other functional attribute (Galaburda, 1994; but see Ojemann, 1983 for evidence of a negative correlation between size and function).

Early morphological asymmetries in the human brain (Witelson, 1983) negate the previously held belief that the two cerebral hemispheres are morphologically identical in infancy (see Dennis & Whitaker, 1977 for a historical review). Accordingly, without having to concede that size is an index of any functional characteristic, one can conclude that the anatomical asymmetries that characterize the adult brain are neither absent nor less marked in the infant brain. The same conclusion applies to the anticlockwise torque that is manifested as a wider frontal lobe on the right side and a greater anterior extension of the right frontal lobe, coupled with a wider posterior left hemisphere and a greater posterior extension of the left hemisphere. Irrespective of its functional significance, if any, in the adult human brain, a similar torque is observed in the immature human brain as well as in the ape brain (LeMay, 1984).

24-1.1.2. Maturational Gradients

Corballis and Morgan (1978) marshaled evidence from diverse sources to support their assertion that the left cerebral hemisphere develops more rapidly than the right hemisphere. Although Corballis and Morgan's arguments stimulated interest in the concept of a lateral gradient of maturation, the left-to-right direction of the gradient was contradicted by a report that the right cerebral cortex in fact develops more rapidly than the left (Dooling, Chi, & Gilles, 1983). Subsequent accounts of neocortical maturation (e.g., Best, 1988; Geschwind & Galaburda, 1985) have accepted the premise that the right hemisphere matures more quickly than the left hemisphere.

A lateral gradient, however, is but one aspect of brain development. The lateral axis may be less important than the other two dimensions of a three-dimensional Cartesian system of coordinates, that is, the neuraxis from brainstem to neocortex and the anterior–posterior axis (Kinsbourne & Hiscock, 1983). Found within the cerebral hemispheres are regional patterns in the development of myelination, in neuronal, dendritic, and axonal density, and in the width of the different cortical layers (Campbell & Whitaker, 1986). In an attempt to summarize all available data on the rate at which different brain regions mature, Best (1988) proposed a growth vector representing the resultant of four developmental gradients: right-to-left, anterior-to-posterior; primary-to-secondary-to-tertiary, and basal-to-cortical.

24-1.2. Electrophysiological Evidence

A substantial body of evidence indicates that electrophysiological asymmetries are present in infancy. This evidence has been reviewed by Molfese and Betz (1988) and by Segalowitz and Berge (1994).

Not all of the infant asymmetries are readily interpretable in terms of the pattern of hemispheric specialization thought to characterize the adult brain. For instance, the unilateral photic driving found in infants may reflect differential maturation of the left and right sides of the neonatal brain rather than functional specialization. However, numerous studies, particularly studies by Molfese and his colleagues (see Molfese & Betz, 1988), have shown consistent differences between the infant's hemispheres in electrophysiological responses to specific speech cues. Event-related potentials (ERPs) indicate bilateral, followed by right-hemispheric, responses to stimuli that differ in voice onset time (VOT), and left-hemispheric, followed by bilateral, responses to stimuli that differ in place of articulation (POA). The ERP findings for infants are similar to those for adults, although the bilateral response to POA occurs prior to the left-hemispheric response in the waveform of adults, and bilateral responses appear to be more localized in adults than in infants.

24-1.3. Behavioral Evidence

24-1.3.1. Head Turning and Postural Asymmetry

Whether turning spontaneously or in response to stimulation, most infants turn their heads to the right more often than to the left. This rightward bias is one of the earliest behavioral asymmetries to be manifested by the neonate (Liederman, 1987), and it has been related statistically to parental handedness (Liederman & Kinsbourne, 1980) as well as to subsequent hand preferences of the infant (Coryell, 1985).

When the infant is supine with the head turned to one side, the ipsilateral arm and foot often are extended, and the contralateral arm and foot are flexed. This asymmetric tonic neck reflex (ATNR) occurs in the newborn infant and persists for at least the first 3 months of life, but certain aspects change during that time. For instance, the ATNR is more evident in the legs of newborns than in their arms, whereas the opposite pattern is observed in infants older than 3 weeks (Liederman, 1987). In infants between the ages of 3 and 10 weeks, head turns to the nonpreferred side are more likely to elicit the ATNR than are head turns to the preferred side, but this may not apply to neonates (Liederman, 1987). Whether the predominant direction of the ATNR predicts subsequent handedness is a matter of dispute (Michel, 1983), although, as noted previously, early head turning by itself seems to bear a relationship to subsequent manual asymmetries.

24-1.3.2. Motor Activity

Studies of arm and hand preferences in infancy have yielded diverse results. The sizable literature on this topic is replete with unresolved conceptual and methodological issues and seemingly inconsistent findings (see Young, Segalowitz, Corter, & Trehub, 1983). Nonetheless, considerable evidence supports Liederman's (1983) contention that most infant behavior is dominated by the left hemisphere and right hand. For instance, it has been observed repeatedly—though not invariably—that most infants hold objects for a longer time with the right hand than with the left hand (see Provins, 1992 for a summary). The right hand typically is preferred over the left hand for a variety of target-related actions that are performed during the first 4 months of life (see Young, Segalowitz, Misek, Alp, & Boulet, 1983 for a review). It has been reported that a hand preference for unimanual manipulation of objects develops between the ages of 5 and 7 months, and that a hand preference in tasks requiring bimanual manipulation develops by the age of 1 year (Ramsay, 1983).

In a critical review of the evidence regarding motor asymmetries during infancy, Provins (1992) emphasized the potential importance of prenatal factors (e.g., intrauterine position of the fetus) and postnatal factors (e.g., head position during feeding) in determining motor and postural asymmetries in the infant. Provins contended that the available evidence fails to establish that motor asymmetries in infancy are either early manifestations or precursors of later handedness. Thus, in addition to lingering doubts about the authenticity or generality of some of the observed asymmetries, there is uncertainty regarding the genesis and the implications of early motor asymmetries.

24-1.3.3. Perception

Studies of auditory perception provide some of the most convincing evidence of early functional asymmetries. This evidence has been summarized by Best (1988). An initial dichotic listening study with infants between the ages of 22 and 140 days (Entus, 1977) yielded a right-ear advantage (REA) for detection of transitions between consonants (e.g., /ma/ to /da/), and a left-ear advantage (LEA) for transitions in musical timbre (e.g, cello to bassoon). Detection of a transition at either ear was indicated by an event-related dishabituation of the infant's nonnutritive sucking. Best and her colleagues (see Best, 1988), using cardiac deceleration to indicate that a stimulus transition had been detected, confirmed both the REA for speech syllables and the LEA for musical stimuli in infants 3 months of age and older. Although an LEA for musical stimuli has been found in 2-month-old infants, a corresponding REA for speech perception has not been reported in infants below the age of 3 months.

A study based on a different behavioral method suggests that a speech-related brain asymmetry is present even in short-gestation infants who are not yet as mature as the typical newborn. Using limb movements as a measure of immaturity, Segalowitz and Chapman (1980) found that repeated exposure to speech, but not to music, caused a disproportionate reduction of right-arm tremor in infants with an average gestational age of 36 weeks, thus implying that speech affected the left side of the brain more than the right side.

Other investigators have found that neonates turn more often to the right than to the left when exposed to speech sounds (Hammer, 1977; Young & Gagnon, 1990). MacKain, Studdert-Kennedy, Spieker, and Stern (1983) reported that 6-month-old infants detect the synchronization of visual (articulatory) and aural components of adults' speech, but only when the adult is positioned to the infant's right. These findings suggest that orientation is biased to the right side of space in the presence of linguistic stimuli, presumably because the left side of the infant's brain is more responsive than the right side to speech-specific activation. This phasic asymmetry appears to modulate the tonic left-hemisphere prepotency that biases orientation to the right.

24-1.3.4. Childhood Laterality

In a review of auditory, visual, tactual, and dual-task laterality studies involving children between the ages of 2 to 12 years, Hiscock (1988) found no consistent evidence of age-related increases in laterality. Irrespective of the modality tested or the method employed, cross-sectional studies typically reveal the expected asymmetries in the youngest children tested, and those asymmetries are comparable in magnitude to the asymmetries found in older children.

Although few longitudinal studies have been published, results of those few studies are similar to results from the cross-sectional studies. When an age-related increase (or decrease) in laterality is observed, the change in laterality seems to reflect either the unreliability of the method or extraneous factors that covary with age. For example, a large-scale longitudinal study by Morris, Bakker, Satz, and Van der Vlugt (1984) yielded different developmental patterns of ear asymmetry for different subsamples. The authors attributed this variability to a deficiency of experimental control that is inherent in the free-report dichotic listening method. Alternative methods, in which the child's attention is focused on one ear or on a single target, may reduce the influence of extraneous variables while introducing other sources of age-related change in ear asymmetry. Geffen and Wale (1979), for instance, found a larger REA in younger children than in older children on a task that required multiple rapid shifts of attention. Unable to reallocate attention quickly, the younger children apparently showed a strong REA by default. Differences in the ability to focus attention selectively also may account for age-related differences in the ability to overcome the REA when asked to attend to the left ear (e.g., Hugdahl & Andersson, 1986).

Even if laterality remains invariant across the childhood years, a quantitative difference conceivably could be found between the asymmetry of children and of adults. However, direct comparisons of children and adults on free-report dichotic listening (e.g., Schulman-Galambos, 1977) as well as selective listening tasks (e.g., Hugdahl & Andersson, 1986) have failed to reveal differences between groups in magnitude of the REA.

24-1.4. Clinical Evidence

The literature on childhood aphasia, hemispherectomy, and recovery of function has been reviewed by a number of authors (e.g., Aram & Eisele, 1992; Aram & Whitaker, 1988; Hiscock & Kinsbourne, 1994; Kinsbourne & Hiscock, 1987; Satz & Bullard-Bates, 1981; Spreen, Risser, & Edgell, 1995; Woods & Teuber, 1978). Even though the hypothesis of progressive lateralization (Lenneberg, 1967) has drawn much of its support from cases of aphasia in children following right-hemispheric lesions, the preponderance of evidence now suggests that childhood aphasia consequent to unilateral right-hemisphere damage is as infrequent in children as in adults.

Studies using quantitative measures of cognitive functioning following unilateral lesions are limited by a reliance on IQ and academic achievement tests, which constitute neither sensitive nor comprehensive measures of linguistic and visuoperceptual functioning. Such measures consequently are not optimal for differentiating between left- and right-sided lesions. Despite this handicap, several studies do show associations between left-sided damage and verbal impairments, and between right-sided damage and nonverbal impairments. Some studies suggest that the selectivity of unilateral lesion effects is less marked when the damage is prenatal or perinatal than when it occurs later in development (cf. Woods, 1980). Nonetheless, there is ample evidence that unilateral brain lesions of prenatal or perinatal origin can lead to differential impairment of verbal and nonverbal functions, especially when those functions are assessed using methods other than IQ tests.

The implications of hemispherectomy performed at different ages are difficult to specify, mainly because of extraneous factors that confound comparisons between hemispherectomy in children and hemispherectomy in adults (St. James-Roberts, 1981) but also because of the inferential limitations of small-sample and single-case studies (Bishop, 1983). If any conclusion is justified by the available hemispherectomy evidence, it is that the outcome of hemispherectomy performed at different ages— which traditionally has been used to support the concept of left- and right-hemisphere equipotentiality—now seems to suggest the opposite conclusion, namely, that specialization of the cerebral hemispheres is manifest early in life (Spreen et al., 1995).

24-2. EVIDENCE REGARDING CHANGES ASSOCIATED WITH AGING

24-2.1. Normal Aging

24-2.1.1. Physiological Changes

Autopsy studies reveal no obvious predominance of neuropathology in either the left or the right hemisphere of the elderly (Esiri, 1994). Studies based on positron emission tomography (PET scanning) report nearly identical levels of glucose metabolism in the left and right cerebral cortices and in the left and right basal ganglia of healthy elderly subjects (Duara et al., 1984).

24-2.1.2. Behavioral Changes

Various psychometric data, typically cross-sectional, establish that certain kinds of tasks are more prone than others to show age-related declines in average performance. For instance, it is evident from normative data that absolute scores on the Performance subtests of the Wechsler Adult Intelligence Scale-Revised (WAIS-R) decline more rapidly with increasing age than do scores on the Verbal subtests. The differential decrement of Performance and Verbal scores is confirmed by longitudinal data (Mortensen & Kleven, 1993). Similarly, tests of "fluid" intelligence—the ability to solve problems requiring novel information and strategies—are more likely to show age-related changes than are tests of "crystallized" intelligence—the ability to utilize previously acquired knowledge and strategies (Horn & Cattell, 1967). Age-related decline in average performance on some tests is quite marked. On Raven's Standard Progressive Matrices, an untimed test that contains seemingly novel problems, an average score for 18-year-olds falls at the 95th percentile for 65-year-olds (Lezak, 1995).

Psychometric data, including results from neuropsychological testing, sometimes have been construed as evidence that right-hemisphere functions are more vulnerable than left-hemisphere functions to deterioration with advancing age. This conclusion, however, is not supported with any consistency by findings from laterality tests. Although two dichotic listening studies (Clark & Knowles, 1973; Johnson et al., 1979) yielded larger right-ear advantages (REAs) in elderly people than in younger subjects—an outcome that is consistent with deterioration of the right hemisphere—the results are obfuscated by the use of multiple pairs of stimuli per trial. Given that subjects tend to report right-ear signals prior to left-ear signals, a larger REA may reflect nothing more than diminished overall performance.

Two of the most salient cognitive deficits observed among the elderly are reduced speed of processing (Van Gorp, Satz, & Mitrushina, 1990) and reduced ability to perform complex or difficult tasks (Crossley & Hiscock, 1992). It is uncertain whether these deficits represent separate limitations—e.g., slowness of processing versus insufficiency of processing resources—or different manifestations of the same limitation (Salthouse, 1988). Nonetheless, since reaction time (RT) and dual-task paradigms are known to yield consistent age effects, these two paradigms are appropriate potential sources of evidence regarding lateralized impairment in the elderly. Should future use of these paradigms yield evidence of age-related changes in laterality, care must be taken to ascertain that the findings reflect differential deterioration of the left and right hemispheres rather than changes in the functioning of the corpus callosum (Hoptman, Davidson, Gudmundsson, Schreiber, & Ershler, 1996).

24-2.2. Clinical Evidence

24-2.2.1. Dementia

Alzheimer's disease (AD), especially early-onset AD, often affects object naming, spontaneous speech, and praxic functions (Nicholas, Obler, Albert, & Helm-Estabrooks, 1985). Although one may be tempted to attribute these symptoms to predominantly left-hemispheric pathology, such symptoms simply may be more noticeable than deficits associated with right-hemispheric disease. Moreover, right-hemispheric symptoms such as spatial confusion are seen frequently in patients with AD, and postmortem examinations of the brain typically reveal diffuse bilateral pathology (Braak & Braak, 1992; Esiri, 1994; Wischik, Harrington, & Mukaetova-Ladinska, 1994).

Even if AD shows no general predilection for one or the other cerebral hemisphere, individual patients with AD sometimes exhibit a cognitive profile suggestive of left- or right-hemisphere dysfunction. Sometimes the cognitive profile can be related to regional differences in cerebral metabolism (Chawluk et al., 1990; Martin, 1990). Using PET to measure glucose utilization in the left and right hemispheres, Martin et al. (1986) found predominantly left-sided hypometabolism in four patients with relatively selective word-finding deficits and predominantly right-sided hypometabolism in five patients with relatively selective visuoconstructional impairment. Massman and Doody (1996) reported that fewer than 40% of 104 patients with AD had finger-tapping asymmetry scores falling within normal limits. Twenty-seven patients showed an exaggerated right-hand advantage, and 38 patients showed reversed asymmetry. The deviant subgroups differed from each other, not only in cognitive performance, but also in educational level. Patients with an exaggerated right-hand advantage had better verbal skills and more years of education than did patients with reversed asymmetry. Massman and Doody suggest that a relatively high level of premorbid left-hemisphere functioning may reduce an individual's vulnerability to the behavioral effects of AD.

24-2.2.2. Aphasia

The relative incidence of different aphasia types appears to shift as a function of age (Brown & Jaffe, 1975; Brown & Grober, 1983; Miceli et al., 1981; Obler, Albert, Goodglass, & Benson, 1978). Brown and Grober (1983) found that expressive disorders and mixed aphasias (i.e., nonfluent speech with moderately impaired comprehension) predominate throughout the life span, but the preponderance of these nonfluent aphasias is most striking during the first 3 decades of life. The incidence of global aphasia begins to rise in the 4th decade, and sensory (fluent) aphasias become relatively common during the 7th decade. Brown and Grober attributed these age-related shifts to a gradually increasing degree of lateralization and focal brain organization: an initially bilateral and diffuse language substrate becomes unilateral and focal over time, with expressive functions lateralizing before receptive functions. It is not clear, however, that the changing incidence reflects shifts in lateralization rather than regional changes in vulnerability to cerebral disease, changes in compensatory ability (i.e., in plasticity), or other variables.

24-3. THE MAJOR ISSUES

24-3.1. What Is Lateralized?

As Kinsbourne (1984) has pointed out with regard to localization of function, neuropsychology has been more successful in specifying where functions are localized than in specifying the functions that are localized. The dichotomy between "what" and "where" underlies many of the most difficult questions concerning life-span development of hemispheric specialization. A misplaced emphasis on the question of "where" has led to disagreement about the lateralization of functions that are only vaguely defined. Porter and Berlin (1975), for instance, pointed out that some of the dichotic listening tasks used with children tap processes that are primarily acoustic or phonetic, whereas other dichotic listening tasks involve semantic and mnemonic processing. As a rule, however, investigators have been content to assume that a particular category of stimuli, such as digit names, represents a much broader and rather ill-defined range of linguistic material.

Luria (1973b) addressed this problem when he emphasized that the development of a new perceptual or motor skill entails first consolidating isolated elements of the function into an integrated and automatized series of elements and then linking the integrated elements into a network of superordinate functions. Luria made it clear that lesion studies cannot reveal the localization of a mental process unless the structure of the process is understood. Much of the research on age-related change in functional lateralization violates this basic Lurian caveat: investigators have attempted to lateralize a function or set of functions without any independent knowledge as to how the functions are organized.

24-3.2. What Do Early Electrophysiological Asymmetries Imply?

In a longitudinal study of children from birth until the age of 3 years, Molfese and Molfese (1986) found that individual differences in ERP recorded in the neonate could predict performance on language tests 3 years later. The best predictor was a left-lateralized consonant-discrimination component of the neonatal ERP. Yet, ironically, lateralized ERPs to speech cues have been recorded in dogs and rhesus monkeys, prompting Molfese and Betz (1988) to suggest that the critical difference between humans and nonhumans with respect to speech discrimination is reflected not by lateralized ERP components but instead by bilateral components, which have been not been found in monkeys and dogs.

24-3.3. Pitfalls in the Interpretation of Age-Related Change in Behavioral Asymmetries

As the selective-listening method becomes more prominent in dichotic listening research, the question of age-related change in ear asymmetry during childhood is being resurrected in the form of a controversy concerning children's ability or inability to overcome the REA when asked to attend to the left ear in the presence of dichotic competition (see Hiscock & Beckie, 1993). The ability to attend selectively to the left ear may be regarded either as a skill that increases developmentally (Hugdahl & Andersson, 1986) or as a manifestation of deficient left-hemisphere language representation (Obrzut, Obrzut, Bryden, & Bartels, 1985). Unless the multiple factors that influence selective attention are controlled, an age-related difference in ability to overcome the REA does not necessarily indicate an age-related difference in hemispheric specialization.

24-3.4. Does Differential Loss of Skills Imply Differential Deterioration of the Cerebral Hemispheres?

The available evidence concerning cognitive changes during aging establishes only that certain measures—especially RT and dual-task performance—are more prone than others to show age-related declines in average performance. Even the widely accepted principle that fluid intelligence deteriorates more rapidly than crystallized intelligence remains open to question (Van Gorp et al., 1990). Longitudinal data reveal a marked variability among individuals in rate of cognitive decline (Schaie, 1993). At this time, inducing a general principle regarding differential decline of left- and right-hemisphere abilities does not appear to be feasible.

24-4. CONCLUSIONS

Brain functions change over the life span as a consequence of maturation, experience, and disease. Various sources of evidence indicate that functional changes often are associated with reorganization of the neural substrate. Yet, the preponderance of evidence suggests that lateralization of function generally does not change: left-hemispheric functions tend to remain left-hemispheric and right-hemispheric functions tend to remain right-hemispheric. Apparent developmental changes in lateralization usually can be attributed to changing plasticity, changing ability levels, or other confounding variables. However, until the various functions of interest are delineated more clearly and measured more adequately, the developmental invariance of lateralization may be regarded as the most plausible hypothesis.

24-5. FUTURE OUTLOOK

The concept of brain lateralization is an abstraction. One may discuss the concept of lateralization but no one can observe lateralization per se. Only specific manifestations of lateralization—hand preferences, asymmetric ERPs, aphasia symptoms following left- and right-sided strokes, and so forth—may be observed. Consequently, there is no critical experiment that will resolve the central question concerning the course of lateralization during the human life span. The definitive answer, if there is to be a definitive answer, will have to be synthesized from a multitude of studies designed to address more specific questions. In some instances, information about lateralization may be incidental to other data, for example, data describing recovery of function.

To interpret accurately the various findings of relevance to the life-span lateralization question as they materialize, it will be necessary to respect certain distinctions. Among them are the following:

1. Change in lateralization versus change in compensatory ability. It is essential to distinguish between decreasing bilaterality of function (evidence of which potentially could be found in normal children) and decreasing equipotentiality (which would be manifested only as differential outcomes following focal brain injury sustained at different ages). As pointed out by Satz, Strauss, and Whitaker (1990), failures to make that distinction have added confusion to the literature on childhood lateralization. The same problem is encountered when attempting to interpret behavioral changes at the other end of the life span. A developmental shift toward bilateralization of a previously unilateral function is different in principle from deterioration of the unilateral function.

2. Change in laterality of activation versus rearrangement of specialized cortical processors. Only in the ideal case would the pattern of regional brain activation match precisely an individual's cortical localization scheme (Hiscock & Kinsbourne, 1987). If laterality of activation is distinguished from lateralized representation of various cortical functions (Segalowitz & Berge, 1995), it follows that age-related changes in one aspect of lateralization may be dissociated from changes in the other aspect. For example, markedly asymmetric activation in a patient with Alzheimer's disease (Martin et al., 1986) need not imply a corresponding shift of cortical functions to the more activated side.

3. Change in lateralization versus change in the microstructure of an activity. New components of a task may materialize as a consequence of maturation or practice, or existing components may become automatized. In addition, increasing proficiency might entail restructuring a task, as when reading is enhanced by the application of material-appropriate strategies. Other changes in task structure—compensatory as well as deleterious—presumably occur in senescence. Whenever an age-related change in the lateralization of an activity is observed, it will be important to establish whether

that change reflects a shifting brain basis for a fixed set of task components or a shifting set of task components.

4. Individual versus population-level asymmetries. If clinical evidence continues to support the existence of hemisphere-selective deficits in some patients with AD, it will be important to know whether the disease process itself is asymmetrical in those patients (at least, in the early stages of the disease), or whether the behavioral consequences of an invariably bilateral disease depend on certain premorbid characteristics of the patient. Either answer would contribute significantly to our understanding of Alzheimer's disease, but neither answer would support a claim that the left and right hemispheres—at the level of the population—are differentially vulnerable.

CHAPTER 25

Language in the Right Hemisphere Following Callosal Disconnection

Eran Zaidel

Department of Psychology, University of California at Los Angeles, Los Angeles, California 90095

The received view is that the disconnected right hemisphere (RH) has a rich semantic system, limited phonology, and poor syntax. It has a large auditory vocabulary and a smaller reading vocabulary but little or no writing and speech. Further, recent studies show that when resources are taxed, word recognition in the disconnected RH can be sensitive to wordness, word length, word frequency, and word concreteness but not to regularity of grapheme–phoneme correspondence. But even though it has no *assembled* phonology, the disconnected RH can sometimes demonstrate *addressed* phonological access to speech. In fact some aphasic symptoms are attributable to RH contribution when left-hemisphere (LH) inhibition is removed, notably semantic paralexias in deep dyslexia, miming without naming in optic aphasia, and covert reading in pure alexia. Converging evidence suggests that language performance in the normal RH can borrow resources from the LH and can surpass the capacity of the disconnected RH, including some competence for grammar and for assembled phonology. Language performance in the disconnected RH can, in turn, borrow resources from the disconnected LH and can surpass the capacity of the aphasic's RH. Word recognition in either normal hemisphere appears to progress in parallel through both a nonlexical and a non–lexical route and the two routes interact.

There is general consensus that the capacity for human language is innate and wired into the left cerebral hemisphere (LH) of most right-handed adults. Then why study language in the right hemisphere (RH)? Because RH language offers a unique perspective on the nativist position about the innateness and modularity of natural language by demonstrating the range of language abilities that can be supported by a more general cognitive system that learns language by experience. Why then study language in the *disconnected* RH instead of studying residual language in aphasia? Because language in the disconnected RH following complete cerebral commissurotomy permits assessment of positive language competence, presumably free of callosally mediated inhibitory effects of left aphasiogenic lesions. But isn't language in the disconnected RH atypical, reflecting reorganization due to early epileptogenic lesions? We find no evidence, either developmental, using clinical observation prior to disconnection, or late, following surgery, using behavioral lateralized testing, for language reorganization in the RH due to early epileptogenic lesions in our key split-brain patients from the Los Angeles series.

25-1. DISCONNECTION SYNDROME

Right-handed patients who had a single-stage complete cerebral commissurotomy, including the corpus callosum, anterior commissure, and hippocampal commissure, demonstrate a loss of normal communication between the two cerebral hemispheres. They cannot compare stimuli across the midline, and they cannot verbally identify objects palpated out of view by the left hand or pictures flashed briefly in the left visual hemifield (LVF). At the same time, these patients demonstrate nonverbally good perception and awareness of the meaning of stimuli restricted to the RH, by appropriate manipulation of the stimuli and reliable retrieval from a collection. Within a few months after the operation, patients appear normal in casual social situations. They perform normally in clinical language tests although they tend to exhibit some pragmatic deficits in conversation, including inappropriate or exaggerated politeness, an occasional tendency to confabulate, and alexithymia, an impoverished verbal description of strong emotional personal experiences (Bogen, 1993). The patients fail to sustain reading of extended texts and do not read for enjoyment.

25-2. SYNOPSIS OF EARLIER FINDINGS

In order to compare linguistic performance across tests, patients, and hemispheres, developmental age norms on standardized tests for language acquisition can be used. The resulting profile does not correspond uniformly to any stage in the ontogenesis of first language (Zaidel, 1990). This means that the RH is not arrested uniformly at some stage of language development when LH specialization starts developing.

The disconnected RHs of patients in the Los Angeles series have substantial auditory language comprehension, more limited reading, little or no spontaneous speech, and no writing. Comprehension of sentences is restricted to short phrases. The disconnected RH has a severely limited short-term verbal memory of 3 ± 1 items, as against 7 ± 2 in the disconnected LH or in the normal brain.

The RH can read words in different sizes, cases, and typefaces (Zaidel, 1982). It reads concrete content words better than abstract function words and it cannot translate graphemes into phonemes (Zaidel & Peters, 1981). Reading comprehension in the disconnected RH, whether for words or for phrases, is consistently worse than auditory comprehension, and the reading lexicon is a proper subpart of the auditory lexicon. Unlike LB and NG, the RHs of JW and DR from the Hanover series are said to possess equal auditory and visual lexicons (Tramo et al., 1995).

The disconnected RHs of four patients with complete commissurotomy from the Los Angeles series (LB, NG, RY and AA) could all perform above chance lateralized lexical decision of concrete nouns and of orthographically regular nonwords (Zaidel, 1983). Other studies confirm that there is receptive language competence in the disconnected RHs of all six patients from the California series who were tested regularly at Caltech. Those studies include auditory language comprehension by AA (Nebes & Sperry, 1971), execution of verbal commands by RY (Gordon, 1980), phonological encoding in CC (Levy & Trevarthen, 1977), lexical auditory comprehension and reading in NG, LB, AA, and RY (Hamilton, Nargeot, Vermeire, & Bogen, 1986), and auditory comprehension in NW (Bogen, 1979). Similar abilities have been demonstrated by the RHs of four commissurotomy patients from the Hanover (PS, JW, DR) and Toledo (VP) series (Baynes, Tramo, & Gazzaniga, 1992).

The disconnected RH has a rich lexical and pictorial semantics. It recognizes linguistic and nonlinguistic references to people and events and it has access to episodic (personal) and semantic knowledge (about the world). It can comprehend not only nouns, verbs, and adjectives but also a variety of grammatical and some simple syntactic structures, ranging from functors and tense markers to transformations such as the passive or the negative. There follows a summary of findings since 1990.

25-3. WORD RECOGNITION IN THE RIGHT HEMISPHERE

25-3.1. Dual-Route Analysis

The dual-route model of naming printed words distinguishes (a) a Visual Analysis System that determines the identity and position of each letter in the input string and assigns the letter an abstract representation that is visually invariant; (b) a lexical route that (i) locates an entry for the word in a Visual Input Lexicon and from it (ii) accesses the appropriate part of the Semantic System; the entry in the Semantic System, in turn, (iii) provides (addresses) an Output Phonological Representation for the word, which is next translated into (iv) an Articulatory Program and subsequently leads to (v) Speech; (c) a parallel nonlexical route that (i) Translates Graphemes into Phonemes and thereby provides (ii) an assembled Output Phonological Representation for the word (e.g., Patterson, Marshall, & Coltheart, 1986). The Visual Analysis System may be characterized by a length effect when resources are taxed. This effect is therefore common to both routes. However, the nonlexical route should be particularly sensitive to string length, relative to the lexical route. By definition, the lexical route should show a strong wordness effect, with a selective advantage for processing words, whereas the nonlexical route can process words and nonwords equally. The lexical route is characterized by a frequency effect, and the nonlexical route is characterized by an orthographic regularity effect. Finally, it is believed that the lexical route is faster than the nonlexical route so that it wins in a horse race. It is also supposed that dual-route variable effects (length, wordness, frequency, regularity) are mandatory rather than merely possible.

The hemispheric version of the dual-route model proposes *(a)* that each hemisphere can store abstract letter identities, *(b)* that each hemisphere has access to "its own" lexical route, but *(c)* that only the LH has access to a nonlexical route. Assume that the two routes operate in parallel without interacting in the LH, and that the lexical route is faster than the nonlexical route. Suppose also that word recognition in the two (normal) hemispheres proceeds in parallel, without interactions. Then word recognition in both VFs should show the effects of a lexical route. This predicts that a lateralized word-recognition or naming experiment should show length effects in both VFs, a frequency effect in both VFs, and no regularity effect in either VF. We have carried out two lateralized lexical decision experiments with LB and with normal controls to test these predictions. Previous experiments had found that the auditory vocabularies of both disconnected hemispheres of LB and NG show similar frequency effects (Zaidel, 1978b), that both hemispheres of LB and NG as well as of normals show concreteness effects (Eviatar, Menn, & Zaidel, 1990), and that the disconnected RH cannot match words with different end spellings for rhyming, that is, that it has no grapheme–phoneme translation and thus no assembled phonology (Zaidel & Peters, 1981). More recently, Eviatar and Zaidel (1994) showed that, in general, both hemispheres of NG, LB, and AA from the Los Angeles series could match letters for both shape and name (Posner–Mitchell paradigm).

The results of two lexical decision experiments with LB and with normal controls are shown in Table 1. Also shown are the results of a similar experiment with congenital phonological dyslexics and with normal children, as well as with deep dyslexia patient RW (see Section 25-5.3). The results do not support either the assumptions or the predictions of the hemispheric dual-route model. First, it is clear that all effects (length, wordness, frequency, and regularity) are possible but none is necessary in any condition. This suggests that an effect occurs in a particular condition only when resources are limited (cf. Eviatar & Zaidel, 1991 concerning length effects). Second, regularity effects do occur in both VFs, suggesting *(a)* that the two routes interact, and *(b)* that the RH has access to a nonlexical route! Third, the prediction that length effects are stronger in the LH because they reflect selectively the contribution of the nonlexical route, which is active predominantly in the LH, is not supported. Rather, length effects can occur either for words or for nonwords in either VF. However, if they occur for words, then they also occur for nonwords, and if they occur in the

TABLE 1

Effects of Dual-Route Variables in Several Subject Groups/Patients

	Normal adults				Normal children				Dyslexic children				RW				LB			
	LVF		RVF		LVF		RVF		LVF		RVF		LVF		RVF		LVF		RVF	
	W+	W−	W+	W−	W+	W−	W+	W−	W+	W−	W+	W−	W+	W−	W+	W−	W+	W−	W+	W−
List I																				
Length	+	+	−	−			+	+					−		−			+		+
Wordness		+	+			+		−						+		+	+		+	
Frequency	+		+		+		+						+		−		+		+	
Regularity	−		−		−		−						−		−		−		−	
List II																				
Length	+	+	−	+		+	+	+	−	+	−	+					+	+		+
Wordness		+	+		~+					−							+		+	
Frequency	+		+		+		+		+		+						−		−	
Regularity	+		+		−				+		−						−		−	

Note. LVF = left visual field; W+ = words, W− = nonwords; + = presence of an effect, − = absence of an effect, ~+ = level of significance .07 > p > .05.

RVF, then they also occur in the LVF. This ordering presumably reflects the order of hemispheric competencies and thus resource availability for the particular function (lexical decision). Fourth, dyslexic children with phonological difficulties show some evidence for ambiguous laterality: there is a selective wordness effect, indexing the lexical route, in the RVF, and there is a regularity effect, indexing the nonlexical route, in both VFs.

25-3.2. Right-Hemisphere Speech?

Patient LB from the Los Angeles series has consistently been able to name some pictures and words in the LVF without being able to do same–different matching of the same stimuli across the two VFs (e.g., Johnson, 1984). This could reflect *(a)* RH speech through addressed rather than assembled phonology. Alternatively, it need not reflect RH speech but could rather reflect *(b)* cross-cuing between the perceiving RH and the verbalizing LH, especially when the choice set is small and overlearned. LB often uses a verbal cross-cuing strategy where the LH seems to guess in turn each letter making up the name of the stimulus by going through the alphabet with the RH apparently signaling when the correct one is reached (D.W. Zaidel, 1988). LVF naming could also reflect *(c)* subcallosal phonological and semantic but not visual transfer from the RH to the LH. Finally, LVF naming can reflect *(d)* ipsilateral projection of sensory information from the LVF to the LH, say, via the extrageniculo-striate system (retina → R superior colliculus → intercollicular commissure → L superior colliculus → LH), although that system is believed to mediate perception without verbal awareness, as in blind sight, rather than verbal identification without perception (same–different matching), as observed here.

25-3.2.1. Patient JW

Baynes, Wessinger, Fendrich, and Gazzaniga (1995) observed similar partial naming of LVF stimuli in callosotomy patient JW from the Hanover series. He could name about 25% of stabilized pictures and words shown in the LVF but naming was sensitive to the perceptual quality of the stimuli. This is consistent with superior-collicular mediation of projecting LVF stimuli to the LH because changes in the perceptual quality of the stimuli (e.g., contrast, eccentricity) may be expected to affect ipsilateral projection more than cross-cuing or subcallosal transfer. There were similar error patterns in the LVF and RVF and this may argue against RH speech, assuming different lexical organizations in JW's two hemispheres. An ingenious attempt to exclude RH-to-LH cross-cuing, by demonstrating naming of an unexpected digit in the LVF, failed, and the error pattern suggested cross-cuing in LVF digit naming. Another ingenious manipulation involved naming of "secret stimuli": "Any time you see an 'X' (where X appears lateralized to the LVF or to the RVF) say 'Y'." The results were ambiguous due to small numbers of trials and low sensitivities (signal detection d'). In sum, the data of Baynes et al. do not strongly support RH speech and are equally consistent with subcallosal transfer.

25-3.2.2. Patient LB

We examined LB's ability to name LVF words as a function of the presence and type of distractors in the RVF. We reasoned that if his LH controlled LVF naming, then LVF naming performance should drop when the LH is simultaneously occupied by a distractor in the RVF. Further, the drop should be greater for word than for figural distractors because then the distractor would engage more of the resources necessary for naming the LVF stimulus. Conversely, if LB's RH controlled LVF naming, then the presence of distractors in the RVF should have no effect or it could even improve LVF naming by removing LH inhibition.

In one experiment, we presented two lists of 96 LVF words each, half with and half without graphic distractors, consisting of nonsense geometric shapes, in the RVF. The first list showed significantly better prompt and correct naming of LVF words without distractors (33.3%) than with distractors (20.8%) (Figure 1). This pattern is consistent with LH naming of LVF words. The second list, however, showed no difference between naming of LVF words with (40%) and without (35%) distractors. This could mean that the LH has enough resources to name the LVF words inspite of the RVF distractors, but the data are more consistent with RH speech.

In another experiment, we presented 12 lists of 96 words each. Half of each list consisted of target words (underlined) in the LVF and half of targets in the RVF. Half of the targets were unilateral and half had distractors in the other VF. Half of the distractors were words and half were nonwords. This experiment was administered twice, 4 months apart. In the first administration, there was no difference in prompt and correct naming of LVF targets with (33%) and without (32%) distractors (Figure 1). Again, this could mean that the LH has enough resources to name LVF targets

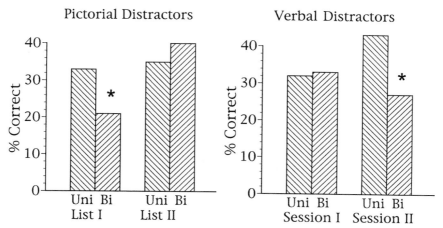

FIGURE 1 Percentage of correct and prompt naming by commissurotomy patient LB of words in the left hemifield without distractors (unilateral) and with distractors (bilateral).
    * = Statistically significant difference.

without interference from verbal distractors in the RVF. In that case, LH resources must fluctuate widely so that it is sometimes not affected by verbal distractors and other times it is affected by nonverbal distractors. This is unlikely. The observed pattern is more consistent with RH naming of LVF targets. In the second administration of the experiment, there was significantly better naming of LVF targets without (43%) than with (27%) distractors (F (1,11) = 25.4, p = .0004). This supports the LH naming hypothesis. Furthermore, both administrations disclosed a frequency effect but only the second administration disclosed a regularity effect. This is consistent with the observation that LB's disconnected LH has access to a nonlexical route whereas his RH does not (Zaidel, 1990).

In sum, there is a great variability in LB's LVF naming strategies both within and between sessions. Naming of LVF stimuli frequently fails. When it succeeds it sometimes reflects LH naming via cross-cuing, ipsilateral sensory projection, and subcallosal cognitive transfer. On other occasions it does appear to reflect RH naming.

25-3.3. Dynamic Language Competence in the Partially Disconnected Right Hemisphere

Faure and Blanc-Garin (1994) describe a provocative case of a patient with a left fronto-occipital white matter lesion sparing the cortex and with a posterior (half) callosal lesion, resulting in temporary global aphasia and a persistent disconnection syndrome. Hemifield tachistoscopic semantic decisions (Is this word an animal name?) following language recovery disclosed good performance in the RVF but chance performance in the LVF. However, LVF performance improved *(a)* with prior RH priming and *(b)* with concurrent LH loading. RH priming was accomplished by requiring the patient to solve silently a maze presented centrally in free vision and point to the solution with the left hand before each lateralized trial in the semantic decision task. LH overloading was accomplished by requiring the patient to remember up to four words presented centrally before each lateralized trial in the semantic decision task and to recall it after the trial was completed. This was assumed to release LH inhibition of RH competence. The authors concluded that the linguistic performance of the RH varies with the balance of interhemispheric activation. However, the roles of prior aphasia and of partial callosal section in these results remain unclear. The authors suggest that smaller fluctuations can be expected in normal subjects because their balance of interhemispheric activation is more even. Thus, posterior callosal section is assumed to be responsible for disrupting the normal balance.

We failed to show comparable priming or removal of inhibition of lateralized lexical decision in the LVF in either the normal or commissurotomized brains by including letter string distractors in the nontargeted VF. Instead, normals showed a loss due to the presentation of distractors and the loss was greater for LVF targets (Iacoboni & Zaidel, 1996). By contrast, LB showed no loss or benefit due to distractors. Taken together, this suggests *(a)* that the anterior callosum mediates inhibitory channels that can be modulated by attention, and *(b)* that posterior channels permit the LH to automatically share resources with the normal (and independent) RH.

═══════   25-4. INTERHEMISPHERIC RELATIONS

25-4.1. Explicit Transfer: A Double Dissociation

The two classical symptoms of the disconnection syndrome, namely, LVF anomia and failure of cross integration across the midline, can show double dissociation in the chronic condition. On the one hand, LB can name some LVF pictures, letters, or words without being able to make same–different judgments about the same stimuli across the vertical meridian. On the other hand, NG can make accurate same–different judgments about visual stimuli in the two visual hemifields, without being able to name the same stimuli in the LVF. This double dissociation between perceptual integration and verbal awareness can serve as an anatomical model of a double dissociation between implicit and explicit knowledge. RH speech in LB does not present a challenge to the disconnection account, but the alternative, semantic transfer, in the absence of sensory transfer, demonstrates that consciousness can be based on higher stimulus codes without access to lower codes.

Thus, whereas LB could match letters for shape and name in either VF, he could not match for either shape or name across the vertical meridian (Eviatar & Zaidel, 1994). By contrast, NG could not only match by shape and name in either VF but she could also match by shape, though not by name, across the two VFs (Eviatar & Zaidel, 1994). (Both LB and NG showed RH superiority for shape matches and LH superiority for name matches.) In fact, NG could match meaningless geometric shapes with poor association value between the disconnected hemispheres (cf. Zaidel, 1994). This makes it unlikely that she matches high-level semantic subcallosal codes, and suggests instead that she transfers a low-level visual representation sufficient for same–different matching but not for conscious identification. The transfer is presumably mediated by the extrageniculo-striate pathway, retina → superior colliculus → pulvinar → extrastriate cortex. Finally, in a control condition for a lateralized Stroop experiment, NG was (barely) able to match colors across the two VFs but she could not name LVF colors. In contrast, LB could not match colors across the two VFs but he could name LVF color patches (Zaidel, 1994). Thus, NG and LB may show selective subcallosal transfer, suggesting that the two patients differ from each other in the channels that are available for transfer. A verbal version of the Stroop experiment with spatially separate color patches and color words, in the same or opposite VF, showed good naming of LVF patches by LB, sometimes consistent with RH naming and other times with LH naming (sensitivity to the color word in the RVF).

Intriligator, Hanaff, and Michel (1995) reported a pattern similar to LB in AC, a patient who had a lesion of the posterior third of his corpus callosum. He could name or identify verbally unilateral or bilateral stimuli in both VFs, ranging in complexity from sine wave gratings to faces and objects. He could not, however, compare (same/different) the same stimuli across the vertical meridian, although he could compare them within either VF. This implies that interhemispheric transfer for visual comparisons and for naming is carried out by different callosal channels, the former more posterior than the latter. Both of these channels appear to have subcallosal analogues.

Visual subcallosal transfer is better developed in NG, whereas linguistic subcallosal transfer may be better developed in LB.

Sergent (1987) found that LB and NG could decide whether four-letter strings straddling the midline were words or not. The stimulus set was small and although patients were tested for, and were shown incapable of, verbalizing the words, it was not clear whether either hemisphere "knew" the meaning of the word. Further, Corballis and Trudel (1993) could not replicate this finding.

25-4.2. Implicit Transfer

Define interhemispheric transfer in the split brain as implicit if both verbalization of LVF stimuli and cross-matching on demand fail, but there is nonetheless some automatic influence of the unattended stimulus in one VF (the distractor) on a conscious decision of an attended stimulus in the other (the target). Given LB's and JW's ability to name some LVF pictures or words and NG's ability to compare shapes across the vertical meridian, the preconditions for implicit transfer may not be satisfied and need to be assessed for each task and patient on a case-by-case basis. Further, what if naming of an LVF word fails but another high level task, say, semantic classification, can be successfully applied to it? Does that indicate conscious explicit transfer or unconscious, but high-level, implicit transfer? Needed is a taxonomy of tasks that demarcates conscious, attentional processing from unconscious, unattentional processing.

In any case, the canonical case for implicit priming in the split brain would require *(a)* evidence for failure of explicit transfer, *(b)* information about priming within each disconnected hemisphere, and *(c)* evidence for significant priming between the disconnected hemispheres. We may even require converging evidence from the normal brain. The most common pattern would include significant, if unequal, priming in each VF with ipsilateral hand responses, found both in the split and the normal brains, as well as smaller priming effects between the hemispheres than within the hemispheres, more so in the split than in the normal brain. But a priming effect that occurs only between the hemispheres would be of special interest because it may *require* interhemispheric interaction.

In an attempt to demonstrate implicit transfer in the split brain, Zaidel (1983b) used a lexical decision task with lateralized targets and lateralized associated primes but found no priming between the hemispheres. Similarly, Reuter-Lorenz and Baynes (1992) found no evidence for letter priming across the disconnected VFs. On the other hand, Lambert (1991, 1993) showed presumed negative priming of an RVF target by an LVF prime in a lexical categorization task both in normals and in LB. However, this result may have been artifactual because the targets that showed inhibition were different from those that did not. Moreover, various control conditions were not reported.

25-4.2.1. Lexicality Priming and Lexical Resource Sharing
between the Hemispheres

Consider a lateralized lexical decision task with bilateral stimulus strings where unilateral targets are cued peripherally. We have observed three distinct distractor effects of a letter string in the unattended field on the decision of the target in the attended field in normal subjects. First, there is a lexicality priming effect, such that word distractors enhance decision of word targets relative to nonword distractors, especially in the LVF (Iacoboni & Zaidel, 1996). This persists in the split brain (LB), although it is mediated by different, subcallosal, channels (Iacoboni, Rayman, & Zaidel, in preparation). Second, there is a bilateral loss that is selective to words, that is, lexical decision of unilateral word targets is more accurate than of word targets accompanied by *different* distractors (Iacoboni & Zaidel, 1996). This effect is absent or dramatically reduced in the split brain (LB) (Iacoboni, Rayman, & Zaidel, in preparation). Third, bilateral copies of the *same* target speed up RVF decisions of word targets ("bilateral gain") in normal subjects (Mohr, Pulvermüller, & Zaidel, 1994) but not in LB (Mohr, Pulvermüller, Rayman, & Zaidel, 1994). We interpret these results to mean *(a)* that there are separate modules for processing words and nonwords in each hemisphere, *(b)* that there is normally resource sharing between the word processing modules but not between the nonword processing modules in the two hemispheres, and *(c)* that the resource sharing is mediated by the corpus callosum. Consequently, *(a)* differences in the lexical competence of the two hemispheres are accentuated in the split brain, and *(b)* the linguistic competence of the disconnected RH underestimates the linguistic competence of the normal RH.

25-5. RELATION OF LANGUAGE IN THE RIGHT HEMISPHERE TO ACQUIRED ALEXIA

25-5.1. Pure Alexia

These patients have a selective inability to read words visually. Their auditory language comprehension and speech are intact and they can also write fluently if no visual feedback is provided. Patients can only read by laboriously sounding out the letters one at a time ("letter-by-letter reading"). The hallmark of letter-by-letter reading is a massive length effect in reading aloud. The standard psycholinguistic account of it is in terms of impaired access to an abstract orthographic code, that is, failure of the Visual Analysis System in the dual-route model (e.g., Shallice, 1988). The standard anatomical account of pure alexia is in terms of a lesion to primary visual cortex in the LH, usually resulting in a right homonymous hemianopia, and a splenial lesion resulting in posterior interhemispheric disconnection. The splenial lesion prevents transfer of visual shape information but permits transfer of abstract letter identities.

Some letter-by-letter readers demonstrate covert or implicit reading. Thus, they may be able to perform correct forced choice pointing to a related picture in a multiple choice array, they may be able to make lexical decision, or they may be able to make semantic judgments, all without being able to identify the word or even being aware of it (Bub & Arguin, 1995). It is natural to interpret the paradoxical dissociation between overt and covert reading in pure alexia or letter-by-letter alexia in terms of a dual-process theory. It is particularly appealing to argue that overt reading reflects control by the damaged LH, whereas covert reading reflects temporary RH contribution because of the classic anatomical disconnection account of pure alexia (Landis, Regard, & Serrat, 1980; Coslett & Saffran, 1989). Coslett and Saffran propose that both overt and covert reading in pure alexia reflect input from the RH. They posit that each hemisphere has an Early Visual Analysis System, an Object Recognition System, a Visual Analysis (Letter Identification) System, a Visual Input Lexicon, and a Semantic System. The model also assumes that there are callosal connections between phylogenetically early modules, including Early Visual Processing, Object Recognition, and Semantics, but not between "recent" modules, such as the Visual Input Lexicons. Splenial disconnection interrupts interhemispheric transfer during early visual processing but permits interhemispheric transfer of letters-as-objects between the two Object Recognition Systems. This leads to letter-by-letter reading in the LH (Coslett & Saffran, 1994). Implicit reading is then due to semantic access without verbalization in the RH. Coslett and Monsul (1994) provided more direct evidence for RH involvement in covert reading in pure alexia by showing that it was disrupted by right but not left temporal-parietal transcortical magnetic stimulation. In this view, pure alexia generally reflects a maladaptive maintenance of control in the diseased LH and inhibition of language in the intact RH, leading to letter-by-letter reading. RH control can occur when LH control is relinquished and this only occurs with brief presentations and other conditions when LH strategies cannot cope with the stimuli.

Arguin (personal communication, Montreal, April 14, 1996) also argues for a dual-process theory of letter-by-letter reading but opts for an opposite hemispheric interpretation. He distinguishes two pathways for reading, *(a)* the normal efficient abstract orthographic pathway, and *(b)* a compensatory "token" or shape-specific pathway essential for overt word recognition. Letter-by-letter readers use the token pathway but also preserve rapid access to an abstract orthographic pathway that controls covert reading tasks. Further, Arguin believes that the token pathway is supported by normal RH reading, whereas implicit reading in pure alexics is supported by the residual LH system. This is based on the assumption that the RH has no abstract letter identities. The evidence is from *(a)* Reuter-Lorenz and Baynes (1992), who showed a length effect in a variant of the word-superiority effect in the LVF of JW, and from *(b)* Marsolek, Kosslyn, and Squire (1992), who show selective same-case stem completion priming in the LVF of normal subjects. However, the relevance of these tasks to reading is questionable, and there is ample evidence, reviewed earlier, that the normal and disconnected RHs can form abstract letter identities. Circumstantial support for Arguin's position comes from *(a)* a SPECT study of a patient with pure alexia showing RH activation during letter-by-letter reading (Iacoboni & Lenzi, personal communication, Los Angeles, CA, September 20, 1996) as well as from *(b)* a case study of an LH stroke patient with pure alexia who had a subsequent RH stroke, abolishing letter-by-letter reading (Bartolomeo, Bachoud-Levi, Degos, & Boller, in press).

25-5.2. Optic Aphasia

These patients cannot name visual objects or words but can demonstrate their use by miming. They can name auditory or tactile stimuli. The standard account is in terms of a lesion to primary visual cortex together with a disruption of the splenium and more anterior portions of the corpus callosum. Coslett and Saffran (1994) see optic aphasia as a more severe form of pure alexia, where the callosal lesion extends further anteriorly than the splenium and disconnects the homotopic callosal channels that interconnect the Object Recognition Systems in the two hemispheres. This prevents the patient from transmitting individual letters as familiar visual forms from the RH to the LH so that even letter-by-letter reading is no longer possible. Schnider, Benson, and Scharre (1994) argue further that when the callosal damage extends even more anteriorly to disconnect the callosal channels that interconnect the semantic systems of the two hemispheres, the resulting symptoms are of visual agnosia. Then semantic access from vision reflects control by the damaged LH rather than by the intact RH.

Common to these models is the assumption that each hemisphere has a separate Visual Analysis System, Object Recognition System, and Semantic System. This assumption is completely consistent with data from the disconnected hemispheres. The assumption challenges Warrington's model of visual object perception according to which the first stage of object perception, initial sensory analysis, occurs separately and in parallel in early visual centers in both hemispheres; the second stage, perceptual classification, including structural descriptions, is specialized in the posterior RH, whereas the third stage, semantic categorization, is specialized in the posterior LH (Shallice, 1988). Others who, like Warrington, resist positing a separate semantic system in the RH explain optic aphasia by a purely visual disconnection (e.g., Hillis & Caramazza, 1994).

25-5.3. Deep Dyslexia

These patients have large LH lesions with variable aphasia but their deficit in reading aloud shows three prominent symptoms: *(a)* semantic paralexias (opposites, superordinates, subordinates, coordinates, synonyms, and associates), *(b)* a concreteness / imageability and part-of-speech effect, where concrete words are read better than abstract words and nouns are read better than function words, and *(c)* failure of grapheme–phoneme translation as required in reading nonsense words aloud or matching strings with different end-spellings for rhyming. This profile is consistent with the reading profile of the disconnected RH and led to the "right hemisphere hypothesis of deep dyslexia" (e.g., Coltheart, 1983). Our version of this hypothesis (Schweiger, Zaidel, Field, & Dobkin, 1989) posits that the nonlexical route in the LH of the deep dyslexic is destroyed and that the lexical route in the LH is impaired. Consequently, when LH reading fails, control is shifted to the lexical route in the intact RH, with its loosely organized lexical semantics. Thus, the semantic paralexias, part of speech effect and concreteness effect all reflect lexical access by the RH. However, output phonology and articulation can still be controlled by the LH. Furthermore, the residual reading system in the LH is able to read aloud many words and sentences not within the competence of the RH. RH takeover in deep dyslexia is therefore dynamic and adaptive.

Saffran, Bogyo, Schwartz, and Marin (1980) tested several deep dyslexic patients, and reported some LVF advantage in bilateral tachistoscopic presentations of lexical decision, but also an LVF advantage for the control tasks. Subsequently, Schweiger et al. (1989) described a deep dyslexic patient, RW, with intact fields and deep dysgraphia and Broca's aphasia. She showed an LVF advantage for lexical decision but not for length decision (does the string have more than four letters?) with the same stimuli, and she had a prevalence of semantic errors in reading aloud with LVF compared to RVF presentations. Ten years after trauma, RW has evolved into a phonological dyslexic (13% of words attempted were semantic paralexias, 5% self-corrected). However, we assumed that RW's earlier semantic paralexias reflected the stable lexical semantic organization of her RH, which should remain consistent following her recovery. We reasoned that if RW read a stimulus word P as the semantically related word T, then P should serve as a selectively potent semantic prime for lateralized lexical decision of the target T in her LVF. And if the lexical semantic organization in RW's RH characterizes that of the normal RH, then P should selectively prime T in lateralized lexical decision in the LVF of normal subjects as well. Against expectation, RW had significant priming of the semantic paralexias only in the RVF, whereas normals had significant priming bilaterally (Langley & Zaidel, 1990; see Table 2). Thus, there is no evidence that RW's semantic paralexias reflected the unique semantic organization in her or the normals' RH.

Another lateralized lexical decision experiment with RW used the same list of words (courtesy of P. H. K. Seymour) varying in frequency and regularity of grapheme–phoneme correspondence used earlier with LB and with normals (see Table 1, Section 25-3.1). RW's accuracy did reveal a word Frequency by VF interaction, with a frequency effect in the LVF but not in the RVF and no regularity effect in either VF (Table 1). This is what we would expect if her nonlexical route is inoperative due to the left brain damage.

If our arguments are correct, then evidence from aphasia generally underestimates the language competence of the normal as well as the disconnected RH. The intact RH is generally free of the dramatic language deficits that follow LH damage, such

TABLE 2
Presence (+) or Absence (−) of Priming of RW's Semantic Paralexias by the Words that Triggered Them

	RT		ACC	
	LVF	RVF	LVF	RVF
RW	−	+	−	−
Controls	+	+	+	−

Note. Priming in a given VF was defined as the following score difference: (Unrelated prime/target pairs minus targets only) minus (Related prime/target pairs minus targets only).

as word deafness, word blindness, or global aphasia. Thus, the psycholinguistic profile of an LH-damaged patient is determined not only by the lost functions of the LH and by the residual competence of the RH but also critically by the balance of activation in a system of control that incorporates several levels of facilitatory and inhibitory circuits. Analysis of these control structures and their response to brain damage is an important topic for further research.

25-6. CONCLUSIONS

Coordinated laterality experiments in split-brain patients and in normal subjects suggest that the disconnected RH underestimates the language competence of the normal RH. The reason is that even when the two normal hemispheres draw on separate lexical representations and exhibit independent strategies, there occur dynamic sharing of resources and variable automatic priming effects across the commissures, which effectively increase RH language competence and range. Even the split brain permits some subcallosal linguistic interhemispheric interaction, including transfer of resources from the disconnected LH to the RH, so that the disconnected RH may overestimate the linguistic competence of the RH following surgical removal or anesthesia of the LH in adulthood.

Lexical variables that show independent effects on word recognition in each normal hemisphere include length (orthographic), concreteness, emotionality, part-of-speech, associative priming, and diverse semantic congruity effects (semantics), derivational morphology (semantic/ phonological), grapheme–phoneme regularity (phonological), and grammatic priming (syntactic).

The new emerging theory from converging clinical and experimental evidence is that RH language competence can assume any of a set of progressive degrees of independence from LH language. With enough resources, the normal RH can even perform grapheme–phoneme correspondence, perhaps also speak. This argues against a strictly modular view of natural language competence. The ability of the normal RH to perform a task then depends not only on its basic competence but also on its concurrent cognitive load, freedom from inhibition, and access to LH resources. Indeed, a similar account may also hold for the normal LH! This is not to argue that RH language competence is as wide as LH competence, but it does suggest greater potential RH competence for online natural language processing than is commonly assumed.

Acknowledgment

Supported by NIH grant 20187.

CHAPTER 26

The Right Hemisphere and Recovery From Aphasia

Marcel Kinsbourne

New School for Social Research, New York, New York 10011

Although language function is overwhelmingly left lateralized, significant recovery from aphasia is possible even after the left language territory is extensively damaged. Some of this recovery is mediated by a transfer of language representation from the lesioned left to the intact right hemisphere. This phenomenon qualifies the view that language is the product of a unique left-hemisphere language acquisition device or module. It may also have broader significance for the mechanisms of cognitive recovery after focal cerebral injury.

Although cognitive functions are distinctively localized in the cerebrum, the cognitive deficits that result from focal cerebral lesions often recover, in part or altogether. This recovery cannot be attributed to regeneration, of which the central nervous system is virtually incapable (Kolb, 1995). Some early recovery is due to a shrinking penumbra of neurons that are in suspended animation surrounding an area destroyed by an acute insult such as a stroke; that is, parts of the compromised area resume function (Furlan, Marchal, Viader, Derlon, & Baron, 1996). But even excision of a whole cerebral hemisphere can be followed by a degree of recovery of that hemisphere's specialized functions. Brain areas that do not normally control a given function must therefore assume control when the specialized area defaults (Finger & Stein, 1982). I address two questions nested in this generalization, using aphasia as the case in point (for a broader review, see Cappa, this volume). Is the territory that assumes control some-times in the opposite hemisphere, rather than ipsilateral, adjacent to the damaged area? If so, when the right hemisphere assumes control of language, does it program lan-guage, in the same way as the left hemisphere did, or in a different way, for instance, one that reflects its distinctive specialization for simultaneous/relational rather than sequential/analytic processing strategies (Kinsbourne, 1982)? Answers to these ques-tions have implications that transcend the specific issues addressed.

1. They bear on the view that the human left hemisphere houses a unique language acquisition device (LAD) in modular form (Fodor, 1983; Lenneberg, 1967) that is structurally and functionally discontinuous with cortex dedicated to other cognitive functions.
2. They bear on more general questions about recovery of function in the human nervous system. Does it involve the compensatory intervention of territories not previously used to control language function and if so, does it deploy novel strategies (Luria, 1963), or does it more simply involve the same organization of verbal skill emanating from a different cortical territory?

I shall return to these general issues after I have discussed the existing data base on the right hemisphere's role in language recovery.

26-1. LANGUAGE LATERALIZATION

The questions as framed assume that language is left lateralized in the intact brain. This generalization is almost correct. It holds for 99% of right-handed individuals, who constitute some 90% of all people. The incidence of crossed aphasia, that is, right-sided lesions causing aphasia, is estimated at 1% (Russell & Espir, 1961). Among the nonright-handed minority, up to 60% have bilateral language control, according to Loring et al. (1990). Loring et al.'s data is particularly persuasive; they based their conclusions on bilateral intracarotid amytal testing, as well as direct mapping of lan-guage cortex in the awake individual. Their method is more definitive than the lower estimate offered by Rasmussen and Milner (1977). It follows that the right hemisphere is to some extent involved in language control in the majority of nonright-handers. Interestingly, right-hemisphere dominance for language is comparatively rare, even in nonright-handers (Loring et al., 1990). One may then ask, when nonright-handers use language do they in fact use distinctive strategies, or identical ones, only the anatom-ical localization differing?

No evidence exists that left-handers whose right hemispheres participate in lan-guage functioning use language at all differently from the right-handed majority. Whether underlying processes differ in this subgroup has not been experimentally addressed. However, Hecaen, De Agostini, and Monzon-Montes (1981) emphasize that the frequency distribution of aphasic symptoms in left-handers after right-hemisphere damage does not noticeably differ from that of right-handers after left-hemisphere damage. According to them, left-handers differ with respect to the wider dispersion and lesser focalization of language circuitry, but not with respect to what that circuitry can accomplish. This leaves open the possibility that the reorganization that takes place after injury of the language areas, may, if it includes the right hemisphere, approximate its role to that which it normally plays in many nonright-handed individuals. Also, crossed aphasics, who premorbidly presumably processed language in their right hemispheres, are not thought to have had premorbid language processes that were at all deviant. I exclude left-handers from further discussion, and only consider patients who became aphasic after left-hemisphere injury.

26-2. THE LEFT-TO-RIGHT TRANSFER HYPOTHESIS

No doubt, many aphasics base their degree of recovery on residual intact cortex in the lesioned left hemisphere. Recovery based on the left hemisphere is not possible, however, after left hemispherectomy, and yet partial recovery has been repeatedly documented (Burklund & Smith, 1977; Gott, 1973; Ogden, 1988; Rossing, 1975; Smith, 1966). The patient of Cummings, Benson, Walsh, and Levine (1979) whose left language area was devastated, made quite impressive strides, especially in comprehension (see also Willmes & Poeck, 1993). I now explore the generality of such contralateral compensation.

At the dawn of aphasiology, Broca (1865b), struck by extensive recovery from aphasia even after very large left-hemisphere lesions, suggested a role for the right-hemisphere in such recovery. Since then, right-hemisphere compensation in aphasia (the "left-to-right transfer hypothesis") has been inferred from:

1. Cases in which a left-sided lesion causes aphasia, the aphasia recovers, and a right-hemisphere lesion reinstates it.
2. Intracarotid amytal suppression of aphasic language after right injection, rather than, as is usually observed, after left injection.
3. Left rather than the normative right ear and hemifield advantage on verbal laterality testing of aphasics.
4. Greater right than left event-related activation in aphasics during language tasks.

The evidence from these sources will now be critically discussed in sequence.

26-3. BILATERAL LESION EFFECTS

1. Aphasics whose language recovery after left hemisphere lesions was subsequently reversed by a second, right-sided insult have been reported since Gowers (1887) inferred that recovery was due to supplemental action of the corresponding areas in the right hemisphere "in patients in whom speech has again been lost when a fresh lesion occurred in this part of the right hemisphere" (p. 132). In early reports

the aphasia recovery was not thoroughly documented. But the recovery from and relapse of language deficit was comprehensively documented in the case study of Lee, Nakada, Deal, Lynn, and Kwee (1984), in two cases each reported by Cambier, Elghozi, Signoret, and Henin (1983) and Basso, Gardelli, Grasso, and Marriotti (1989), and most recently in a case study by Cappa, Miozzo, and Frugoni (1994). The obvious inference is that the right hemisphere was the site of recovered language (see review by Cappa & Vallar, 1992). However, Gainotti (1993) offered the following criticisms of this interpretation of the lesion data:

a. The right-sided lesion may have impaired recovered language in the left hemisphere through transcallosal diaschisis (Andrews, 1991). Diaschisis (Von Monakow, 1914/1969) is a hypothesized adverse effect of a dysfunctional area of brain on a distant area to which it projects.

However, it is inherent in the notion of diaschisis that this recedes over time. The concept is therefore not applicable to the majority of cases in question, in whom the relapse in aphasia proved to be permanent.

b. If the right hemisphere were the only backup system, then after right-sided injury there should be no recovery. But recovery had been reported (Lee et al., 1984, one case of Basso et al., 1989).

However, the concept of right-hemisphere compensation does not exclude compensation by yet other structures (such as cortical and subcortical areas adjacent to the right-sided lesion).

c. The right-sided insult complicated the language deficit by inducing cortical deafness (bitemporal) and buccolingual dyspraxia (bifrontal), rather than by compromising language processing proper.

But the described language deficits cannot all be shoehorned into these two categories.

2. Intracarotid amytal injections, which normally inactivate speech expression when left-sided, in aphasics have done so when right-sided. Kinsbourne (1971) reported three such cases, in two of whom left-sided anesthesia was also done and left the patients' speech unchanged. Similar observations were recorded by Czopf (1972) in the 19 more severe and long-standing cases of his series of 25 patients with aphasia, also studied with intracarotid amytal. These observations render it beyond question that the right hemisphere can assume control of expressive speech after left-hemisphere injury (particularly if it is extensive). It appears that the right hemisphere assumes its expressive verbal role gradually rather than all at once. This could either reflect a learning process or the gradual lifting of inhibition or diaschisis emanating from the left-sided lesion.

The situation for receptive speech is less clear, as this is, at least at a simple level, within the usual range of right-hemisphere abilities, as demonstrated in callosally sectioned individuals, and nonaphasic patients who undergo left-sided amytal anesthesia (Hart et al., 1991). Receptive abilities after left-hemisphere compromise could therefore reflect a continuing right-sided capability, rather than true compensation.

The evidence marshaled under headings (1) and (2) contrasts interestingly with split-brain evidence that the right hemisphere lacks the ability to use or produce either phonology or syntax (Zaidel, 1983b). Either this is not true of all right hemispheres (Gazzaniga & Smylie, 1984) or the right hemisphere does have syntactic and phonological potential, but which is only released when the left language area is massively inactivated by lesion that cause aphasia, rather than simply disconnected from the right hemisphere by callosal section.

On the other hand, it cannot be that all aphasics recover by right-hemisphere compensation. As shown by the intracarotid amytal findings, some recovery is based on the left brain, and also, some unilaterally lesioned aphasics recover little if at all. In fact, the majority of children with left cortical dysplasia retain left-sided language lateralization (Duchovny et al., 1996). After postnatal injury, left-to-right transfer is thought to vary in frequency inversely with age (Varga-Khadem, O'Gorman, & Walters, 1985).

26-4. LATERALITY IN APHASIA

3. Laterality studies of aphasics have often shown an attentuation or even a reversal of the expected right-sided dichotic or hemifield advantage for verbal material consistent with right-hemisphere control of language (Johnson, Summers, & Weidner, 1977; Moore & Weidner, 1974, 1975). But insofar as laterality reflects an attentional bias contralateral to the active hemisphere (Kinsbourne, 1970a), this finding might be nonspecific. It is confounded by the left-sided lesion itself. Lateral attending is implemented by the reciprocal interaction of opponent processors that respectively direct attention to the right and to the left (Kinsbourne, 1974). A left-sided lesion might disinhibit leftward attending nonspecifically, resulting in a leftward bias of attention irrespective of the processing domain. A left-ear or hemifield advantage for verbal material might result, even if the left hemisphere remains in control of language.

Longitudinal studies of aphasics have sometimes shown an increasing left-ear bias over time, consistent with the gradual assumption of a verbal role by the right hemisphere (Petit & Noll, 1979). In the data of others this has been true only for a nonfluent subset of patients (Castro-Caldas & Botelho, 1980) or not true at all (Niccum & Speaks, 1986).

4. Event-related electrophysiological measures during verbal tasks evidence more right than left cerebral activation (Moore 1989; Papaniculaou et al., 1988). However, it may be that the lesion disrupts and obscures such changes on the left. Regional blood flow measurements find greater right activation (Demeurisse & Cappon, 1987; Knopman, Rubens, Selnes, Klassen, & Meyer, 1984; Meyer, Sakai, Yamaguchi, Yamamoto, & Shaw, 1980; Yamaguchi, Meyer, Sakai, & Yamamoto, 1980). This right-sided improvement appears to increase over time. It is generally most clearly demonstrable in Broca's aphasics. Positron emission tomography has either revealed bilateral increments in activation over time (Cappa et al., 1991), or highlighted the importance of the left temporoparietal area (Metter et al., 1981). But most recently, clear evidence for right-sided activation contralateral to the language area has been reported in Wernicke's aphasics engaged in nonword repetition and verb generation (Weiller et al.,

1995). Transcranial Doppler methodology has also provided evidence that recovered aphasics enlist their right hemisphere when they perform word-fluency tasks (Silvestrini, Troisi, Matteis, Cupini, & Caltagirone, 1995).

26-5. DEGREE OF LANGUAGE COMPENSATION

After left-to-right transfer, does the right hemisphere subserve language to the extent that it has been shown capable of doing so in other studies, notably split-brain, as well as laterality? Or does the reorganization open up a broader capability, even coextensive with normal language use? Specifically, is the compensation limited to the lexical–semantic domain (reviewed by Joanette, Goulet, & Hannequin, 1990), or does it also extend to syntax and phonology, generally thought to necessitate the unique specialization of the language hemisphere? This fundamental issue has not been rigorously studied. Clinical reports of right hemisphere-based language recovery are usually insufficiently detailed, not least because we are not yet able to identify reliably the controlling hemisphere in individual aphasic patients on clinical grounds, and we cannot anticipate which very few cases will suffer relapse due to subsequent right-hemisphere infarction. The intracarotid amytal procedure offers the potential for thorough testing of individuals identified as right hemisphere-supported for language, but it is rarely applied to aphasics, and such a study has not been reported. We are left with circumstantial evidence pointing in opposite directions. On the one hand, the apparent nonexistence of a right hemisphere-compensated aphasic syndrome—for example, one in which phonology and syntax are inaccessible, but lexical–semantic function is preserved—suggests that the right hemisphere can support various patterns of partial (or even complete?) recovery. On the other hand, conservatism is in order, given the strong evidence from other sources of the putatively incomplete nature of any recovery from focal cortical deficit, a principle first articulated by J. Hughlings Jackson (1879–1880), who posited hierarchical representations of cognitive function with lower levels, when disinhibited by selective injury of higher levels, showing a sparser and more rigidly constrained repertoire for the function in question. On such a view, the right hemispheres might be considered "lower" for language purposes and not capable of fully restoring the previous level. Given the importance in principle of this question, and the dramatically increasing repertoire of psychophysiological techniques adaptable to this issue, it could rightly be regarded as a research priority to define the limits of right language compensation in identified patients with left-to-right transfer.

26-6. MECHANISM OF TRANSFER

It would be easier to validate right hemisphere use in aphasia, and identify it in the individual case, if it accounted for a specific subset of aphasic syndromes. If the right hemisphere's productions betrayed its bent toward relational processing, then a "right-hemisphere aphasia" would have long been recognized. Contrary to such expectation, it appears that any of the major left-hemisphere aphasia syndromes can derive from uneven compensation by the right hemisphere. This is mystifying, in view of the anatomical relationships that have been assumed to obtain between lesion locus in the left hemisphere and the type of the aphasic handicap (e.g., Geschwind, 1970). It is hard to envisage an effect of the location of a left-sided lesion on the nature of a processing limitation in the undamaged right hemisphere (except possibly by diaschisis). The notoriously high incidence of exceptions to the classical functional anatomy may even largely be due to dilution of the brain-behavior relationships by left-to-right transfer aphasias. In any case, the incidence of right-hemisphere compensation, and the functional limitations inherent in such compensation, are obviously subject to massive individual variation. It is intriguing that the verbal capabilities of the disconnected right hemisphere are also widely variable (Gazzaniga, Smylie, Baynes, McCleary, & Hurst, 1984).

By what physiological mechanism is transfer effected? If by a dissipation of left-to-right cross inhibition (e.g., Kinsbourne, 1970b; Moscovitch, 1973), that must be a gradual process. Or reorganization may be less specific, simply taking advantage of what intact and usable neural network is available, in either hemisphere. This may in turn interact with intersubject variability in premorbid, albeit latent, right-hemisphere verbal skills. Finally, any proposed mechanism will have to explain why many severely aphasic left stroke patients make *no* significant recovery at all. Why not if left lesions disinhibit, or if reorganization uses an intact neural network?

The sources of variability of the language potential of the right hemisphere are unknown. It would not be surprising, however, if a part of the brain that is engaged in an activity other than that for which it is primarily specialized shows interindividual variability. The language potential of the right hemisphere is perhaps genetically underspecified, and therefore subject to great variability based on fortuitous factors in embryogenesis (and even postnatal experience).

26-7. IMPLICATIONS OF RIGHT-HEMISPHERE COMPENSATION

For neurolinguistic theory, left-to-right transfer might engender some skepticism about the uniqueness of human language circuitry and its allegedly modular organization. The human right hemisphere, compared unfavorably with the cerebrum of the chimpanzee by Gazzaniga and Smylie (1984), can acquire useful language nonetheless. Certainly, vast tracts of the cerebral neural network afford elements of language potential. When nonlanguage cortex assumes a language function, it may even conform to the same constraints as language cortex. The networks of the classical language area, and other networks of the brain, may differ in degree only in how amenable they are to the language function. Any "language acquisition device" must be co-extensive with much usually nonlanguage cerebral cortex. Such a "continuity" view of language in the brain would be congenial to those who look for continuity between the preverbal and the verbal infant, and who suspect the existence of language precursors in nonhuman primates.

Outstanding questions about the right hemispere's role might prove pertinent also to the practical issues of aphasia remediation (see Code, 1987). If the right hemisphere reorganizes so as to duplicate the left hemisphere's manner of controlling language, then perhaps no modification in remedial methods is called for. But if it is limited (e.g., to lexical–semantic processing), then perhaps patients should be taught to use their lexical–semantic skills to the extent possible, bypassing the phonological–syntactic routes. Embedding verbal material in tasks that are known to be congruent with right-hemisphere specialization (e.g., musical intonation therapy) might assist this hemisphere in assuming its newer, verbal role. Finally, nonspecific stimulation of the right hemisphere (e.g., analogous with vestibular stimulation of the right hemisphere in cases of left neglect of space and person) is worth a try.

26-8. CONCLUDING COMMENTS

Although many specifics remain to be determined, the evidence is overwhelming that the right hemisphere *sometimes* participates in recovery from aphasia. Should we conclude that the human right hemisphere has latent language equipment, evolved as a backup system in case of left-sided injury? An alternative to this implausible post hoc assumption would be desirable.

Cerebral anatomy constrains us into dichotomous thinking: right versus left. But more probably the cortical gray matter functions as a continuous integrated (as well as differentiated) network. In the left-to-right transfer phenomena, we may be observing a special case of a much more general attribute—the multipotentiality of the cerebral network. There may be no difference in principle between the ipsilateral and the contralateral rearrangements of the network that enable recovery in many aphasic patients. Left-to-right transfer may be a test case for the broader task of identifying the mechanism of recovery after any focalized function is impaired.

CHAPTER 27

Anatomical Asymmetries in Language-Related Cortex and Their Relation to Callosal Function

Francisco Aboitiz and Andrés Ide

Programa de Morfología, Instituto de Ciencias Biomedicas, Facultad de Medicina, Universidad de Chile, Santiago, Chile

Lateralization of linguistic and visuospatial functions has anatomical correlates that can be observed in gross brain anatomy, especially in the size and position of the planum temporale. Furthermore, the Sylvian fissure has qualitatively distinct morphologies that are unequally distributed across the hemispheres. These correlates may reflect the differential development of specific architectonic areas in the two hemispheres, which are of direct relevance for lateralized function. Additionally, interhemispheric connectivity is inversely related to anatomical asymmetry in males more than in females, indicating that there may be sex differences in the interhemispheric organization for language. We discuss these findings in the context of the development of hemispheric asymmetry and its relation to interhemispheric interaction, both from an ontogenetic and a phylogenetic viewpoint.

Anatomical asymmetry in language-related cortex has been found at the levels of gross anatomy, size of cytoarchitectonic areas, and neuronal dendritic patterns (for recent reviews, see Witelson, 1995; Galaburda, 1995). To avoid confusion, in this chapter we will speak of lateralization to mean functional laterality, and of asymmetry when speaking of structural differences between the hemispheres. Across subjects, the degree and direction of anatomical asymmetry have been found to correlate with functional lateralization—especially handedness and linguistic skills—and more recently a relation between these asymmetries and interhemispheric communication has been detected. The two lateralized hemispheres must interact with each other through the corpus callosum in order to generate coherent behavior. Therefore, callosal function is of direct relevance for the appropriate workings of a lateralized brain.

The main problem to be addressed in this chapter is how callosal connectivity relates to hemispheric asymmetry. In a lateralized or asymmetric brain, is there an increase in interhemispheric relations to compensate for the differences in hemispheric capacity? Or does callosal connectivity decrease, thus progressively isolating the two hemispheres? Or is it the other way around, with changes in callosal connections inducing the generation of asymmetry in the brain? Partial answers and lines of research oriented to elucidate these questions will be suggested here, from anatomical and embryological viewpoints. In the first part of the chapter, recent developments on structural asymmetry in the human brain will be reviewed, followed in the second part with studies on callosal anatomy and the relation between these two variables.

27-1. ANATOMICAL ASYMMETRY IN PERISYLVIAN REGIONS

Only after Geschwind and Levitsky's (1968) classic paper, did the concept that the human brain is structurally asymmetrical become widely accepted. In an analysis of 100 human brains, Geschwind and Levitsky found that the planum temporale, a planar cortical field located in the posterior floor of the Sylvian fissure (see Figure 1), was larger on the left in 65% of the cases and larger on the right in 11%, while symmetric plana were found in 24% of the specimens. In the left hemisphere, the planum temporale is located in a region that overlaps with Wernicke's area, the posterior language region. This asymmetry is associated with a longer Sylvian fissure on the left side (Witelson, 1995; Galaburda, 1995). The planum temporale corresponds to a cytoarchitectonic area denominated as Tpt by Galaburda and Sanides (1980), which is strongly asymmetric toward the left side (having a volume up to seven times larger in the left than in the right; Galaburda, Le May, Kemper, & Geschwind, 1978). Asymmetries in the Sylvian fissure can be observed in 31-week fetuses or even before, and newborns already have the distribution observed in adults (see Witelson, 1995). Regions related to Broca's language area have also been found to be asymmetric (Albanese, Merlo, Albanese, & Gómez., 1989; Foundas, Leonard, & Heilman, 1995; Foundas, Leonard, Gilmore, Fennell, & Heilman, 1996), but they will not be reviewed here since they have not been analyzed in relation to interhemispheric communication.

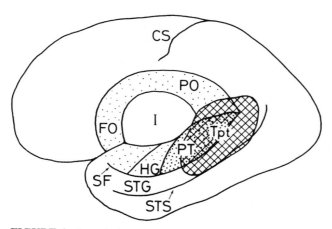

FIGURE 1 Lateral view of a left hemisphere, with the Sylvian fissure (SF) opened so as to expose the insula (I) and the floor of the Sylvian fissure. In the latter, the planum temporale (PT, densely dotted) is located behind Heschl's gyrus (HG). Cytoarchitectonic area Tpt is indicated in cross-hatching. CS, central sulcus; FO, frontal operculum; PO, parietal operculum; STG, superior temporal gyrus; STS, superior temporal sulcus. As in all figures, anterior is to the left side.

27-1.1. *In Vivo* Studies Relating Anatomical Asymmetry and Functional Laterality

In vivo imaging techniques such as carotid angiography and computerized tomography (CT) were initially fundamental to establish a relation between behavioral lateralization and anatomical asymmetry (Damasio & Geschwind, 1984). More recently, magnetic resonance imaging (MRI) studies of cerebral asymmetries in the language areas of living subjects have been performed, confirming the concept of a relation between anatomical asymmetry and functional laterality in both Wernicke's area (Kertesz, Polk, Black, & Howell, 1992; Bergvall, Habib, Jiddane, Rumeau, & Salamon, 1986; Steinmetz, Volkmann, Jäncke, & Freund, 1991; Foundas, Leonard, Gilmore, Fennell, & Heilman, 1994; Foundas, Leonard, & Heilman, 1995; Jäncke, Schlaug, Huang, & Steinmetz, 1995; Habib, Robichon, Lévrier, Khalil, & Salamon, 1995) and Broca's area (Foundas et al., 1995, 1996). In general, the strongest behavioral indicators of anatomical asymmetry in language regions are handedness and linguistic lateralization (as determined by dichotic listening, visual field advantage, or the sodium amytal test). Furthermore, there is a good correlation between right-handedness and left-language dominance (Witelson, 1995), and perhaps the latter evolved from an already lateralized brain with regard to handedness and certain visuospatial functions. An interesting finding is that, among left-handers, those who have inverted writing tend to show leftward asymmetry of the planum temporale, while those having noninverted writing have a rightward asymmetry (Foundas et al., 1995). Using MRI, Habib et al. (1995) found two leftward asymmetric parameters: the size of the planum temporale, and the distance between the central sulcus and the end of the Sylvian fissure. These two asymmetries were not correlated. Although each of the two asymmetry measures correlates with handedness, the two asymmetries combined showed a much stronger correlation to manual preference. This indicates that the overall morphological pattern rather than that of a specific measure may yield better estimates of lateralization of function. Anomalous patterns of asymmetry have been reported in the brains of dyslexics (see Galaburda, 1993). Interestingly, Schlaug, Jäncke, Huang, & Steinmetz (1995) found that musicians gifted with perfect pitch have a highly asymmetric planum temporale (left larger than right).

27-1.2. Sex Differences

Recent reports on postmortem material (Witelson & Kigar, 1992; Aboitiz, Scheibel, & Zaidel, 1992; Ide, Rodríguez, Zaidel, & Aboitiz, 1996) indicate that sex differences in Sylvian asymmetries are not significant, although using MRI Kulynych, Vladar, Jones, and Weinberger (1994) reported asymmetry of the planum temporale in males but not females. Using CT, Mussolino and Dellatolas (1991) suggest that functional laterality has a stronger structural correlate in males than in females; and in a postmortem study Witelson and Kigar (1992) determined that the horizontal segment of the Sylvian fissure (see the next section) tended to be bilaterally larger in right-handed than in nonconsistent right-handed men (in females there were no differences). In general, many studies have determined that when an anatomical relation exists between functional laterality and structural asymmetry, it is more significant in males than in females. However, this difference may be marginally significant or simply not robust, resulting in discrepant findings across studies when sample sizes are at the limit of statistical power.

27-1.3. Morphology of the Sylvian Fissure and the Planum Temporale and Their Variability

Since the macroscopic morphology of the human brain is now accessible *in vivo*, in order to determine reliable structure–function neurolinguistic correlations it is of the greatest interest to establish the precise arrangements of cortical areas in different sulci and gyri, especially for regions like the planum temporale and cytoarchitectonic area Tpt. This is of importance also because there is an immense individual variability in the morphology of the Sylvian fissure. A qualitative and quantitative analysis of the variability of this region has been performed Ide et al., (1996). Anatomically, the Sylvian fissure divides into an anterior segment that ends posteriorly in the gyrus of Heschl (containing the primary and secondary auditory areas); behind the gyrus of Heschl there is usually a horizontal segment containing the lateral extension of the planum temporale. More posteriorly, the Sylvian fissure bifurcates into two, usually more superficial ascending and descending rami (see Figure 2). There are two main types of Sylvian fissure. In the standard type, the ascending ramus is larger and deeper than the descending one (Figure 2A). Two subtypes can be distinguished within the

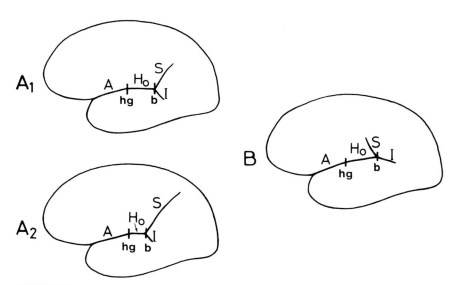

FIGURE 2   Three common types of Sylvian fissure: **A,** the superior or ascending branch (S) is longer than the inferior. **A₁,** the horizontal ramus (H$_o$) is relatively long (27.5% of the cases); **A₂,** the horizontal ramus (H$_o$) is rather short, but the ascending branch (S) is very long and deep (33.75% of the cases). **B,** the superior ramus (S) is distorted and directed forward (21.25% of the cases). The remaining 17.5% of the cases belong to different types.

standard type, according to the relative length of the horizontal segment (compare types A1 and A2 in Figure 2); type A2 being more common in the right hemisphere (Table 1). Ide et al. (1996) have also described a type of fissure (Figure 2B) in which a sometimes small, but other times relatively long fissure, directed upward and forward (instead of upward and backward) exists, that is more common in the left hemisphere (Table 1). This type of fissure corresponds in many cases to what other authors have described as the absence of an ascending ramus (Steinmetz, Ebeling, Huang, & Kahn, 1990; Witelson, & Kigar, 1992). Ide et al's. (1996) interpretation is that the forward-directed ramus is a distorted and sometimes vestigial superior branch.

The distinct fissurization patterns may result from differences in the relative growth of specific cortical regions, which in turn may be directly associated with divergent modes of cortical processing. In particular, the position of the planum temporale and area Tpt may differ in the three fissure types reported here, an issue of relevance for studies that incorporate imaging analyses. For example, Witelson and Kigar (1992) suggest that in some cases (in their "V-type" of fissure) the planum temporale runs into the ascending branch of the Sylvian fissure. Ide et al. (1996) assert that Witelson and Kigar's "V-type" corresponds to their type A2 (Figure 2) and agree with the latter regarding the position of the planum in this type. However, in the other cases (especially type B) it appears that the planum temporale is restricted to the horizontal segment of the Sylvian fissure. Since type A2 is more common on the right,

TABLE 1
Distribution of Major Sylvian Fissure Types According to Hemispheres and Sex

Fissure type	Number of cases in left hemisphere	Number of cases in right hemisphere	Total
A1	10	12	22
A2	09	18	27
B	13	04	17
Other types	*08*	*06*	*14*
Total	40	40	80

Note: Values in each cell indicate number of cases. (Data from Ide et al., 1996. N = 40, 20 males, 20 females; X^2 = 8.22; P<0.05). No significant sex differences were found in this distribution.

this position of the planum may reflect the development of cortical areas related to right-hemisphere skills or the reduction in the size of areas related to left-hemisphere skills, while the type B fissure that is more common on the left may be related to the development of cortical areas involved in linguistic skills. In this context, the issue of the location of the planum temporale may not be as important as the location and size of asymmetric areas such as Tpt. For instance, it could be that Tpt closely corresponds to the planum in some cases but not in others. Further studies are needed to settle the question of a correspondence between variability at the level of gross anatomy and variability at the level of cytoarchitectonics.

27-2. THE CORPUS CALLOSUM

27-2.1. Anatomy of Callosal Connections

Another anatomical structure that appears implicated in lateralization is the corpus callosum, a massive tract containing some 200,000,000 fibers (Aboitiz, Scheibel, Fisher, & Zaidel, 1992a) connecting the two cerebral hemispheres. A commonly used and straightforward method to subdivide the corpus callosum is the partition proposed by Witelson (1985), in which the corpus callosum has been arbitrarily divided into three regions according to maximal straight length. The anterior third (genu) is a rather bulbous region that contains fibers connecting prefrontal cortices. The mid-third is the mid-body of the corpus callosum, a slender region that contains projections from motor, somatosensory, and auditory cortices. The posterior third is divided into the posterior fifth (the splenium), which contains temporal, parietal, and occipital (visual) fibers; and the isthmus, a region between the mid-body (mid-third) and the splenium (posterior fifth), which is believed to contain fibers connecting superior temporal and parietal regions (perisylvian areas).

27-2.2. Fiber Types and Regional Differentiation of the Corpus Callosum

In the human, callosal fiber diameters may range from 0.4 to 15 mm in diameter, but the most common fiber diameters are found between 0.6 and 1 mm; unmyelinated fibers are scarce as seen in light microscopy (about 5% of the total, except in the genu, where they may compose about 16%; Aboitiz et al., 1992a). The fiber composition of the corpus callosum reflects the topographic arrangement of cortical areas, as there are regional differences in terms of the proportions and densities of fibers of distinct diameter (Aboitiz et al., 1992a; see Figure 3). Fast-conducting, large-diameter fibers (larger than 3 mm in diameter) tend to be concentrated in the posterior body and the posterior splenium, which may correspond to the representation of primary and secondary sensorimotor areas, particularly auditory and somatosensory (posterior mid-body) and visual (posterior splenium). On the other hand, callosal regions representing higher-order areas are mainly composed of small-diameter, slower-conducting fibers. This is of interest, as callosal cells connecting primary and secondary sensorimotor areas are relatively few, restricted to a narrow strip in the border of the area that usually represents the sensory or motor midline, while in higher-order areas callosal cells can be found all along the respective regions (see Innocenti, 1986). In

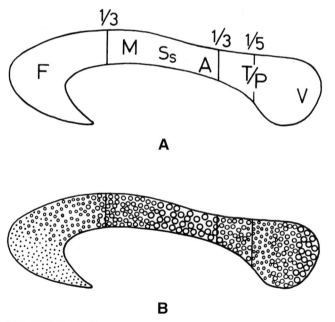

FIGURE 3 **A,** diagram of the corpus callosum, indicating the regions through which fibers connecting different cortical areas pass. F, frontal; M, motor; Ss, somatosensory; A, auditory; T/P, temporo-parietal; V, visual. **B,** distribution of fiber diameters in different callosal regions.

primary and secondary visual and somatosensory cortices, the role of fast-conducting, gigantic callosal fibers may have to do with fusing the two hemirepresentations in both hemispheres, while in auditory areas they may represent an additional stage of fast bilateral interaction in the auditory pathway that serves to localize sounds in space (Aboitiz et al., 1992a).

27-2.3. Individual Differences in the Corpus Callosum

The gross morphology of the corpus callosum shows a large degree of individual diversity. Differences in the cross-sagital size of the corpus callosum may have functional significance, as a larger total or partial callosal size reflects an increased number of small-diameter fibers (Aboitiz et al., 1992a), although the sparse large-caliber fibers of more than 3 mm are so variable in density that they fail to show a relation with callosal size. An additional factor contributing to callosal size is myelin deposition, which also has functional correlates since increased myelination is associated with higher conduction velocity and hence shorter interhemispheric transmission times.

27-2.4. Sex Differences

Some studies indicate a more bulbous splenium in females than in males, while additional studies reported a sex difference in the size of the isthmus in favor of females (for recent reports, see Allen, Richey, Chai, & Gorski, 1991; Witelson, 1989; Clarke, & Zaidel, 1994). Other findings suggest no sex differences in callosal morphology or fiber composition (for a review, see Aboitiz et al., 1992, 1992a; see also Steinmetz, Staiger, Schlaug, Huang, & Jäncke, 1995).

Taking all the conflicting reports on callosal sex differences, probably the safest conclusion at this point is that if they exist, they are not robust. This contrasts with the striking difference in brain size between the sexes (Aboitiz et al., 1992a; Steinmetz et al., 1995; although Rauch & Jinkins, 1994 found no sex differences in callosal size relative to brain volume). Peters (1988) has argued that sex differences in brain weight relate to cortical areas connected with sensorimotor surfaces, but not to higher-order areas involved in cognition, which make up the bulk of callosal fibers (see earlier in this section), but there is no strong evidence in his support. Despite differences in overall size, it is still unclear whether the male brain actually has more cortical cells than the female. Haug (1987) and Witelson, Glezer, and Kigar (1995) have reported that cortical cell density is larger in females than in males, which may result in similar numbers of cortical cells in the two sexes (Haug, 1987). If this is the case, and if the proportion of cells projecting to the corpus callosum does not differ, it would not be surprising that callosal size and fiber numbers are no different in females and males.

27-3. RELATION OF CALLOSAL STRUCTURE AND BRAIN ASYMMETRY: IMPLICATIONS FOR LANGUAGE LATERALIZATION

There is evidence that callosal area (as seen in MRI) is inversely related to behavioral laterality measures (Hines, Chiu, McAdams, Bentler, & Lipcamon, 1992; Yazgan, Wexler, Kinsbourne, Peterson, & Leckman, 1995), supporting the concept that hemispheric specialization is associated with a decreased communication between the hemispheres. Clarke and Zaidel (1994) reported that an inverse relation between isthmus size and laterality holds only for males. Regarding handedness, Witelson (1985, 1989) found that callosal measures are larger in non-consistent right-handers than in consistent right-handers. Subsequently, Witelson and Goldsmith (1991) reported that, in males, the callosal isthmus was larger in nonconsistent right-handers than in consistent right-handers, although among females there were no differences.

In an apparently discrepant finding, Steinmetz et al., (1992) found no differences in the callosal isthmus among left- and right-handers, but they classified their subjects as self-described dextrals and sinistrals instead of consistent or nonconsistent right-handers. Habib et al. (1991) found that the total size, and the size of the anterior half of the corpus callosum, was larger in nonconsistent than in consistent right-handed males; and the posterior mid-body was larger in consistent than in nonconsistent right-handed females. These results partially support our findings, although the possibility of a larger posterior mid-body in consistent right-handed females needs to be investigated further.

Aboitiz et al. (1992, 1992b) investigated the relations between callosal connectivity and hemispheric specialization, as indexed by the degree of perisylvian asymmetry. The Sylvian fissure was measured from Heschl's gyrus to the end of the ascending branch, and asymmetries in length were calculated. An inverse relation was found between the magnitude (i.e., regardless of direction) of these asymmetries and the size and fiber numbers of a specific callosal segment—the isthmus. As mentioned, fibers from perisylvian regions are believed to cross through the callosal isthmus (Witelson, 1989).

When separating sexes, some differences emerged. First, the inverse correlation between isthmus size and asymmetry was concentrated in males, a result reminiscent of Witelson and Goldsmith (1991) and Clarke and Zaidel (1994) in that the relation between asymmetry/lateralization and callosal morphology is stronger in males than in females, and more robust in the callosal isthmus. Second, at the level of fiber composition, only the isthmus of males had a significant correlation with the numbers of small, medium-sized, and moderately large (but not very large) callosal fibers, confirming the gross-morphology findings. However, in a small segment immediately posterior to the isthmus (the anterior third of the splenium), small and medium-sized fibers (and only them) had a significant negative relation to asymmetry levels *only in females*. This may indicate either a differential mapping of asymmetric regions in the corpus callosum of males and females, or that distinct cortical regions are asymmetric in the two sexes. A third alternative is that both the isthmus and anterior splenium include fibers connecting perisylvian areas in the two sexes, the observed difference resulting from sampling error due to a limited number of subjects (10 of each sex) and to the fact that fibers connecting perisylvian areas are only a fraction of all the fibers crossing through these two regions. Another intriguing result is that only in males, moderately large fibers (between 1 and 3 mm) correlated with asymmetry, perhaps indicating that in males, but not females, relatively fast interhemispheric transfer is affected by the asymmetry of language-related areas. When testing the correlation between the asymmetry levels of the different components of the Sylvian fissure and callosal connectivity, no significant relation was found, presumably because the components involve smaller cortical regions whose contribution to isthmus connectivity is too small to be detected.

These findings indicate that increasing asymmetry (and hence increasing hemispheric specialization for language) is associated with a decreased interhemispheric communication, and that perhaps there are sex differences in these relationships. If this hypothesis is correct, the interplay between callosal connectivity and hemispheric asymmetry may be of great importance during the development of language lateralization (recall that anomalous cerebral dominance may be associated with learning deficits such as developmental dyslexia; Galaburda, 1995). In the remainder of this chapter, some recent studies on developmental correlations between interhemispheric connectivity and hemispheric asymmetry will be reviewed.

27-4. DEVELOPMENTAL AND EVOLUTIONARY ASPECTS

It has been postulated that since adult callosal projections tend to be homotopical—that is, connecting equivalent regions in both hemispheres (Innocenti, 1986, 1995)—in asymmetric areas there will be fewer points of homotopy, which will produce a higher than normal process of callosal fiber retraction during development. This would result in a smaller amount of callosal fibers than would be the case for the same regions in symmetric brains (Aboitiz et al., 1992a, 1992b). Alternatively, an increased retraction of callosal terminals may induce the generation of hemispheric asymmetry and lateralization (Witelson, & Nowakowski, 1991; Witelson, 1995), presumably due to the emphasis on intrahemispheric processing and progressive isolation of the two hemispheres (Galaburda, Rosen, & Sherman, 1990; Ringo, Doty, Demeter, & Simard, 1994). Witelson and Nowakowski (1991) found that very premature infants show a high proportion of left-handedness, which is interpreted as that the normal course of callosal axon loss has been interfered with, resulting in loss of laterality and of asymmetry. Supporting this view, Lassonde, Bryden, and Demers (1990) reported that subjects with callosal agenesis tend to be more lateralized than callosotomized and control subjects. However, in this proposed mechanism there is no explanation of why language should become lateralized to the left instead of the right; and furthermore, anatomical asymmetries may appear earlier in development (see section 27.1.) than the period of callosal axon retraction, which apparently takes place between the 35th gestational week and the first postnatal month (Clarke, Kraftsik, van der Loos, & Innocenti, 1989). A likely possibility is that the two factors may reinforce each other; that is, after early generated anatomical asymmetries have induced an increased retraction of axons in the corpus callosum, the reduced corpus callosum itself plays a role enhancing the incipient functional lateralization by constraining the communication between the two hemispheres. In this way, the development of language dominance would depend on the interplay between anatomically asymmetric structures and their respective degree of interhemispheric communication.

One interesting and recurrent finding is that the relations between callosal connectivity and hemispheric asymmetry tend to be stronger in males. This might imply that, in females, the process of perinatal reaccommodation of connectivity may not be as intense as in males. For example, if asymmetry determines an increased retraction of terminals in males, for some reason in females there might be less competition between axons, resulting in a lack of negative correlation between asymmetry and callosal connectivity, which in turn would imply that at least for some processes, females might be able to tolerate a higher level of interhemispheric interaction with increasing hemispheric specialization than males. Perhaps sex hormones have an effect on the process of terminal retraction in late neural development, thus causing the proposed developmental differences between sexes.

Ringo et al. (1994) proposed that in phylogeny, a longer interhemispheric delay due to increased brain size might produce an emphasis on local processing within each hemisphere, resulting in hemispheric isolation, which in turn would facilitate hemispheric asymmetry and lateralization. Thus, hemispheric lateralization for language might have partially become a by-product of having a larger brain. Jerison (1991) reported no differences in average callosal fiber diameter between the mouse and the macaque, suggesting that conduction velocity remains constant despite an increased interhemispheric distance, thereby effectively increasing interhemispheric delay in larger brains. On the other hand, preliminary findings by one of us (F.A.) and collaborators suggest that callosal fibers tend to be thicker (and hence faster-conducting) in large-brained species like the cow compared to a small-brained species like the rat, thus compensating at least to some extent for the longer interhemispheric distance. Nonetheless, in Jerison's (1991) study there was a small population of large-diameter fibers that also increased in size and numbers from the mouse to the macaque. Perhaps this population of large, fast-conducting fibers tends to be concentrated in sensory regions that, as said, have the most stringent timing requirements (Aboitiz et al., 1992a). Comparing the sizes of axons connecting visual areas 17/18 in the mouse and the cat, Innocenti (1995) reports an increase in fiber diameter with brain size. Furthermore, across species no relation was found between interhemispheric conduction delay in the 17/18 border and brain size when comparing studies performed by different authors (four species were compared: mouse, rabbit, cat, and monkey; Innocenti, 1995). Interestingly, the shortest interhemispheric delay (antidromic stimulation) is found in the cat (between 2 and 3 msec), and the monkey has a delay comparable to the mouse (7 to 8 msec) despite the approximately tenfold difference in interhemispheric distance. The rabbit has an unusually long interhemispheric delay (about 17 msec). The relatively short interhemispheric delays in cat and monkey may relate to the fact that both are frontally eyed species with a high degree of binocularity and depth perception, for which fast interhemispheric conduction may be especially useful. This suggests that besides brain size, callosal fiber composition may also depend on behavioral and perceptual specializations such as the degree of stereopsis.

27-5. SUMMARY AND CONCLUSION

There is consensus that there are anatomical asymmetries in the human brain, and that these correlate with functional lateralization (especially handedness and language-related functions). Although sex differences in anatomical asymmetry and callosal structure seem to be small if they exist, in males there are apparently better structural correlates of lateralization as seen in anatomical asymmetry or callosal connectivity. Furthermore, there are indications of an inverse relation between interhemispheric connections and cerebral asymmetry/laterality that is more robust in males. More details about the asymmetric regions in the Sylvian fissure need to be worked out in order to determine if there is a relation between gross brain morphology and the arrangement of cytoarchitectonic areas, which will be of unquestionable utility for neurolinguistic studies that incorporate brain-imaging techniques. In order to analyze the interplay between the development of language lateralization and interhemispheric communication, studies on the development of ipsilateral/contralateral connectivity in animal models need to be performed, and in the human precise developmental chronologies of asymmetry and callosal structure must be established. The issue of sex differences in these relationships and the role of steroid and other hormones is also worth tackling. Furthermore, studies on individual and interspecific variation in callosal fiber composition are necessary in order to establish functional and phylogenetic correlates of callosal structure and its relation to brain asymmetry. In short, this is a newly developing field that will bring important information for neurolinguistics and more generally for the anatomical and neurobiological bases of hemispheric specialization and interaction.

Acknowledgments

We are grateful to Eran Zaidel, Ximena Rojas, and the editors for their comments during the preparation of the manuscript. Supported by FONDECYT grant 194050 (Chile), and DTI grant M3604 9633 (Chile).

CHAPTER 28

Unity of Language
and Communication
Interhemispheric Interaction
in the Lateralized Brain

Joseph B. Hellige

Department of Psychology, University of Southern California, Los Angeles, California, 90089

The left and right cerebral hemispheres are functionally asymmetric for certain aspects of language and communication. Despite the fact that the two hemispheres take the lead for different aspects of language and communication, there is unity in language processing. This chapter provides a brief overview of hemispheric superiority for selected aspects of language, consideration of why so many of these hemispheric superiorities seem to complement each other, consideration of how complementary hemispheric superiorities might emerge over the course of time, and discussion of various mechanisms of interhemispheric collaboration that serve to unify the processing of the two hemispheres.

There is a great deal of evidence that the left and right cerebral hemispheres are functionally asymmetric for certain aspects of language and communication. In this chapter, I note ways in which left-hemisphere and right-hemisphere dominance for different aspects of language and communication complement each other. For example, left-hemisphere dominance for speech production, phonology, and syntax is complemented by right-hemisphere dominance for intonation, emotional tone, and pragmatics.

Despite the fact that the two hemispheres take the lead for different aspects of language and communication, there is unity in language processing. Consequently, I also consider how it is that the two hemispheres, with their different abilities and propensities, collaborate with each other to permit efficient, unified processing related to language and communication. With these goals in mind, I begin with a brief overview of hemispheric superiority for selected aspects of language. I then consider why it is that these different hemispheric superiorities seem to complement each other and how complementary hemispheric superiorities might emerge over the course of time. This is followed by a consideration of various mechanisms of interhemispheric collaboration.

28-1. EXAMPLES OF HEMISPHERIC SUPERIORITY

It is commonly stated that the left hemisphere of humans is dominant or specialized for linguistic, language-related processing. Although there is some truth to this assertion, it has become clear that the extent of left-hemisphere superiority is not uniform for all language-related functions. Indeed, there is growing evidence that the right hemisphere makes important contributions to the processing of language and may even be superior to the left for some language-related functions. In this section, I review some of the hemispheric asymmetries for which there is the greatest amount of converging evidence.

Perhaps the most dramatic example of language-related laterality is left-hemisphere dominance for the production of overt speech, a dominance that characterizes upwards of 95% of right-handed individuals and 65-70% of left-handed individuals (for reviews, see Hellige, 1993a). In fact, this left-hemisphere dominance is so striking that in most cases the right hemisphere is incapable of generating overt speech (though it may sometimes learn to do so after left-hemisphere injury; see Kinsbourne, this volume; Zaidel, this volume). Overt speech is such a striking aspect of language that it is little wonder that the left hemisphere is seen as dominant for language more generally.

Evidence from individuals with unilateral brain injury, from split-brain patients, and from neurologically intact individuals has also identified additional aspects of left-hemisphere dominance for language, though in most of these other cases the right hemisphere is not completely without ability. The left hemisphere is typically better than the right for the perception of speech and for other tasks that require phonetic processing. For example, the left hemisphere outperforms the right in tasks that require the ability to rhyme and in such things as naming printed words and pronounceable nonwords (e.g., Banich, 1995a, 1995b; Hellige, 1993a; Hellige & Cowin, 1996; Hellige, Cowin, & Eng, 1995; Rayman & Zaidel, 1991; Zaidel, this volume). Left-hemisphere dominance has also been demonstrated for determining whether a string of letters spells a word (that is, responding to words in a lexical decision task), though, as we shall see, certain kinds of lexical priming are greater when stimuli are presented to the right hemisphere. The left hemisphere is also superior to the right for dealing with syntactic information, both in the production and in the perception/understanding of language (see Hellige, 1993a).

When linguistic processing is viewed in the larger context of communication, the right hemisphere makes significant contributions. In fact, the right hemisphere appears to be superior to the left for a variety of communication-related factors. Studies with brain-injured and intact individuals have demonstrated right-hemisphere superiority for the production and perception of prosody in speech (e.g., Buhlman-Flemming & Bryden, 1994; Van Lancker & Sidtis, 1992). Thus, the speech of patients with right-hemisphere injury is often said to be characterized by a monotone, lack of inflection, or flat affect. In addition, dichotic listening studies show a left-ear (right-hemisphere) advantage for identifying such things as the emotion conveyed by a speaker's tone of voice (e.g., Buhlman-Flemming & Bryden, 1994). The right hemisphere also plays a role in processing what might be called the pragmatic aspects of language. For example, right-hemisphere damage interferes with the ability to understand certain kinds of jokes that depend on the ability to build a context across sentences that serves to make the punch line funny (e.g., Brownell, Michel, Powelson, & Gardner, 1983) and to process the metaphoric meanings of words such as *warm* and *cold* (which can refer to feelings as well as to temperature) (Brownell, Simpson, Bihrle, Potter, & Gardner, 1990).

For certain aspects of language, the hemispheric asymmetries can be thought of as complementary. For example, it has been suggested that the left-hemisphere dominance for recognizing stop consonants (e.g., *b, d, g, p, t, k*) comes about because the left hemisphere is well adapted for processing the rapid acoustic changes over time that differentiate one such consonant from another (e.g., for high temporal frequency). By way of contrast, right-hemisphere dominance for recognizing emotional tone may come about because the right hemisphere is well-adapted for processing acoustic changes over a longer temporal interval (i.e., low temporal frequency). For review of these and other frequency-based asymmetries in audition, see Ivry and Robertson (1997). A second example comes from recent lexical priming studies, which show that priming has different characteristics in the two hemispheres. A particularly interesting suggestion is that, when a word is presented, the left hemisphere very quickly restricts access to one possible meaning (either the dominant meaning or the one most consistent with the preceding words), whereas the right hemisphere maintains activation of many possible meanings for a longer period of time (e.g., Chiarello, 1995). Note that the left-hemisphere propensity would permit a rapid focus on the relevant meaning most of the time, whereas the right-hemisphere propensity would be useful for getting processing back on track if the meaning settled on by the left hemisphere turned out to be inappropriate in view of succeeding context.

This brief review is intended to be illustrative rather than exhaustive. Among other things, it illustrates that both hemispheres contribute to the use of language and communication. Furthermore, there is a sense in which at least some of these asymmetries complement each other. I now consider why it might make sense for the two hemispheres to develop complementary abilities.

28-2. WHY IS HEMISPHERIC SUPERIORITY COMPLEMENTARY?

In recent years, examples of hemispheric asymmetry in a number of domains suggest that left- and right-hemisphere superiorities complement one another. For example, in terms of visual pattern recognition, the right hemisphere appears to be dominant for the processing of global characteristics of a visual pattern that are carried by visual channels tuned to relatively low visual spatial frequencies. By way of contrast, the left hemisphere appears to be dominant for the processing of local details carried by visual channels tuned to relatively high visual spatial frequencies (for review, see Hellige, 1993a, 1995, 1996; Ivry & Robertson, 1997; Kitterle, Christman, & Conesa, 1995; Robertson, 1995). In terms of localizing visual stimuli in space, the right hemisphere appears to be dominant for processing information about metric distance or what has been termed coordinate spatial relations (e.g., determining whether two stimuli are within 3 cm of each other), whereas the left hemisphere may be dominant for processing information about categorical spatial relations (e.g., whether a dot is above or below a line) (for review, see Hellige, 1995, 1996). In terms of motor control, the right hemisphere is hypothesized to be dominant for postural control, the performance of closed-loop movements, and for performing movements of low temporal and spatial frequencies. By way of contrast, the left hemisphere is hypothesized to be dominant for the performance of skilled activities, the performance of open-loop movements, and for performing movements of high temporal and spatial frequency (for review, see Hellige 1993a). In these cases as well as others, the neural computations that are well suited for one hemisphere's preferred activity may be poorly suited for the other hemisphere's preferred activity.

In view of so many examples of complementary aspects of hemispheric asymmetry, it is important to consider what advantages might result from the assignment of complementary processing to the two cerebral hemispheres. From this perspective, it is interesting to consider the results from dual-task studies, which indicate that two unrelated tasks performed simultaneously interfere with each other more when they require the specialized abilities of the same hemisphere than when the processing load can be spread more evenly across both hemispheres (e.g., Friedman, Polson, & Dafoe, 1988; Hellige & Longstreth, 1981; Kinsbourne & Hiscock, 1983). Such results have led to a variety of hypotheses about each hemisphere containing processing resources that are at least somewhat independent of the resources of the other hemisphere. According to one such view, two tasks interact with each other to the extent that the "functional cerebral distance" between the two is small (e.g., Kinsbourne & Hiscock, 1983). When coupled with the view that the functional cerebral distance is typically greater across the two hemispheres than within a single hemisphere (perhaps because there are many more intrahemispheric cortical connections than there are interhemispheric connections), this accounts for the observation that two tasks interfere with each other more when they both require processing from within the same hemisphere. If this is the case, then there would seem to be some benefit from having the two hemispheres take the lead for different aspects of a complex activity (e.g., global versus local aspects of visual processing or pragmatic cues to understanding language versus syntactic and phonological cues); that is, the assignment of complementary processing to the two hemispheres would seem to have the advantage of allowing each type of processing to proceed with little interference from the other.

The seeds of functional hemispheric asymmetry seem to be sown long before an individual's birth and may even date back to the ontogenetic formation of the first neural structures, to asymmetries of the ovum, or even to various molecular asymmetries. From these early origins, functional hemispheric asymmetries are shaped by the interaction of many biological and environmental factors. Asymmetries that are very small and subtle when they first emerge can eventually have profound consequences for functional hemispheric asymmetry via a kind of snowball effect. Although a detailed review of various developmental scenarios of this sort is beyond the scope of the present chapter, it is instructive to consider a few examples that have implications for why it is that the two hemispheres become specialized for complementary aspects of language and communication.

During the course of fetal development, certain areas of the right hemisphere develop somewhat earlier than homologous areas of the left hemisphere (see Geschwind & Galaburda, 1987; Hellige, 1993a, 1994; Hiscock, this volume). Scenarios have been suggested as to how certain functional asymmetries might arise from the interaction of these maturational gradients and the changing nature of environmental stimulation that reaches the brain. With respect to hemispheric asymmetry for speech, it has been suggested that, because of its greater maturity, the right hemisphere is more influenced than the left hemisphere by the impoverished auditory information that is presented to the brain as early as the second trimester of pregnancy (digestive sounds, the sounds of the mother's heartbeat, and so forth). As a result of this early influence, the right hemisphere is predisposed to become dominant for a variety of nonlinguistic sounds—including, perhaps, intonation and prosody. By way of contrast, the later developing left hemisphere is in some sense saved to become dominant for the more detailed information that is presented only later, as a result of the acoustic transmission properties of the uterus as it expands (e.g., linguistic sounds from the mother's voice during the last few weeks of pregnancy). As a result, the left hemisphere may be predisposed to become dominant for the perception as well as for the production of speech (for more discussion, see Hellige, 1993a, 1994; Turkewitz, 1988).

Additional prenatal asymmetries have also been proposed to influence later functional hemispheric asymmetry for linguistic and nonlinguistic processes. For example, certain cranio-facial asymmetries that begin to appear during the first trimester of pregnancy may constitute important prenatal origins of left-hemisphere dominance for perceiving speech by increasing the sensitivity of the right ear to sounds in the frequency range that is typical of human speech (e.g., Previc, 1991, 1994).

Although there is debate about the merit of these various scenarios, there is sufficient circumstantial evidence to make them worthy of additional study. In addition, these scenarios are consistent with the fact that neither hemisphere is completely dominant for all aspects of language and communication and that, instead, the two hemispheres play complementary roles across so many domains. Such scenarios are also consistent with the fact that, to the extent it can be measured, hemispheric asymmetry is present in newborns and seems to change little over the course of childhood development—except as a by-product of more general cognitive development (see Hiscock, this volume).

28-3. MECHANISMS OF INTERHEMISPHERIC COLLABORATION

In view of the fact that there is no such thing as complete dominance of one hemisphere for all aspects of language and communication, the two hemispheres must collaborate to produce the full range of language processing. This being the case, it is important to consider how it is that the two hemispheres, with their different processing biases and propensities, collaborate with each other in an efficient manner. With this in mind, this section provides a brief review of the mechanisms of interhemispheric collaboration.

Investigations of the biological mechanisms that underlie interhemispheric collaboration have focused on the roles of the corpus callosum and various subcortical structures. The corpus callosum is the largest fiber tract that connects the two hemispheres. Although there are no anatomical landmarks that divide the corpus callosum into discrete regions, anterior portions of the corpus callosum connect premotor and frontal regions of the two hemispheres, middle portions of the corpus callosum connect motor and somatosensory regions of the two hemispheres, and posterior regions of the corpus callosum connect temporal, postparietal and peristriate regions of the two hemispheres (e.g., Aboitiz & Ide, this volume; Witelson, 1995). Given this topographic structure, it is not surprising that the corpus callosum plays a significant role in the transfer of information from one hemisphere to the other (see Aboitiz & Ide, this volume; Zaidel, this volume). The integrity of the corpus callosum is particularly important for the transfer of explicit, conscious information about the identity of stimuli presented to one hemisphere, perhaps because the corpus callosum permits interhemispheric synchronization that is critical for binding the dimensions of objects and experiences into a single, coherent event memory (for discussion, see Liederman, 1995). The corpus callosum has also been hypothesized to regulate the state of asymmetric activation or arousal between the two hemispheres or to serve as a kind of barrier between the hemispheres, minimizing potentially maladaptive cross talk between the processes for which each hemisphere is dominant (for discussion, see Hellige, 1993a; Kinsbourne & Hiscock, 1983). Recent studies that examine the relationship between functional hemispheric asymmetry and the size of regions of the corpus callosum are consistent with the view that the corpus callosum plays an important role in both the transfer of information from one hemisphere to the other and in maintaining some degree of isolation between the two hemispheres (e.g., Aboitiz & Ide, this volume; Clarke, Lufkin, & Zaidel, 1993; Zaidel, Aboitiz, Clarke, Kaiser, & Matteson, 1995). However, the mechanisms by which the corpus callosum accomplishes these two very different things are not understood and it may be difficult to do so without explicit computational models of hemispheric asymmetry and callosal activity (e.g., Chiarello & Maxfield, 1996).

Studies with split-brain patients, whose hemispheres can no longer communicate via the surgically severed corpus callosum, indicate that a good deal of interhemispheric interaction can take place subcortically. Although explicit information about the identity of a stimulus cannot be transmitted subcortically, other types of information can: implicit (unconscious) information about stimulus identity, information about the location of objects in space, information about the categories to which a stimulus belongs and contextual information about an object (e.g., Liederman, 1995; Sergent, 1990, 1991; see also Zaidel, this volume). Subcortical structures also play a role in permitting each hemisphere to receive information about decisions made by the other hemisphere. In this way, subcortical structures may play an essential role in coordinating the activity of the two specialized hemispheres.

With respect to language and communication, this system of callosal and subcortical connections permits the two hemispheres to collaborate in a variety of ways. As noted earlier in this chapter, the two hemispheres are dominant for different and often complementary aspects of language and communication. Under such conditions, each hemisphere is likely to take the lead for those components of processing that it handles best. So, for example, when listening to a fanciful tale of a neurolinguist who dreams about being abducted by space aliens, it is likely that the left hemisphere takes the lead for phonetic, syntactic, and certain semantic aspects of processing, whereas the right hemisphere takes the lead for processing intonation, prosodic cues to emotion, and pragmatic clues to meaning. Coordination of these activities requires that the various dimensions of the event be bound into a coherent experience and, at a functional level, involves the transfer of relevant information and decisions from one hemisphere to the other as well as the insulation of certain hemisphere-specific processes from each other so that those processes can proceed in parallel.

The two cerebral hemispheres are capable of sharing many types of information, ranging from sensory input to complex decisions. However, cooperation at all levels does not necessarily take place all of the time. For example, when a stimulus is projected to only one hemisphere, it is sometimes the case that the hemisphere that receives the information directly carries out all of the processing. One language-related example of this "direct access" possibility comes from experiments that require observers to indicate whether a string of visually presented letters spells a word (e.g., Zaidel et al., 1995). In fact, a variety of lexical decision experiments suggest that the only thing the two hemispheres may share at a functional level is the final decision about whether the stimulus was a word or a nonword (e.g., Hardyck, 1991). In view of results such as this, it is important to consider when the benefits of distributing processing across both hemispheres might be outweighed by the costs of interhemispheric transfer.

With this in mind, recent studies have examined the benefits and costs of interhemispheric collaboration. In such studies, it has been useful to compare performance of the same task under conditions that demand interhemispheric collaboration by presenting each hemisphere with only a portion of the total information needed to perform the task (the between-hemisphere condition) and conditions that present all of the relevant information to a single hemisphere (the within-hemisphere condition). For example, the task might require observers to indicate whether two simultaneously presented letters are identical to each other, with the two letters sometimes presented to the same hemisphere (the within-hemisphere condition) and sometimes presented to different hemispheres (the between-hemisphere condition). One important conclusion to arise from such studies is that distributing information across both hemispheres becomes more beneficial as the task becomes more demanding (for discussion, see Banich, 1995a, 1995b; Hellige, 1993a, 1993b; Liederman, 1995); that is, when processing demands are very simple (e.g., indicating whether two letters are physically identical), there is often a within-hemisphere advantage, suggesting that the costs associated with interhemispheric transfer are sufficiently detrimental to offset the potential benefits of interhemispheric collaboration. However, when processing demands are sufficiently great (e.g., indicating whether two letters of different case have the same name), there is typically a between-hemisphere advantage. Dividing the relevant input between the two hemispheres is also advantageous if it permits the hemispheres to engage in mutually exclusive perceptual processes (e.g., one hemisphere restricts processing to upright letters while the other hemisphere restricts processing to inverted letters). Such results suggest why it may be advantageous to have the two hemispheres take the lead for aspects of language and communication that are at least complementary, if not actually mutually exclusive.

Thus far, I have considered interhemispheric collaboration in which each hemisphere takes the lead for different components of language processing and communication. Recent research suggests that the two hemispheres can collaborate in other ways as well, differing from each other with respect to how much of the interhemispheric interaction can be predicted from the processing performed by each hemisphere alone.

An experimental paradigm that has proven useful in the study of interhemispheric interaction compares performance on unilateral visual half-field trials with performance on redundant bilateral trials, on which exactly the same information is presented to both hemispheres. In a number of experiments, the pattern of performance on redundant bilateral trials has been very similar to the pattern obtained for one of the two unilateral visual fields (see Hellige, 1993a, 1993b). For example, when observers attempt to identify consonant-vowel-consonant (CVC) nonword stimuli, there is a consistent right visual field (left-hemisphere) advantage, as would be expected for a task that requires phonetic processing. In addition, the pattern of errors is different for the two visual fields (and hemispheres). Specifically, on left visual field (right-hemisphere) trials, there are many more third-letter errors than first-letter errors. On right visual field (left-hemisphere) trials, the difference between first-letter and third-letter errors is reduced (even when the error types are normalized for each visual field to compensate for the right visual field advantage). In view of the fact that there is a left-hemisphere advantage for CVC identification, one might expect the error pattern on redundant bilateral trials to be identical to the pattern obtained on right visual field (left hemisphere) trials. Despite the intuitive nature of this expectation, the error pattern on redundant bilateral trials is actually more similar to the pattern obtained on left visual field (right-hemisphere) trials (e.g., Cherry, Hellige, & McDowd, 1995; Hellige & Cowin, 1996; Hellige, Taylor, & Eng, 1989; Hellige et al., 1995; Luh & Levy, 1995). In fact, in several studies, the error pattern on redundant bilateral trials is not significantly different from the pattern obtained on left visual field (right-hemisphere) trials.

It is not clear why the error pattern obtained on redundant bilateral trials in CVC identification is different from that obtained on right visual field (left-hemisphere) trials. It has been suggested that the error pattern obtained on right visual field (left-hemisphere) trials reflects the operation of a phonetic mechanism that cannot be used by the right hemisphere. Accordingly, the processing strategy used on redundant bilateral trials may be restricted to a nonphonetic strategy that can be used by both hemispheres and that is ordinarily used by the right hemisphere (see Hellige & Cowin, 1996). It has also been suggested that transmission time is faster from the right hemisphere to the left hemisphere than vice versa and for visual information carried by low spatial frequency channels than for visual information carried by high spatial frequency channels (e.g., Brown, Larson, & Jeeves, 1994; Kitterle et al., 1995). Both of these properties of interhemispheric transmission might contribute to increased involvement of the right hemisphere on redundant bilateral trials. At the same time, however, it must be noted that there are two (albeit, redundant) stimuli on bilateral trials and only one stimulus on each unilateral trial, and it remains to be determined how the number of stimuli per se might influence such things as the qualitative error patterns.

Despite the fact that, in the CVC identification experiments, the results on redundant bilateral trials are somewhat counterintuitive, they can be derived from knowledge of the unilateral results. This is not always the case. For example, in a series of experiments that involved rhyming judgments, Banich and Karol (1992) found certain effects of letter font and letter case on bilateral trials that could not be predicted at all from the absence of such effects on unilateral trials. As Banich (1995a, 1995b) argues, such results indicate that interhemispheric processing can also be emergent, in the sense that it cannot be deduced from the sum of the parts provided by the single hemispheres. It will be important in future studies to examine how these emergent aspects of interhemispheric processing contribute to the unity of language and communication.

28-4. FUTURE OUTLOOK

A complete understanding of the neural mechanisms that underlie language and communication must take into account the fact that the left and right cerebral hemispheres make different contributions. Clearly, more research is needed to refine our understanding of the nature of these hemispheric asymmetries and to consider how it is that complementary specializations unfold over the course of ontogenetic development. It will be particularly important to investigate the manner in which functional hemispheric asymmetries are shaped by the interaction of biological and environmental factors, beginning even before an individual's birth. At the same time, it will prove necessary to learn more about the various ways in which the two hemispheres, with their different processing biases and propensities, collaborate with each other to provide a fundamental unity of language and communication. This will involve further studies of both the biological and the functional mechanisms of interhemispheric interaction, as well as studies of the relationship between structure and function.

Acknowledgments

Preparation of this chapter was supported in part by Grant SBR-9507924 from the National Science Foundation to the author.

PART IV

Clinical Neurolinguistics

CHAPTER 29

Language and Communication in Multilinguals

Michel Paradis

Department of Linguistics, McGill University, Montreal, Quebec, Canada H3A 1G5

The various recovery patterns of bilingual aphasic patients may be accounted for in terms of interference with the normal distribution of inhibitory resources. The differences between unilinguals and bilinguals may be only quantitative, reflecting the extended reliance on metalinguistic knowledge and/or pragmatic competence in the use of their weaker language. Also, one language (or certain items within a language) may be more readily available, depending on the frequency of their use.

Distinctions between the linguistic and neurolinguistic levels of description, lexical and conceptual representations, language and verbal communication, implicit linguistic competence, and explicit metalinguistic knowledge will help our investigation of the complex problems posed by bilingual aphasia and the representation of two languages in the same brain.

The first systematic scientific study of aphasia in bilinguals and polyglots appeared just over one hundred years ago (Pitres, 1895). In this monograph, Pitres reported that, contrary to expectation, some polyglot patients did not recover all of their languages to the same extent. Sometimes one was recovered much sooner and/or better than the other, and sometimes one remained totally unavailable. Over the half century that followed, discussions centered on the characteristics of the languages that might be responsible for the various types of nonparallel recovery, such as primacy (the native language should recover first—Ribot, 1891), familiarity (the language most familiar at the time of insult is recovered first—Pitres, 1895), or affect (the language with the greatest affective load is recovered first—Minkowski, 1928). None of these proposals was very successful (Paradis, 1977).

One of Pitres' proposals that had unanimously been taken for granted (and often vigorously reaffirmed) for more than 80 years (that is, that the languages of a bilingual speaker are not each represented in a different cortical area) has more recently been challenged by the differential lateralization hypothesis (Vaid & Hall, 1991). As we shall see, the available clinical evidence is incompatible with such a hypothesis.

Other proposals, which had gone essentially unnoticed, are being revived (for example, the notion of inhibition—Green, 1986; modified in terms of activation thresholds—Paradis, 1993; and neurofunctional modularity—Paradis, 1987a). Moreover, a number of investigations have explored the neurobiological bases of multiple language acquisition, storage, or processing (Jacobs, 1988; Jacobs & Schumann, 1992; Danesi, 1994; Pulvermüller & Schumann, 1994; Paradis, 1994; Schumann, 1994).

29-1. RECOVERY PATTERNS

Pitres (1895) had identified three types of recovery: parallel, as it came to be known, when both languages are recovered simultaneously and to the same extent; selective, when one language is never recovered; and successive, when one is recovered long before the other(s). He also mentioned selective aphasia, though without giving examples, and none in fact are found in the literature until Paradis and Goldblum (1989); in such cases, aphasia affects only some of the languages of a polyglot. Since then, cases of antagonistic recovery have been described, in which the language that is recovered first is eventually replaced by the language subsequently recovered, so that, as one language improves, the other proportionately regresses and ceases to be available. There have also been cases of blending, where the patient mixes the two languages (sometimes at the monomorphemic word level) and is unable to speak only one language at a time, and cases of differential recovery, in which one language is recovered to a much better extent than the other, regardless of their relative premorbid fluency.

In the 1970s, cases of differential aphasia, for example, Broca's aphasia in one language and Wernicke's aphasia in the other were reported (Albert & Obler, 1978; Silverberg & Gordon, 1979). However, after careful examination of the data, it seems that both languages exhibited agrammatism, except that in one language (Hebrew) it was instantiated by substitution errors, rather than omissions as the theory would have predicted at the time. It is now known that the symptoms of agrammatism will vary in accordance with the structure of each language, and that inflectional morphemes cannot be omitted in Hebrew. In addition, the patients in question probably had not mastered Hebrew quite as fully as their respective native languages before their aphasia, which would result in lower scores on comprehension after insult. This would account for two of the three patients described. As to the third, reported as having global aphasia in Hebrew and conduction aphasia in Russian, it turns out that the patient had very little Hebrew at his command to begin with, and exhibited a not-too-counterintuitive selective recovery of Russian (with symptoms of conduction aphasia).

Cases of alternating antagonistic recovery have also been reported (Paradis, Gold-blum, & Abidi, 1982; Nilipour & Ashayeri, 1989), where patients have access to only one language for alternating periods of time ranging from 1 day to 3 weeks. As well, cases of paradoxical translation behavior, where patients are able to translate into a language that, at the time, they are unable to speak spontaneously or use to name the simplest objects, but cannot do the reverse (that is, they are unable to translate from a language they understand into a language they are nevertheless able to speak spon-taneously and use to name objects), have been documented (Paradis et al., 1982; De Vreese, Motta, & Toschi, 1988). This paradoxical directional translation ability has also been reported in patients who were able to translate from their native to their second or third language, but not vice versa (Fabbro & Paradis, 1995).

Three distinct questions can be raised with respect to patterns of recovery: (1) What makes these various patterns possible in the first place? That is, how are lan-guages represented and processed in the brain in such a manner that these various phenomena may occur? Then, assuming that the reason why the brain is capable of giving rise to all these different patterns has been identified, (2) what mechanism determines which of the many possible patterns will actually materialize in any given patient? And finally, assuming that we know what mechanism implements selective rather than differential or successive recovery, (3) what is responsible for English being recovered rather than Japanese, or French rather than Gujarati?

In answer to the first question, Pitres argued at length against the notion of different cerebral locations for each of the different languages spoken by the patient. Instead, he proposed that, when a language is not available, it is not because its cerebral substrate has been physically destroyed, but because it has been weakened. In some form, this solution has continued to be widely accepted to this day. The weakening of the system can be conceived of in terms of increased inhibition, that is, of a raised activation threshold for the affected system. In the normal speaker, one might expect that the activation threshold of the unselected language is raised so as to avoid inter-ference. This mechanism, however, can be hindered by pathology. Thus, various lan-guages can be differentially inhibited, giving rise to differential recovery, temporarily inhibited, as in successive recovery, or permanently inhibited, as in selective recovery. When both languages are inhibited to the same degree, the patient exhibits parallel recovery. When inhibition affects one language for a period of time, and then the other, the patient exhibits antagonistic recovery. When one of the languages cannot be selectively inhibited, blending occurs. Selective aphasia (one language unimpaired, the other impaired) and selective recovery (one language impaired, the other un-available) are thus seen as two poles of a differential inhibition continuum.

The second question has never really been addressed as such. However, an answer might be conceived of in terms of impairment to the control mechanism that distributes resources among the various language subsystems (that is, individual languages), of the type proposed by Green (1986). Depending on the nature of the disruption of the control system, resources are evenly distributed among the various languages (parallel recovery), or skewed toward one of the languages (differential recovery), or supplied to only one of the languages (selective recovery). It may also be the case that the supply of resources is normal for one language but reduced for the other (selective aphasia), or the control mechanism is unable to deactivate one of the systems (blending recovery), or the control mechanism that partially supplied one language system eventually supplies the other one as well (successive recovery), and if over time this operation is reversed, it gives rise to antagonistic recovery. The nature of the control mechanism and hence of its possible breakdown will likely be the next focus of inquiry.

The third question is the one that has been debated by far the most over the past century and it continues to be a matter of active investigation. Several factors have been proposed, although none has been able to account for most of the facts. Neither the native language, nor the most familiar to the patient at the time of insult, nor the most useful, the most affectively loaded, the language of the environment, or the last one used before insult, is necessarily recovered first or best (Paradis, 1977). Nor does it seem to be a matter of whether the two languages were acquired and used in the same context as opposed to acquired and/or used in different contexts, at different times of development. Junqué, Vendrell, and Vendrell (1995) report on Catalan-Spanish bilingual aphasic patients who exhibit differential recovery even though they had acquired both languages from the crib and were using both equally every day before insult, compared to other patients who exhibited parallel recovery even though they had acquired one language long after the other. Sasanuma and Park (1995) also describe patients who underwent parallel recovery even though their languages were learned in quite different environments. Moreover, two languages that are structurally closely related (for example, Catalan and Spanish, or Korean and Japanese) can be recovered differentially (Junqué et al., 1995; Sasanuma & Park, 1995), whereas languages structurally distant (Azari and Farsi, or English and Farsi) can be recovered in a parallel fashion (Nilipour, 1988).

Given the lack of correlation between any single factor and the recovery pattern, researchers tended to favor a "multiple factor" view, whereby several factors might contribute to the recovery of one particular language over another. However, one reason why the answer to this question has been so elusive may be that researchers were looking in the wrong direction. It might be that these various factors have little to do with the preferential recovery of a particular language, but could be a direct result of the nature of the disruption of the control system (that is, the system that controls the distribution of resources). The activation threshold of a particular system (or subsystem) may be raised simply because the distributive system, for reasons independent of the actual language that happens to be subserved, no longer has access to that system (or has only reduced access).

29-2. LATERALIZATION

The hypothesis that language is less asymmetrically represented in the cerebral hemispheres of bilingual speakers can be interpreted as a form of differential localization of the two languages. If one of the languages is more extensively represented in the right hemisphere (RH), then it is at least partially represented in a different locus. This has been suggested as a possible explanation for certain nonparallel recovery patterns. In cases of selective or successive recovery, the recovering language might be more bilaterally represented than the nonrecovering language and thus be less severely impaired and recovered sooner (Albert & Obler, 1978). However, all the clinical data collected since this hypothesis was formulated contradict this possibility. There is no greater incidence of crossed aphasia in bilingual than in unilingual populations (Chary, 1986; Karanth & Rangamani, 1988; Rangamani, 1989) and the evidence from Wada testing, electrical stimulation of the brain, and PET scans confirms that both languages are represented in the left hemisphere (LH) in bilingual individuals (Rapport, Tan, & Whitaker, 1983; Klein, Zatorre, Milner, Meyer, & Evans, 1995).

Yet this has been one of the most controversial and most discussed issues of the past 2 decades. At first it was proposed that the languages of bilinguals (in general) were less lateralized than the language of unilinguals. Since studies kept failing to show any difference between the lateralization pattern of bilinguals and unilinguals, the hypothesis was gradually restricted to smaller and smaller subpopulations of bilinguals: only bilinguals who have not acquired their languages concurrently; only in the initial stages of second language acquisition; only in the initial stages, provided that the second language is acquired informally. Not only were none of these findings replicated, but some experimenters found differential lateralization in the direction opposite to the predictions of these age, stage, and manner hypotheses, for example, at the later stages of the formal learning of a second language. Meanwhile, about half the studies continued to find language to be represented no less asymmetrically in bilinguals than in unilingual controls.

These discrepancies have mostly been attributed to subtle differences in method, task, or stimulus characteristics (Vaid & Hall, 1991). This line of argument implies that some specific procedures will yield results indicative of greater RH participation whereas others will not. One is then left with the problem of deciding which of these subtle methodological differences are actually tapping into what they purport to be tapping into, and which are not, since different but equally valid procedures would not be expected to yield incompatible results. In the absence of a consensus on the matter, and given that the same technique (for example, dichotic listening or tachistoscopic visual half-field presentation) can produce contradictory results—which, incidentally, as we have seen, run counter to all the available clinical evidence—it is more likely that the experimental paradigms employed to measure language lateralization in individuals and subpopulations are not measuring what they purport to measure. And, in point of fact, this had already been demonstrated at the time these studies began to be popular (Satz, 1977). The acknowledgment that their experiments are "based on impoverished measures," that their stimuli are "hardly representative" of what they attempt to measure (that is, language), that their technique is "subject to many criticisms," and that "no suitable monolingual control group" can be found for their bilingual experimental sample does not, however, prevent some experimenters from continuing to conduct and publish such experiments (Richardson & Wuillemin, 1995).

The contradictions among experimental studies may be intensified by the fact that experimenters fail to specify what they mean by "language," that is, what it is that they suspect of being less lateralized in bilinguals. If they mean implicit linguistic competence, they are demonstrably wrong. All the evidence gathered to date points to both languages being represented in the language areas of the LH, in the same proportion as in unilinguals. If they mean communicative competence, including pragmatics, they are probably right, but the experimental procedures they use, by their very rationale, could not possibly speak to the issue. Indeed, it is difficult to see how a subject's response to the tachistoscopic presentation of single four-letter words or the dichotic presentation of digits could possibly be indicative of the use of pragmatic strategies in the processing of utterances in situational contexts.

On the basis of studies of communication deficits subsequent to RH lesions (Brownell, Gardner, Prather, & Martino, 1995), one can safely assume that the RH is crucially involved in the processing of pragmatic aspects of language use. Given that second language speakers, like children, are likely to compensate for the gaps in their implicit linguistic competence in their weaker language by increasing their reliance on pragmatic inferences, it is to be expected that they will involve their right hemispheres to a correspondingly greater extent. Increased RH involvement in these cases, however, does not reflect the representation or processing of the language system (implicit linguistic competence) but, on the contrary, of whatever nonlinguistic competence is substituted for it.

29-3. BILINGUAL VERSUS UNILINGUAL REPRESENTATION AND PROCESSING

When one language has not been fully mastered, the speaker may resort to compensatory mechanisms in order to interpret and formulate sentences. These may be of at least two sorts: metalinguistic knowledge and pragmatic competence. The differences between unilinguals and bilinguals, or between the stronger and the weaker language of a bilingual, may be quantitative, simply reflecting the extended reliance on either metalinguistic knowledge or pragmatic competence, or both, in the use of the weaker language. Also, one language (or certain items within a language) may be more readily available, depending on the frequency of use, just as some items are more or less readily available to unilinguals, for the same reason.

There is no verbal function available to the bilingual speaker that does not have its homologue in the unilingual speaker. Bilinguals switch languages, unilinguals switch registers (which implies a change of vocabulary, morphosyntax, and in some cases even phonology). Bilinguals mix their languages intrasententially and borrow words from the other language, unilinguals mix their registers and borrow words from other registers (either in jest or as a result of interference, just as bilinguals do). Bilinguals are able to translate from one of their languages into the other, unilinguals may paraphrase an utterance from one register into another, or even within the same register (that is, say more or less the same thing with different words and different syntactic structures). There is therefore no need to postulate neural mechanisms specific to bilinguals.

In our comparisons between bilingual and unilingual speakers, we must carefully distinguish between the contents of representations (that is, what is represented, the language system with its phonological, morphosyntactic, and lexical-semantic components, as well as pragmatic aspects, including sociolinguistic registers and the appropriateness of using specific items in particular circumstances, also proverbs, conventional metaphors, and indirect speech acts) on the one hand, and the cerebral organization of representations on the other. The contents necessarily differ from one language to the next (obviously, implicit linguistic competence in Korean is not the same as implicit competence in English). The contents even differ among bilinguals speaking the "same" pair of languages. (Of course, to the extent that their grammar deviates from the native pattern, one might argue that they do not really speak the language [Paradis, 1993b], but that is not important here.) Given that the grammar (including the lexicon) of few, if any, bilingual speakers is identical in all respects with that of unilingual speakers of the respective languages, and that the deviances from the unilingual norm are different in different individuals as a result of the context of acquisition/learning and/or use, the actual contents of the representations of the languages will differ among bilinguals, according to their degree of mastery of each aspect of grammar and pragmatic competence for each language.

The way the two grammars (and all other aspects of verbal communication) are organized in the brain, however, need not depend on the internal structure of each grammar. In other words, whether or not (and to what extent) one or both languages are affected by what is linguistically described as interference in their contents is independent of whether the two language systems are represented in the brain as one extended system for language; as two distinct systems; as one common system for whatever the two languages have in common (whether legitimately or not), and separate systems for what is unique to each language (different phonemes, stress patterns, morphosyntax, lexical-semantic constraints, etc.); as two language-specific subsystems within the language system (Paradis, 1987b); or any other manner in which two languages may happen to be represented in the human brain.

Although no qualitative differences need exist between unilinguals and bilinguals and between various types of bilinguals with regard to how languages are processed, extensive quantitative differences in the use of the various cerebral mechanisms involved in verbal communication may nevertheless result from (1) differences in the degree of mastery of the two languages, (2) the age at which the second language has been acquired, and (3) the degree of motivation in acquiring the second language.

To the extent that linguistic competence in one language is incomplete, speakers, like unilingual children during their first years of language acquisition, have to rely on (presumably right hemisphere-based) pragmatic processing in order to derive an interpretation for utterances in that language whose automatic linguistic processing has not yet been fully internalized. (Note that this increased RH participation could not, by definition, be revealed by procedures explicitly devised to measure lateralization of the language system [that is, implicit linguistic competence, the grammar]. If these procedures [dichotic listening, tachistoscopic visual half-field presentations, dihaptic tasks] were sensitive to pragmatic processing, it would be a demonstration that they were measuring something other than what they purport to measure—in other words, that they were not valid.)

Independently, second language learners may also obtain extensive metalinguistic knowledge about their second language in the form of explicit pedagogical grammar rules. Like any native speaker who has obtained such knowledge at school, they may consciously process sentences, though to a much greater extent than in their native language and than unilinguals, in order to compensate for their lack of automatic implicit competence in their weaker language. To the extent that they control their production, they engage their declarative memory system (when their procedural memory for language fails them), and thereby engage a particular neural substrate to a greater extent (Paradis, 1994).

There are also indications that speakers who have acquired their second languages after a certain age (that is, about 7) process functional categories (closed-class words) in the same way as lexical categories, whereas unilinguals and early bilinguals process them differently (Weber-Fox & Neville, 1996). This might be due to a decrease in the capacity of procedural memory to acquire implicit linguistic competence with age, and learners may consequently have to rely to a greater extent on declarative aspects (vocabulary being one of them—Paradis, 1994). Function words may then be learned as lexical items, with a sound and a particular meaning, rather than as a structural form within the implicit grammar. This process is probably not very different from the way genetic dysphasics learn inflectional morphology in their native language (Paradis & Gopnik, 1994). We may thus eventually have to distinguish between (early) bilinguals and fluent second language speakers. The ERP evidence (Weber-Fox & Neville, 1996) points to a neurofunctional processing difference. If our speculation that it may reflect a difference in the type of memory system involved turns out to be correct, then we may also have a difference at the level of the neural substrates involved in unilinguals and early bilinguals, on the one hand, and fluent second language speakers (or late bilinguals), on the other.

29-4. FUTURE DIRECTIONS IN THE NEUROPSYCHOLOGY OF BILINGUALISM: THE NEED FOR DISTINCTIONS

Too often, investigators of the organization of two or more languages in one brain in general, and of aphasia in bilinguals and polyglots in particular, have suffered the consequences of a failure to make a number of distinctions, namely, between (*a*) the linguistic and neurolinguistic levels of description (or domains of discourse); *(b)* lexical and conceptual representations; *(c)* language qua grammar (or system or code) and language qua verbal communication (the grammar plus pragmatics); *(d)* implicit linguistic competence and explicit metalinguistic knowledge; and *(e)* quantitative versus qualitative differences. Once these distinctions are taken into consideration, one realizes that the many controversies can be resolved. This goes a long way toward ending the state of confusion that has led to numerous contradictory claims concerning language laterality in bilinguals, the number of lexical stores, and cerebral representation and processing in various types of bilingual speakers, to mention only a few of the issues that have mobilized the energies of psychologists and neurologists investigating bilingualism over the last two decades.

29-4.1. Linguistic and Neurolinguistic Levels of Description

The organization of the neural substrate has often been mistakenly supposed to be modified in some way as a consequence of the differential organization of language structure. Compound organization of an individual's linguistic system, whereby elements of the two languages are blended or fused (a linguistic construct), has thus been assumed to necessarily entail a cerebral organization (a neurolinguistic construct) that would differ from that of individuals with a system in which elements of the two languages retain their respective unilingual integrity. Although the *contents* of the grammar may differ from one language to the next and from one bilingual speaker to another (according to the degree and kind of deviation from the unilingual norm in each language), there is nevertheless no necessary effect on the neural underpinnings of the grammar, and so far none has been documented. Although the internal structure of the blended grammar may differ, there is no reason why it should not be subserved by the brain as a grammar, irrespective of its specific contents and hence irrespective of the number of illicit borrowings or blends incorporated into it.

As we have seen, there is no evidence from aphasia to support the hypothesis that bilinguals who have acquired both their languages concurrently, in the same environment, store their languages in ways that are more neurofunctionally similar, whereas those who have learned their second language in a different context from the first store their two languages in more neurofunctionally separate ways.

One linguistic consequence of early bilingualism in some individuals may be a systematic deviation from the unilingual norm at some or all levels of language representation (for example, some degree of compoundness in voice onset time or vowel quality, or the idiosyncratic use of a particular preposition in some contexts) without any consequence for the manner in which the grammar is subserved by its neural substratum. There is no reason to believe that the brain should process legitimate elements in one way and illegitimate ones in another. Blends are processed as if they were legitimate elements, as long as they are treated as such by the speakers and have been incorporated into their implicit linguistic competence. The legitimacy of an element is often a mere historical accident. For example, the actual number of elements shared between English and French is a result of such historical accidents—it is not necessary and could easily have been different. In fact, it continues to change in our own lifetime.

Linguistic and neurolinguistic domains of discourse bear on objects that are of a different nature, and hence the particular form of the linguistic elements has no bearing on the neural principles that govern the related substrate and vice versa. Future research in language sciences will have to explicitly state whether the hypothetical constructs that are proposed and the claims that are made concern the language system (the linguistic level of description), the use of mental representations in real time (the psycholinguistic level), or the underlying neural mechanisms and structures (the neurolinguistic level). Any claim that the content/form of elements at one level of representation has a direct influence on the structure and operations of another level will have to be logically or empirically demonstrated.

29-4.2. Lexical and Conceptual Representations

Another source of confusion has been a neglect to distinguish between lexical semantic constraints and nonlinguistic mental representations, or concepts. For the last four decades, experimental psychologists have investigated whether bilingual speakers possess two linguistic memory stores or one.

The various experimental studies attempting to ascertain the number of linguistic memory stores were all based on the following general rationale. If subjects responded to translation equivalents in the same way as they responded to the repetition of the stimulus word, results were interpreted as supportive of the one-store hypothesis. If they responded to translation equivalents in the same way as they responded to altogether different words in the same language as the stimuli, results were interpreted as supporting the two-store hypothesis. Some experiments claimed to support the two-store hypothesis, others were construed as supporting the one-store hypothesis, while some supported neither, as the subjects' responses were too different for a one-store explanation but not different enough for a two-store account. However, these contradictory results are uninterpretable because these studies suffered from a triple lack of distinctions: (1) between different types of semantic organization in their bilingual subjects' lexicon; (2) between stimuli that inherently have considerable overlap in meaning between languages and those that only have minimal degree of overlap; and (3) between semantic and conceptual levels of representation (lexical-semantic representations versus nonlinguistic mental representations).

The results were thus partly a function of linguistic rather than psycholinguistic criteria: they varied according to the degree of overlap in meaning between the stimulus words and their translation equivalent. The more extensive the overlap in meaning, the greater the chances that the translation equivalent would trigger the same response as a repetition of the initial stimulus; the less extensive the overlap, the greater the chances that it would trigger a different response. These differences can be exacerbated by an increased overlap in meaning in the lexicon of subjects with a compound and subordinate set of lexical meanings. This is also true in the case of priming in current lexical decision studies. Results tend to vary based on subjects' degree of fluency (generally correlated to a great extent with subordinate organization of the lexicon) and structural distance between languages (that is, the degree of inherent overlap).

One of the major problems, with many ramifications, has been the failure to distinguish between the meaning of words and nonlinguistic mental representations. The semantic field of each word is determined by language-specific constraints on its possible uses. Words share some but not all of the semantic features of their translation equivalents and will therefore not denote all of the same referents. The mental representation that corresponds to a word will thus differ to some extent from the mental representation corresponding to its translation equivalent. But clinical evidence has shown that the speaker has a third, language-independent system that contains conceptual representations. During language comprehension or production, the mental representations are organized (that is, conceptual features are grouped together) in accordance with the lexical semantic constraints peculiar to the selected language system. This third cognitive system, phylogenetically and ontogenetically anterior to the language system(s), is independent of language and hence of the bilingual's two languages, and remains available to the aphasic patient (Lecours & Joanette, 1980). Even though an aphasic patient may have lost access to the words *cupcake, brioche,* and *muffin,* that person may nevertheless go to the store and buy a muffin, not a cupcake (that is, the person has the concept, whether or not he or she is able to verbalize it).

We should thus distinguish between the lexical meanings of words, a part of the speaker's linguistic competence (a component of the lexical item, like its syntactic features and phonological form) and hence vulnerable to aphasia, and conceptual representations that are outside of implicit linguistic competence and thus are not vulnerable to aphasia (though they are vulnerable to other forms of mental disruption). The conceptual system, where messages are elaborated before they are verbalized in the course of the encoding process, and where a mental representation is attained at the end of the decoding process, remains independent and isolable from the language systems.

The form of any aspect of the grammar (phonology, morphosyntax, lexical-semantic features—a linguistic construct) may be the result of language contact, but once it has been internalized as implicit linguistic competence, there is no indication that it is not represented in the mind (a psycholinguistic construct) or in the brain (a neurolinguistic construct) in the same manner as any other linguistic system. The context of acquisition may influence the content of a grammar and the context of use may influence the availability of some of its elements, but they have no influence on the principles by which brain mechanisms store and process these elements. Future research will likely concentrate on identifying, in bilingual speakers, the cerebral areas and mechanisms involved in the processing of lexical items versus their conceptual representation, teasing apart what is language-specific and hence part of the linguistic system, as opposed to what is conceptual and hence independent of the linguistic system, possibly along the lines of research developed by Damasio (1989).

29-4.3. Grammar and Pragmatics

Before the claim can be made that the RH is involved to a greater extent in the language functions of bilinguals than of unilinguals, it is necessary to distinguish between implicit linguistic competence (that is, phonology, morphology, syntax, and the lexicon) on the one hand, and pragmatic aspects of language use (for example, reliance on inferences from situational context, general knowledge, emotional prosody, facial expressions, and so forth) on the other. All clinical studies to date unambiguously point to the fact that implicit linguistic competence is subserved by areas of the LH of right-handed bilinguals in the same proportion as in unilinguals. As we have seen, there is also increasing evidence from RH lesion studies to the effect that pragmatic and paralinguistic aspects of language use are subserved by areas of the RH.

The failure to differentiate between grammar and pragmatics has resulted in a series of controversies as to whether there is greater participation by the RH in processing language in bilinguals than in unilinguals. The answer seems to be an unqualified "yes" or "no," depending on what one means by "language." Some bilinguals, like some unilinguals, may process some of their grammar in their RH and show subtle morphosyntactic and/or phonological deficits subsequent to a RH lesion (Joanette, Lecours, Lepage, & Lamoureux, 1983). But there is no clinical evidence (such as a greater incidence of crossed aphasia) or neuropsychological evidence (for example, from Wada testing or electrical stimulation of the brain) that would indicate *greater* participation. If, however, "language" refers to the normal use of language, including pragmatics, then the answer is a qualified "yes"—only in the case of the use of a weaker language.

As far as the experimental evidence is concerned, a meta-analysis (Vaid & Hall, 1991) has failed to come up with a difference between unilinguals and bilinguals and most researchers have finally abandoned this fruitless search for a differential cerebral asymmetry. This does not mean that bilinguals cannot rely to a greater extent on RH-based pragmatic features when using their weaker language. Future research is likely to examine the extent of pragmatic compensation in the use of any weaker language by bilingual speakers with methods inspired by studies of RH involvement in language use in unilinguals.

29-4.4. Linguistic Competence and Metalinguistic Knowledge

It is equally important to distinguish between implicit linguistic competence and metalinguistic knowledge. The former is acquired incidentally, is stored in the form of procedural know-how without conscious knowledge of its contents, and is used automatically. The latter is learned consciously, is available for conscious recall, and is applied to the production (and comprehension) of language in a controlled manner. Implicit linguistic competence is acquired through interaction with speakers of the language in situational contexts. Metalinguistic knowledge is usually learned through formal instruction. The extent of metalinguistic knowledge about one's native language is by and large proportional to one's degree of education. Very often, a foreign language is learned almost exclusively through metalinguistic knowledge and whatever linguistic competence develops subsequently does so through practice of the language in communicative situations. If implicit linguistic competence is what is affected by aphasia, given that explicit knowledge is ordinarily not impaired in aphasic patients, it is not unreasonable to expect that metalinguistic knowledge, together with all other items of episodic and encyclopedic declarative memory, should remain available to aphasic patients. Some patients may have been exposed to much more explicit metalinguistic knowledge during the learning of a foreign language than during the acquisition of their native language. Some may, in fact, have been schooled entirely in a second language, while continuing to use the first language only in the home, thus remaining illiterate in, and having little metalinguistic knowledge about, that language.

As a result, in such individuals, metalinguistic knowledge about their second and possibly weaker language may be more extensive than about their native language. Subsequent to aphasia, they may lose access to their implicit linguistic competence equally in both languages but nevertheless retain full access to their metalinguistic knowledge which, being more extensive in their nonnative language, may give the impression of a preferential recovery of the language that was the least fluent before insult. In reality, however, even though implicit linguistic competence is equally impaired in both languages, these patients simply have more metalinguistic knowledge to fall back on in their weaker language. The fact that they speak slowly, control their production consciously, and keep asking their interlocutor to slow down and/or repeat may go unnoticed, since the patients, being aphasic, are expected to speak more effortfully anyhow. Future research is likely to investigate this compensatory mechanism and its impact on language recovery and rehabilitation.

29-5. CONCLUSIONS

The neuropsychology of bilingualism over the past two decades has concentrated on various attempts to explain the diversity of recovery patterns in terms of models of language representation and processing. Two issues in particular have monopolized much of the literature: differential hemispheric asymmetry and differential localization within the dominant hemisphere. The ontological status of the claims investigators make must be clarified, namely, whether they are intended as descriptive metaphors (like the various characterizations of implicit linguistic competence) or as descriptions of the actual state of affairs in the real world in real time; or, in other words, whether these claims are about the linguistic or neurolinguistic level of description. We must also distinguish between the language-specific meaning of words and nonlinguistic mental representations. Finally, we need to distinguish between implicit linguistic competence and pragmatic aspects of language use, and between explicit metalinguistic knowledge and implicit linguistic competence. Once that has been accomplished, we will be in a better position to examine the complex problems posed by aphasia in bilinguals and polyglots and by the representation and processing of two languages in the same brain. The systematic collection of data from bilingual aphasic patients with instruments that are equivalent in each of their languages will greatly facilitate the process. New imaging techniques, in particular fMRI and ERP studies of bilingual populations, should also help us identify the cerebral mechanisms involved in the various language functions of bilingual speakers.

CHAPTER 30

The Role of Subcortical Structures in Linguistic Processes
Recent Developments

Bruce Crosson and Stephen E. Nadeau

Department of Clinical and Health Psychology, University of Florida Health Science Center, Gainesville, Florida, 32610

With infarction or hemorrhage of the dominant basal ganglia, cortical dysfunction due to hemodynamic factors accounts for the more prominent language symptoms. At best, cortical symptoms obscure a minor role for the basal ganglia in language. If this role exists, it is a supportive, as opposed to a primarily linguistic, one. On the other hand, dominant thalamic nuclei appear to play a role at the lexical-semantic interface, explaining semantic paraphasias commonly seen in thalamic aphasia. The frontal lobes, inferior thalamic peduncle, nucleus reticularis, and perhaps centromedian nucleus may be involved in selective engagement of cortical mechanisms that subserve this interface. Other phenomena after dominant thalamic lesion include category-specific naming deficits and neglect dyslexia. Such findings raise the possibility that the thalamus has multiple functions that support language.

Interest in the role of subcortical structures in language has been evident for more than a century. Wallesch and Papagno (1988) traced the concept of subcortical aphasia as far back as Broadbent (1872), who believed that the basal ganglia were the structures in which words were generated as motor acts. As in more recent times, this position was not held by all. Kussmaul (1877) assigned a strictly motor function to the basal ganglia. Even Carl Wernicke (1874), who, along with Broca and Lichtheim, can be considered a founder of modern aphasiology, discussed the function of the basal ganglia in language. Marie and his students (Marie, 1906, cited by Démonet, 1987; Moutier, 1908, cited by Wallesch & Papagno, 1988) believed the basal ganglia were involved in language. Regarding the thalamus, Penfield and Roberts (1959) suggested that it played an integrative role in language. Schuell, Jenkins, and Jimenez-Pabon (1965) speculated that the thalamus was involved in preverbal feedback about the adequacy of formulated responses. Thus, some of the more notable names in the history of aphasiology have endeavored to discern the role of subcortical structures in language.

However, it was not until the advent of the X-ray computerized tomography (CT) scan, and later magnetic resonance imaging (MRI), that serious consideration of subcortical structures in language emerged. These imaging techniques have allowed us to visualize and localize subcortical lesions. It was discovered that when located in the dominant hemisphere, subcortical lesions were frequently accompanied by aphasia. Stereotactic surgery studies, mostly on Parkinson's disease patients, have yielded stimulation and lesion data that have played a lesser, but important, role in fueling theories of subcortical functions in language. In the last decade, significant advances have been made that are contributing to our understanding of the role of subcortical functions in language. In this chapter, we will endeavor to address these advances, first for the basal ganglia and then for the thalamus. For each of these anatomic regions, we will provide a brief theoretical history, then discuss recent advances, and finally raise issues for future consideration. Our focus will be primarily theoretical, and because of space limitations, our review will be somewhat selective. For those readers wishing further detail, we cite reviews that can be perused.

30-1. ROLE OF THE BASAL GANGLIA IN LANGUAGE

30-1.1. Historical Perspective on Theory

Numerous studies in the 1960s, 1970s, and 1980s implicated the basal ganglia in language (see Crosson, 1992). Although studies of large hemorrhages into the basal ganglia were flawed because they failed to take into account the potential pressure effects on surrounding structures, numerous studies of infarction in the basal ganglia of the dominant hemisphere also described aphasia. This phenomenon encouraged speculation about the role of the basal ganglia in language. Nonetheless, studies of infarction were not flawless, in part because they typically involved surrounding white matter structures.

Divergent theoretical positions have appeared regarding the role of the basal ganglia in language. Alexander, Naeser, and Palumbo (1987) suggested that the basal ganglia were not involved in langauge. They explained the aphasia accompanying basal ganglia infarction as due to damage to white matter pathways surrounding the basal ganglia that interconnect important language-related structures. Crosson (1985, 1992) proposed a different explanation for aphasia after dominant basal ganglia lesions. He thought that language segments were formulated anteriorly and held in a buffer until they could be semantically verified by posterior mechanisms used in language decoding. The role of the basal ganglia was to provide a signal initiating the release of the formulated language segment for motor programming once it had been semantically verified. Notice that this proposal does not posit that the basal ganglia process linguistic features in order to accomplish this role. Their role is simply to trigger the release of language segments from the buffer for the purpose of motor programming. Finally, Wallesch and Papagno (1988) suggested that the basal ganglia were involved in monitoring multiple cortically generated lexical alternatives in order to select the one best meeting semantic and motivational constraints. Recent evidence, however, suggests a different origin of language problems accompanying dominant basal ganglia lesions.

30-1.2. Recent Developments

30-1.2.1. Relationship of Aphasia after Dominant Basal Ganglia Infarction to Hemodynamic Factors

The problem with all these hypotheses relates to the variability in aphasia syndromes observed after infarction of dominant basal ganglia structures. Crosson (1992) reviewed numerous studies of infarction of various portions of the dominant basal ganglia. No matter which specific parts of the basal ganglia were involved in the lesion, no definable aphasia syndrome emerged. To some degree, this analysis was flawed because, in most of the studies, lesions involved multiple basal ganglia structures, as well as surrounding white matter pathways. Nadeau and Crosson (1997), therefore, took a slightly different approach to analyzing this problem. Their review of cases from the literature addressed only dominant striatocapsular infarctions, which have a characteristic comma-shaped appearance and invariably involve the head of the caudate nucleus, the putamen, the anterior limb of the internal capsule, and perhaps some of the globus pallidus. Because the anatomic structures involved are relatively consistent, some consistency of aphasic symptoms would be expected if any of these structures are involved in specific language processes. No such consistent pattern of language symptoms could be found. Aphasias after this type of lesion could be fluent or nonfluent, could involve significant comprehension deficits or no comprehension deficits, might or might not involve impairment of repetition, and so on. Further, there were a significant number of cases in which dominant striatocapsular infarction was accompanied by no language symptoms. This evidence mitigates against a specific role in language for these structures, and an alternative explanation for aphasia after dominant basal ganglia lesions must be sought.

A series of studies by investigators in cerebrovascular disease, most notably that of Weiller et al. (1993), have shed light on this issue. These authors found that striatocapsular lesions were caused by occlusion of the initial segment of the dominant middle cerebral artery (MCA), or occasionally by occlusion of the internal carotid artery (ICA). The presence of significant aphasia after dominant basal ganglia infarction was determined by two factors: (1) how early the MCA recanalized and (2) adequacy of anastomotic circulation. Significant aphasia was found primarily in cases of late recanalization and inadequate anastomotic circulation. Thus, these findings suggest that circulatory dynamics play a crucial role in language symptoms after dominant basal ganglia infarction.

Nadeau and Crosson (1997) have explained these circulatory dynamics in some detail (Figure 1). The reader is referred to this review for the specifics, but the essentials are as follows: Because striatocapsular infarction is caused primarily by occlusion of the initial segment of the MCA or occasionally by occlusion of the ICA, the entire circulation of the MCA is at risk. This circulation includes the perisylvian cortex essential for language functions. The adequacy of anastomotic circulation to various areas of cortex determines the impact of MCA or ICA occlusion on cortex in the

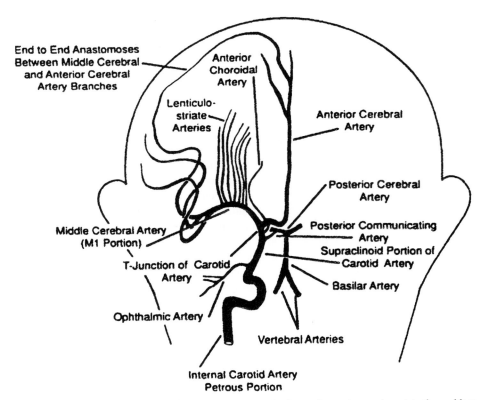

FIGURE 1   This semischematic diagram represents cerebral vascular anatomy relevant to the problem of aphasia that accompanies dominant basal ganglia infarction. In this view, the anterior to posterior dimension has been compressed into a single coronal plane. (From S. E. Nadeau & B. Crosson [1997], Subcortical aphasia. *Brain and Language.* Reprinted by permission of Academic Press.)

MCA circulation; however, the position of the occlusion frequently prevents anastomotic circulation from supplying the infarcted subcortical territory. When anastomotic circulation for a particular cortical region is grossly inadequate, cystic infarction of the cortex as well as the basal ganglia occurs. In this case, the cortical damage is clearly visualized on CT or MRI scans. However, inadequate anastomotic circulation can cause two other states that will impair function but will not be visualized on structural brain imaging. First, blood flow may not be so compromised that cystic infarction occurs, but it may be compromised enough that ischemic neuronal dropout takes place. Second, circulation may not be so compromised as to cause ischemic neuronal dropout, but it may be compromised enough that normal tissue function is impossible. These conditions will have the effect of permanently or temporarily causing dysfunction of the affected cortical regions. The nature of the language symptoms is dependent on which areas of language cortex are compromised in this way. Which cortical regions are compromised is dependent on the adequacy of their anastomotic circulation. In basal ganglia hemorrhage, Nadeau and Crosson (1997) have also noted circulatory dynamics that affect the cortex and thereby play a role in the presenting symptoms. In such instances, the authors noted that pressure-induced ischemia causes cortical dysfunction, which results in aphasic symptoms.

Thus, Nadeau and Crosson (1997) concluded that any role of the dominant basal ganglia in language is at best obscured by cortical dysfunction in cases of basal ganglia infarction or hemorrhage, and probably is minimal or nonexistent. Other reviewers have reached similar conclusions regarding the lack of basal ganglia functions in language. For example, Bhatia and Marsden (1994) reviewed 240 cases of basal ganglia lesions. In instances where lesions appeared confined to a single structure of the basal ganglia, significant speech or language deficit was very infrequently found for the caudate nucleus (2 of 43), putamen (1 of 20), or globus pallidus (1 of 17). Thus, much recent evidence suggests the basal ganglia play no major, direct role in language processes.

30-1.2.2. Syntax Comprehension Deficits in Parkinson's Disease

One possible exception to this conclusion is the complex syntax comprehension deficits noted by Grossman and his colleagues, as well as others, in Parkinson's disease (PD) patients (for example, Grossman et al., 1991; Grossman, Carvell, Stern, Gollomp, & Hurtig, 1992; Lieberman, et al., 1992). Major impairment of mesostriatal dopamine pathways in PD significantly affects putamen and caudate nucleus functions. Grossman et al. (1992) noted that a variety of grammatic factors were correlated with overall sentence comprehension problems in PD patients, but performance on sentences requiring greater working memory was also correlated with general sentence comprehension. Lieberman et al. (1992) found that PD patients with voice onset timing deficits had more syntactic comprehension errors than PD patients without voice onset timing deficits in speech. At this time, it is still uncertain whether syntax processing deficits in PD patients represent a primary linguistic processing deficit or are due to other underlying cognitive dysfunction, such as working memory or procedural memory. Further, it must be acknowledged that cortical pathology exists in these patients (for example, de la Monte, Wells, Hedley-Whyte, & Growdon, 1989), which could account for these deficits.

30-1.3. Future Directions

In spite of evidence that cortical dysfunction plays a major role in language deficits after basal ganglia stroke, it is possible that this cortical dysfunction obscures a minor, supportive role for the basal ganglia in language. Alexander (1992) tested language in patients with infarcts in the dominant basal ganglia and surrounding white matter. In some instances, the lesions were small. Motor speech deficits were prominent. Even in cases where the lesions were small, Alexander noted some word-finding deficits, semantic substitutions, and impairment in complex comprehension. He concluded that this deficit pattern had little "instrumental specificity"; that is, it did not implicate any particular linguistic function. One explanation he offered for the language symptoms was that a "facilitatory," "activating," or "enabling" system affecting multiple language functions was damaged. Raymer, Moberg, Crosson, Nadeau, and Rothi (1997) have suggested that giving sensitive tests of language function to patients with lacunar infarcts limited to the striatum would be a way of resolving this issue. Lacunar infarcts are unlikely to reflect occlusion of large cerebral vessels (Poirier, Gray, & Escourolle, 1990) and, therefore, may avoid the confounding effects of circulatory dynamics present in larger lesions.

The work of Malapani, Pillon, Dubois, and Agid (1994) offers what might be a more promising approach to this question. They administered auditory and visual choice reaction-time tasks to Parkinson's disease (PD) patients either separately or concurrently in both the on and the off (dopamine-depleted) states. They found impairment in the concurrent but not the separate administration in the dopamine-depleted state. Since this state has a major impact on striatal function, it is likely that differences in cognitive function during the off state were related to basal ganglia function. Nonetheless, it must be conceded that mesocortical or mesolimbic dopamine pathways could impact cognition in the off state. Still, this approach also could be used with language functions, including syntactic comprehension. However, we are not optimistic that a significant indirect role for the basal ganglia in language will be defined.

30-2. THE ROLE OF THE THALAMUS IN LANGUAGE

30-2.1. Historical Perspective on Theory

As with the basal ganglia, numerous studies from the 1970s on have demonstrated aphasia after dominant thalamic lesions. However, the picture is more coherent than for the basal ganglia. First, thalamic lesions are usually caused by small vessel disease, and therefore, less likely to be accompanied by aberrant cortical circulation. Aphasia is almost invariably found after lesions in the dominant tuberothalamic (polar) artery territory. Not infrequently, aphasia also is found after a lesion in the dominant inter-peduncular profundus (paramedian) artery territory (Crosson, 1992). Aphasia after thalamic lesions is almost always characterized by normal or minimally impaired repetition. Word finding with semantic paraphasia is common and may deteriorate into jargon. Neologisms are sometimes apparent. Comprehension is generally less impaired than output. Aphasia lasts months or years in some patients (Crosson, 1984, 1992). These findings led some theorists to speculate about a role for the thalamus in language.

A number of authors suggested that this role involves arousal mechanisms (Horenstein, Chung, & Brenner, 1978; McFarling, Rothi, & Heilman, 1982; Riklan & Cooper, 1975). Others have suggested that thalamic nuclei direct activation important for the modulation and integration of language (for example, Cooper et al., 1968; Samra et al., 1969). Ojemann (1983) carried this reasoning a step further, suggesting that "thalamic alerting processes selectively activate the specific discrete cortical areas (mosaics) appropriate to the particular language processing" (p. 203). Others have suggested an integrative role for the thalamus in language (for example, Penfield & Roberts, 1959; Schuell et al., 1965). Along these lines, Crosson (1985, 1992) had suggested that the thalamus was involved in semantic verification of formulated language segments prior to their release for motor programming. In the scheme of Wallesch and Papagno (1988), the role of the thalamus in language was primarily related to its position in cortico-striato-pallido-thalamo-cortical loops (see earlier discussion on the basal ganglia).

30-2.2. Recent Developments

30-2.2.1. Selective Engagement of Cortical Mechanisms as a Role for Thalamic Nuclei in Language

Nadeau and Crosson (1997) examined lesions in four cases of thalamic aphasia in order to ascertain the precise nuclei involved. In two cases, the lesion was in the territory of the tuberothalamic artery, and, among other structures, the ventral anterior portion of the nucleus reticularis was damaged. The nucleus reticularis (NR) is a thin shell of neurons surrounding most of the thalamus that has inhibitory (GABAergic) output to most thalamic nuclei. The ventral anterior portion of this nucleus has dense synaptic connections with and is pierced by a large bundle of fibers that connects the frontal lobes with the thalamus, the inferior thalamic peduncle. In the other two cases, the lesion was more posterior, in the territory of the interpeduncular profundus artery. While these lesions did not impact the ventral anterior portion of the NR, they did damage the centromedian nucleus of the thalamus (CM). Two of these patients, one with a lesion affecting the ventral anterior NR and one with a lesion affecting the CM, received extensive language testing by our group (Raymer et al., 1997). Both had similar naming deficits that compromised both written and oral naming and affected naming to definition as well as visual confrontation naming. Other lexical and semantic functions were intact. These patients consistently had greater difficulty with low- as opposed to high-frequency words, and errors were primarily semantically related to the target. Raymer et al. concluded that these patients had an impairment affecting the lexical–semantic interface. What would cause such an impairment at the lexical–semantic interface?

Nadeau and Crosson (1997) reasoned that the similarity in these cases was due to the involvement of different portions of the same neural system. The hypothesized system (Figure 2) involved the frontal lobes, the inferior thalamic peduncle (ITP), the nucleus reticularis (NR), and the centromedian nucleus of the thalamus (CM). According to the authors, the system is involved in "selective enhancement of thalamic output relevant to intentionally guided attention and behavior." In addition to their analysis of lesion location in thalamic aphasia, Nadeau and Crosson's hypothesis was based on a large body of physiological evidence. These studies included evidence that thalamic neurons reside in one of two states: a burst-firing mode in which there is little correspondence between input and output, and a single-spike mode in which there is good correspondence between input and output (McCormick & Feeser, 1990; Steriade, Jones, & Llinas, 1990). Steriade, Domich, and Oakson (1986) suggested that input from the nucleus reticularis serves to put thalamic neurons into the single-spike firing mode whereby information can be transferred between afferent sources of the thalamus and targets of thalamic projections. In addition, the work of Yingling and Skinner (1977) indicated that the frontal-ITP-NR-CM system could enhance or suppress sensory signals traveling via the thalamus to the cortex. Nadeau and Crosson (1997) reasoned that the action of the frontal-ITP-NR-CM system could affect not only the transmission of sensory information (that is, attention) but also could serve to selectively engage cortical systems necessary to perform various cognitive functions.

One way to conceptualize the impact of selective engagement on lexical selection processes is in terms of levels of activation for individual lexical items (similar to the concept used in parallel distributed processing [PDP] models [Rumelhart, McClelland, and the PDP research group, 1986]). Selective engagement helps to amplify the difference in activation level between the target item and competing items. Without selective engagement, the difference in activation levels between the target lexical item and semantically related, competing items is reduced, and the probability of misselection of a semantically related item increases dramatically. This explanation would explain the pattern of errors in the cases of Raymer et al. (1997).

Some aspects of this selective engagement theory remain to be worked out. For example, the precise role of the CM in selective engagement is not clear. The CM and related intralaminar thalamic nuclei have diffuse projections to most areas of the cerebral cortex (Jones, 1985). It is possible that these connections are the means by which different cortical mechanisms are engaged, but the connections may be too sparse to support such a function. The NR has a consistent relationship with most thalamic nuclei; portions of the NR are related to specific thalamic nuclei, sending inhibitory projections to their targets (Jones, 1985). Selective engagement may occur

FRONTAL CORTEX

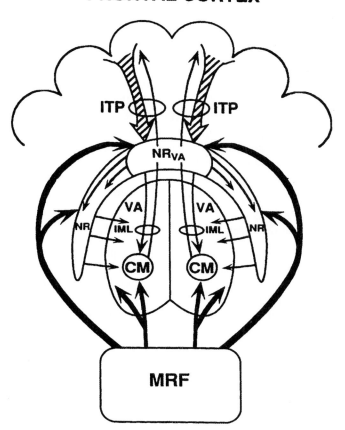

FIGURE 2   This schematic diagram represents the anatomical system that has been proposed by Nadeau and Crosson (1997) to regulate selective engagement of cortical mechanisms. CM = centromedian nucleus of thalamus; IML = internal medullary lamina; ITP = inferior thalamic peduncle; MRF = midbrain reticular formation; NR = nucleus reticularis of thalamus; NR$_{VA}$ = ventral anterior portion of the nucleus reticularis; VA = ventral anterior nucleus of thalamus. For the sake of clarity in the figure, diffuse projections from the cerebral cortex to the thalamus, projections from thalamic nuclei to the cortex, and collateral projections from cortex and thalamus to NR have not been diagrammed. (From S. E. Nadeau & B. Crosson [1997], Subcortical aphasia. *Brain and Lannguage.* Reprinted by permission of Academic Press.)

through the influence of the frontal lobes via the ITP on the NR, which acts to gate output from most thalamic nuclei. Since the CM and other intralaminar nuclei have almost no projections to other thalamic nuclei, it is unlikely that they influence other thalamic nuclei and this gating directly. However, connections from the CM to the frontal lobes may provide a means by which the CM impacts frontal functions, which in turn influence the NR's gating of various other thalamic nuclei. Because of its connections with perisylvian language cortex, the pulvinar is a prime candidate for a nucleus whose output could be gated to selectively engage language mechanisms. This is an area where good research in animal models could impact our knowledge about subcortical systems that impact cognition. Studies that follow up on the CM component of Skinner and Yingling's (1977) model would be welcome in this regard. The reader wishing further details is referred to Nadeau and Crosson (1997).

In addition to the selective engagement theory, two other recent developments raise interesting questions regarding the role of the thalamus in linguistic processing. Cases of category-specific naming deficits (that is, semantically bounded anomia) after thalamic lesion suggest that thalamic nuclei are involved in processing information in semantic or related realms that in turn impacts naming of specific categories of items. Neglect dyslexia accompanying visual neglect in a case of thalamic lesion indicates that thalamic mechanisms impact reading early in visual processing. We will address the topic of category-specific naming deficits first.

30-2.2.2. Category-Specific Naming Deficits after Thalamic Lesion

Cortical lesions have been reported to cause category-specific naming deficits for such items as body parts (Dennis, 1976), fruits and vegetables (Farah & Wallace, 1992; Hart, Berndt, & Caramazza, 1985), letters (Goodglass, Wingfield, Hyde, & Theurkauf, 1986), country names (McKenna & Warrington, 1978), indoor objects (Yamadori & Albert, 1973), animals (Hart & Gordon, 1992; Hillis & Caramazza, 1991; Stewart, Parkin, & Hunkin, 1992; Temple, 1986), medical items (Weinstein & Kahn, 1955), proper names (Semenza, & Zettin, 1988), and artifactual objects (Sacchett & Humphreys, 1992). Explanations for these deficits have ranged from dysfunction in the semantic system (for example, Sacchett & Humphreys, 1992) to effects of frequency, familiarity, and/or complexity (for example, Stewart et al., 1992; Funnell & Sheridan, 1992) to inability to map semantic information onto items in the output lexicon (Farah & Wallace, 1992). Thus far, the literature has not suggested any one mechanism that can account for all cases. Category-specific naming deficits have offered a challenge to our models about how information is represented in the brain because some authors have suggested that specific lexical, semantic, or lexical–semantic knowledge about the involved items is stored in specific regions of cortex (see Farah & Wallace, 1992). This issue has been hotly debated. It is also possible that category-specific naming deficits represent an anomalous organization that is present in a significant minority of cases.

Lucchelli and De Renzi (1992) described a case of anomia for proper names (primarily the names of people) in a case of infarction in the territory of the dominant tuberothalamic artery, affecting the ventral anterior nucleus, the ventral lateral nucleus, the mammillothalamic tract, and the internal capsule. The deficit was present whether naming from photographs or from description. Semantic knowledge was intact because the patient nearly always could give details about the persons he was unable to name, and phonemic cuing was less effective than giving the first name to evoke the surname. Further, the patient was unable to learn associations between new names and faces, though he could learn to associate occupations with faces. Three years later, Moreaud, Pellat, Charnallet, Carbonnel, and Brennen (1995) also reported a case of anomia for proper names that included both names of persons and geographic locations. Again, the infarction was in the territory of the tuberothalamic artery. The deficit was present whether naming from picture or description. We have also reported a patient with a category-specific deficit for naming medical items from definition (Crosson, Moberg, Boone, Rothi, & Ramer, in press), which was established by comparing his performance on experimental lists to that of normal controls and that of anomic patients. However, he had a hemorrhagic lesion in the posterior thalamus (including portions of the pulvinar and the centromedian nuclei) and the internal capsule. Like other patients with category-specific naming disorders, the items he missed were consistent across different test sessions. Similar to the case of Lucchelli and De Renzi (1992), phonemic cuing did not assist with retrieval. Familiarity, imageability, and abstractness did not account for his deficit.

Because it is highly unlikely that the thalamus is the site of storage for representations relevant for converting semantic information to a specific lexical item, the idea that the relevant stored representations (see Farah & Wallace, 1992) are destroyed in category-specific naming deficits can be ruled out, at least for these cases. An alternative explanation is that thalamic nuclei are engaged in neural nets that translate semantic information into a specific lexical output. Lucchelli and de Renzi (1992) suggested how this might happen in one case. In addition to his anomia for proper names, their patient had trouble learning both arbitrary associations between colors and numbers and associations between names and faces. Yet he was able to learn associations between faces and occupations much better. They emphasized a difficulty associating labels (that is, names) that carry little descriptive value for the object they represent. This difficulty in associating arbitrary labels or associations occurred for well-learned information (that is, proper narmes) as well as for new information. The findings of Lucchelli and De Renzi (1992) indicate that a principle more fundamental than semantic category may underlie many, if not most, category-specific naming deficits. Unfortunately, such principles may remain opaque to investigational methods in many instances, including our own case (Crosson et al., 1997).

The anatomical implications for these category-specific naming deficits are not clear. The anterior thalamus was involved in the two cases of proper name anomia (Lucchelli & De Renzi, 1992; Moreaud et al., 1995), but the posterior thalamus was involved in difficulty naming medical items (Crosson et al., 1997). Similar systems may be involved in all cases and include anterior and posterior thalamic structures (see earlier discussion on selective engagement). On the other hand, different thalamic nuclei may be involved in specific processes that will affect different substrates for naming.

30-2.2.3. Neglect Dyslexia after Thalamic Lesion

Some cases of neglect dyslexia have implicated deficits early in visual processing. In neglect dyslexia (Ellis, Flude, & Young, 1987; Hillis & Caramazza, 1990), some patients perform much better when words are spelled orally for them than when they are actually reading, and some cases do not show this pattern. For their case of neglect dyslexia who performed well with oral spelling, Ellis et al. (1987) suggested that the deficit was not in the patient's visual image of the word (that is, visual input lexicon) because it could be accessed through oral spelling. Rather, the deficit was earlier in visual processing and affected the ability to encode letter identity but not position. In such cases, top-down influences from the visual input lexicon would influence how the neglected portion of the word was completed, resulting in frequent errors.

We reanalyzed the reading errors of a previously reported case of thalamic aphasia (Crosson et al. 1986) and discovered that they could be attributed to neglect dyslexia (Ellis, Flude, & Young, 1987; Hillis & Caramazza, 1990). This patient had a hemorrhagic infarction of the lateral nucleus of the thalamus that extended into the pulvinar. Although his language output was characterized by semantic paraphasias, similar to other cases of thalamic aphasia, his oral reading was characterized by the substitution of words similar in phonology to the target response. He exhibited evidence of right-sided spatial neglect. Upon further examination, it was discovered that the words he substituted in reading frequently had the same letters on the left-hand side as the target word, but different letters on the right-hand side of the word; that is, he showed neglect dyslexia. (This phenomenon has been seen somewhat more frequently with right-hemisphere lesion where the right side of the word is the same and the left side of the substituted word is different from the target word.) Similar to the patient of Ellis et al. (1987), our patient's performance improved with oral spelling. Following Ellis et al. (1987), it is suggested that impairment was in early visual processing of words, not in the visual image contained in the visual input lexicon. This finding underscores the need for careful evaluation of reading functions in cases of thalamic lesion.

30-2.3. Future Directions

There are reasons to believe that the thalamus plays an important role in language, though it may be one of supporting linguistic functions as opposed to one of primary linguistic processing. First, unlike striatocapsular infarctions, a syndrome of language impairment emerges from reports of thalamic aphasia (Crosson, 1992): Repetition is almost always unimpaired or minimally impaired. Semantic paraphasia usually predominates over phonemic paraphasia and neologism; paraphasia may be so severe as to deteriorate into jargon. Although patients may have difficulty initiating language in the initial stages after the cerebrovascular accident, language usually evolves into a relatively fluent output with intact grammar. Second, isolated thalamic infarctions are due to small vessel disease, which does not implicate cortical function in the same way that major basal ganglia infarcts do. Third, there is some consistency in symptomatology and lesion location, with infarcts in the dominant tuberothalamic territory almost always producing language deficits (Crosson, 1992). For these reasons, the role of the thalamus in language processes is not in question in the way that the role of the basal ganglia is. The relevant question regarding the thalamus is, How does it play a direct or supportive role in language?

In this regard, we have proposed a selective engagement mechanism that increases the efficiency of structures needed to process language in such a way that appropriate lexical choices are more reliably differentiated from semantically related but inaccurate alternatives. Several questions need to be answered about this selective engagement theory. We have noted that animal studies could be useful in determining the role of the CM, that is, whether it directly affects cortical processing or whether it indirectly affects the NR's gating of other thalamic nuclei thereby affecting cortical processing. Although selective engagement accounts for the type of anomia seen in the cases of Raymer et al. (1997), it is not clear how well it could account for cases of category-specific naming deficit. The responses of Raymer et al.'s cases of general naming deficit differed in some important ways from Crosson et al.'s (1997) case of category-specific naming deficit. Most notably, in the category-specific naming deficit, no frequency effect was seen and errors were generally nonresponses as opposed to semantic errors. These differences may be related to the underlying cognitive process that is interrupted. More careful analysis of category-specific deficits in thalamic lesion would be helpful in resolving such issues. Similarly, the role of thalamic gating and neglect dyslexia in reading deficits after thalamic lesions deserves further attention.

Another potential area for investigation involves the distinction between knowledge of procedures that can be applied without conscious attention and knowledge of facts and relationships that is consciously declared. We have suggested elsewhere (Crosson, 1992; Nadeau & Crosson, 1997) that grammar may fall into the category of procedural knowledge while lexical and semantic knowledge falls under the declarative rubric. In this regard, it is of interest that Raymer et al.'s (1997) cases of thalamic anomia showed no impairment of syntax, suggesting that the damaged thalamic mechanisms affected processes within the declarative realm. Further exploration regarding the dissociation of grammatical and lexical–semantic functions after thalamic lesion would be useful. For example, is knowledge of other well-learned procedures unaffected by thalamic lesion, and how does this compare to cases of basal ganglia dysfunction like PD where new procedural learning may be affected (see Crosson, 1992 for review)?

Although it is likely that analysis of patients with thalamic lesions will continue to be an important source of information concerning thalamic functions in language, several problems with this lesion methodology exist. One problem is the relative rarity of thalamic lesion cases. The lack of sizable groups of patients with thalamic lesions means that studies are almost always case studies, and a large group study is virtually impossible. This problem leaves us vulnerable to case reports where symptoms may reflect premorbid anomalous brain organization as well as the results of the lesion. A second problem is that lesions most often do not encompass only the structure of interest. For example, it would be difficult on the basis of lesion data to distinguish between the contribution of the ventral anterior nucleus from that of the ventral anterior nucleus reticularis because both structures are likely to be damaged in tuberothalamic artery infarctions. A third problem is that it is difficult to distinguish the contribution of the lesioned structure from the optimal compensatory operation of cognitive system of interest without the structure. For example, do semantic errors in thalamic aphasia reflect disruption of the semantic system or compensation by the semantic system for some other type of retrieval problem?

Thus, it would be optimal to have other methods to investigate the role of the thalamus in language. One potential alternative is functional neuroimaging. The potential resolution of functional magnetic resonance imaging (fMRI) is high enough to allow us to distinguish between regions of the thalamus. For example, in an fMRI study of working memory conducted at the Medical College of Wisconsin (Crosson et al., 1996), we found an area of increased activity in the anterior thalamus, a separate area of increased activity in the posterior thalamus, and an area of increased activity in the body of the caudate nucleus during a semantic working memory task. This task required subjects to detect membership of stimulus words in remembered categories. Increased activity of the anterior thalamus was also detected on a phonological working memory task that required subjects to detect if stimulus words rhymed with remembered words. In an orthographic working memory task, in which subjects attended to parts of words rather than whole words, no thalamic activity was seen. Nadeau and Crosson (1997) have suggested that the construct of selective engagement could be applied to lexical–semantic working memory: the transient selective engagement of one or more neural nets in a particular way instantiates the working memory of a particular semantic construct or lexical representation. The activity increases seen in the anterior thalamus by Crosson et al. (1996) during working memory tasks are at the extreme margin of the anterior thalamus and may include the portions of the nucleus reticularis implicated by Nadeau and Crosson (1997) in selective engagement. Positron emission tomography has also been used to study thalamic activity in cognitive tasks such as those requiring attention (for example, LaBerge & Buchsbaum, 1990). Thus, functional neuroimaging of normal subjects can overcome some of the problems involved in studying thalamic lesion cases, including rarity of thalamic lesion cases and inability to study normal operation of a cognitive system after damage to one of its components. Although there are many caveats about the use of functional neuroimaging in cognitive studies (Nadeau & Crosson, 1995), the techniques have the potential to contribute to our knowledge concerning the role of the thalamus in language.

30-3. CONCLUSIONS

Ultimately, the goal of neurolinguistics is to understand how the brain operates as a system to process and generate linguistic communications. One area of research in this field endeavors to determine the components of neural systems involved in language and the roles they play in language processing. It is generally true of all forms of cognition and behavior, including language, that we must understand the role of subcortical structures if we hope to have a complete understanding of how brain systems process cognitive information. In some instances, the answer at which we arrive may be that a particular subcortical structure is not involved in language. Based on our review, we believe this may be the case for the basal ganglia. Future research should focus on this question. In other cases, subcortical structures may play important supporting roles in cognitive processes, and in some instances perhaps even crucial roles. Our research and assessment of the literature indicate that the thalamus plays a significant role in language. We have suggested that this role includes involvement in selective engagement of cortical mechanisms necessary to carry out a cognitive task; however, further research will be needed to precisely define the role or roles of the thalamus in language.

CHAPTER 31

Language and Communication in Non-Alzheimer's Dementias

Monique M. Cherrier,[1] Mario F. Mendez,[2] Jeffrey L. Cummings,[3] and D. Frank Benson[4]

[1]Department of Psychiatry and Biobehavioral Sciences, University of California at Los Angeles School of Medicine, and West Los Angeles Veterans Affairs Medical Center; [2]West Los Angeles Veterans Affairs Medical Center and Department of Neurology, University of California at Los Angeles School of Medicine; [3]Department of Psychiatry and Biobehavioral Sciences, University of California at Los Angeles School of Medicine, and West Los Angeles Veterans Affairs Medical Center, and Department of Neurology, University of California at Los Angeles School of Medicine; [4]Department of Neurology, University of California at Los Angeles School of Medicine, Los Angeles, California

Disturbances in language and communication are prominent symptoms in dementia. Many dementing illnesses lead to aphasia, others primarily impair verbal fluency or result in motor speech disturbances such as dysarthria. This chapter discusses language and communication disturbances that occur in non-Alzheimer's dementias, including related abnormalities in cognition, aspects of neuropathology, and neuroimaging. The language deficits in vascular dementia (VaD), frontotemporal dementia (FTD) (e.g., Pick's disease), Huntington's disease (HD), progressive supranuclear palsy (PSP), Creutzfeldt-Jakob disease (CJD), and primary progessive aphasia (PPA) are included. Implications for future research are also noted.

This chapter reviews language and communication disorders due to cerebrobvascular disease (vascular dementia), degeneration of the frontal lobes and the anterior temporal lobes (frontotemporal dementias), and other neurodegenerative disorders such as Huntington's disease, progressive supranuclear palsy, and Creutzfeldt-Jakob disease. Most patients with the unique entity of primary progressive aphasia eventually progress to dementia, and this disorder is also discussed. Because these are dementing disorders, language deficits occur within the context of other cognitive deficits, and this relationship will be reviewed.

31-1. VASCULAR DEMENTIA

31-1.1. Clinical Characteristics and Etiology

Studies suggest that vascular disease accounts for between 9% and 35% of dementia patients among elderly subjects (Kase, Wolf, Backman, Linn, & Cupples, 1989). Vascular dementia (VaD) is more common among men than among women, and increases in prevalence with increasing age. Autopsy studies indicate that 25–33% of persons who die with dementia have vascular disease as a major or sole contributing cause (Chui et al., 1992).

The concept of vascular dementia has undergone many changes in the last few years. Previous studies of VaD have used the diagnostic and statistical manual of mental disorders, 3rd edition revised (*DSM-III-R,* American Psychiatric Association, 1987) or a rating on the Hachinski Ischemia Scale to classify patients (Hachinski, Lassen, & Marshall, 1974). Recent improvements in our understanding of the disease process have led to new guidelines for diagnosis (Roman et al., 1993). According to the new National Institute of Neurological Disorders and Stroke and Association Internationale pour la Recherche et l'Enseignement en Neurosciences (NINDS-AIREN) diagnostic criteria for VaD, three things must be included: presence of dementia, evidence of cerebrovascular disease, and a temporal relationship between these two disorders. The subtypes or cerebrovascular lesions associated with VaD include major vessel multi-infarct dementia, strategic single infarct dementia, small vessel ischemic disease with dementia, multiple lacunar strokes, hypoperfusion, hemorrhagic changes, and others. VaD often progresses in a stepwise fashion and depression, anxiety, and apathy are common (Cummings, 1988; Sultzer, Levin, Mahler, High, & Cummings, 1993). Sometimes focal neurologic changes are present such as paresis, gait abnormalities, incontinence, and bradykinesia. Gait problems and bradykinesia may be prominent even when the dementia is not severe (Thal, Grundman, & Klauber, 1988). Although the new classification system (Roman et al., 1993) is likely to improve future research findings by ensuring a more precise subject selection, most studies to date reflect old classification approaches (Loeb & Gandolfo, 1988) and may include patients with various disease states including VaD, Alzheimer's disease (AD), and mixed dementia.

31-1.2. Cognitive Changes

The most common cognitive impairments in VaD are bradyphrenia, decreased initiation and spontaneity, and executive function deficits such as difficulties with set shifting and verbal fluency (Wolfe, Linn, Babikian, Knoefel, & Albert, 1990). These changes occur irrespective of the type of VaD and may relate to the common subfrontal and periventricular ischemic lesions. Moreover, executive function abilities discriminate between VaD and AD (Almkvist, 1994; Kertesz & Clydesdale, 1994; Mendez & Mendez, 1991). Otherwise, the most common neuropsychological pattern is one of heterogeneity of performance depending on the location of the strokes (LaRue, 1992).

VaD with strokes in the left hemisphere temporal and parietal cortex often results in aphasia, alexia with agraphia, acalculia, or apraxia. VaD with occipital lobe strokes often results in visual field deficits or hemianopsias. Vascular damage to the hippocampi and adjacent structures frequently results in a profound anterograde amnesia.

31-1.3. Overall Speech and Language Changes

Language and speech findings reported in VaD, although sometimes contradictory, reflect both the focal neurological motor signs and the heterogeneity of lesion locations. Nevertheless, most studies have shown shared features such as reduced verbal fluency in VaD in both free speech and formal verbal fluency tasks (Hier, Hagenlocker, & Shindler, 1985; Mendez & Mendez, 1991; Powell, Cummings, Hill, & Benson, 1988). The decreased verbal fluency in VaD patients is likely due to bilateral subcortical lesions in the frontal regions (Ishii, Nishihara, & Imamura, 1986).

Other common language findings can be attributed to more focal neurological lesions. For example, Cummings and Benson (1992) noted that one of the most common occlusions occurs in the middle cerebral artery stem or branches and aphasias commonly follow when the left hemisphere is lesioned (Benson & Geschwind, 1975; Geschwind, 1975). Infarctions anterior to the rolandic fissure often produce nonfluent aphasias, and infarctions in the posterior temporal or parietal regions usually produce fluent aphasias. Specific lesions in the left angular gyrus region may produce alexia and the Gerstmann syndrome (acalculia, right-left disorientation, agraphia, and finger agnosia) (Cummings & Benson, 1992). In general, these patients often complain of word-finding problems and may have a semantic anomia with inability to recognize the meaning of words. Refer to Table 1 for a summary of reported deficit findings in VaD.

31-1.3.1. Verbal Communication

Powell et al. (1988) examined speech and language functions in VaD compared to AD patients with comparable dementia severity. They found that VaD patients exhibited more abnormalities of motor aspects of speech than the AD patients, including poor articulation and rate; VaD patients had shorter and less grammatically complex phrases but communicated more information. Other investigators have reported similar findings of reduced verbal output, shorter phrase lengths, and simplified syntax (e.g., fewer subordinate clauses, fewer words, and fewer unique words) (Hier et al., 1985). In addition, VaD patients may have perseverations and intrusions in their speech (Shindler, Caplan, & Hier, 1984).

31-1.3.2. Comprehension

In several studies, impaired comprehension was present in nearly all of the patients with VaD (Almkvist, 1994; Kertesz & Clydesdale, 1994; Kertesz, Polk, & Carr, 1990; Powell et al., 1988). Villardita (1993) found deficits on the Token Test and the reporter's test. Relative to AD patients, VaD patients perform better on Luria and Western Aphasia Battery tasks concerned with understanding complex grammatical structures (Kontiola, Laaksonen, Sulkava, & Erkinjuntti, 1990; Powell et al., 1988).

31-1.3.3. Confrontation Naming and Repetition

Confrontation naming deficits are often present (Kertesz & Clydesdale, 1994; Kontiola et al., 1990), although they are usually not as severe as those in AD (Bayles & Tomoeda, 1983; Powell et al., 1988; Villardita, 1993). The reports are conflicting with regard to greater repetition difficulty in VaD than AD (Kertesz & Clydesdale, 1994; Kontiola et al., 1990).

31-1.3.4. Reading and Writing

Subtle word recognition deficits may occur in VaD patients (Kontiola et al., 1990), and prominent alexia can occur with lesions in the left angular gyrus. Kertesz and Clydesdale (1994) found that VaD patients had more dysgraphia than AD patients. The VaD group had greater difficulty writing letters presented individually in dictation, had significantly poorer copying of words in sentences, and obtained little benefit from the opportunity to copy previously dictated material. These language and speech findings likely reflect focal neurological motor signs found in VaD patients along with heterogeneity of these signs resulting in contradictory results between studies.

31-1.4. Gaps in the Literature

All of the studies reviewed and summarized in Table 1 specifically examined language functions in VaD patients in comparison to AD patients. While this comparison can

TABLE 1
Vascular Dementia

Areas of impairment	Almkvist (1994)	Hier et al. (1985)	Kertesz & Clydesdale (1994)	Kertesz et al. (1990)	Kontiola et al. (1990)	Mendez & Mendez (1991)	Powell et al. (1988)	Villardita (1993)
Articulation & prosody							X	
Perseveration		X						
Verbal fluency	X	X				X	X	
Phonemic structure							X	
Syntactic structure		X			X			
Comprehension	X		X	X			X	X
Repetition					X			
Confrontation naming	X	X	X		X			X
Writing			X				X	
Reading					X			

be useful, what is lacking is an examination of language impairments in VaD in a longitudinal manner. Verbal fluency and confrontation naming may be less useful in discriminating patients with VaD and AD than an analysis of narrative discourse (LaRue, 1992). As noted earlier, studies to date reflect patient groups of various etiologies due to the use of different diagnostic classification criteria. In fact, investigators have often excluded patients with large lesions or lesions in Wernicke's or Broca's areas (Hier et al., 1985; Kertesz & Clydesdale, 1994). This type of exclusion, while useful for some types of research questions, obscures important information concerning the manifestations of various VaD subtypes. The examination of language in VaD can profit greatly from the increasing availability of sophisticated imaging technology along with the recent development of diagnostic guidelines.

31-2. FRONTOTEMPORAL DEMENTIAS

31-2.1. Clinical Characteristics and Etiology

The frontotemporal dementias (FTD) make up about 10% of all dementias. The FTDs are composed of frontal lobe degeneration without Pick bodies and with or without motor neuron dysfunction. Although much of the literature is devoted to Pick's disease, it has been shown that Pick's disease is usually not clinically distinguishable from the less specific frontal lobe degeneration. Pick's disease makes up about 2-3% of all dementias and 20-25% of FTDs (Gustafson, 1993; Mendez, Selwood, Mastri, & Frey, 1993).

FTD is a progressive dementia, characterized by early personality changes and frontal or frontotemporal atrophy on neuroimaging (Brun, 1993; Knopman et al., 1989; Lund & Manchester, 1994; Mendez et al., 1993; Miller et al., 1991). Neuropathological changes include prominent lobar atrophy of gray and white matter. Pick's disease is distinguished by the additional presence of argentophilic intranuclear inclusions known as Pick's bodies (Baldwin & Forstl, 1993). Males and females are equally affected. Symptoms include prominent apathy, decreased initiative, disinhibition, and impulsivity. Poor personal hygiene, socially inappropriate behaviors, and sexual disinhibition are common. Emotional lability frequently occurs with depression, mania, anger, and irritability (Mendez et al., 1993; Miller et al., 1991). If bilateral temporal lobe degeneration and amygdala damage are also present, FTD patients are predisposed to the Klüver-Bucy syndrome (Cummings & Duchen, 1981; Lilly, Cummings, Benson, & Frankl, 1983; Mendez et al., 1993). This syndrome includes hypermetamorphosis, hyperorality, hypersexuality, visual agnosia, and blunted emotional reactivity.

31-2.2. Cognitive Changes

The neuropsychological findings of FTD include prominent executive function disturbances, even early in the disease. There are deficits on planning and follow-through, set shifting, sequencing, and judgment. Memory is less impaired than executive functions although there are eventual retrieval deficits for verbal information and episodic information. Occipitoparietal functions are relatively preserved, and visuospatial functions remain intact for most patients (Hodges & Gurd, 1994; Johanson & Hagberg, 1989).

31-2.3. Overall Speech and Language Changes

Speech changes include a progressive reduction of speech (economy of utterances), stereotypy of speech (repetition of same words, phrases, themes), echolalia, and perseveration (Knopman et al., 1989; Mendez, Selwood, Mastri, & Frey, 1993; Wechsler, Verity, Rosenschein, Fried, Schrebel, 1982). At later stages of the disease, mutism is common (Cummings & Benson, 1992; Mendez et al., 1993). A comparison of Pick's and AD patients also found mutism along with preserved receptive language to be more common in Pick's disease (Gustafson, Brun, & Risberg, 1990). Holland, McBurney, Moossy, and Reinmuth (1985) reported a well-studied patient with autopsy-confirmed Pick's disease. The patient first evidenced slight errors in his speech, along with slowed pace and long pauses. He began to substitute lower-frequency words for higher-frequency words and evidenced word-finding difficulties despite intact confrontation naming and intact reading and writing skills. The patient commonly used generic terms such as "thing" and relied on particular words to convey several meanings. Writing samples also demonstrated word-finding difficulties, semantic errors, reduction of outputs, and perseverations. The patient continued to deteriorate to the point of complete mutism but still retained the ability to read and write despite impaired auditory comprehension. A summary of reported impairments in frontotemporal dementias is provided in Table 2.

Cummings and Duchen (1981) examined five cases of autopsy-confirmed Pick's disease for the presence of Klüver-Bucy syndrome. They found language disturbances

TABLE 2
Pick's Disease

Areas of impairment	Cummings & Duchen (1981)	Hodges & Gurd (1994)	Holland et al. (1985)	Knopman et al. (1989)	Mendez et al. (1993)	Wechsler et al. (1982)
Articulation & prosody					x	x
Perseveration	x			x	x	x
Semantic structure			x			
Phonemic structure			x			
Syntactic structure						
Comprehension	x		x			
Repetition						
Confrontation naming	x	x		x		
Writing			x			
Reading	x		x			

in all five cases. Three of the five had early word-finding difficulties in spontaneous speech and failed tests of confrontation naming. All five eventually had reduced vocabulary, excessive use of verbal stereotypes, echolalia, or complete loss of intelligible speech.

31-3. HUNTINGTON'S DISEASE

31-3.1. Clinical Characteristics and Etiology

Huntington's disease (HD) is a neuropsychiatric disorder distinguished by choreatheosis, caudate atrophy, and an autosomal dominant inheritance (Mendez, 1994). The gene responsible is on the terminal short arm of chromosome 4 and has nearly 100% penetrance, with no phenotypic differences between heterozygotes and homozygotes (Martin, 1984). Recently, the genetic defect has proven to be an abnormal, excessive repetition of the triplet nucleotide of cytosine and guanine. Equally common in men and women, HD affects 3 to 10 per 100,000 people with a mean age of onset around 41 years (Martin, 1984).

The diagnosis of HD is largely dependent on the recognition of choreoathetosis, a movement disorder characterized by jerky movements and slower writhing motions often leading to dystonic posturing. Clinicians may mistake these movements for nervousness, mannerisms, or intentional acts (Mendez, 1994). Eventually, choreoathetosis decreases as dystonia and akinesia-rigidity supervenes. Other diagnostic features include ocular motor changes and psychiatric symptoms such as depression, mania, personality disorders, suicides, and hyper- or hyposexuality in the later stages of the disease (Mendez, 1994). Computerized Tomography (CT) and Magnetic Resonance Imaging (MRI) demonstrate caudate atrophy in moderate or advanced HD.

31-3.2. Cognitive Changes

All HD patients eventually develop cognitive deficits. Changes in cognition are not strongly correlated with choreoathetosis; however, once choreoathetosis is present, HD patients have notable difficulties on timed tasks (e.g., performance measures on the WAIS) (Baxter et al., 1992; Josiassen, Curry, & Mancall, 1983). Although they usually have better explicit memory than AD patients, HD patients evidence procedural learning deficits. In addition, they usually develop executive functioning problems, including managing and maintaining a cognitive set, abstraction, judgment, reasoning, sequencing, and planning, along with mental flexibility (Mendez, 1994).

31-3.3. Overall Speech and Language Changes

Compared to other cognitive functions, language in HD is relatively preserved. HD patients do not develop aphasic syndromes and are better at object naming than AD patients (Brandt, Folstein, & Folstein, 1988; Fisher, Kennedy, Caine, & Shoulson, 1983). The language changes that occur are more common in the later stages of the disease and include dysarthria, decreased verbal fluency and initiation, decreased verbal output, and terminal mutism. Table 3 outlines reported deficit findings for HD patients.

31-3.3.1. Verbal Communication

Early in HD, there is a decrease in spontaneous speech with single-word or short-phrase responses, and frequent silent intervals or pauses. The loss of conversational initiative substantially contributes to the patient's communicative disability. For example, on a 4-minute speech sample, HD patients demonstrated a 25% decrease in the number of words compared to normal controls (Gordon & Illes, 1987). The decreased verbal output may gradually progress to mutism.

The impact of dysarthria has a great effect on articulation and prosody in HD. Podoll, Caspary, Lange, and Noth (1988) found that motor speech was impaired in 13% of early HD patients, 50% of middle-stage patients, and 87% of later-stage patients. Spontaneous speech difficulties are most notable for the middle- and late-stage patients and include articulation and prosody errors (Podoll et al., 1988). Semantic and phonemic structure of free speech is usually well preserved into the late stages, however, the dysarthria may become so severe as to render the patient unintelligible (Gordon & Illes, 1987; Podoll et al., 1988). An in-depth examination of HD patients using the Montreal-Stanford Neurolinguistic Protocol investigated both the temporal and the lexical forms of spontaneous speech (Illes, 1989). Early stage HD patients evidenced short, rapid language segments, which declined with increasing disease severity. The HD patients also manifested a reduction in syntactic complexity, which was the most distinctive difference in comparison to healthy controls. Controversy remains over the existence of primary language disturbances in HD independent of motor speech changes. In an extensive review of language changes in HD, Podoll et al. (1988) failed to find primary language changes in HD. They found that the only language changes were neurological or neuropsychologically based. Rosser and Hodges (1994) examined verbal fluency tasks (letter and category) in AD, HD, and progressive supranuclear palsy (PSP) patients. They suggest that the poor letter fluency of HD and PSP, as opposed to category fluency, results from initiation and retrieval problems secondary to disruption of frontostriatal circuits. Thus, on a task of verbal fluency, it appears that frontostriatal degeneration found in HD may particularly affect verbal search abilities.

HD patients produce significantly more self-corrections (whole-word repetitions), aborted phrases, and vocal interruptions, in comparison to a Parkinson's group. Illes (1989) suggested that these particular errors were representative of a self-corrective, compensatory strategy. As disease severity worsened, an increase in aborted phrases was found indicating failure of the self-corrective strategy. In general, perseverations and verbal stereotypes are uncommon and may occur only in HD patients with severe dementia. Podoll et al. (1988) found the semantic structure of spontaneous speech was intact with no evidence of semantic paraphasias or semantic jargon. Although some of the demented HD patients evidenced poverty of ideas, coherence of expression was preserved. They also found that syntactic structure was intact for all of the early stage patients, 71% of the middle-stage patient and 43% of the late stage patients.

31-3.3.2. Comprehension

Impaired comprehension in HD correlates highly with overall cognitive impairment. As with other language functions, comprehension is relatively intact in the early stages and is progressively impaired as the disease advances and cognitive impairments become pronounced. Podoll et al. (1988) found that 50% of the middle-stage patients and 74% of the late-stage patients scored below the normal range on a measure of verbal comprehension. Performance correlated highly with an overall cognitive impairment measure. Speedie, Brake, Folstein, Bowers, and Heilman (1990) found that HD patients were impaired in the ability to understand prosody. Two types of prosody were assessed—affective prosody, when the patient was asked whether the tone of a recorded voice was happy, angry, or sad, and propositional prosody, which determines whether an utterence was a question or not. HD patients were impaired for both types of prosody, which may further hamper communication ability.

31-3.3.3. Confrontation Naming and Repetition

Confrontation naming may show little or no decline (Bayles & Tomoeda, 1983; Butters, Sax, Montgomery, & Tarlow, 1978). Generally, early stage HD patients demonstrate intact ability whereas middle- and late-stage HD patients have naming errors (Podoll et al., 1988). Bayles and Tomoeda (1983) suggested that HD patients are impaired due to a central semantic deficit that occurs at the word-search stage of naming because many confrontation naming errors were semantically related to the item. However, Hodges, Salmon, and Butters (1991) examined a large group of AD, HD, and control patients and found that the naming deficits of HD patients could be attributed to visual or other perceptual deficits, whereas AD deficits were attributed to a breakdown in the organization of semantic knowledge. Repetition may be impaired due to the dysarthric motor speech disorder. Repetition of consonants is more likely to be impaired than is the pronunciation of vowels (Podoll et al., 1988). Errors in repetition generally include the omission of words or the substitution of words or parts of words.

31-3.3.4. Reading and Writing

Podoll et al. (1988) found that 14% of middle-stage and 65% of severe-stage HD patients were compromised in their ability to read aloud due to dysarthria. Errors observed included hesitations, self-corrections, omissions, substitutions, and additions of letters and words. However, as with naming errors, reading difficulties may be due to a disturbance in visual registration or visual dyslexia (Marshall & Newcombe, 1973). HD writing is slow, effortful, and characterized by sudden stops and strokes,

TABLE 3
Huntington's Disease

Areas of impairment	Podoll et al. (1988) early	Podoll et al. (1988) middle late	Speedie et al. (1990)	Hodges et al. (1991)	Illes (1989)	Rosser & Hodges (1994)
Articulation & prosody	x	x				
Perseveration		x			x	
Semantic structure						
Phonemic structure	x	x				
Syntactic structure	x	x			x	
Comprehension	x	x	x			x
Repetition	x	x				
Confrontation naming	x	x		x		
Writing	x	x				
Reading	x	x				

omissions, and perseveration. Some patients with severe chorea show an enlargement of script described as "choreatic macrographia" (Hochheimer, 1936).

31-3.4. Gaps in the Literature

Future investigations may clarify the question of whether language disturbances in HD are primary or secondary to motor and cognitive changes of the disease. Longitudinal studies of language changes in HD would be particularly useful in defining the nature of progression of these changes.

31-4. PROGRESSIVE SUPRANUCLEAR PALSY

31-4.1. Clinical Characteristics and Etiology

Progressive Supranuclear Palsy (PSP) is a neurodegenerative disorder resembling Parkinson's disease (Steele, Richardson, & Olszewski, 1964). PSP is characterized by extrapyramidal rigidity, unsteady gait, supranuclear paralysis of eye movements, and prominent dementia. Age of onset is generally around age 55 (range, 45 to 75 years) and the clinical picture is often fully developed within 2 to 3 years. In addition to rigidity and ophthalomoplegia, pseudobulbar palsy with emotional lability and axial dystonia in extension also occur (Agid et al., 1986). Due to the similarities and overlap with Parkinson's disease, the actual prevalence of this disease is difficult to determine. A rough estimate is that 18% of patients initially diagnosed with PD are found to have neuropathological signs of PSP at autopsy (Agid et al., 1986). While the initial presenting picture of PSP may mimic Parkinson's disease, these patients do not respond to anti-Parkinson's medications. Neuropathological changes include neuronal cell loss, gliosis, and neurofibrillary tangles in the brain stem, basal ganglia, and cerebellar nuclei (Steele et al., 1964).

31-4.2. Cognitive Changes

Early cognitive impairments are major aspects of PSP (Kimura, Barnett, & Burkhart, 1981; Pillon et al., 1995; Steele et al., 1964). Albert, Feldman, and Willis (1974) originally described the cognitive impairments of PSP as characteristic of a "subcortical dementia." Rosser and Hodges (1994b) also found that in comparison to AD and HD, PSP patients were less impaired on memory and attention measures and more impaired on verbal fluency and abstraction. Maher, Smith, and Lees (1985) and Pillon et al. (1995) found that PSP patients have difficulty performing executive function tasks such as verbal fluency. Cambier, Masson, Viader, Limodier, and Strube (1985) found frontal executive and related deficits that included mental slowing, impaired attention, reduced verbal fluency, poor abstract thinking and reasoning, dynamic apraxia, motor impersistence, and imitation and utilization behaviors. Although neuropathological findings of PSP do not indicate lesions in the frontal cortex, PET studies have demonstrated a significant reduction in cerebral glucose utilization in the frontal lobes (D'Antona et al., 1985), likely due to decreased activation from subcortical neuronal afferents (Agid et al., 1986).

31-4.3. Overall Speech and Language Changes

Dysarthria is one of the most striking features of PSP. It is among the earliest symptoms and may progress to complete anarthria (Steele, 1972). Other changes include decreased verbal fluency, decreased comprehension of complex commands, and disturbed reading and writing.

31-4.3.1. Verbal Communication

In PSP, communicative disability results primarily from the general mental slowing and dysarthria (Podoll, Schwarz, & Noth, 1991). PSP patients may be loquacious and rambling in their speech in the early stages (David, Mackey, & Smith, 1968; Messert & Nuis, 1966); however, much more characteristic of PSP is a slow rate, a low volume, and a flat prosody. The latencies between questions and the patients' responses may be inordinately long (Albert et al., 1974; David et al., 1968). Palilalia and other reiterative speech changes may occur (Rafal & Grimm, 1981). The dysarthria of PSP may be so severe that speech utterances are only poorly intelligible, and in the later stages of the disease, most patients are mute (Albert et al., 1974; David et al., 1968; Janati & Appel, 1984). Occasionally, patients have been able to communicate by writing or blinking or movements of the extremities when they have been rendered completely mute (David et al., 1968). Syntactic structure, semantic understanding, and lexical access appear intact.

31-4.3.2. Comprehension

Podoll et al. (1991) found that PSP patients had impaired comprehension for complex visual or verbal information. Their performance on these tasks correlated significantly with impairment on the Comprehension subtest of the WAIS. Simple, one-step comprehension may be preserved. Rosser & Hodges (1994b) found that only 20% of PSP patients demonstrated impaired performance in simple comprehension items on a dementia rating scale.

31-4.3.3. Confrontation Naming and Repetition

Several but not all studies report confrontation naming and repetition to be preserved in PSP patients (Alpert, Rosen, Welkowitz, & Lieberman, 1990; Maher et al., 1985; Podoll et al., 1991). In one study, naming errors were visually related to the item, suggesting that the naming deficit results from an impairment at the encoding stage of naming where the processes of visual perception are involved (Kimura et al., 1981; Maher et al., 1985; Rafal & Grimm, 1981).

31-4.3.4. Reading and Writing

In PSP, reading aloud is slow with pronunciation errors due to dysarthria. Some patients read dysfluently, deciphering words in a hesitant, disconnected manner (David et al., 1968; Dix, Harrison, & Lewis, 1971; Podoll et al., 1991). PSP patients may also have difficulty reading due to a disturbance in visual registration during reading, possibly related to the occulomotor disorder (Dix et al., 1971). Writing is also slow due to the loss of fine motor movement. Abnormal slanting and micrographia are evident. Dysgraphic errors included omissions or perseverations of letters and words. Copying words is compromised by difficulties in shifting gaze from model to copy and the paralysis of downward gaze makes writing particularly difficult.

31-4.4. Gaps in the Literature

In general, the literature regarding language and cognitive changes for PSP is sparse. The diagnosis of PSP is difficult to make early in the disease and a confident diagnosis may not emerge until the patient is no longer able to withstand testing. Postmortem confirmation and retrospective studies will help shed light on the early signs of language deficits in PSP.

31-5. CREUTZFELDT–JAKOB DISEASE

31-5.1. Clinical Characteristics and Etiology

Creutzfeldt-Jakob disease (CJD) is a rare and rapidly progressive dementia caused by an infectious agent (Prusiner, 1987). Death usually occurs 5-6 months after onset, although rare cases last years (Mendez, 1994). In the early stages of CJD, the diagnosis may be difficult due to the variation and complexity of the presenting features (Drobny, Krajnak, Svalekova, & Pithova, 1991; Van Rossum, 1972). In addition to cognitive decline, commonly occurring initial signs include fatigue, emotional lability, mood disorders, and even psychosis. CJD is also characterized by early abnormalities in the pyramidal, extrapyramidal, cerebellar, oculomotor, or other neurologial systems. In the middle to late stages, muscular rigidity, spasticity, loss of consciousness, and advanced pyramidal or extrapyramidal motor dysfunction occur. The presence of startle myoclonus and periodic complexes on EEG are the most helpful for the clinical diagnosis of CJD (Drobny et al., 1991).

31-5.2. Cognitive Changes

Memory and cognition decline early in the disease in a pattern suggestive of frontal-subcortical involvement (Drobny et al., 1991). There is a decline in sustained and divided attention and concentration. Confusion and disorientation are common along with bradyphrenia and confabulation (Van Rossum, 1972). In a review of 33 CJD patients, a pattern of deficits further suggestive of frontal systems dysfunction included impaired digit span and poor interpretation of similarities (Maher et al., 1985).

31-5.3. Overall Speech and Language Changes

Dysarthria is the most common speech symptom and may be of bulbar, pseudobulbar, cerebellar, or extrapyramidal origin (Van Rossum, 1972). The combination of rigidity affecting the speech muscles along with apathy and loss of initiative often results in mutism. Echolalia also occurs possibly as a consequence of frontal-subcortical circuit dysfunction. In a retrospective study of 12 CJD patients, five developed language difficulties indicating the variability of language changes (Drobny et al., 1991).

Investigators have reported an absence of primary language impairment among patients with CJD (Maher et al., 1985); however, McPherson et al. (1994) documented a case of CJD who presented primarily with mixed transcortical aphasia including nonfluent spontaneous speech and impaired comprehension. Moreover, there was relative preservation of repetition and grammatical rules. In addition, Mandell, Alexander, and Carpenter (1989) reported a case of CJD presenting as a primary progressive (fluent) aphasia. Subsequent clinicopathologic studies of primary progressive aphasia indicate that CJD is a potential cause of an early, isolated, and insidious aphasia.

31-6. PRIMARY PROGRESSIVE APHASIA

31-6.1. Clinical Characteristics and Etiology

In primary progressive aphasia (PPA), a language disorder develops in relative isolation to other cognitive changes until later stages. Mesulam (1982) originally described six patients with primary progressive aphasia that did not evidence intellectual declines until very late stages of the disease. In fact, many of his patients continued to work at part-time jobs or remained active in civic duties and managed their personal finances. Reasoning and judgment abilities remained intact for many of these patients despite language deterioration to the point of muteness. Some patients, however, had early evidence of apraxia or acalculia (Poeck & Luzzatti, 1988).

PPA manifests as either a dysfluency with decreased verbal fluency or a semantic anomia. Patients with primary progressive aphasia are generally able to convey information effectively, if not by speech, then by written messages, gestures, and even sign language. Because this condition is rare, incidence and prevalence rates are difficult to assess; however, on neuropathology many if not all of these patients prove to have frontotemporal dementia and some have Alzheimer's disease or CJD or some other dementing illness. Neuroimaging often discloses focal degeneration of the left perisylvian region.

31-6.2. Overall Speech and Language Changes

In the dysfluent subtype of PPA, spontaneous speech is slow, hesitant, and sometimes agrammatic (Karbe, Kertesz, & Polk, 1993). Patients typically complain of difficulties in expressing themselves and finding the right word. As a result, lengthy descriptions or responses are difficult and generally avoided. In the advanced stages of aphasia, speech is characterized by logopenia, long word-finding pauses, circumlocution, rare paraphasias, and dysarthria (Mesulam, 1982; Scholten, Kneebone, Denson, Fields, & Blumbergs, 1995). Phonemic paraphasias and verbal stereotypes may be present (Karbe et al., 1993). In the semantic anomia subtype of PPA, naming deficits are the most prominent symptom along with word-finding difficulties in spontaneous speech (Karbe et al., 1993; Scholten et al., 1995). These patients have difficulty in comprehending the meaning of words. In general, PPA is of specific interest because of the relatively isolated nature of language difficulty, at least early in the disorder. PPA most probably represents an atypical presentation of AD, CJD, VaD, and Pick's disease, and later speech and language changes vary depending on the nature of the underlying disease.

31-7. SUMMARY AND CONCLUSIONS

In summary, disturbances in language and communication are often prominent symptoms in dementing illnesses. Many dementing illnesses lead to aphasia, others primarily impair verbal fluency or result in motor speech disturbances such as dysarthria. The language deficits in vascular dementia (VaD) usually include poor verbal fluency and decreased confrontation naming. These deficits generally reflect the disturbances of ischemic insults in VaD. The frontotemporal dementias (FTD) such as Pick's disease produce a progressive reduction of speech, stereotypy of speech, echolalia, and perseveration. Among patients with Huntington's disease (HD) and progressive supranuclear palsy (PSP), motor impairments impact speech, resulting in dysarthria and sometimes complete communicative disability (anarthria). Similarly, Creutzfeldt-Jakob disease (CJD) produces dysarthria of bulbar, pseudobulbar, cerebellar, or extrapyramidal type along with aphasia in the middle to late stages. Patients with primary progressive aphasia (PPA) demonstrate a wide variety of aphasic symptoms, with relative sparing of other cognitive abilities. Although substantial information is available regarding language changes in dementia, controversies remain regarding the extent and nature of language deficits, and further research is needed with well-characterized patient groups, standardized testing procedures, and longitudinal follow-up measures to more fully understand the complex communicative manifestations of dementing illnesses.

CHAPTER 32

Language and Communication Disorders in Dementia of the Alzheimer Type

Paulo Caramelli,[1] Leticia Lessa Mansur,[2] and Ricardo Nitrini[3]
[1]Cognitive and Behavioral Neurology Unit, Neurology Division, Hospital das Clínicas of the University of São Paulo School of Medicine, São Paulo, Brazil; [2]Department of Internal Medicine (Speech Pathology), University of São Paulo School of Medicine, São Paulo, Brazil; [3]Cognitive and Behavioral Neurology Unit, Neurology Division, Hospital das Clínicas of the University of São Paulo School of Medicine, São Paulo, Brazil

Language disturbances may appear early in the course of dementia of the Alzheimer type (DAT) and become frequent as the disease progresses. Moreover, the presence of language impairment seems to be associated with a faster cognitive and functional decline. The profile of the linguistic changes depends on the stage of the disease, with anomia and impairment in verbal fluency tests being early and prominent features. In the initial and intermediate phases of the disease there is a marked lexical and discourse deficit, characterizing a semantic impairment, with preservation of syntactic and phonological abilities. This semantic deficit is probably secondary to pathological involvement of the temporal neocortex, an anatomical region related to lexical retrieval. In later stages, all language components become involved.

Dementia of the Alzheimer type (DAT) is the leading cause of dementia in most countries, accounting for more than half of all dementia cases (Canadian Study of

Paulo Caramelli held a scholarship from CAPES (Coordenação de Aprimoramento de Pessoal de Nível Superior), Brazil, from October 1994 to March 1996.

Health and Aging Working Group, 1994; Nitrini et al., 1995). Its prevalence ranges from 1% among individuals 65 to 74 years old to 26% at age 85 and older (Canadian Study of Health and Aging Working Group, 1994). Clinically, DAT is characterized by progressive memory loss associated with decline in other cognitive domains, such as attention, language, visuospatial, and constructional abilities, as well as behavioral changes. Memory disorders are among the first symptoms of the disease.

Language disturbances may occur as early symptoms in DAT and become more frequent as the disease progresses. A wide scope of research has been developed, especially in the last decade, aiming at determining not only the frequency of language deficits in DAT, but also characterizing and interpreting the clinical picture of these linguistic changes.

The goal of this chapter is to present and discuss the current data on language and communication disorders in DAT. The profile of the linguistic changes and the evolution of these deficits over the course of the disease are presented first, followed by a brief discussion about the specificity of the linguistic changes. The relationships between language impairment and subgroups of DAT patients are then outlined. Finally, the neuropathological correlates of language impairment are shown and some current research trends in this field are pointed out. Emphasis will be placed on recent studies and ideas, and also on classic works that constitute the basis of contemporary research.

32-1. CHARACTERIZATION OF THE LINGUISTIC CHANGES IN DAT

Most of the studies focusing on language disturbances in DAT have compared demented patients with brain-damaged subjects (especially stroke and head trauma) and elderly controls. The language profile of DAT presented in the literature has been mainly traced from the instruments usually employed in aphasia diagnosis, such as the Western Aphasia Battery and the Boston Diagnostic Aphasia Examination (Cummings, Benson, Hill, & Read, 1985; Cummings, Darkins, Mendez, Hill, & Benson, 1988; Kertesz, Appell, & Fisman, 1986). Tests usually include examination of verbal fluency, naming, auditory and reading comprehension, repetition, reading aloud, and discourse abilities. Another method of investigation is based on assessment of performance in communication situations. The seminal work of Irigaray (1973), using language batteries and conversation, drew attention to some special characteristics of language in dementia and set the basis for many of the later studies in this area.

The frequency of language disturbances in DAT seems to depend on the severity of dementia, ranging from 36% to 100%, respectively, in mild and severe cases (Faber-Langendoen 1988). The linguistic changes found in DAT, in language comprehension and production, are described in the following sections.

32-1.1. Language Comprehension

32-1.1.1. Auditory Skills

Impairment in auditory comprehension in DAT is progressively greater the higher the complexity of information. Hence, patients display preserved comprehension of simple commands and questions, with trouble in understanding complex sentences that involve inferences, comparisons, or causal relationships (Appel, Kertesz, & Fisman, 1982; Cummings et al., 1985).

Hart (1988) evaluated the performance of DAT patients in the Token Test and found a correlation between the results in this test and those obtained by the same patients in the Wechsler Memory Scale, suggesting that the memory deficits may interfere in auditory comprehension tests. However, the patients showed difficulties in the final part of the test, which has similar requirements of short-term verbal memory to the previous sections, but is of greater linguistic complexity.

Additional factors must also be considered in interpreting these data, such as the extension of the material to be understood (number of propositions) and the role of the postinterpretive stage of sentence comprehension. Postinterpretation refers to the organization of verbal material, which entails the integration with nonlinguistic capacities, such as visual and memory abilities. Waters, Caplan, and Rochon (1995) have studied 14 mild to moderate DAT patients, showing that the postinterpretive stage is affected in the disease.

32-1.1.2. Reading Skills

Reading comprehension is affected early in the course of DAT and is considered to be more seriously affected than auditory comprehension (Hart, 1988). Faber-Langendoen et al. (1988) report that the deficits in reading comprehension seem to be more sensitive indicators of the language disturbance in DAT than anomia. Bayles, Tomoeda, and Trosset (1992), however, examining patients at different stages of the disease, observed that reading comprehension is relatively stable in mild and moderate cases. However, these studies have analyzed only the comprehension of single written words, which clearly limits the interpretation of these data.

Reading aloud is relatively preserved in DAT (Cummings et al., 1985), although in many cases patients are able to read but without understanding (Hart, 1988). The published data indicate that reading comprehension is affected more than auditory comprehension and reading aloud in DAT.

32-1.2. Language Production

32-1.2.1. Oral Skills

Oral production is the linguistic aspect of DAT that has received the greatest attention in recent studies. For this reason, this section will be divided by different linguistic levels: phonology, syntax, and semantics, the latter being addressed in lexical tasks and discourse.

32-1.2.1.1. Phonological and Syntactic Abilities

In virtually all studies on language disturbances in DAT, phonological abilities have been found to be unimpaired, a feature that allows these patients to be classified as fluent (Bayles, 1994). This characteristic has favored the development of increasing research on the semantic and syntactic aspects of language in DAT.

Syntactic abilities are relatively well preserved, at least in early and intermediate stages (Bayles, 1982; Kempler, Curtiss, & Jackson, 1987). Bayles (1982) reports that DAT patients are able to repeat totally incoherent sentences, and to correct sentences containing syntactic errors. However, it has been argued that syntactic competence is limited to routine contexts and that deficits may emerge in situations demanding the use of propositional syntax, that is, requiring a higher level of voluntary generation (Patel & Satz, 1994). Furthermore, the employment of correct syntax could be merely a reflection of a stereotypical and automatic procedure (Cardebat, Démonet, Puel, Nespoulous, & Rascol, 1991).

The relative preservation of phonological and syntactic abilities stands in sharp opposition to the semantic impairment, one of the most characteristic language traits in DAT.

32-1.2.1.2. Semantic Abilities

Semantic abilities are markedly impaired in DAT, contrasting with preserved phonological/syntactic capacities. This dissociation was highlighted by the work of Irigaray (1973). Many later studies confirmed that the semantic disturbances constitute a main feature of the linguistic changes in DAT (Bayles, 1982; Bayles & Tomoeda, 1983; Huff, Corkin, & Growdon, 1986). In these studies, semantic abilities are basically assessed through naming and verbal fluency tasks, as well as in discourse.

32-1.2.1.3. Naming Tasks

Lexico-semantic disturbances are clearly evident in confrontation naming tests. Anomia constitutes one of the most common and earliest linguistic changes in DAT (Bayles et al., 1992). In mild stages, confrontation naming errors consist mainly in production of superordinate labels and semantic-related attributes (Bayles, Tomoeda, & Trosset, 1990; Hodges, Salmon, & Butters, 1991). Circumlocutions also occur. These features indicate a loss of detailed knowledge, that is, the attributes of a specific item within a category, which is a major putative cause for anomia in DAT, and contrasts with the preservation of superordinate knowledge of semantic categorization (Chertkow & Bub, 1990; Martin & Fedio, 1983). As dementia progresses, semantic-related errors become more frequent and circumlocutions decrease, suggesting a reduction in the specificity of the semantic information available (Hodges et al., 1991). These deficits have a clear impact on discourse (Glosser & Deser, 1990).

32-1.2.1.4. Verbal Fluency Tasks

Verbal fluency, assessed through the production in a limited time of a list of words belonging to the same semantic category, is usually reduced in DAT. Some investigations suggest that this is a useful screening test for the diagnosis (Monsch et al., 1992; Nitrini et al., 1994), as well as for the prediction, of dementia (Masur, Sliwinski, Lipton, Blau, & Crystal, 1994).

Another interesting feature disclosed by some studies is that patients present a weaker performance in tests of verbal fluency with semantic criteria when compared to letter criteria, that is, words initiated by the same letter (Monsch et al., 1992; Pasquier, Lebert, Grymonprez, & Petit, 1995). This finding suggests that the lexico-semantic impairment may, in part, explain the reduction in verbal fluency. However, it must be considered that these verbal fluency tasks are time-constrained and thus could be evaluating not only fluency (and semantics) but also other aspects of cognition (Faber-Langendoen et al., 1988).

32-1.2.1.5. Nature of the Lexico-Semantic Deficit

Two possibilities must be considered regarding the nature of the lexico-semantic deficit: an access deficit to the semantic system or a disintegration of the system itself, with loss of previously established concepts.

Several studies addressed this question. Huff et al. (1986) selected a group of DAT patients without evidence of visuoperceptual deficits and submitted them to a naming test. Subsequently, they compared the anomia of these patients with the results obtained in a comprehension test including the same items used in the naming task. The authors observed a close correlation between the results in the two tasks, as the patients presented difficulties in comprehending the same items that they were unable to name.

Additional data came from the study of Chertkow and Bub (1990). These authors selected 10 patients with DAT, without any evidence of visuoperceptual deficits, and compared them to 10 matched elderly controls in different tasks assessing semantic memory: picture naming, word-to-picture matching (using the same words of the picture-naming task), verbal fluency, semantic cueing, superordinate knowledge, and semantic priming). The patients showed consistency of errors between the picture-naming and the matching tasks, loss of semantic cueing, and preservation of superordinate knowledge with marked deficit of detailed knowledge.

Therefore, both studies suggest that there is a destruction of the semantic system in DAT. Further investigations also support this notion (Henderson, Mack, Freed, Kempler, & Andersen, 1990; Hodges, Salmon, & Butters, 1992). However, since most of these studies have included mild and moderate cases, the occurrence of an access deficit in very mild cases (i.e., in the early *pathological* stages of the disease) is still not ruled out.

32-1.2.1.6. Discourse

The analysis of communication possibilities in DAT has been based on models that take into account linguistic and extralinguistic aspects. There are only a few studies addressing the semantic–pragmatic aspects related to nonlinguistic contexts. Most of these investigations have been conducted through assessment of narrative and conversation, comparing speech samples from DAT subjects to Wernicke's aphasics and elderly controls.

The studies of the pragmatic–conceptual aspects have reinforced the notion that semantic disintegration is at the root of the linguistic changes in DAT. These pragmatic–conceptual aspects were studied in relation to language formulation and to referential coherence and relevance, which support the global organization of the oral text.

Blanken, Dittmann, Haas, and Wallesch (1987), and Glosser and Deser (1990) found that DAT patients have preserved language formulation and thus are able to deal with isolated propositions, but have difficulty in organizing them as a whole. Further, Glosser and Deser verified that patients have difficulty in employing the markers of referential coherence. Blanken et al. suggest that the deficits lie in the conceptual structure of the intended speech act.

Fromm and Holland (1989) further investigated this aspect of communication. They analyzed speech acts produced by mild and moderate DAT subjects, Wernicke's aphasics, and elderly depressed in a role-playing situation. The comparison of DAT patients with depressed subjects allowed these authors to exclude the influence of motivational aspects on the impairment of speech acts production. Performance of both subgroups of DAT patients was impaired in the more complex cognitive processing (divergencies, utilization of context, reading, calculating, and sequential relations), while overlearned communicative acts, such as social conventions, remained relatively preserved.

More recently, Marcie, Roudier, and Boller (1994), and Mentis, Briggs-Whittaker, and Gramigna (1995) evaluated oral texts produced by DAT patients and detected irrelevant and incomplete units, as well as units unrelated to the topic under discussion, leading to a rupture in the development of the topic. Moreover, Mentis et al. (1995) also analyzed the ability to introduce and change topics and observed a difficulty in introducing and shifting topics in an active and coherent way during conversation, which increased the responsibility of the conversational partner.

Although these studies have examined orally produced texts with different methods, they generally suggest that in DAT there is a loss of competence in managing the totality of semantic units, and a difficulty in integrating cognitive and linguistic components during development of the discourse.

This inability to deal with the discourse as a whole may be considered secondary to memory troubles, an aspect addressed by the work of Ska and Guénard (1993). They analyzed the basic semantic structure of narrative production (narrative schema) in early stage DAT patients in comparison to controls. The two groups were submitted to three distinct tasks: the first without image support (to tell a well-known children's story), and the remaining two involving presentation of images (single or sequenced pictures). The patients produced fewer schema components than the controls in all three situations. They made more errors in sequencing the reported events and produced a greater number of irrelevant propositions. Even in visually supported tasks, the performance of patients was affected, thus excluding episodic memory disturbance as the primary cause for the narrative production impairment in the disease. Further studies, however, are still necessary to clarify the relationships between language and cognition in DAT.

32-1.2.2. Writing Skills

Writing skills have not been studied as much as oral production in DAT, and to date most of the studies are mainly descriptive. Horner, Heyman, Dawson, and Rogers (1988) submitted 20 DAT patients to a test of describing an image in writing. They observed poor organization, perseverations, intrusions, verbal paragraphias, grammatical and orthographic errors, and also spatial agraphia associated with difficulty in letter formation. Although this study evaluates writing skills in a textual situation, the authors' analysis pertains only to the word and sentence levels and does not discuss the global structure of the text.

Henderson, Buckwalter, Sobel, Freed, and Diz (1992) evaluated narrative writing material from 33 DAT patients and compared them to 41 matched-controls, and found that the patients had lower writing scores, wrote fewer words, mentioned fewer categories of information, and made significantly more writing errors. The poor performance in the narrative writing task correlated with severity of dementia.

In another study, Croisile, Adeleine, Carmoi, Aimard, and Trillet (1995) examined 42 DAT patients and 30 controls by employing a protocol based on writing regular and irregular words, as well as neologisms, from dictation. The patients performed worse than the control group in the dictation task for all three word categories, with more difficulty in writing irregular words, a finding similar to the one reported by Rapcsak, Arthur, Bliklen, and Rubens (1989). The spelling errors correlated with overall language deficits and also with the severity of dementia. Furthermore, the authors observed a word-frequency effect in the writing task. These features suggest that a lexical deficit occurs at the orthographic level in DAT.

Hence, writing disturbances are frequent in DAT, being present even in early phases of the disease and correlating with severity of dementia. These data assign a clear diagnostic potential for writing evaluation in DAT patients. A study by Snowdon et al., (1996) added another dimension to the importance of semantic and syntactic writing skills. These authors analyzed autobiographic texts written by 93 Catholic nuns when they were young. They found that measures of low idea density and low grammatical complexity were strongly related to a diagnosis of DAT in late life. Although this study may be somewhat controversial in its subjacent theory of DAT, it suggests that reduced writing skills in early life may have a potential predictive value for the development of dementia in old age.

32-2. EVOLUTION OF THE LINGUISTIC CHANGES IN DAT

According to several works, a relatively typical language course occurs in DAT, which is generally characterized by three phases (Cardebat et al., 1991; Bayles, 1994).

In the early stages of the disease, communication is already impaired: the patient may have difficulty in initiating conversation and in understanding humor, sarcasm, verbal analogies, and indirect propositions. Anomia is the most striking trait, associated with circumlocutions and rare verbal paraphasias. Furthermore, oral expression shows a qualitative output reduction, with omission of items and the use of somewhat generic and imprecise terms at the lexical level, as well as in sentence generation. The phonetic, phonemic, and syntactic aspects of language are preserved. Reading comprehension and writing show signs of impairment. Auditory comprehension is relatively unaffected, except for new information and situations demanding abstraction. Repetition and reading aloud are preserved. Some authors identify this clinical picture with that of anomic aphasia.

In a more advanced stage, semantic abilities are severely impaired, with paucity of ideas leading to frequent repetitions. Phonetic and phonological stability is still evident. However, at the syntactic level, speech production is fragmented. Automatisms are used as compensatory strategies. Anomia becomes even more frequent. Verbal and semantic paraphasias, occasionally associated with neologisms, emerge. There is considerable worsening of writing as well as of reading and auditory comprehension, especially for complex segments. Repetition, and even reading aloud (although to a lesser degree), are usually preserved. Therefore, the clinical profile is sometimes compared to transcortical sensory aphasia (Cummings et al., 1985), although patients present less paraphasias and echolalias than the aphasics with cerebrovascular etiology and with worse performance than the latter in overlearned tasks like recitation.

In the final stage of the disease, all linguistic functions are impaired, with marked reduction in oral expression and severe problems of comprehension. Only some automatisms and verbal perseverations may persist. In a few patients repetition is still possible, leading to echolalia. This feature has been described as resembling global aphasia.

32-3. SPECIFICITY OF THE LINGUISTIC CHANGES

The diagnosis of DAT is essentially based on widely accepted clinical criteria, but without objective markers. An important issue is to discuss how far the analysis of language could contribute to the diagnosis, that is, if there are specific linguistic markers of DAT. In this sense, a clear limitation is that DAT manifests itself at an age of natural neurobiological and cognitive decline and of increased susceptibility to the development of language impairment secondary to other neurological diseases. Hence, overlapping diseases, or even aging, could be affecting the performance.

The analysis of symptoms detected in language evaluations of elderly individuals, aphasics, right brain-damaged and demented patients reveals marked similarities. All linguistic symptoms may be common to these populations, namely, anomia, difficulties in textual and complex syntactic comprehension, reduction of informational content, increased discourse tangentiality, and referential errors (Chapman & Ulatowska, 1994).

It is clear, therefore, that the search for the nature of these deficits is of central importance for the differential diagnosis. In the case of DAT, it implies a need to identify the relationships between different language components and between language and other aspects of cognition. In this sense, discourse, in all modalities, is the ideal situation for this analysis, since its formulation involves comprehension and arrangement of the information in a meaningful way (Chapman & Ulatowska, 1994). However, to date there have been very few attempts to make use of discourse material in order to identify specific linguistic and cognitive markers of DAT.

32-4. INFLUENCE OF NON-LINGUISTIC DEFICITS ON LANGUAGE IMPAIRMENT

Different clinical factors during the course of DAT may interact with language, making it difficult to interpret the linguistic data. Tasks that imply extensive auditory input, for example, may be affected by the global cognitive decline, especially by memory. In this sense, visual support can rule out the interference of memory functions; however, it introduces a new bias in the case of the occurrence of visual agnosias.

32-4.1. Visuoperceptual Problems

As already discussed, several works supporting the involvement of the semantic aspects of language in DAT were based on the poor performance of patients in confrontation naming of objects and figures. Thus, visuoperceptual problems could have an effect on these tasks. Appel et al. (1982) have shown that DAT patients improve their performance on naming tasks through tactile stimulation. In another study, Shuttleworth and Huber (1988) observed that performance in naming real objects was better than in confrontation naming of line drawings. Other studies, however, do not corroborate these data (Bayles, 1982; Bayles & Tomoeda, 1983).

Differences in the profiles of cognitive impairment may explain the discrepancies found in these studies. Neuropsychological manifestations of DAT are heterogeneous. Major subgroups may be identified according to the prominence of some specific cognitive deficits, such as visuoperceptual impairment (Neary et al., 1986). It is clear that in these patients the visual deficit might influence their performance in language tests, especially in confrontation naming.

32-4.2. Memory Problems

Language and verbal communication depend very much on knowledge of the world, on linguistic knowledge, and on expertise in manipulating the rules of linguistic usage, which pertain to the domain of semantic memory. On the other hand, they also depend on short-term verbal memory capacities. Short-term and long-term (semantic and episodic) memory are significantly impaired in DAT. Recent works have investigated the degree of interdependence of short-term memory and language in DAT, although there are no definite conclusions (Patel & Satz, 1994; Waters et al., 1995).

Episodic memory relates to autobiographical experience and thus represents an obvious prerequisite for communicative activities. Bayles and Tomoeda have proposed a specific test for language evaluation in dementia, the Arizona Battery for Communication Disorders of Dementia (ABCD; Bayles & Tomoeda, 1991 as cited in Bayles, 1994). Besides the examination of linguistic comprehension and expression, the ABCD includes additional tasks for cognitive components that clearly interact with language, such as verbal episodic memory. The use of such a battery in further studies may help to clarify the relationship between memory functions and language in DAT.

32-5. LANGUAGE IMPAIRMENT IN DAT: A MARKER FOR CLINICAL SUBGROUPS?

The presence and the severity of linguistic changes have been considered in some studies as possible indicators of particular clinical subtypes of DAT, although some data are still controversial. It has been postulated that there are significant differences between presenile (onset before 65 years) and senile (onset after 65 years) forms of DAT. Accordingly, several authors have found more frequent and even more prominent language disturbances in presenile DAT (Chui, Teng, Henderson, & Moy, 1985; Koss, et al.; 1996; Lawlor, Ryan, Schmeidler, Mohs, & Davis, 1994).

Other studies found a relationship between language disturbances and family history of dementia, suggesting that these features could be an indicator of familial DAT (Duara et al., 1993; Folstein & Breitner, 1981). However, Chui et al. (1985) and, more recently, Swearer, O'Donnell, Drachman, and Woodward (1992) did not observe a higher frequency, nor a higher severity, of language disturbances in this particular group of patients. A less controversial finding is the positive correlation between the presence of linguistic changes and a faster progression of dementia (Boller et al., 1991; Bracco et al., 1994; Faber-Langendoen et al., 1988).

32-6. NEUROPATHOLOGICAL CORRELATES OF LINGUISTIC CHANGES

Classically, linguistic changes in DAT have been ascribed to the involvement of the left-temporo-parieto-occipital junction area (Cummings et al., 1985; Delay & Brion, 1962). However, in the initial stages of the disease, the neuropathological process affects the medial temporal lobe, especially the entorhinal cortex and the hippocampus (Braak & Braak, 1991). The temporal lobe bears the brunt of the disease as the pathological process advances. Neocortical areas of the temporal lobe, such as the temporal pole and the middle and inferior temporal gyria, are also affected, and this probably occurs before the involvement of the temporo-parieto-occipital junction.

Recent studies reveal that damage to the left temporal neocortex, particularly the middle and inferior temporal gyria, and to the temporal pole, severely impairs the ability to retrieve words, but is not accompanied by any grammatical or phonemic defects (Damasio, Grabowski, Tranel, Hichwa, & Damasio 1996; Semenza & Zettin, 1989). It is likely that anomia and reduction in verbal fluency tests seen in DAT depend on the involvement of these areas.

As already mentioned, syntactic and phonological abilities are relatively preserved in the initial phases of DAT. Patel and Satz (1994) state that the perisylvian/peri-Rolandic cortex, which are the neuroanatomical structures responsible for these language mechanisms, must be intact at this stage.

As the disease progresses, the temporo-parieto-occipital junction area is also affected, and it is likely that the involvement of this area is responsible for the clinical profile similar to transcortical sensory aphasia usually described in the moderate stages of the disease. In the final stages of the disease, extension of the neuropathological process to the frontal regions may contribute to the marked reduction of spontaneous speech production seen in severe DAT patients.

32-7. CONCLUSIONS

Language disturbances may occur as early symptoms in DAT and are characterized by poor performance in naming and verbal fluency tests. As the disease progresses, linguistic changes become more frequent, and semantic problems predominate over syntactic and phonological problems. Efforts have been made to investigate the nature of these linguistic deficits, although some questions still remain open. One major issue has to do with the specificity of the linguistic changes since virtually no language feature seems to be exclusive to DAT, given the test instruments employed so far.

Future studies on language function in DAT will have to include longitudinal data in order to better characterize the evolution of the linguistic changes, particularly in relation to the neuropsychological heterogeneity that occurs with the disease and that may probably arise as well in the language domain. This approach would also allow the determination of whether the language disturbances are consistently related to specific subgroups of DAT. Furthermore, evaluation of very mild DAT cases might help to ascertain whether a deficit in semantic access could be present in the very early stages of the disease.

The systematic use of neuroimaging methods, in particular magnetic resonance imaging (MRI), functional MRI, and Positron Emission Tomography, allied with carefully elaborated language tests, might substantially advance the comprehension of the anatomic-functional correlates of language function in DAT. In relation to language tests, it is important to develop highly sensitive batteries that are specific to the language functions likely to be impaired in dementia. For this reason, assessment of discourse abilities is of great interest, since it is an ideal set for the analysis of language and communication abilities, and also for the investigation of their relationships with cognition, and especially memory.

CHAPTER 33

Language Impairment in Parkinson's Disease

Henri Cohen

Laboratoire de Neuroscience de la Cognition et Département de Psychologie, Université du Québec à Montréal, Montréal, Québec, Canada, H3C 3P8

Deficits in verbal fluency and naming, memory and comprehension of verbal information, as well impairment in verbal and logical reasoning are frequently associated with linguistic performance in Parkinson's disease, a neurodegenerative disorder primarily involving subcortical structures and the depletion of dopaminergic neurons in the substantia nigra. Impairment is more pronounced in more complex aspects of language processing as in sentence comprehension, understanding of relational terms and relative clauses, and in aspects of logical reasoning. It is believed that constraints in accessing the cognitive resources needed to implement the processing of syntactic and semantic information, as well as the disruption of reciprocating neural connections between the basal ganglia with the prefrontal cortex, may explain these observed deficits.

Parkinsonism is a symptom complex that occurs in a variety of disorders of the central nervous system (CNS). The principal locus affected is the pigmented neuronal system of the brain stem, with damage to the substantia nigra (SN) as the major site of pathological changes. There are a variety of causes and variants of the Parkinsonian syndrome, including idiopathic and other multisystem degenerations. The most common variant is Parkinson's disease (PD) or idiopathic parkinsonism (i.e., of unknown etiology; about 70%; Jellinger, 1986). PD is featured by unilateral or symmetrical degenerative changes involving mainly the basal ganglia, the central and caudal parts of the zona compacta of the SN, the locus coeruleus, and the nucleus basalis of Meynert. Frequently associated with this focal neuronal loss is the presence of Lewy bodies (Gibb, Scott, & Lees, 1991) or various types of neurofibrillary degeneration. This neuropathologic alteration is responsible for the conspicuous motor disorders characteristic of PD, which include akinesia, bradykinesia, muscular rigidity, alterations of posture, and tremor at rest (Koller & Hubble, 1992). The diagnosis is based on these clinical observations.

33-1. STATEMENT OF THE PROBLEM

In addition to these deficits in motor function, there is a growing consensus that cognitive deficits may be a part of the complex manifestations of PD. The traditional view of cognition in PD, perhaps because of James Parkinson's strong denial of intellectual changes in the disease that now bears his name, was that the senses and intellect are uninjured (Parkinson, 1817). Until recently, impaired cognitive functioning was ascribed to "bradyphrenia" (i.e., slow thinking) and few studies had addressed empirically the nature of the cognitive changes in PD. Evidence from many studies now suggests that at least subtle alterations and deficits in attention (Brown & Marsden, 1988), memory (Sagar, Sullivan, Gabrieli, Corkin, & Growdon, 1988) and procedural learning (Saint-Cyr, Taylor, & Lang, 1988; Allain, Lieury, Quemener, & Thomas, 1995) are present in PD. Moreover, similarities between the cognitive disorders of PD and those found in patients with frontal cortex lesions (Pillon, Dubois, Lhermitte, & Agid, 1986; Taylor, Saint-Cyr, & Lang, 1986) suggest that deficits in so-called executive functions are part of the clinical picture of PD.

There is also good clinical and experimental evidence of impairment in motor aspects of speech (e.g., dysprosody, hypophonia, dysarthria; Darley, Aronson, & Brown, 1975; Cohen, Laframboise, Labelle, & Bouchard, 1993) and, until recently, it was believed that language abilities were relatively well preserved and showed only gradual deterioration with the progression of the disease (Levin & Tomer, 1992; Levin & Katzen, 1995; Mahurin, Feher, Nance, Levy, & Pirozzolo, 1992). The first investigations of language in PD, using either standard aphasia batteries or WAIS subtests (e.g., Bentin, Silverberg, & Gordon, 1981), had generally concluded that language is relatively well spared. Mounting evidence, however, strongly suggests that some aspects of linguistic processing are affected in PD, the nature of which still remains to be defined. For instance, aphasic symptoms have been reported following lesions of the thalamus and the basal ganglia (a structure affected by the depletion of dopaminergic neurons in PD); this suggests that the involvement—direct or indirect—of subcortical structures affected in PD may not be confined to speech impairment and may encompass language processes as well (Crosson, 1992). At issue, then, is the question of whether language deficits are a clear manifestation of PD.

33-2. LANGUAGE DEFICITS IN PD

It is well known that PD patients represent a heterogeneous group, with some developing dementia or showing some intellectual impairment, and some remaining intellectually stable or normal. Often, depressive symptoms complement the clinical profile of PD, especially in the early stages. The studies that have been conducted on the language abilities of PD patients can be grouped into the following categories: verbal fluency including naming, organization of verbal information in memory, and comprehension.

33-2.1. Verbal Fluency and Naming Deficits

Studies of verbal fluency in PD—measured both by initial letter (phonological fluency) and by category retrieval (semantic fluency)—have yielded contradictory results. Deficits of initial-letter fluency have been reported (Gurd & Ward, 1989; Gurd, Ward, & Hodges, 1990) but these results must be interpreted with caution since it is not clear whether the patients in these studies were screened for dementia and depression. It has been shown that demented or depressed PD patients perform significantly worse than controls on initial-letter fluency tasks (Cummings, Darkins, Mendez, Hill, & Benson, 1988; Huber, Shuttleworth, & Freidenberg, 1989; Starkstein, Preziosi, Berthier, Bolduc, Mayberg, & Robinson, 1989; Bayles, Trosset, Tomoeda, Montgomery, & Wilson, 1993) and that naming and word fluency are correlated with severity of dementia (Fisher, Gatterer, & Danielczyk, 1988). The majority of studies, however, show an absence of deficits in initial-letter fluency with nondemented and nondepressed PD patients (Miller, 1985; Beatty, Staton, Weir, Monson, & Whitaker, 1989; Hanley, Dewick, Davies, Playfer, & Turnbull, 1990; Raskin, Sliwinski, & Borod, 1992; Auriacombe et al., 1992; Cohen, Bouchard, Scherzer, & Whitaker, 1994).

Deficits in category fluency, on the other hand, are more frequently associated with PD (e.g., Beatty & Monson, 1989; Raskin et al., 1992; Auriacombe et al., 1992) but a substantial number of studies also fail to find any impairment in semantic fluency (e.g., Hanley et al., 1990; Levin, Llabre, & Weiner, 1989; Cohen et al., 1994). In the Raskin et al. (1992) study, for example, PD subjects produced fewer semantic clusters in cued semantic category tasks than did the control subjects. The authors interpret their finding to suggest that this deficit is characteristic of an executive dysfunction that interferes with the activation of semantic networks (see also Tweedy, Langer, & McDowell, 1982). Hines and Volpe (1985), on the other hand, had previously found that PD subjects were unimpaired in their ability to activate semantic memory, as revealed by faster reaction times to words following a semantically associated prime. Beatty, Monson, and Goodkin (1989) also found that PD subjects who performed normally on the Boston Naming Test generated as many specific exemplars and category labels as did controls. Nondemented PD subjects in Beatty and Monson's (1989) study, however, showed selective language deficits, suggesting that naming in particular can be compromised in subcortical disease. The clinical implications of their observations are, however, unclear. In another study, Matison, Mayeux, Rosen, and Fahn (1982) have suggested that impaired category fluency, coupled with the difficulties observed in confrontation naming, may represent a form of cognitive anomia that shares the clinical characteristics of the tip-of-the-tongue phenomenon seen in some aphasics. Again, others have reported mild or no naming deficits in nondemented PD patients (e.g., Pirozzolo, Hansch, Mortimer, Webster, & Kuskowski, 1982; Bayles & Tomoeda, 1983; Freedman, Rivoria, Butters, Sax, & Feldman, 1984; Levin et al., 1989) or in PD patients who had undergone sterotactic surgery a decade earlier (Gamsu, 1986). Furthermore, Pillon et al. (1986) failed to find naming deficits in PD patients matched for impairment with Alzheimer patients on Raven's matrices and subtests of the WAIS-R.

Two general neuroanatomic hypotheses have been put forward to account for some of these observed fluency deficits. In one, they may be attributable to lesions in subcortical structures (basal ganglia and thalamus), and investigation of fluency deficits in patients with progressive supranuclear palsy, a disease with subcortical degeneration, adds support to such a notion (e.g., Agid et al., 1986). In the other, lesions to the frontal cortical structures are responsible for such fluency deficits (e.g., Ramier & Hecaen, 1970), a view that finds agreement with Eslinger and Grattan's (1993) position that the frontal lobes play a strategic role in "spontaneous flexibility," a notion similar to set shifting discussed in relation to the frontal lobe syndrome. It has been argued that fluency tasks rely on the internal control of attention, which is presumed to task the limited attentional resources of PD patients. One test of this hypothesis, however, does not suggest that this is the case; rather, PD may be associated with a deficit in inhibitory attentional resources and an impairment in the maintenance of those internal representations that control action (see Downes, Sharp, Costall, Sagar, & Howe, 1993).

33-2.2. Organization of Verbal Information in Memory

Memory and learning deficits with verbal material have frequently been reported in PD and the California Verbal Learning Test (CVLT; Delis, Kramer, Kaplan, & Ober, 1987) is the instrument commonly used to evaluate various aspects of verbal learning and memory. Briefly, the test consists of three lists of words including a recognition list. The first list includes 16 items from four semantic categories; the second list has the same structure, with two semantic categories being the same as in the first list. The recognition list includes 16 target items from the first list, plus distractors that share or do not share semantic or phonological membership with that list. Taylor, Saint-Cyr, and Lang (1990) found, in a study using the CVLT, that short-term memory was impaired in both free recall and clustering in PD subjects. In a German study using CVLT-like stimuli and procedures, overall performance and semantic clustering scores were lower for PD subjects (Karamat, Ilmberger, Poewe, & Gerstenbrand, 1991); the authors suggest that the observed deficit is comparable to that seen in normal aging or in subjects with frontal lobe lesions. Buytenhuijs et al. (1994) studied the differential influence of external versus internal control on recall performance in PD, in an experiment using a Dutch version of the CVLT. Recall according to the semantic categories was considered to be the result of unprompted, internally generated strategy, and recall according to the sequence in which the lists were read by the experimenter was viewed as an externally offered strategy. Their results showed that, unlike controls who appeared to rely mainly on an internally generated semantic organization, PD subjects appeared to adhere more to the externally imposed serial sequence. In a study using a French version of the CVLT, deficits were also observed on all measures of free recall and source memory, suggesting that the memory impairment in PD may also include long-term memory (Cohen et al., 1994). Source memory is the ability to remember the context within which chunks of information have been learned (Taylor et al., 1990). In the CVLT, however, source memory and memory for the order of presentation may be confounded; therefore, these results must be interpreted with caution. It should also be noted that, in the Cohen et al. (1994) study, PD subjects did not differ from control subjects in measures of semantic or serial clustering, suggesting that organization of verbal information in memory, in PD, presents a complex profile.

Short-term memory impairment in PD appears to affect the semantic organization of words as well as of drawings and faces in PD. In a study by Raoul, Lieury, Decombe, Chauvel, and Allain (1992) with PD and control subjects, results of interest showed that deficits were material-specific: both overall verbal recall and semantic organization were lower for PD subjects. Moreover, except for recognition of drawings, which was similar in all groups, the young controls performed better than the aged controls and the PD subjects in all tests. To explain their results, Raoul and colleagues suggest that encoding and organization processes, which depend on the limited resources and capacity of short-term memory, are negatively affected in PD.

There seems to be general agreement that the frontal lobe in part mediates the organization of verbal information in memory, and this aspect of frontal lobe function is implicated by the pattern of memory impairments observed in PD. Impairment in verbal recency discrimination and content recognition have been recognized in both depressed and demented PD patients (Sagar et al., 1988) as well as nondepressed and nondemented PD patients (Fisher et al., 1988). Moreover, proactive interference, which is the progressive decline in recall performance subsequent to intrusion from previously presented material, seems to be preserved in PD despite impaired recall (Sagar, Sullivan, Cooper, & Jordan, 1991). This would suggest difficulties in effortful processing, as is the case with explicit memory, and preservation of more automatic processes in PD.

Strategic processes involved in sequencing may also be impaired in PD, as revealed by deficits in reconstructing the temporal ordering of words and drawings (e.g., Vriezen & Moscovitch, 1990). An alternate view of the cause of temporal ordering deficits in PD implicates effects of disease chronicity and medication. Cooper, Sagar, and Sullivan (1993) found that memory deficits were greater in chronically medicated patients, at least in the early stages. In remote memory, a specific dating impairment seems to be present despite a good recognition memory for the content of both public and autobiographical episodes (Friedman & Wilkins, 1985).

33-2.3. Language Comprehension

Impaired comprehension of verbal material is also a feature of the cognitive loss encountered by PD subjects. Comprehension in PD, although grossly intact, may be affected by frontal-lobe mediated attentional deficits. Grossman, Carvell, Stern, Gollomp, and Hurtig (1991) examined the ability of nondemented PD subjects to interpret grammatical aspects of sentences and found that impairment in some patients on tests of sentence comprehension was correlated with reduced mesial frontal lobe glucose metabolism (as measured by PET imaging), suggesting an association with frontal-lobe attentional mechanisms. In another PET study, Grossman et al. (1991) implicated the anterior cingulate cortex, suggesting that a defect in this area contributes to the specific linguistic impairments the authors observed in PD. Grossman, Carvell, and Peltzer (1993) also observed a significant correlation between bilateral mesial frontal cortex and comprehension of sentences with center-embedded subordinate phrases. In another study, the authors found that a majority of PD patients were compromised in their ability to answer simple questions about sentences such as "The Eagle chased the hawk that was fast. Which bird was chased?" (Grossman, Carvell, Gollomp, et al., 1991). These impairments were greater when sentences increased in syntactic complexity. In general, it was found that sentence comprehension deficits were similar in PD and in frontal lobe insults. In a more recent study, nondemented PD subjects were exposed to a new verb and the grammatical and semantic information that they learned about the verb was probed. Grossman and colleagues observed that most of the PD subjects demonstrated a language-sensitivity deficit in appreciating the grammatical information and that a small number of patients responded randomly to the probes. This difficulty in appreciating grammatical information may be, following Grossman, Stern, & Gollomp, (1994), one of the factors implicated in the language impairments of PD patients.

"Greek PD subjects were also tested for their competence to process relative clauses *(that)* in sentences without semantic constraints. Relative clauses, as a form of pronoun assignment, present a concise picture of syntactic, semantic, and pragmatic function at the level of sentence grammar. The findings showed a clear language deficit in PD (Natsopoulos, Katsarou et al., 1991). Additional evidence in the study also showed that PD patients' performance was similar to that of Grade 1 children. The authors propose a "regression hypothesis" to explain their results; they also suggest that PD subjects processed sentences with complex semantic reversibility on a heuristic and not on an algorithmic basis. Syntax comprehension has also been studied, using the Rhode Island Test of Language Structure, in PD with and without dementia (Lieberman et al., 1992). It was found that most demented PD patients had high comprehension error rates for sentences that had moderately complex syntax. In contrast, the error rates of the nondemented subjects were low and similar to those of control subjects. The authors propose, in line with other formulations to explain general cognitive deficits in PD, that destruction of the midbrain regions that stimulate the frontal cortex may be responsible for these syntactic comprehension deficits. Pathology of the neostriatum has also been implicated in disruption of syntactic organization in spontaneous language production (Illes, 1989).

Understanding of relational terms, such as *before* and *after* denoting succession, is impaired in PD patients relative to control subjects matched for age, sex, education, and socioeconomic status (Natsopoulos, Mentenopoulos et al., 1991). Data from the study further reveal that *before* is better understood than *after,* and that order of event mention may be a predominant language strategy in PD patients. Appreciation of mass and count quantifiers in PD appears also to be affected. Grossman et al. (1993), for example, found that patients often erred by pointing to the incorrect mass or count type of a substance and made more detecting errors in the agreement between a noun and a quantifying adjective. These deficits point to the multifactorial and complex nature of comprehension impairments in PD and suggest that PD patients have difficulty in the integration of surface structure information conveyed by syntax, semantics, pragmatics, and the interrelation of causal elements.

Attention has recently been directed to more complex linguistic processes such as verbal reasoning. In a study by Cohen et al. (1994), results from a battery of tests including the similarities subtest of the WAIS, logical reasoning and invited inference were obtained from nondemented PD patients and education-matched controls. The invited inference test was made up of pairs of sentences such as "The boy is coming out of the kitchen; there is an empty glass on the counter. What do you think has occurred?" The verbal logical reasoning test (VLRT; Whitaker, Markovits, Savary, Grou, & Braun, 1991) was made up of logical problems of the form "Suppose it is true that: If p then q." For each problem, subjects had to choose between a nonlogical conclusion, a definite conclusion, or an uncertain conclusion that conformed to or was contrary to the premises. An example of such a problem is presented in Table 1. Significant impairments were revealed in the similarities test showing an inability, on the part of some patients, to generate more than one instance of similarity between the items in a pair. It is as yet unclear whether the poorer performance of PD subjects is due to underlying conceptual or flexibility difficulties, or confining one's responses to a restricted semantic field or representation; the results, however, can be interpreted in the context of a deficient ideational fluency (see also Wilson & Gilley, 1992). In

TABLE 1

Example of a Counterfactual Problem With an Affirmation of the Consequent

Suppose it is true that:

If we put an object in the snow, it will become hot.

An object is hot. Therefore:

a. It is certain that the object is heavy.
b. It is certain that the object was not put in the snow.
c. It is impossible to be sure that the object was put in the snow or not.
d. It is certain that the object was put in the snow.

the VLRT, PD subjects more frequently selected conclusions that were contrary to the premises, a response probably attributed to a problem in the comprehension of syntactically complex material. There was no impairment in the invited inference test, suggesting that "common sense" appeared to be well preserved in that sample of PD subjects.

33-3. CONCLUSION

Most of the studies reviewed in this chapter point to specific language deficits in PD. They suggest that PD subjects have some difficulty with category fluency and more pronounced impairment with the temporal organization of verbal information, with learning and memory, and with the processing and comprehension of syntactic information. In addition, it was found that complex verbal reasoning deficits were also a feature of PD.

It is tempting to conclude that neurolinguistic disorders are part and parcel of the Parkinsonian syndrome. It remains, however, to be confirmed whether disorders of the neostriatum or executive-type deficits cannot adequately (and more parsimoniously) explain the nature of the linguistic impairments observed in these studies. Parkinson's disease is considered by many to be a "frontal" disorder in which executive deficits occur and are important for understanding the clinical symptoms, mainly because of the extensive reciprocal connections between the basal ganglia with the prefrontal cortex and the limbic system. Indirect evidence points to the disruption of a fronto-striatal network following degradation of the dopaminergic (DA) projection system to explain the physiological basis of the linguistic deficits in PD. So far, it appears that simpler aspects of language processing are performed normally whereas PD patients experience difficulty in accessing the cognitive resources needed for the treatment of more complex linguistic information. Related issues are whether PD affects equally language and nonlinguistic functions, and whether the dementia in PD affects linguistic performance differently than the dementia in Alzheimer's disease (see also Bayles, 1993).

Moreover, there is a need for more sophisticated statistical treatment of comparative data in investigations of linguistic performance in PD. It is often the case that univariate analyses of comparative data, between PD and adequately matched control subjects, fail to reveal significant differences between these groups—although the performance of PD subjects is always lower. Trend analyses and appropriate evaluation of effects (see Bernstein, Garbin, & Teng, 1988) would most probably reveal meaningful differences that are not otherwise evident. A complementary procedure, which now appears necessary given the number of studies addressing the question of linguistic processing and impairment in PD, is the use of meta-analytical approaches to determine the extent and robustness of the trends and differences observed in the literature.

Finally, side effects of drug treatment of parkinsonism on cognitive or linguistic function have seldom been taken into account (Taylor, Saint-Cyr, & Lang, 1987).

Pharmacological treatment of PD typically includes dopaminergic substances such as direct dopamine agonists (e.g., bromocriptine, pergolide) or dopamine reuptake inhibitors (e.g., deprenyl). Anticholinergic agents are also often prescribed with older subjects (e.g., trihexyphenidyl). Although they have been extensively studied, the consequences of chronic PD medication on cognition are not well known. Data from recent studies, however, show that dopaminergic treatment (deprenyl excepted) may negatively affect cognitive performance in tasks that require important attentional resources such as memory and language (Malapanis, Pillon, Dubois, & Agid, 1994). With younger patients without cognitive deficits, minimal therapeutic doses of anticholinergic medication have been shown to induce a "frontal syndrome" for the duration of the treatment. Also, in large doses, these neuroleptic treatments may induce confusional states and, in older patients, may give rise to irreversible dementia (Dubois & Pillon, 1992; Bédard, Pillon, Dubois, Masson, & Agid [in press]). Almost none of the studies reported here have examined the impact of PD medication on cognitive performance. Future efforts should also carefully consider these treatment variables.

CHAPTER 34

Communication and Language Disturbances Following Traumatic Brain Injury

Skye McDonald

School of Psychology, University of New South Wales, Sydney, Australia

Severe traumatic brain injury (TBI) results in aphasic disturbances in a minority and impaired communication competence in many more. Different assessment approaches have revealed problems in maintaining coherence across extended discourse, difficulties in the production and comprehension of conversational inference, and clumsy participation in conversation, including poor topic maintenance and failure to cater to the needs of a conversational partner. TBI produces cognitive impairments that vary in both severity and nature. The particular cognitive deficits experienced may have direct ramifications for impaired communication. In particular, loss of attention, slowed information processing, and loss of executive control may underlie much of poor communicative ability in TBI.

Traumatic brain injury, frequently incurred as a result of motor vehicle accidents, leads to diffuse, microscopic, axonal injury as well as multifocal damage throughout the cerebrum. Lesions are commonly concentrated in the temporal and medial-orbital zones of the frontal lobes due to rapid acceleration–deceleration forces during impact, crushing these areas against the bony fossas of the cranium in which they are cradled (Walsh, 1986). Recovery from severe TBI is variable but there is almost always some residual impairment as frequently psychosocial as sensorimotor (Tate, Lulham, Broe, Strettles, & Pfaff, 1989). Communication disturbances are common following

TBI and have been considered one of the major impediments to successful rehabilitation in a patient group that is typically composed of young adults with a full life span ahead. This chapter will review the nature of these disturbances and some approaches to their measurement, and then consider some explanations for them.

Until fairly recently, descriptions of language impairment following TBI focused on aphasic symptomatology. The actual incidence of aphasia has been estimated from between 2% (1,544 cases: Arseni, Constantinovici, Iliescu, Dobrota, & Gagea, 1970; 750 cases: Heilman, Safran, & Geschwind, 1971) to roughly 30% (125 patients: Sarno, 1988), with anomic aphasia the most prevalent in adults (Heilman et al., 1971; Levin, Grossman, & Kelly, 1976; Thomsen, 1975) and nonfluent aphasias (Global and Broca-like) more prevalent in children and adolescents (Basso & Scarpa, 1990). For a proportion of these patients, such aphasic deficits resolve over the ensuing months (Grosswasser, Mendelson, Stern, Schecter, & Najenson, 1977; Thomsen, 1975, 1984) although few completely regain premorbid language abilities (Basso & Scarpa, 1990; Thomsen, 1984). Although not classified as aphasic, many other TBI patients perform lower than expectations on specific language subtests, for example, confrontation naming, word finding, or verbal associative tasks, as well as structured tests of comprehension such as the Token Test (Gruen, Frankle, & Schwartz, 1990; Levin et.al., 1976; Levin, Grossman, Rose, & Teasedale, 1979; Sarno, 1988). Opinion has differed as to whether such deficits reflect a "subclinical aphasia" not apparent in casual conversation (Sarno, 1988) or, as currently more widely accepted, attentional and memory disorders rather than linguistic impairment per se (Sohlberg & Mateer, 1989; Holland, 1984).

Although frank aphasic features in conversational speech are thus relatively uncommon, profound communication difficulties are not. TBI patients have been found to have a range of difficulties in everyday communication that fall into several broad categories. First, they have been described as overtalkative (Hagan, 1984; Milton, Prutting, & Binder, 1984; Milton & Wertz, 1986) but inefficient (Hartley & Jensen, 1992), drifting from topic to topic (Snow, Lambier, Parson, Mooney, Couch, & Russell, 1987) and making tangential and irrelevant comments (Prigatano, Roueche, & Fordyce, 1986). Alternatively, some patients are impoverished in the amount and variety of language produced (Chapman et al., 1992; Hartley & Jensen, 1991, 1992; Ehrlich, 1988), their conversational style characterized by slow, frequently incomplete responses, numerous pauses, and a reliance on set expressions (Thomsen, 1975). A third category of discourse characterized by confused, inaccurate, and confabulatory verbal behavior has also been advocated (Hartley & Jensen, 1992), and other combinations of these features can occur in individual patients (Hartley & Jensen, 1992; McDonald, 1992a).

Conventional approaches inadequately characterize the breadth and pervasiveness of these communication disorders as evidenced by a lack of correlation between aphasia language batteries and other measures of communicative competence (Hartley & Jensen, 1991; Liles, Coelho, Duffy, & Zalagens, 1989; McDonald & van Sommers, 1993). As an alternative, advances in linguistic theory have proven fruitful in capturing the nature of some of the problems faced by TBI patients and can be discussed under three headings: discourse analysis, pragmatic analysis, and conversational analysis.

34-1. DISCOURSE ANALYSIS

Where conventional language tests analyze language at no greater complexity than the sentence, discourse analyses are concerned with the discourse as a whole, its continuity, semantic organization, and the relation between structure and function (see Patry & Nespoulous, 1990 for review). Discourse analysis in TBI has focused on monologues such as telling a story or relating a procedure and the resultant text has been analyzed to determine whether it meets certain functional requirements, in particular, whether it is efficient and coherent.

34-1.1. Efficiency

Efficiency of discourse has been characterized as the rate of speech or amount of information imparted in the words produced. Adult and adolescent TBI patients have been found to speak more slowly (Hartley & Jensen, 1991; Wychoff, 1984), produce fewer meaningful words overall (Hartley & Jensen, 1991; Chapman et al., 1992; Wychoff, 1984), more incomplete or ambiguous utterances (Hartley & Jensen, 1991; Wychoff, 1984), as well as shorter informational units (C-units) (Hartley & Jensen, 1991) and less information per minute (Ehrlich, 1988) across both narrative and procedural discourse tasks. In some subjects, such findings may reflect the presence of motor speech problems (Hartley & Jensen, 1991; Wychoff, 1984), but not all subjects exhibiting decreased efficiency have dysarthria (e.g. Ehrlich, 1988). Alternatively, reduced verbal retrieval as demonstrated on tests of fluency or naming may be associated with low productivity in some subjects (e.g. Levin, Grossman, Sarwar, & Meyers, 1981), but again, this is not always the case (Hartley & Jensen, 1991).

34-1.2. Coherence

The coherence of discourse relies on the semantic continuity of the text (Patry & Nespoulous, 1990). Discourse may be coherent at a "local" level, that is, there are appropriate relations between adjacent propositions, or at a "global" or macrostructure level, that is, there is development and maintenance of an underlying discourse plan (Coelho, Liles, & Duffy, 1991c; Patry & Nespoulous, 1990). Both notions of coherence have been investigated in TBI subjects.

34-1.2.1. Local Coherence

One measure of local coherence is provided by an estimate of textual cohesion created by the interdependence of linguistic items occurring in separate clauses in a text. A cohesive link occurs wherever one linguistic item relies on another for its interpretation, via reiteration of semantically linked items ("she picked up her *shawl* and put the *wrap* over her shoulder"), or grammatical links such as that formed via pronominal reference ("he had *a plan of action* but would not divulge *it*") (Halliday, 1985; Halliday & Hasan, 1985; Hasan, 1985).

Because TBI discourse is frequently tangential and disorganized, it has been of interest to determine whether this reflects problems developing and maintaining textual cohesion. In support of this notion, some TBI subjects have been found to use less cohesive ties than non-brain-damaged control subjects (Hartley & Jensen, 1991; Mentis & Prutting, 1987; Wychoff, 1984), and in virtually all studies of cohesion, TBI subjects make more incomplete references where the source for interpretation of a given linguistic unit is missing or ambiguous (Hartley & Jensen, 1991; Liles et al., 1989; McDonald, 1993b; Mentis & Prutting, 1987). Also, some TBI subjects appear to use certain types of cohesion differently from controls (Liles et al., 1989; Hartley & Jensen, 1991; McDonald, 1993b; Mentis & Prutting, 1987), although in general they remain sensitive to the demands of different discourse requirements, for example, procedural versus narrative discourse, varying the amount and type of lexicogram-matical cohesion accordingly, sometimes in a pattern similar to matched controls (Hartley & Jensen, 1991) and sometimes differently (Liles et al., 1989; Mentis & Prutting, 1987).

Although cohesion analyses have thus yielded some consistent findings and also proven sensitive to changes in discourse function over time (Coelho et al., 1991b, 1991c) there is also considerable variability between studies, including null findings (Jordan, Murdoch, & Buttsworth, 1991). This may reflect the fact that TBI subjects are not homogeneous with respect to communication impairments and will not, there-fore, demonstrate a uniform deficit in the maintenance of cohesive relations. Even so, further work is needed to improve the reliability and stability of the measures (Strong & Shaver, 1991) and to provide better characterization of cohesion patterns of normal speakers across tasks. In addition, the validity of cohesion measures requires explo-ration, given that two studies that attempted to match cohesion to subjective impres-sions of the coherence of the discourse yielded nonsignificant associations (Glosser & Deser, 1990; McDonald, 1993b).

34-1.2.2. Global Coherence

Global coherence has been gauged by identifying the nature and sequence of the propositional content, propositions being roughly equivalent to a predicate with one or more arguments (Kintsch & van Dijk, 1978; van Dijk & Kintsch, 1983). In narrative tasks, the propositional structure of TBI discourse has been characterized using story grammar analysis, which focuses on the nature of the story constituents as well as their interrelationships within the story structure (Fayol & Lemaire, 1993). According to story grammar analysis, stories consist of key as well as optional story components (e.g., Chapman et al., 1992) organized appropriately, that is, into episodes defined as sequences of events with defined beginnings, middles, and ends (Chapman et al., 1992; Jordan et al., 1991; Liles et al., 1989). Using this approach, TBI adults reportedly produced as many complete episodes in a story retelling task as controls (Liles et al., 1989), but fewer complete and more incomplete episodes on a story generation task (Liles et al., 1989; Coelho, Liles, & Duffy, 1995). In contrast, Chapman et al. (1992) reported that their adolescent TBI subjects produced less-complete episodes in a story retelling task and also less of the essential story elements. In another variation, Jordan et al. (1991) failed to find any differences between their TBI children and controls in the number of complete episodes produced in a story generation task.

More consistent results have been yielded by investigating the propositional structure of procedural discourse produced by nonaphasic TBI subjects with clinically defined communication problems (McDonald, 1993b; McDonald & Pearce, 1995; Turkstra, McDonald, & Kaufman, 1995). In each of these studies, the subject was asked to explain how to play a novel board game, "the dice game," to a naive listener. The explanations were then evaluated as to the number, type, and sequence of propositions produced. In the first of these studies (McDonald 1993b), the procedural texts produced by two TBI subjects were also the subject of cohesion analysis (see the preceding section) and independent raters' evaluations. According to the propositional analysis, one subject provided fewer different propositions than the controls, while the other repeated more propositions. These features were consistent with raters' impressions that there were too few details and too much repetition, respectively. Furthermore, both subjects made discrete errors on sequencing, failing to present information in the same order of priority as the controls and focusing on irrelevant propositions at the beginning of their explanations. The independent ratings of the texts as both confusing and disorganized validated these sequence characteristics as genuinely disruptive. In subsequent studies it has been repeatedly demonstrated that TBI subjects, both adults and adolescents, omit essential information, provide a disrupted sequence of explanation, and include irrelevant and ambiguous material (McDonald & Pearce, 1995; Turkstra et al., 1995).

34-2. PRAGMATIC ANALYSIS

A second approach to clinical communication disorders has developed on the basis of pragmatic theory, which addresses the relationship between the surface structure of an utterance, the context in which it occurs, and the information imparted via inference (see Levinson, 1983 for review). Of particular interest is the notion that any given message may be communicated via a variety of surface forms. Choice of a particular form will be guided by politeness or other cultural conventions (Brown & Levinson, 1987) and in many situations, such as advertising, will be made in order to communicate a number of messages simultaneously (Nippold, Cuyler, & Braunbeck-Price, 1988). Successful communication requires the ability to understand and use pragmatic inference, and pragmatic theory has emerged as a fruitful framework with which to discuss a variety of nonaphasic communication disorders, for example, those associated with right-hemisphere lesions (Molloy, Brownell, & Gardner, 1990;

Stemmer, 1994; Stemmer, Giroux, & Joanette, 1994). Studies of TBI subjects have also benefited from this approach.

34-2.1. Comprehension

A number of studies have demonstrated impaired comprehension of pragmatic inference in TBI subjects. TBI subjects selected on clinical grounds for their poor communication skills were found to be inferior to matched controls in their capacity to derive alternative meanings in ambiguous advertisements (e.g., "Fascinating things happen on 'Impulse,'" "Impulse" being a brand of deodorant) (Pearce, McDonald, & Coltheart, 1995), conventional indirect speech acts (e.g., "Can you pass the salt") (McDonald & van Sommers, 1993), or sarcastic remarks in which the intended meaning is the opposite of the literal meaning ("What a *great* football game") (McDonald, 1992b; McDonald & Pearce, 1996). In each case they preferred the literal meaning, regardless of the context.

34-2.2. Production

TBI subjects also demonstrate differential impairment in their use of pragmatic strategies when making requests. A sample of such subjects have been shown to be sensitive to the relative politeness of different forms of simple requests and able to moderate the formality of requests in different contexts (McDonald & van Sommers, 1993), and yet to be ineffective in the formulation of more complex requests designed to overcome listener reluctance (e.g., borrowing a car when the owner needs it herself) (McDonald & Pearce, in press). They produced less elaborate requests than controls that were unlikely to address the obstacle to listener compliance and more likely to encompass counterproductive comments that would discourage the listener from complying. It has also been demonstrated that TBI can compromise the ability to produce nonconventionally indirect requests, that is, hints, that allude to the request by inference alone (McDonald & van Sommers, 1993). Similar deficits in comprehension and production have been found in adolescent TBI subjects (Dennis & Barnes, 1990; Turkstra et al., 1995; Wiig & Secord, 1989).

34-3. CONVERSATIONAL ANALYSIS

Conversational analysis focuses on how verbal interaction between two or more participants is managed during spontaneous conversation (Patry & Nespoulous, 1990), and in TBI research has incorporated a variety of behavioral and linguistic measures. Checklists and rating scales have been used that range in focus from global measures of verbal and nonverbal behavior, for example, "social performance" (Newton & Johnson, 1985; Spence, Godfrey, Knight, & Bishara, 1993) to measures of specific and discrete attributes of communication, for example, "frequency of questions" (Godfrey, Knight, Marsh, Moroney, & Bishara, 1989). Using these it has been repeatedly demonstrated that severe TBI subjects have poor topic maintenance (Ehrlich

& Barry, 1989; Milton, Prutting, & Binder, 1984; Snow et al., 1987), poor initiation (Ehrlich & Barry, 1989), are unskilled in their response to questions from their conversational partner (Spence, Godfrey, Knight, & Bishara, 1993), egocentric in their discussion, fail to actively involve their conversational partner, for example by asking questions or supporting them with the use of verbal reinforcers (Marsh & Knight, 1991a; Flanagan, McDonald, & Togher, 1995), and elicit high rates of facilitative behavior in their conversational partner when engaged in discussing a problem (Godfrey, Knight, & Bishara, 1991).

Linguistic analyses of these kinds of conversational behaviors in TBI are only just emerging but tend to support the subjective impressions described earlier. For example, Mentis and Prutting (1991) segmented and classified spontaneous conversation as well as monologues in order to characterize topic maintenance. They reported that, relative to the non-brain-injured control subject, their TBI subject changed topic in an incoherent manner without signaling his intention, made ambiguous statements due to a failure to establish clear reference, produced less novel information to develop a topic and more of this in response to specific questions. Coelho et al. (1991a) categorized utterances as to who initiated them, whether they were responded to, and the degree to which they were appropriate. TBI subjects, relative to controls, had more conversational turns, elicited more prompts from their partner, and differed in the perceived appropriateness of their conversation. Finally, conversational analysis has not only been directed toward uncovering the nature of impairments in the conversational skills of TBI subjects, but also the manner in which conversation is altered in reaction to these impairments. Compensatory strategies used by TBI subjects and their conversational partners have been identified and analyzed (Penn & Cleary, 1988), as well as the manner in which linguistic markers of social dominance alter in the conversational style of speakers talking to TBI versus normal adults (Togher, Hand, & Code, 1995).

34-4. EXPLANATIONS FOR COMMUNICATION DISTURBANCES IN TBI

Explanations for TBI communication disturbances need to be considered within the context of prevalent patterns of neuropathology and coexisting cognitive deficits. As mentioned in the introductory section, TBI commonly (but not universally) results in diffuse damage and multifocal lesions concentrated in the temporal and frontomedial lobes of the brain. Different constellations of language impairment are apparent in TBI (see the introductory section) and particular characteristics may reflect the specific effects of these lesions. Diffuse injury, associated with attentional problems, slowed information processing, and verbal retrieval difficulties (Lezak, 1995), has implications for communication efficiency and coherence. The finding that immediate memory span (Digits Forward, from the Wechsler Memory Scale) is associated with productivity measures (Hartley & Jensen, 1991) corroborates a role for attention in linguistic efficiency, although, surprisingly, information-processing speed has not been found to be related (Godfrey et al., 1989). It has also been queried whether verbal retrieval deficits underlie reduced cohesion, in particular, underutilization of lexical cohesion

(Mentis & Prutting, 1987), although an attempt to establish this empirically was unsuccessful (Hartley & Jensen, 1991).

Frontal lobe injury, exacerbated by diffuse damage, leads to impairment of the executive control of other cognitive activity, resulting in inertia, perseveration, rigidity, stimulus-bound thinking and poor abstract thought, reduced problem-solving skills, a failure to plan ahead and/or monitor behavior in order to achieve a goal, and failure to inhibit inappropriate or unadaptive responses (Walsh, 1986; Lezak, 1995). Disturbances in the control of cognition and behavior have obvious ramifications for the ability to engage in socially-effective verbal interaction. Furthermore, many of the characteristics of TBI communication skills bear a striking resemblance to descriptions of nonaphasic language impairment following focal lesions to the frontal lobes (for review, see Alexander, Benson, & Stuss, 1989; McDonald, 1993a).[1]

Work is just beginning, however, to link executive deficits to specific communicative disturbances. First, general problems of stimulus-bound behavior and poor abstraction skills may result in inability to comprehend pragmatic inference. In support of this, TBI patients who were poor at understanding pragmatic inference in conventional and nonconventional indirect speech acts (indirect requests and sarcasm, respectively) were also poor on standard neuropsychological measures of concept formation (McDonald & van Sommers, 1993, McDonald & Pearce, in press-a), although not all efforts to find such correlations have proven successful (Turkstra et al., 1995). Within discourse analysis it has been speculated that patients who display differential reliance on lexical cohesion in storytelling have a pathological fixation with the concrete, visible attributes of the stimuli used to generate stories (Liles et al., 1989). This interpretation is feasible but awaits empirical confirmation.

Second, problems in the planning and monitoring of behavior may lead to difficulty adhering to conventional discourse macrostructure. For example, when relating a story, an association has been reported between the frequency of incomplete episodes and perseverative errors on the Wisconsin Card Sorting Test (Coelho et al., 1995), suggesting that perseveration leads to a pathological failure to terminate episodes when developing the narrative. Similar problems in regulating behavior appear to influence procedural discourse performance. A TBI subject with predominantly impaired impulse control on neuropsychological testing produced a disorganized and tangential procedural explanation, while, conversely, a TBI subject suffering predominantly from inertia on formal tests was laborious and repetitive in the development of his discourse

---

[1]Descriptions of TBI language disturbances also bear a close relationship to RH language (for comparison, see Fredriksen & Stemmer, 1993; Joanette & Goulet, 1990; Molloy, Brownell, & Gardner, 1990). There is no a priori reason to suspect differential RH impairment in TBI, but the two groups do appear related in terms of the type of cognitive deficit associated with the communication disturbance. For example, in one of the few studies to relate neuropsychological measures to language performance, it was revealed that abilities traditionally attributed to the RH such as visuospatial skills bore little relation, while skills in planning, monitoring, and integrating were linked (Stemmer, Giroux, & Joanette, 1994). These latter abilities are often compromised in TBI, have typically been termed executive control, and attributed to the frontal lobes of the brain, left and right (Lezak, 1995). It may be that RH language impairment thus reflects executive impairment secondary to anterior or more generalized cognitive impairment.

(McDonald, 1993b). Despite these qualitative similarities, the relationship between specific executive cognitive disorders and failure to achieve global coherence in procedural tasks has not yet been established empirically (McDonald & Pearce, 1995).

Third, disinhibition, another facet of executive impairment, has further ramifications for the production of socially appropriate communication. TBI patients, who had difficulty producing nonconventional indirect speech acts, not only found the use of inference difficult, but were also unable to refrain from stating their true intention, thus failing in their attempt to be indirect (McDonald & van Sommers, 1993). Direct requests likewise suffered, but in this case, because the subjects failed to inhibit tangential remarks that were not conducive to listener compliance. In the latter study, such behaviors appeared to be associated with neuropsychological measures of disinhibition (McDonald & Pearce, in press).

Finally, poor abstract reasoning, poor inferential skills, and poor planning and monitoring of verbal behavior are likely to impact upon the ability to converse in an interactive manner that is sensitive to the needs of a conversational partner. In one of the few attempts to relate conversational skills to neuropsychological deficits, it has been verified that poor performance on a verbal fluency task, taken as a measure of inflexibility, was positively correlated to broad ratings of social competence in spontaneous conversation (Marsh & Knight, 1991b). Interestingly, numerous measures of new learning and memory were not correlated to estimates of social competence. Thus, even though memory impairments reflecting temporal lobe injury are as common following TBI as executive impairments (Tate, Fenelon, Manning, & Hunter, 1991), these do not appear to have a direct impact on social communication.

In summary, although the likelihood that many communication disturbances after TBI are a manifestation of coexisting deficits in attention, information processing, and executive control is supported by the emergence of these few correlational studies, further work is required. The heuristic value of considering higher-level language disturbance in terms of coexisting cognitive impairment is also enhanced by advances in cognitive neuropsychological theory. In particular, executive control is being specified in more precise terms, for example, as a supervisory attentional system to override routine behavior (Shallice & Burgess, 1991), or as a weakening of associations between goals, environmental stimuli, and stored knowledge in working memory (Goldman-Rakic, 1987; Kimberg & Farah, 1993). This precision helps to generate hypotheses regarding the role of executive functions in producing smooth and effective discourse. For example, independent measures of working memory capacity (repetition from the Western Aphasia Battery, Digits Backward, Wechsler Memory Scale) have been associated with linguistic measures of cohesion in TBI (Hartley & Jensen, 1991). The achievement of cohesion in extended discourse, for example, by ensuring that pronominal references have a clear source, requires on-line processing and monitoring of verbal output, clearly a working memory task. The capacity to generate inferences from stories has also been associated with working memory performance in TBI patients (Dennis & Barnes, 1990), and may represent an advance on the specification of how TBI impairs the ability to comprehend inferential links in language. Thus, continued refinement of models of cognitive dysfunction in TBI provide an opportunity to facilitate understanding of the communication problems seen.

34-5. CONCLUSIONS

In conclusion, recent advances in linguistic theory and methodology have been fruitful in delineating the nature of communication disturbance(s) experienced in particular TBI subjects. Although there has been some variability between studies using similar methods, there have also been some consistent findings, and in this chapter three approaches have been discussed. First, discourse analyses have revealed problems in maintenance of coherence from one utterance to the next, as well as problems in the ability to prioritize and sequence information according to knowledge of conventional discourse structure. Second, using pragmatic theory as a framework, it has been demonstrated that TBI can impair sensitivity to inference impeding comprehension and production of socially effective communication. Finally, conversational analysis has focused on spontaneous conversation, using a range of measures, from behavioral and linguistic theoretical backgrounds and demonstrated difficulty monitoring topic, taking into account the cooperative nature of conversation, and meeting the needs of a conversational partner. TBI does not result in uniform cerebral pathology and, not unexpectedly, a number of difficulties in communication have been observed, which appear to fall into different profiles. However, attempts to discriminate subgroups of TBI communication disturbance on empirical grounds are only just emerging (e.g., Hartley & Jenson, 1992; McDonald, 1993b). Many studies discussed have focused on small groups or single cases in order to circumvent problems of heterogeneity. This is sensible within a given study, but it may also explain some of the variability of findings produced by different research groups using similar methods. Study of communication disorders in TBI would benefit from clearer criteria with which to classify subtypes of communicative difficulty, and this task is made more feasible by the advent of some of the research techniques reviewed here. Although there is inherent variability in the TBI population, some characteristics of communication difficulties are extremely frequent and are likely to reflect common functional consequences of the trauma. In particular, coexisting cognitive impairments such as attentional, information-processing, and executive disorders appear to have an important role in producing many of the disturbances in communication seen. Efforts to demonstrate such associations using specific neuropsychological measures and indices of discourse failure have had limited success to date. This is not entirely unexpected because standard neuropsychological tests and discourse tasks differ substantially in the nature of scores yielded. Furthermore, it is entirely possible that particular manifestations of cognitive and executive dysfunction have differential impact on discourse versus other kinds of goal-directed tasks. Ongoing improvements in the sophistication of linguistic measures in TBI, along with advances in cognitive neuropsychological theory, should enable better specification of the impact of impaired cognitive functions on sustained and effective social communication in traumatic brain injury.

CHAPTER 35

Language Abnormalities in Psychosis: Evidence for the Interaction between Cognitive and Linguistic Mechanisms

Joseph I. Tracy
Allegheny University of the Health Sciences, Medical College of Pennsylvania/Hahnemann School of Medicine and the Norristown State Hospital Clinical Research Center, Philadelphia, Pennsylvania 19129

Frameworks that organize language and communication research in psychosis are described and the two major patterns of language deficits emerging from current research are reviewed (complexity-related deficits in oral speech and poor lexical cohesion/unclear reference). An integrated cognitive/neurolinguistic approach to language research in psychiatry is advocated. This approach argues that many linguistic acts have cognitive requirements and the way in which cognitive deficits interact with and disrupt linguistic processing needs to be specified through a cognitive/neurolinguistic model.

Oddities of language and communication have always been hallmarks of major psychiatric disorders, playing key roles in their clinical picture and, in some cases, their diagnostic criteria. The term "disorganized speech" in the current DSM-IV (American Psychiatric Association, 1994) criteria has been salient to definitions of schizophrenia going back to its original describers (Kraepelin, 1971/1919; Bleuler, 1911). "Pressured" or "expansive" speech and "flight of ideas" have always been crucial to the definition of mania. In spite of this, language dysfunction is rarely listed as a "characterizing" deficit in neuropsychological reviews of these disorders (Randolph, Goldberg, & Weinberger, 1993; Gray, Feldon, Rawlins, Hemsley, & Smith, 1991). This does not reflect a lack of good empirical data, as many studies have shown that schizophrenia and other psychotic patients differ from normals on a host of language measures. Instead, it reflects a common view that these language problems are caused by something else, namely, a disturbance in underlying thought structure or content (i.e., a thought disorder) or a failure in cognitive processing (e.g., working memory, attention).

I will briefly describe the two major research perspectives on language/communication abnormalities in psychosis that continue to frame current language research in psychiatry. Next, I will describe two patterns of results that emerge from recent research in psychosis and describe the cognitive factors used to explain them. Finally, I will advocate an integrated cognitive/neurolinguistic approach to language research. This approach argues that most linguistic acts have cognitive requirements, and that although a class of automatic linguistic processes may exist and need to be specified, most often cognitive and linguistic processes interact in fundamental ways that are quite difficult, perhaps even artificial, to disentangle. Throughout I will emphasize the major psychoses (schizophrenia and mania) because the disruptions in language and communication are most striking for these disorders and most language-oriented studies in psychiatry have had this focus.

35-1. MAJOR RESEARCH PERSPECTIVES

35-1.1. The Speech—"Thought"—Disorder of Schizophrenia and Mania

Thought disorder in psychiatric disorders was originally described by Bleuler (1911). By the "Bleulerian" view, speech anomalies in schizophrenia patients were spurned by abnormal conceptual structures or associative processes directly related to the psychotic state (e.g, bizarre mental content, autistic thought, tangentiality, poverty of speech, non sequiturs, derailment, neologisms, and "word salads"). The thought disorder of mania differed in that it involved more prominent flight of ideas and circumstantial speech. By this view, the mechanics of phonology, syntax, and semantics, when examined in isolation, were intact in the psychoses and placed no constraints on speech or comprehension.

Psychiatry, in an attempt to operationalize the "Bleulerian" position and despite evidence that thought and language are not isomorphic (see Rieber & Vetter, 1994), developed the tradition of measuring thought disorder through verbal behavior, that is, speech, making it impossible to empirically segregate true linguistic failures from disturbances in thought content or structure, or cognitive processes (e.g., The Scale for the Assessment of Thought, Language, and Communication Disorders, TLC; Andreasen, 1982).[1] Research in the "Bleulerian" tradition has sought to determine if "thought disorder" symptoms reflected disturbances in the underlying structure (actual semantic linkages) or function (speed or spread of activation) of lexical networks. Most studies have failed to demonstrate the former, that is, schizophrenics are not more prone to unusual responses on word association tasks (Laffal, 1965; see Cohen, 1978). Recently, however, Aloia, Gourovitch, Weinberger, and Goldberg (1996) reported results implying abnormality in the structure of semantic networks of chronic schizophrenia patients. Applying multidimensional scaling techniques to category fluency data, they found that, compared to normals, patients were less likely to group exemplars into subordinate clusters and produced category exemplars that "did not follow any logical ordering in two-dimensional space" (p. 270).

Maher's work (1972) reflected the "Bleulerian" position and investigated the function of semantic networks. For instance, Manschreck, Maher, Celada, Schneyer, and Fernandez (1991) reported a higher ratio of objects to subjects in thought-disordered individuals (object chaining, i.e., listing objects at the end of a sentence) and attributed this to an attentional deficit that promotes associative intrusions. Research using lexical priming has built considerable evidence suggesting that schizophrenia thought intrusions and derailments in discourse arise from the hyperactive functioning of semantic networks. For instance, Kwapil, Hegley, Chapman, and Chapman (1990) used a degraded-word recognition task and found increased semantic priming in schizophrenia patients compared to bipolars and normal controls. This effect has been delineated by Spitzer et al., (1994), who utilized both semantic and phonologic priming tasks. Spitzer et al. found more pronounced semantic priming in thought-disordered schizophrenia patients, and suggested that information spreads more quickly and farther through the semantic networks of these individuals. They also observed phonologic inhibition in normals (to permit unimpeded articulation), but not in thought-disordered patients. Spitzer et al. concluded that thought disorder was associated with more highly activated semantic and more disinhibited phonemic networks. Spitzer's work is an advance, specifying well the associational disturbance discussed by Bleuler and Maher, but leaves unclear the exact type of psychotic speech such abnormal activation causes and how this abnormality interacts with normal language operations.

---

[1]Note, the terms "thought-disordered" and "speech-disordered" are, unfortunately, used interchangeably in the literature, reflecting the underlying assumption that the problem is one of disturbed thought manifested through abnormal speech. The term "formal thought disorder" has been used to identify structural failures in speech unique to schizophrenia (although there is no consensus on its definition; see Jampala, Taylor, & Abrams, 1989) and to distinguish it from both mania and the phenomena of disturbed thought content. Use of "psychotic speech disorder" has been advocated by some, and is the recommendation of this author, as it is the less presumptive term.

35-1.2. Schizophrenic or Manic Speech as Aphasia

Kraepelin (1971/1919), so struck by the similarity with aphasia, used the term "schizophasia" to describe schizophrenic speech. Indeed, the at least superficial similarity of schizophrenic to fluent aphasic speech prompted many comparisons, with the hope that they could be understood by the same etiologic mechanism. The hypothesis of this perspective is that the speech disorder found in some psychotic patients constitutes a deficit in primary language mechanisms. Prominently representing this position, Chaika (1974, 1982) utilized taped interviews of a schizophrenia patient to identify characteristics that "suggest a disruption in the ability to apply those rules which organize linguistic elements, such as phonemes, words, and sentences, into corresponding meaningful structures, namely words, sentences, and discourse" (1974, p. 275). She suggested that schizophrenic speech represented an "intermittent aphasia," distinct from disordered thought that correlated with acute psychotic states, and was more prominent in the nonmedicated (for criticism see Fromkin, 1975).

Several empirical studies have distinguished schizophrenia from aphasia (Gerson, Benson, & Frazier, 1977). Faber et al., (1983) found that thought-disordered schizophrenia patients had better auditory comprehension and use of complex phrases and polysyllabic words than aphasics, but more private word usage. Faber et al. (1983) argued that no classic aphasia syndrome existed for schizophrenia despite sharing features with fluent aphasics (fluent/spontaneous speech, idiosyncratic word usage, paraphasic responses, empty speech, and so forth). Landre, Taylor, and Kearns (1992) found that speech-disordered schizophrenia patients and fluent aphasics were identical in language comprehension, single-word naming, repetition, and spontaneous speech (e.g., semantic paraphasias, unclear reference). Landre et al. also found that an index of general intelligence was strongly associated with language performance in the schizophrenia patients, leading the authors to attribute the problems to a general cognitive deficit.

Methodological issues plague comparative work with aphasia. Seldom are lesion locations matched within the aphasic sample. Regarding psychosis, standardized diagnostic procedures are not always used, and patients are often medicated and chronically institutionalized, making unclear the attribution of deficits to a primary speech/language disorder. There are many reasons to differentiate schizophrenia from aphasia. Psychodynamically oriented writers such as Sass (1992) have noted that the intermittent nature of schizophrenic speech indicates a dependence on social contexts such as stress, and perhaps a motivation to avoid focus on threatening material. Others have noted that the speech of psychotic patients does not show a recovery curve as aphasics do, and, unlike aphasia, appears resistant to speech therapy. Finally, no clear dissociations of language functions has been shown in psychotic speech, as might be expected if it was an aphasia (intact fluency with impaired comprehension). Many workers would accept Faber et al.'s (1983) conclusion that similarities with aphasia exist, yet few would argue that psychosis brings with it an aphasic disorder. Interestingly, even those that liken schizophrenia to aphasia, such as Chaika, consider the etiologic mechanisms of it and aphasia to be different.

35-2. PATTERNS IN LANGUAGE/ COMMUNICATION RESEARCH

35-2.1. Evidence for Complexity-Related Deficits in Oral Speech

Several studies have noted lower syntactic complexity in schizophrenic speech and suggested that more basic linguistic processes are intact (Fraser, King, Thomas, & Kendell, 1986; Morice & McNicol, 1986). Morice and Ingram (1982) found that complexity properties (frequency and depth of embedded clauses), fluency (pause-fillers, false starts, repeats), and integrity of speech (syntactic, semantic errors) distinguished schizophrenia, mania patients, and normal controls. Later, Morice and McNicol (1986) compared manic and schizophrenic speech and found that the latter used shorter sentences with fewer embedded clauses, more word repeating, fewer reduced relative clauses (elision), and more semantic deviancy. These authors also reported that as syntactic complexity increased, so did semantic deviance. Morice and McNicol held that basic receptive and expressive language functions such as naming, repetition, and simple syntax were intact in schizophrenia, but that deficits were present at larger, more complex units (i.e., the level of organizing sequential, coherent discourse). They concluded that schizophrenic speech is less syntactically complex than that of manics and healthy controls, although its stability and relation to clinical state or general cognitive deficits remain unclear.

Hoffman, Hogben, Smith, and Calhoun (1985) compared schizophrenia patients and controls using written output and found that errors were particularly frequent in representing input propositions when these propositions were written in the more complex passive voice. Later studies by Thomas, King, Fraser, and Kendall (1990; also King, Fraser, Thomas, & Kendell, 1990) found that chronic schizophrenia patients were more impaired on complexity measures (e.g., length of utterance, number of embedded clauses, levels of clause embedding) than acute patients and normal controls, and suggested that these problems were associated with negative symptoms, worsened with chronicity, but were unrelated to medication, institutionalization, or acute state. Recent work by Goldfarb, Stocker, Eisenson, and DeSanti (1994) used measures of communicative responsibility (production of comprehensible, usable speech) and found that schizophrenia patients had trouble generating a story in response to a picture (producing more irrelevant responses), whereas aphasics had trouble specifying the semantic features of a picture due to the naming demands. However, they found no differences between the groups at smaller, less complex units of linguistic analysis (e.g., specifying object properties or antonyms).

This complexity finding is not without challenges. Some research has suggested that not all basic functions (e.g., fluency, naming, comprehension, repetition) are intact in schizophrenia. Barr, Bilder, Goldberg, Kaplan, and Mukherjee (1989) found that semantic paraphasias during confrontation naming were a common error in their sample of chronic schizophrenics. Seidman, Cassens, Kremens, and Pepple (1992) reported poor sentence repetition in thought-disordered schizophrenia. Kolb and Whishaw (1983) observed poor oral word fluency. Anand, Wales, Jackson, and

Copolov (1994) studied syntax, semantics, cohesion, and metaphor in an early psychosis sample and found that semantic comprehension best discriminated their psychotic patients from controls. A study by Thomas, Leudar, Newby, and Johnston (1993) using a writing-output paradigm found that schizophrenia and mania patients' written products were equal in syntactic complexity (sentence length, subordinate clauses) to normal controls. This finding suggested a dissociation between oral and written speech, and thus looms important for future efforts at determining whether the "complexity" finding in psychotic speech is based on a general problem of competence.

Problematic is that some aphasics (Miyake, Carpenter, & Just, 1994) and other neurologic groups (right hemisphere and frontal lobe patients: Gordon, 1980; Lezak, 1995) also have difficulty processing complex linguistic input and giving complex output. Thus, it is unlikely that these deficits in complex linguistic processing are unique to psychosis.

35-2.2. Evidence for Poor Lexical Cohesion and Unclear Reference

Perhaps the most widely reported flaws in the speech of psychotic patients, not highly indicative of aphasic speakers, are poor cohesion between discourse elements and unclear reference for pronouns, phrases, and clauses. Because of these failures, the speech of these patients becomes more and more incomprehensible to the listener as it becomes more distant from initial referents and context.

Rochester and Martin's seminal study (1979) utilized interviews of thought-disordered and non-thought-disordered schizophrenia patients and established two widely cited aspects of discourse failure that distinguished the groups: (a) reliance on lexical or prosodic features to achieve cohesion between clauses, sentences, or prior context, and (b) poor introduction of nonshared, new information (i.e., unclear references for nominal groups, "predicative speech"). Harvey (1983; also Harvey & Serper, 1990) studied coherence and reference failures and showed that these failures are specific to the speech termed thought-disordered. Harvey (1983) found that the speech of thought-disordered patients, regardless of diagnosis, was more poorly integrated with fewer, less effective cohesion strategies and more incompetent references than both non-thought-disordered patients and controls. He also demonstrated that reference failures occurred in children of acute patients as well as in remitted patients, suggesting that such failures were not due to acute state. More recently, Harvey and Serper (1990), using auditory distraction, reality monitoring, and word-span/encoding tasks, found that incompetent reference was less severe in mania than in schizophrenia, and that in schizophrenia it was predicted by distractibility and poor reality monitoring. Ragin and Oltmanns (1986) analyzed lexical cohesion while controlling for verbosity and found that schizophrenia patients produced fewer within-clause lexical cohesion devices than manics, and that for manics this measure improved with treatment. More recently, Docherty, Sledge, and Wexler (1994) studied linguistic references in schizophrenia outpatients and their families during discussion of negative and positive memories. They found that reference performance deteriorated under the negative memories condition and that poor reference performance in parents predicted both performance and positive symptoms in their patient offspring. Importantly, both Harvey's and Docherty's data raise the possibility that reference failure represents a genetic vulnerability marker for schizophrenia.

The specificity of reference and cohesion failures to schizophrenia is uncertain, however, as mania patients appear to have similar difficulties (Miklowitz et al., 1991). For instance, Hoffman, Stopek, and Andreasen (1986) used discourse analysis (counts of phrase units in a semantic hierarchy that capture the truth suppositions of prior statements) to study the hypothesis that speech in schizophrenia and in mania both reflect cohesion problems but arise from different mechanisms. They found that mania patients used complex and well-organized plans but inappropriately shifted from one discourse plan to another ("structural shifts," i.e., a shift in topic), yet produced intact links between propositions in terms of logic and presupposition. Schizophrenia patients, in contrast, showed a deficiency in forming plans ("structural deficiency," i.e., poorly organized propositions within one topic). Hoffman et al. (1986) also found that a subgroup of "manic-like" schizophrenia patients with accelerated speech, and so forth, were quite similar in discourse to mania patients, suggesting that symptomatology, not diagnosis, best related to the linguistic process (see Taylor, Reed, & Berenbaum, 1994 for further comparison of the speech of schizophrenia and mania patients).

Based on the above, it is clear that providing unclear referents and poor cohesion are major findings for both schizophrenia and mania; however, several issues remain unsettled: the role of neuroleptic medication, chronicity, heterogeneity of the disorders, and the power of specific symptoms versus diagnosis to predict the speech disturbance. Some studies have identified a positive impact for medication on cohesion problems. For instance, Clark, Harvey, and Alpert (1994) found that medicated schizophrenia patients produced a greater number of clear references, and fewer episodes of bizarre, "interruptive" speech than nonmedicated patients. Also, chronicity may be a factor as syntactic and cohesion deficits appear most common in well-established schizophrenia (Morice & Ingram, 1982; Hoffman & Sledge, 1988; King et al., 1990). Again, the specificity problem exists as aphasic patients have also been reported to have difficulty reactivating antecedents, although perhaps for different reasons (Swinney, Zurif, Prather, & Love, 1994). Regarding the etiology of cohesion and reference failures, many of these researchers lean toward a cognitively oriented explanation, with the implication that the linguistic failures, while very real, are not the primary problem.

35-3. COGNITIVE EXPLANATIONS OF LANGUAGE/ COMMUNICATION ABNORMALITIES

Theorists who held that abnormal language was a mirror of disordered thought utilized broad cognitive explanations in hopes of explaining not just disturbed thought but other schizophrenic symptoms such as hallucinations and delusions (see Maher, 1972). Recent explanations of abnormal speech in psychosis have provided more specific cognitive models, drawn clearer links to specific linguistic failures, and, in some cases, proposed underlying biological substrates. Three cognitive functions have been implicated, and, generally, the dysfunction is viewed as disrupting both expressive and receptive language. The first posits a dysfunction in an attentional filtering mechanism. Schwartz (1982) held that schizophrenia and mania patients displayed a flawed filtering mechanism (pigeonholing) such that messages cannot be distinguished (and behavior cannot be guided) on the basis of prior knowledge (response sets), but only on the basis of immediately available, salient stimulus features as reflected in the tendency to repeat recently heard words, respond to strong, not weak, word meanings, and produce phonetic "clang" responses. As schizophrenia patients can discriminate phonemes, recognize and form words, and so forth, Schwartz concluded that there is adequate processing and filtering of basic linguistic features, and thus, schizophrenia is a cognitively based disorder of "speech" not "language" competence.

The second cognitive function implicated involves a failure to hold information over the short term and keep it available for manipulation during speech or comprehension, that is, a working memory failure. For instance, Harvey and Serper (1990) argued for a combined distractibility and short-term memory explanation of abnormalities such as unclear reference, suggesting that the "inability to discriminate the origin of information in short-term memory could lead to a situation in which discourse plans are confused for prior discourse, causing communication problems by leading to speech that refers to information never presented, only planned" (p. 492). The link between specific linguistic performance and working memory was well described by Thomas et al. (1993) in their analysis of written speech. They found that schizophrenia patients left out more propositions as syntactic complexity increased, and these errors correlated with working memory. Thomas suggested that working memory deficits placed limitations on the ability to organize and keep accessible all the input propositions needed for production of a final syntactic framework. Docherty, Hawkins, et al., (1996) found that reference performance was related to measures of working memory and sustained attention in nonacute schizophrenia patients, but related to concept formation and verbal fluency in nonacute bipolar patients. Another study from this group suggested that both manic and schizophrenic inpatients produced much higher frequencies of reference failures than normal controls, with indications that manics and schizophrenia patients show similar reference failures though to varying degrees (Docherty, DeRosa, & Andreasen, 1996). Finally, a study by Niznikiewicz et al. (1996) provided a good demonstration of the role that deficient working memory may play in certain linguistic findings. Schizophrenia patients showed an increase in N400 amplitude to "congruent" words at the end of a sentence (in normals this appears only for "incongruent" words), suggesting that patients were not using context to constrain associations to oncoming words.

The third cognitive dysfunction that has been proposed emphasizes a failure in planning or self-monitoring (executive functions). Morice (1986) suggested that prefrontal-based executive deficits underlie the reductions in syntactic complexity he had observed in prior studies. Barr et al. (1989) concluded that schizophrenia expressive deficits, particularly in object naming, are secondary to executive function/prefrontal-based deficits in self-monitoring, planning, and perseveration. Hoffman (1986) proposed a breakdown in the voluntary control of language, whereby schizophrenia patients fail to maintain a discourse plan. Specifically, he suggested that at the level of syntactic discourse potential structures strongly compete in schizophrenia but not in normals. Thus, plans unintentionally generated during unconscious thought, which are inconsistent with conscious goals, can intrude into ongoing speech or be misattributed as having an external source (hallucinations). Finally, a theory of flawed monitoring and planning with clear neuroanatomical foci was offered by Crosson and Hughes (1987). They noted that thalamic lesions in normals can result in a fluent aphasia that presents as a formal thought disorder (paraphasia, empty speech, poor turn taking). They suggest that a medial thalamic–basal ganglia–prefrontal cortex network that mediates language monitoring is disrupted in schizophrenia, producing compromised semantic accuracy in the context of intact motor aspects of speech.

35-4. AN INTEGRATED COGNITIVE/ NEUROLINGUISTIC APPROACH

Although cognitive explanations of communication breakdowns in psychosis may be fruitful and even correct, unclear reference, poor cohesion, and other language problems, for example, neologisms, still need to be explained at a linguistic level; that is, even if schizophrenia cognitive deficits do cause linguistic failures, the way in which the cognitive deficits interact with and disrupt linguistic processing still needs to be specified through a cognitive/neurolinguistic model. All but the most automatic of linguistic acts depend on the highly integrated functioning of multiple cognitive (e.g., attention, memory) and linguistic (e.g., lexical, semantic, phonologic, syntactic, pragmatic) systems that occur under social demands. For instance, selective attention may be needed to parse heard sentences, working memory to comprehend extended text, and executive functions to plan, organize, and monitor speech output consistent with social communication requirements. The empirical studies cited in the preceding section showing that linguistic performance covaries with extent or type of cognitive deficit (see Docherty, Hawkins et al., 1996) highlight the possibility that psychosis is an instance where cognitive and linguistic mechanisms interact.

Tracy, Glosser, and DellaPietra (1996) utilized the single-word retrieval model of Goodglass (1993) to identify the specific breakdown in linguistic processes that could account for a subset of naming errors in schizophrenia and integrated this with the well-established cognitive deficits in schizophrenia to show how the breakdown likely occurred. They reported two patients who produced combined phonemic/semantic errors on the Boston Naming Test, that is, responses containing correct phonological elements and a semantic relation to the target (e.g., *re-recorate* for *wreath*). In Goodglass's model the "interactive" track relies on monitoring (e.g., evaluating phonologic precursors, organizing output), selecting (e.g., selecting among early, undifferentiated semantic associates), filtering (e.g., deleting off-target phonemic or semantic inputs), and attentional switching (moving focus between phonemic and semantic tracks). This model permitted specification of where in the process of word retrieval (the "inter-active track") schizophrenia deficits in monitoring, selecting, filtering, and switching might exert their disruptive effects. This preliminary work provides an example of the way cognitive and linguistic mechanisms might be integrated to explain specific speech abnormalities in psychiatry.

35-5. FUTURE OUTLOOK

Psychosis is a clinical phenomenon in which both cognitive dysfunction and disordered speech are evident. It may be a particularly fertile ground to study ways in which cognitive and linguistic mechanisms could interact to produce flawed language and communication. Studies that combine measures of cognitive processes, linguistic mechanisms, and psychotic speech should prove most useful. Toward this end, two research approaches can be united. First, attempts to dissociate cognitive (e.g., attention resource limitations) and linguistic factors (e.g., syntactic processing) are common now in the aphasia literature (Martin, 1995; Swinney, Zurif, Prather, & Love, 1996). It is through such precise decoupling of linguistic and cognitive factors that we will be able to identify whether certain linguistic processes are truly automatic (require no cognitive resources) or the product of a cognitive/linguistic interaction. As measures highly sensitive to linguistic processing make their way into language studies of psychosis (e.g., the "P600" event-related potential response to syntactic anomalies; the "N400" response to sentential semantic anomalies—Osterhout & Holcomb, 1992; Osterhout, 1994), it will be possible to determine if psychotic patients display normal brain responses to linguistic stimuli and whether such responses vary as a function of cognitive load. Second, attempts to link specific linguistic failures to specific types of psychotic speech are emerging (see Anand & Wales, 1996, for discussion). Along this line, Barch and Berenbaum (1996) demonstrated several language production/communication disorder associations (e.g., monitoring speech related to derailment, non sequiturs).

Further testament that psychosis may be productive ground for studying the way cognitive and linguistic mechanisms interact is the observation that the morphologic regions implicated in schizophrenia—left frontal, left temporolimbic, basal ganglia, and thalamic structures—are also known to be involved in both language and cognitive processing (Roberts, 1991). For instance, the anterior frontal region considered important for word generation and sequencing (Nobre & McCarthy, 1995; Petersen, Fox, Posner, Mintun, & Raichle, 1988) has also been implicated in some of the cognitive deficits of schizophrenia (Weinberger, 1987). Basal ganglia and thalamic dysfunction have been implicated in schizophrenia (Gray et al., 1991; Andreasen et al. 1994), are considered important in attention (Posner & Petersen, 1990), and the left basal ganglia and left thalamus are integral parts of the subcortical, "habit learning" route connecting language-concept mediation areas of the occipitotemporal axis to anterior speech motor areas (Damasio & Damasio, 1992). Research on the overlap in the pathophysiology of schizophrenia, cognitive, and language disorders is in its infancy and constitutes an important area of inquiry. Unfortunately, we do not yet have models to explain how such shared anatomic substrates might, under certain conditions, translate into the flawed cognitive, linguistic, and symptomatic processes that we observe in psychosis.

Regarding language/communication abnormalities in psychosis, a host of questions remain: *(a)* Are they limited to certain subgroups of patients? *(b)* Do environmental events trigger them (e.g., negative affect; Docherty, Evans, Sledge, Seibyl, & Krystal, 1994)? *(c)* Are problems limited to acute exacerbations of the illness? *(d)* Are they mediated by the same neurochemical or neuroanatomical substrates that subserve cognitive deficits or psychotic symptoms such as hallucinations? *(e)* Are they influenced by psychotropic medication, chronicity, or institutionalization? *(f)* Do they imply abnormal language acquisition and support the possibility that some psychoses, that is, schizophrenia, are a developmental disorder? *(g)* Do they imply abnormal brain activation patterns during language processing? The role of medication is important as contrary reports emerge. Some data suggest that enhanced priming effects in schizophrenia are tied to antipsychotic medication (Barch et al., 1995). Others argue that antipsychotics most likely reduce language production disturbances (Barch & Berenbaum, 1996), and that medication and other factors (e.g., chronicity) are not key factors in language disturbances such as complexity deficits (King et al., 1990). Disorder subtypes may be important as some suggest that semantic priming effects are less reliable in paranoid schizophrenics compared to nonparanoids (Ober, Shenaut, & Vinogradov, 1995). Also, other patient characteristics may affect linguistic processing (e.g., cognitive deterioration). Factors such as medication, chronicity, and patient subtyping usually work as confounds to research findings, but need not if they are properly built into experimental designs.

In the case of psychiatric disorders, brain abnormality is inferred from the presence of impaired and quite multidetermined behavior; in aphasia, the job is different, characterizing the behavioral deficits of known brain lesions. Language research in psychiatry can take advantage of its multidetermined nature and less clear grounding in brain-behavior relations by recognizing that psychoses may provide naturalistic examples of how an abnormality arises from the interaction between multiple brain systems (linguistic, cognitive, and psychotic mechanisms). Such an interactionist perspective will allow us to move beyond dichotomies regarding language and thought, and debates about the primacy of linguistic versus cognitive impairment. Such a perspective will also be crucial for mapping out the complex neuroanatomy and neurochemistry that drive the abnormal speech and language seen in the psychoses.

Acknowledgments

This project was supported, in part, by NIMH Grants # MH53397 and # MH52690 awarded to the author.

CHAPTER 36

Landau–Kleffner Syndrome
Clinical and Linguistic Aspects

Gianfranco Denes

Department of Neurological and Psychiatric Sciences, University of Padova, Padova, Italy

Clinical neuropsychological and linguistic aspects of the Landau–Kleffner syndrome (LKS, acquired childhood epileptic aphasia) are reviewed. Unlike child aphasia following focal lesions, LKS is characterized by severe auditory comprehension disorders and by a strong relationship between age of onset and outcome: the older the child at onset, the better the outcome.

Alternative hypotheses underlying comprehension disorders (purely linguistic versus sensory higher-order auditory processing deficit) are discussed.

A specific impairment of language development can occur in two forms: a deficient acquisition, usually named *Developmental Dysphasia* (D.D.), or a loss of previously acquired language skills, *Acquired Aphasia* (A.A.) (Van Hout, 1992).

Among the latter, of particular importance both in terms of frequency (more than 160 cases reported from 1957 to 1990; Paquier, Van Dongen, & Loonen, 1992) and severity, is the symptom complex of "acquired aphasia and convulsive disorders," first described by Landau and Kleffner in 1957 (Landau–Kleffner Syndrome, LKS).

The lack of structural abnormalities, the prevalence of auditory comprehension deficits, and the poor outcome, often leading to a permanent loss of language production and comprehension, differentiate LKS from other forms of child aphasia following focal lesion (Van Hout, 1992).

The aim of this chapter is to briefly describe the clinical characteristics, to review the linguistic deficits, and to speculate on the nature of the underlying auditory comprehension deficit, if primarily linguistic or secondary to an impairment of a sensory/cognitive mechanism necessary for language development and use.

36-1. CLINICAL PICTURE

LKS occurs in children who achieved developmental milestones at an appropriate age. The onset varies from 18 months to late infancy (Bishop, 1985; Paquier et al., 1992; Gerard, Dugas, Valdois, Franc, & Lecendraux, 1993), with the peak incidence between 2 and 5 years, affecting subjects of both genders who had already developed age-appropriate speech. Exceptionally, LKS is seen in subjects with a history of developmental language disorder (Marien, Saerens, Verslegers, Borggreve, & De Deyn, 1993).

Etiologies such as neurocysticercosis (Otero, Cordova, Diaz, Garcia-Teruel, & Del Brutto, 1989), cerebral arteritis (Pascual-Castroviejo, Lopez-Martin, Martinez-Bermejo, & Perez-Higueras, 1992), and inflammatory demyelinating disease (Perniola et al., 1993) have been discussed in some cases, but so far they have not been confirmed in the majority of cases.

The onset may be subacute or "stuttering" and the first clinical manifestation of the language disturbance is an auditory inattention to both verbal and nonverbal stimuli. The symptom may progress up to a pattern of cortical or word deafness (verbal auditory agnosia, VAA; Rapin, Mattis, Rowan, & Golden, 1977). At the same time, expressive language deteriorates with the appearance of phonological paraphasias and articulatory disorders. Eventually, the child may become completely mute, expressing himself or herself through unarticulated sounds or by the use of appropriate gestures. Rarely, fluent aphasia (Lerman, Lerman-Sagie, & Kivity, 1991) and jargon have been observed (Landau & Kleffner, 1957; Rapin et al., 1977; Paetau et al., 1991).

The second landmark of the syndrome is epilepsy, present in more than 80% of the cases (Dugas, Masson, Le Henzey, & Régnier, 1982; Beaumanoir, 1985). The seizures are heterogeneous in both their manifestations and their frequency. Generalized seizures are of the motor type and consist of eye blinking or brief ocular deviations.

In contrast to the language impairment, seizures have a benign course, are responsive to common antiepileptic standard treatment, and, by the age of adolescence, generally subside (Mantovani & Landau, 1980).

The spectrum of EEG abnormalities reported in LKS may vary. The most common abnormality is represented by bilateral independent or synchronous temporal or temporoparietal spikes, bilateral 1-to-3-Hz spike and wave activity centered over temporal regions or generalized sharp waves. An EEG abnormality recorded during slow-wave stages of sleep and consisting of continuous spikes and waves (Electrical Status Epilepticus, ESES) has been described in a number of LKS patients (Patry, Lyagoubi, & Tassinari, 1971; Tassinari, Bureau, Dravet, Dalla Bernardina, & Roger, 1985) and related to the severity of their language impairment (see Figure 1).

EEG recordings taken at the earliest stages of illness, or waking recordings at any stage can show the presence of unilateral, focal, posterior-temporal discharges.

Neurological examination, pure tone audiograms, CSF examination, and neuroimaging studies (CT scan and MRI) are within normal limits. In the few metabolic studies

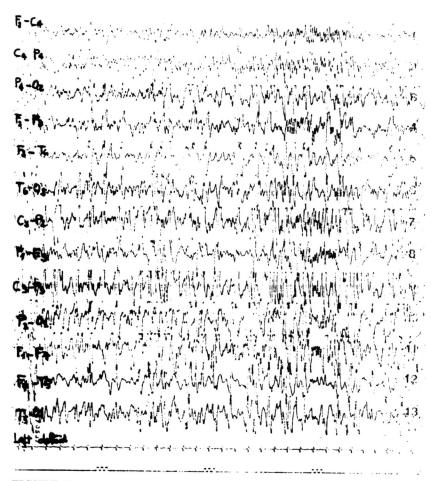

FIGURE 1   Paroxismal status epilepticus recorded during sleep in an LKS patient.

TABLE 1
Landau–Kleffner Syndrome: Summary of the Clinical Findings

1. Subacute (or sometimes stuttering) onset of progressive comprehension and production language disorders in a 2–5-year-old normal child who has already developed age-appropriate language competence.
2. Predominance of auditory comprehension disorders, including auditory agnosia in the initial phase.
3. Absence of psychotic traits and neuropsychological evidence of preservation of nonverbal cognitive abilities.
4. Mild to pronounced behavioral abnormalities (hyperactivity, aggressiveness, depression).
5. Absence of associated neurological signs. Normality of neuroimaging study (CT scan, MRI, cerebral angiography).
6. Clinical seizures usually but not invariably present. Heterogeneous in character, often subtle and involving only eye movements. Good response to anticonvulsant drugs.
7. Unilateral or bilateral independent, temporal or temporoparietal foci, bilateral spikes and waves activity. Continuous spikes and waves activity during slow sleep (ESES).

Note: From Morrell et al., 1995, modified.

performed (SPECT and PET), a reduced brain glucose utilization, as well as an asymmetrical hypoperfusion predominantly over the temporal lobes, were found (Maquet et al., 1990).

The language impairment cannot be regarded as the most evident symptom of a generalized cognitive deficit, since nonverbal abilities such as memory, orientation, and praxis are well within normal limits. Only Papagno and Basso (1993) described a persistent mathematical impairment well after the resolution of aphasic symptomatology.

The presence of psychotic traits is exceptional (Dugas et al., 1982; Gordon, 1990; Hirsch et al., 1990; Zivi, Broussaud, Daymas, Hazard, & Sicard, 1990), while behavioral disorders ranging in severity from mild to pronounced and characterized by hyperactivity, sometimes coupled with aggressive traits, are often reported (White & Sreenivasan, 1987; Sawhney, Suresh, Dhand, & Chopra, 1988).

Table 1 summarizes the usual clinical findings.

36-2. PROGNOSIS AND TREATMENT

Although epileptic seizures have a benign course, generally subsiding at the time of adolescence, the case is unfortunately different as far as the language deficit is concerned: the rate of recovery is, in most cases, poor, sometimes characterized by remissions and exacerbations that parallel the EEG course.

Corticosteroid treatment has been found to have a beneficial effect on language disorders with a concomitant disappearance of EEG abnormalities; this effect is, however, transient, with recurrence of electrical abnormalities and language disturbances when the drug is tapered off (Marescaux et al., 1990).

A dramatic improvement of language function up to a total reacquisition of language after 2 years of mutism was reported by Morrell et al. (1995) in a subset of LKS children, following the surgical elimination of the epileptogenic discharge from the perisylvian region of one hemisphere by multiple subpial intracortical resections.

A close relation was found by several authors (Toso, Moschini, Cagnini, & Antoni, 1981; Dulac, Billard, & Arthuis, 1983) between age at onset and linguistic accomplishment: the younger the child, the poorer the outlook. In a metanalysis of the published cases, Bishop (1985) extended and confirmed the conclusions of previous authors and attempted an intepretation of this age-related clinical course: If, as suggested by different sources (see Section 36-4.1), the nature of the language impairment is a high-level auditory processing disorder with the effect of blocking the auditory path to language areas, the process of language acquisition in a child whose language is not yet fully developed will be blocked. On the other hand, the auditory deprivation effect will be less severe in older children, whose language acquisition has already been accomplished, even if not to its full extent.

The auditory processing disorder hypothesis nicely fits with the finding of the poor recovery of language following conventional speech therapy, while a substantial improvement has been achieved through the visual modality. Rapin et al. (1977) first noted the retained ability of some of these children to acquire language through the visual modality. Fostering communication with use of sign language, a method that was found very effective in restoring communication in a case of pure word deafness following early bilateral temporal damage (Arias et al., 1995), has not been employed in LKS children up to now, perhaps because of the reluctance of professionals and parents to resort to a method of developing sign language in a nondeaf child (Deonna, Beaumanoir, Gaillard, & Assal, 1977).

The LKS boy reported by Denes, Balliello, Volterra, and Pellegrini (1986) was thought, when 6 years old, to match objects to their corresponding written word. The boy, whose auditory comprehension and oral production were nil, was able to learn reading and writing to such an extent that he could successfully attend primary school.

36-3. DESCRIPTION OF LINGUISTIC IMPAIRMENT

In the most severe cases, children are unresponsive to both verbal and nonverbal sounds (cortical deafness), followed by a recovery of the ability to respond to nonverbal sounds (Dugas et al., 1982) or, in some cases, to correctly identify the meaning of a nonverbal sound, as shown by their correct performance in matching a nonverbal meaningful sound with the corresponding picture (Denes et al., 1986). Discrimination and identification of vowel and consonant sounds are severely impaired (Deonna et al., 1977; Denes et al., 1986). Equally impaired is performance on auditory comprehension tasks, such as the Peabody Picture Vocabulary Test or the Token Test (Klein et al., 1995). Production disorders are equally severe up to almost total disappearance of speech, oral production being limited to nonverbal sounds. Singing also can be absent.

In less severe cases, oral production, although defective, is present and a detailed analysis has been reported by Van De Sandt-Koenderman, Smit, Van Dongen, and Van Hest (1984) in an LKS child who showed a fluctuating course. Total number of words and mean length of utterences matched the clinically observed fluctuations of aphasia. Analysis of paraphasias showed that most of them could be classified as literal paraphasias. Semantic paraphasias were absent, while neologisms, present in the stage of language breakdown, tended to disappear very fast during the recovery phase.

In contrast to the severity of the language impairment in the oral–auditory modality, the ability to learn and use written language can be significantly superior. Few experimental studies have examined the extent of language achievement through the visual modality. C.S., the LKS patient described by Denes et al. (1986), whose oral language was totally disrupted, showed a discrepancy between lexico-semantic and morphosyntactic aspects of visual language processing. His performance was flawless in lexical decision tasks, as well in the written naming subtest of BDAE (Goodglass & Kaplan, 1972) and in written naming, following semantic cuing. On the other hand, in contrast to age-matched controls, he showed a marked difficulty in handling free morphemes (articles, prepositions), whereas he did not show any particular difficulty in using bound morphemes that distinguish number, gender, and person. This discrepancy seems to support Caramazza, Berndt, and Basili's (1981) hypothesis that producing and understanding function words demands some form of phonological coding and decoding.

36-4. HYPOTHESIS ON THE UNDERLYING DEFICITS

36-4.1. Neurophysiological Findings

Experimental data pointing to a disruption of a specific neurological pathway as the basis of the language comprehension deficit in LKS patients was provided by Gascon, Victor, Lombroso, and Goodglass (1973) and Denes et al. (1986), who studied evoked responses to auditory stimuli. In both studies, the early components of evoked potentials (EPs) were normal, while long latency event-related potentials (ERPs), such as the P3, were abnormally long in latency and of reduced amplitude. These data point to a normal conduction of acoustic stimuli at the peripheral and brainstem level up to the cerebral cortex, with a selective difficulty when acoustic stimuli are further processed.

From a neurolinguistic point of view, the application of the EP and ERP method can help to disentangle the problem on the nature of auditory comprehension deficit, if primarily acoustic in nature or subsequent to a phonemic decoding disorder, leading to a selective inability to extract phonetic information from the acoustic speech signal.

In a recent study, Klein et al. (1995) examined six young adults who had presented a pattern of verbal auditory agnosia in childhood and exhibited some residual speech processing impairment of varying degrees at the time of testing. Four of them had bilateral EEG abnormalities of epileptic type and had presented epileptic seizures, thus confirming the diagnostic criteria of LKS. Auditory brain stem responses (ABRs), middle latency responses (MLRs), and ERPs to nonverbal and verbal (synthetic consonant-vowel syllables) stimuli were recorded. ABRs and MLRs were normal, while ERPs showed an abnormal delay in the N1 component recorded over the lateral temporal cortex. This abnormal slowing of processing for speech and nonspeech stimuli seems to suggest a global deficit of auditory processing rather than a language specific deficit, supported by a structural, although not detectable, abnormality at the level of secondary auditory cortex.

36-4.2. Behavioral Data

Tallal and Piercy (1974, 1975) were able to demonstrate a specific deficit in language impaired children in discriminating and indentifying rapidly presented acoustic events with a resulting deficit in speech perception. In other studies, Tallal and Stark (1987) and Tallal, Sainburg and Jernigan (1991) showed that language-delayed children have a specific difficulty in both perceiving and producing those speech sounds and syllables that incorporate rapidly changing acoustic spectra.

Given the severity of the central acoustic deficit, most LKS patients were unable to perform the test. Only two children with visual–auditory agnosie (VAA) and impaired production were given the task and it was found that their performance improved when the duration of formant transition was artificially lenghtened (Frumkin & Rapin, 1980). If, as suggested by previous studies, the deficit is prephonemic, all linguistic operations not involving the auditory channel should be performed at a normal level. If, on the contrary, the lack of phonemic input has prevented the acquisition of cerebral phonological abilities, all cognitive processes that require phonological mediation should be equally affected.

This question was experimentally investigated by Denes et al. (1986) by submitting their LKS patient to a series of tests using the visual modality that requires phonological mediation. Following Posner's seminal work (Posner, 1978; Posner & Mitchell, 1967), it is well established that when two letters are visually presented and matched for physical identity (e.g., AA), for nomimal identity (e.g., Aa), or for nonidentity (e.g., AB), a reaction time (RT) pattern emerges in which physically identical pairs are responded to faster than either of the other two types.

On the basis of this evidence, Posner has suggested that "AA" pairs are matched using a visual code, which is generated rapidly and does not involve linguistic analysis, while "Aa" pairs are matched using a phonetic or name code, which involves the phonetic representation of the letter names. The availability of the two proposed codes was investigarted in C.S., and his performance was compared with that of a group of 24 children of the same age and school level. C.S.'s performance did not significantly differ from that of the normal controls, suggesting an unimpaired availability of phonetic or name code. It must be pointed out, however, that Boles and Douglas (1983) challanged the existence of the phonetic mediation in name identity tasks. In normal subjects, short-term memory (STM) for sequences of phonologically similar items (strings of letters or acoustically similar words) is worse than for dissimilar items. This effect does not depend on the mode of presentation (visual or auditory). This and other evidence (Conrad & Hull, 1964; Conrad, 1971; Baddeley, 1976) has suggested a critical role of phonological mediation in verbal STM. C.S.'s mean score was 4, for both similar and dissimilar strings of letters, thus showing no effect of phonological similarity.

36-5. CONCLUSIONS

Four decades after it was first recognized as an autonomous entity, many aspects of LKS remain unresolved: although the hypothesis of the central auditory processing disorder as the basis of the language comprehension disorders is suggested by converging evidence from behavioral and neurophysiological studies, it is not clear why, at least in some cases, a regression of written language abilities in some previously literate children should appear (Deonna et al., 1977; Mantovani & Landau, 1980).

Equally thorny is the question why LKS children should, in the absence of demonstrable structural lesions, have a prognosis significantly worse than children with aphasia following focal lesions. Morrell et al. (1995) proposed that the presence of an epileptogenic focus in language-specific cortex during development would activate elements in the language circuitry that would not normally be active and thereby perpetuate and sustain synaptic arrangements that are functionally inappropriate. Only the cessation of epileptiform discharges through surgery performed before the end of the critical period of language acquisition will allow restoration of functionally appropriate connections within language-specific circuitry.

Finally, the finding that, following surgical treatment LKS subjects who were language deprived for years can recover at normal level (Morrel et al., 1995) seems to show that the language system can survive an almost total deprivation, without any major effect on future development.

Acknowledgment

Part of the work was supported by an MPI 60% grant. Thanks are due to Dr. G. Casara, Department of Pediatrics, University of Padua, for performing the EEG recording.

CHAPTER 37

The Development of Language in Some Neurological Diseases

Shirin Sarkari, Arlene A. Tan, and Dennis L. Molfese
Department of Psychology, Southern Illinois University, Carbondale, Illinois 62901

The development of infants and children with neurological disorders such as Down's syndrome, Williams syndrome, Turner syndrome, Persistent Hyperinsulineric Hypoglycemia of Infancy, Sturge–Weber syndrome, and Porencephaly is reviewed in this chapter. The clinical and neuroanatomical features of each syndrome are provided and studies investigating deficits in language development and in visuospatial skills specific to each disorder are discussed.

The development of language is intricately interwoven with neuroanatomical development (Jernigan, Bellugi, Sowell, Doherty, & Hesselink, 1993), development of cognition (Piaget & Inhelder, 1969), articulatory abilities (Mahoney, Glover, & Finger, 1981), receptive abilities (Kumin, Councill, & Goodman, 1994), and social development (Williams, 1994). While various deficits associated with any of these factors may result in linguistic delays, others may impair the development of visuospatial abilities, while preserving language skills. Moreover, some neuroanatomical deficits associated with these syndromes may selectively affect the development of language, dissociating between different aspects of linguistics. This chapter provides a review of specific neurological disorders, focusing on the related language and cognitive impairments that are associated with each disease.

37-1. DOWN'S SYNDROME

37-1.1. Neuroanatomical Features

Down's syndrome (DS) is a congenital condition derived from trisomy of chromosome 21 in 95% of cases, and is characterized by typical facies and mental retardation (Smith & Jones, 1982). Neuroanatomical features include reduced overall cerebral volume with relative preservation of the basal ganglia. The frontal cortex is disproportionately reduced in volume. Moreover, a dramatic reduction of limbic structures of the temporal lobes (including the uncus, hippocampus, amygdala, parahippocampal gyrus, etc.) is usually present. There is a striking larger asymmetry in the temporal limbic region (Jernigan et al., 1993).

37-1.2. Language in Down's Syndrome

A review of research on language development indicates that the pattern and sequence of development of language in DS children is similar to that of nonretarded children (babbling, developmental patterns of consonant production, etc.) (see Pruess, Vadasy, & Fewell, 1987 for a review). However, it has been postulated that deficits in expressive language development do occur, possibly due to auditory deficits, articulation disorders, cognitive delays, and environmental factors. For example, when DS and normal children were chronologically age-matched, DS children were found to show a distinct developmental lag in speech production (Leifer & Lewis, 1984). When matched with normal controls by mental age, however, DS children were found to be superior in conversational response abilities. Leifer and Lewis (1984) concluded that during the one-word stage of development, DS children may be more advanced than controls, but stay at this stage longer, resulting in delayed language development.

Evidence indicates that DS children are at risk for malformations of the auditory system, such as small external ears and a narrowed external auditory canal. These malformations frequently lead to an increased incidence of otitis media in this population. Whiteman, Simpson, and Compton (1986) report that the recurrence of early otitis media may be a contributing factor to language deficits often associated with DS children. Although a direct causal relationship between hearing loss and language deficits has not yet been established, it is imperative that children with DS receive comprehensive auditory assessments regularly in the first few years of life (Kumin et al., 1994; Pruess et al., 1987; Strome, 1981; Whiteman et al., 1986).

Another consideration in the cause of language delays observed in DS children is the presence of articulatory deficits due to hypotonicity of the orofacial musculature. Mahoney et al. (1981) examined the relationship between sensorimotor and language development in DS and non-retarded children with matched developmental ages of 17 months. They determined that the deficit in language ability was related to poor vocal imitation skills. However, Mahoney and Snow (1983) concluded that cognitive and sensorimotor factors were significantly correlated with expressive language delays. The relationship between cognitive abilities and language development has been investigated by various researchers (Hill & McCune-Nicolich, 1981; Moore & Meltzoff, 1978; Piaget & Inhelder, 1969; Smith & von Tetzchner, 1986). The development of object permanence and object identity has been found to be a crucial aspect of language formation. Hence both cognition and articulatory ability are implicated in the deficits in language seen in the DS population.

An additional aspect to the evaluation of expressive language development is the assessment of phoneme acquisition. Normative data previously reported require a mastery level based on achieving 75–90% accuracy in phonemic production (e.g., Smit, Hand, Freilinger, Bernthal, & Bird, 1990). More recently, Kumin et al. (1994) postulate a change in approach to the evaluation process in DS children. Rather than mastery of specific phonemes as measured by formalized articulation testing, the authors suggest looking at the emergence of sounds, and the use of these emergent phonemes in daily communication.

Other theories have been introduced in an attempt to account for the language deficits observed in DS children. For example, Bird and Chapman (1994) tested the hypothesis that DS patients experience most difficulty with language tasks due to a "pervasive sequencing" deficit, where recalling the order of information is problematic (Rosin, Swift, Bless, & Kluppel Vetter, 1988). According to Bird and Chapman (1994), neither a deficit in sequential processing nor specific difficulties in recalling order of information were found. Alternatively, these authors suggest that deficits in auditory working memory span and a rate-based limitation to the phonological loop could account for the problems in language production observed in DS individuals. However, these possibilities need to be validated by future research.

Mundy, Kasari, Sigman, and Ruskin (1995) proposed that the acquisition and development of nonverbal communication skills impact development of language. These skills reflect the development of cognitive processes (e.g., capacity for representational thought, executive functions such as selective inhibition of responses, and planned action sequences). Mundy et al. (1995) found that DS children were impaired in nonverbal requesting regardless of developmental level. In addition, nonverbal requesting was positively correlated with language outcome. These findings supported those of Mundy, Sigman, Kasari, and Yirmiya (1988) and Beeghly, Weiss-Perry, and Cicchetti (1990).

Few studies have investigated language comprehension in DS individuals. Chapman, Schwartz, and Bird (1991) found that the differences between lexical and syntactic skills differed with increasing age in DS children. Although the differences are distinct, the source of the discrepancy remains unresolved. Chapman et al. (1991) suggest that the differences may be either the result of a lexical advantage or due to a syntactic deficiency.

The studies cited above provide evidence for many contributing factors, but further research in psycholinguistics and the acquisition of language is needed in many areas before the language deficiency associated with DS can be fully understood. Causes of language delays in DS children are still debatable, with problems with hearing and cognitive difficulties being contributors, but not necessarily causal factors. Instructional strategies need to be highly structured and a clear "best way" has not been found. Language intervention should occur early, with families and school being very involved in the process so that the best use can be made of experiences occurring naturally.

37-2. WILLIAMS SYNDROME

37-2.1. Neuroanatomical and Clinical Features

Williams Syndrome (WS) is a rare autosomal dominant disease that may have an idiopathic or a genetic etiology. It occurs in around 1 in 25,000 live births (Udwin & Yule, 1991). First described in 1961 by Williams, Barratt-Boyes, and Lowe, the distinguishing physical defect is a supravalvular aortic stenosis that impedes blood flow from the aorta. This disorder is found predominantly (90%) in males. An abnormal regulation of calcitonin may result in an infantile hypercalcemia, which causes an abnormal density of the bones. This often causes cranial sutures to close early, giving rise to micro-encephaly in some cases (Culler, Jones, & Deftos, 1985). Increased intracranial pressures, hernia, and cardiovascular anomalies are also present in severe cases. Other features include an elfin-like facies and mild to moderate mental retardation (with IQ scores in the 40–60 range), but with relative preservation of linguistic and affective skills (Wang & Bellugi, 1994).

In early infancy, irritability, failure to thrive, and feeding difficulties have been described. Common features of WS include hyperacusis, hyperactivity, deficient social skills, and inappropriate behaviors. Commonly, oral-cavity malformations, including microdontia, hypoplasia of the enamel, caries, and missing teeth have been described. Occlusal problems such as micrognathia and prognathia are also found in many cases of WS (Myerson & Frank, 1987).

Cerebral MRIs of WS individuals have shown that cerebral volume is significantly decreased, while neocerebellar vermal lobules are increased in size when compared to DS children (Wang, Hesselink, Jernigan, Doherty, & Bellugi, 1992).

37-2.2. Language in Williams Syndrome

One distinctive feature of WS is the dissociation of language and cognitive functions. WS individuals display superior language abilities in contrast to impaired cognitive functions (Wang & Bellugi, 1993; Wang, Doherty, Hesselink, & Bellugi, 1992). According to Reilly, Klima, and Bellugi (1991), adolescents with WS are not capable of resolving conservation tasks; that is, in a standard Piagetian situation, the children are unable to determine whether one row of objects contains more or fewer elements if the overall length of the two rows is the same. However, these same children excel in receptive and expressive vocabulary tasks. WS individuals also possess a remarkable command of grammar and vocal prosody.

In a longitudinal study of early lexical development, WS, DS, and normal children, aged 13–26 months, were followed for 15–26 months (Mervis & Bertrand, 1995).

Children with WS were found to do significantly better in language items than in cognitive items, whereas DS children displayed the opposite pattern. Normal children passed equal numbers of language and cognitive items.

Wang and Bellugi (1993) tested a group of 10 adolescents with WS, comparing their performance with DS subjects who were matched for age, full scale IQ (FSIQ) score (41–59), and educational background. A wide battery of tests including tests for cognitive functioning, language, and neurophysiological measures was administered. Both WS and DS subjects displayed general impairments in cognition. Subjects were also tested separately on each subdomain of language—grammar, affective prosody, semantics, and narrative abilities. Subjects with WS performed above expected levels relative to their cognitive abilities. They had large vocabularies and used significantly more uncommon words than either the DS subjects or normal controls. WS subjects also produced more syntactically and morphologically complex and well-formed sentences than children of equal mental age. In addition, these subjects showed good paralinguistic skills, including narration and affect.

In contrast to studies finding a high degree of loquacity and verbal skill in WS subjects, other researchers have found low FSIQs (between 41 and 72) with no significant verbal and performance differences (Bennett, LaVeck, & Sells, 1978; Arnold, Yule, & Martin, 1985). Strang and Rourke (1985) have identified one group of WS children as resembling what they describe as "a nonverbal perceptual organization-output disability" (NPOOD). These children are characterized as being overly talkative, with no notable difficulty with syntax, but having a tendency to use clichés often, and with an unfocused quality to their speech.

Kataria, Goldstein, and Kushnik (1984) tested seven children with WS, 18–71 months old, (M = 48 months). All were from lower-middle-class to middle-class backgrounds. All had borderline to severe mental retardation. The five youngest children were tested with the Mental Scale from the Bayley Scales of Infant Development (Bayley, 1969). The two older children were tested with the Stanford-Binet Intelligence Scale (Terman & Merrill, 1973). The Receptive-Expressive Emergent Language Scale (Bzoch & League, 1978) and the Peabody Picture Vocabulary Test (PPVT) (Dunn, 1959) were used to test specific language development in the older and younger children, respectively. In addition, motor development was tested. Follow-up testing of intelligence and language development was completed 14–42 months after the first evaluation. All seven children tested in the mild to severe retarded range (IQs 28–64 [M = 50.6]). Five of the seven children had relatively high verbal skills when compared to their performance subtests. None, however, reached the statistical .05 level of significance. The average range for language scores was not even close, as would have been expected from earlier reports. This is an important finding as remedial early intervention can be attempted to maximize whatever potential these children have.

Although it may appear that children with WS are verbally strong and possess superior articulatory skills, these very attributes could mask more debilitating cognitive deficits. Continued research in this area, with children and adult patients, would contribute to more skillful interventions. Further, cross-sectional studies with larger sample sizes would help determine the justifiability of differing findings in this population.

37-3. TURNER SYNDROME

37-3.1. Anatomical Features

Turner Syndrome (TS) affects 1 in 2,500 newborn female children. The anatomical characteristics of children with TS include a chromosomal variation of chromosome 45, X (XO syndrome), with all or part of one X chromosome missing; gonadal dysgenesis; and congenital renal and cardiac anomalies. Other physical features include sexual infantilism, webbed neck, elbow deformations, small stature, and high-frequency hearing loss (50%).

37-3.2. Cognitive Impairments

TS has been associated with a high incidence of neurocognitive impairment. This is specific to mental processing and sequencing of rotational transformations of shapes in the spatial dimensions of left/right, up/down, front/back (Money, 1993; Williams, Richman, & Yarbrough, 1992). The syndrome does not affect verbal intelligence but may severely affect nonverbal functioning (Williams, 1994). A marked and significant disparity between Verbal IQ (VIQ) and Performance IQ (PIQ) may be >30 points with a VIQ of >130 (Money, 1993).

In daily living, the deficit in spatial processing may prove a handicap in such tasks as map reading, following travel directions, deciphering sounds stereophonically, and recognition of facial expressions (Money, 1993). There is no handicap in learning to read with a rotational transformation disability—in fact, there is an advantage as these children do not mix their *p/q, b/d,* and so on.

Lewandowski, Costenbader, and Richman (1984) tested 10 females with TS between 9 and 33 years of age. The researchers found that the subjects scored significantly lower in PIQ in comparison to controls. There was no difference in VIQ scores between TS subjects and controls. TS subjects performed significantly poorer than controls on the Raven Color Progressive Matrices, providing further evidence for the hypothesis that TS subjects have a visuoperceptual deficit independent of verbal, memory, and motor skills.

Shucard, Shucard, Clopper, and Schacter (1992) found that TS females are deficient in visual–spatial processing tasks. Again, VIQ scores between TS subjects and controls did not differ significantly, although the comprehension and arithmetic subtests showed lower scores in TS subjects.

Williams (1994) tested 10 children with TS on various measures including behavioral and verbal abilities. The controls were 10 children with no learning disability history, who were acting as controls in another endocrinology study. Ten other children diagnosed with a nonverbal learning disability (NLD) were also recruited into the study. No overall significant main effect for age or VIQ was found (p > .05).

37-3.3. Language in Turner Syndrome

Language acquisition in TS subjects does not appear to be impaired. TS children tend to have verbal and language abilities with the average range, although spatial relations and poor perceptual motor organization skills tend to pull down nonverbal scores. Visual–spatial functioning and nonverbal learning disabilities are the areas in which they show deficits (Netley, 1983; Williams, 1994). In addition to nonverbal learning disabilities, Williams (1994) describes these children with TS as having problems with peer relationships, behavior, and social functioning.

37-4. PERSISTENT HYPERINSULINERIC HYPOGLYCEMIA OF INFANCY

37-4.1. Clinical Features

Persistent Hyperinsulineric Hypoglycemia of Infancy (PHHI) is a condition wherein significant and rapidly occurring low glucose levels in the blood may result in seizure activity and permanent CNS damage. Diagnostic criteria include:

1. Serum glucose levels of <40mg/dL. during the first 72 hours, with detectable circulating insulin levels;
2. Low levels of free fatty acids and ketones during hypoglycemic episodes;
3. A glucose requirement of >15mg/kg/min; and
4. A glycemic response to glucagon despite hypoglycemia (Aynsley-Green, Polak, & Bloom, 1981).

Hypoglycemia in infancy may lead to irreversible damage in the central nervous system. The brain damage noted is similar to that found with ischemia (Auer, Wieloch, Olsson, & Siesjo, 1984). In the autopsies of three children with hypoglycemia, extensive degeneration of neurons throughout the CNS was noted. Basal ganglia and occipital lobes showed the most degeneration (Anderson, Milner, & Strich, 1967).

37-4.2. Language in PHHI

In a study by Gross-Tsur, Shalev, Wertman-Elad, Landau, and Amir (1994), seven school-age PHHI patients and six healthy controls matched for age, sex, IQ, and socioeconomic status (SES) underwent a battery of tests. IQs of six out of the seven were in the normal range (FSIQ scores 91–123; the sixth had an FSIQ of 65). However, writing skills and serial-word learning of PHHI patients were found to be significantly impaired. This was demonstrated by primary phonetic errors in the spelling of words, and missing letters and words. Frontal lobe functioning appeared normal even though the Trail Making Test showed slowed times. This could have been due in one case to the medication (Phenobarbital) given for control of the epilepsy. Gross-Tsur et al. (1994) describe the neuropsychological profile of the children in their study as suggesting bilateral posterior hemispheric dysfunction.

A number of studies with PHHI children have shown a pattern of normal development (Gough, 1984; Hirsch et al.,1977; Horev, Ipp, Levey, & Daneman, 1991). Although mental retardation and severe neurological defects are not common with good postnatal care, seizures, learning disabilities, and motor incoordination are seen. Overall, it appears that the more aggressively the PHHI infants were treated to stabilize their glucose levels and seizure activity early in development, the better their neurological outcomes.

37-5. STURGE–WEBER SYNDROME

Sturge–Weber Syndrome (SWS) patients are characterized by the presence of facial angioma, a portwine stain or nevus, involving the upper face, periorbital region, and ipsilateral leptomeningeal angiomatosis. The angioma can involve the mucous membranes, nasopharynx, and ocular choroidal membrane, resulting in glaucoma in about 25% of patients. A hemiparesis is present in around 25% of patients with SWS. Hemianopia occurs in about 25% of patients, but is difficult to assess in very young children.

SWS can cause partial or generalized seizures, infantile spasms, and myoclonic seizures, with a relentless progression of frequency and severity of seizure activity. Intracranial calcifications typically assume a linear-parallel configuration or a convolutional pattern, most often in the parietal or parieto-occipital regions. MRI scans show a thickened cortex, decreased convolutions, and areas of abnormal white matter.

A study by Chugani, Mazziotta, and Phelps (1989) looked at PET scans of patients with SWS and found that hypometabolism and hypermetabolism in the brain were common. Chiron et al. (1989) studied this population with a regional cerebral blood flow (rCBF) paradigm, and in all six patients in the group saw an rCBF decrease in a large cortical and subcortical area in the affected hemisphere. The decrease of rCBF was remarkable—ranging from 32 to 75%.

A study from the Mayo Clinic showed that 47% of patients had a subnormal intelligence (Gomez & Bebin, 1987). Mental retardation was more likely in patients with seizure disorders. Hemispherectomy or lobectomy is considered if seizures prove intractable in order to preserve mental function.

37-5.1. Language in Sturge–Weber Syndrome

Studies involving patients with SWS who underwent a hemidecortication in infancy show that though written language acquisition is possible, competency is dependent on the hemisphere that remains. Generally, the isolated left hemisphere appears to be involved in the grammatical structures of English (Dennis, Lovett, & Wiegel-Crump, 1981). Lovett, Dennis, and Newman (1986) found that in hemidecorticate adolescents who had SWS the use of referential language was again dependent on the remaining hemisphere.

═══════════ **37-6. PORENCEPHALY**

37-6.1. Clinical Features

Porencephaly is a congenital anomaly described as a focal defect of the cerebral mantle, where the "pore" connects the subarachnoid space with the ventricles. The region is often served by the major cerebral arteries (Levine, Fisher, & Caveness, 1974). Although the definite cause is not known, it is probably caused by an intrauterine insult during the second half of pregnancy.

Children with porencephaly may develop raised intracranial pressure, seizures, hemisensory deficits, hemiparesis, visual field defects, and hydrocephalus. Destruction of the brain tissue occurs in the area of the brain where the defect is found. If the anomaly involves the left hemisphere, then problems with language are more likely.

However, in a case study, Frazen, Tishelman, Seime, and Friedman (1986) describe a female congenital porencephaly individual who was studied at age 42. Although she had a congenital massive LH anomaly, part of her left hemisphere missing, her language development was adequate. She reported that she had had no special education and graduated from high school without problems, although she often felt slower than her peers, and thought she had to work a lot harder. She had similar experiences in college but received her M.A. degree just 1 year late. She held a job in a technical library and at this point self-referred for a neuropsychological exam. She was experiencing problems with making quick decisions, with memory, and in logical thinking. She was evaluated with a comprehensive battery of tests.

CT scan localized a porencephalic cyst in the LH. The results show a pattern of spatial task weaknesses and adequate skills in verbal subtests. She received the lowest score in Block Design, usually considered an RH task, an area of the brain in which the scan showed no anomalies. These results are consistent with other findings, which show that in the presence of a considerable LH or RH lesion the functions will transfer to the opposite hemisphere (Bigler & Naugle, 1985).

37-6.2. Language in Porencephaly

The spatial skills are more likely to be impaired than verbal skills in presence of a lesion in either the LH or the RH. As noted earlier, a LH lesion is more likely to result in a language impairment.

═══════════ **37-7. CONCLUSION**

The review in this chapter provides a brief insight into the various deficits associated with neurological diseases. Although in some cases a severe language problem is clear, as in the case of Down's syndrome, other diseases such as Williams syndrome more severely impact cognitive abilities. Overall, it is obvious that each of these neuroanatomically related problems do impact some aspect of language functioning in the individual. Hence, it is important that such problems are detected early in life. With early detection and the implementation of intervention programs, along with special personalized care and training, infants with such neurological diseases may be able to develop to optimal levels, given that the brain is more resilient and plastic during the critical periods of development early in life. Although it is clear that many language deficits are associated with the various syndromes reviewed in this chapter, detailed information continues to be lacking with respect to porencephaly, PHHI, SWS, and TS. Further research into language processes is needed in these areas. With the availability of added information, more specialized intervention techniques may be designed, wherein infants diagnosed with any of these syndromes may be provided with better language experiences, enabling them to maximize their linguistic and cognitive development.

CHAPTER 38

Language and Communication Disorders in Autism and Asperger's Syndrome

Francesca Happé

Social, Genetic, and Developmental Psychiatry Research Centre, Institute of Psychiatry, London, UK

Communication problems form one of the key diagnostic criteria for autism. As with the other diagnostic criteria (social impairments, restricted interests) there is a wide variety of manifestations. The challenge is to understand such diverse problems as Kanner-type mutism and Asperger-type loquaciousness in terms of a unifying cognitive theory. The theory that autistic individuals are unable to represent mental states can shed light on the nature of the whole range of communication impairments, while still allowing for the presence of islets of ability in other areas. This theory predicts that the specific deficit lies in the use of language to affect other minds. Normal language acquisition appears to build on the ability to recognize and orient toward ostensive behavior. Children with autism fail to share attention and appear insensitive to the speaker's intentions. Such deficits may be sufficient to explain the almost universal prevalence of language delay in children with autism, without postulating additional language impairments. Autism, then, provides a model for studying the important distinction between language and communication, and demonstrates the vital part that "mind reading" plays in normal verbal and nonverbal interaction.

This chapter reviews current work on language and communication in autism, and explores what this disorder can tell us about the cognitive capacities that underlie normal communicative functioning and language acquisition. I will focus in particular on one cognitive theory of autism; although there are many current hypotheses of the core cognitive impairment, one in particular has wide-ranging implications for our understanding of autistic language and communication. Before this, however, it is necessary to take a step back and look at what it means to communicate.

38-1. LANGUAGE AND COMMUNICATION

As many authors, including Grice and more recently Sperber and Wilson (1986), have made clear, language and communication are distinct processes. Although we commonly use language (a grammar-governed representational system) to communicate, words and sentences are just one type of tool that can be employed. Communication has to do with conveying our intended meanings, and we can convey our intentions in any number of different ways: we can use gestures, pantomime, gaze, and so forth to let another person "read our mind." The precedence of intended thought over spoken word is evident in our ability to understand nonliteral language such as indirect requests, metaphor, and other "loose usage."

It seems that young children too are more interested in intentions than words. In fact, it may be this bias that leads preschoolers to make errors in message-evaluation tasks—apparently believing that they know the speaker's intended meaning even when the surface form of the utterance is in fact uninformative. This "blindness" to literal meaning should, perhaps, be seen not as an immaturity but as a normal feature of the precedence of communication over language. After all, this precedence remains in adults too; think of the case of asking, "Would you mind telling me what time you open?" to which a common answer begins, "Yes, of course. . . ."

What underlies the extraordinary flexibility of communication? We can begin to answer this question by exploring the possible dissociations between language and communication: communication is possible without language, but communication can also fail in the presence of apparently good language skills. This brings us to the case of autism.

38-2. AUTISTIC LANGUAGE AND COMMUNICATION

Autism is a profound developmental disorder, with biological origin, that affects around 1 in 1,000 children born, and lasts throughout life (Frith, 1989b). Recent years have brought the recognition that autism can have a range of manifestations, from mild to severe, with or without additional handicaps and mental retardation. What is common, and definitional, across this spectrum of disorders is a constellation of impairments in socialization, communication, and imagination, with a restricted repertoire of interests and activities.

In current practice, impairments in verbal and nonverbal communication form an important part of the diagnostic criteria for autism (e.g., DSM-IV; American Psychiatric Association, 1994). What do we know about autistic communication? Perhaps the most striking finding is the great variety of problems seen. Consider the clinical picture presented by three individuals with autism. At one extreme is the child with no language and, strikingly, no compensating gesture or sign language. Such a child may be "aloof" and, because of a failure to orient to speech (Klin, 1991), deafness may be suspected (though subsequently ruled out). The child may be mute, but need not be silent; odd vocalizations, not resembling babbling, may be made, and can sometimes be decoded by parents as signals of the child's mood or needs. These cries, however, are idiosyncratic and are not understood by parents of other autistic children (Ricks & Wing, 1976). The child may show evidence of good auditory memory; for example, singing a complex melody heard only once. The child may lack gestures, but need not be motionless; odd stereotypical movements (hand flapping, toe walking) vie with surprisingly agile purposeful actions (spinning coins, climbing roofs).

A second type of child, particularly well described by Kanner (1943), has some language, while still showing a striking inability to communicate. Echolalia, either immediate or delayed, may serve in some cases a self-stimulatory function, and in others may be used in an instrumental or code-like way—for example, after receiving a gift, saying "You say 'Thank you!' ". Well known is the example of saying "Do you want a biscuit?" to mean "I want a biscuit." Such wholesale parroting is probably the source of much of the pronoun reversal ("you" for "I") common in autism. It is remarkable that in these echoed phrases the speaker's original intonation is often well preserved—in stark contrast to the lack of normal modulation in spontaneous speech (Fay & Schuler, 1980). The same child may use single words in a simple, associative way, so that "Apple" always means "Give me apple." The single words acquired are often esoteric (e.g., "Beethoven") and not like the first words of a normally developing preschooler. Neologisms (e.g., "bawcet" for "bossy"; Volden & Lord, 1991), or familiar words with special meanings ("yes" meaning "carry me on your shoulders"; Kanner, 1943), also reflect the very concrete association of word and object. The child may memorize complex verbal material, which commonly has little meaning for him/her (e.g., an encyclopedia index page, a French lullaby). It is also common for such a child to start reading words, even apparently "teaching himself to read."

The third type of child may have very advanced, adult-like language skills. Several such children were described by Asperger (1944/1991) as sounding "like little professors." Vocabulary in particular may be extensive, and syntax is more formally correct than is usual in everyday speech. However, content and use of language are often bizarre. It is, typically, impossible to sustain a conversation with such a child; either yes/no answers are given, or the topic is hijacked to the child's own special interest—at which point a monologue on train times, pylons, or beetles ensues. Such a child usually shows overliteral understanding of communication—for example, asking earnestly for glue when told, "stick your coat down over there." Although single words are correctly used, discourse and narrative are disjointed, hard to follow, and lacking in coherence. No longer aloof, this child may seek social contact, even with strangers, for example, with repetitive questioning. Voice, prosody, intonation, and timing all tend to be odd; monotone, singsong, too soft, too fast, or stressing unimportant parts of the utterance (Fine, Bartolucci, Ginsberg, & Szatmari, (1991). Odd, wooden, or overblown gestures may accompany speech, typically lacking coordination with verbal content. Eye gaze may be staring, rather than averted as in the aloof child. Such individuals are likely to receive the diagnosis of Asperger's syndrome, which is currently used for individuals at the able end of the autism spectrum (Frith, 1991).

These three children may, in fact, be one and the same child at different ages, or three members of the same family (Frith, 1991). There is good reason, then, to think of the three pictures as different manifestations of a similar underlying handicap. What is that handicap?

38-3. COGNITIVE THEORIES: UNDERSTANDING THE MIND IN AUTISM

The puzzling constellation of symptoms in autism has provoked a large number of theories concerning the underlying cognitive impairment (see Happé, 1994 for a review). At the present time, accounts split into two broad types: those that posit a primary impairment in social development, and those that suggest a more or less widespread nonsocial deficit with secondary social consequences. In this latter group is the hypothesis that autism is characterized by impaired "executive functions" (e.g., Ozonoff, Rogers, & Pennington, 1991). Difficulties in higher cognitive functions, such as inhibition of prepotent responses, shifting set, planning, and monitoring action, are suggested to underlie the restricted interests and insistence on sameness in autism, and to cause social difficulties as a result. An alternative, though compatible, hypothesis of nonsocial impairment in autism comes from the anatomical investigations of Courchesne, which find evidence of cerebellar damage and parietal lobe abnormalities in many individuals with autism (Courchesne et al., 1994). An impairment in rapid shifting of attention is postulated, which might disrupt social functioning from an early age by interfering with the child's development of shared attention and social referencing. As we will see, the major challenge to these nonsocial theories is to explain how the very specific pattern of impaired and intact abilities in autism could be caused by early deficits in these domain-general processes.

Cognitive theories suggesting a primary social impairment in autism are numerous, and in recent years have converged on many common ideas. A number of researchers have shown that older children with autism have difficulty reading facial expressions, sharing attention, and so forth (work reviewed in Baron-Cohen, Tager-Flusberg, & Cohen, 1993). What is still in dispute is the causal and primary status of such deficits; for example, Hobson suggests that autism is the result of an early lack of "personal relatedness" or emotional connection to others which he claims is the foundation for social cognition. Sigman and colleagues give early joint attention deficits primary causal status in autism, while Gopnik and Meltzoff, for example, hypothesize a damaging lack of neonatal imitation (see chapters in Baron-Cohen et al., 1993). To date we lack the data to decide among these different accounts of the primary impairment in autism because this disorder is rarely diagnosed before the third year. Although the primary impairment causing social difficulties is still uncertain, there is more agreement concerning the nature of the social impairment in children and adults with autism.

38-4. THE THEORY OF MIND HYPOTHESIS

Perhaps the most influential theory in recent years has proposed that children and adults with autism are impaired in "theory of mind" or "mentalizing"—that is, the ability to attribute independent mental states to self and others in order to explain and predict behavior (Frith, 1989b). This hypothesis has been successful in predicting impairments in the ability to understand false beliefs, deception, ignorance, and knowledge (work reviewed in Baron-Cohen et al., 1993; Happé, 1994). It also provides a good explanation for the observed lack of pretend play, social understanding, and empathy (Frith, 1989b). It fits the observed range of manifestations: a child without theory of mind may be aloof, passive, or socially active in odd ways. The theory makes sense of otherwise puzzling behavioral observations; for instance, children with autism, when playing a simple game, show pleasure at winning but do not coordinate their smiling with eye contact to communicate their glee to others (Kasari, Sigman, Baumgartner, & Stipek, 1993).

Experimental work prompted by the theory of mind hypothesis has shown that individuals with autism are most clearly impaired where the listener's thoughts and feelings must be taken into account (Frith, 1989a; Tager-Flusberg, 1993). The first demonstration used a false belief task, in which the child must predict where Sally will look for her ball, having been out of the room when Ann moved it. Normal 4-year-olds understand that Sally will look in the old, now-empty location because she acts on her mistaken belief, but even able individuals with autism expect Sally to look where the ball really is—failing to take her mental state into account. This impairment in mentalizing leads to a specific pattern of intact and deficient performance. Individuals with autism can comprehend and use pointing to direct action, but not to direct and share attention; they can use gestures to modulate behavior but not to change thoughts and feelings; they can supply information but do not take account of listeners' needs in supplying the relevant missing facts; they can conceal objects but fail to conceal informative clues in a penny-hiding game; they often score well on the Information subtest but badly on the Comprehension subtest of the Wechsler scales; on vocabulary tests, lacunae are shown for words to do with feelings and mental states (work reviewed by Happé 1994, and in Baron-Cohen et al., 1993). These "fine cuts," between apparently similar behaviors that differ only in their requirement for mentalizing, are shown only in individuals with autism. Nonsocial deficit theories of autism have not, to date, risen to the challenge of explaining these findings, and although there is considerable debate among social deficit accounts concerning, for example, the possible precursors to theory of mind (Baron-Cohen et al., 1993), there is widespread agreement that the idea of "mind blindness" captures something important about the nature of autism.

Even those autistic individuals who do pass simple theory of mind tasks (requiring attribution of a false belief to a story character) seem to show impairments on higher-order theory of mind tasks, failing to attribute more complex mental states (e.g., intention to deceive) to speakers in short scenarios (Happé, 1994). Rather than disproving the mentalizing hypothesis, the existence of these individuals has important implications for refining this theory. Level of theory of mind performance seems to relate closely to the ability to understand similes, metaphors, and irony (as predicted by relevance theory; Sperber & Wilson, 1986) in individuals with autism and in normally developing young children (Happé, 1993).

38-5. DEVELOPMENTAL EFFECTS OF MENTALIZING DEFICITS

The experimental studies summarized above, like most to date, explore the moment-to-moment effects of theory of mind, and its impairment, on communication; that is, studies of protodeclarative pointing, gesture, and so on, tap the individual's ability to understand the audience's mental state, and gear his/her communication to affect that state. Lack of ability to attribute mental states in moment-to-moment processing time will thwart communicative interaction in both production and comprehension, leading to difficulties in understanding indirect requests and other nonliteral language. Direct effects of lack of mentalizing, such as these, may be rather different from the long-term or developmental effects of this cognitive deficit. For example, one can imagine an adult car-crash survivor with damage to the neurological system underlying mentalizing, who can no longer imagine the mental states of others on-line, but who has, nonetheless, all the accumulated routines for insightful social and communicative behavior intact. One group in which on-line mentalizing may be disrupted is patients with acquired right-hemisphere damage, who are known to show pragmatic impairments (e.g., failure to comprehend indirect requests, jokes, metaphors; Foldi, 1987). Indeed, it has been suggested that autism, and especially Asperger's syndrome, resembles right-hemisphere disorder in cognitive profile (Ellis, Ellis, Frazer, & Deb, 1994; Klin, Volkmar, Sparrow, Cicchetti, & Rourke, 1995). Brain scans have given direct evidence of right-hemisphere involvement in three cases of Asperger's syndrome (McKelvey, Lambert, Mottron, & Shevell, 1995), although the causal status of the abnormalities found, and the generality of these findings, are as yet uncertain.

What might be the developmental effects on communication of deficits in mentalizing from birth? In particular, what functions does mentalizing serve in the normal development of language? Although language acquisition could not even begin without the existence of innate, dedicated cognitive systems, it will be facilitated by other processes of development, including social development. Language acquisition appears to make great use of the normal infant's tendency to orient to ostensive behavior and to follow and share another's focus of attention. Recent work has highlighted the role of joint attention in lexical acquisition. For example, Baldwin (1995) has asked how it is that mapping errors are avoided in normal word learning. For example, what happens when infants hear a new word (spoken by an adult talking on the phone, perhaps) while they just happen to be focused on an incorrect novel object? What happens when the baby's attention is momentarily distracted by another object, while the parent is labeling something on which the baby had previously been focused? Baldwin's work has shown that such unintended object–word pairings do not, typically, lead to incorrect lexical acquisition, precisely because infants appear to be sensitive to the speaker's focus of attention. In such cases, children of 18 months typically look to the adult and follow his/her line of regard. Without this sensitivity to the speaker's focus of attention, it is hard to see how ostension could serve as the useful aid to word learning that it, in fact, seems to be.

Autistic children do not orient preferentially to ostension, nor share attention—and this may be one of the earliest signs of the disorder (Baron-Cohen, Allen, & Gillberg, 1992). In the light of Baldwin's work, it is fascinating to speculate, then, on the source of the idiosyncratic word use that is reported in many individuals with autism (Volden & Lord, 1991). For example, Kanner (1943) reports the case of Paul, whose habit of exclaiming "Peter eater" whenever he saw a saucepan was traced back to when his mother dropped a saucepan while reciting to him the well-known nursery rhyme. This appears to be an example of a mapping error, of precisely the type that does not occur in normal language learning.

It is possible, then, that the developmental effects of theory of mind impairment may account for the autistic child's problems in acquiring language. Perhaps a quarter of such children remain mute, and almost without exception language delays are reported (Frith 1988b). Severe mental handicap may explain some cases of muteness, and it is possible that additional, superimposed language problems may occur in some children, just as a few autistic children have additional sensory or motor handicaps. However, it seems plausible that a lack of mentalizing might have as one consequence a failure to orient to and use human language. That is to say, communication may be necessary for normal language development. The autistic child's failure to orient preferentially to speech, to share attention, and to use eye gaze to disambiguate an adult's intention (Baron-Cohen et al., 1993) would be a particularly damaging combination of factors for the acquisition of words.

Although morphology, phonology, semantics, and syntax in autism have not received as much attention in recent research as might be wished, it appears that the underlying cognitive substrates for language are, as a rule, intact in individuals with autism (Tager-Flusberg, 1993). Therefore, alternative routes to word learning can be successful, if slow. Thus some children with autism acquire a limited vocabulary used to achieve concrete ends, and others acquire extensive sight vocabularies of words not fully understood. Written language vocabulary, interestingly, is not normally acquired through ostension—and this may be why many autistic children do not suffer significant delay in learning to read and spell. Reading is sometimes reported to precede speech, reading accuracy typically exceeds comprehension, and hyperlexia is particularly common in autism (Frith & Snowling, 1983).

The phonological and syntactic skills that the verbal autistic child demonstrates in reading are also manifest in the delight often taken in puns and word games (Van Bourgondien & Mesibov, 1987). One able autistic boy composed a whole "joke" book, starting with an overheard joke: "Where does a sick wasp go? A Waspital"; followed by "Where does a sick Alex go? An Alexpital," and so on, repetitively, through a long list of names.

Thus, although it used to be thought that autism was at heart a language disorder (a problem of symbolic abstract thought), the present argument proposes that language may be delayed and peculiar due to a lack of insight into minds, and the resulting inability to enter into normal ostensive–inferential communication. This is a long-range developmental effect. In theory, then, the effects of a deficit in mentalizing ability early in development might remain (as do the effects of, e.g., early visual deprivation), even when theory of mind is gained later on and can then be used moment to moment.

38-6. RELATIONSHIP OF LANGUAGE TO THEORY OF MIND

To date there is no evidence against the idea that all individuals with autism suffer at least a delay in developing the ability to represent mental states. However, in some this ability does seem to emerge eventually, typically in adolescence. These individuals, often diagnosed as having Asperger syndrome, show more insightful communication and socialization in their everyday lives (Happé, 1994). People with autism who pass theory of mind tests also tend to be more verbally able than those who fail. Interestingly, individuals with autism appear to require a much higher verbal mental age in order to pass theory of mind tests than do either normal or mentally handicapped individuals (Happé, 1995). Language may play a different role in theory of mind test performance for autistic and control individuals; verbal mental age is highly correlated with mentalizing performance in autism, but not in other mentally handicapped groups.

It is not that autistic people who fail theory of mind tests do so due to lack of verbal comprehension (after all, they pass control questions)—rather, in order to pass theory of mind tests, individuals with autism may be relying on advanced language skills. Could it be that language somehow allows the autistic child to circumvent his/her theory of mind impairment—that language can become an artificial route to the representation of mental states? Tager-Flusberg and Sullivan (1994) have suggested that some individuals with autism use their competence with complex embedded syntactic forms (e.g., sentential complements) to develop an understanding of the embedded nature of propositions in mental states.

38-7. ASPERGER'S SYNDROME

Exploration of the links between better language and better social understanding (based on mentalizing) brings us naturally to focus on Asperger's syndrome. Although this new label is still somewhat controversial, what is generally agreed is that it applies to those individuals on the autistic spectrum who have rather better social and communication skills. Recent research has suggested that people diagnosed as having Asperger's syndrome do not show the striking failure on theory of mind tasks typical of other autistic individuals (Ozonoff et al., 1991). Participants in these studies are typically adolescents, with near-normal IQ. Their good performance on laboratory tests suggests that theory of mind is working for them at that moment—at least in these simple structured tasks. These individuals are still impaired in everyday life, however, and show the characteristic restricted interests and odd stereotypes. It remains to be seen whether their mentalizing mechanism was functioning earlier in development. It is possible that these people have deficits due to early impairments in theory of mind that have irreversibly affected the developmental course.

One way to find out whether individuals with Asperger's syndrome have early deficits in theory of mind is by exploring their acquisition of language. Language development, and specifically lexical acquisition, may be a barometer of early sensitivity to mental states (orientation to ostension, recognition of attentional focus, etc.). Therefore, a strong prediction would be that children with deficits in these early theory of mind functions will have abnormal word learning. By contrast, a child with absolutely normal acquisition of words through ostension (pointing and naming), and an absence of mapping errors, would have to be credited with a normally functioning theory of mind mechanism. Interestingly, among children with autistic disorders, the development of some communicative speech by age 5 is a marker for a good prognosis and, apart from IQ, language features appear to provide the best predictors of psychosocial outcome (Frith, 1989b).

Somewhat controversially, the new diagnostic criteria for Asperger's syndrome (American Psychiatric Association, 1994) distinguish this disorder from other forms of autism by specifying that there must be no significant language or cognitive delay. As yet there have been no follow-up studies to demonstrate that language delay is the important discriminating factor. However, in terms of the theory proposed in this chapter, it would clearly be interesting if some children on the autism spectrum (those fitting the Asperger picture) were not subject to the typical autistic language delay and abnormality. Indeed, it would suggest that for these individuals theory of mind is intact even early in life—and hence a different explanation for the triad of impairments would be needed. Looking at early language history, then, may aid us in identifying cognitively distinct subtypes within the autism spectrum, and may have predictive value for many other areas of functioning.

38-8. CONCLUSIONS AND FUTURE OUTLOOK

The case of autism reminds us vividly that language and communication are distinct domains. Communication appears to be intimately intertwined with our human ability to attribute mental states to ourselves and others. In contrast, language (syntax and phonology) is a self-contained module that can be intact even though the ability to think about thoughts is impaired. Autism reminds us, also, that development is a complex process of interaction. Even a child with intact language abilities may have problems acquiring the agreed names for things in the normal socially mediated manner. On the one hand, normal language acquisition appears to rely importantly on the existence of communication; on the other hand, communication (including to the self) is well served by an external, flexible, abstractly mapped code such as speech. Asperger's syndrome promises to be an important and informative subgroup for future research on the relationship between language development and theory of mind. Recent advances in early diagnosis (Baron-Cohen et al., 1992), and the recognition of a genetic component in autism, render possible large-scale prospective studies of the development of language, communication, and social understanding in autism and Asperger's syndrome. Through studying these disorders, we may gain a better understanding of the role of normal "mind reading" in language acquisition and everyday communication.

Acknowledgments

I am indebted to Uta Frith, with whom many of the ideas presented here were developed while preparing a joint Royal Society lecture on this topic.

CHAPTER 39

Spontaneous Recovery from Aphasia

Stefano F. Cappa

Laboratorio di Neuropsicologia, Università di Brescia, Brescia, and Dipartimento di Scienze Cognitive, Scientific Institute H S. Raffaele, Milan, Italy

Studies of spontaneous recovery from aphasia indicate that only a limited set of variables (severity, lesion size, time post-onset) have proven to be reliable prognostic factors, while others, such as age, gender, or handedness, have a less predictable relationship with clinical outcome. Limited evidence is available about the possible role of other potentially relevant patients characteristics (education, language background, motivational and intellectual level). The neurological mechanisms underlying functional recovery are still largely unknown. Evidence from cerebral blood flow and metabolism studies indicates that regression of distance effects (diaschisis) in areas undamaged by the lesion may underlie recuperation in the first few months after an acute stroke. Takeover of selective aspects of language function by the healthy hemisphere may be the most relevant mechanism underlying recovery in later stages.

Disorders of neurological function due to an acute, nonprogressive pathological involvement of the nervous system (as in the case of ischemic or hemorrhagic stroke, or trauma due to physical agents), if not extreme in severity, undergo a variable degree of recuperation in the period following injury.

A preliminary problem for the scientific investigation of functional recovery lies in the definitions used to define the consequences of disease. The widely used World

Health Organization classification (World Health Organization, 1980) underlines the distinction between impairment, disability, and handicap. In order to apply this terminology to aphasic disorders, we refer, respectively, to the direct consequence of the lesion on a measurable linguistic function (such as, for example, auditory comprehension of single words and sentences), to its functional consequences (in this case the inability to engage in useful communicative exchanges), and to the final social outcome (loss of job, social isolation). Recovery can be assessed in relation to temporal modifications of each of these three aspects, which do not necessarily proceed in parallel.

Most studies of recovery in aphasia have addressed the "impairment" level. This is hardly surprising, considering that the usual end point of recovery studies is the scores on aphasia tests, which, with few exceptions, yield profiles of linguistic impairment. This point should be kept in mind, because, at the clinical level, what of course matters for the individual patient is the recovery from disability and the prevention of the handicap.

Another general problem is what can be considered as "spontaneous" and as "treatment induced" recovery. There is probably no such thing as spontaneous recovery: the patient is always engaged in some form of "treatment," which can vary from totally aspecific, to highly specialized. In this chapter, "spontaneous" is used in a very broad sense: although some of the studies that are referred to have tried to assess the role of therapeutic intervention, the problem of the efficacy of specific forms of rehabilitation will not be discussed here.

39-1. CLINICAL STUDIES OF SPONTANEOUS RECOVERY

Several studies of recovery from aphasia due to vascular lesions are available in the neurological literature. Some of them have been guided by the clinical necessity to formulate a prognosis for the individual patient, and to provide a baseline for the evaluation of treatment effects. These investigations are extremely heterogeneous. Most of the studies are population-based, and the general statistical framework is correlational or regression analysis. The size and the characteristics of the sample vary widely, however, from small groups of selected patients (Nicholas, Helm-Eastabrooks, Ward-Lonergan, & Morgan, 1993) to large consecutive series (Pedersen, Joergensen, Nakayama, Raaschou, & Skyhoj Olsen, 1995); the methods of assessment of aphasia range from clinical scales (Pedersen et al., 1995) to standardized aphasia tests (Pickersgill & Lincoln, 1983); the number of independent variables that are entered into the outcome analysis also varies, and is not always appropriate to sample size.

This classical epidemiological approach has been supplemented, in recent years, by detailed longitudinal studies of single cases, which make use of the tools of cognitive neurolinguistics (see, for example, Simmons & Buckingham, 1992; Trojano, Balbi, Russo, & Elefante, 1994). This distinct approach is another important source of information about language recovery.

The following, selective discussion of the variables affecting recovery is organized according to the usual practice of clinical medicine, subdividing the variables into patient-related and disease-related variables. Most of the evidence has been reviewed in detail by Basso (1992) to which the reader is referred for a complete bibliography.

39-1.1. Patient-Related Variables

39-1.1.1. Age

Increasing age is associated with inferior recovery in some (Marshall, Thompkins, & Phillips, 1982) but not in all studies (Basso, Capitani, & Vignolo, 1979). For example, a large-scale study by Pedersen et al. (1995) found a "minimal" effect of age on recovery at 6 months. It must be remarked that age is associated with a number of phenomena that might play a detrimental role on recovery: in particular, many age-associated medical and neurological diseases, such as hypertension, diabetes, and de-menting conditions, can affect the brain, and thus interfere with the neurological modifications associated with functional recovery (comorbidity). When these considerations are kept in mind, what can be concluded from the available evidence is that old age per se does not appear to be a major negative prognostic factor for aphasia recovery. This is a point worth considering, given that economical constraints in the provision of rehabilitation services in most countries tend to penalize older patients.

On the other hand, the effect of age on recovery is unequivocal in the case of childhood aphasia: a fast and relatively complete recovery can be expected in children with acquired aphasia (Martins, Castro-Caldas, Van Dongen, & Van Hout, 1991); moreover, within this population there is a trend for a better prognosis being associated with earlier lesion onset (Martins & Ferro, 1992).

39-1.1.2. Gender

The subtle gender-related differences observed in neuropsychological performance have been suggested to indicate different patterns of cerebral organization, possibly related to the effects of sexual hormones on neural differentiation (Fitch & Denenberg, 1995). The idea of a different, less lateralized pattern of hemispheric specialization in females (McGlone, 1977), which has received some support by a recent functional imaging study (Shaywitz et al., 1995), has led to the prediction of better recovery in females. Some empirical support has been obtained for this hypothesis (Basso, Capitani, & Moraschini, 1982; Pizzamiglio, Mammucari, & Razzano, 1985), although the evidence remains limited and controversial.

39-1.1.3. Education

Education has usually not been included in prognostic analyses. Clinical experience suggests that education does indeed play a role, but it is unclear if the positive effects of a higher cultural level are independent from those of other potentially related variables, such as socioeconomic status and general intelligence (no effects of IQ on recovery were found, however, by David & Skilbeck, 1984). Given that studies of aphasia in illiterates have suggested that the acquisition of reading skills modulates the clinical picture of aphasia (Lecours, 1989), an effect on the pattern of recovery could also be expected.

39-1.1.4. Linguistic Background

Several interesting observations have been reported about differential recovery in patients who know more than one language. No clear-cut pattern emerges from the literature; in particular, the idea that the first acquired language recovers first has been challenged by several observations (Aglioti & Fabbro, 1993; Paradis, this volume).

39-1.1.5. Handedness

The main sources of evidence for an effect of handedness on recovery from aphasia have been studies in left-handed patients (Basso, Farabola, Grassi, Laiacona, & Zanobio, 1990) and in patients with "atypical" hemispheric language dominance, such as "crossed" aphasics (Basso, Capitani, Laiacona, & Zanobio, 1985). Both these conditions have been associated with better, or faster, recovery. However, superior recovery does not seem to be the rule in all crossed aphasics (see Table 1).

The in-depth analysis of patterns of recovery in these patients with "atypical" language lateralization (crossed right-handed aphasics and left-handers, with both crossed and uncrossed aphasias) might provide important clues about the cerebral organization of linguistic functions. These patients often show unusual dissociations of linguistic performance, which can be associated with specific patterns of recovery (Berndt, Mitchum, & Price, 1991; Trojano et al., 1994).

39-1.1.6. Hemispheric Asymmetries

This anatomical variable, as assessed *in vivo* on the basis of CT- or MRI-detected asymmetries, is related to handedness (Steinmetz, Volkmann, Jänke, & Freund, 1991) and language lateralization (Foundas, Leonard, Gilmore, Fennel, & Heilman, 1994).

TABLE 1
Rate and Degree of Recovery in Crossed Aphasia

	"Fast" recovery	"Expected" recovery	Total
Cortical lesion	4	4	8
Subcortical lesions	2	2	4
Total	6	6	12

Note: "Fast" or "expected" recovery is a clinical judgment, based on the amount of recovery expected in noncrossed aphasics (i.e., right-handers with left-hemispheric lesions) with lesions of comparable site and extent. From Basso et al., 1985; Cappa et al., 1993; Perani, Papagno, Cappa, Gerundini, & Fazio, 1988.

Some studies have suggested a role of hemispheric asymmetries in predicting outcome. The original observation of better recovery in global aphasics with "atypical" hemispheric asymmetry (i.e., right temporal wider than left) has not been confirmed in other studies; in particular, Burke, Yeo, Delaney, and Conner (1993) have reported the reverse association.

39-1.1.7. Motivational and Social Factors

Despite their obvious clinical significance, these aspects have been largely neglected. There is evidence that depression, which may be particularly severe in patients with anterior lesions, interacts negatively with recovery (Robinson, Starr, & Price, 1984).

39-1.2. Disease-Related Variables

39-1.2.1. Etiology

It is obvious that the main negative prognostic factor for aphasia is a nonstatic etiology, that is, its association with a progressive disease of the nervous system, such as a malignant tumor or a degenerative condition. Leaving these general considerations aside, there is evidence that aphasia due to traumatic brain injury recovers better than aphasia due to cerebrovascular lesion (Basso & Scarpa, 1990); this effect may be dependent, however, on the younger age of the patients and on the generally milder clinical picture. Concerning aphasia due to vascular lesions, hemorrhagic strokes tend to be associated with a more severe clinical picture, but with better outcome in comparison with ischemic strokes (Basso, 1992; Nicholas et al., 1993); this may be related to the less destructive effects of a hematoma on brain tissue, and/or to the less frequent presence of diffuse ischemic changes in the brain of patients with hemorrhagic stroke.

39-1.2.2. Lesion Size

Many studies performed after the introduction of CT scan have assessed the size of vascular lesions in aphasic patients. The size of the cerebral lesion, together with the closely associated variable of aphasia severity, seems to be the strongest negative predictor of recovery in all studies (for examples, see Goldenberg & Spatt, 1994; Mazzoni et al., 1992).

39-1.2.3. Lesion Site

The role of lesion site in determining recovery has of course attracted considerable interest, given the information it may give on the neural organization of language. Within this framework, the negative influence of a lesion localized to a specific cerebral area on the recuperation of a definite aspect of language function has been taken to indicate a relevant functional contribution of that area. A remarkable series of papers by Knopman and his colleagues has addressed the question of the role of lesion site on the recovery of several aspects of language function, such as speech fluency or auditory comprehension, as assessed by traditional language tests. Cortico-subcortical lesions of the left precentral gyrus were found in patients who remained nonfluent 6 months after onset (Knopman et al., 1983). Other investigations have suggested that the extension of the lesion toward the basal ganglia (Ludlow et al., 1986) or mesial frontal white matter (Naeser, Palumbo, Helm-Eastabrooks, Stiassny-Eder, & Albert, 1989) is associated with a negative prognosis for fluency. Naming disorders persisted at 6 months only in patients with lesions of the posterior temporoparietal and insulo-lenticular region (Knopman, Selnes, Niccum, & Rubens, 1984), while repetition disorders were long-lasting in the case of lesions involving Wernicke's area (Selnes, Knopman, Niccum, & Rubens, 1985). The recovery of single-word comprehension disorders was related only to lesion size (Selnes, Niccum, Knopman, & Rubens, 1984), while the lesion extent in Wernicke's area and suprasylvian parietal regions predicted the recovery of sentence comprehension (Selnes, Knopman, Niccum, & Rubens, 1983). Subsequent studies have addressed this point. According to Naeser and colleagues (Naeser, Helm-Eastabrooks, Haas, Auerbach, & Srinivasan, 1987), only the lesion extent in the temporal lobe correlated with severity and recovery of comprehension disorder. A different finding has been reported by Kertesz, Lau, and Polk (1993): in their study, it was lesion extent in the inferior parietal area (supramarginal and angular gyrus) that predicted comprehension recovery. In an investigation of global aphasia, patients were more likely to recover auditory comprehension if the lesion were centered on the temporal isthmus, rather than on Wernicke's area (Naeser, Gaddie, Palumbo, & Stiassny-Eder, 1990).

Aphasias associated with subcortical lesions are associated often, but not invariably, with a fast and complete recovery (Vallar et al., 1988), leaving the patients only with a mild residual semantic–lexical disorder (Kennedy & Murdoch, 1993).

It has been suggested that pathological involvement of the temporobasal regions has a negative effect on rehabilitation-induced, but not on spontaneous, recovery (Goldenberg & Spatt, 1994); this interesting finding deserves replication in a larger patient sample.

39-1.2.4. Clinical Picture

Several studies have addressed the question of the relationship of aphasic syndromes, according to the traditional taxonomy, with recovery. The results are clearly confounded by severity effects, as some syndromes, such as global aphasia, are intrinsically more severe than others. Moreover, the assignment of a patient to a taxonomic category is based on a cluster of symptoms defined in terms of performance in different language modalities (typically, oral expression, auditory comprehension, and repetition). This approach is not suitable for the acute stage, when the patients differ mainly along the severity dimension (Wallesch, Bak, & Schulte-Mönting, 1992). Differences in the rate of recovery for different aspects of linguistic function further complicate the effort to assess the influence of aphasia type: auditory comprehension usually recovers first (Vignolo, 1964), followed by oral expression and written language. This means that many patients evolve from one clinical syndrome to another in the acute period (Pashek & Holland, 1988).

All studies concur in indicating that initial severity, defined according to global impairment measures, appears to be the strongest predictor of recovery (Goldenberg & Spatt, 1994; Mazzoni et al., 1992; Pedersen et al., 1995).

39-1.2.5. Time Post-Onset

The rate of spontaneous recovery is maximal in the first 6 months after stroke, with a very steep curve in the first 6 weeks (Pedersen et al., 1995; Wade, Hewer, David, & Enderby, 1986). This finding, which is typically derived from population studies on an unselected patient sample using very broad assessment procedures, must not obscure the relevant fact that significant improvement can be observed in severely aphasic patients up to 2 years post-onset (Hanson, Metter, & Riege, 1989; Nicholas et al., 1993).

To summarize, there is unequivocal evidence that clinical severity, lesion size, and time post-onset are negative prognostic factors. Lesion site has predictive value for specific aspects of linguistic recovery. The role of other variables, such as gender, age, handedness, and linguistic background, deserves further investigation.

39-2. THE NEUROLOGICAL BASIS OF RECOVERY

The mechanisms underlying recovery of aphasia (and of neurological impairment in general) remain largely unknown.

An amelioration of communicative abilities can sometimes be wholly accounted for at the behavioral level, as the consequence of "strategic" reorganization, both spontaneous and induced by specific training. However, recovery of basic aspects of linguistic impairment, such as disorders of phonemic discrimination or of lexical decision, suggests that modifications take place not only at the behavioral, but also at the neural, level. From this point of view, there appear to be substantial differences between aphasia and other neuropsychological syndromes. In the case of amnesia, for example, most signs of recovery can be expected at the functional level, through the use of memory aids and other "substitution" approaches, while measures of memory impairment, such as the free recall of word lists, are usually very resistant to treatment (Wilson, 1987).

It is well known that the central nervous system has a limited potential for regeneration: damaged axons of mature neurons fail to show spontaneous regrowth, unless they are presented with peripheral nerve grafts (for a comprehensive introduction to the neurobiology of recovery, see Kolb, 1995). The increasing body of knowledge about neurobiological mechanisms of recovery at the synaptic and cellular level is still far from providing an explanation for functional recuperation of complex functions, such as language. Two hypotheses have been advanced at the macroscopic level of brain organization:

1. Regression of diaschisis: In the early period following an acute brain lesion, a functional impairment is present in structurally unaffected brain regions connected to the damaged area, both in the ipsilateral and in the contralateral hemisphere. This phenomenon is often called diaschisis, following von Monakow (1914) (for a discussion of the diaschisis concept, see Feeney & Baron, 1986).

2. "Takeover" of function: In the case of a lateralized disorder such as aphasia, this takes the form of the hypothesis, originally formulated by Gowers (1895), of a takeover of linguistic functions by the contralateral, undamaged hemisphere (right-hemisphere theory of recovery). Within a Jacksonian framework (York & Steinberg, 1995), this is often considered as an "unmasking" of a preexisting functional commitment, inhibited by the left hemisphere (Moscovitch, 1977). Another possibility is that the reorganization of function takes place in undamaged areas within the same hemisphere.

Much of the empirical evidence for these two hypotheses has been reviewed (Cappa & Vallar, 1992), and will be briefly summarized and updated here.

39-2.1. Regression of Diaschisis

The presence of diaschisis has been related to the severity of the clinical picture in the acute period after a stroke. A regression of diaschisis is usually, although not invariably, found in the following months; this finding may be related to the clinical recovery that is observed in the early period after a stroke. Knopman et al. (1984), using the Xe-133 method for the measurement of cerebral blood flow, found a "re-activation" of left temporoparietal areas in patients with good recovery of auditory comprehension. In a study with single photon emission computerized tomography (SPECT) of patients with aphasia or neglect after a stroke involving the subcortical areas, the regression of hypoperfusion in the ipsilesional, structurally unaffected cortical areas has been found to parallel the recovery of neuropsychological disorders (Vallar et al., 1988).

Related mechanisms seem to be at work also in patients with aphasia due to cortical lesions. In a series of studies devoted to the assessment of regional cerebral metabolic abnormalities in structurally affected and unaffected brain areas in aphasic patients, Metter and coworkers (Metter, Jackson, Kempler, & Hanson, 1992) found a significant positive correlation between changes of left and right temporoparietal glucose metabolism and the change in the auditory comprehension score of the Western Aphasia Battery. Heiss and coworkers (Heiss, Kessler, Karbe, Fink, & Pawlik, 1993) found that the value of left-hemispheric glucose metabolism outside the infarcted area in the acute stage after left-hemispheric lesion was the best predictor of recovery of auditory comprehension measured on Token test scores after 4 months, suggesting an important role of intrahemispheric diaschisis in determining the severity of the clinical picture in the acute stage and its recovery. The day following the first examination, a sub-sample of patients with mild aphasia was submitted to a second PET examination with (^{18}F) FDG in what the authors considered as an "activated" state, that is, while they were engaged in an open conversation. The metabolic values in the "infarcted" region, in its contralateral mirror area, and in left Broca's area during activation were highly predictive of the recovery of auditory comprehension, indicating that the possibility of activating an extensive, bihemispheric neural network was crucial for recovery. A further study (Karbe, Kessler, Herholz, Fink, & Heiss, 1995) indicated that at a 2-year follow-up the post-stroke metabolic rates of language-relevant regions (left superior temporal and left prefrontal cortex) were still the best predictors of language comprehension and word fluency, respectively.

In a recent study, using quantitative assessment of regional glucose metabolism with positron emission tomography (PET), Cappa et al. (1997) looked for a similar phenomenon (regression of functional deactivation in brain regions remote but connected to the primarily injured areas) in patients with aphasia due to lesions of the cortical language areas. The study focused on the early period after an acute stroke involving the cerebral cortex of the left hemisphere; only patients with lesions of limited size and mild aphasia, where significant spontaneous recovery can be expected, were included. The patients were studied for the first time 2 weeks after the stroke, that is, when the clinical picture of aphasia can be considered "stable." The follow-up study was performed 6 months later, when extensive recovery can be expected in this patient group. This study confirmed the presence of extensive metabolic depression in structurally unaffected areas in patients with acute stroke. The reduction in glucose metabolism was not limited to the ipsilateral hemisphere, but extended contralaterally. An association was found between regression of functional deactivation in structurally intact regions connected with the area of anatomical damage and spontaneous recovery in the first 3 months after a stroke.

Taken together, these findings suggest that the regression of intrahemispheric and transhemispheric diaschisis may be associated with the recovery of a function, such as language, that is subserved by an extensive network of interconnected regions in both hemispheres, at least in the first 6 months after stroke.

39-2.2. Functional Takeover

Several case studies of patients with bilateral lesions have reported the worsening of language function in a recovered patient after a second stroke in the right hemisphere (for a recent example, see Cappa, Miozzo, & Frugoni, 1994); these observations concur with the results of pharmacological inactivation of the right hemisphere in aphasics (Kinsbourne, 1971), and suggest a role of the right hemisphere in mediating residual or recovered linguistic performance (see also Kinsbourne, this volume). This hypothesis has been supported by the results of some noninvasive studies of hemispheric language dominance in aphasics. Most dichotic listening studies have shown a reversed ear advantage for linguistic stimuli (i.e., superiority of the left ear). This finding, however, cannot be univocally interpreted as the result of a reversal of language dominance. The presence of a lesion in the left hemisphere can be expected to interfere by itself with right-ear superiority if it involves the auditory cortex (Niccum & Speaks,

1991). The interpretation of the results of divided visual field studies is hampered by the same problem, that is, lesion effects. Neurophysiological studies, using the recording of evoked responses to linguistic stimuli, have shown a reversed asymmetry in the amplitude of the responses in recovered aphasics (Papanicolaou, Moore, Deutsch, Levin, & Eisenberg, 1988).

The measurement of regional cerebral blood flow in aphasics using the xenon-133 method (Demeurisse & Capon, 1987) has also provided evidence of an increased right-hemispheric contribution (Demeurisse & Capon, 1991). The investigation of regional cerebral activations in recovered and nonrecovered aphasic patients, studied with PET while engaged in linguistic tasks, allows a more direct investigation of the functional reorganization of the brain after acute damage. Activation methods have been applied to the study of recovery of motor function after striatocapsular ischemic lesions (Chollet et al., 1991; Weiller, Chollet, Friston, Wise, & Frackowiak, 1992). An extensive functional reorganization, including both ipsilateral and contralateral regions, has been observed in recovered patients performing a motor task. Several methodological problems, which only recently have been superseded by technical developments, and which are associated in particular with the necessity to perform group studies, have hampered the application of language activation paradigms to the study of aphasia. An example of this approach is a report by Weiller et al. (1995) concerning a group of partially recovered Wernicke's aphasics studied with PET while engaged in two linguistic tasks: nonword repetition and verb generation. The results of this study indicate a significant recruitment of frontal and temporal areas in the right hemisphere, mirroring the left-hemispheric areas that were activated in control subjects. Buckner, Corbetta, Schatz, Raichle, & Petersen (1996) reported a PET activation study in an aphasic patient with a left inferior frontal infarction. During a word-stem completion task, he showed a right lateralized instead of a left lateralized prefrontal response, suggesting recruitment of a compensatory brain pathway. A recent transcranial Doppler study (Silvestrini, Troisi, Matteis, Cupini, & Caltagirone, 1995) provides further evidence for the contribution of the right hemisphere to the word-fluency task in recovered aphasics.

However, there is evidence that some aspects of language function do not migrate to the opposite hemisphere. For example, Vallar and colleagues (1988) used an interference task (articulation in normal subjects interferes with visual reaction time performance only when the stimuli are presented in the right hemispace, that is, to the left hemisphere). Recovered aphasic subjects showed the same effect as normal controls, suggesting that the left hemisphere was still in charge for articulatory programming. Similarly, the normal right-sided asymmetry in mouth opening during speech has been shown to be preserved in most aphasic patients (Graves & Landis, 1985).

39-3. CONCLUSIONS

In conclusion, there is considerable evidence that both intrahemispheric and interhemispheric reorganization play a role in the recovery process. It may be hypothesized that in the first months after a stroke, when the recovery proceeds at a fast rate, the regression of functional depression in ipsi- and controlateral areas is the main mechanism underlying recovery. At the clinical level, this period is usually associated with a prevalent improvement of auditory comprehension. The subsequent phase, characterized by a much less steep recovery function, might be related to the process of functional reorganization; the relative contribution of undamaged regions of the left hemisphere and of the healthy right hemisphere remains to be assessed. For example, the available evidence suggests that different aspects of expressive function may be subserved by a specific pattern of reorganization: a contribution of the healthy hemisphere is probable for lexical retrieval, while speech programming aspects seem to remain strongly lateralized to the left hemisphere.

Knowledge about spontaneous recovery and its underlying biological mechanisms provides the foundations for the study of therapeutic intervention. A relatively large amount of data is available about the natural history of aphasia and the influence of several variables on its course; what is still missing is an understanding of its neurological underpinnings. PET appears to be a promising tool of investigation, addressing the crucial issues of cerebral plasticity and reorganization of function. There are several clinically relevant aspects of the recovery process that have never been investigated, and that deserve particular attention. For example, there is now sufficient evidence that rehabilitation has a positive effect on recovery: specific predictions about the neurological correlates of different treatment methods can now be put to test using functional imaging. Another important aspect is the effect of drugs, and their possible interaction with behavioral methods. The time is ripe for more precise studies, based on linguistically and neurologically grounded hypotheses, taking advantage of the potentials offered by the current advancement of functionally imaging methods, such as 3-D data collection with PET, which allows single-subject studies, and functional MRI.

CHAPTER 40

Recovery from Language Disorders
Interactions between Brain and Rehabilitation

Leo Blomert

Department of Psychology, University of Maastricht, Maastricht, The Netherlands

Most aphasic patients recover to some degree from acute language impairments. Improvements are often considered the result of ''spontaneous'' recovery, followed by changes due to language treatment. Recent neurobiological findings show that the mechanisms underlying brain reorganizations after injury are also involved in normal learning processes. Language relearning and neuroplasticity work in concert to modify brain organization over considerably longer periods of time than is often assumed. The appreciation of this symbiosis between language and brain repair may enrich recent developments in rehabilitation research. It is argued that longitudinal studies of structured learning paradigms combined with simultaneous drug treatments are needed to study and remediate the restitution and substitution of language functions.

Brains can recover from injury. Patients can recover from language and communication disorders. How might the two be related?

Aphasia rehabilitation methods often formulate the goals of treatment in terms of restoration or substitution (Holland & Forbes, 1993; Méthe, Huber, & Paradis, 1993) implying that treatment ultimately results in some specific neurobiological changes. However, brain repair theories are rarely included in models of rehabilitation (but see Holland, 1989; Weniger & Sarno, 1990). Is there anything to learn from the way language experience interacts with a reorganizing and learning brain? Recent findings in neuroscience about the mechanisms underlying brain repair and learning and memory do not leave much doubt about the answer. Aphasia treatment constitutes a learning situation and learning does change the brain, especially a brain that is energetically trying to cope with and repair the results of injury.

These developments in the neurobiology of recovery are paralleled by hotly debated new developments in the rehabilitation of language disorders; see the special journal issues *Aphasiology,* 1994, 8(5); *Brain and Language,* 1996, 52(1); *European Journal of Disorders of Communication, 1995, 30(3); Neuropsychological Rehabilitation,* 1995, 5(1–2). As the debate centers around the cognitive neuropsychological approach to rehabilitation, it seems appropriate to start the quest about language and brain interactions by investigating the contributions of this approach to an understanding of recovery from language and communication disorders.

40-1. THE COGNITIVE NEUROPSYCHOLOGICAL APPROACH TO RECOVERY

The development of cognitive neuropsychological approaches to rehabilitation (Seron & Deloche, 1989; Humphreys & Riddoch, 1994) has first and for all led to theory-guided deficit interpretations.

Model-driven cognitive neuropsychological treatment is focused on "fixing" impairments that are interpreted to reflect specific defects in processing models. The problems inherent in handling a cognitive theory of normal processing as a theory of cognitive rehabilitation have been sufficiently discussed elsewhere (e.g., Basso, 1989; Caramazza, 1989; Wilson & Patterson, 1990; Holland, 1994; Mitchum & Berndt, 1995). The blurring of the distinction between the explanation and treatment of language disorders (e.g., Howard & Hatfield, 1987; Lesser & Milroy, 1993) and the subsequent disappearance of the distinction between deficit assessment and treatment methods and goals (Edmundson & McIntosh, 1995) seems to set a natural limit to this purely model-driven rehabilitation approach.

Other cognitive rehabilitation methods try to account for the fact that models of normal language processing are still clearly underspecified and that patients show heterogeneous deficit patterns, by developing focused and explicit intervention methodologies. For example, Schwarz, Fink, and Saffran (1995) propose a modular approach, consisting of component skill treatments sensitive to different causes for specific aphasic symptoms. Springer and Willmes (1993) are among the few who explicitly emphasize the learning component in treatment procedures and advocate the use of crossover designs to compare the effects of different structured learning paradigms within patients. It is noteworthy that both approaches focus on symptom-specific small groups with special attention for individual differences.

The emphasis on theory-guided assessments, explicit descriptions of materials (Weniger, Springer, & Poeck, 1987; Byng, 1995) and training methods (e.g. Springer & Willmes, 1993; Schwarz, Saffran, Fink, Myers, & Martin, 1994; Schlenck, Schlenck, & Springer, 1995), constitutes a methodologically critical ingredient for the study of language recovery after brain damage. However, it is evident that the effects of behavioral intervention on recovery processes are intimately related to brain changes. Unfortunately, cognitive neuropsychological approaches do not differ from most other rehabilitation approaches in that they sometimes acknowledge, but seldom address, the consequences of brain repair for language rehabilitation and the effects of language learning on brain function. This neglect contrasts sharply with the firm roots early aphasia rehabilitation had in neurobiological and learning theory (Schuell, Jenkins, & Jimenez-Pabon, 1964; Luria, Naydin, Tsvetkova, & Vinarskaya, 1969; Weigl, 1979). In the following I will explore the question of whether it is worthwhile to restore or to substitute these roots.

40-2. RESTORATION AND SUBSTITUTION OF FUNCTION

Concepts like restoration and substitution of function are not only central to neuro-biological theories of recovery after brain damage (e.g., Almli & Finger, 1988), but also to cognitive theories of recovery from language disorders. For example, Howard and Patterson (1989) described the improvements of three cases (patient P.S.: de Partz, 1986; B.R.B.: Byng, 1988; B.B.: Jones, 1986) in terms of a psycholinguistic model of normal language processing. The fact that it was possible to fit these data in such a model constituted the sole reason to interpret this improvement as a *restitution* of the original function. Weniger and Bertoni (1993) challenged this restoration interpretation by pointing out that the improvements were linguistically limited and context-dependent. They interpreted the reported improvements as the learning of a strategy to adapt to a processing impairment and, therefore, as a partially successful *substitution* for the original function. This example illustrates that one cannot distinguish between restitution and substitution of function without a careful analysis of what the patient actually did learn. Interpreting changes after training is at least as dependent on the availability of a theory of normal processing as it is on an analysis of learning paradigms and thus training methods: "A theory of remediation without a model of learning is a vehicle without an engine" (Baddeley, 1993b, p. 235).

Neuroscience is confronted with questions of whether brain repair mechanisms are necessary for functional improvement to occur (Finger & Almli, 1985) and sufficient to sustain recovery (Stein, Brailowsky, & Will, 1995). Although the questions asked in neuroscience differ in content from questions in cognitive neuropsychology, both research areas nonetheless may profit from the same set of behavioral data. Longitudinal intervention studies with an emphasis on explicit learning paradigms are needed to clarify the mechanisms involved in the restoration and substitution of function. Such studies will also naturally address the how-to-treat-what-and-when issue so prominent in aphasia rehabilitation.

Before embarking on the relations between brain and language repair, relevant neurobiological concepts and mechanisms of recovery will be briefly introduced.

40-3. NEUROPLASTICITY AND RECOVERY

The main neurobiological mechanisms traditionally thought to mediate recovery were vicariation, redundancy, and diaschisis (for a critical review, see Finger, LeVere, Almli, & Stein, 1988). *Vicariation* is the idea that another area of the brain, not previously involved in a particular function, takes over the function of the damaged area. *Redundancy* assumes that uninjured neurons in the damaged area function as spare systems that can compensate for those that are nonfunctional. *Diaschisis* asserts that brain injury to a certain area will inhibit connected noninjured areas and recovery consists of the removal of this inhibition and therefore of a return to normal function (see also Cappa, this volume).

Experimental evidence for these theories is mainly negative or equivocal (Marshall, 1985), and known neurobiological mechanisms of recovery do not relate to these theories (e.g., Slavin, Laurence, & Stein, 1988; Stein & Glazier, 1992). However, this critique does not appreciate that the effects of stroke (Zivin & Choi, 1991) deviate from the effects of experimental lesions. After all, long standing lowered metabolic rates in areas remote from the lesion site, predicted by diaschisis theory have been demonstrated (Metter, 1987).

The traditional theories shared a common origin: they offered explanations for the reappearance of "lost" skills given a finite brain with rigidly localized functions. However, neuroscientists have established that a brain responds to damage by producing trophic factors and neuromodulators, stimulating regeneration processes and changing synaptic sensitivity. "This means that when damage occurs it triggers a complex cascade of processes that take place over *potentially very long periods of time* (days, weeks, months and perhaps even years) and produce changes *throughout the entire nervous system* which are not limited to the injury site alone" (Stein & Glazier, 1992, p. 8; my emphasis). The relative merits of these neuroplasticity mechanisms for cognitive improvements after brain damage are still controversial (see also Section 7.3 in this chapter). However, the finding that the mechanisms underlying brain reorganizations are also involved in normal learning processes (e.g., Weinberger, 1995; Kolb, 1995) indicates a strong relation between neuroplasticity and recovery.

40-4. NEUROPLASTICITY AND LEARNING

The brain "has the capacity for continuously changing its structure and ultimately its function, throughout a lifetime" (Kolb, 1995, p. 4) and thus allows us to adapt to changes in our internal and external environments. Recent research revealed evidence for the functional reorganization of cortical representational maps not as a consequence of damage, but as a consequence of sensory stimulation (Jenkins, Merzenich, & Recanzone, 1990). Weinberger et al. (1990) have demonstrated the possibility of re-tuning representations in auditory cortex: changes in receptive field sizes could be attributed to an associative learning process. This plasticity of cortical representational maps as a consequence of learning has been replicated in different species and different sensory modalities (Recanzone, Schreiner, & Merzenich, 1993; Kaas, 1995; Weinberger, 1995). Now that it has been shown that learning changes representational maps in primary sensory cortices, it would be surprising if representations of cognitive functions are not dramatically dynamic in nature and sensitive to learning experience.

A complete understanding of these neuroplastic mechanisms is not necessary to appreciate that the concept of a continuously changing brain sensitive to behavioral manipulation poses serious challenges for core beliefs about the rehabilitation of cognitive impairments, and language and communication disorders in particular.

40-5. LIMITS TO NEUROPLASTICITY AND RECOVERY

Although representational changes due to neuroplasticity may occur at any time, it will come as no surprise that there are limits to recovery. The size and site of the lesion play a significant role in level of outcome (e.g., Mazzoni et al., 1992; Goldenberg & Spatt, 1994). Marshall (1985) estimated that 10–15% of the tissue of a given structure must be spared to allow at least some recovery to occur.

An example of a limit to recovery is the globally impaired aphasic patient. The widely held belief that these patients have little chance to recover (e.g., Schuell et al., 1964; Mark, Thomas, & Berndt, 1992) probably rests on the assumption that the damage to their brains is beyond repair. An accepted rule of thumb—". . . the lower the level of performance at onset, the more limited the amount of recovery" (Basso, 1992, p. 342)—is based on the results of large aphasia recovery studies (e.g. Kertesz & McCabe, 1977). However, it is likely that the use of standard aphasia batteries as well as limited follow-up periods have restricted the possibility of measuring recovery in globally impaired patients. For example, Sarno and Levita (1981) reported significant functional communication improvement in seven out of seven globally aphasic patients, although none of them improved on standard neurolinguistic measures. A Dutch study followed first-time stroke patients with unilateral damage in the left hemisphere over the first year post-onset (Blomert, 1994). It was shown that half of the 76 patients who were initially very severely impaired on a measure of purely *verbal* communicative efficacy (ANELT: Blomert, Kean, Koster, & Schokker, 1994) did show improvements: 17% just significant, 28% good, and 9% excellent recovery. Most patients still improved significantly in the second half of the first year. Economic pressures on health-care systems may deprive such patients of access to researchers and therapists long before recovery starts to be measurable or even to have occurred.

40-6. NEUROCHEMICAL THRESHOLDS AND COGNITIVE PROCESSING LIMITATIONS

Adequate cognitive functioning requires a relatively high level of neurochemical capacity. Functional (metabolic) "lesions" may represent depressed levels of neurotransmitters and/or receptor populations and thus form the biological basis of many processing limitations. Because the quality and quantity of receptor populations is a dynamic property, long-standing functional lesions may signal the potential restitution of impairments.

Processing limitations may be a primary cause of impairments at central levels of ongoing language activity (Friederici, 1995). Linebarger, Schwartz, and Saffran (1983) showed that agrammatic patients were able to detect syntactic anomalies, but were not able to use this knowledge on-line to produce well-formed sentences. Broca's aphasic patients did detect syntactic violations in spoken language if sentences were simple, but not if they were complex (Haarman & Kolk, 1994). Martin, Dell, Saffran, and Schwartz (1994) studied a patient over a period of 6 years post-onset. His overall improvement in naming performance as well as change in error pattern was interpreted as a *normalization* of the decay time of activated lexical items. This demonstration of a *restitution* of function after long-standing specific processing limitations is as elegant as it is rare. Adequate lexical retrieval no doubt requires sufficient processing resources to occur. A restitution of processing capacity very likely indicates a change in neurochemical thresholds. This change may have been brought about indirectly, by brain reorganizations unrelated to the function under study, or more directly by specific behavioral and/or pharmacological treatments.

Luria et al. (1969) were among the first to appreciate the dynamic nature of cognitive deficits. They used a combination of pharmacological and behavioral treatment for seemingly permanent disorders and showed significant behavioral improvements by boosting levels of the neurotransmitter acetylcholine. It was further reported that the drug effects were stronger if combined with behavioral training. This beneficial effect of drugs as potential adjuvants to behavioral treatment was recently confirmed in a study reporting significant improvements in chronic aphasic patients receiving systematic language training in combination with a "cognitive enhancer" such as piracetam (Huber, Willmes, Poeck, Van Vleymen, & Deberdt, 1997).

The concept of biochemical thresholds for the recovery of different functions was investigated by Russell, Smith, Booth, Jenden, and Waite (1986). They injected into rodents a compound binding *irreversibly* to cholinergic receptors in the brain and reducing the receptor population to approximately 10% of its normal volume. The results showed that recovery, from an almost complete neurochemical lesion, occurred in a strict hierarchical and temporal order: physiological functions reappeared first and cognitive functions last. This hierarchical recovery matched the increase of a newly synthesized receptor population. Cognitive functions were observed only after the receptor population had reached 90% of its normal level. These studies indicate that biochemical thresholds and balances in damaged areas and their connections may constitute key concepts for understanding *(a)* cognitive processing impairments in the absence of extensive structural damage, *(b)* improvements without extensive structural changes, and *(c)* late recovery phenomena. Training programs may sometimes not be successful because they inappropriately tax the available processing capacity. Pharmacological and/or stimulation treatment preceding structured training may sufficiently alter neurochemical activity levels to make improvements possible and enhance the learning process. It has recently been shown that it is possible to significantly improve verbal memory of aged adults, if the subjects were administered a drug, sensitive to memory encoding, prior to the learning task (Lynch et al., 1997). These drugs (ampakines) promote the induction of long-term potentiation (Staubli et al., 1994) and thus change functional biochemical thresholds. The authors suggest that the drugs used may be particularly effective in the context of reduced memory processing. There probably is more potential for recovery than is often assumed in the field of cognitive rehabilitation.

40-7. THE INTRICATE RELATION OF LANGUAGE REHABILITATION AND BRAIN REPAIR

Let us now consider the relevance of some neurobiological research findings for recovery and rehabilitation.

40-7.1. "Spontaneous" Recovery versus Rehabilitation

The following is a recurrent statement in studies of the efficacy of treatment: the patient was x months post-onset and therefore improvements cannot be attributed to "spontaneous" recovery. Kertesz (1993) discriminated two recovery stages. During the first stage, absorption of cellular debris and edema are paralleled by attempts to reestablish circulation. This period may take anywhere from a few days to a month. After this, a second phase dominated by brain repair mechanisms may continue for months, and even for years. The period of "spontaneous" recovery often referred to in aphasia rehabilitation research presumably coincides with the start of this second stage and is typically assumed to end after 3 to 6 months. This assumed limit is based on large-scale heterogeneous group studies using compound average scores over a range of language modalities and is contradicted by evidence from single-case long-term recovery studies (e.g., Schlenk & Springer, 1989).

Theories of the neurobiology of recovery as well as learning and memory, cast serious doubt on a categorical separation of "spontaneous" recovery and rehabilitation effects: brain damage results in behavioral changes and these in turn produce changes in the brain. The patient under investigation may change by the mere act of being studied. "Spontaneous" recovery and (re)learning can only be separated at a conceptual level. The actual outcome of recovery processes reflects changes within an *integrated organism* that are determined by structural, neurochemical, physiological, and behavioral interactions (Russell, 1992).

40-7.2. The Time Course of Recovery

The controversy over the time course of recovery is mainly due to considerable differences in assessment methods and patient selection criteria. Estimates range from a few weeks or until "the lesion stabilizes" (Delwaide & Young, 1992), to months (Basso, 1992) or even years (Sarno, 1991). It has also been observed that recovery from aphasia may start to occur after many years (Geschwind, 1974, 1985; Luria et al., 1969).

A time course analysis of brain plasticity revealed the partial overlap of initial de- and regeneration processes; glial scar tissue is already forming when sprouting and reactive synaptogenesis have barely started (Cotman & Nieto-Sampedro, 1985).

This illustrates that the different repair processes cannot be separated in time and that "there is no principle of neuronal organization that demands that functional recovery occurs immediately or not at all" (Stein & Glazier, 1992, p. 15). Treatment in an early period of massive biological change in a complex system may be as crucial as in seemingly stable stages of recovery. There is convincing evidence that recovery often does not occur "spontaneously" or will not reach a functional optimum without appropriate behavioral and/or pharmacological treatment. Recovery of stable, long-standing deficits has been reported following specifically designed training for complex motor (Bach-y-Rita, 1990, 1992), visual-perceptual (Zihl & von Cramon, 1979), and syntactic production deficits (Jones, 1986; Springer, Willmes, & Haag, 1993). Success of intervention may be dependent on the time post-onset when the treatment was given, "windows of opportunity" (Kolb, 1995), but interactions between time course and type of intervention is still terra incognita. The potential for long-lasting as well as late recovery, sometimes only to be uncovered by rehabilitation, indicates that the results of treatment efficacy studies using "chronic" or "stable" patients may need a reinterpretation.

40-7.3. Are Changes Always Improvements?

The main problem that confronts neuroscientists is not *if* the brain changes after injury, but *whether* these reparations are beneficial or detrimental (Stein et al., 1995). A comparable problem faces aphasia rehabilitation. If a patient fails a given criterion after considerable training, it is not reasonable to conclude that nothing was learned. This is not a prominent issue in discussions about rehabilitation (but see Baddeley, 1993b; Weniger & Bertoni, 1993; Holland, 1994; Byng & Black, 1995). The hidden premise in rehabilitation seems to be: if it doesn't help, it doesn't hurt. This may be misleading. "It is likely that many changes, perhaps the majority, are unfavorable, serving to prevent the expression of latent restorative processes. Studies of these unfavorable mechanisms may be even more important than investigation of spontaneous recovery in leading to improved methods of therapy" (Geschwind, 1985, p. 2). The large number of permanently impaired patients is advanced as evidence for these maladaptations.

Brain repair mechanisms seem not quite prepared for the task set for them (Finger & Almli, 1985). The reason may be that potential recovery mechanisms have been sacrificed in evolution to favor other specializations, as, for example, functional lateralization instead of duplicate hemispheres. Inappropriate rewiring or synaptic reorganization could potentially be more detrimental than neuronal loss (Galaburda, 1990). "Bad brain is far worse than no brain" (Kolb, 1992, p. 176). Cotman, Cummings, and Pike (1993) argue that a disruption of the balance between "adaptive and pathological plasticity" may in time prevent a restoration of function, especially in an aging brain subject to multiple injuries and compensations. Therefore, it seems reasonable to assume that incomplete recovery may result from the damage per se and/or the neurobiological maladaptations. Furthermore, since learning is one of the modulators of brain (re)organization, it is necessary to find out which training methods improve and which actually may prevent a restoration of function. An evaluation of changes over time may identify beneficial and pathological adaptations and controlled learning paradigms may clarify the impact of treatment at different times post-onset. Reporting null effects of treatment is, therefore, not only useful for the evaluation of method, but certainly contributes to the building of an integrated theory of recovery.

40-8. CAN SYNTAX BE (RE)LEARNED?

What constitutes restitution and what substitution in the process of relearning language? Insights may be gained from the study of language properties that need inherent brain capacities to develop and thus possibly constrain the degree to which this property can be (re)learned. Syntactic proficiency requires an innate computational device and a critical period of exposure to spoken language. For example: children and adults deprived of early language experience never really mastered complex syntactic structures, such as *wh*-sentences (e.g., Pinker, 1994). Ludlow (1973) investigated when and how syntactic structures not essential for the production of simple declarative sentences would reappear in the speech of fluent and nonfluent aphasic patients. She showed that the order of recovery of the investigated structures did not differ between individuals or aphasic groups and mainly correlated with the frequency of occurrence of these structures in normal speech. Given these potential constraints on the degree to which syntax can be (re)learned, what are the mechanisms underlying syntactic improvements in severely agrammatic patients?

It has been shown that agrammatic patients did not generate untrained *wh*-structures despite elaborate training consisting of isolated sentences (Thompson & McReynolds, 1986). Thompson and Shapiro (1995) based their treatments on the assumption that the training of a given sentence type will generalize to sentences subject to similar syntactic rules and principles, supposedly using the same "mentalistic operations." If sentences similar in structure to the target picture/sentence were primed, enhanced use of these same structures was measured in restricted sentence production and discourse tasks. However, no generalizations across sentence types were observed.

Springer et al. (1993) used a two-period crossover design combining a linguistically specific learning approach and a prosodic stimulation approach. Remarkable improvements in the use of *wh*-questions were observed in dyadic conversations. The linguistic learning approach was significantly more effective than the stimulation approach. Interestingly, it was also found that prosodic stimulation followed by specific language training proved even more advantageous. The authors concluded that these improvements in chronic patients indicate that "it is not just deblocking of temporarily affected abilities but controlled (re)learning that takes place in linguistically oriented therapy." Blomert and de Roo (1995) analyzed the evolution of syntax in the first year poststroke. Twenty-one patients were presented with the same standardized daily life situations (ANELT: Blomert et al., 1994) 1 and 13 months post-onset. All patients received some form of therapy during the year, but nobody received specific syntactic training. The results showed a clear relation between initial and outcome level of syntactic complexity and well-formedness within patients, irrespective of aphasia type. These findings suggested that, without specific training, patients evolve to a level of syntactic proficiency mainly determined by the pattern of initial impairment and the degree of biologically conditioned "spontaneous" recovery. Interestingly, some agrammatic Broca patients showed significant *verbal communicative* improvements without showing improvements on neurolinguistic measures. A syntactic analysis revealed that these patients preferably produced bare VP (verb phrase) structures, which are syntactically simple, but potentially rich in information (Blomert & de Roo, 1992). This latter finding emphasizes that the results of syntactic elicitation procedures are not only sensitive to the context of the task, but also to the cognitive processes involved in formulating a communicative intention. Findings show that the ability to structure information at the propositional level (Levelt, 1989) may change dramatically over time (personal observation) and may be more important in determining the form and content of aphasic verbal utterances than is often assumed (Blomert, 1990).

The studies above show that the relearning of syntactically complex structures, often thought of as not relearnable in principle, may depend on the use of structured learning paradigms. It was also shown that the nature of the sentence elicitation procedures influences the observed syntactic variability. The context of the task and therefore the relevance of the communicative intent help shape syntactic performance. The study of syntactic impairments in a meaningful context may help disentangle the mechanisms involved in the recovery from language and communication disorders.

40-9. CONCLUSIONS

In this chapter, I have argued that theories of recovery after brain damage should approach brain and language repair as a Siamese twin. Language relearning and neuroplasticity work in concert to modify brain organization. It is suggested that longitudinal studies of symptom-specific small groups receiving structured learning tasks probably best probe the how-to-treat-what-and-when issue. Systematically varied combinations of different intervention approaches and simultaneous pharmacological treatments are promising. Future rehabilitation programs may not only concentrate on enhancing recovery but also on preventing disabling adaptations. The occurrence of these maladaptations deserves more attention in an era in which evidence from single cases is highly valued. Given the many new directions in recovery and rehabilitation research, it seems worthwhile to stress that it is critical to show the functionality of an improvement, independent of the approach, method, and language function chosen for treatment. Aphasic patients, like most of us, use language mainly as a communicative device.

Acknowledgments

The research by the author and his colleagues was supported in part by The Dutch "Praeventiefonds," grant 28-1530-1. The chapter has benefited greatly from discussions with colleagues in the Center for the Neurobiology of Learning and Memory during my affiliation with the University of California Irvine.

CHAPTER 41

Recovery and Treatment of Acquired Reading and Spelling Disorders

Nadine Martin

Center for Cognitive Neuroscience, Temple University School of Medicine, Philadelphia, Pennsylvania 19140

Studies of recovery and treatment of acquired language disorders are not new, but the application of such investigations to the development of neuropsychological models of language is a relatively recent development. Patterns of recovery and effects of treatment provide insight into the relations of the functional components of the language system and dynamics of processes that mediate verbal and written language. This chapter aims to acquaint the reader with recent advances in the study of recovery and treatment of reading and spelling impairments resulting from acquired brain damage.

This chapter focuses on the study of recovery and treatment of acquired dyslexias and dysgraphias—reading and spelling disorders. Although this enterprise is relatively young, it holds promise as a window on the dynamic relations among components of the reading system that are not readily apparent in cases studied at one period in time. We will begin with a brief overview of processes involved in reading and spelling and a description of neurologically based disorders of these processes. This will be followed by a review of recent investigations of recovery and rehabilitation of reading and spelling. Most of the work in this area has been carried out on disorders of central reading and spelling mechanisms that process orthographic representations of language, as opposed to peripheral systems that process perceptual aspects of visual stimuli. Accordingly, this chapter will limit the discussion to those reading and spelling disorders that are due to disruption of linguistic processing of visual stimuli. As the review will indicate, these studies lead to some new insights into old questions, as well as many new questions that set the stage for future research in this area.

41-1. PATHWAYS TO READING

Peripheral visual systems are responsible for the analysis of the visual stimulus and identification of the visual form of letters within the visual context in which the letters are viewed. Linguistic processing of visual stimuli begins when the visual forms of letters are mapped onto abstract letter identities. These representations then map onto a word form and its corresponding semantic representations to mediate comprehension of the letter string. Pronunciation of the letter string, according to most models, can occur by a mapping of the semantic representations onto output phonological units. Reading words aloud, however, is not limited to this semantically mediated route. In fact, we can pronounce words that have no meaning and novel words whose meaning is unfamiliar via a direct conversion of the abstract letter representations extracted from the printed word to the phonological units that guide production processes. The conversion process entails segmenting the letter string into pronounceable units (syllables, morphological units, phonological segments), a process that takes time. For this reason, many models of reading assume a buffer that maintains phonological information as the input is processed. Ellis (1984) terms these two routes as "reading by ear" (orthographic to phonological conversion) and "reading by eye" (reading via access to the orthographic word-form store). It is likely that most readers use both routes when reading—sometimes relying on one over another depending on the nature of the stimulus. Impairment to these systems "fractionates" the pathways so that a dyslexic reader is forced to rely on one route over another.

41-2. READING DISORDERS

41-2.1. Pure Alexia or Alexia without Agraphia

It is generally agreed that pure alexia reflects a failure in transmission of sensory feature information onto word units, but it remains a point of dispute whether the syndrome is due to a central or a peripheral visual processing disturbance. Patients with pure alexia are often referred to as letter-by-letter readers because they tend to laboriously read each letter of a printed word aloud in order to access its pronunciation. This syndrome was once attributed to a partial or complete impairment of the word-form system (Warrington & Shallice, 1980). Subsequent studies indicated, however, that pure alexic patients gain access to some higher-level information via the printed word (Bub & Arguin, 1995; Coslett, Saffran, Greenbaum, & Schwartz, 1993). For example, they can perform lexical decision and semantic categorization tasks without explicit report. Additionally, reading performance improves with rapid presentation of words, presumably forcing a more visually based direct activation of lexical–semantic information. This evidence does not disprove the involvement of an impaired word-form system, but does indicate that such impairment can only be partial (Shallice & Saffran, 1986).

There are numerous competing hypotheses regarding the fundamental impairment underlying pure alexia. Friedman and Alexander (1984) propose that the locus of the impairment is more peripheral at the early stages of visual analysis. Others have argued that the problem impairs the ability to synthesize multiple forms, a consequence of a more general visual deficit termed simultagnosia in which separate visual forms cannot be perceived simultaneously (Coslett & Saffran, 1989). Rapp and Caramazza (1991) propose that the deficit affects the efficiency of an early stage of processing in which attention is deployed across a spatial map of the visual stimulus. Finally, Behrmann and Shallice (1995) claim that the syndrome is rooted in a letter activation deficit in which rapid and efficient processing of single letters is disrupted. They argue that impairments described by previous accounts as the source of letter-by-letter reading are present in some, but not all, pure alexics. Letter processing is disrupted in most pure alexics, however, and this fact argues for the view that deficient letter processing leads to the letter-by-letter reading strategy.

In addition to understanding the deficit underlying pure alexia, theorists seek an account of mechanisms underlying the residual reading abilities in this syndrome. How do pure alexics gain access to lexical and semantic information about words they cannot pronounce? One hypothesis is that their reading is mediated by access to a right-hemisphere lexicon that is known to be limited in the range of words it can process. This account is similar to that which has been proposed for reading patterns in deep dyslexia (see Section 41-3), although the impairments underlying the two syndromes are not the same.

41-2.2. Surface Dyslexia

Patients with this reading disorder rely on use of the phonological route that maps input orthography (abstract letter identities) onto output phonological representations (reading by "ear" in Ellis's [1984] terms). Dependence on this route for reading stems from a difficulty in mapping the abstract letter identities onto the input orthographic lexicon. Although the direct route is adequate for reading many words, it cannot be used to access the pronunciation of irregularly spelled words (spellings that do not correspond directly with the sounds used in pronunciation, for example, *yacht*). In order to access the pronunciation of irregularly spelled words, the reader must first access the word-level representation in the input orthographic lexicon and then use this representation to gain access to output phonology.

Three characteristics of surface dyslexia reflect use of the orthographic-phonological route to reading. First, regularity effects are observed in reading; words whose spellings do not convert transparently to pronunciation will be mispronounced in a manner that forces the direct orthographic-phonological conversion. The word *pint*, for example, will be pronounced as /pInt/ and *blood* will be read as /blud/. High-frequency irregularly spelled words may be preserved in these patients, suggesting that deployment of orthographic-phonological mappings may be influenced by the frequency of letter–sound associations such that more frequent mappings (for example, *oo* → /u/) replace those that are relatively rare (for example, *oo* → /U/). Nevertheless, the inability to read irregularly spelled words serves as evidence of a reading disorder that impairs access to the orthographic lexicon via written input. Second, reading-comprehension errors tend to reflect the reader's would-be pronunciation of the word. For example, the written word *bear* is defined as "something you would drink" (/bir/ —*beer*). Third, these patients have a preserved ability to read nonwords with regular spellings (a task mediated by the orthographic-phonological conversion route).

41-2.3. Phonological Dyslexia

In this disorder we see a pattern somewhat opposite to that of surface dyslexia. There is no regularity effect and known words are read better than nonwords or unknown words. Word-class effects and morphological errors are also common in the reading pattern of this syndrome. Nouns are read better than function words and sometimes more accurately than verbs. Affix substitutions are common (for example, *farming* → *farmer*) and typically occur with inflectional affixes (for example, *directed*) more than derivational affixes (for example, *direction*). Inflectional affixes do not change meaning or grammatical class of word and are probably computed. Derivational affixes, on the other hand, are stored as whole words and are presumed to have their own lexical entry. Finally, concreteness effects (concrete words are read more accurately than abstract words) are sometimes present in these patients. This overall error pattern indicates that, to read aloud, the subject is relying on a reading pathway by which graphemic input maps onto representations in the orthographic lexicon, then onto representations in the phonological lexicon before phonological output systems are activated.

41-3. DEEP DYSLEXIA

This reading disorder shares many features with phonological dyslexia and has been characterized by some as a more extreme form of that syndrome (for example, Glosser & Friedman, 1990). The ability to read nonwords is severely impaired, indicating damage to the system that translates grapheme input directly to phonological output. Also, as in phonological dyslexia, errors occur on word inflections and function words. Visually based errors *(kind → king; desire → desert)* are common but tend to occur more often on abstract words. Two features that distinguish this syndrome from phonological dyslexia are the production of semantic errors in reading single words (for example, *tulip* → *crocus; applaud* → *clap*) and the consistent presence of concreteness effects in single-word reading.

Most accounts of deep dyslexia postulate two functional impairments: one that disrupts the orthographic-phonological conversion route, and a second partial impairment of processes that disrupts reading by the lexical–semantic route (for example, Shallice, 1988). Of those theorists who view deep dyslexia as a severe form of phonological dyslexia, Glosser and Friedman (1990) propose that it is the severity of the semantic impairment that determines whether symptoms will be more in keeping with deep or phonological dyslexia. Evidence to support this hypothesized continuum comes primarily from investigations of recovery and will be addressed in more detail in the discussion of recovery studies.

A well-known account of reading in deep dyslexia is that it is mediated by the unaffected right hemisphere. Patients with deep dyslexia often have very large lesions affecting language abilities other than reading. According to this account, access to the left hemisphere is rendered impossible by the impairment but does not preclude access to the limited language stored in the right hemisphere. Lexical entries accessed in the right hemisphere are used as input to output lexical systems in the left hemisphere (Coltheart, 1980).

41-3.1. Nonsemantic Reading

This reading impairment affects the later stages of processing written information. Words and sentences can be read, but are not understood. Nonwords can be read, and so in that sense, the syndrome resembles surface dyslexia. However, unlike that syndrome, words with irregular spelling (that cannot be mapped directly to phonology; for example, *leopard*) can be read aloud. Schwartz, Saffran, and Marin (1980) first described this syndrome in a case report of a patient, W.L.P, who suffered from presenile dementia. W.L.P. showed a remarkable dissociation between her ability to read aloud both regularly and irregularly spelled words and her ability to comprehend those words. This finding provided strong evidence for a third route to reading aloud that makes use of word-form representations but not semantic information.

41-4. PATHWAYS TO SPELLING

When evaluating disorders of spelling, we must first consider that writing is not simply the reverse of reading in the same way that speaking is not simply the reverse of hearing and comprehending. There are parallels, however, in our taxonomy of spelling disorders. As in reading, the pathways to spelling vary depending on the task. Writing thoughts to compose a letter involves a different combination of processes than writing a dictated letter or copying lecture notes from a chalkboard. Thus, it is conceivable that acquired brain damage might affect spelling ability via one route but not another.

Spontaneous generation of written language (for example, writing a letter or story) is mediated by a mapping of conceptual–semantic representations onto word forms and their corresponding orthographic representations. Spelling words that are spoken can proceed by one of three pathways (Margolin & Goodman-Schulman, 1992). One route involves a direct mapping of nonlexical phonology onto individual orthographic segments. This route enables us to spell nonwords or unfamiliar words. Two other routes by which spoken words are spelled access output orthographic word forms by way of lexical phonology or semantics. Finally, the process of copying a written stimulus (word, nonword, letter) can be carried out by mapping orthographic units onto output orthographic representations or copying visual forms of letters directly without any access to the abstract letter identities they represent. In the latter case, writing is purely dependent on peripheral input and output mechanisms.

41-5. SPELLING DISORDERS

41-5.1. Surface (Lexical) Dysgraphia

This disorder parallels surface dyslexia and is characterized by difficulty in accessing lexical–orthographic representations of words. Thus, spelling must be carried out primarily by relying on phoneme–grapheme correspondences. The clearest indication of a surface dysgraphia can be observed when a subject is asked to write irregularly spelled words. Regularization errors will be observed in this spelling-to-dictation task, that is, irregularly spelled words will be misspelled in a way that corresponds to their pronunciation (for example, *broad* will be written as *brawd*).

41-5.2. Phonological Agraphia

This disorder involves an inability to associate input phonological representations with output orthographic representations, thus forcing a reliance on orthographic word-form representations stored in lexical memory in order to write. A key symptom of phonological agraphia is a preserved ability to write words but not nonwords. Additionally, the ability to write irregularly spelled words is preserved. Other problems usually present in this syndrome include errors in retrieving affixes (because they are not stored lexically) and in accessing the spelling of low-frequency words (a consequence of relying on lexical information to mediate the spelling process).

41-5.3. Deep Dysgraphia

Symptoms of this disorder overlap with those of phonological agraphia, in the same way that deep and phonological dyslexia are related. Nonword spelling is impaired, and there are numerous errors in spelling grammatical words and inflectional morphemes. The ability to write concrete words is invariably better than the ability to write abstract words (a problem sometimes present in phonological dysgraphia), and semantic errors (for example, *star → moon, lilac → orchid*) as well as visual errors (for example, *rabbit → raffit*) are produced in spelling to dictation. Finally, written output tends to be agrammatic in style, indicating that it is difficult for these subjects to access orthographic representations of grammatical/syntactic morphemes.

41-6. STUDIES OF RECOVERY AND TREATMENT

We now turn to the focus of this chapter: recovery and rehabilitation of reading and spelling disorders. What can be gained from an examination of changes in language performance that accompany recovery or treatment? Early neuropsychological investigations of reading and spelling disorders aimed to identify symptom complexes that dissociated one process from another. These dissociations revealed the possible routes by which a task could be mediated but shed less light on the dynamics of information processing. A primary goal of recovery studies is to observe and account for change in language performance over time: How do once inaccessible representations become accessible? Treatment research is also essentially concerned with dynamics of processing in that a primary goal is to develop better means to effect change in language performance. Thus, recovery and treatment studies require theories that account for the dynamic aspects of information processing, but at the same time, they will provide an important testing ground for such theories. In this sense, studies of treatment and recovery have the potential to make an important contribution to neuropsychological theories.

Apart from their theoretical contributions, recovery and treatment studies have obvious practical applications. Recovery studies serve as a useful guide to treatment of acquired language disorders (Behrmann, Black, & Bub, 1990) by outlining sequences of changes over the course of recovery. Treatment studies identify stimuli, tasks, and modalities that maximize immediate and long-term improvements as well as generalization of these effects to untreated items. Of equal practical value, treatment studies identify stimuli and methods that do not effect change in a particular disorder.

Currently, model-based investigations of recovery and treatment in the area of reading and spelling disorders are relatively few in number and sporadic with respect to the amount of attention paid to some syndromes over others. Nevertheless, they have made an important contribution to our understanding of the dynamic relations among processes and stages of representation in reading and spelling, and have led to a number of questions yet to be answered in future studies. We will now review critical studies of recovery and treatment of disorders of central processes of reading and spelling.

41-6.1. Studies of Recovery

The most informative recovery studies to date are those that have followed the evolution of deep dyslexia to phonological dyslexia. Recall that deep and phonological dyslexia share many features but differ in severity and in the presence of one specific error type, the semantic error (the defining feature of deep dyslexia). Early conceptions of deep dyslexia postulated two separate impairments affecting the two routes to reading—one more severely than the other. Friedman (1996) reviewed five cases of deep alexia whose symptoms evolved into the pattern associated with phonological dyslexia (Laine, Niemi, & Marttila, 1990; Glosser & Friedman, 1990 [two cases]; Job & Sartori, 1984; Klein, Behrmann, & Doctor, 1994). That is, the semantic errors that were present in the early stages of their disorder dropped out of the symptom complex as recovery ensued. In one case (Laine et al., 1990), the evolution was documented from 2 weeks post-onset, at which time the patient was unable to read a single word, to 4 years post-onset, at which time reading was near normal. Assessments at intervals over the course of recovery revealed stages at which the subject's reading was characteristic of deep dyslexia and then phonological dyslexia. Friedman (1996) has further identified a succession of symptoms along this evolution from deep to phonological dyslexia.

This evolution of error patterns suggests that deep and phonological dyslexia represent a single impairment whose symptoms vary along a continuum of severity. Klein et al. (1994) suggest that a double deficit remains a viable account of the disorder and propose that with recovery, only one reading route, the semantic route shows any recovery. In another version of this account, Friedman (1996) proposes that the distinction between deep and phonological alexia lies in the severity of the semantic deficit. Thus, interpretation of this recovery pattern has, for the most part, adhered to the idea that two deficits underlie the more severe syndrome, deep dyslexia, and that recovery reflects changes in one of those routes.

The double-deficit account of deep dyslexia and its relation to phonological dyslexia developed largely on the basis of information-processing models that allow descriptions of impairments in terms of the locus within a model of language processing and the level of representation affected by the impairment. A single impairment that affects phonological and semantic reading processes is conceivable in computational models (interactive activation and parallel distributed processing models) that postulate a highly interactive system in which representations processed at one stage are not independent of representations processed at another earlier or later stage. Computational models allow exploration of the dynamics of processing, an element critical to our understanding of changes in performance that accompany recovery and treatment. Two studies of recovery carried out in the framework of computational models are relevant to the relation of deep and phonological dyslexia. Martin, Saffran, and Dell (1996) proposed a single deficit account of the recovery patterns observed in deep dysphasia (the auditory–spoken counterpart to deep dyslexia in which semantic errors are produced in repetition). This account defined the deficit in terms of processing characteristics (activation spread rate and activation decay rate) rather than locus of transmission route affected. Martin et al.'s (1996) patient was determined to have an impairment that affected maintenance of activated representations due to an abnormally high decay rate. Recovery, in this model, took the form of resolution of that decay rate toward the "normal" premorbid rate. A similar account of the resolution of deep dyslexia to phonological dyslexia is conceivable in such a model.

In another recent study, Plaut (1996) attempted to simulate resolution of pure alexia to deep dyslexia and then to phonological dyslexia in a connectionist network. Although the model successfully reproduced the early shift from few responses to the deep dyslexic pattern, it failed to produce the dropout of semantic errors that characterizes the evolution from deep to phonological dyslexia. This was the case when lesions were applied to connections between intermediate and semantic representations and also between orthographic and intermediate representations. This finding led Plaut (1996) to conclude that the evolution from deep to phonological dyslexia actually reflects improvements in both phonological and semantic routes of reading. Case studies of recovery patterns echo this conclusion in that many patients show a recovery pattern in which the ability to read nonwords (a barometer of phonological function) improves in conjunction with the reduction of semantic errors.

41-6.1.1. Recovery Patterns in Deep Alexia

Recovery patterns in this rather severe impairment of reading have been documented by Behrmann et al. (1990). Their subject, D.S., was able to read words accurately with a letter-by-letter strategy, but only after considerable delay. Thus, Behrmann et al. (1990) based their investigation of recovery on changes in response latencies in conjunction with word-length effects rather than changes in error patterns. Letter-by-letter readers show increased reading times as a consequence of their laborious attempts to read words by reading each individual letter and this latency increases as word length increases. Behrmann et al. (1990) followed their subject from the 1st week post-onset and periodically over the following 12-month period. Initially, this patient showed all the classic symptoms of a severe pure alexia: intact visual and peripheral analytic abilities, letter-by-letter reading, and reading times that were positively associated with word length. Additionally, D.S. displayed some ability to gain access to higher-level information about written words, but only with unlimited exposure time. Thus, the underlying impairment, difficulty in accessing the word-form system, was successfully circumvented by this patient with the letter-by-letter strategy. Behrmann et al. (1990) found that reading times improved dramatically with recovery, although they still remained below normal. Nevertheless, subsequent testing of the underlying deficit (access to the word form) revealed no evidence of improvement. Behrmann and her colleagues concluded that improved reading times resulted from an improvement in D.S.'s compensatory reading strategy rather than any resolution of the underlying deficit.

This case study emphasizes that recovery can take two forms and illustrates the relevance of recovery patterns to treatment research. Changes in reading ability associated with recovery from pure alexia can occur because impaired mechanisms recover or because compensatory strategies become more efficient (see also Tuomainen & Laine, 1991). Thus, a therapist can determine which approach coincides with a subject's natural course of recovery and treat the deficit accordingly.

41-6.2. Studies of Treatment

41-6.2.1. Deep and Phonological Dyslexia

Most treatment studies of deep dyslexia and its near relative, phonological dyslexia, have focused on direct retraining or reestablishment of the impaired orthographic–phonological route. An important early study of treatment for acquired deep dyslexia (de Partz, 1986) illustrates how early models of the functional organization of reading processes served as a useful guide to an organized program of treatment. De Partz (1986) treated the reading deficit in a deep dyslexic subject by systematically retraining the impaired orthographic–phonological conversion route. Initially, the subject was taught cue words whose initial segments held direct grapheme–phoneme correspondences to associate with each letter of the written alphabet. Following this, the subject was taught to blend the grapheme–phoneme correspondences into words. The first part of the program targeted only those correspondences in which a single letter and sound are associated. The final part of the program involved training correspondences in which a single phoneme was represented by a sequence of graphemes (for example, /u/ $\leftrightarrow$ *ou*). Although de Partz demonstrated substantial improvement in this subject's reading, it was not entirely clear that the changes were due to the treatment alone since the training program coincided with the period of natural recovery.

In a more recent study, Nickels (1992) applied the treatment program developed by de Partz (1986) to a patient, T.C., with a more long-standing (11 months post-CVA) reading impairment. Using link words as cues to the identities of grapheme–phoneme correspondences, Nickels demonstrated improvement in her subject's ability to link graphemes with corresponding phonemes. However, T.C. remained unable to blend the grapheme–phoneme correspondences into words (except high-imageability words) and nonwords. Patterson (1994) notes that these two studies illustrate how a particular therapy technique may not have identical effects in all subjects with a common reading deficit (in this example, deep dyslexia) because more than one underlying impairment may give rise to a set of symptoms associated with that deficit. This necessitates that treatment studies provide accounts for the deficits being treated as well as the changes that occur in response to treatment (Patterson, 1994).

Treatment programs for deep dysgraphia (which frequently co-occurs with deep dyslexia) have also made use of cues to link known words with words that cannot be spelled. Hatfield (1983) pioneered the use of a link-word technique to tackle the problem of deep dysgraphics' difficulty in spelling closed-class words. She trained subjects to link a content word that could be spelled by the subject with an orthographically similar closed-class word (for example, INN $\rightarrow$ IN). This technique continues to be applied in more recent studies of treatment in deep dysgraphia (for example, Carlomagno & Parlato, 1989). The associative cues provide a means by which access can be facilitated. More specifically, the content word cues (as were used in Hatfield's study) supply a link with a semantic representation that is richer and more accessible than any semantic level representation normally accorded a closed-class word. If deep dyslexia and dysgraphia do indeed reflect reading by the semantic route (or reliance on semantic representations to gain access to grapheme representations), then this interpretation is plausible.

41-6.2.2. Treatment of Pure Alexia

Behrmann et al.'s (1990) study of recovery patterns in this syndrome (see Section 41-6.1.1) demonstrated two paths of recovery that would lead to a reduction in the letter-by-letter reading strategy observed in pure alexics. Other studies confirm that effects of treatment strategies differ depending on whether they address compensatory strategies or the impairment itself (for example, Coslett & Saffran, 1989; Rothi et al., 1985). Whereas the subject of Behrmann et al. (1990) did not show improved access to the word-form system with the improved compensatory strategy, other subjects have shown such improvement with treatments that emphasize direct access to the word-form system, even after natural recovery processes have occurred (Rothi et al., 1985). These investigations of pure alexia have two important implications for treatment. First, Rothi et al.'s (1985) subject showed reading improvement 4 years after onset of the disorder and in spite of a firmly established letter-by-letter reading strategy. This indicates that the subject shifted reading strategies and that direct training led to that shift (Coslett & Saffran, 1989). Thus, although access to the word-form system might improve naturally, it can also be "trained" with some possibility of success. Second, like many treatment studies targeting other language impairments, improvements in function are noted long after spontaneous recovery has occurred, indicating that continued treatment after neurophysiological recovery is justified.

41-6.2.3. Treatment of Surface Dyslexia and Dysgraphia

A central goal in many treatments of surface dyslexia (which affects access to the meaning of written words) is to facilitate direct or indirect access to semantic information about words that are being read; that is, treatments tend to include some means of forcing the subject to learn a written word's meaning. One subject, E.E., studied by Coltheart and Byng (1989), showed the typical surface dyslexic pattern indicating an inability to derive semantics from orthography. Coltheart & Byng directed training specifically toward relearning the pronunciation of words with highly irregular spellings. Rather than just drilling E.E. on these pronunciations, however, they included pictures that corresponded to the irregularly spelled words. Combining the semantic representation with the word-pronunciation drill resulted in improved performance on both treated and untreated words. In another study, Scott & Byng (1988) incorporated a semantic task into treatment of reading in a surface dyslexic, J.B., who made errors on homophones (words that sound similar but are spelled differently and have different meanings). They used a sentence completion task in which J.B. had to choose the target word among other words. Here, the subject is asked to consider a written word's meaning and is cued not by a picture but by the circumstances described by the sentence to be completed.

The importance of (re)establishing links between written words and their meanings is also apparent in treatment studies of surface dysgraphia. Behrmann (1987) used training tasks that paired a written word with its meaning (sentence completion, picture matching) in a treatment program for a surface dysgraphic 53-year-old patient, C.C.M. Training was carried out in two stages, the first to train spellings of homophonic word pairs and the second to train spellings of nonhomophonic irregular words. The pictures were used to distinguish the meanings of the homophonic words. Of interest in this study was the finding that training generalized to untrained words with irregular spellings, but not to untrained homophonic pairs. In a more recent study, de Partz, Seron, and Van der Linden (1992) used an imagery technique to facilitate access to spellings of words with ambiguous or irregular spellings. Specifically, they trained the subject to associate a visual image semantically related to the whole word's meaning but also visually similar to the letters in the misspelled portion of the word. The purpose of the visual–semantic cues was to facilitate memorization of the ambiguous and irregular spellings. They were found to be effective compared to no training and to more classic training approaches using repetitive presentation of the ambiguous and irregularly spelled words.

The positive effects of combined semantic and phonological stimulation have been emphasized by a number of authors in the treatment of anomia (for example, Howard, Patterson, Franklin, Orchard-Lisle, & Morton, 1985; Nettleton & Lesser, 1991), a disorder that parallels surface dysgraphia in that it involves an inability to link word meanings with the phonological form. This combined approach has been shown to have a more profound effect on the endurance of treatment effects as well as generalization of those effects to untrained items. Generalization of effects is notoriously difficult to obtain, and there are few strong clues as to what should be effective in this regard. Thus, the success of studies employing combined semantic–phonological or semantic–orthographic stimulation to access written or spoken word forms represents a significant contribution to the treatment literature.

Generalization of treatment effects remains an important area of investigation, and despite the success of some approaches, others have met with little success. It is equally important to understand why one treatment is effective and why another treatment is ineffective. This is particularly true when a theory predicts success of a particular treatment strategy. The next few studies report training programs in which improvement was restricted to the items that were trained.

Hillis (1993) studied immediate effects and generalization of treatment on a subject, P.S., who had multiple impairments affecting different sites within the reading system. Treatments were tailored to the levels of representation affected. For example, in order to improve function of the phonological route, treatment involved direct training of grapheme–phoneme correspondences. To improve access to lexical–orthographic representations, a lexical decision training task was used. And finally, to improve reading by the semantic route, homophonic pairs were used in tasks that connect a written word with some representation of its meaning (definition, picture, sentence completion). Although all treatments were effective in improving performance on targeted items, generalization was minimal.

Behrmann and Lieberthal (1989) investigated treatment of the reading abilities of C. H., a 57-year-old globally aphasic subject who demonstrated, among other things, severe semantic impairment. Therapy was directed toward improving the association of words and their meanings. Treatment generalization was tested within semantic categories; that is, generalization was anticipated from treated words to untreated words within the same semantic category. Some generalization was obtained between treated and untreated items, but the amount varied considerably across semantic categories.

As these studies indicate, the mechanisms that maximize generalization to untrained items are poorly understood. Clearly, incorporation of semantic processing in treatment is an important aspect, but even this is not always effective. The Behrmann and Lieberthal (1989) study suggests that we should anticipate generalization primarily to semantically related untrained items, although category effects may lead to inconsistent generalization. Plaut (1996) provides a useful insight into this issue in a simulation study of treatment changes in reading within a connectionist network model of reading. He notes that generalization of retraining after the network is lesioned is influenced by the structure of mapping between orthography and semantics and that the structure of that mapping is related to the extent to which word units are semantically organized. Thus, generalization should depend in part on the degree to which a set of treated words encompasses an estimate of the semantic characteristics of other words within that semantic category. Plaut (1996) proposed that using a set of words representing atypical members of a category would provide the estimate needed to achieve generalization to both atypical and prototypical members of a category. This is because relatively atypical members contain (collectively) more features of all members of a category than prototypical members and thus would provide more information about the overall structure of that category. Plaut demonstrated this principle in a simulation of reading in which trained words varied in the typicality of their semantic representations. This study is an excellent demonstration of the ways in which connectionist models can be used to demonstrate effects that depend on dynamics of processing.

41-7. SUMMARY AND INDICATIONS FOR FUTURE RESEARCH

In this review of studies that focus on processes of change that accompany recovery and treatment of reading and spelling disorders, I have tried to illustrate the major issues targeted in such investigations and some current approaches employed in studies of recovery. There are practical issues to resolve such as documenting the course of recovery, identifying stimuli and methods that result in long-term effects of treatment, and maximizing generalization of treatment from trained to untrained items. Additionally, there are theoretical issues of importance such as the relationship among components and processes that compose the language system and the nature of processes underlying facilitation effects obtained in treatment.

The practical concerns of treatment studies have been greatly aided by the development of theoretical models that identify possible routes to mediate a particular task such as reading. These models provide a framework within which a reading impairment can be diagnosed in terms of its functional characteristics. In turn, the model-based diagnosis is readily translatable into an appropriate set of treatment tasks that consider the characteristics of the words to be treated (for example, semantic versus phonological characteristics) and processes that need to be reestablished (for example, phoneme–grapheme correspondences). The success of these studies in recent years has established the use of a psychological model of cognition to devise a treatment plan as the standard approach to investigating treatment issues. This enterprise has been further aided by the introduction of connectionist models of language. Such models enable a description of the dynamics of processing written and spoken language that are central to our understanding of recovery and treatment effects.

Although our understanding of processes underlying treatment and recovery is in its infancy, there is no better time for this enterprise to proceed. Our knowledge of the functional components of language systems is well established, and in recent years our knowledge of the dynamic relations among those components has increased greatly. Advances in computer implementations of cognitive models provide a means to make and test predictions about dynamics of processing that previously was not available. It is likely that treatment and recovery studies will grow in number and quality in the next decade, in part because the means to pursue critical questions about these processes are available, but also because the endeavor has a direct application to theoretical issues as well as to treatment of neuropsychological impairments such as reading and spelling.

Acknowledgments

This chapter was prepared with the support of a grant from the National Institutes of Health, DC 360564-193 to the author. Thanks go to Brigitte Stemmer and Harry Whitaker for helpful comments on an earlier version of this chapter.

CHAPTER 42

Neurolinguistic Issues in the Treatment of Childhood Literacy Disorders

Philip H. K. Seymour

Department of Psychology, University of Dundee, Dundee, Scotland

Childhood dyslexia is a disorder affecting the acquisition of written language. Approaches to treatment can be discussed in relation to questions about the causes of the difficulty, and the cognitive architecture and development of the reading process. According to peripheralist theories, treatment should be directed toward a cause located in the transient component of the visual or auditory systems, or in higher-level visual orthographic or phonological processes. The alternative view places the difficulty in one or more of the components of a central orthographic system. The important questions for treatment concern the linkage between orthographic and phonological segments, the size and nature of these segments, and their interaction with stages of reading development.

The term "dyslexia" refers to disorders that affect the use of written language. It is well known that reading and spelling processes can be disturbed by damage to the mature brain, producing an *acquired dyslexia* or dysgraphia (Ellis & Young, 1988). Such damage generally involves the left (language) hemisphere. The disorder can also occur in a childhood form, producing a *developmental dyslexia* in which the learning of the basic elements of reading and writing is affected (Critchley, 1970). Possible connections with developmental pathology of the left hemisphere have been suggested

(Galaburda, Sherman, Rosen, Aboitiz, & Geschwind, 1985), and also a dependence on a genetic etiology (De Fries, 1991).

This chapter is concerned mainly with childhood dyslexia. The disorder is manifest as a failure to advance in a normal age-related way when appropriate instruction in reading and spelling is provided in school. It is held that the difficulty may be dissociated from general intelligence. Opinions differ as to whether it should be viewed as an isolated disorder, specifically affecting written language processes, or whether it is part of a larger syndrome that includes more general disturbances of language processing (Miles, 1983). A view that is currently widely favored states that the reading and spelling difficulty is commonly associated with a disturbance of *phonological processing* that constitutes a proximal cause of the dyslexia, and that may affect other aspects of language, such as syntactic processing and comprehension (Stanovich, 1988; Pennington, 1991; Shankweiler & Crain, 1987).

The objective in this chapter is to examine questions concerning the most appropriate forms of *treatment* of dyslexic disorders. In particular, the aim is to draw on current neuropsychological research and theory in order to consider whether particular recommendations can be supported. To do this, it will be necessary to explore a number of fundamental issues, including the following:

1. Does dyslexia have an identifiable *cause?* If so, at what level (physiological, sensory, cognitive) should the cause be defined? Can the disorder be treated by removal or amelioration of the cause?
2. What is the structure of the *cognitive system* that supports reading and spelling? Can different forms of dyslexia be identified, depending on which components are most affected? If so, do such differences have implications for the design of treatment programs?
3. Is there a normal sequence in reading and spelling *development?* If so, should treatment programs mimic the normal sequence?

42-1. CAUSAL ARCHITECTURE

The second of the questions above refers to the *cognitive architecture* of the processes involved in reading and spelling. This has typically been represented as an arrangement of specialized processors (or "modules") that are linked by pathways along which information can be transferred (the "box-and-arrow" models). These models have proved valuable in the study of the acquired dyslexias because they have made it possible to formulate and test hypotheses regarding the particular components or pathways that have been compromised by neurological damage (Ellis & Young, 1988). This information is held to be relevant for the design of treatments in that the identification of the damaged functions provides a rationale for the direction of remediation (Coltheart, Bates, & Castles, 1994).

The cognitive architecture models represent the systems involved in the processing of information. As such, they are well adapted for the description of the failures to perform standard tasks (such as reading aloud of words or nonwords) that are characteristic of acquired dyslexia. However, the models do not address the first issue, the question of causation, or the third issue, the course of reading development. On these grounds, it can be argued that a theoretical model that represents both the architecture of the cognitive system and causal influences on the development of processes may be needed as a first step in a discussion of the treatment of developmental dyslexia. A schematic format for such a model is presented in Figure 1. The framework incorporates the distinction between *central* and *peripheral* processes that has previously been applied in the studies of the acquired dyslexias (Shallice, 1988). Thus, it is suggested that reading and spelling depend on the integrity of a set of "central orthographic processes." There are, in addition, various peripheral processes that deal with the reception and categorization of visual and auditory input.

The diagram in Figure 1 is intended to illustrate two general hypotheses about the causes of developmental reading disorders. One hypothesis states that the cause is intrinsic to the *central* orthographic processes themselves. The other ascribes the difficulty to a *peripheral* process that is functionally adjacent to the central processes and that provides them with essential data. If the first hypothesis is correct, it would be appropriate to direct treatment toward the construction of the central orthographic processes. Further, if the process has an internal modular structure, treatment should be directed toward the subcomponent that is most severely affected. If the second hypothesis is correct, it would become important to identify the peripheral process that was impaired so that remediation could, perhaps, be directed toward that process. This peripheralist hypothesis will be considered first.

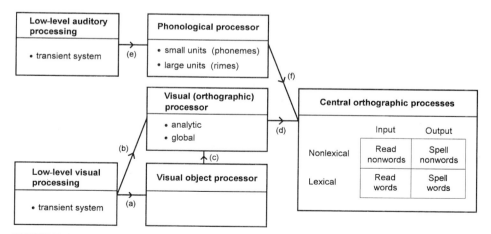

FIGURE 1   Schematic representation of the cognitive architecture of peripheral and central processes in reading, including possible causal influences (marked by arrows).

42-2. PERIPHERALIST HYPOTHESES

Figure 1 suggests the existence of a number of modular processing systems in addition to the central orthographic processes. In the diagram, the arrows represent *causal influences* of an adverse or beneficial nature rather than information flow. The additional components can be viewed as peripheral visual and auditory processes, which can be defined at two levels. The first level concerns the functioning of the sensory pathways that are involved in the direct acquisition of information from the environment. These are referred to as "low-level" processes in the diagram. The second level is cognitive and deals with the internal representation of incoming information, including, for example, the possibility of distinct processors for categorization of visual objects and written language.

42-2.1. Low-Level Visual Processes

The first of the peripheralist hypotheses holds that central dyslexic difficulties result from an impairment located in the *visual* processing pathways. At the level of physiology and basic sensory functioning, the visual system is held to be divisible into two pathways, which are referred to as the parvocellular (P) or "sustained" channel, and the magnocellular (M) or "transient" channel (Lehmkuhle, 1993). The pathways may be traced from the retina through the lateral geniculate nuclei into the visual cortex and are associated with differing processing characteristics. In particular, the P pathway is specialized for perception of color and detail (high spatial frequencies) over an extended period of processing time, whereas the M pathway responds rapidly and briefly to stimulus change and deals with global features of patterns (low spatial frequencies). There is accumulating evidence that dyslexia may be associated with abnormally slow functioning in the M (transient) pathway. This comes from anatomical studies (Livingstone, Rosen, Drislane, & Galaburda, 1991) and from perceptual studies of evoked potentials and stimulus judgments (Lovegrove, Martin, & Slaghuis, 1986) in which the spatial and temporal frequencies of simple patterns (gratings or checkerboards) have been systematically manipulated.

Figure 1 suggests that an adverse influence of an impaired visual transient system on central reading and spelling processes presupposes an intervening effect on the visual cognitive analyzers. One possibility is that the transient system defect influences visual object recognition (link [a]) which in turn affects the establishment of a specialized visual (orthographic) processor (link [c]) which in its turn undermines the development of the central processes (link [d]). A second is that the transient system defect directly affects the development of the visual (orthographic) processor (link [b]). At the present time, the studies needed to trace these possible connections have not been carried out. Hence, although there is good evidence of a co-occurrence of transient system defects and central reading difficulties, the underlying causal mechanism has not been adequately documented. Indeed, the main interpretation of the effect has focused on eye movements in reading, and the role of the transient system in clearing the sustained channel prior to reception of new input following a saccade (Breitmeyer, 1993).

If the transient system is the origin of difficulty, then an implication is that treatments that improve transient system functioning might be effective. These could include manipulations of text format, such as reduction in line length (to reduce the involvement of saccadic eye movements) or changes in the color of letters or background that are calculated to enhance transient system responsiveness (Lovegrove & Williams, 1993). A convincing case for the applicability of this type of treatment would require *(a)* a demonstration of the effect on transient system processing; *(b)* confirmation of improved functioning at the visual (orthographic) level; and *(c)* an analysis of central reading processes that showed changes in specific components. In the absence of studies of this type, the value of peripheral visual treatments of dyslexia remains uncertain.

42-2.2. Visual Orthographic Processes

A common element in the visual hypothesis is the suggestion that the central reading impairment depends on a dysfunction in the visual (orthographic) processor (link [d]). This could occur, independently of the status of the low-level processes, if the neural substrate of the processor was underresourced. It is possible to test for deficiencies located within the processor by using reaction time methods, including same–different judgments of letter arrays and analyses of effects of distortion of format (e.g., zigzag or vertical arrangements of letters). Studies of this type indicate that a proportion of dyslexic children (maybe 50% or more) exhibit entirely normal processing at this level (Seymour & Evans, 1993). According to the model in Figure 1, this means that the visual hypothesis cannot be true of all cases of dyslexia, but may be appropriate for a subset of cases. This could be connected with the proposed differentiation within the central orthographic processes (see Figure 1), that is, with the distinction between lexical and nonlexical processes. Thus, it might be that an impairment of the visual (orthographic) processor undermines only one aspect of the central process, for example, the lexical pathways that are required for word identification and spelling. If this was true, it could follow that treatment directed at the efficiency of visual (orthographic) processes might be appropriate only for cases having a predominant impairment of the lexical processes.

In order to support this more differentiated hypothesis, one would wish to establish that impaired visual (orthographic) processing accompanied a central dyslexia of the lexical (or "surface") type. In this context, it will be important to recognize that the functions of the peripheral processor are, in themselves, quite complex, and may include aspects such as *(a)* the internalization of sequential regularities in legal letter sequences; *(b)* a capability for "analytic" (letter-by-letter) processing; and *(c)* a capability for "global" processing of entire words or segments of words (morphemes, syllables, and so forth). There is evidence that "surface dyslexia" is particularly associated with impairment of the "global" functioning of the visual processor

(Seymour & Evans, 1993). This is because surface dyslexic individuals appear unable to recognize words as familiar wholes and characteristically resort to an analytic letter-by-letter mode of processing. Nonetheless, studies that have sought to establish a link between specific impairments of the visual processor and particular forms of central dyslexia have not produced particularly clear outcomes. Seymour (1986) found some cases who exhibited large effects in visual (orthographic) processing that were not associated with comparable central effects. Nor has it proved possible to establish a secure association between impaired holistic visual processing of letter arrays and "surface dyslexia" (Seymour & Evans, 1993). Hence, although there is a good prima facie case for attempting to treat "surface dyslexia" by training the global function of the visual (orthographic) processor, there is currently a lack of evidence that the connection is a real one.

Also, it remains unclear what the best method of training might be. Seymour and Bunce (1994) reported a longitudinal treatment study of a surface dyslexic boy, R.C. Despite prolonged training, which was effective in improving lexical reading and spelling, the underlying approach remained strongly sequential (indexed by the effects of variations in word length on the reading reaction time). It is worth noting that other individuals show a contrasting pattern that suggests an impairment of the "analytic" function of the visual processor. This is shown by an overreliance on global processing that is effective in recognition of whole words but that results in numerous errors when attempts are made to read unfamiliar nonwords. Seymour and Bunce (1994) described a second case, D.K., of this type. In this instance, training that was directed toward the adoption of a more analytic approach to reading was effective in improving accuracy of response to nonwords. The training also produced a change in processing strategy, including the adoption of a more deliberate and analytic procedure when attempting to read unfamiliar forms (indexed by changes in reaction time distributions).

42-2.3. Auditory Processes

The second main hypothesis represented in Figure 1 is that dyslexia results from a peripheral cause that is located in the auditory system. The arguments here are very similar to those that have been mounted in respect of the visual hypothesis. It is held that the processing of incoming speech stimuli takes place within an auditory pathway that includes a "transient" component sensitive to the rapid shifts in acoustic characteristics that convey information necessary for the categorization of speech segments. There is evidence from anatomical studies of the medial geniculate nucleus (Galaburda, Menard, & Rosen, 1994) and from psychophysical procedures (Tallal, 1980) that this auditory transient system may be defective in cases of dyslexia.

The contention that a peripheral disturbance of speech processing might be causally related to dyslexia raises questions about the existence of the connections illustrated in Figure 1. It would be desirable to be able to show that the auditory transient system disturbance was associated with deficiencies in a "phonological processor" in which speech segments may be represented and manipulated (link [e]), and that this disturbance was, in its turn, associated with a specific impairment of the central reading processes (link [f]). Alternatively, it would be necessary to show that training that improved auditory processing exerted a consequent effect on the activities of the phonological processor, and that this eventually led to improved reading and spelling. Tallal et al. (1996) have indicated that this might be done by modifying the duration or intensity of the critical segments of the speech signal. However, it is not known whether the expected effects on phonological processing and reading would follow from the application of such treatments.

42-2.4. Phonological Processes

At the present time, the most widely acknowledged hypothesis regarding the causation of dyslexia postulates an influence of "phonological processing" on the central reading processes (link [f]). This could arise independently of the integrity of low-level auditory processes as a consequence of neural underresourcing of the phonological processor. A large amount of research has been directed toward the evaluation of this hypothesis. In essence, this research has aimed *(a)* to develop methods of measuring the functioning of the phonological processor; *(b)* to show that an association exists between impaired phonological functioning and impaired reading; and *(c)* to demonstrate that the direction of training toward the phonological processor will exert a beneficial effect on reading development. Various techniques for assessment of "phonological awareness" have been devised, the common feature being a requirement to isolate and manipulate segments of speech (by counting, addition or deletion of elements, inversion of sequence, and so forth). Ability to perform these speech manipulation tasks is strongly and reliably correlated with reading progress (Goswami & Bryant, 1990).

This research supports the reality of link (f) in Figure 1. However, there are some complications. One of these concerns the size and nature of the phonological units that are held to be related to reading acquisition. In the past, it has been maintained that small segments of speech, the *phonemes* of the language, are crucial, because these units stand in an approximate correspondence with the letters of the alphabet and form the basis of the relationship between written and spoken language (Gough & Hillinger, 1980). But the studies suggest that preliterate children are not aware of the phonemic structure of speech and that it is only after formal instruction in alphabetic reading that children begin to succeed in phoneme manipulation tasks (Goswami & Bryant, 1990). This raises a question about the directionality of link (f). It looks as though reading acquisition influences phonological processing, implying the reverse of the direction of influence, or at least the possibility of bidirectionality (Morais, Alegria, & Content, 1987).

In addition, it has been noted that preliterate children may be able to demonstrate awareness of some segments of speech, particularly if the units involved are large ones, such as whole syllables or rhymes. This observation has been discussed in the context of linguistic analyses of the internal structure of the syllable, particularly the suggestion that an intermediate level of organization into an *onset* group (the initial consonant or consonants) and a *rime* group (the vowel + final consonants) can be identified (Treiman, 1992). The ability of preliterate children to perform tasks such as oddity detection with sets of words that are distinguished in terms of onsets or rimes has been taken to imply that the phonological processor supports an awareness of these segments (Goswami & Bryant, 1990). As a corollary, it is argued that phonological processes influence reading development by supporting a classification of words in terms of onsets and rimes and the development of a capacity to read new words by looking for rime analogies with known words (Goswami, 1993).

It follows that the implications of the phonological hypothesis for the remediation of dyslexic difficulties are somewhat ambiguous. It is not clear whether teachers should emphasize large units (rimes) or whether they should emphasize small units (phonemes). Further, the uncertainty about the directionality of link (f) means that teachers cannot be confident that a deficiency within the phonological processor can be identified as a cause of the reading difficulty that might be corrected by the direction of training toward the processor.

An effective test of this possibility requires that training in speech manipulation should be shown to be successful (i.e., to result in a clear improvement in phonological skill) and to result in consequent gains in reading. Some positive results have been obtained by Lundberg and his group in Scandanavia. Lundberg, Frost, and Petersen (1988) trained preliterate kindergartners in a range of phonological activities that had large effects on phonological skill and produced advantages in later reading and spelling relative to the results obtained from a control group. Lundberg (1994) subsequently reported results for a subset of children who had performed poorly on phonological tasks at the outset of the study and who could, accordingly, be seen as being "at risk" for the development of subsequent reading difficulties. The comparison with similar control cases suggested that phonological training had had the effect of offsetting the risk of reading difficulty. However, a comparable study by Bradley and Bryant (1983), directed toward children who had shown poor preliterate skills in rime oddity detection, did not demonstrate significant gains in reading following two years of phonological training (carried out concurrently with normal instruction in reading).

Hatcher, Hulme, and Ellis (1994) investigated the effect of intensive phonological training on 7-year-old children with documented reading difficulties. This was effective in producing improvements in phonological skill, but no consequent gain in reading and spelling occurred. Hence, this study calls into question the assumptions (a) that there is a unidirectional causal link from phonological processing to reading, and (b) that direction of treatment toward the phonological processor is an effective method of improving reading difficulties.

42-3. CENTRALIST HYPOTHESES

It is assumed (Figure 1) that the literate brain contains a set of "orthographic processes" that support information about written word forms, print–sound correspondences, and links with semantics. Damage to the neural basis of these processes will produce a "central dyslexia." In the diagram, it is suggested that the central orthographic process may have its own internal modular structure, which is organized around the distinctions between input and output (reading and spelling), and between lexical (word-specific) and nonlexical (letter–sound) processes. These submodules may themselves be vulnerable to damage, allowing for the possibility that central dyslexias may take different forms, depending in particular on whether the predominant effect is on the lexical processes or on the nonlexical processes. Such effects are by now well documented and support a distinction between two categories of central dyslexia: (1) "surface dyslexia," which reflects damage to the lexical processes, and (2) "phonological dyslexia," which results from damage to the nonlexical processes (Shallice, 1988; Ellis, & Young, 1988). A similar contrast can be observed in spelling ("surface dysgraphia" versus "phonological dysgraphia").

With regard to developmental disorders, the critical question is whether an analogous pattern of contrasting difficulties can be observed in children who are having difficulty in learning to read. Castles and Coltheart (1993) have maintained that developmental forms of "surface dyslexia" and "phonological dyslexia" occur. In their study, the reading of words of irregular spelling was treated as an index of the lexical process and the reading of nonwords as an index of the nonlexical process. Results obtained from severely dyslexic children were compared with results of normal readers of the same age (chronological age, or CA match) in a regression analysis. Castles and Coltheart (1993) reported that there were some dyslexic children who exhibited impaired irregular word reading combined with "normal" (within chronological age range) nonword reading, and others who displayed the opposite pattern (impaired nonword reading combined with normal irregular word reading). They argued that dyslexic children present a dissociation between surface and phonological patterns of dyslexia that is comparable to that found in adults. Their conclusion was that the dual process (lexical, nonlexical) model of the cognitive architecture of the central orthographic processes was applicable to developmental disorders as well as to adult disorders.

According to the centralist hypothesis, the basis of developmental dyslexia is an underresourcing of the neural substrate of the central orthographic processes. One implication is that, to be effective, remedial training should be directed toward the orthographic processes themselves and not toward a peripheral process. This implication is supported by the intervention studies. Bradley and Bryant (1983) gave one of their groups of "at risk" children training that included both practice in classifying spoken words according to their sounds and demonstrations of the way in which these distinctions were reflected in the letters making up the words. This dual approach resulted in substantial gains in reading and spelling. Hatcher et al. (1994) treated one of their groups of 7-year-old dyslexic readers in a similar way. A group that received phonological training combined with reference to letters and orthography improved their reading and spelling relative to an untreated control and to groups receiving training in phonology alone or (nonphonological) remedial reading. The authors concluded that the key requirement for successful remediation was the teaching of *linkage* between phonological segments and orthography. The question, raised earlier, of the size and nature of the segments was not addressed in this study. Thus, it remains unclear whether the linkage should be established for small units (phonemes) or for large units (onsets and rimes).

A further issue that was not considered by Hatcher et al. (1994) concerns the possibility of individual variations, particularly the distinction between "surface dyslexia" and "phonological dyslexia." Should this distinction be taken into account in designing remedial programs, with different approaches, targeting either the lexical or the nonlexical process, being required for different individuals? Or is there a single approach that is relevant for all dyslexic cases, notwithstanding the occurrence of individual variations?

Treatment studies based on the modular standpoint have been reported by Broom and Doctor (1995a, 1995b). In the first study, an 11-year-old dyslexic boy, D.F., who exhibited the features of "surface dyslexia" (reliance on letter–sound processing, difficulty with irregular words, homophone confusions, regularization errors, phonetically plausible spelling errors), was given training (using the method of "simultaneous oral spelling") designed to establish visual representations of unknown irregular words. An individual case multiple baseline design was used to assess the effectiveness and generality of the training. Acquisition of specific words occurred, but there was little evidence of generalization to other words or tasks. The second study concerned another boy, S.P., who exhibited the features of "phonological dyslexia" (errors and response delays in nonword reading, occurrence of visual and derivational errors). An attempt was made to direct treatment toward the nonlexical process by training the sounds attached to complex graphemes contained in unfamiliar regular words. The treatment was successful in establishing grapheme–phoneme correspondences and there was evidence of generalization to the reading of new words and nonwords.

Broom and Doctor's studies were based on the assumption that central orthographic processes possess the modular structure shown in Figure 1. Other investigations have produced somewhat different outcomes. For example, Seymour and Bunce (1994) reported a study of two dyslexic boys, aged about 10 years. R.C., who has already been mentioned in connection with peripheral visual accounts of dyslexia, presented the features of "surface dyslexia," including regularity effects, phonetic spelling, and reaction time anomalies (slowness of response, letter-by-letter reading). D.K. was considered to be a case of "phonological dyslexia," showing a very large dissociation between word reading (9% error rate) and nonword reading (94% error rate). Both cases participated in an extended treatment study during which an attempt was made to establish the basis of orthographic structure. This involved highlighting of letter groups at two levels of complexity. In both cases, the interventions were successful in producing changes in reading and spelling, although the underlying contrast between the cases (i.e., the "surface" and "phonological" dyslexia) remained.

In the present context, the important point concerns the generalization of the effects of the treatments. The approach adopted acknowledged the proposed modular structure of the orthographic system and attempted the selective training of the four components identified in Figure 1. This was done by devising instructional programs that emphasized recognition (reading) or production (spelling), and that used nonwords (nonlexical process) or real words (lexical process) as exemplars. The emphasis on orthographic segments produced a large amount of generalization, which extended beyond the specific vocabularies that were used, and beyond the system that was targeted. Thus, teaching that emphasized reading of words produced effects on, for example, the spelling of nonwords.

These results suggest that an important issue in the evaluation of teaching approaches concerns the degree to which the effects *generalize* across vocabularies and functions. This principle is at variance with the modular conception of the orthographic process. It might, indeed, be seen as more in the spirit of the connectionist formulations (Seidenberg & McClelland, 1989) in which lexical and nonlexical regularities are held to be encoded within a single network.

42-4. ORTHOGRAPHIC DEVELOPMENT

The analysis presented so far suggests that the remediation of childhood dyslexic disorders might most usefully focus on the central orthographic processes, and might aim to promote generalized learning by emphasizing orthographic segments that are related to phonology. However, this is a very general prescription that provides little guidance regarding either the content or the sequence of instruction.

Several models of literacy acquisition have been formulated. They typically represent development as a progression through a succession of "stages" in which new strategies or processes emerge. A common view is that reading starts with reliance on a visual "logographic" strategy but that, with the development of phonological awareness, an "alphabetic" strategy emerges that supports the capacity to read unfamiliar forms (i.e., the nonlexical process), while at the same time providing a phonological basis for word recognition (the lexical process) (Gough & Hillinger, 1980; Ehri, 1992).

One way of looking at this is to make a distinction between an early *foundation* level in reading development and a subsequent *orthographic* level (Seymour, in press). The foundation level involves acquisition of basic letter–sound knowledge and the formation of two primitive functions: (1) a *logographic* process of global word recognition and storage; and (2) an *alphabetic* process of sequential letter–sound translation. Later development depends on the formation of a central structure called the orthographic framework. This is built up step-by-step by a process of internal phonologically motivated redescription of data held in the logographic system. According to this view, the approach to treatment will differ according to level, focusing on specific components and phonemic awareness at the "foundation level," and on the construction of generalizable knowledge, perhaps based on rime segments, at the "orthographic level."

42-5. CONCLUSIONS

This chapter has explored implications of cognitive neuropsychological research for treatment of childhood literacy disorders. One possibility is that treatment might be directed toward the supposed *cause* of the difficulty. However, despite large research investments, the causal connections needed to justify peripheralist approaches to treatment are not yet established. The second possibility is that treatments should be directed toward the *central* orthographic processes themselves. These processes may have an internal modular structure, implying a need for *(a)* identification of the weakened process, and *(b)* direction of treatment toward that process. However, it is uncertain whether this degree of selectivity is possible or desirable. A conclusion is that progress may depend on the availability of a *developmental* account of the steps involved in construction of a central orthographic system, especially the achievement of *linkage* between orthographic and phonological segments.

CHAPTER 43

The Role of Computers
in Aphasia Rehabilitation

Volkbert M. Roth[1] and Richard C. Katz[2]
[1]Kliniken Schmieder, Allensbach, Germany and Universität Konstanz, Germany; [2]Department of Veterans
Affairs Medical Center, Phoenix, Arizona 85012 and Arizona State University, Tempe, Arizona 85287

**The computer can be an effective clinical tool by incorporating what we know
about aphasia treatment and computer programming. Computers can administer
activities designed by clinicians, vary stimulus characteristics, adjust response
requirements, present cues, and select tasks, all in response to patient performance. Many supporters of computers as a clinical tool focus on relevant issues
such as cost effectiveness and operational efficiency, but ignore the important
issue of the efficacy of the treatment provided. This chapter reviews major factors
that influence the development and use of computer therapy programs in aphasia
rehabilitation. Attention is given to theories of aphasia and limitations inherent
in the computer medium. The structure, content, and efficacy of aphasia treatment software as well as the role and responsibilities of the clinician are also
discussed.**

Computers can be powerful clinical tools. They can administer activities designed by
clinicians, vary stimulus characteristics, adjust response requirements, present cues,
and select tasks, all in response to patient performance. Computers can increase the
amount of time patients are involved in treatment and provide more treatment at lower
cost. Performance data aggregated across patients and tasks can be used to create a
database from which aphasia and the effects of therapy can be abstracted. Most

advocates of computers focus on their "appealing" features—cost effectiveness and operational efficiency—all but ignoring the important issue of the efficacy of treatment provided (Loverso, 1987).

Many factors influence the development and use of computer therapy programs in rehabilitation (Kaasgaard & Lauritsen, 1995). Software for treating aphasia, however, is shaped primarily by *(a)* theories of aphasia, *(b)* limitations inherent in the computer medium, and *(c)* effectiveness of treatment software.

43-1. MODELS OF APHASIA

Effective therapy is individualized for each patient's abilities and disabilities. Explicit models of aphasia treatment (e.g., stimulation-facilitation, linguistic, modality, processing, minor hemisphere, functional) should guide therapy structure and content (Chapey, 1994; Horner, Loverso, & Gonzalez Rothi, 1994). For example, treatment software based on the *stimulation-facilitation model* (Duffy, 1994) is repetitive and intensive, uses many different and meaningful stimuli, and elicits responses that are unforced and not corrected (Katz & Wertz, 1997). Treatment software developed from the *linguistic model* (Lesser, 1978) attempts to restore language by organizing stimuli according to linguistic systems (Guyard, Masson, & Quiniou, 1990; Loverso, Prescott, Selinger, Wheeler, & Smith, 1985). Treatment software developed from the *modality model* (Luria, 1973b; Weigl & Bierwisch, 1970) attempts to reorganize language and communication through intra- and intersystemic stimulation by pairing weak modalities with strong ones to "deblock" impaired performance (Scott & Byng, 1989). Horner and Loverso (1991) provided four major reasons for describing aphasia treatment in terms of explicit models: (1) facilitate the translation of linguistic, psycholinguistic, neurolinguistic, and cognitive psychology literature into meaningful and rational treatment techniques; (2) articulate the rationale of treatment; (3) help clinicians to evaluate the specific therapeutic technique; and (4) accumulate data from model-driven scientific investigations of therapy efficacy to help determine the validity of various theoretical models. Many modality-specific tasks and modality combinations can be reconstructed using modern multimedia computers to address the therapeutic needs of aphasic patients (Kosa, 1994; Roth, 1992a, 1996).

43-2. LIMITATIONS OF COMPUTERS IN APHASIA THERAPY

Contemporary computer treatment programs consist of a finite set of rules that are stated explicitly. From these rules, an *algorithm* (i.e., finite series of steps) is derived that specifies actions at any particular point during therapy. However, unlike most computer programs, aphasia therapy is governed by an evolving set of assumptions derived from a diverse body of research. Treatment provided by a computer is quantifiable and qualitatively different from treatment provided by a clinician. Although many nonlinguistic communication behaviors are omitted (e.g., facial expressions, gestures, postures, spatial positions, clothing styles), the information generally exchanged through computer-controlled modalities (visual and auditory, and, to a certain extent, kinesthetic) is often sufficient to support effective (though rarely comprehensive) language therapy. Although many large and complex computer treatment programs provide systematic and reliable auditory and visual stimulation and accept and evaluate single-key, "pointing," and typing responses, a clinician can do much more by generating an infinite number of novel and relevant stimuli and by recognizing, evaluating, and modifying treatment in response to previously unacknowledged associations and unanticipated responses. In this last aspect alone, computerized *training* will always be a subset of speech and language *therapy* provided by clinicians.

43-2.1. Properties of Computer-Provided Treatment

Bolter (1984) provided four properties of computers and computer programming in general that help clinicians describe limitations inherent in the application of computers to aphasia treatment: discrete, conventional, finite, and isolated.

43-2.1.1. Discrete

Because computers are discrete (i.e., digital), qualitative description and decisions are difficult for them. Events must first be separated into distinct, unconnected elements before they can be acted upon. However, treatment of communication disorders, such as aphasia, is recognized as a multilevel, simultaneously interactive behavioral exchange. Face-to-face communication is a complex act described as our "oldest and highest bandwidth technology" (Rheingold, 1991, p. 216). In summarizing his research on competing messages (i.e., the listener's perception of the meaning of words versus the message's emotion) during face-to-face communication, Mehrabian (1968) estimated that 55% of a message's affect (i.e., the emotional content) is communicated via facial cues, 38% is communicated vocally, but only 7% is communicated by the actual words. Although many individual language tasks can be programmed, applying computer technology appropriately to aphasia rehabilitation requires appreciation for the intricacies and interdependence of verbal (i.e., language) and nonverbal (e.g., kinesics, proxemics) channels of communication.

43-2.1.2. Conventional

Most computer programs apply predetermined rules to symbols that have no effect on the rules. While changing values of the symbols directs the outcome of the program, the rules themselves never change. However, although a consensus exists for some fundamental guidelines of aphasia treatment, all the rules are not known and those that are may not always be right (Rosenbek, 1979). Conventional computer-provided therapy programs, even those utilizing complex branching algorithms (e.g., Katz & Wertz, 1992), do not follow the clinical cycle of *(a)* administer treatment, *(b)* analyze performance, *(c)* modify treatment, and *(d)* readminister treatment, and therefore are not adequately responsive to the dynamics of patient performance. Programs utilizing artificial intelligence techniques, such as Guyard et al. (1990), more closely approached the process employed by aphasia clinicians.

43-2.1.3. Finite

Programs incorporating dynamic rules or "fuzzy logic" to create additional rules of therapy as new data are received are unfortunately rare in aphasia rehabilitation. Although impressive for their flexibility and responsivness, these "expert systems" are generally limited in scope in the same manner as more conventional therapy programs. For example, Guyard et al. (1990) describe a prototype program that utilizes a linguistic model, the "mediation theory" (Gagnepain, 1982), to guide artificial intelligence techniques to provide patients with appropriately selected stimuli and detailed feedback. Each patient is presented with a unique sequence of exercises based on a diagnostic "patient portrait." However, the goal of the exercise described is deliberately restricted to improve accurate selection of the French language articles *le* and *la* by determining the gender of a list of nouns. Little if any research has described programs using artificial intelligence to treat areas of rehabilitation that are more complex or functional.

Most currently available aphasia therapy software does not incorporate artificial intelligence programming. The rules and symbols that control these conventional computer-provided treatment programs are limited to those defined within the programming code. Unforeseen problems and associations do not result in creation of new rules and symbols. Therapy presents the opposite case. Not all therapeutically relevant behaviors are identified; those that have often vary in importance between patients and situations. Computer-assisted learning commonly defers decisions to software designers and programmers who are not physically present during the session, but must plan in advance how to handle the intervention and code these steps into a computer program (Odor, 1988).

43-2.1.4. Isolated

Problems and their solutions presented by computer exist within the computer's own parameters, apart from the real world. Problems are stated in a way that symbols can be manipulated to solve the problem by following an algorithm. This lack of "world knowledge" is perhaps the most significant impediment to comprehensive computer-provided treatment. Language tasks as presented by computers essentially differ from the meaningful, pragmatic setting in which communication occurs among people. Computers, therefore, only consider problems in which all the variables and rules are known ahead of time, and can be solved in a step-by-step procedure with a finite number of steps, much like a game of chess. Learning to solve linguistic riddles on a computer program may lead to better scores in a standardized aphasia test; improving communication, however, is quite another matter.

43-2.2. Characteristics of Aphasic Patients

Patients vary according to age, gender, handedness, education, premorbid learning style, site of lesion, etiology, time post-onset, severity of aphasia, type of aphasia, coexisting communication problems, referral source, inpatient/outpatient status, family support, and other characteristics that have been determined to influence potential for recovery. For some authors (Katz, 1990; Kotten, 1989), the ability of aphasic people to use computers is also influenced by elements that are cognitive (e.g., attention, vigilance, memory, resource allocation), cybernetic (e.g., slow rise time, noise buildup, intermittent imperception), behavioral (e.g., discriminatory stimuli, chaining, extinction), pragmatic (e.g., functionality, social status), and emotional (e.g., interest, relevance, novelty, enjoyment). In addition, pathologic characteristics commonly associated with brain damage, such as perseveration and abstract attitude, can interfere with use of treatment software. These problems may affect motivation, independence, pragmatics, and interpersonal relations, ultimately diminishing potential for recovery, success of treatment, and quality of life for aphasic patients. Computer treatment programs rarely attend to these factors.

43-2.3. Task Structure

Language therapy presented on computer is usually organized in one of four nonexclusive formats: stimulation, drill and practice, simulation, and tutorial (Katz, 1995).

43-2.3.1. Stimulation

As mentioned earlier, stimulation-facilitation activities permit the patient to respond quickly and usually correctly over time for the purpose of achieving and stabilizing the underlying processes or skills, rather than learning a new set of responses. The process, not a specific response, is the focus of the task. Stimuli are selected not for informational content, but for characteristics affecting their processing, such as length, complexity, and number of critical elements. Stimulation tasks can be programmed on computers rather easily as intervention can be minimal (Mills, 1982).

43-2.3.2. Drill and Practice

The goal of drill-and-practice tasks generally is to learn a specific set of (functional) responses. Stimuli are particular to each patient (e.g., family names, favorite foods, home address), so convenient authoring or editing capability is required for this format of computer therapy (Roth, 1992c). Patients commonly experience a higher error rate with drill-and-practice tasks than with stimulation tasks. Since computer programs typically offer a limited variety of cues, a clinician must be prepared to intervene, especially during complex tasks. Because specific responses are the focus, drill-and-practice programs are usually convergent tasks (Seron, Deloche, Moulard, & Rouselle, 1980).

43-2.3.3. Simulation

Simulation tasks are also called, "microworlds." They present a structured environment with an intentional problem. The patient must often use elements within the simulated environment to solve the problem. Microworlds provide the opportunity to design divergent therapy tasks with various solutions to real-life problems. Although microworlds require complex programming, the skills patients acquire from using them seem more likely to generalize to real-world situations than gains achieved through more traditional formats as microworlds more closely approximate variables found in the real world. The focus of simulations is to develop an effective (problem-solving) strategy (Crerar, Ellis, & Dean, 1996).

43-2.3.4. Tutorial

Simply "teaching" aphasic patients in the same manner as new information is taught in a classroom is rarely successful. However, early work by Eisenson (1973) and others suggests that aphasic patients are best served by modifying their communicative environment. Computerized tutorials can provide important information, such as techniques to facilitate communication (Pulvermüller & Roth, 1991) or explanations to foster greater empathy and patience (Brumfitt, 1995) to family members, friends, and others who influence the aphasic person's world to help improve the aphasic patient's independence and satisfaction. Non-computer-based tasks, such as *PACE* (Steiner, 1993) and *PACT* (Roth, 1989), are suitable therapeutic follow-up activities to such tutorials.

43-3. EFFICACY OF TREATMENT SOFTWARE

Treatment software can be viewed along a continuum of complexity that represents salient features of therapy, including those discussed earlier (i.e., aphasia models, program properties, patient characteristics, task structure). At one end of the continuum, simple, isolated drills are utilized in the same manner as therapy workbooks to improve specific language-related skills, such as identifying pictures of objects named verbally by the clinician (Mills, 1982). Toward the far end of the continuum, series of integrated activities utilize complex branching algorithms (Katz & Wertz, 1997, 1992), multiple tasks, or artificial intelligence (Guyard et al., 1990) to provide a comprehensive approach to improving language ability and communication (Pulvermüller & Roth, 1993; Roth, 1992b).

Although general discussions of efficacy of aphasia rehabilitation are available elsewhere (Caprez, 1992; Wertz, 1995), the issue is central to the examination of the validity of computer-provided aphasia treatment and deserves mention. Aphasia treatment is the systematic application of conditions designed to improve language ability and communicative competence. Treatment efficacy considers whether outcomes are improved as a result of a specific intervention (McGlynn, 1996). According to Rosenbek (1995), *efficacy* is improvement resulting from treatment applied in a rigidly controlled design when treatment and no-treatment conditions are compared. Wolfe (1987) found that early reports of computerized aphasia treatment did not provide explicit models of rehabilitation (e.g., linguistic models) from which the software could be evaluated. In an extensive review of the literature, Robinson (1990) reported that the efficacy of computerized treatment for aphasia as well as for other cognitive disorders had not been demonstrated. The research studies reviewed suffered from inappropriate experimental designs, insufficient statistical analysis, and other deficiencies. Since Robinson's critique, a number of studies have reported the effect of particular computerized interventions (Crerar et al., 1996; Katz & Wertz, in press, 1992; Loverso, Prescott, & Selinger, 1992). The following is a description of five areas illustrating the development of computer-provided aphasia treatment.

43-3.1. Reading and Writing

Even before the availability of multimedia computer systems, computers were viewed as an appropriate medium for treating reading and writing problems. Most aphasic people have these problems. As communicative acts, reading and writing are usually done alone, and having greater interpersonal distance, are socially appropriate to practice in isolation. Reading requires minimal response by the patient, thereby simplifying software development and hardware requirements. The independent use of computers to provide reading and writing treatment outside of the traditional treatment session can increase the total time a patient is involved in therapeutic activities. The clinician, no longer directly providing reading and writing treatment, can then use the time freed during traditional treatment sessions to focus on auditory comprehension, verbal expression, and other components of face-to-face communication (Pulvermüller & Roth, 1991, 1993).

43-3.1.1. Reading Comprehension

Early reports of reading treatment software included mostly simple drills converted for the computer. Scott & Byng (1989) tested the effectiveness of a program designed to improve reading comprehension of similar sounding words (homophones) for a 24-year-old woman who suffered traumatic head injury and underwent subsequent left temporal lobe surgery that resulted in aphasic symptoms as well as surface dyslexia and surface dysgraphia. The patient demonstrated particular difficulty with homophones. The computer program designed to treat this problem was based on an information-processing model. After using the program over 10 weeks, the subject improved in recognition and comprehension of treated ($p < .001$) and untreated ($p < .002$) homophones used in sentences. Improvement was also demonstrated on recognition of isolated homophones that were treated ($p < .05$) and on defining isolated treated ($p < .03$) and untreated homophones ($p < .02$). Recognition of isolated untreated homophones and spelling of irregular words showed no improvement.

Katz & Wertz (1997) conducted a longitudinal group study based on earlier research (Katz & Wertz, 1992) to investigate the effects of computer-provided reading comprehension activities on language performance for chronic aphasic adults. Fifty-five aphasic adults were assigned randomly to one of three conditions: 78 hours of computer reading treatment, 78 hours of computer stimulation, or no treatment. Computer stimulation software consisted of nonverbal games and cognitive rehabilitation tasks. Standardized language measures were administered to all subjects at entry and after 3 and 6 months. Significant improvement ($p < .01$) over the 26 weeks occurred on five language measures for the computer reading treatment group, on one language measure for the computer stimulation group, and on none of the language measures for the no-treatment group. Results suggest that computerized reading treatment can *(a)* be administered with minimal assistance from a clinician, *(b)* generalize to non-computer language performance, and *(c)* be efficacious for chronic aphasic patients. Additionally, improvement resulted from the language content of the software and not the stimulation provided by using computers.

43-3.1.2. Writing

Most computer programs designed to improve writing substitute typing for writing. Selinger, Prescott, and Katz (1987) found no significant differences on standardized graphic tasks when comparing writing and typing performance for seven left hemisphere-damaged subjects. Several investigators have incorporated complex branching algorithms in computerized writing programs to provide multilevel intervention for spelling (Seron et al., 1980), written confrontation naming (Katz, Wertz, Davidoff, Shubitowski, & Devitt, 1989), and written sentence completion (Glisky, Schlacter, & Tulving, 1986). Deloche, Dordain, and Kremin (1993) developed software to treat oral and written modality differences in confrontation naming for two aphasic subjects, a surface dysgraphic and a conduction aphasic. The intervention focused on written naming from the keyboard. Both subjects maintained improvements 1 year following therapy.

43-3.2. Verbal Output

A number of early efforts demonstrated how computers could be used to assist aphasic patients experiencing dysnomia and other verbal problems. Van de Sandt-Koenderman (1994) described a computer program, "Multicue," that is designed to help aphasic patients identify and select self-cuing and compensatory strategies for word-finding problems. A picture is presented in the upper-left quarter of the computer screen and the patient's goal is to find the appropriate name. If the patient does not know the name, various cues can be selected (e.g., word meaning, word form, sentence completion) to help the patient retrieve the word. The goal of the program is to allow the patient to evaluate and practice several word-finding strategies. Further study is under way to assess the effectiveness of Multicue.

Early attempts to develop a computerized "speech prosthetic device" for people with aphasia met with limited success due in part to the existing state of computer technology (e.g., Colby, Christinaz, Parkison, Graham, & Karpf (1981). Recent attempts appear more hopeful. *Lingraphica* is an integrated, computerized communication system that combines spoken words, pictures (icons), and text processing. It is an extension of the C-VIC system (Computerized Visual Communication) developed as an alternative mode of communication for severely impaired aphasic people. The empirical evidence to support the efficacy of C-VIC was collected in a series of single-case studies (Steele, Weinrich, Kleczewska, Wertz, & Carlson, 1987; Weinrich et al., 1989). Lingraphica and C-VIC are the current results of these efforts. Aphasic patients use a mouse device to select one of several icons, each of which represents a general category. The selected icon then "opens up" to reveal pictures of the items within the selected category. After selecting the desired item, the picture is added to a sequence of other selected pictures; this "string" of pictures represents the message. The message can be read via the sequence of icons, words printed below the sequence, or, in some cases, heard through digitized speech. Much attention is given to the selection of icons. Weinrich et al. (1989) reported that concrete icons were learned and generalized faster than abstract icons, but neither type of icon generalized well to new situations. Steele et al. (1987) reported improvement on expressive and receptive tasks for a globally impaired aphasic subject using C-VIC, although communication through more traditional modes (e.g., speech) remained unchanged. Conversely, Weinrich, McCall, Weber, Thomas, and Thornburg (1995) trained two Broca's aphasic subjects in the production of locative prepositional phrases and S-V-O sentences on C-VIC and reported that their verbal ability improved considerably. Clearly, further single-subject and group research utilizing standardized measurements is needed to test the effectiveness of new computer technology adapted to the communicative needs of aphasic patients.

43-3.3. Training at Home and in a Self-Help Context

Athough not intended specifically as a treatment efficacy study, Petheram (1996) assessed the feasibility and acceptability of computers for therapy by patients in their homes. Computer systems were installed in the homes of 10 aphasic patients. Therapy software automatically varied the level of difficulty of the treatment material for each patient. Although patients demonstrated a wide variation in performance, the author reported that their extensive use of the computer supports the notion of therapeutic application of home computer systems. In a similar fashion, multimedia computers have been used in self-help groups of aphasic people (Roth, 1996). In Germany, for example, the first two language-training computers (with digital speech input and output) were donated in 1990. Self-help groups in metropolitan and provincial areas had an early opportunity to gain experience using computers outside of rehabilitation centers. For example, Roth and Schönle (1992) followed a "systemic" approach. Sessions incorporating a PC were arranged for an aphasic father (39 years old, polytrauma following a car accident) and his family at their home. In addition to a version of the Stachowiak program (1992), a second program, *WEGE* (Roth, 1992a), was used that focused on understanding everyday language. The authoring system allowed for the introduction of new material designed to measure performance during pleasant, nonstressful family sessions at the computer. Role-playing activities and drawings from the sessions were stored on the multimedia computer. The aphasic father participated (minimally) in the overall production, but he could run through the stored activities repeatedly when they were part of exercises such as matching audible and written language. The gains made during half a year of daily training facilitated initially by his wife and two students, supervised by a professional speech-and-family therapist, were significant (Roth & Schönle, 1992). Other aphasic people occasionally participated, working with the aphasic father, who in turn responded to tasks presented by the language training program.

Many people suffering the mild residual effects of aphasia will no longer be able to work productively in a paying job, but still want to improve their ability to communicate. Training with a speech-language therapy computer in a self-help group can be a way to provide this care efficiently. Programs commonly used by German self-help aphasia groups include LingWare/STACH (Stachowiak, 1993) and NeueWEGE/MODAKT (Zechner & Roth, 1996). An evaluation study in five German self-help groups (of an estimated total of 20 computer-user groups) is being prepared by Tollkühn at the University of Leipzig (section Förderpädagogik) concentrating on the aspect of user acceptability for different programs and types of tasks that use familiar pictures and voices from the patient's own environment. Integrating significant others in the rehabilitation process has been advocated by many authors. Today's multimedia PC technology offers new ways to do this. In the last few years, the self-help movement has started to offer patients and their families a chance to see for themselves how technology can help them communicate (Roth, 1994).

In Sweden, Magnusson (1996) described a "telematic pilot study" at the University of Karlstad funded by the Foundation for Communication Research in which "video-telephony" is used as a therapy tool for communicatively impaired patients who live in remote areas of Sweden where speech therapy is not readily available. The project (called "VITSI") includes a 3-year project to provide remote therapy using both visual and auditory signals ("videoconferencing") to people with aphasia and to evaluate their attitudes about this form of support. Twenty-six aphasic adults are participating in the study, all of whom have had prior conventional therapy and most of whom have prior experience with computer-based therapy. Sessions are 45 minutes long and are scheduled once every 2 weeks in order to accommodate all 26 participants. In addition to individualized therapy programs, surveys and interviews will be used to evaluate the response of the participants to this mode of treatment.

43-3.4. Artificial Intelligence

Guyard et al. (1990) introduced a new stage in the development of treatment software that integrates artificial intelligence (AI) programming and computer-assisted instruction (CAI) for the rehabilitation of aphasic patients. Barr and Geigenbaum (1982) described a union between AI and CAI as "Intelligent CAI" (ICAI). ICAI can expand the scope, responsiveness, and flexibility of aphasia therapy software so that a computer program would determine the type, sequence, and rate of stimuli presented based on evaluation of the patient's responses. Although valued in education, ICAI has met limited success in aphasia rehabilitation primarily due to two factors: the heterogeneity of the aphasic population and the complexity of aphasia therapy (Katz, 1990).

43-3.5. Microworlds

Roth (1992b, 1995) and Gadler and Zechner (1992) each described computer-simulated worlds (NeueWEGE and AUSWEGE, respectively) in which patients, guided by their clinicians, explore a microworld, making decisions, "traveling," and taking chances without any real physical or interpersonal risks. Messmer and Roth (1996) and Norosel and Roth (1997) developed a WEGE version running under Windows 95, opening aphasia therapy software up to current technological possibilities such as PC to PC linking via ISDN, integration of video. Crerar and Ellis (1995) described the "Microworld Project," a computer system based on sound neuropsychologic and psycholinguistic theory and designed to treat sentence processing impairments. Concepts such as agent, action, object, and spatial relations were manipulated to improve sentence comprehension. A series of experiments (Crerar & Ellis, 1995; Crerar et al., 1996) demonstrated improvement in chronic aphasic subjects after a relatively short duration of treatment. Other attempts at developing microworlds and other therapeutic virtual reality environments are described by Brodin and Magnusson (1993) and others. Much is expected from this approach as computer and communication technologies continue to develop and interact.

43-4. CONCLUSION

We reviewed three factors central to the development of effective computer-provided aphasia treatment: aphasia models, technological limitations, and published research. These three factors influence the way clinicians perceive the role of computers in treatment. Computers can be effective clinical tools for clinicians who incorporate what is known about aphasia, computers, and the state of the art. Those who are concerned that researchers are advocating computers in place of clinicians are missing our point. In the studies described in this chapter, clinicians selected and tested the patients, designed treatment plans, designed and modified treatment tasks, trained the patients to use the computers, and measured treatment efficacy. Computers and treatment software, like all tools, should extend the abilities of the clinician, allowing clinicians to intervene when skills, experience, and flexibility are required. Rather than emphasize what computers can or cannot do better than clinicians, our focus should be on an intelligent division of labor between computers and clinicians, a combination that can do more than either alone. For example, Stachowiak (1993) reported that daily multimedia computer-provided supplemental treatment led to additional gains in therapy for aphasic patients receiving a minimum of 3 hours a week of conventional treatment.

A real danger comes from a failure to appreciate the scope and depth of clinical work. How many symbols and algorithms do we need to represent aphasia therapy? A program that accurately represents clinician-provided therapy would quickly exceed the capacity of most computers. Treatment software will always be an imperfect reflection of clinician-provided therapy, but, by improving the software, clinicians and programmers will learn more about how and why treatment works. In the best tradition of scientific and rehabilitative efforts, aphasiologists can work together to shape this new tool of technology for the development of their professions and the benefit of all patients. Controlled treatment studies can assist clinicians in developing treatment models and tasks that translate to the computer medium by identifying the salient features of treatment and the strengths and limitations of the computer.

PART V

Resources in Clinical and Experimental Neurolinguistics and Related Fields

CHAPTER 44

Computational Transcript Analysis and Language Disorders

Brian MacWhinney

Department of Psychology, Carnegie Mellon University, Pittsburgh, Pennsylvania 15213

The errors and omissions found in aphasic language production are a rich source of information about how language is processed in the brain. However, in order to fully exploit this type of data, we need a consistent methodology for elicitation, recording, transcription, and analysis. One such framework is provided by tools developed for the CHILDES (Child Language Data Exchange System) Project. This paper examines those tools in the light of the development of research methodology from the precomputer period into the current period of connectivity and exploratory reality. Although these tools were originally developed for the analysis of language acquisition data, they can be readily adapted to the study of language disorders.

When asked to describe simple pictures or to recite simple narratives, aphasics often illustrate a wide variety of paraphasias, word-finding difficulties, phonological disfluencies, and grammatical errors. These errors and omissions provide us with two important windows into the functioning of language in the brain. First, by studying the various types of errors and omissions in psychological and linguistic terms and by comparing their relative frequencies, we can learn a great deal about how aphasia affects the basic mechanisms of language processing. For example, the study of omissions of markers like the plural on the noun or the past tense on the verb has been useful in developing our understanding of production processes in aphasia (MacWhinney & Osman-Sági, 1991; Menn & Obler, 1990).

Error patterns also offer us potential information about clinical groupings. By looking at differential patterns of errors and omissions, we can distinguish the telegraphic production patterns found in agrammatism or Broca's aphasia, the more verbose and error-prone patterns characterizing Wernicke's aphasia, the exclusively lexical error patterns characterizing anomia, and the more exclusively phonetic error patterns found in dysarthria and apraxia of speech. These patterns in aphasia can also be compared with possibly similar language patterns in the speech of people with schizophrenia, right-hemisphere damage, frontal lobe damage, Alzheimer's disease, and other neural disorders. We can then couple this information with additional information from comprehension, neural imaging, and other methodologies to advance claims regarding the structuring of language in the brain.

Unfortunately, the study of production errors and clinical patterns in production has often been a rather hit-or-miss endeavor. Because there is no standardized reference database of production data, it is difficult to evaluate the relative position of new clinical samples and data from single subjects. There are a number of standard diagnostic tests available for the study of aphasia, but the actual transcripts from these tests have not been collected in a publicly available repository and organized in a fashion that permits easy comparison between individual subjects and the larger database.

There are some fairly good reasons why this database has not yet been developed. Although it is extremely easy to collect production data, it is much more difficult to analyze these data in a scientifically consistent manner. Many laboratories have dozens and dozens of transcripts of disordered speech sitting in paper repositories or stored on computer disks. It is easy to turn on a tape recorder or videotape recorder and build up a huge library of hundreds of hours of tapes. However, transcribing, coding, and analyzing hours upon hours of recordings involves an enormous time commitment. If the work spent doing this transcription is to be meaningful, we need to set standards to guarantee the comparability of data across subjects, laboratories, protocols, and transcription formats.

44-1. THE CHILD LANGUAGE DATA EXCHANGE (CHILDES) SYSTEM

Fortunately, there already is a well-developed framework for the process of database formation and analysis that can be directly applied to the study of aphasia. This is the system of programs and codes developed by the Child Language Data Exchange (CHILDES) Project (Higginson & MacWhinney, 1990, 1994; MacWhinney, 1991a, 1991b, 1993, 1994b, 1994c, 1995, in press; MacWhinney & Snow, 1985; 1990). The CHILDES project has been directed by the author in collaboration with Catherine Snow of Harvard University and has been supported since 1987 by the National Institute of Child Health and Human Development (NICHD).

It is important to realize that, although the CHILDES system was originally developed for use in the study of child language data, we took care early on to make sure that the system would also be applicable to the analysis of aphasic language data. We did this by emphasizing the development of tools for the analysis of speech errors and by including data from both children and adults with language disorders. At this point, the acronym CHILDES is largely a historical relic, since the system is being used not only by child language researchers, but also by discourse analysts, sociolinguists, speech pathologists, aphasiologists, computer scientists, and applied linguists studying second language acquisition.

Sometimes researchers who are unfamiliar with CHILDES think of it only as a database. However, in order to develop a useful and consistent database, we had to construct a system based on three integrated components:

1. **CHAT** is the system for discourse notation and coding. This system includes detailed conventions for marking all sorts of conversational features, such as false starts, drawling, overlaps, interruptions, errors, and so on. This system was developed over the course of 6 years with continual input from language researchers. This standard transcription system is used for all the data in the database.

2. **CLAN** is the set of computer programs for searching and manipulating the database. Rather than focusing on canned analyses or rigid clinical packages, these programs provide the user with a tool kit of analytic possibilities that can be combined to fit a specific research agenda. Most recently, the programs have been extended to provide tools for linking transcripts to digitized audio and video records.

3. **Database.** Finally, the system includes the database itself, with data donated to the language community from more than 40 major projects in English and additional data from Cantonese, Danish, Dutch, French, German, Greek, Hebrew, Hungarian, Italian, Japanese, Mambila, Mandarin, Polish, Portuguese, Russian, Swedish, Tamil, Turkish, and Ukrainian. Along with data from normally developing children, there are data from children with language disorders, adult aphasics, second language learners, and early childhood bilinguals. In essence, the database is simply a set of standard text files of transcripts of conversational interactions. In a few cases, the computerized transcript is accompanied by digitized audio and even video, but the vast majority of the corpora have only transcripts without audio or video.

The system has been used as the basis of nearly 400 published research studies in the areas of language disorders, aphasia, second language learning, computational linguistics, literacy development, narrative structures, formal linguistic theory, and adult sociolinguistics.

There are two major modes in which researchers have used the CHILDES system. The first mode focuses on the examination of patterns in the existing database. Researchers operating in this first mode need to learn the basic functions of the CLAN programs for searching across corpora. However, they are mostly interested in understanding the shape of the database and the nature of the various existing corpora. They may be interested in studying the development of specific syntactic constructions or parts of speech, such as questions, prepositions, plurals, or demonstratives. To study these issues, they typically use the basic search and tabulation programs in CLAN. Because there are fewer data on child language disorders and even fewer still on adult aphasics, this mode of research is somewhat less attractive currently for the areas of developmental language disorders and aphasiology.

The second mode of research uses the CLAN programs and the CHAT transcript format to transcribe and analyze new data. Workers operating in this second mode usually develop their own coding schemes and analysis routines designed to address project-specific questions. When researchers have completed their work, they then contribute their transcripts as new corpora for the database. Researchers operating in this mode are particularly interested in understanding the ways in which the various CLAN programs can help them address their current research needs. In order to maximize their use of the CLAN programs, they also need to understand the various alternative ways in which one can use the CHAT transcription system.

Each of the three components of the CHILDES system was designed from the beginning to be useful across languages. The crosslinguistic focus of the CHILDES system is not just an optional methodological nicety; it is conceptually central. And this centrality holds equally well for both child language and aphasiology. In both of these fields, certain prominent theories make strong claims about universals of grammar (Chomsky, 1965), conceptual structure (Bickerton, 1984; Slobin, 1985), or sentence processing (Frazier, 1987). Proponents of these universalist theories often argue that these abilities are located in specific brain modules that can be damaged in specific ways (Friederici & Frazier, 1992; Grodzinsky, 1990; Warrington & McCarthy, 1987). In the simplest case, these accounts would claim that all language is organized in the same basic way and that the patterns of dissociation we find across languages (Paradis & Lebben, 1987) should be basically the same. An alternative view stresses the importance of language differences in determining patterns of errors and omissions in aphasia (Bates, Wulfeck, & MacWhinney, 1991). These between-language differences are then attributed to variation in cue distribution (MacWhinney & Bates, 1989) or social interaction (Schieffelin & Ochs, 1987). In order to understand how both universals and particulars interact in language learning and language loss, researchers must adopt a crosslinguistic perspective. Ideally, we will also want to maximize comparability across languages by settling on a standardized set of pictures and other tasks for the elicitation of language production.

Before we begin a more detailed examination of the current status of the CHILDES system, it may be useful to step back a bit to look at the ways in which the methodology for the study of spontaneous language production has evolved historically. This historical overview can help us gain some perspective on the status of our current methodological advantages and the ways in which changes in methodology are linked to changes in theory. Starting in ancient times and continuing up through the present, we can distinguish five major periods. In each of these periods, our understanding of the nature of language has been closely linked to the nature of the methodology that has been available for studying language performance. During each of these periods, the methodology used for the study of language acquisition has been essentially the same as the methodology used for the study of language disorders.

44-2. FIVE METHODOLOGICAL PERIODS

44-2.1. Period 1: Naive Speculation

The first attempt to understand the process of language development appears in a remarkable passage from the *Confessions* of Saint Augustine (Augustine, 397). In this passage, Augustine actually claims that he remembered how he had learned language:

> This I remember; and have since observed how I learned to speak. It was not that my elders taught me words (as, soon after, other learning) in any set method; but I, longing by cries and broken accents and various motions of my limbs to express my thoughts, that so I might have my will, and yet unable to express all I willed or to whom I willed, did myself, by the understanding which Thou, my God, gavest me, practise the sounds in my memory. When they named anything, and as they spoke turned towards it, I saw and remembered that they called what they would point out by the name they uttered. And that they meant this thing, and no other, was plain from the motion of their body, the natural language, as it were, of all nations, expressed by the countenance, glances of the eye, gestures of the limbs, and tones of the voice, indicating the affections of the mind as it pursues, possesses, rejects, or shuns. And thus by constantly hearing words, as they occurred in various sentences, I collected gradually for what they stood; and, having broken in my mouth to these signs, I thereby gave utterance to my will. Thus I exchanged with those about me these current signs of our wills, and so launched deeper into the stormy intercourse of human life, yet depending on parental authority and the beck of elders (p. 4).

Augustine's fanciful recollection of his own language acquisition remained the high-water mark for child language studies through the Middle Ages and even the Enlightenment. However, Augustine's recollection technique is no longer of much interest to us, since few of us believe in the accuracy of recollections from infancy, even if they come from saints.

44-2.2 Period 2: Diaries and Biographies

The second major technique for the study of language production was pioneered by Charles Darwin. Using note cards and field books to track the distribution of hundreds of species and subspecies in places like the Galapagos and Indonesia, Darwin was able to collect an impressive body of naturalistic data in support of his views on natural selection and evolution. In his study of gestural development in his son, Darwin (1877) showed how these same tools for naturalistic observation could be adapted to the study of human development. By taking detailed daily notes, Darwin showed how researchers could build diaries that could then be converted into biographies documenting virtually any aspect of human development. Following Darwin's lead, scholars such as Ament, Preyer, Gvozdev, Szuman, Stern, Ponyori, Kenyeres, and

Leopold created monumental biographies detailing the language development of their own children.

Darwin's biographical technique also had its effects on the study of adult aphasia. Following this tradition, studies of the language of particular patients have been presented by Low (1931), Pick (1913, 1971), Wernicke (1874), and many others.

44-2.3. Period 3: Transcripts

The limits of the diary technique were always quite apparent. Even the most highly trained observer could not keep pace with the rapid flow of normal speech production. The emergence of the tape recorder in the 1950s provided a way around these limitations and ushered in the third period of observational studies. This period was characterized by projects in which groups of investigators collected large data sets of tape recordings from several subjects across a period of 2 or 3 years. As long as there was sufficient funding available, these tapes were transcribed either by hand or by typewriter. Typewritten copies were reproduced by ditto master, stencil, or mimeograph. Comments and tallies were written into the margins of these copies and new, even less legible, copies were then made by thermal production of new ditto masters. Each investigator devised a project-specific system of transcription and project-specific codes. As we began to compare handwritten and typewritten transcripts, problems in transcription methodology, coding schemes, and cross-investigator reliability became more apparent.

44-2.4. Period 4: Computers

Just as these new problems were coming to light, a major technological opportunity was emerging in the shape of the powerful, affordable microcomputer. Microcomputer word-processing systems and database programs allowed researchers to enter transcript data into computer files, which could then be easily duplicated, edited, and analyzed by standard data-processing techniques. In 1981, when the CHILDES Project was first conceived, researchers basically thought of these computer systems as large notepads. Although researchers were aware of the ways in which databases could be searched and tabulated, the full analytic and comparative power of the computer systems themselves was not yet fully understood.

44-2.5. Period 5: Connectivity and Exploratory Reality

Since 1981, the world of computers has gone through a series of remarkable revolutions, each introducing new opportunities and challenges. The processing power of the home computer now dwarfs the power of the mainframe of the 1980s, new machines are now shipped with built-in audiovisual capabilities, and devices such as CD-ROMs, DAT tapes, and optical disks offer enormous storage capacity at reasonable prices. This new hardware has opened up the possibility for multimedia access to transcripts of aphasic language production. In effect, a transcript is now the starting point for a new Exploratory Reality in which the whole interaction is accessible through the transcript in terms of both full audio and video images. For those who are just now becoming familiar with this new technology, Table 1 summarizes some of the relevant pieces of hardware and software options.

Most recently, microcomputers all across the world have become interconnected through a global high-speed network called the Internet that supports the movement of all sorts of information, including text, sound, and video. This connectivity between computers is matched by an increasing interactivity between the operating system and individual programs. The user can record a sound in one program, take it immediately to another for detailed acoustic analysis, and then to a third for database storage. Together, these new hardware and software developments have led to an enormous increase in interconnectivity between computers, between programs, and between researchers. We are just now beginning to understand the potential consequences of this connectivity for researchers.

From this quick survey of the development of tools for language analysis, we see that the possibilities for careful, detailed analysis of production data have markedly widened in the last few years. The methodological tools that are now available far exceed those of previous eras. What we lack now in the field of aphasiology are not the conceptual or computational tools, but the organizational commitment that will be

TABLE 1
Some Computer Terminology

Term	Explanation
Audiovisual/AV	Computer that can control sound and video
CD-ROM	Removable disk that gives access to huge amounts of nonerasable data
CHAT	CHILDES transcription and coding format
CLAN	CHILDES programs for data analysis
DAT tape	Inexpensive way of archiving large amounts of data
digitized speech	Storage of sound in a form that can be played by the computer
electronic bulletin board	A forum for the discussion of issues through computer mail For CHILDES this is info-childes@andrew.cmu.edu
E-mail	Electronic mail that operates over the Internet
FTP	File transfer protocol—a program for moving data between computers
hard drive	Built-in device that gives access to large amounts of erasable data
Internet	System of electronic links that allows computers to transfer data
Macintosh	An operating system designed to machine user-friendliness
MS-DOS	A common and easily controlled operating system for microcomputers
optical disk	Removable disk that gives access to huge amounts of erasable data
poppy.psy.cmu.edu	The machine that makes CHILDES data and programs available by FTP
TAR	A program that puts many files into one (like Zip or Compactor)
UNIX	A powerful, but sometimes difficult, operating system
World Wide Web	Software that facilitates use of the Internet for conceptual connections

needed to push forward the development of a standardized database. In order to envision the possibilities that are open for constructing a database for aphasia, let us look at the shape of the current database for first and second language development and for child language disorders.

44-3. THE DATABASE

The first major tool in the CHILDES workbench is the database itself. Through CD-ROM or FTP, researchers now have access to the results of nearly a hundred major research projects in 20 languages. Using this database, a researcher can test a vast range of empirical hypotheses directly against either the whole database or some logically defined subset. The database includes a wide variety of language samples from a wide range of ages and situations. Almost all of the data represent real spontaneous interactions in natural contexts, rather than some simple list of sentences or test results. Although more than half of the data come from English speakers, there is also a significant component of non-English data.

Until 1989, nearly all of the data in the CHILDES database were from normally developing children. However, recent additions to the database have included several major corpora from children with language disorders. These include data from Down's syndrome contributed by Nahid Hooshyar, Jean Rondal, and Helen Tager-Flusberg; data from autistic children contributed by Helen Tager-Flusberg; data from SLI (Specific Language Impairment) contributed by Lynn Bliss, Patricia Hargrove, Gina Conti-Ramsden, and Larry Leonard; and data from children with articulatory disorders contributed by Susan Fosnot-Meyers and the Ulm University Clinic.

In the area of adult aphasia, the database includes two large corpora. The first is a set of conversational interviews with 42 aphasic patients during the period of recovery from stroke donated by Audrey Holland. The second is a collection of interview and picture description data from aphasic speakers of English, German, Hungarian, Chinese, and Italian donated by Elizabeth Bates and her colleagues. One of the major priorities for the CHILDES project is the inclusion of additional data on both childhood language disorders and aphasia during the coming years. We are aware of a variety of additional computerized corpora in the area of adult aphasia (Menn & Obler, 1990; Paradis & Lebben, 1987) and we hope to be able to convince researchers in aphasiology of the importance of making these data sets publicly available.

All of the major corpora have been formatted into the CHAT standard and have been checked for syntactic accuracy. The total size of the database is now approximately 180 million characters (180 MB). The corpora are divided into six major directories: English, non-English, narratives, books, language impairments, and bilingual acquisition. In addition to the basic texts on language acquisition, there is a database from the Communicative Development Inventory (Dale, Bates, Reznick, & Morisset, 1989) and a bibliographic database for Child Language studies (Higginson & MacWhinney, 1990).

Membership in CHILDES is open. Members are listed in a standard database and receive electronic messages through the info-childes@andrew.cmu.edu electronic bulletin board. In order to be officially included in the info-childes electronic mailing list and database, researchers should send E-mail to childes@cmu.edu with their computer address, postal address, affiliations, and phone number. Users are asked to abide by the rules of the system. In particular, they should abide by the stated wishes of the contributors of the data. Any article that uses the data from a particular corpus must cite a reference from the contributor of that corpus. The exact reference is given in the CHILDES manual (MacWhinney, 1991b). In addition, researchers should cite the 1991 version of the manual, since this allows us to track references in the literature.

All of the CHILDES materials can be obtained without charge by anonymous FTP to childes.psy.cmu.edu in Pittsburgh and atila-ftp.uia.ac.be in Antwerp. Our address on the World Wide Web is http://childes.psy.cmu.edu. For users without access to the Internet, as well as for those who want a convenient way of storing the database, we have published (MacWhinney, 1994a) a CD-ROM that can be read by Macintosh, UNIX, and MS-DOS machines that have a CD-ROM reader. The disk contains the database, the programs, and the CHILDES/BIB system. One directory contains the materials in Macintosh format and the other contains the materials in UNIX/DOS format. The CD-ROM, the printed manual, and the research guide are available at nominal cost through Lawrence Erlbaum Associates.

44-4. CHAT

All of the files in the database use a standard transcription format called CHAT. This system is designed to accommodate a large variety of levels of analysis, while still permitting a bare-bones form of transcription for those research projects in which additional levels of detail are not needed. Here is a brief example of segment of a transcript from a Broca's aphasic transcribed in CHAT. The file begins with these 10 lines of identifying material, or "headers."

```
@Begin
@Participants:    PAT Patient, INV Investigator
@Age of PAT:      47;0.
@Sex of PAT:      male
@SES of PAT:      middle
@Date:            22-MAY-1978
@Comment:         Group is Broca
@Filename:        B72
@Coder:           JMF
@Situation:       Given/New task
```

After the headers, the actual transcript begins. This is a picture description task and each picture is identified with an @g marker to facilitate later retrieval. In the first three @g segments, the patient is describing a set of three pictures used in Bates, Hamby, and Zurif (1983) and MacWhinney and Bates (1978). In this first set, various animals are all eating bananas. In its "raw" form, what the patient said was simply, "rabbits, squirrel, monkeys." Here is how this is transcribed:

@g:	3c = bunny is eating banana
*PAT:	rabbits [*].
%mor:	DET|0 N|rabbit-*PL
%err:	rabbits = rabbit $SUB
@g:	3b = squirrel eating banana
*PAT:	squirrel.
%mor:	DET|0 N|squirrel
@g:	3a = monkey eating banana
*PAT:	monkeys [*].
%mor:	DET|0 N|monkey-*PL.
%err:	monkeys = monkey $SUB

Here, the *PAT line conveys the simple shape of the patients description of the three pictures—"rabbits, squirrel, monkeys." We can notice several things about this transcription. First, the err or "error" lines code the fact that plurals are used for two of the pictures when, in fact, only a single animal appears in each. The locus of these errors is marked in the main line or *PAT line with the symbol [*]. The %mor line is designed to indicate the morphological shape of the words on the main line. This line is used to study the use of different parts of speech and syntactic constructions. In this example, the %mor also provides a backup to the %err line, since both lines code for errors of omission and commission. The %mor line is intended to have a one-to-one correspondence with the main line, but when an item is marked as missing on the %mor line, it does not need to be present on the main line. For example, the code "DET\|0" indicates that the determiner is missing on the main line. The code "N\|monkey-*PL" indicates that the patient used the noun *monkey* in the plural, but that the use of the plural was an error in this case. The advantage of the elaborate coding on the %mor line is that it provides a more systematic structure for search programs that tabulate missing items by part of speech.

Let us look at one more segment from the same patient in the same study. Here the picture involves the dative verb *give*. It is "raw" form, what the patient said was simply, "boy, girl, school, rat, boy no girl, girl truck girl." Here is how this is transcribed:

@g:	8a = lady giving present to girl
*PAT:	boy [*] [//] girl # school [*].
%mor:	DET|0 N|girl N|*xxx.
%err:	boy = girl $SUB ; school = [?] $SUB
@g:	8c = lady giving mouse to girl
*PAT:	rat .
%mor:	DET|0 N|rat
@g:	8b = lady giving truck to girl
*PAT:	<boy [*] no>[//] girl [/] girl truck # girl +...
%mor:	DET|0 N|girl N|truck N|girl.
%err:	boy = girl $SUB

In this example, we see several additional features. In description for picture 8a, the self-correction or retracing of *boy* by *girl* is marked by [//]. The repetition of the word *girl* is marked by [/]. Pauses are marked by # and the trailing off of the last sentence for picture 8b is marked by +... In the description for picture 8c, there is no %err line, since the characterization of the *mouse* as a *rat* is not judged to be so far off the mark as to constitute an error.

These two examples illustrate only a few of the many symbols and conventions available in the CHAT system. The system provides many options, but the transcriber only needs to select out those options that are relevant to the particular case. The simpler the transcription, the better, as long as it still captures the important aspects of the aphasic production.

The examples we have looked at illustrate some of the basic principles of the CHAT transcription system. Three of the most fundamental aspects of the system are the following:

1. Each utterance is transcribed as a separate entry. Even in cases when a speaker continues for several utterances, each new utterance must begin a new entry.
2. Coding information is separated out from the basic transcription and placed on separate "dependent tiers" below the main line. The CHILDES manual presents coding systems for phonology, speech acts, speech errors, morphology, and syntax. The user can create additional coding systems to serve special needs.
3. On the main line, transcription is designed to enter a set of standard language word forms that correspond as directly as possible to the forms produced by the learner. Of course, learner forms differ from the standard language in many ways and there are a variety of techniques in the CHAT system for notating these divergences, while still maintaining the listing of word forms to facilitate computer retrieval.

For full examples of the coding system and its many options, the reader should consult the CHILDES manual.

44-5. CLAN

The main emphasis of new developments in the CHILDES system has been on the writing of new computer programs. Currently, there are two major components of the CHILDES programs. The first is the set of programs for searching and string comparison called CLAN (Child Language Analysis). The second is a set of facilities built up around an editor called CED (CHILDES Editor).

The CLAN programs have been designed to support four basic types of linguistic analysis (Crystal, 1982; Crystal, Fletcher, & Garman, 1989): lexical analysis, morphosyntactic analysis, discourse analysis, and phonological analysis. In addition, there are programs for file display, automation of coding, measure computation, and additional utilities. Table 2 lists the full set of programs by type.

44-5.1. Lexical Analyses

The programs for lexical analysis like FREQ and KWAL focus on ways of searching for particular strings. The strings to be located can be entered in a command line, one

TABLE 2
CLAN Programs and Their Function

Group	Program	Description
Lexical search	FREQ	Tracks the frequency of each word used
	FREQMERG	Merges outputs from several runs of FREQ
	KWAL	Searches for a specific word or group of words
	STATFREQ	Sends the output of FREQ to a statistical program
Block search	GEM	Searches for premarked blocks of interaction
	GEMFREQ	Does a FREQ analysis on a particular block type
	GEMLIST	Profiles the types of blocks found in a file
Discourse/Interaction	CHAINS	Displays "runs" or "chains" of speech acts
	CHIP	Tracks imitations, repetitions, lexical overlap
	DIST	Tracks the distance between particular codes
	KEYMAP	Looks at the variety of speech acts following a given act
	TIMEDUR	Computes overlap and pause duration
	PAUSE	Computes speaking, pause, and overlap times
Morphosyntax	COMBO	Searches for combinations of words or types of words
	COOCCUR	Tabulates pairwise co-occurrence frequency
	KWAL	Searches for a specific word or group of words
	MOR	Performs a full morphological analysis using rules
	POSFREQ	Does a FREQ analysis by sentence position
Phonology	MODREP	Matches phonological forms to their corresponding words
	PHONFREQ	Tabulates the frequency of each phoneme or cluster
	Sonic CHAT	Uses the CED editor to link the transcript to actual sound
Coding tools	CED	A multipurpose editor for CHAT files
	RELY	Compares two sets of codes to compute reliability
Measures	CDI DB	A database of early maternal reports on lexical growth
	DSS	Computes the Developmental Sentence Score
	MAXWD	Lists the longest words and longest utterances in a file
	MLU	Computes mean length of utterance
	MLT	Computes mean length of turn
	FREQ	Includes computation of the type–token ratio
	WDLEN	A frequency distribution by word and sentence length
File display	COLUMNS	Displays CHAT files in the old "column" format
	FLO	Removes complex codes from a CHAT file
	LINES	Adds line numbers to a CHAT file
	SALTIN	Converts data from SALT to CHAT
	SLIDE	Puts a file onto one line that can be scrolled horizontally
Utilities	CHIBIB	A bibliographic access system with 14,000 references
	CHECK	Examines CHAT files for syntactic accuracy
	CHSTRING	Converts strings
	DATES	Computes a child's age for a given date
	TEXTIN	Takes simple unmarked text data and outputs a CHAT file

at a time, or put together in a master file. The strings can contain wild cards and words can be combined using Boolean operators such as *and, not,* and *or.* Together, these various capabilities give the user virtually complete control over the nature of the patterns to be located, the files to be searched, and the way in which the results of the search should be combined into files or even reduced into data for statistical analysis. Scores of studies have appeared in the published literature using these techniques to track the development of lexical fields, such as morality, kinship, gender terminology, mental states, causative verbs, and modal auxiliaries. It is also possible to track the use of words of a given length or a given lexical frequency. FREQ outputs a complete frequency analysis for a single file or for groups of files. Here is an example of a FREQ frequency count for a single small file with only the Mother's utterances being analyzed.

```
freq sid.cha +f +t*MOT
Sun Jul 16 01:31:13 1995
freq (21-NOV-94) is conducting analyses on:
    ONLY speaker main tiers matching: *MOT;

*******************************************
From file <sid.cha> to file <sid.fr0>
13 a
2 about
1 ah
4 all
1 all+right
1 ambulance
7 and
7 are
1 are-'nt
2 back
2 be
1 because
1 bet
3 big
1 bought
3 boy
1 bring-ing
1 build
1 building
1 can
2 clever
2 come
1 crash
1 daddy
1 dear
1 did
7 do
5 do-'nt
```

In this analysis we see that the Mother used the word *big* three times. If we want to look more closely at these usages, we can use KWAL and we will get this output:

```
kwal +t*MOT +sbig sid.cha
Sun Jul 16 01:33:11 1995
kwal (21-NOV-94) is conducting analyses on:
     ONLY speaker main tiers matching: *MOT;
*****************************************
From file <sid.cha>
----------------------------------------
*** File sid.cha. Line 336. Keyword: big
*MOT: is it go-ing to be a big ship ?
----------------------------------------
*** File sid.cha. Line 344. Keyword: big
*MOT: and that-'is go-ing to be a big ship .
----------------------------------------
*** File sid.cha. Line 379. Keyword: big
*MOT: that-'is <all the small lego> [//] all the big lego@ you-'ve got .
```

Each of these programs has many options that can allow the user to vary the shape of the input, the shape of the output, and the type of analysis that is being conducted.

44-5.2. Morphosyntactic Analyses

Many of the most important questions in child language require the detailed study of specific morphosyntactic features and constructions. Typically, this type of analysis can be supported by the coding of a complete %mor line in accord with the guidelines specified in Chapter 14 of the CHILDES Manual. Once a complete %mor tier is available, a vast range of morphological and syntactic analyses becomes possible. However, hand-coding of a %mor tier for the entire CHILDES database would require perhaps 20 years of work and would be extremely error-prone and noncorrectable. If the standards for morphological coding changed in the middle of this project, the coders would have to start over again from the beginning. It would be difficult to imagine a more tedious and frustrating task—the hand-coder's equivalent of Sisyphus and his stone.

To address this problem, we have built an automatic coding program for CHAT files, called MOR. Although the system is designed to be transportable to all languages, it is currently only fully elaborated for English, Japanese, Dutch, and German. The language-independent part of MOR is the core processing engine. All of the language-specific aspects of the systems are built into files that can be modified by the user. In the remarks that follow, we will first focus on ways in which a user can apply the system for English. The MOR program takes a CHAT main line and automatically inserts a %mor line together with the appropriate morphological codes for each word on the main line. Although you can run MOR on any CLAN file, in order to get a well-formed %mor line, you often need to engage in significant extra work. In particular, users of MOR will often need to spend a great deal of time engaging in the processes of lexicon building and ambiguity resolution. To facilitate lexicon building, there are several options in MOR to check for unrecognized lexemes and to add new items. To facilitate ambiguity resolution, we have integrated a system for sense selection into the CED editor.

Construction of a full %mor line using MOR also makes possible several additional forms of analysis. One is the automatic running of the DSS program, which computes the Developmental Sentence Score profile of Lee (1974). Parallel systems of analysis will eventually be developed for systems such as IPSYN (Scarborough, 1990) or LARSP (Crystal et al., 1989). The %mor line can also be used as the basis for CLAN programs such as COOCCUR, which examines local syntactic structures, and CHIP; which examines recasts, imitations, and structural reductions.

Because of the importance of agrammatism in the study of aphasia, it would seem that the MOR program would be of particular interest to aphasiologists. However, the presence of large numbers of lexical, phonological, and syntactic errors in aphasic speech makes automatic application of the MOR program more difficult. Despite these difficulties, this is an area of great potential interest for work on language disorders.

44-5.3. Discourse and Narrative

The most important CLAN tool for discourse analysis is the system for data coding inside the CED editor. CED provides the user with not only a complete text editor, but also a systematic way of entering user-determined codes into dependent tiers in CHAT files. In the coding mode, CED allows the user to establish a predetermined set of codes and then to march through the file line by line making simple keystroke movements that enter the correct codes for each utterance selected.

Once a file has been fully coded in CED, a variety of additional analyses become possible. The standard search tools of FREQ, KWAL, and COMBO can be used to trace frequencies of particular codes. However, it is also possible to use the CHAINS, DIST, and KEYMAP programs to track sequences of particular codes. For example, KEYMAP will create a contingency table for all the types of codes that follow some specified code or group of codes. It can be used, for example, to trace the extent to which a mother's question is followed by an answer from the child, as opposed to some irrelevant utterance or no response at all. DIST lists the average distances between words or codes. CHAINS looks at sequences of codes across utterances. Typically, the chains being tracked are between and within speaker sequences of speech acts, reference types, or topics. The output is a table that maps, for example, chains in which there is no shift of topic and places where the topic shifts. Wolf, Moreton, and Camp (1994) apply CHAINS to transcripts that have been coded for discourse units. Yet another perspective on the shape of the discourse can be computed by using the MLT program that computes the mean length of the turn for each speaker.

44-5.4. Phonological Analyses

Currently, phonological analysis is a bit of a stepchild in CLAN, but we have plans to correct this situation. These plans involve two types of developments. One is the amplification of standard programs for inventory analysis, phonological process analysis, model-and-replica analysis, and other standard frameworks for phonological investigation. Currently, the two programs adapted to phonological analysis are PHONFREQ, which computes the frequencies of various segments, separating out consonants and vowels by their various syllable positions, and MODREP, which matches %pho tier symbols with the corresponding main line text. For more precise control of MODREP, it is possible to create a separate %mod line in which each segment on the %pho corresponds to exactly one segment on the %mod line.

The second set of plans for improving our ability to do phonological analysis focuses on the use of digitized sound within the CED editor. On the Macintosh, the CED editor allows the transcriber direct access to digitized audio records that have been stored using an application such as Sound Edit 16. We hope to implement a similar utility for the Windows platform. Using this system, which we call "sonic CHAT," one can simply double-click on an utterance and it will play back in full CD-quality audio. Moreover, the exact beginning and end points of the utterance are coded in milliseconds and the PAUSE program can use these data to compute total speaker time, time in pausing between utterances, and overlap duration time. A sample of a file coded in sonic CHAT with a waveform displayed at the bottom of the window

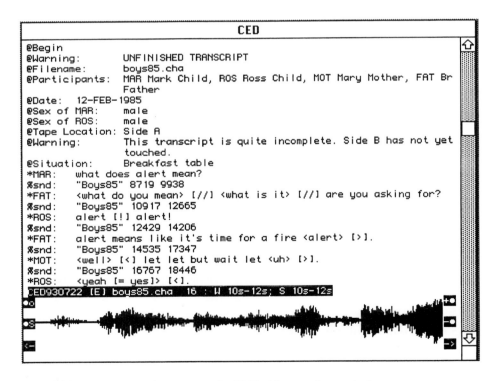

FIGURE 1   A sample file displayed in sonic CHAT with a waveform at the bottom.

is given in Figure 1. In this file, the numbers on the %snd tier refer to absolute time in milliseconds from the beginning to the end of a particular utterance.

The basic CLAN programs like FREQ and KWAL are extremely easy to use and understand. They work on a simple MS-DOS type command line and one can often get the basic answers to important research questions without an understanding of any of the more arcane uses of some of the less common CLAN programs. In addition, users can rely on a well-tested manual that is now in its second edition; and there are additional support resources available over the Internet.

44-6. CONCLUSION

Earlier we looked at four periods in the growth of observational studies of language development. We are now entering the fifth period of methodological development. Our plans for the future development of the CHILDES system are based on the view of the fifth stage of observational research as being the period of electronic connectivity and exploratory reality. Our first priority for this period is to make full use of the facilities of the World Wide Web (WWW) to provide multimedia access to the database, the bibliographic system, and the manual. Using currently available tools such as Netscape, Macintosh AV facilities, and HTML formatting programs, it is now possible for a user to use a sequence of mouse clicks to open up pages of the CHILDES manual, search for particular files in particular corpora, open up those files, and hear the sounds in each. It is even possible to have pictures of the children and parents accessible over the net.

Equally important is the growth of connectivity between programs on a single computer. An example of the type of development we are currently supporting is the linkage of the CED editor to high-level speech analysis tools such as Signalyze on the Macintosh or WAVES on UNIX. We also plan to have access to a reference database of IPA sounds, as well as audio examples of specific uses of CHAT symbols and codes.

44-6.1. The Glossome

The emergent connectivity of the Internet has opened up an exciting prospect that few researchers have yet considered. This is the potential for the establishment of a Glossome Database. Much like the Human Genome Database, the Glossome Database would be supported by data entry over the Internet. The creation of a set of standards for data transcription and transmission will allow us to store and access a wide variety of data from a wide variety of normal and disordered populations.

In order to make successful use of these new opportunities, we will need to develop a higher level of consciousness in both the adult aphasia research community and the child language disorders research community. In each of these areas, the strong commitment to patients' rights must be protected and encouraged. However, researchers often cite patients' rights as a motivation for not sharing their data with the broader research community. This interpretation of personal rights does damage to the progress of the very field that is dedicated to improving the condition of the aphasic patient. The only way to counter these protectionist sentiments is for major figures in the field to lead by contributing new data to the database and by encouraging younger researchers to follow their lead.

Currently, we have had much more success in convincing students of child language disorders than students of adult aphasia to enter their data into CHILDES. Given the fairly advanced state of methodology in the CHILDES system and the small amount of aphasic data currently in CHAT format, it may now make more sense to focus our efforts on collecting new sets of well-transcribed data that are accompanied with full digitized audio records that could be accessed directly over the Internet. Ideally, we would like to see a large body of consistently transcribed data for comparable tasks, which could provide us with a consistent basis for comparison. Although the transcription standards and analytic programs are already in place, there must be a period of further dialogue regarding elicitation tasks and related issues. We would like to work together with workers in the field of adult aphasia to build a solid empirical database for studies of disordered language production in both adults and children.

CHAPTER 45

Neurolinguistic and Related Assessment and Rehabilitation Software: A Listing

Brigitte Stemmer[1] and Brigitta Gahl[2]

[1]Centre de Recherche du Centre Hospitalier Côte-des-Neiges, Montréal, Québec, Canada H3W 1W5, and Lurija Institute for Rehabilitation and Health Sciences at the University of Konstanz, Kliniken Schmieder, Allensbach, Germany; [2]Kliniken Schmieder, Allensbach, Germany

Assessment and rehabilitation tools play an important part in clinical neurolinguistics and in neurolinguistic research. Whereas information on nonelectronic tools and material is readily available, it is still laborious to obtain information on what is available in electronic form. The information listed in this chapter was obtained by placing a ''want ad'' in major Internet discussion lists, by performing Internet, database, and library catalog searches on the subject, by contacting publishers and authors, and last but not least by ''word-of-mouth'' information from researchers in the field. Despite our efforts, we received very little feedback concerning software aimed at non-English or non-German-speaking target groups. Consequently, our search may not have reached all publishers and authors, and available sources may have been overlooked. In order to improve this listing in the next edition of the *Handbook*, we would like to encourage readers to send us any information that seems suitable for inclusion in the next edition. In this context, we would like to point out that no evaluation of the content of the cited material has been made. Hence, neither the editors nor the publisher can endorse specific programs. Furthermore, most software was furnished us on a demo disk; thus we cannot attest to the functionality of the complete program.

The chapter is divided into two main sections. The first will refer the reader to addresses or references for further information such as software providers, software listings, and useful software addresses on the Internet. The second will list individual programs, that is, the name of the program, the authors (if this information was available), information on the software (including the type of stimuli, the language the stimuli are presented in, and the tasks/activities), minimal hardware requirements, whether a demo disk is available, publications on the software (if available), and where to obtain the software. Some of the software listed is of a more general neuropsychological nature but includes modules aimed at neurolinguistic issues. We list software used for clinical evaluation and therapy as well as for neurolinguistic research. Excluded from this list is scoring software and report generating software. For information on such programs, the reader is referred to the addresses given in the "Obtaining Information" section that follows.

45-1. OBTAINING INFORMATION ON ASSESSMENT AND REHABILITATION MATERIAL

1. Information on the Internet
Cog & Psy Sci: Software
http://yul.yu.edu/fgs/lists/software.htm
Cognitive & Psychological Science: Software
http://matia.stanford.edu/cogsci/software.html
COMPSYCH (Computerized software information service for psychologists)
http://www.awa.com/psy/compsych.html
CTI Centre for Psychology
http://www.york.ac.uk/inst/ctipsych/web/MainMenu.html
    The CTI Directory of Psychology Software
    http:/www.york.ac.uk/inst/ctipsych/web/CTI/DirTxt/contents.html
    Reviews of Software from Psychology Software News
    http:/www.york.ac.uk/inst/ctipsych/web/CTI/DirTxt/reviews/rcontents.html
CyberPsychLink
http://cctr.umkc.edu/user/dmartin/psych2.html
Online Psych—Psychological Software
http://www.onlinepsych.com/tour/software.html
Mental Health Net
http://www.cmhc.com
Neuroscience software
http://ivory.lm.com:80/~nab/neurosoft.html
Psychguide-Software List
http://www.designers-int.com/Psychguide/lib/soft.html
PSYCH SITE: Software Sites
http://www.unipissing.ca/psyc/soft.htm

2. Other sources of information
AGS (American Guidance Service) Clinical Catalogue
4201 Woodland Road
Circle Pines, Minnesota 55014-1796
USA
Tel: (USA) 1-800-328-2560
Apparatezentrum der Testzentrale, Hogrefe-Verlag GmbH & Co. KG
Rohnsweg 25
Postfach 3751

D-37027 Göttingen
Germany
Tel: (+49) (0)551-49609-37/38
Fax: (+49) (0)551-49609-88
Dr. G. Schuhfried Ges.m.b.H
Hyrtlstr. 45
A-2340 Mödling
Austria
Tel: (+43) (0)2236-42315-0 or 43680-0
Fax: (+43) (0)2236-46597
Kuratorium ZNS
Humboldstr. 30
53115 Bonn
Germany
Tel: (+49) (0)228-631153
Fax: (+49) (0)228-630397
The ZNS Kuratorium has published a catalog that is regularly updated and that provides information on more than 200 (neuro)psychological (including neurolinguistic) software programs designed for native speakers of German.
NFER-Nelson
Tests and Assessment
Clinical Psychology
Darville House
2, Oxford Road East
Windsor
Berkshire SL4 1DF
UK
Tel: (+44) (0) 1753-858961
Fax: (+44) (0) 1753-856830
E-mail: claire.gordon@nfer-nelson.co.uk
PAR—Psychological Assessment Resources
P.O. Box 998
Odessa, FL 33556
Tel: (USA) (800) 331-8378
Fax: (USA) (800) 727-9329
Psychological Software Services
6555, Carrollton Ave.
Indianapolis, IN 46220
USA
Tel: (USA) (317) 257-9672
Fax: (317) 257-9674
Software sourcebooks
Krug, Samuel E. (1993). *Psychware sourcebook: A reference guide to computer-based products for assessment in psychology, education and business.* (Available at: MetriTech, Inc., 111 N. Market, Champaign, IL 61820, Tel: (USA) (800) 747-4868.)
Spitczok von Brisinski, Ingo (1996). *Sofwareführer Psychologie, Psychotherapie, Psychiatrie, Sonderpädagogik.* Heidelberg: Asanger Verlag.
The Psychological Corporation
Harcourt Brace & Company

555 Academic Court
San Antonio, TX 78204-2498
USA
Tel: (800) 228-0752
Fax: (800) 232-1223
The catalog lists tests and related products for (neuro)psychological assessment; including neurolinguistics.

45-2. SOFTWARE PROGRAMS

The software is listed alphabetically, either by citing the name of the software (in most cases) or by citing the name of the provider of the software. A bullet denotes a major software name or provider. Particular software modules which belong in this category are underlined.

• *BrainTrainMedia: PicTo 200 and Comic 200* (language: German)

Software series for all ages that can be used for medical rehabilitation as well as geriatrics. The multimedia programs are aimed at training word and text comprehension, concept formation, naming and matching, and form and color perception. The therapist has to select the task material and the way it is presented (such as time limit, on-line help, and acoustic feedback). The exercise files contain pictures, sounds (language, noise) and text (words, sentences) to choose from.

Hardware requirements: IBM-compatible PC 486, sound card, color graphic card, 2MB RAM, HD 40MB, mouse, or touch screen. Demo disk available:

Provider:
l.e.b.e.n H. & A. Eickhorst
Lotter Str. 77
49078 Osnabrück
Germany
Tel: (+49) (0) 541-41029

• *CogniSpeed* (version 1.2, version 2.0 in development) (languages: Finnish, Swedish, English in development)

(Antti Tevonsuo and Raija Porton)

CogniSpeed is a software tool in clinical diagnostics and rehabilitation that offers a small lab of experimental cognitive psychology tasks for the clinician or researcher, and that measures different aspects and phenomena of human information processing with a particular focus on automatic and controlled processing. The output of the tests can be to screen, paper, and disk and is transferable to Excel for statistical analyses. The current version is limited to the use of letters, digits, and words as stimuli. Tasks include *(a)* simple and 10-choice reaction time, *(b)* mental arithmetic, *(c)* sentence verification tasks, *(d)* perceptual recognition threshold for single number, letter and word stimuli, *(e)* visual vigilance tasks, *(f)* Stroop-like tasks, and *(g)* lexical decision tasks.

A new version 2.0 is in preparation and is planned to include features such as *(a)* the combination of behavioral tasks with simultaneous recording of brain activity, *(b)* a reference database, and *(c)* use of pictorial stimuli.

Hardware requirements: IBM-compatible PC 286 or higher, EGA or VGA graphic card, sound card optional. Demo disk available.

For scientific information and publications contact:

Dr. Antti Revonsuo
Center for Cognitive Neuroscience
University of Turku
FIN-20014 Turku
Finland
Tel: (FIN) - 358-2-333-6340
Fax: (FIN) - 358-2-333-6270
E-mail: revonsuo@sara.utu.fi
Software is commercially available from
AboaTech Ltd.
Tykistokatu 4A
FIN-20520 Turku
Finland
E-mail: lmakela@utu.fi

• *Cognitive Rehabilitation Software: Language* (language: English)

No detailed information on activities or stimulus type is given in the catalog. The language software includes comprehension, abstract thinking, word finding, memory retrieval, quantitative thinking, logical thinking, text comprehension and memory, error detection in texts, fluency of retrieval, tachistoscopic reading, listen and type, and read and say.

Hardware requirements: IBM-PC compatible, CGA graphics or better, Apple II+ or Macintosh. Demo disk available.

Provider:
Life Science Associates
One Fenimore Road
Bayport, NY 11705-2115
USA
Tel: (USA) (516) 472-2111
Fax: (USA) (516) 472-8146
E-mail: franklsa@aol.com

• *COMLES-Familie Lernprogramme* (language: German)

The software consists of five programs especially designed to support children (ages 6–9) with reading and writing disabilities. All programs are designed as creative and interactive games with animation pictures, acoustic and optical signals, and open program architecture to allow for a choice of the learning content, the level of difficulty, and learning speed.

Abenteuer mit Rudi Wieselwurm (Adventures with Rudi Wieselwurm)
The program develops the construction of words from their individual components.

Mano mit dem Lesepfeil (Mano with the reading arrow)
A text-processing program designed specially for children to teach segmentation of words, to remember the building blocks of words, and to write words.

Mit Lalipur in der Schatzkammer (With Lalipur in the treasure room)
Program to develop, in a playful way, the reading and writing of a basic vocabulary. The individual game events also require the development and application of planning strategies.

Lokführerin Lilli in der Buchstabenfabrik (Engine driver Lilli in the letter factory)
Program teaches writing spoken words and building words from syllables.

Reise mit Käpten Tom (Journey with Captain Tom)
Program to develop fast recognition of letters, building blocks of words and sentences. Words or sentences can be presented as moving over the screen.

Hardware requirements: IBM-compatible PC 386 or higher, MS-DOS, 1MB RAM, 40MB

HD, floppy drive, sound card, speakers, earphones, microphone, mouse. Demo disk available.
Literature:
Baumann-Geldern, Irene (1995). *Das Projekt „Computerunterstützter Schriftspracherwerb in sonderpädagogischen Diagnose- und Förderklassen". Der Einsatz von Lernprogrammen der „COMLES-Familie" bei Schülerinnen und Schülern mit sonderpädagogischem Förderbedarf.* München. [Report] (Report can be obtained from Cornelsen)
Provider:
Cornelsen Sofware
Mecklenburgische Str. 53
14197 Berlin
Germany
Tel: (+49) (0) 30-89785-434
Fax: (+49) (0) 30-89785-599

- *Cool Spring Software (language: English)*

The Naming Test

Assessment tool for naming ability. On-line help is provided by cues in hierarchical fashion (associated sound stimulus, semantic cues, phonemic cues, and multiple choice).

Williams Picture Vocabulary Test (WPVT)

Assessment tool for vocabulary. A word is presented in natural, digitized speech along with four color photographs. The subject is requested to indicate the word. Norms are currently collected for the test and users are invited to participate in the study.

The Category Test

A test for the assessment of cognitive styles and neuropsychological syndromes. Stimulus sets must be organized according to rules invented by and consistently used by the person.

Auditory Perception

Tool to examine auditory neglect and cerebral dominance for auditory information. The program allows the user to perform the standard audiological neglect studies, test auditory acuity for amplitude and pitch using pure tones, perform dichotic listening studies with pure tones and speech, and examine auditory agnosia using environmental sounds. Users may also enter their own auditory stimuli using an inexpensive sound digitizer.

Rhythm Perception

Tool to create a series of randomly generated musical rhythm patterns and present them over headphones. The pattern lengths and presentation order are set by the user. The program presents the target pattern and then requests the subject to identify the target when presented with one distractor pattern of the same length and sound pitch. Although the program will create pattern sets, the user has complete editing control and may alter any set or create new sets by typing in rhythm patterns, or even melodies, using conventional musical notation.

Similar Sounds

Program for the assessment of auditory agnosia. The program presents common environmental sounds and requests the subject to identify and name them. The program includes a section that requires the subject to name similar, popular musical tunes.
Hardware requirements: Most programs are available for IBM-compatible PC (Windows) and Macintosh. No further details on hardware requirements were given.
Provider:
Cool Spring Software
P.O. Box 130
Woodsboro, MD 21798
USA

Tel: (USA) 301-845-8719

E-mail: CoolSpring@aol.com

Internet: http://users.aol.com/CoolSpring/CSpring.html

• ***COPIA and LETRAS: Global and analytical reading*** (language: Spanish)

The program COPIA trains global word reading by tachistoscopically displaying the entire word. The program LETRAS presents a word tachistoscopically letter by letter.

Hardware requirements: IBM-compatible PC 8086 or higher.

Information:

Josep M. Vendress

Escola de Patologia del Llenguatge

Laboratori d'Informàtica

Pavelló Santa Victòria, 2ª planta

Hospital de la Santa Creu i Sant Pau

Avgda. St. Antoni Maria Claret, 167

08025 Barcelona

Spain

Tel: (+93) 291 90 79

Fax: (+93) 291 90 78

• ***COPROF*** (Computerunterstützte Profilanalyse) (language: German)

(D. Hansen & H. Clahsen)

COPROF is an interactive diagnostic instrument to describe and evaluate linguistic impairments particularly at the grammatical level during child language acquisition.

Hardware requirements: IBM-compatible PC 386 or higher, MS-DOS 3.3 or higher.

Provider:

FOCUS

Postfach 410302

50863 Köln

Germany

Information also from Dr. D. Hansen, Universität Hannover, Fachbereich Erziehungswissenschaften I, Bismarckstr. 2, 30173 Hannover, Germany (E-mail: hansen@fbez1.uni-hannover.de)

• ***CURE-Software*** (language: German)

The software focuses on training particular cognitive functions and has been designed for traumatic brain-injured individuals as well as mentally disabled people. The program includes functions such as attention, visual-spatial memory, mental shifting, calculation, language (semantic discrimination, verbal memory), and visual recognition. The therapist can choose the number of exercises, size of font, type of presentation, degree of difficulty, and time of presentation. Special hardware components such as special keyboards and input devices are available.

Hardware requirements: IBM compatible PC 386, 4MB RAM, 4MB HD, Windows 3.1, VGA card. Demo disk available.

Provider:

Siemens Nixdorf Informationssysteme AG

Projekt „Computer helfen heilen"

Vorgebirgsstr. 49

D-53119 Bonn

Germany

Tel: (+49) (0)228-9825-195

Fax: (+49) (0)228-9825-193

• *Dyslexia Screening Instrument* (language: English) (Kathryn B. Coon, Mary Jo Polk, & Melissa McCoy Waguespack)

Scoring program for teachers, counselors, psychologists, and clinicians to assess individuals (6–21 years of age) who have reading, spelling, writing, or language processing problems. Classifications are based on observations and ratings.

Hardware requirements: DOS-based PC, DOS 3.0 or higher.

Distributed by The Psychological Corporation (for mailing address see Section 1).

• *EXDEN:* semantically and phonologically oriented rehabilitation (languages: Danish, French, German, English, Dutch, Portuguese, Spanish)

Two programs to teach word retrieval, either by phonological or by semantic cuing. There are four different levels of cuing for each program.

Hardware requirements: IBM-compatible PC 386 or higher, Super-VGA card, HD, mouse, a special speech card is provided with the program, amplifiers, and loudspeakers.

Information:
Josep M. Vendress
Escola de Patologia del Llenguatge
Laboratori d'Informàtica
Pavelló Santa Victòria, 2ª planta
Hospital de la Santa Creu i Sant Pau
Avgda. St. Antoni Maria Claret, 167
08025 Barcelona
Spain
Tel: (+93) 291 90 79
Fax: (+93) 291 90 78

• *Exler and Fonex* (language: Spanish)

Program to train naming to language-impaired children and adults. The program includes exercises such as noun exploration (introduction of vocabulary), oral comprehension (the computer speaks a word that has to be matched to a picture/scene), written comprehension (a written word has to be matched to a picture/scene), dictation (computer dictates word), written naming (the name of a presented item has to be written down), and semantic fields (group words into categories and add new words).

An additional program, Fonex, focuses on phonological aspects and groups the words according to phonological criteria. Exercises are similar to the ones mentioned previously.

Hardware requirements: Version 1.60: IBM-compatible PC 286 or higher, MS-DOS, graphic card, mouse, speakers; Version 2.00 for Windows: IBM-compatible PV 386 or higher, 2MB RAM, VGA graphic card, Sound Blaster compatible sound card, mouse. Demo disk available.

Provider:
Escola de Patologia del Llenguatge
Laboratori d'Informàtica
Pavelló Santa Victòria, 2ª planta
Hospital de la Santa Creu i Sant Pau
Avgda. St. Antoni Maria Claret, 167
08025 Barcelona
Spain
Tel: (+93) 291 90 79
Fax: (+93) 291 90 78

• *GrundSchulSoftware (GSS) Jansen* (language: German)

Various programs developed to improve reading, writing, and calculation skills of people with impaired reading, writing, and calculation abilities.

1. Programs focusing on reading and writing skills: *Gleitzeile I and Gleitzeile II* (moving lines) Program designed for intensive reading support at the beginners' level. Target group is students with reading and writing disabilities, juvenile and adult illiterates, and vision-impaired individuals. The program presents large-font script that moves over the screen. The font size, moving speed, letter distance, and level of difficulty are adjustable. Self-designed texts and exercise material can be added to the program. Exercises are based on syllables, words, groups of words, sentences and texts, and fragmented texts. Tachistoscopic presentation is possible.

Fehlerfeind (error enemy)

Program to support orthographic abilities. Target group is elementary school students with orthographic and attention problems. Three different types of exercises, which can be presented as a competitive game between two children, are available.

Lesegeister (reading ghosts)

Program designed to train attention and improve reading and writing skills. Target group is students with writing disabilities. Exercises are copying of words presented in different ways, and putting syllables into a correct order.

Wortbilder (word pictures)

Program to train reading and writing skills and acquire a basic vocabulary. Target group is 6- to 7-year-old elementary school children and children with writing and reading disabilities. Various files containing a total of 400 word and pictures are presented using 20 different tasks to support, for example, recognition of beginning and ending sounds and syllables, word–picture matching, copying from pictures with a help function, and so forth.

Signale I

Program to support synthetic and analytic processes in beginners' language training, and to associate words with similar orthography. Exercises are designed as games such as "letter rain" with letters falling from the sky and a decision has to be made which of the letters is necessary to write a particular word. Target group is elementary school children and children with reading and writing disabilities.

2. Programs focusing on calculation skills: *Prisma B, SI, SII*

All programs have been designed for elementary school children but can also be used for children and adults with calculation disabilities. The programs are made up of a variety of different tasks to train elementary mathematical skills, support memory skills, concentration, imagery, logical reasoning, and creative abilities.

Hardware requirements: IBM-compatible PC, DOS-compatible 386 upward (no further information given). Demo disk available.

Provider:

GrundSchulSoftware (GSS) Jansen

Talstr. 13

41844 Wegberg

Germany

Tel/Fax: (+49) (0) 2434-1236

• *GUS-Multimedia Speech System* (language: English)

A set of software programs and hardware components aimed at improving speech output of patients with acquired speech impairments.

Information:

Gus Communications, Inc.

E-mail: gus@gusinc.com

• *IBM PS SprechSpiegel* (languages: German, English, French, Spanish; a total of 13 language options are available)

SprechSpiegel is a diagnostic and treatment tool for auditory- and speech-impaired children and adults consisting of 15 therapeutic exercises and six modules for organization and administration. Focus is on training sound elements, voice and articulation, and patterns of sentences, words, and sounds. The program provides optical feedback to voice and articulation exercises. Administrative functions make it possible to define specific patient characteristics (e.g., pitch, language, feedback), objectives for sound characteristics, syllables, and words.

Hardware requirements: IBM-compatible PC 386 or higher, DOS 4.01 or higher, or OS/2 2.0 or higher, 2MB RAM (DOS), 8MB (OS/2), VGA card, color monitor, ACPA-adapter, microphone, speakers.

Provider:

IBM Beratungszentrum

Informationstechnik für Menschen mit Behinderungen

Postfach

70548 Stuttgart

Germany

Tel: (+49) (0) 711-785-3471

Fax: (+49) (0) 711-785-3698

• *INTACT* (language: English)

Program specifically designed to supplement speech and language therapy in the clinic or independently at home. The program contains a resource library of more than 700 texts and picture- and sound-based exercises. An authoring facility enables the therapist to create his or her own exercises.

Hardware requirements: IBM-compatible PC 486, 4MB RAM, SVGA graphics adapter and color monitor, 25MB HD space, 16 bit sound blaster card, DOS 5.0 or higher.

Provider:

Aphasia Computer Team

Speech and Language Therapy Research Unit

Frenchay Hospital, Bristol BS16 1LE

UK

Tel: (GB) (0) 117-970 1212 Ext. 2291

Fax: (GB) (0) 117-970-1119

• *ITS-Integriertes Therapie System* (institute, clinic, and outpatient versions) (language: German)

Multimedia authoring and therapy system focusing on clinical applications (target groups are aphasics, learning-disabled children, neuropsychologically impaired patients), as well as for research purposes. The authoring system includes databases for words, sentences, drawings, spoken words, sounds, texts, and film scenes that can be extended, modified, and combined. The therapy system includes the administration of exercises and the patient database. Structure, layout, and content of each task, as well as the form and order of presentation, can be designed individually. Therapeutic modules focusing on particular impairments (e.g., text-processing impairments, agraphia, alexia, acalculia, speech apraxia) aim at improving, for example, self-monitoring abilities, nonverbal signaling, word finding, lexical semantics, morphosyntax, sound formation, auditory comprehension, reading comprehension, and action planing. Reaction time recording is possible. The data can be exported for further statistical processing.

Hardware requirements (depending on the type of version purchased): IBM-compatible PC with multimedia capabilities, touch screen, mouse, Windows 95/NT.

Provider:

NEUROsoft

Technologiezentrum Muensingen

Rudolf Diesel Str. 3

72525 Muensingen

Germany

E-mail: ufschwarz@aol.com

• *LeMo* (Lexikon and Morphologie) (language: German; other languages are being developed) (R. de Bleser, J. Cholewa, N. Stadie, & S. Tabatabaie)

LeMo is an expert system for the assessment of linguistic impairments with respect to monomorphemic (testpart "lexicon") and polymorphemic (testpart "morphology") word processing to be used by speech language pathologists, neurolinguists, and clinical linguists. The system offers largely automated test presentation and on-line scoring and provides special programs for the off-line linguistic analysis of phonemic and graphemic errors.

Hardware requirements: IBM-compatible PC 386 or higher, 4-8MB RAM, Windows 3.0 or higher; Macintosh: 4MB RAM, system 6.0 or higher.

Literature:

Stadie, N., Cholewa, J., De Bleser, R., & Tabatabaie, S. (in press). LeMo, an expert system for theory-based single-case assessment of aphasia. *Neuropsychological Rehabilitation*.

For further information contact:

Nicole Stadie

University of Potsdam

Institute of Linguistics

Dept. of Neurolinguistics

Postfach 60 15 53

14415 Potsdam

Germany

E-mail: nstadie@rz.uni-potsdam.de

• *Laureate Learning Systems* (languages: English and Spanish, unless specified otherwise)

A number of programs mostly designed for children (but can also be used with adults) with language learning disabilities, developmental disabilities, visual impairments, hearing impairments, and autism. All programs combine natural-sounding speech, colorful graphics, and animation and can be accessed by a variety of interfaces, such as the keyboard, single switch, touch screen, and mouse. Most programs can be customized by the teacher to meet the needs of the users. For several programs, monographs describing the background and applications are available.

Hardware requirements: Most programs can be run on Apple IIe, Apple IIGS, DOS, Windows, and Macintosh. All systems need speech capabilities and for some a CD-ROM drive is required (for details, consult the company's catalog).

The Language Activities of Daily Living Series (Barbara Couse Adams)

Programs for children and adults to improve the ability to understand and express the language necessary to perform everyday activities within the home ("My house"), community ("My town"), and school ("My school"). Each of these programs contains six scenes that can be chosen to work in. There are four activity options: In "Discover Names" and "Discover Functions or Descriptions," the computer speaks the name of a selected item, or its function or description, respectively. In "Identify Names" and "Identify Functions and

Descriptions," the computer verbally prompts the user to find items according to their names, or their functions and descriptions, respectively.

The Early Vocabulary Development Series: First Words, First Words II, First Verbs (Mary Sweig Wilson & Bernard J. Fox)

The programs aim at building core vocabulary and improving vocabulary comprehension. The "Exploring Early Vocabulary Series" reinforces and expands the vocabulary from First Word, First Words II, and First Verbs. Activities include explore nouns/verbs (the computer gives the name of a selected item or shows the verb action), explore nouns and descriptions (the computer gives the name of a selected item and a short description), and identify nouns/verbs.

Talking Nouns I, Talking Nouns II, and Talking Verbs (Mary Sweig Wilson and Bernard J. Fox)

Programs for children and adults to train expressive use of words and phrases, and promote conversational abilities. Activities include interactive communication, picture matching, picture identification, and nouns by category.

First Categories (Mary Sweig Wilson & Bernard J. Fox)

Program to train categorization skills. Activities include reviewing categories, and including and excluding words into a particular category.

Talk Time with Tucker

Program to train vocalization skills and experiment with the duration, pitch, and volume of the utterance.

Let's Go to the Circus

Program to introduce and reinforce the concept of opposites. Activities include simple object identification and the identification of opposites.

Simple Sentence Structure (Mary Sweig Wilson & Bernard J. Fox)

Program to train subject-verb-object word order in simple sentences. The roles of subjects, verbs, and objects are taught first, and then the user has to select the picture that represents a particular sentence.

Micro-LADS (Microcomputer Language Assessment and Development System)(languages: English and French)

Program to train syntactic structures. Different modules are available: plurals and noun-verb agreement, verb forms, prepositions, pronouns, negatives, wh-questions, passive, and deictic expressions, and prepositions II.

The Words & Concepts Series

Programs to develop vocabulary and concepts in six integrated units: vocabulary, categorization, word identification by function, word association, and the concepts of same and different.

The Following Directions Series (Eleanor Semel)

Program to develop left–right discrimination and increase the ability to follow one- and two-level commands.

Twenty Categories (Mary Sweig Wilson & Bernard J. Fox)

Program to develop advanced categorization skills. Activities include word selection that fits a category, category selection for a given word, and word selection that does not belong to a category.

The Sentence Master (Marion Blank)

Program to teach reading. Words are individually trained through word recognition, spelling, and sentence completion tasks. Stories that contain the trained words are presented visually and auditorily. Reinforcement is also given through a printout version of the story. The program is available for four levels of reading instruction.

Demo CD-ROM available.

Readable Stories (Donna J. Crowley & Philip J. Grise)
Program to increase reading comprehension ability. Comprehension questions cover the concepts of "main idea," "sequence of events," and "who, what, where, when, why, and how."
Provider:
Laureate Learning Systems
110 East Spring Street
Winooski, VT 05404-1898
USA
Tel: 1 (800) 562-6801 (USA and Canada)
(USA) (802) 655-4755 (International)
Fax: (USA) (802) 655-4757
Web: http://www.LLSys.com

• *LernReha Programme* (language: German) (Dr. Erich Kasten)

Ratewort (guessing a word): Training program for anomia and dyslexia. Words are presented on screen and missing letters have to be supplied.

Wissen (knowledge): Training program (high level of difficulty) for mild aphasics, memory impaired and dyslexic individuals. Words have to be guessed in crossword puzzle fashion.

Leseprogramme (reading program): Program designed for step-by-step training of reading ability. Part One focuses on recognition of letters/numbers, syllables, and words of different length. Part Two trains discrimination of individual letters.

Bild-Begriff-Funktion (picture-word-function): Program combining reading and writing training and developed for 7–10-year-old children. May be used with brain-damaged adults. The program consists of four differently designed tasks that basically require matching pictures, written words, and the function of objects.

Rechtschreibeprogramme (orthography program): Training program developed for dyslexia. Focus is on teaching rules and application of these rules using a variety of tasks.

Hardware requirements: IBM-compatible PC, DOS-compatible 286, 386, 486, color monitor, VGA graphic card.

Demo discs are available for the programs Bild-Begriff-Funktion, Rechtschreibung, Wissen, Gesichter, and Sehtra.
Provider:
NEUROsoft
Karlheinz Siegmund
Burladinger Str. 10
72393 Burladingen
Germany
Tel: (+49) (0) 7126-1329
Fax: (+49) (0) 7126-1425

• *Lernsoftware* (Teaching/rehabilitation software) (language: predominantly German, some English modules integrated)

Each software program uses a variety of creatively and pedagogically designed methods to train various language components.

Universelles Worttraining (universal word training): 10 different task modules have been designed to diagnosis and train German orthography. A self-designed vocabulary list can be added to the dictionary that is provided by the program. Some of the programs include animation and allow for more than one individual to work on the task at the same time, thus creating "competitive play."

Alphabet: 11 different modules train various aspects of the alphabet. Can be "played" by two individuals in a competitive fashion.

Lesen + Schreiben lernen (learning how to read and write): 12 programs summarized in three blocks training different levels of competencies. Block 1 operates at the letter level, block 2 at the syllable and word level, and block 3 at the word and sentence level.

Bild > Wort PRO: Teaching and training of reading and writing through a variety of different tasks using clipart pictures of different categories and including a painting program. The program is suitable for therapy with aphasic patients.

Die Wortbaustelle (words as a construction site) (languages: German, English)
Syllables, morphemes, and word stems are trained using 11 different programs.

Plättchenrechnen (calculation with tokens) (language: German)
Training program for elementary school and dyscalculia. The various programs include calculation with the help of tokens, a calculator, or a clock, and train addition/subtraction, simple multiplication, and perception and logic (using picture arrangement).

Die Zahlenwaage (number balance) (language: German)
Training program for elementary school and dyscalculia. The programs use various aids, such as a balance, sticks, thermometer, and mice, to practice calculation. Includes animation.

Hardware requirements: IBM-compatible PC 286/386/486, VGA card, color monitor, MS-DOS or Windows 3.0 or higher. Demo disk available.

Provider:
Eugen Traeger
Lernsoftware-Verlag
Bramkamp 39
49076 Osnabrück
Germany
Tel: (+49) (0) 541-128288
Fax: (+49) (0) 541-128288

• *Lexical Access Model* (Gary Dell & Dan Foygel)

The program instantiates the model of normal and disordered lexical access developed by Dell, Schwartz, Martin, Saffran, and Gagnon. The model specifies the probabilities of correct responses, semantic errors, formal errors, neologisms, mixed semantic/formal errors, and unrelated word errors as a function of model parameters in a picture-naming task. For example, if a patient's naming errors are categorized in a particular way, the program will find the best-fitting parameters for that patient, according to the authors' theory.

The authors are planning to make the program available on the WWW. For further information, contact:

Gary Dell, E-mail: gdell@s.psych.uiuc.edu
Dan Foygel (foygel@uiuc.edu)

• *LF, LFPRON, and LFPREP: Reading Comprehension* (language: Spanish)

Three programs designed for children and adults with reading disabilities to train reading comprehension: LF (sentence comprehension), LFPRON (pronoun comprehension), and LFPREP (preposition comprehension). The user has to select from three sentences the one that best matches the displayed scene.

Hardware requirements: IBM-compatible PC 286 or higher, HD, CGA graphic card.

Provider:
Escola de Patologia del Llenguatge
Laboratori d'Informàtica
Pavelló Santa Victòria, 2ª planta
Hospital de la Santa Creu i Sant Pau
Avgda. St. Antoni Maria Claret, 167

08025 Barcelona
Spain
Tel: (+93) 291 90 79
Fax: (+93) 291 90 78

• *Lies mit* (language: German) (Birgit Königes, Thomas Störmer, Rita Völker, Franz Huber, & Friedrich Haberkom)

The program is distributed to teachers only; commercial use is not allowed.

The program is designed as a training program for children with reading and writing disabilities. Stimuli are written or spoken words, texts, and pictures, and the tasks include picture–word assignment, analysis and synthesis of letters in a word or words in a sentence, syllable tasks, semantic tasks, and morpheme tasks.

Hardware requirements: 386, 512KB RAM, 3MB storage, VGA color graphic card, color monitor, audiocard 300E, printer, joystick. Demo disk available.

Provider:
Bayerische Landesschule für Körperbehinderte (Ed.)
Staatsinstitut für Schulpädagogik und Bildungsforschung
Information for teachers:
Zentralstelle für Computer im Unterricht
Schertlinstr. 9
D-86159 Augsburg
Germany
Tel: (+49) (0)821-573011
Fax: (+49) (0)821-2589095
Information for nonteachers:
ADULO
Frank-Klüpfel-Szarny BGB
Steigerfurtweg 42
D-8700 Würzburg-Heidingsfeld
Germany
Tel: (+49) (0)931-612686
Literature:

Thomas Störmer (1991). *Handbuch zum Programm „Lies mit".* München: Zentralstelle für Computer im Unterricht, Staatsinstitut für Schulpädagogik und Bildungsforschung.

• *Lingraphica System* (languages: English and Spanish) (Tolfa Corporation)

Lingraphica is a portable visual-language prosthesis used as a therapy tool for recovering and chronic aphasic patients. It is based on the Visual Communication (VIC) approach reported on in 1976 by Howard Gardner and his VA-based group in Boston and later adapted and extended by researchers at the Palo Alto VA Medical Center using microcomputers, the so-called C-VIC (Computerized visual communication) system. Everyday items and activities are displayed as picture images that can, upon "touch," be enlarged, animated, spoken, or displayed in written form. Images can be copied to a storyboard to create complete expression. The therapist can select from a large amount of activities. The system is self-documenting and patient reports are automatically compiled. The system can be used in a clinical setting or at the patient's residence.

Hardware: Lingraphica is regulated by the FDA as a medical device and the portable unit is dispensed by physician prescription only.

For information and publications contact:
LingraphiCARE Regional
Aphasia Rehabilitation Center

3600 West Bayshore Road
Suite 202
Palo Alto, CA 94303
USA
Tel: (USA) (415) 842-7444
Fax: (USA) (415) 842-7447

• *LingWare Therapiesysteme* (language: German) (F. J. Stachowiak)

LingWare is a neurolinguistic therapy program containing different modules that address various aspects of language such as naming, comprehension, categories, minimal pairs, diction, syntax, word formation, and text. Presentation form is auditory or visual display. For some tasks on-line help is available. Topics and scenes that are displayed are about activities of daily living. Activities include word and sentence completion, multiple-choice tasks, auditory discrimination, and dictation.

Hardware requirements: IBM-compatible PC 386 or higher, VGA card, sound card, speakers, microphone, mouse, or keyboard.

Literature:

Stachowiak, F. J. (1993). Computer-based aphasia therapy with the Lingware/STACH system. In F. J. Stachowiak (Ed.), *Developments in the assessment and rehabilitation of brain-damaged patients* (pp. 353–380). Tübingen: Gunter Narr Verlag.

Provider:
Phoenix Software GmbH
Küdinghovener Str. 98
53227 Bonn
Germany
Tel: (+49) (0) 228-975840
Fax: (+49) (0) 228-9758418

• *Lübecker Lernprogramme* (language: German) (M. Barmwoldt)

The programs are designed for use in elementary school and for reading- and writing-disabled children.

Alphalex 2000 is a program to train alphabetical sequencing.

Buchstaben-Specht 2000 and *Krypto 2000* are programs that train the ability to combine letters to form words.

Silben 2000 and *Silben 3000* are programs that focus on syllable identification, syllable copying, finding syllables in a word, and combining syllables to form words.

Diktat 3000 and *Duo 3000* are programs that train grammar, dictation, and text comprehension.

Wort-Mixer 3000 is a program that trains the combination of letters and syllables to form words.

Hardware requirements: IBM-compatible PC 386 or higher, EGA/VGA graphic card, mouse.
Demo disks available.

Provider:
l.e.b.e.n H. & A. Eickhorst
Lotter Str. 77
49078 Osnabrück
Germany
Tel: (+49) (0) 541-41029

• *MicroCog:* Assessment of Cognitive Functioning (version 2.4, 1993; language: English) (Douglas Powell, Edith Kaplan, Dean Whitla, Sandra Weintraub, Randolph Catlin, & Harris Funkenstein)

Screening instrument or diagnostic test including 18 subtests for nine areas of cognitive functioning: attention/mental control, memory, reasoning/calculation, spatial processing, reaction time, information processing accuracy, information processing speed, cognitive functioning, cognitive proficiency. Age-specific norms and education level-adjusted norms are provided. Can be used with PsyTest Assessment Management Systems for computer scoring.

Hardware requirements: DOS-based PC 80286 or higher, DOS version 5 or higher, 640K RAM, HD at least 3MB free space, EGA or color graphics monitor, specific hardware add-ons, mouse. Demo disk available.

Developed in cooperation with the Risk Management Foundation of the Harvard Medical Institution. Distributed by The Psychological Corporation (for mailing address, see Section 1).

• *Neurop–Neuropsychologische Batterie* (language: German) (Dr. Laco Gaál)

Collection of various neuropsychological test and training programs, which can be combined with a database system ("Badok") for evaluation and statistical analyses. One of the modules ("Kombination") is directed toward training of simple arithmetic (addition and subtraction). Another module ("Kommunikationshilfe") provides communication aids for severely language- and motor-impaired patients.

Various modules are currently revised. The program "Logot" has been developed for (multilingual) aphasia training. No further details are available.

Hardware add-on required. Other hardware requirements not specified.

Provider:
NEUROsoft
Karlheinz Siegmund
Burladinger Str. 10
72393 Burladingen
Germany
Tel: (+49) (0) 7126-1329
Fax: (+49) (0) 7126-1425

• *Paced Auditory Serial Attention Test (PaSaT)*

Assessment of concentration and speed of information processing, mental calculating abilities, and ability to serial-track numbers. Includes automated test presentation and extensive data analyses in graphic, tabular, and text formats.

Hardware requirements: IBM-compatible 386 or higher, 20MB free HD space, DOS 5.0 or higher, Windows 3.1 or higher.

Distributed by The Psychological Corporation (for mailing address, see Section 1).

• *Parrot Software for Communication Disorders* (language: English)

Parrot software offers a large variety of programs to train verbal expressive and receptive skills. A list of this software follows (for details and other related software the reader is referred to the catalog and the demo disks).

Hardware requirements for the following listed programs are PC-compatible 386 or higher, Windows 3.× or higher, 4MB RAM, graphics 640×480×16. Some programs also require a sound card, a microphone, and a mouse. (Some programs also exist as an Apple II-compatible version. This will be indicated).

Multi-Sensory Words (Frederick F. Weiner)

Program designed to practice word-naming skills. A picture is displayed on the screen and the users can cue themselves with the first letter, the spelling, a recording of the word, or a list containing the word. They can record their own production and play it back and compare it to the recording of the actual name.

Picture Categories (Frederick F. Weiner)
Program to train vocabulary, word retrieval, and reading comprehension. The user has to select the appropriate category of a picture displayed on the screen.

Categorization, Association, and Sequencing Plus (Frederick F. Weiner)
Program to train vocabulary, word retrieval, reading comprehension, sequencing, thought organization, and deductive reasoning. Tasks include word finding, word identification, word association, and word sequencing.

Listening Skills Plus (Frederick F. Weiner)
Program to train listening skills. The user has to follow verbal instructions such as assigning colors to geometric forms.

Conditional Statements Plus (Frederick F. Weiner)
Program to train deductive reasoning, memory, logical thinking, and thought organization. The user has to follow a conditional statement presented either auditory or visually.

Auditory Perception (Frederick F. Weiner)
Program to train auditory perception and auditory memory. The user has to remember or identify environmental sounds.

Verbal Picture Naming Plus (Frederick F. Weiner)
Program to practice word-naming skills. The user has to name photo-realist pictures. The response is evaluated by the computer and feedback is given.

Using Propositional Speech (Frederick F. Weiner)
Program to train expressive language skills. The user must move pictures to corresponding positions in a grid by giving verbal instructions.

Auditory and Visual Picture Recognition (Frederick F. Weiner)
Program to train making associations between printed or verbal words and presented pictures. Pictures are presented on the screen and the user is asked to identify one of the pictures.

Memory for Animated Sequences (Frederick F. Weiner)
Program to train attention, short-term memory, receptive language, and expressive language. Users watch an animated sequence of events and then choose from a list to indicate whether the statements correspond to what they have seen or not.

Verbal Analogies (Frederick F. Weiner)
Program to improve naming and vocabulary. The user has to choose the correct analogy according to a previously presented analogy example. (An Apple II-compatible version is available.)

Inferential Naming (Lorie Houston Dillon & Frederick F. Weiner)
Program to improve naming and reasoning skills. The user has to try to determine the identity of a concealed word. Clues from five semantic features will help the user finding the word. (An Apple II-compatible version is available.)

Word Order (Frederick F. Weiner)
Program to improve syntactic skills. The user has to construct a sentence from a scrambled set of words. (An Apple II-compatible version is available.)

Category Discrimination and Reasoning (Lorie Houston Dillon & Frederick F. Weiner)
Program to train thought organization, memory, deductive reasoning, vocabulary, semantics, category discrimination, analysis of category attributes, and category reasoning. The user has to identify the word that does not belong to a group of presented words, select a reason why it does not belong to the group, and choose a new word that fits into the word list. A "lesson creation option" provides therapists with the possibility of creating their own lesson. (An Apple II-compatible version is available.)

Sorting by Category (Frederick F. Weiner)
Program to train categorization and reasoning. The user has to sort words into categories. (An Apple II-compatible version is available.)

Auditory and Visual Instructions (Frederick F. Weiner)
Program to train visual and auditory receptive deficits and attention deficits. The user must identify geometric forms that fit a previously given auditory or visually presented description.

Word Association (Frederick F. Weiner)
Program to train word retrieval and vocabulary. The user has to recognize words that are associated with other words or are parts of other words.

Opposites and Similarities (Frederick F. Weiner)
Program to train vocabulary, word retrieval, and reading comprehension. The user has to select opposite or similar word pairs.

Categories (Frederick F. Weiner)
Program to train word retrieval and categorization, to curb perseverative behavior, and to increase vocabulary. The user has to decide within a given time frame whether a presented word belongs to a specific category or not.

Definitions (Frederick F. Weiner)
Program to train word retrieval, vocabulary, and reading comprehension. The user must select the most appropriate definition of a word. Therapists can create their own lessons to adapt the vocabulary to the user's interests.

Sentence Completion (Frederick F. Weiner)
Program to train word retrieval, vocabulary, reading, and semantics. The user has to select appropriate words or phrases to complete a sentence.

Multiple Meaning Words 1 and Word 2 (Lorie Houston Dillon & Frederick F. Weiner)
Program to train word retrieval and develop vocabulary. The user has to find appropriate homophones or homographs to complete a sentence. (An Apple II version of these programs is available.)

Picture Naming (Frederick F. Weiner)
Program to train word retrieval, spelling, vocabulary, and reading comprehension. The user must find or type the name of a presented real-life picture. The program can be adjusted to accept approximate spellings of words.

Picture Identification (Jonathon Snavely)
Program to train word retrieval and vocabulary. The user must select the word that best describes a presented real-life picture.

Semantique 1: Antonyms, Synonyms, and Multiple Meanings: A Learning Game (Elisabeth H Wiig & Erik Wiig)
Program to develop word comprehension and production. Users can compete with one another in providing antonyms, synonyms, or definitions of words. Therapists can expand the database answers to adapt to the users' needs.

Minimal Contrast Stories (Christine Heinbaugh, Thomas Evans, & Frederick F. Weiner)
Program to train articulation. The user's articulatory response governs the outcome of a story.

Multimedia Reading Comprehension (Frederick F. Weiner & Courtney E. Weiner)
Program to improve reading comprehension. The user has to read a story on the screen and, by using help functions that provide auditory or visual cues, answers questions on the story or on particular vocabulary.

Automatic Articulation Analysis (Frederick F. Weiner)
Program to provide a phonological articulation analysis. The user has to respond to pictures presented on the screen. The response is then analyzed by the program.

Visual Confrontation Naming (Lorie Houston Dillon & Frederick F. Weiner)
Program to train word retrieval, spelling, vocabulary, and reading comprehension. The user has to type the name of a real-life picture presented on the screen. Help is available by auditory or visual cues.

Functional vocabulary (Frederick F. Weiner)
Program to train word retrieval, spelling, vocabulary, and reading comprehension. The user has to select the best description of a picture displayed on the screen.
Hardware requirements for the following programs are either an IBM-compatible PC with VGA graphics and a mouse or Apple II-compatible computers (Macintosh LC with a IIE card):

Aphasia 1: Noun Association; Aphasia 2: Opposites and Similarities (Frederick F. Weiner)
Two programs to train word retrieval, word association, reading comprehension, and vocabulary. The user has to match a presented word with the appropriate word from a word list.

Aphasia 3: Categories (Frederick F. Weiner)
Program to train word retrieval and categorization. The user has to decide whether a presented word is a member of a target category.

Aphasia 4: Reading Comprehension (Cathy A. Pelletier & Frederick F. Weiner)
Program to train reading comprehension, inferential reasoning, and vocabulary. The user has to answer comprehension questions that require either a "literal" or an "inferred" response. A help function provides the user with a series of logical statements if the answer was incorrect.

Aphasia 5: Sentence Completion; Aphasia 6: Definitions (Frederick F. Weiner)
Two programs to train word retrieval, reading comprehension, and vocabulary. The user must select the correct phrase to finish an incomplete sentence, or a definition of the presented word. An authoring option allows the therapists to create their own lessons.

Aphasia 7: Picture Identification, Aphasia 8: Picture Naming (Frederick F. Weiner)
Program to train word retrieval, spelling and vocabulary. The user has to type the name of a real-life picture displayed on the screen, or match the picture with its appropriate name from a presented choice of three words.

Aphasia 9: Initial Phoneme Cueing for Aphasics (Alan Gallaher & Frederick F. Weiner)
Program to train word retrieval and vocabulary. A picture is displayed and the user has to type the first letters of a word. Screen cues, which can be faded according to the level of difficulty, aid the user.

Aphasia 10: Orientation for Aphasia and Cognitive Disorders: Mastering Personal Information (Alan Gallaher & Frederick F. Weiner)
The program helps the user recall personal identifying information. The user has to type in highly familiar written materials.

Aphasia 11: Understanding Attributes (Frederick F. Weiner)
Program to train syntax, grammar, reading comprehension, and vocabulary. The user has to find an attribute in a sentence and replace it by another attribute that can be chosen from a list of three.

Categories: Completion from Partial Information (Frederick F. Weiner)
Program to train word retrieval, phonemics, and vocabulary. The user must complete a word of which the category and some parts are given. The therapists can add self-designed lessons.

Antonyms and Synonyms (Lorie Houston Dillon & Frederick F. Weiner)
Program to train semantics and vocabulary. The user has to either select the antonym or the synonym of a target word, or decide whether two words are antonyms or synonyms.

Reading Comprehension and Picture Association (Frederick F. Weiner)
Program to train reading comprehension abilities of children and adults. The user has to read a displayed story and then answer questions by selecting the appropriate pictures.

Preposition Fill-ins and Verb Fill-ins (Alice Kamin, Merle Joblin, & Frederick F. Weiner) Two programs that train prepositions, verb forms, grammar, and syntax. The user has to provide the correct preposition or verb out of a list of two to complete a sentence.

Function Pictures, Association Pictures, and Rhyming Pictures (Edna Carter Young & Frederick F. Weiner) The programs are sold separately. Programs to teach naming, vocabulary, and semantics. Depending on which program is used, based on visually displayed instructions, the user has to reveal the correct picture, which is hidden under a box.

Provider:
Parrot Software, P.O. Box 250755
West Bloomfield, MI 48325-0755
USA
Tel: (800) 727-7681 (USA and Canada)
(810) 788-3223 (International)
E-mail: parrot@mcs.net
Web: http://www.parrotsoft.com

• ***PHONO*** (language: German) (R. De Bleser, J. Cholewa, N. Stadie, & S. Tabatabaie) Phono is a program that allows automatic quantitative and qualitative analysis of phonemic errors in oral naming, repetition, and reading aloud. The stimuli to which the patient has to respond are selected and implemented by the examiner. The program analyzes phonemic errors and determines segmental-phonological error processes, the frequency of phoneme substitutions and omissions, and the frequency of lexicalizations and neologizations. Furthermore, the degree of correspondence between a produced item and the target is calculated, and a phonemic feature analysis for phoneme substitutions and omissions is performed.

Hardware requirements: IBM-compatible PC 386 or higher, 4-8MB RAM, Windows 3.0 or higher; Macintosh: 4MB RAM, system 6.0 or higher, sound card, microphone, speakers.

Literature:

Cholewa, J., Tabatabaie, S., Stadie, N., & De Bleser, R. (1994). Das Programm PHONO: Ein automatisiertes Verfahren zur Analyse expressiver phonologischer Fehlleistungen. *Neurolinguistik,* 1, 27–40.

For information contact:
Nicole Stadie
University of Potsdam
Institute of Linguistics
Dept. of Neurolinguistics
Postfach 60 15 53
14415 Potsdam
Germany
E-mail: nstadie@rz.uni-potsdam.de

• ***PsychLab*** (version 1.099, last update July 1996) (Teren Gum) Application software to be used by researchers to set up different experimental paradigms. The experiments are specified by a control panel. The user clicks on the various pop-up menus, buttons, or boxes to set different parameters defining the events in the experiment, then chooses an input file and runs the experiment on-line. An experiment can have a number of trials and each trial can contain a number of stimuli. Simple stimuli can be words, sentences, tones, sound, or pictures. Compound stimuli consist of a number of simple stimuli presented successively or simultaneously. Timing parameters, interstimulus intervals, and response type can be specified for all kinds of stimuli that PsychLab is capable of handling. Responses are measured in units of 1 ms and can be obtained from any single key from the keyboard, a number of keys in sequence, or no response at all. Any font can be chosen for presentation of word and sentence stimuli. An unlimited number of display positions can be specified. A viewing window can be resized for each experiment. A trial randomization option is included. The program will save the responses and reaction times, and means and standard deviations for responses are computed.

Hardware requirements: All non-Power Macs, 2MB RAM minimum, 4MB RAM recommended. (A version for Power Macs is being developed.) Demo disk available.

For information contact:
Teren Gum, E-mail: teren@sprynet.com
PsychLab mailing address:
PsychLab
3175 Toupin Boulevard
St.-Laurent, Québec
Canada H4K 1YP

- **RehaCom**

A training program for cognitive impairment that can be adapted to the patient's abilities. The system consists of modules to train attention, vigilance, topological memory, verbal memory, visual memory, reaction time, visual–motor and visual–constructive abilities. Tools to develop individualized software (to be programmed in Turbo Pascal) are also provided.

Hardware requirements: PC 386/25, 4MB RAM, 200MB HD, 5¼ or 3½ floppy drive, serial port for the specially designed patient desk, SVGA graphic card, mouse, keyboard, 16 bit sound card, MS-DOS 5.0 or higher, VGA 15″ monitor, printer, patient desk.

Providers: Dr. G. Schuhfried Ges.m.b.H (for Austria) and Hogrefe (for Germany). (For mailing address, see Section 1).

- **Suvalino** (version 2.0) (language: German)(G. Caprez)

Program to train different neuropsychological areas. The language part contains the module "Textaufgaben," which trains text comprehension and memory for text. Activities include completing fragmented texts, finding words in a text, finding differences in two similar texts, finding and learning to remember main text ideas.

Information:
Rehabilitationsklinik Bellikon
Sekretariat Neurorehabilitation
Postfach
5454 Bellikon
Switzerland
Tel: (CH) (0) 56-485 5111
Fax: (CH) (0) 485-5444

- **Text & Co.** (language: German) (H. Spitzer and U. Behrends)

Program to improve text comprehension. The texts vary in length and level of difficulty. Text recall operates on different levels of difficulty and the nature of the information to be recalled falls into specific categories (names, features, dates and numbers, events, general world knowledge). Activities include recall of individual information, text completion, explaining words, check on statements, wh-questions, highlighted cue words, sequencing text lines, and text recall. Depending on the type of task, different forms of feedback are provided: simple correct–incorrect feedback, acceptance of a variety of answers, comparison between user answer and program answer (requires the user's ability to discover mistakes). The program can be customized by the therapist to meet the users' needs. The program provides several "tasks boxes" that contain tasks suitable for particular language impairment syndromes (such as Wernicke's aphasia and Broca's aphasia). Therapists can add their own texts to the program and adjust the program in such a way as to allow the patient to work independently with the program.

Hardware requirements: IBM-compatible PC 386 or higher, 4MB RAM, 2MG HD space, VGA color monitor, VGA graphic card, Windows 3.1 or higher, CD-ROM drive, sound card, speakers. Demo disk available.

Provider:
Reha-Service Petra Rigling
Bahnhofstr. 13d
76337 Waldbronn
Germany
Tel: (+49) (0) 7243-68859
Fax: (+49) (0) 7243-65702

TheDiaS, TUS, PhonX (language: German) (Clinical Neuropsychology Research Group, City Hospital Bogenhausen, Munich, Germany, in cooperation with the Institute for Communication Engineering, Federal Armed Forces University, Munich, Germany).

TheDiaS (Therapy and Diagnostics of Speech) has been developed for therapy and diagnosis of central speech disorders. Individual components of speaking can be objectively differentiated and evaluated. Specially designed training procedures have been developed to improve the impaired speech–motor components.

PhonX (Phonetic eXperimentation) (J. Teiwes & W. Ziegler) provides an experimental toolbox to investigate the different processes hypothesized to be relevant in current models of production and comprehension of spoken language. The "Listening" component contains tools for devising perceptual experiments in the acoustic domain. The "Speaking" component enables users to design repetition, reading, and naming tasks. PhonX can be used for experimental purposes as well as in standard clinical examinations of aphasia, apraxia of speech, dysarthria, or dementia.

The ***TUS*** (Testing Unit for Speech Disorders) (T. Ahrndt & W. Ziegler) is a tool to evaluate central speech impairments and to train patients with speech disorders. Speech can be recorded and replayed with a focus on standard clinical applications such as intelligibility testing, auditory rating scales of dysarthria, and syllable repetition tasks.

The patients' responses are saved in a database, which allows further statistical processing and graphical output.

Hardware requirements: IBM-compatible PC 486/66, PC card with digital signal processor, AD/DA converting unit, minimum of 4 RAM, 500 MB HD. PhonX and TheDiaS run under Windows, TUS under DOS.

Provider:
esq electronic und software GmbH
Fasanenstr. 106
D-82008 Unterhaching
Germany
Tel/Fax: (+49) (0)89-6116211

Literature:
Ziegler, Wolfram, Vogel, Mathias, Teiwes, Jürgen, Ahrndt, Thomas (in press). Microcomputer-based experimentation, assessment and treatment. In C. Code & M. J. Ball (Eds.), *Instrumental clinical phonetics.* London: Whurr Publishers.

• ***The Computerized Boston*** (language: English) (Chris Code, Manjit Heer, & Matt Schofield)

Software scoring program for the Boston Diagnostic Aphasia Examination (BDAE). It allows assessment results and provides profiles, classifications, and performance printouts. The program provides the possibility of exporting and importing data for use with other databases.

Hardware requirements: DOS-based PC, 640K RAM, monochrome or color monitor.

Provider: The Psychological Corporation (for mailing address, see Section 1).

• *Vigil Continuous Performance Test* (language: English)

Assessment of attention including learning disabilities and dyslexia, for ages 6–90. The test is computer-administered and -scored and can be modified by the clinician and the researcher.

Hardware requirements: IBM-compatible PC, XT, AT or higher; 640K RAM, DOS 3.1 or higher.

Distributed by The Psychological Corporation (for mailing address, see Section 1).

• *Western Aphasia Battery Scoring Assistant* (language: English) (Andrew Kertesz)

A software program that converts subtest scores into the Aphasia Quotient and the Cortical Quotient and reports results as a table, graph, or narrative.

Hardware requirements: DOS-based PC, VGA monitor, 640K RAM, DOS 4.0 or higher.

Provider: The Psychological Corporation (for mailing address, see Section 1).

• *Wiener Test System* (basically language-independent as it is mostly nonverbal; the testing procedures have been translated into 14 languages, including some with special characters such as Japanese, Korean, Russian, Hebrew, and Arabic).

The Wiener Test System is a computer-supported (neuro)psychological diagnostic system including test batteries that measure general and intellectual efficiency (including language comprehension), neuropsychological tests, personality tests, and so on. This system also includes an authoring system that enables the user to design and apply his or her own testing and scoring procedures. Furthermore, hardware such as a test subject's control panel, specially designed keyboards, and other input devices (light pen, etc.) can be obtained.

Hardware requirements: IBM-compatible PC 486/33, 8MB RAM, color monitor, Windows 3.1, additional hardware components such as the subject's control panel. Demo disks for Windows available.

Providers: Dr. G. Schuhfried Ges.m.b.H (for Austria) or Hogrefe (for Germany) (for mailing addresses, see Section 1).

• *WinWege 1996* (language: German)

WinWege is a software system designed for neuropsychological rehabilitation, especially aphasia therapy. Focus is on the four language skills: reading, writing, speaking, and listening. Tasks include vocabulary work such as semantic field matching, syntax exercises, fragmented texts, dialogue routines, dialogues (such as interactive dialogue sequences with scripts), and tasks relating to activities of daily life. The program is designed for multimedia use including audio and video sequences. Self-designed material and exercises can be integrated into the program.

Hardware requirements: Pentium PC100 Mhz with 16MB RAM running MS-Windows 95 or Windows NT 3.51 and Excel 7.0, quadruple CD ROM drive, 17″ color monitor, 32 bit VGA graphic card, 2MB video memory, audio card, HD 1GigaByte, scanner recommended.

Literature: V. M. Roth (Ed.) (1992). *Computer in der Sprachtherapie—Neue WEGE.* Tübingen: Gunter Narr Verlag.

Provider:

CRYSTALS, Computersysteme & Expertensystemberatung

Bruderturmgasse 3

78462 Konstanz

Germany

Tel and Fax: (+49) (0) 7531-15583

CHAPTER 46

Neurolinguistic and Related Journal and Book Resources: A Listing

Brigitte Stemmer[1] and Sieglinde Lacher[2]

[1]Centre de Recherche du Centre Hospitalier Côte-des-Neiges, Montréal, Québec, Canada H3W 1W5 and Lurija Institute for Rehabilitation and Health Sciences at the University of Konstanz, Kliniken Schmieder, Allensbach, Germany; [2]Kliniken Schmieder, Allensbach, Germany

The objective of this chapter is to provide reference material for those new to the field of neurolinguistics. We list journals that regularly publish articles on neurolinguistic issues and books that are directly or indirectly related to neurolinguistics. Due to the large amount of material published and space constraints, the list is not exhaustive. For similar reasons, no evaluations of the quality of the contents is given. In keeping with our policy of providing a "state-of-the-art" handbook, we have restricted ourselves to listing books that were first published since 1990. This does not mean, however, that "older" books are of no relevance. On the contrary, there certainly are contributions prior to this date that are important and will continue to influence current thinking in the field. Here we trust the reader's own initiative to locate these books. Finally, we have included sourcebooks that either are directly related to neurolinguistics or, in a somewhat broader perspective, point the reader to other areas of interest that have an impact on neurolinguistics.

The listing is based on searches in databases such as *PsycLit, Psyndex, Social Sciences Citation Index SSCI, Medline, Science Citation Index,* and *PsychInfo.* University catalogs and publishers' catalogs were also consulted.

Despite our efforts to provide an up-to-date listing, references may have slipped by and new material may have been in press at the time this listing was compiled. We thus encourage the reader to provide us with information on such material, which will be included in a revised version of the *Handbook of Neurolinguistics.*

46-1. JOURNALS

Aphasie und Schlaganfall. Zeitschrift für Rehabilitation und Selbsthilfe. Bundesverband für die Rehabilitation der Aphasiker e.V. (BRA) (Ed.). Wesseling: BRA.

The journal focuses on rehabilitation and addresses those having acquired a speech language impairment, their caregivers, and professionals dealing with the rehabilitation of aphasia.

Aphasiology. Chris Code, Dave Müller, & Robert Marshall (Eds.). London: Taylor and Francis.

The journal is concerned with all aspects of language impairment and related disorders resulting from brain damage. Submissions are encouraged on theoretical, empirical, and clinical topics from any disciplinary perspective. The manuscripts can be experimental, and clinical research papers, reviews, theoretical notes, comments, and critiques are accepted.

Brain. A journal of neurology. W. J. McDonald (Ed.). Oxford: Oxford University Press.

The journal publishes papers on neurology and related clinical disciplines, and on basic neuroscience, including neuropsychology.

Brain and Language. Harry A. Whitaker (Ed.). San Diego: Academic Press.

The journal publishes original research articles, case histories, theoretical articles, critical reviews, historical studies, and scholarly notes relevant to human language or communication and any aspect related to the brain or brain function.

Clinical Aphasiology, Margaret L. Lemme, Jennifer Horner, Michael L. Kimbarow, Linda A. Nicholas, Richard K. Peach, & Connie A. Tompkins (Eds.). Austin: Pro-Ed.

Clinical Aphasiology publishes the yearly proceedings of the Clinical Aphasiology Conference.

Cognitive Neuropsychiatry. Anthony S. David & Peter W. Halligan (Eds.). Hove, UK: Psychology Press, an imprint of Erlbaum (UK) Taylor & Francis.

The journal encourages the exploration, integration, and application of theories, methods, and research findings from related fields of clinical psychiatry, behavioral neurology, and cognitive neuropsychology.

Cortex. Ennio de Renzi (Ed.). Milan: Masson Italia Periodici.

A journal devoted to the study of the interrelations of the nervous system and behavior, particularly as these are reflected in the effects of brain lesions on mental functions.

Developmental Neuropsychology. Dennis L. Molfese (Ed.). Mahwah, NJ: Lawrence Erlbaum.

The journal explores the relationships that exist between brain and behavior across the life span and publishes scholarly papers on the appearance and development of behavioral functions such as language, perception, and cognitive processes as they relate to brain functions and structures.

Human Brain Mapping. Peter T. Fox (Ed.). New York: Wiley-Liss.

The journal focuses on clinical, technical, and theoretical research in the interdisciplinary field of human brain mapping and features research derived from noninvasive brain-imaging modalities used to explore the spatial and temporal organization of the neural systems supporting human behavior.

Journal of Clinical and Experimental Neuropsychology. Louis Costa & Byron P. Rourke (Eds.). Lisse, The Netherlands: Swets & Zeitlinger.

The journal publishes research on the neuropsychological consequences of brain disease, disorder, and dysfunction, and promotes the integration of theories, methods, and research findings in clinical and experimental neuropsychology.

Journal of Communication Disorders. Theodore J. Glattke (Ed.). New York: Elsevier Science.

The journal publishes original articles on topics related to disorders of speech, language, and hearing. The submission of reports of experimental or descriptive investigations, theoretical or tutorial papers, case reports, or brief communications is encouraged.

Journal of Medical Speech-Language Pathology. Leonard L. LaPointe (Ed.). San Diego: Singular Publishing Group.

Focus is on clinical and research articles, notes, tutorials, dialogue, and letters that are relevant to clinicians and researchers interested in human communication and its disorders as it is studied and practiced in a health-care or medical orientation.

Journal of Memory and Language. Gregory L. Murphy (Ed.). San Diego: Academic Press.

The journal publishes original research (empirical and theoretical) in the areas of human memory and language processing.

Journal of Neurolinguistics. John Marshall, & Michel Paradis (Eds.). Amsterdam, New York: Elsevier Science.

The journal is an international forum for the integration of the language sciences and the neurosciences. Contributions from all disciplines from linguistics and cognitive psychology to neurology and neuropsychiatry are invited.

Journal of Psycholinguistic Research. R. W. Rieber (Ed.). New York: Plenum Publishing Corporation.

Papers from disciplines engaged in psycholinguistic research and the study of communicative processes are accepted, including papers dealing with the psychopathology of language and cognition and the neuropsychology of language.

Journal of Speech and Hearing Research (JSHR). Sandra Gordon-Salant, Richard G. Schwartz, & John W. Folkins (Eds.). Rockville, MD: American Speech-Language-Hearing Association.

The journal pertains broadly to studies of the processes and disorders of speech, hearing, and language, and to the diagnosis and treatment of such disorders. Manuscripts can be in the form of reports of original research, single-subject experiments, theoretical, tutorial, or review articles, research notes, and letters to the editor.

Language and Cognitive Processes. Lorraine K. Tyler, Gerry T. Altmann, Kathryn Bock, François Grosjean, & William Marslen-Wilson (Eds.). Hove, UK: Lawrence Erlbaum.

The journal publishes theoretical and experimental research into the mental processes and representations involved in language use. Research papers in experimental and developmental psychology, as well as work derived from linguistics, philosophy, computer science and artificial intelligence, are encouraged. Experimental and observational studies, theoretical discussions, short notes and replies, and review articles are accepted.

Laterality. Philip Bryden, Michael Corballis, & Chris McManus (Eds.). East Sussex, UK: Lawrence Erlbaum.

The journal's principal interest is in the psychological, behavioral, and neurological correlates of lateralization. Contributions from any discipline that can illuminate the general problems of the evolution of biological and neural asymmetry, including the linguistic consequences of lateral asymmetry, are also accepted.

Linguistische Berichte. Günter Grewendorf & Arnim von Stechow (Eds.). Opladen: Westdeutscher Verlag GmbH.

The journal publishes manuscripts dealing with theoretical and applied linguistics, including neurolinguisitics, and publishes in German, English, and French.

Neurocase. Ian H. Robertson, John R. Hodges, & H. Branch Coslett. Oxford: Oxford University Press.

The journal is a rapid-publication journal of adult and child case studies in neuropsychology, neuropsychiatry, and behavioral neurology. The journal considers novel reports, replication reports, and reviews for publication.

NeuroImage. Arthur W. Toga, Richard S. J. Frachowiak, & John C. Mazziotta (Eds.). San Diego: Academic Press.

The journal publishes reports of imaging and mapping strategies to study brain structure and function, from the whole brain to the tissue.

Neurolinguistik. Zeitschrift für Aphasieforschung und -therapie. G. Blanken & E. G. de Langen (Eds.). Freiburg: HochschulVerlag.

The journal focuses on all aspects of aphasia research and therapy.

Neuropsychologia. G. Berlucchi (Ed.). Oxford: Elsevier Science.

The journal accepts research reports and reviews dealing with the neural bases of cognition and behavior.

Neuropsychologia Latina. Jordi Peña-Casanova, André Roch Lecours, María Alice de Mattos Pimenta, Alfredo Ardila, & Gabriele Miceli (Eds.). Madrid: Editorial Garsi (Grupo Masson).

The journal is concerned with all aspects relevant to clinical and theoretical neuropsychology and publishes in Portuguese, Spanish, and French.

Neuropsychology. Laird S. Cermak (Ed.). Washington: American Psychological Association.

The journal focuses on (a) basic research, (b) basic and applied research, and (c) improved practice in the field of neuropsychology. Submissions of human experimental, cognitive, and behavioral research with implications for neuropsychological theory and practice are encouraged.

Revue de Neuropsychologie. Michel Habib & Eric Sieroff. Marseille: ADRSC.

The journal publishes papers relevant to all aspects of neuropsychology.

Zeitschrift für Neuropsychologie. (Organ der Gesellschaft für Neuropsychologie). W. Hartje (Ed.). Bern: Hans Huber.

The journal publishes review articles and empirical and clinical research articles dealing with all aspects of neuropsychology, including experimental and clinical neurosciences, neurology, neurophysiology, neuropharmacology, neurobiology, neuroanatomy, and neurolinguistics.

46-2. BOOKS

Altmann, Gerry (Ed.). (1990). *Cognitive models of speech processing: Psycholinguistic and computational perspectives.* Cambridge: MIT Press.

The book's 23 chapters review concerns in speech and language processes. The topics range from lexical access and the recognition of words in continuous speech to syntactic processing and the relationship between syntactic and intonational structure.

Beeman, Mark, & Chiarello, Christine (Eds.) (in press). *Right hemisphere language comprehension: Perspectives from cognitive neuroscience.* Hove, UK: Lawrence Erlbaum.

The book consists of 15 chapters presented in three sections. Section 1 deals with decoding speech sounds and individual words covering aspects such as the neurology of consonant perception, the corpus callosum and language, word recognition, the visual lexicon, and reading and the right hemisphere. Section 2 discusses lexical and sentence-level semantics, including topics such as codes of meaning, evidence of language comprehension from sentence priming, and asymmetries in language comprehension and in high-dimensional space. Section 3 deals with discourse processing and problem solving and covers coarse semantic coding and discourse comprehension, verbal aspects of emotional communication, deficits in inference and social cognition, the interpretation of narrative discourse in a multilevel discourse model, and right-hemisphere contributions to creative problem solving.

Benson, Frank D., & Ardila, Alfredo (1996). *Aphasia: A clinical perspective.* **Oxford: Oxford University Press.**

This book covers 21 chapters organized in four parts dealing with assessment techniques, linguistic analyses, aphasia classification, and related disorders such as alexia, agraphia, acalculia, and anomia. The book ends with a discussion on rehabilitation and recovery.

Blanken, Gerhard (Ed.). (1991). *Einführung in die linguistische Aphasiologie. Theorie und Praxis.* **Freiburg: HochschulVerlag.**

In 11 chapters, this book introduces various aspects of aphasiology, including phonemic disorders in aphasia, apraxia in aphasia, neurolinguistic bases of language automatisms, agrammatism, paragrammatism, impairment of text processing, dysgraphia and dyslexia, nonverbal communication in aphasia, and aphasia therapy.

Blomert, Leo (1994). *Assessment and recovery of verbal communication in aphasia.* **Zutphen: Koninklijke Wöhrman.**

In 6 chapters, the author discusses traditional assessment strategies of verbal communication in aphasia, the Amsterdam-Nijmegen Everyday Language Test and its sensitivity for changes over time, prognosis for recovery from aphasia, initial severity and outcome of recovery, and test validity and interpretation of recovery.

Bloom, Ronald L., Obler, Loraine K., De Santi, Susan, & Ehrlich, Jonathan S. (Eds.). (1994). *Discourse analysis and application. Studies in adult clinical populations.* **Hillsdale, NJ: Lawrence Erlbaum.**

The objective of the 13 contributions in the book are *(a)* to examine current theoretical issues concerning models of discourse processing via study of their breakdown resulting from damage to the central nervous system, *(b)* to investigate what discourse analysis contributes to our understanding of the linguistic, cognitive, and social aspects of disorder in different clinical populations, *(c)* to explore the clinical utility of discourse analysis in different adult populations, and *(d)* to provide readers with a broad range of methodological approaches to evaluate discourse in adults.

Bouton, Charles P. (1991). *Neurolinguistics: Historical and theoretical perspectives.* **(Trans. Terence MacNamee). New York: Plenum Press.**

The book's 12 chapters discuss the relationship between the human body and language, including major sections on the prehistory of linguistics, the methodology of body and mind, from the realm of words to the realm of objects, and the birth of neurolinguistics.

Brownell, Hiram H., & Joanette, Yves (Eds.) (1993). *Narrative discourse in neurologically impaired and normal aging adults.* **San Diego: Singular Publishing Group.**

Fifteen chapters are organized around four topics: (1) ''Theoretical considerations'' covering theoretical aspects of narrative discourse theories; (2) ''Narrative capacities of healthy elderly subjects'' discussing intra-individual changes in text recall, the impact of working memory capacity on narrative tasks, and sociolinguistic and comparative aspects; (3) ''Narrative capacities of patients with focal brain damage,'' including comprehension abilities of aphasic patients, discourse production patterns of various brain-damaged individuals, contextual and thematic influences on narrative comprehension of left- and right-brain-damaged patients, narrative expressive deficits associated with other cognitive abnormalities, and the application of discourse models to narrative discourse production; and (4) ''Narrative capacities of patients with dementia'' reviewing the narrative abilities of demented populations.

Bryan, Karen L., & Maxim, Jane (1996). *Communication disability and psychiatry of old age.* **London: Whurr Publishers.**

The book discusses the management of communication disorders in the psychiatry of old age and describes in detail language change and language assessment in psychiatric disorders associated with old age, especially the dementias.

Caplan, David (1992). *Language: Structure, processing, and disorders.* **Cambridge: MIT Press.**

The book's 10 chapters describe the linguistic and psycholinguistic bases of aphasias that are a result of acquired neurological disease. It discusses topics such as the recognition and production of spoken words, the meaning of words, reading and writing single words, recognizing and producing morphologically complex words, sentence comprehension and production, the comprehension and production of discourse, and notes on diagnosis and treatment of language disorders.

Caramazza, Alfonso (Ed.) (1990). *Cognitive neuropsychology and neurolinguistics: Advances in models of cognitive function and impairment.* **Hillsdale, NJ: Lawrence Erlbaum.**

The 8 chapters in the volume cover cognitive neuropsychological research carried out in the United States and include topics such as number production, phonological deficits in aphasia, sentence parsing, sentence comprehension in aphasia, short-term memory and language processing, image generation, distributed networks and selective attention, and effects of attentional deficits on reading and spelling.

Carbonnel, Serge, Gillet, Patrice, Martory, Marie-Dominique, & Valdois, Sylviane (Eds.) (1996). *Approche cognitive des troubles de la lecture et de l'écriture chez l'enfant et l'adulte.* **Marseilles: Solal Éditeurs.**

The book covers 22 chapters organized in four parts dealing with anatomical-functional aspects of written language, impairments in acquiring written language, dyslexia and dysgraphia in adults, and rehabilitation of written language impairments.

Carlomagno, Sergio (1994). *Pragmatic approaches to aphasia therapy.* **London: Whurr Publishers.**

The book focuses on the use and efficacy of PACE (Promoting Aphasics' Communicative Effectiveness) methodology.

Code, C., & Müller, D. J. (1995). *Treatment of aphasia: From theory to practice.* **London: Whurr Publishers.**

The book presents state-of-the art papers on the treatment of aphasia.

Code, C., & Müller, D. J. (Eds.). (1996). *Forums in clinical aphasiology.* **London: Whurr Publishers.**

The book brings together a selection of the 13 Clinical Forum features from the journal *Aphasiology.*

Code, Chris, Wallesch, Claus-W., Joanette, Yves, & Roch Lecours, André (Eds.). (1996). *Classic cases in neuropsychology.* **Hove, UK: Psychology Press.**

The book discusses classic cases in language, memory, perception, attention, and praxis.

Crosson, Bruce (1992). *Subcortical functions in language and memory.* **New York: Guilford Press.**

Part I of the book deals with subcortical structures in language and the four chapters cover subcortical neuroanatomy and the role of the basal ganglia and the thalamus in language, as well as theories of subcortical functions. Part II deals with subcortical structures in memory covering subcortical neuroanatomy, the role of the diencephalon, and the basal forebrain and the basial ganglia in memory, and ends with a discussion of subcortical functions and memory theories.

Davidson, Richard J., & Hugdahl, Kenneth (1995). *Brain asymmetry.* **Cambridge: MIT Press.**

The principal aim of the book is to survey research on brain asymmetry in its widest context, from sociocultural influences to basic physiology. The book covers 23 chapters divided into nine parts: historical overview; phylogenetic antecedents and anatomic bases; perceptual, cognitive, and motor lateralization; attention and learning; central-autonomic integration; emotional lateralization; interhemispheric interaction; ontogeny and developmental disabilities; and psychopathology.

Eling, Paul (Ed.) (1994). *Reader in the history of aphasia: From [Franz] Gall to [Norman] Geschwind.* **Amsterdam: John Benjamins.**

The book provides access to the most important passages of influential classical papers on aphasiology and aims at presenting the original text in a readable format. The following authors are included in the book: Franz Joseph Gall, Paul Broca, Carl Wernicke, Henry Charlton Bastian, John Hughlings Jackson, Sigmund Freud, Jules Dejerine, Pierre Marie, Arnold Pick, Henry Head, Kurt Goldstein, and Norman Geschwind.

Ellis, Andrew W. (1993). *Reading, writing, and dyslexia.* **Hove, UK: Lawrence Erlbaum.**

The 8 chapters of the book deal with historical aspects and the nature of writing, skilled word recognition, models of word recognition, acquired dyslexias, words in combinations, and writing and spelling.

Ender, Uwe F. (1994). *Sprache und Gehirn. Darstellung und Untersuchung der linguistischen Aspekte des Verhältnisses von Sprache und Gehirn. Patholinguistica* **(Vol. 16). München: Wilhelm Fink Verlag.**

The author discusses aspects of hemispheric asymmetry in relation to phonetics and phonology, lexicon and semantics, syntax and grammar, emotion and consciousness, space and gestalt, and music.

Fasotti, Luciano (1992). *Arithmetical word problem solving after frontal lobe damage: A cognitive neuropsychological approach.* **Amsterdam: Swets & Zeitlinger.**

Using an information-processing model of mathematical word problem solving as a framework, the author distinguishes encoding of the problem and search for a solution in the problem-space as two main solution stages. The author reports on several studies that indicate that the first of these stages can be severely impaired in patients with frontal lobe lesions. Implications of these findings for treatment are discussed.

Gathercole, Susan E., & Baddeley, Alan D. (1993). *Working memory and language.* **Hove, UK: Lawrence Erlbaum.**

The book evaluates the role of working memory in five aspects of language processing: vocabulary acquisition, speech production, reading development, skilled reading, and comprehension.

Goodglass, Harold (1993). *Understanding aphasia.* **San Diego: Academic Press.**

The book is divided into 13 chapters covering the nature and scope of the problem of aphasia, the history of aphasia, the anatomy of language, disorders of motor speech implementation, word retrieval, syntax and morphology, auditory comprehension, repetition, reading and writing, as well as classification of aphasia and the related phenomenon of apraxia, and finally the relation of aphasia to normal language.

Hagoort, Peter (1990). *Tracking the time course of language understanding in aphasia.* **Zutphen, The Netherlands: Koninklijke Wöhrmann.**

The book's six chapters discuss on-line studies and their implications for the investigation of the processing of lexical, syntactic, and semantic meaning in aphasia. The author concludes with a discussion on the nature of the processing deficit observed in aphasia.

Hartley, Leila L. (1994). *Cognitive-communicative abilities following brain injury: A functional approach.* **San Diego: Singular Publishing Group.**

The text focuses on head-injured adults in the later stages of recovery and discusses topics such as a model for understanding social communication following brain injury, functional approaches to assessment and treatment of cognitive-communicative abilities, and rebuilding cognitive–communicative competence and community integration.

Hawellek, Barbara (1992). *Sprachorganisation im Gehirn bei mehrsprachigen Personen (deutsch/polnisch).* **Hamburg: Verlag Dr. Kovac.**

The author discusses the organization of language in multilingual individuals based on empirical investigations of multilingual (German/Polish) aphasics.

Heilman, Kenneth M., & Valenstein, Edward (Eds.). (1993). *Clinical neuropsychology* **(3rd ed.). New York: Oxford University Press.**

This books deals with all aspects of neuropsychology, including neurolinguistics.

Helm-Estabrooks, Nancy, & Albert, Martin L. (1991). *Manual of aphasia therapy.* **Austin, Tx: Pro-Ed.**

The authors describe day-to-day concerns with aphasia in 21 chapters, organized into five sections, dealing with foundations of aphasia rehabilitation, diagnostic processes, implementation of aphasia therapy and measurement of its effects, specific therapy programs, and the impact of aphasia on the patient and family.

Hillert, Dieter (Ed.). (1994). *Linguistics and cognitive neuroscience: Theoretical and empirical studies on language disorders.* **(Linguistische Berichte, Sonderheft 6). Opladen: Westdeutscher Verlag.**

This special issue provides a mixture of state-of-the-art research and new experimental findings investigating grammatical and semantic processing in normal and neurologically impaired speakers/hearers (i.e., fluent and nonfluent aphasics).

Holland, Audrey L., & Forbes, Margaret M. (Eds.). (1993). *Aphasia treatment: World perspectives.* **New York: Chapman & Hall.**

The book investigates approaches to and various aspects of aphasia therapy in a number of different countries. Sociolinguistic and psychosocial aspects, as well as a multidisciplinary approach, are also discussed.

Honeck, Richard P. (Ed.) (1996). *Figurative language and cognitive science.* **Hove, UK: Lawrence Erlbaum. (A Special Issue of** *Metaphor and Symbolic Activity***).**

The articles focus on experimental psychological, neuropsychological, and computer-based approaches to figurative language.

Hulme, Charles, & Snowling, Margaret (Eds.). (1994). *Reading development and dyslexia.* **London: Whurr Publishers.**

The book provides a collection of papers in the field of reading development and dyslexia presented at the Third International Conference of the British Dyslexia Association, April 1994.

Hüttemann, Joachim (1990). *Sprachstörung und Kommunikation aus handlungstheoretischer Sicht. Diagnostische und therapeutische Aspekte eines handlungsorientierten Konzepts zur Patholinguistik.* **Tübingen: Gunter Narr Verlag.**

The author argues for a theoretical framework of speech language pathology that is based on action theory, including aspects of psycholinguistics, the social sciences, and ecological validity. The theoretical framework is applied to an empirical study investigating quantitative and qualitative aspects of therapeutical interaction.

Joanette, Yves, & Brownell, Hiram H. (Eds.) (1990). *Discourse abilities and brain damage: Theoretical and empirical perspectives.* **New York: Springer Verlag.**

The book aims to illustrate in detail the nature, relevance, and importance of the effect of brain damage on discourse ability. A total of 11 chapters are covered in two sections. Section 1 provides a theoretical framework for analyzing levels of discourse in brain-damaged populations, including related fields such as linguistics, pragmatics, philosophy, and cognitive psychology. Section 2 focuses on empirical work and illustrates how neurolinguistic studies have made use of the theoretical issues raised in the first section.

Joanette, Yves, Goulet, Pierre, & Hannequin, Didier (1990). *Right hemisphere and verbal communication.* **New York: Springer Verlag.**

The main objective of the book is to present a critical review of the issues and facts relating to the contribution of the right hemisphere to verbal communication. In seven chapters, issues such as historical questions, methodological and conceptual limitations, aphasia and the right hemisphere, the contributions of the right hemisphere to lexical semantics, right hemisphere and written language, prosodic aspects of speech, and pragmatic issues are discussed.

Joshi, R. Malatesha, & Leong, Che Kan (Eds.). (1993). *Reading disabilities: Diagnosis and component processes.* **Dordrecht: Kluwer Academic Publishers.**

The 18 chapters of the book cover topics such as differential diagnosis of reading disabilities, access to language-related component processes, and reading/spelling strategies.

Kayser, Hortencia (1995). *Bilingual speech-language pathology: An Hispanic focus.* **San Diego: Singular Publishing Group.**

Part I of the book discusses speech and language development, disorders, assessment, and intervention, Part II deals with assessment issues, and Part III draws conclusions and points out research needs.

Lafond, Denise, DeGiovani, René, Joanette, Yves, Ponzio, Jacques, & Taylor Sarno, Martha (Eds.). (1992). *Living with aphasia: Psychosocial issues.* **San Diego: Singular Publishing Group.**

The book covers a broad perspective of the person with aphasia, the physical experience, psychological affects of aphasia, treatment; and aphasia and the family and society.

Lees, Janet (1993). *Children with acquired aphasias.* **London: Whurr Publishers.**

The book focuses on acquired childhood aphasias, its different causes and effects. Part I covers aphasias of traumatic origin and Part II aphasias due to convulsive activity. Details of assessment procedures, recovery, and management strategies are also given.

Lesser, Ruth, & Milroy, Lesley (1993). *Linguistics and aphasia: Psycholinguistic and pragmatic aspects of intervention.* **London: Longman Group.**

The main purpose of the book is to investigate the impact of recent developments in psycholinguistics and pragmatics on aphasia research and on the practical management of aphasia. The book includes 13 chapters, which are divided into three parts: (1) "The Background" focuses on the causes and nature of aphasia, its recovery, and approaches to therapy; (2) "Models and Methods" examines the role of linguistics in aphasia research and therapy, the application of psycholinguistic models to lexical and sentence processing, pragmatic issues and different aspects of discourse; and (3) "Implications and Applications" discusses the application of psycholinguistics and pragmatics in intervention.

Lezak, Muriel Deutsch (1995). *Neuropsychological assessment* **(3rd ed.). New York: Oxford University Press.**

This book discusses in 8 chapters the theory and practice of neuropsychological assessment. Another 12 chapters are devoted to describing individual tests, testing procedures, and assessment techniques for different brain functions.

Lubinski, Rosemary (Ed.) (1995). *Dementia and communication.* **San Diego: Singular Publishing Group.**

The purpose of the book is to provide a state-of the-art perspective on dementia and communication. The book contains 17 chapters, which cover four main topics: (1) "Bases for communicative changes in dementia" focuses on the scope and definition of dementia and its physiologic, cognitive, auditory, and motor-speech characteristics; (2) "Language changes and dementia," explores the impact of aging and dementia on language and discourse abilities in demented individuals; (3) "Social impact of dementia" investigates the impact of dementia on the individual and family; and (4) "Diagnostic and rehabilitative considerations" concerns differential diagnosis and assessment, as well as management and professional services.

Mattingly, Ignatius G., & Studdert-Kennedy, Michael (Eds.). (1991). *Modularity and the motor theory of speech perception: Proceedings of a conference to honor Alvin M. Liberman.* **Hillsdale, NJ: Lawrence Erlbaum.**

The book's contributors discuss in a controversial manner Liberman's views on speech perception and their implications not only for speech perception and production but also for related areas such as the production and perception of sign language, perception in nonhuman animals, lipreading, language acquisition, sentence processing, reading, and learning to read.

McCarthy, Rosaleen A., & Warrington, Elizabeth K. (1990). *Cognitive neuropsychology: A clinical introduction.* **San Diego: Academic Press.**

Besides covering a broad area of neuropsychological issues (object and face recognition, spatial perception, voluntary action, memory, and problem solving), seven chapters are devoted to neurolinguistic issues such as auditory word comprehension, word retrieval, sentence processing, speech production, reading, spelling and writing, and calculation.

Mellies, Rüdiger, Ostermann, Frank, & Winnecken, Andreas (Eds.). (1990). *Beiträge zur interdisziplinären Aphasieforschung. Sprachtherapie 3. Arbeiten zum Workshop „Klinische Linguistik II".* **Tübingen: Gunter Narr Verlag.**

The book contains a collection of 10 articles ranging from investigating ERP studies and language and the study of pragmatic aspects of language to the role of computers in aphasia diagnosis and therapy.

Miller, Joanne L., & Eimas, Peter D. (Eds.). (1995). *Speech, language, and communication.* **San Diego: Academic Press.**

The 10 chapters of the book cover issues of representation in psycholinguistics, speech production and perception, spoken and visual word recognition, sentence production and comprehension, language acquisition, the neurobiology of language, and pragmatics and discourse.

Minifie, Fred D. (Ed.). (1994). *Introduction to communication sciences and disorders.* **San Diego: Singular Publishing Group.**

An introductory text that covers topics such as communication and its disorders in a multicultural society, in children, and in adults; the neurological bases; scientific substrates; voice disorders; fluency and stuttering; and hearing disorders.

Murdoch, Bruce E. (1990). *Acquired speech and language disorders: A neuroanatomical and functional neurological approach.* **London: Chapman & Hall.**

The author aims at providing a link between neuroanatomical and neurological knowledge on the one hand, and specific neurologically based communication disorders on the other hand. The 11 chapters cover subjects such as neuroanatomy, aphasia syndromes, speech-language disorders related to traumatic head injury, right-hemisphere lesions and dementia, neurological disturbances related to aphasia, dysarthrias, and acquired childhood aphasia.

Nespoulous, Jean-Luc, & Villiard, Pierre (Eds.). (1990). *Morphology, phonology, and aphasia.* **New York: Springer Verlag.**

The 16 chapters of the book are the result of the authors' efforts to bring together theoretical constructs of general linguistics and aphasiology with a focus on morphology and phonology, and across languages.

Ohlendorf, Ingeborg M., Pollow, Thomas A., Widdig, Walter, & Linke, Detlef B. (Eds.). (1994). *Sprache und Gehirn. Grundlagenforschung für die Aphasietherapie.* **Freiburg: HochschulVerlag.**

This Festschrift in honor of Anton Leischner's 85th birthday discusses, in 18 contributions, a variety of topics related to aphasiology.

Paradis, Michel (Ed.). (1995). *Aspects of bilingual aphasia.* **Oxford: Elsevier Science.**

Besides practical issues pertinent to the assessment, diagnosis, prognosis, and therapy of bilingual aphasia, the book discusses theoretical issues of the organization of two or more languages in the brain. Case studies are presented and the patients reported on speak a variety of languages.

Pulvermüller, Friedeman (1990). *Aphasische Kommunikation. Grundfragen ihrer Analyse und Therapie.* **Tübingen: Gunter Narr Verlag.**

The author argues for an approach to aphasia guided by communicative and pragmatic aspects. He discusses topics such as brain and communication, the analysis of aphasic communication, traditional aphasia therapy, and bases for a communicative approach to aphasia therapy.

Sloan Berndt, Rita, & Mitchum, Charlotte C. (Eds.). (1995). *Cognitive neuropsychological approaches to the treatment of language disorders.* **(Special issue of** *Neuropsychological Rehabilitation*). **Hove, UK: Lawrence Erlbaum.**

The book reports six original treatment studies, targeting symptoms of word retrieval, sentence comprehension, reading aloud, and written spelling.

Snowling, Margaret, & Stackhouse, Joy (1996). *Dyslexia, speech and language: A practitioner's handbook.* **London: Whurr Publishers.**

The book is divided into four main sections, which cover the theoretical framework, assessment, and treatment; it ends with an overview.

Spitzer, Manfred, Uehlein, Friedrich, Schwartz, Michael A., & Mundt, Christoph (Eds.). (1992). *Phenomenology, language and schizophrenia.* **New York: Springer Verlag.**

The contributions in the book investigate schizophrenia from a clinical and philosophical, as well as cognitive and language, perspective. The nine chapters dealing with language and cognition cover aspects such as *Zerfahrenheit* and incoherence, word associations in experimental psychiatry, language planning and alterations, thought disorders, and some historical and cognitive issues.

Spreen, Ottfried, & Strauss, Esther (1991). *A compendium of neuropsychological tests.* **Oxford: Oxford University Press.**

The book provides an exhaustive listing of assessment procedures used in neuropsychology (and neurolinguistics) and discusses its applications, research findings, and advantages and disadvantages.

Tompkins, Connie A. (1995). *Right hemisphere communication disorders: Theory and management.* **San Diego: Singular Press.**

The book is intended to help readers ascertain what is known and what is not known about right-hemisphere communication disorders and the processes presumed to underlie them. It aims at illustrating how to generate hypotheses about evaluation and management based on theory, data, logic, and patients' communicative needs, and to help clinicians evaluate the literature contributing to these hypotheses, and their own treatment plans and procedures. The book covers eight chapters. Chapter 1 introduces the book and the population, including a discussion on epidemiology, demographics, and etiology; presumed localization and nature of selected neuropsychological and cognitive functions, and the generality of communicative disorders after right-hemisphere damage. Chapter 2 reviews selected symptoms and current thinking about explanations of these symptoms. Chapters 3 and 4 focus on appraisal, evaluation, and diagnosis. Chapter 5 discusses prognosis, recovery, treatment efficacy, and outcome, Chapter 6 treatment principles and considerations, and Chapter 7 treatment approaches and strategies. The last chapter focuses on patient management.

Turner, Martin (1996). *The psychological assessment of dyslexia.* **London: Whurr Publishers.**

The author reviews the major tests in use for children and adults.

Tyler, Lorraine K. (1992). *Spoken language comprehension: An experimental approach to disordered and normal processing.* **Cambridge: MIT Press.**

In 15 chapters, organized into five parts, the author describes the psycholinguistic approach to the study of language disorders and in-depth analyses of individual aphasic patients. The topics covered include a psycholinguistic processing approach to the study of aphasia, the process of contacting form representations, accessing lexical content, constructing higher-level representations, and processing morphologically complex words in utterances.

Vogel, Deanie, & Carter, John E. (1994). *Effects of drugs on communication disorders.* **San Diego: Singular Publishing Group.**

The book discusses causes and symptoms of disorders, agents used to treat the disorder, the desirable and undesirable effects of the drugs, and alternatives to medical management.

Warnke, Andreas (1990). *Legasthenie und Hirnfunktion. Neuropsychologische Befunde zur visuellen Informationsverarbeitung.* **Bern: Verlag Hans Huber.**

The author empirically pursues the question to what degree severe dysgraphia in intellectually normal children influences the processing of visual information.

Willows, Dale M., Kruk, Richard, & Corcos, Evelyne (Eds.). (1993) *Visual processes in reading and reading disabilities* **(2nd ed.). Hove, UK: Lawrence Erlbaum.**

The book brings together a broad range of evidence concerning the role of visual information in reading and reading disabilities. Part I covers background information, Part II the neuropsychological bases of visual processes, Part III visual processes in reading, Part IV visual factors in reading disabilities, Part V parameters affecting visual processing and part VI provides a conclusion and future directions.

46-3. SOURCEBOOKS

Ayd, Frank J., Jr., (1995). *Lexicon of psychiatry, neurology, and the neurosciences.* **Baltimore: Williams & Wilkins.**

Blanken, Gerhard, Dittmann, Jürgen, Grimm, Hannelore, Marshall, John C., & Wallesch, Claus-W. (Eds.). (1993). *Linguistic disorders and pathologies: An international handbook.* **Berlin: Walter de Gruyter.**

The handbook consists of 87 chapters divided into five parts covering language use in normal speakers and its disorders, acquired organic pathologies of language behavior (neurolinguistic and neurophonetic disorders), pathologies of language use in psychiatric disorders, and pathologies and disorders of language development.

Boller, François, & Grafman, Jordan (Eds.). (1988–1995). *Handbook of neuropsychology.* **(Vols. 1–10). Amsterdam: Elsevier Science Publishers.**

Essential reference source to all topics related to neuropsychology. Several volumes contain sections on neurolinguistics: *Language & Aphasia Parts 1 and 2* (Vol. 1 & 3, Section 3, edited by H. Goodglass) and *Child Neuropsychology: Language and Its Disorders* (Vol. 7, Section 10, Part 8, edited by S. J. Segalowitz and I. Rapin).

Dittmann, Jürgen, & Tesak, Jürgen. (1993). *Studienbibliographien Sprachwissenschaft.* **Vol. 8,** *Neurolinguistik.* **Heidelberg: Julius Groos Verlag.**

This reference source contains an introduction to various topics related to neurolinguistics (Part I) (neurolingusitics, language and the brain, aphasia, dementia, schizophrenia, and psychosis), and a bibliography for each of these topics (sorted according to linguistic levels of description and nonlinguistic aspects) is provided in Part II.

Gazzaniga, Michael S. (1995). *The cognitive neurosciences.* **Cambridge: MIT Press.**

The book covers a wide spectrum of the cognitive neurosciences, starting at the molecular level and continuing right up to the problems of human conscious experience. The 92 chapters are divided into 11 parts covering topics such as molecular and cellular plasticity, neural and psychological development, sensory systems, strategies and planning, motor systems, attention, memory, language, thought and imagery, emotion, evolutionary perspectives, and consciousness.

Gernsbacher, Morton Ann (Ed.). (1994). *Handbook of psycholinguistics.* **San Diego: Academic Press.**

In 34 chapters, the handbook covers psycholinguistic aspects of research methods; theoretical models; the representation and processing of words, sentences, figurative language, and discourse; memory and language; mental models; language comprehension and production; first and second language acquisition; reading disabilities; language and the brain; and the neuropsychology of language.

Kirshner, J. S. (Ed.). (1995). *Handbook of neurological speech and language disorder: Neurological disease and therapy.* **New York: Marcel Dekker.**

Menn, Lise, & Obler, Loraine K. (Eds.). (1990). *Agrammatic aphasia: A cross-language narrative sourcebook.* **(Vols. 1–3). Amsterdam: John Benjamins.**

The three volumes provide a rich source of language data of aphasic patients in 14 languages and of control subjects. Volumes 1 and 2 contain three sections: orientation (Chapters 1–3), language findings and data (Chapters 4–17), and language comparisons and conclusions (Chapters 18–21). Volume 3 contains control transcripts, that is, the narratives of all control subjects. The core work is to be found in Chapters 4–17, which contain all narrative texts collected from patients in 14 languages, together with analyses and discussions. Chapters 18–21 consist of comparative studies.

Obrzut, John E., & Hynd, George W. (Eds.) (1996). *Neuropsychological foundations of learning disabilities: A handbook of issues, methods, and practice.* **San Diego: Academic Press.**

In 27 chapters, the book offers a comprehensive overview of advances made in the field of learning disabilities, in the service of diagnosis and remediation. The work includes genetic, electrophysiological, and brain-imaging studies, and presents data on cognitive functions, laterality mechanisms, neuropsychological assessment, and subtyping of learning disabilities. Specific neuropsychological syndromes, in such areas as mathematics, reading, and speech and language are also discussed.

CHAPTER 47

Neurolinguistic and Related Resources on the Internet: A Listing

Brigitte Stemmer[1] and Manfred Hild[2]

[1]Centre de Recherche du Centre Hospitalier Côte-des-Neiges, Montréal, Québec, Canada H3W 1W5 and Lurija Institute for Rehabilitation and Health Sciences at the University of Konstanz, Kliniken Schmieder, Allensbach, Germany; [2]Kliniken Schmieder, Allensbach, Germany

The Internet is a decentralized network that connects networks of computers. In the last few years, information that can be obtained from the Internet has grown tremendously and new information is added continuously. Because of the rapid changes and growth of the Internet, it is impossible to give a comprehensive, up-to-date account of all the current resources that are helpful for the reader interested in neurolinguistics and related fields. The main purpose of this section is to provide a variety of information sources that can be used as a basis for continued browsing. Sites directly and indirectly relevant and related to neurolinguistics (such as linguistics, psychology, neuroscience, neurology, and so forth) will be cited. In this context, a word of caution is necessary: addresses which exist today on the Net may have disappeared by tomorrow. Still, we hope that the listing will help the "wired" reader to move around in neurolinguistic and related cyberspace in an efficient and time-saving way.

Although it is beyond the scope of this section to introduce the reader to all the possibilities for navigating the Internet, a brief explanation as to how the listed addresses (so-called *URLs*—Uniform Resource Locators) were obtained may help the novice to the Internet to update information himself or herself.

In order to gather and access information on the Net, a service called the Global Network Navigator organizes information through its World Wide Web (WWW) by weaving a weblike route among individual information entities. For example, if you do a search using the word *neurolinguistics,* a list of words and phrases will appear on the screen, some of which are highlighted, such as *institutes* and *directories.* If you click on the word *institutes,* a list of neurolinguistic institutes appears on the screen; these again are highlighted and thus can be selected (by clicking on the word or phrase) to obtain further information on a particular institute. To access the Global Network Navigator, Internet access providers offer "web browsers" such as Netscape Navigator or MS Explorer.[1] Each Web destination has its own home address, a "home page." To travel to this home page, you need an address called URL (Uniform Resource Locator). On some servers, the URL is case-sensitive, that is, you need to use the exact spelling. If you do not know the URL, you can use a search engine, which will search for the address and connect you to the home page. The following are some common search engines and their URLs:

alta vista	http://www.altavista.com/
yahoo!	http://www.yahoo.com/
lycos	http://www.lycos.com/
webcrawler	http://wc3.webcrawler.com/

All Internet addresses listed in this chapter were found by using such a search engine. Once you have typed in the item you want to do a search on, the search engine browses the Web and an index of search-related items appears on the screen, which can then be used again as links to information related to that particular item. The information we provide in the listing is mostly restricted to first- and second-level links as space does not permit a listing of all follow-up links. Once the reader has accessed the address, however, s/he can easily follow up on links of her or his own interest.

Throughout the listing, the following notation conventions will be observed: The name of the major site is given first (denoted by a • and underlined). This is followed by its URL.[2] If the home page of the site provides further links that we find helpful, we have listed those links, or a selection thereof. An → will denote that the item following the arrow is a selectable link. Links are separated from each other by |. Deeper-level links will be cited using ⇒. Items that appear in square brackets [ ] usually show menu items displayed by the Web site. These items can be selected just like link items.

The chapter is organized into four main sections: (1) Libraries and Bookstores, (2) Publishers and Journals, (3) Newsgroups and Mailing Lists, and (4) Other useful URLs.

47-1. LIBRARIES AND BOOKSTORES

• *Association of Research Libraries (ARL)*
    http://arl.cni.org/pubscat/index.html (publication catalog index)
    http://arl.cni.org/scomm/edir/ (journals and newsletters)
    http://www.n2h2.com/KOVACS (discussion lists)

---

[1]In order to use Netscape or MS Explorer you will need access to a server via SLIP or PPP. If your site does not support SLIP or PPP, there are nevertheless alternatives to browse the Web and you should inquire at your computer department. On most UNIX servers a program called "lynx" can be used. At the UNIX prompt, type in the word *lynx* follwed by the enter/return key. If you know the URL, type in *lynx URL.* For example: *lynx http://www.excite.com* will call the search engine "Excite," which you can use to browse the WEB in lynx. Another possibility is to obtain "Slipknot," a shareware program that does not need SLIP or PPP and can be used very similarly to Netscape or MS Explorer. Slipknot can be obtained at *http://plaza.interport.et/slipknot/whatsnew.html*

[2]Note that the sometimes complex addresses can be saved as a bookmark and thus only need to be clicked on for retrieval.

The Association of Research Libraries is a not-for-profit membership organization including 120 libraries of North American research institutions. Its mission is to shape and influence forces affecting the future of research libraries in the process of scholarly communication. For further information: Patricia Brennan, Information Services Coordinator, pubs@cni.org.

• *BrainBooks*
    http://www.brainbooks.com/
    → Neurology Books: http://www.brainbooks.com/neurolog.htm
       ⇒ | clinical neurology | clinical neurophysiology | neuroimaging | neurosurgery | neuropharmacology | history of neurology |
    → Neuropsychology Books: http://www.brainbooks.com/neuropsy.htm
       ⇒ | general neuropsychology | consciousness | language |
    → Neuroscience Books: http://www.brainbooks.com/neurosci.htm
       ⇒ | general neuroscience | neuroanatomy | neurochemistry | neurophysiology | sensory physiology |
    → Psychology Books: http://www.brainbooks.com/psycholo.htm
       ⇒ | general psychology | psychological testing | psychopathology | experimental psychology | physiological psychology | perception (incl. sensory physiology) | history of psychology |
    → Brain Literature Search: http://www.brainbooks.com/research.htm

• *Institute for Scientific Information (ISI)*
    http://www.isinet.com
The Institute for Scientific Information is a database publishing company that maintains a comprehensive, multidisciplinary, bibliographic database of research information. The ISI database covers more than 16,000 international journals, books, and proceedings in the sciences, social sciences, and arts and humanities, indexing complete bibliographic data, cited references, and author abstracts for every item it includes.

• *International Digital Electronic Access Library (IDEAL)*
    http://www.apnet.com
The library contains journals published by Academic Press. The user can browse the IDEAL database of journal tables of contents, and, for a limited time, search and browse abstracts free of charge.
    → http://www.apnet.com/www/ap/whatsnew.htm
This page lists the journals that are available on IDEAL as well as the date of the most recent issue and is updated as new material is added to the IDEAL service. An excerpt of on-line issues of journals available as of September 1996 shows: *Brain, Behavior, & Immunity; Brain & Cognition; Brain & Language; Cognitive Psychology; Consciousness & Cognition; Experimental Neurology; Frontiers in Neuroendocrinology; Journal of Experimental Child Psychology; Journal of Memory & Language; Journal of Phonetics; Neurobiology of Learning & Memory; NeuroImage; Seminars in the Neurosciences.*

• *Internet Bookshop Homepage*
    http://www.bookshop.co.uk/
    They claim to be "the largest online bookshop in the world."

• *Internet Grateful Med*
    http://igm.nlm.nih.gov/
The U.S. National Library of Medicine (NLM) offers free access to databases such as MEDLINE, HealthSTAR, PREMEDLINE, AIDSLINE, AIDSDRUGS, AIDSTRIALS, DIRLINE, HISTLINE, HSRPROJ, OLDMEDLINE, SDILINE. When first invoked, Internet

Grateful Med is set to search in MEDLINE.. Select the "Search Other Files" action on the Search Screen to change from MEDLINE to one of the other databases accessible through IGM.

• *World-Wide-Web Virtual Library (WWWVL) URLs: A selected listing*

> WWW VL Academic Publishers
> http://www.edoc.com/jrl-bin/wilma/pac
> WWWVL Applied Linguistics
> http://alt.venus.co.uk/VL/AppLingBBK
> WWWVL Electronic Journals List
> http://www.edoc.com/ejournal/
> The WWWVL Electronic Journals List: Publishing Companies Online
> http://www.edoc.com/ejournal/publishers.html
> WWWVL History of Science, Technology and Medicine—Overview
> http://www.asap.unimelb.edu.au/hstm/hstm_ove.htm
> WWWVL: Libweb—Libraries on the Web
> http://sunsite.berkeley.edu/Libweb
> WWWVL Neuroscience (Biosciences)
> http://neuro.med.cornell.edu/VL/
> | Comprehensive list of Biomedical Sites in WWW Virtual Library | Neuroscience Internet Resource Guide | Neurosciences on the Internet | WWWVL Medicine |
> The Virtual Psychology Library
> http://www-mugc.cc.monash.edu.au/psy/psylinks.html
> The WWWVL Publishers
> http://www.comlab.ox.ac.uk/archive/publishers.html
> | Virtual Library | Electronic Journals | Libraries | Literature | Conferences |
> WWWVL Subject Catalogue
> http://www.w3.org/hypertext/DataSources/bySubject/Overview.html
> Some selected links:
> | Applied Linguistics | Communications | Conferences | Electronic Journals | Languages | Libraries | Linguistics | Medicine | Neurobiology | Philosophy | Publishers | Statistics | Vision | Science | Other virtual libraries |

47-2. PUBLISHERS AND JOURNALS

• *Academic Press—APNet Home Page*

> http://www.apnet.com/
> → | Welcome to APNet | Academic Press | New and Notable | Online Journals | Textbooks | Product Catalogs | Visit AP PROFESSIONAL | What's New at APNet | AP History | Customer Service and Orders | Meeting Place | Other AP Sites | WWW Pocket Directory | Book Updates |
>> [Product Catalogs] [Journal Sites] [IDEAL] [Meeting Place] [New and Notable][Textbooks] [What's New at APNet] [Customer Service] [Registry] [Other AP Sites] [APPNet Home]

• *Cambridge University Press*

> http://www.cup.org/

• *Electronic Journals*

> http://www.livjm.ac.uk/resource/ejournals/ejhome

- *Electronic Journals: Full Text*
  http://www.swan.ac.uk/library/ejnlwww.htm
  Currently available are *(a)* journals provided under the UK Pilot Site Licence (electronic version of printed journals that also exist in hard copy), *(b)* other electronic journals and related material that also exist in hard copy, and *(c)* journals that exist only in electronic form.
  → | How to access electronic journals | What can you access? | Terms and conditions | Downloading and printing |
- *Elsevier*
  http://www.elsevier.com/catalogue/SAC/530/08250/08252/Menu.html
  → | Internet Catalogue Neurolinguistics | Cognitive Science | Neurolinguistics | Journals | Electronic Publications | Books |
  ⇒ Journals
      | Cognition | Journal of Neurolinguistics | Neuroscience Package |
      | Special Issues and Supplements | Cognition (Special Issues) |
      | Speech Communication (Special Issues) |
- *Lawrence Erlbaum Associates (LEA)*
  http://www.erlbaum.com/intro.htm
  http://www.erlbaum.com/inform.htm
  [LEA] [Info Center] [Send E-mail] [Request Catalogs] [Search by Keyword] [Place an Order]
- *Lawrence Erlbaum Associates (LEA) Journals*
  http://www.erlbaum.com/journdis.htm
  Links to journals by disciplines:
  → | Applied psychology | Clinical psychology | Cognitive science | Communication | Computer science | Developmental and lifespan psychology | Experimental psychology | Health and behavioral medicine | Langauge and linguistics | Neuroscience/ neuropsychology |
- *Massachusetts Institute of Technology (MIT)*
  http://www-mitpress.mit.edu
  → | New releases | Books | Journals | Order | Search | Q and A |
  → | About the MIT Press | Staff | News | Electronic books | The MIT Press bookstore | Conference exhibits schedule | Textbook information | Recommended excursions | About this site |
  [New release] [Books 1993–1996] [Journals] [Order] [Search] [Comments/Questions] [About The MIT Press] [Electronic Books] [The MIT Press Bookstore] [Recommended excursions] [Textbook information]
- *Medical & Biological Journal List*
  http://www.dokkyomed.ac.jp/users/dinet/medic/medjrnl.html
  An excerpt of links:
  → | Science | Nature | Nature Japan | American J. Physiology (This Year's Abstracts) | American Physiological Society | Neuronet (Thomson Publishing) | The Federation of American Societies for Experimental Biology | The Journal of Biological Chemistry |
- *Psycoloquy: Psychology—a refereed electronic journal (ISSN 1055-0143)*
  http://www.princeton.edu/~harnad/
  → Science:Psychology:Journals
      | Psycoloquy Journal, Behavioral & Brain Sciences (BBS) |

→ Science:Psychology:Usenet
| sci.psychology.digest–PSYCOLOQUY: Refereed Psychology Journal and Newsletter.
(Moderated) |
- *Singular Publishing Group*
   http://www.singpub.com/
- *Singular Publishing Group: Catalog*
   http://www.singpub.com/cata.html
   → | Speech-Language Pathology |
- *Springer Verlag*
   http://www.springer-ny.com/
   [Home] [Map] [Search] [Comment] [Catalog] [Order] [Customer Service]
   Selected subject areas:
   → | Biology & Biomedicine | Medicine | Psychology | Statistics |
- *Springer Verlag Medicine Page*
   http://www.springer-ny.com/medicine/disc.htm
   [Home] [Map] [Search] [Comment] [Catalog] [Order] [Customer Service]
   → | Books | Journals | Electronic Media | Online Products | Samples & Supplements |
   Instructions for Authors | Upcoming Conferences | Links to Medical Sites |

47-3. NEWSGROUPS AND MAILING LISTS

47-3.1. USENET and Newsgroups

This is a collection of thousands of electronic bulletin boards with topics you can and cannot imagine. These bulletin boards are also known as newsgroups and the discussions range from bizarre encounters to serious academic exchanges. You can just browse these newsgroups, or actively participate in any discussion. Internet service providers offer a "newsreader" and tell you how to use it. If you are using a UNIX system on which PPP or SLIP is not available, usually the command <u>rn</u> or <u>nn</u> at the command prompt will give you access to the newsgroups from which you can choose those you want to read. Because of the large number of newsgroups, we will refrain from citing any here but refer the reader to the newsread browsers.

47-3.2. Mailing Lists—LISTSERV

Unlike newsgroups, which are electronic bulletin boards everybody can just look at and post messages on, mailing lists require that you subscribe to them via a list server in order to get information or participate in a discussion; and some are organized by a moderator.

Information on Listserv
http://www.nova.edu/Inter-Links/listserv/listhlp.html
→ | What is a Listserv? | Subscribing, Unsubscribing, and Posting | Other Listserv
   Commands | Searching List Archives |
Below an excerpt of the deeper-level information on
   ⇒ What is a Listserv, and
   ⇒ Subscribing, unsubscribing, and posting
will be given.

47-3.2.1. What Is a Listserv?

A listserv is a program that maintains one or more mailing lists (i.e., a list server). A listserv automatically distributes an E-mail message from one member of a list to all other members on that list. Listservs maintain thousands of lists in the form of digests, electronic journals, discussion groups, and the like. When you subscribe to a list, your name and E-mail address are automatically added to the list. From that time on, you will receive all mail (postings) sent to the list by its members. You may follow the discussions or join in on them. If you need help on the specific server, you can send an E-mail message to the listserv address. In the text body, type only the word HELP and you will receive an E-mail message that lists commands of and useful information about the specific server.

47-3.2.2. Subscribing, Unsubscribing, and Posting

http://www.nova.edu/Inter-Links/listserv/subscribe.html

47-3.2.2.1. How to Subscribe

To subscribe to a list, send an E-mail message to the listserv address with one line in the body of the letter:
   SUB listname yourfirstname yourlastname
   where listname is the name of the list, and yourfirstname is your firstname and yourlastname is your last name (e.g., SUB PSYCOLOQUY Brigitte Stemmer).
Recently, some list servers have changed the subscription procedure and allow you to simply type the word "subscribe" into the subject or topic header of your E-mail.

47-3.2.2.2. Unsubscribing

To have your name removed from a listserv, send an E-mail message to the listserv address with one line in the body of the letter:
   SIGNOFF listname
   where listname is the name of the list (e.g. SIGNOFF PSYCOLOQUY).

47-3.2.2.3. Posting

If you have an article (comments, questions, etc.) that you wish to distribute to all members of a list, send it as E-mail to the list address for that list. Note that the list address is different from the listserv address.

47-3.3. Searching for Mailing Addresses

Most listserv servers can provide you with up-to-date information on all known mailing lists across the network. The following sites are useful for finding mailing addresses:
   E-mail Discussion Groups: http://www.nova.edu/Inter-Links/listserv.html
   → | Directory of Scholarly E-Lists | Inter-Links Search for Discussion Groups | List Directory of E-mail Discussion Groups | Search the List of Lists | Tile.net List of Internet Discussion Groups | Mail List Manager Commands | Listservs |
      ⇒ Tile.net List of Internet: http://tile.net/listserv/
         | Alphabetical listing by description | Alphabetical listing by name | Alphabetical listing by subject | Grouped by host country |

⇒ Search The List of Lists: http://catalog.com/vivian/interest-group-search.html
If you want to use a file transfer protocol (FTP) to receive an ASCII version of the List of Lists go to ftp://sri.com/netinfo/interest-groups.txt

This directory may also be obtained through E-mail by sending a message to mail-server@sri.com with "SEND interest-groups" in the body of the message. Be aware that this list is very long and will take up a lot of hard disk space.

47-3.4. Selection of Mailing Lists

(For instructions how to subscribe or unsubscribe to these lists, see explanations in the section "Mailing Lists—LISTSERV".)

- **List Name: *ALZHEIMER***
    Subscription Address: MAJORDOMO@WUBIOS.WUSTL.EDU
    Owner: Kathy Mann Koepke <mannkoepkek@neuro.wustl.edu>
    ALZHEIMER is an E-mail discussion group for patients, professional and family caregivers, researchers, public policymakers, students, and anyone with an interest in Alzheimer's or related dementing disorders in older adults.
- **List Name: *AUTINET—Autism & Asperger's Syndrome***
    Subscription Address: AUTINET-request@iol.ie
    Owner: Peter Wise <Autinet.wise@iol.ie>
    AUTINET is an open, unmoderated forum on autism, especially high-function autism (HFA) and Asperger's syndrome.
- **List Name: *BRAIN-L***
    Subscription Address: LISTSERV@listserv.net
    BRAIN-L is a mind-brain discussion group.
- **List Name: *COGSCI***
    Subscription address: LISTSERV@HERAN or LISTSERV@NIC.SURFNET.NL
    COGSCI is an open, unmoderated discussion list about cognitive science.
- **List Name: *DEVELOPMENTAL NEUROPSYCHOLOGY***
    Subscription address: DEVNEUROPSY@DNP.SOMLSII.SIU.EDU
    To subscribe to this list simply place the word
    Subscribe
    in the "Subject" or "Topic" header for your E-mail. The list server will send you back a notice that your message was received as well as a help file.
    Owner of this list: Dennis Molfese
    This list discusses issues in developmental psychology and neuropsychology.
- **List Name: *IAPSY-L@UACSC2.ALBANY.EDU***
    Subscription Address: LISTSERV@UACSC2.ALBANY.EDU
    Owner: Bernardo M. Ferdman <csppbmf@class.org>
    This is the Interamerican Psychologists' List. It is intended to facilitate and encourage communication and collaboration among psychologists throughout the Americas and the Caribbean, and to aid the Interamerican Society for Psychology/Sociedad Interamericana de Psicología in its activities. The languages of the list are English, French, Portuguese, and Spanish (the languages of the ISP).
- **List Name: *Info-Psyling***
    Subscription Address: Info-Psyling@psy.gla.ac.uk
    Owner: Kerry Kilborn <kerry@psy.gla.ac.uk>
- **List Name: *LINGUIST***
    Subscription Address: LISTSERV@TAMVM1.TAMU.EDU

Owner: Anthony Aristar <aristar@tamuts.tamu.edu>
This is a forum for issues relating to the discipline of linguistics and related fields.

- *List Name: Neurology List Server*
  Subscription Address: rivner@emgmhs.mcg.edu
  Owner: Michael Rivner <Rivner@emgmhs.mcg.edu>
  The purpose of this list is to provide a forum for neurologists and other related health-care professionals to discuss topics in neurology.

- *List Name: NEURO1-L*
  Subscription Address: LISTSERV@uicvm.uic.edu
  NEURO1-L is a neuroscience information forum.

- *List Name: NEURON@CATTELL.PSYCH.UPENN.EDU*
  Subscription Address: NEURON-REQUEST@CATTELL.PSYCH.UPENN.EDU
  Owner: Peter Marvit <marvit@cattell.psych.upenn.edu>
  NEURON is a list (in digest form) dealing with all aspects of neural networks (and any type of network or neuromorphic system).

- *List Name: NL-KR@CS.RPI.EDU*
  Subscription Address: NL-KR-REQUEST@CS.RPI.EDU
  Owner: Christopher Welty <weltyc@cs.rpi.edu>
  NL-KR is open to discussion of any topic related to the natural language such as knowledge representation, natural language understanding, discourse understanding, philosophy of language, plan recognition, computational linguistics, cognitive psychology, human perception, linguistics, and so forth.

- *List Name: PSYC@PUCC.PRINCETON.EDU*
  Subscription Address: LISTSERV@PUCC.PRINCETON.EDU
  Owner: Stevan Harnad <psyc@pucc.princeton.edu>
  This list's full name is PSYCOLOQUY. It is a refereed electronic journal sponsored by the American Psychological Association. It contains both newsletter-type materials (announcements, conferences, employment notices, abstracts, queries) and short articles refereed by the Editorial Board, as well as refereed interdisciplinary and international commentaries on the articles "Scholarly Skywriting"). PSYCOLOQUY is also available as the moderated Usenet newsgroup sci.psychology.digest.

- *List Name: PSYGRD-J*
  Subscription Address: LISTSERV@ACADVM1.UOTTAWA.CA
  Owner: Matthew Simpson <054340@acadvm1.uottawa.ca>
  This is a electronic journal in the field of psychology, called The Psychology Graduate Student Journal: The PSYCGRAD Journal (PSYGRD-J). The purpose of the journal is to publish, from the graduate student perspective, professional-level articles in the field of psychology.

- *List Name: PSYCHNEWS INTERNATIONAL (PNI)*
  http://www.cmhc.com/pni/mailing.htm
  Subscription Address: LISTSERV@VM1.NODAK.EDU
  Owner: plaud@badlands.nodak.edu.
  PsychNews International (PNI) is an independent electronic publication and publishes articles on current events in mental health, updates form relevant Internet discussion groups, as well as original research and theoretical articles.

- *List Name: SCHIZ-L@UMAB.UMD.EDU*
  Subscription Address: LISTSERV@UMAB.UMD.EDU
  Owner: Steven Roy Daviss <sdaviss@cosy.ab.umd.edu>
  SCHIZ-L is an unmoderated discussion list devoted to schizophrenia research. The objective

of the list is to provide a forum for communications among researchers and others interested in this mental illness.

- *List Name: SCR-L*

  Subscription Address: LISTSERV@listserv.net

  SCR-L is a discussion group concerned with the study of cognitive rehabilitation and traumatic brain injury.

- *List Name: SEMIOS-L@ULKYVM.LOUISVILLE.EDU*

  Subscription Address: LISTSERV@ULKYVM.LOUISVILLE.EDU

  Owner: Steven Skaggs <S0SKAG01@ULKYVM.LOUISVILLE.EDU>

  SEMIOS-L is a discussion group for those interested in semiotics, verbal and nonverbal communication, language behavior, visual issues, and linguistics.

- *List Name: TBI-SPRT*

  Subscription Address: LISTSERV@listserv.net

  TBI-SPRT is a traumatic brain injury support group list at St. John's University.

47-4. OTHER USEFUL URLS

- *Agenesis of the Corpus Callosum*

  http://www.indiana.edu/pietsch/agenesis01.html

  The site presents a selection of abstracts of scholarly journal articles on the congenital absence of the corpus callosum.

- *BrainMap*

  http://ric.uthscsa.edu/services/

  BrainMap is a software environment for meta-analysis of the human functional brain-mapping literature.

- *Brain Meds*

  http://maui.net/jms/brainuse.html

  The site presents information on brain chemistry.

- *CCS (Center for Cognitive Science) at Buffalo*

  http://www.cs.buffalo.edu/pub/WWW/cogsci/

  → | Institutes | Directories |

  Following up on the Institutes link, the following listing was obtained:

  ⇒ Neurolinguistic Institutes

  Active Window Productions: http://www.actwin.com/NLP/whats-nlp.html

  Carnegie Mellon University:

  http://www-cgi.cs.cmu.edu/afs/cs.cmu.edu/project/cnbc/CNBC/CNBC.html

  Harvard University: http://dem0nmac.mgh.harvard.edu/units/cognitive.html

  John Hopkins Medical Institutions: http://infonet.welch.jhu.edu/research.html

  Massachusetts Institute of Technology: http://broca.mit.edu/LingFac.lst.html

  MIC-KIBIC: Karolinska Institute: http://www.mic.ki.se/Diseases/c10.html

  Northwestern University:

  http://www.psych.nwu.edu/psych/department/psychobiology.html

  Ohio State University: http://www.cog.ohio-state.edu/

  Stanford University: http://www-portfolio.stanford.edu:6380/845

  Sussex University: http://www.cogs.susx.ac.uk/users/mike/rad/mrtutor.html

  The Jackson Laboratory: http://www.jax.org/

University at Buffalo: http://www.nucmed.buffalo.edu/nrlgy1.htm
University of Alberta: http://nshade.uah.ualberta.ca
University of California, Davis: http://neuroscience.ucdavis.edu/
University of California, Los Angeles: http://www.loni.ucla.edu/
University of Groningen: http://www.let.rug.nl/Linguistics/www/staff.html
University of Pennsylvania:
http://www.seas.upenn.edu/~mengwong/add/add_books.txt
University of Rochester: http://www.cvs.rochester.edu/
University of Southern California:
http://www.usc.edu/dept/cs/research_index/neurolinguistics.html
University of Washington: http://weber.u.washington.edu/~wcalvin/bk2.html
University of Waterloo:
http://www.science.uwaterloo.ca/undergraduate_calendar/psych.html
Zürich University Hospital: http://www.unizh.ch/~vor/

Following up on the Directories link, the following listing was obtained:
⇒ Directories ⇒ Neurolinguistic Directories

EINet: http://galaxy.einet.net/galaxy/Medicine/Medical-Specialties/Neurology.html
Human Brain Project, WWW Servers: http://137.131.192.144/HBP_html/HBPsites.html
Journal: Linguist: http://www.ling.rochester.edu/linguist/6-502.html
Med Web: http://www.cc.emory.edu/WHSCL/medweb.neurology.html
MIT Press: http://www-mitpress.mit.edu/jrnls-catalog/science-toc.html
Neurology/Neuroscience: http://www.informatik.uni-rostock.de/HUM-MOLGEN/neurology/
Neurosciences on the Internet: http://ivory.lm.com/~nab/
Current Opinion in NEUROBIOLOGY: http://www.cursci.co.uk/BioMedNet/nrb/
nrb94.html
Yahoo Index: http://www.yahoo.com/Health/Medicine/Neurosciences/

- *Cerebral Hypoxia & Ischemia Web Resources*
  http://sable.ox.ac.uk/~scro0037/guide.htm
- *Clinical Linguistics Bibliography*
  http://www.ims.uni-stuttgart.de/phonetic/joerg/biblio/clinical.html
- *Cognitive Neuroscience Resources on the Internet: Homepages (CNBC)*
  http://www.cs.cmu.edu/afs/cs.cmu.edu/project/cnbc/other/homepages.html
- *Disabilities* (What are the disabilities that affect learning?)
  http://www.fln.vcu.edu/ld/conf1.html
  → | Dyslexia | Dysgraphia | Dyscalculia | Language deficit | Visual perception | Auditory deficits | ADD | ADHD |
- *Info-CHILDES* (The CHILDES language data exchange system; for more information, see MacWhinney, this volume)
  http://poppy.psy.cmu.edu.childes
- *Massachusetts General Hospital Neurovascular News*
  http://neurosurgery.mgh.harvard.edu/vaschome.htm
- *MEDICAL MATRIX 3.6.23: Guide to Internet Clinical Medicine Resources*
  Some selected links:
  → | Neurology | Neurosurgery | Psychiatry | Rehabilitation and Physical Medicine | Major Journals | Electronic Journals and Newsletters | Software Databanks | Telemedicine, Imaging, and Computing | Internet Resource Guides | Introduction to Medical Internet Resources | How to Access Internet Resources | Clinical Document Retrieval Overview | Internet Medicine Resources News |

- *Medical Sites*
  http://www.neurosim.wisc.edu/nsl/nsl/medical.html
- *National Aphasia Association (NAA)*
  http://www.aphasia.org/
  → | Aphasia community groups | NAA regional and area | Representatives | Associations and agencies | Aphasia community group resources |
- *Neurologist Online*
  http://www.eastnc2.coastalnet.com/~cn3877/index.htm
  Adult neurology: migraine headache, stroke, seizures, parkinsonism, back pain, managed care, consultation, brain, spinal cord, nerves. Aids for patients and doctors.
- *Neurology Web-Forum [Mass. General Hospital]*
  http://dem0nmac.mgh.harvard.edu/neurowebforum/neurowebforum.html
  An interactive on-line discussion about various neurology-related topics.
- *Neuropsychology Central*
  http://www.premier.net/~cogito/neuropsy.html
  http://www.premier.net/~cogito/index.html
  → | Neuropsychological Assessment | Brain Imaging | Cognitive Neuropsychology | Developmental Neuropsychology | Forensic Neuropsychology | Geriatric Neuropsychology | Homepages | Laboratories | Mailgroups | Neuropsychology Central Forum | Newsgroups | Professional Organizations | Publications | Training Programs | Treatment and Rehabilitation Software | General Neuroscience | Various Psychology Links |
    ⇒ General Neuroscience Links
       | Neurosciences on the Internet | Neurofeedback Archive | Sleep Medicine WWW Archive | Neuroscape: Journal: Reading List | Massachusetts General Hospital Neurovascular News | Massachusetts General Hospital History of Neurosurgery Homepage | Neuroanatomy Study Slides | Neuroanesthesiology Manual | Neurology Web-Forum [Mass. General Hospital] | Neurosciences Internet Resource Guide | Neurotrophin | Purkinje Park | Shuffle Brain | University of Texas Southwestern Medical Center - Department of Neurology | USC Neurosurgical Information Resource | Wisconsin/Michigan State Brain Collections | Yale Neurosurgery |
    ⇒ Neuropsychology Organizations & Conferences
       | Organizations & Contacts | Organization Directory Listing | Medical Societies Directory | Resources of Scholarly Societies - Health Sciences |
       | American Academy of Neurology Home Page | American Association of Electrodiagnostic Medicine | American Psychological Association | American Psychological Society | American Society of Neuroradiology | Brain Injury Interdiscipinary Special Interest Group of the American Congress of Rehabilitation Medicine (BI-ISIG to ACRM) | National Academy of Neuropsychology | National Coalition for Research in Neurological Disorders | National Foundation for Brain Research | National Institute of Neurological Disorders and Stroke | Society of Behavioral Medicine | Society for Behavioral Neuroendocrinology | Society for Computers in Psychology |
    ⇒ Professional Conferences
       | Biomedical Imaging Conferences | Meetings of the Mind | TENNET Conference Home Page |

⇒ Neuropsychology Central - Various Psychology Links
| American Psychological Association - PsychNet - (Online reprints of APA
Monitor articles and job search listings) | Behavior OnLine - (Behavioral science
resources and reviews) | Psychological Journals & Conference Information |
Grohol Mental Health Page - Ask for the Internet Resource Pointers | Hardin
Meta Directory - Neurology/Neurosciences | Hardin Meta Directory - Psychiatry/
Mental Health | InterPsych Email Groups Homepage | Mental Health Net | Mental
Health Research Institute - Australian psychology resources | PIE Online - A
fairly comprehensive listing of resources for the professional and interested
layperson | PsycGrad Project Gopher - If you are a graduate student of
psychology, you need to investigate this link. | Psych Web | Psychology Self-Help
Resources on the Internet | Psychological Bulletin - APA Journal Psychological
Review - APA Journal | PSYCOLOQUY by topic | Psychiatry On-Line |

- *Neuroscience for Kids*
  shttp://weber.u.washington.edu/~chudler/neurok.html
  Created for elementary and secondary school students and teachers who would like to
  learn more about the nervous system.
- *Neurosciences Internet Resource Guide*
  http://http2.sils.umich.edu/Public/nirg/nirg1.html
- *Neurosciences on the Internet*
  http://www.lm.com/~nab
  [Table of Contents] [Recent Additions] [Best Bets] [Original] [Search]
  [Guides] [Academic: Basic] [Academic: Clinical] [Organizations] [Images]
  [Software] [Journals] [Diseases] [Newsgroups/Web Forums] [Mailing Lists]
  [Biology] [Medicine] [WWW: About] [WWW: Start] [WWW: Indexes] [WWW:
  Newsgroups] [About the Editor] [Feedback]
- *Social Science Links—Psychology*
  http://www.ssn.flinders.edu.au/socsci/links/psych.htm
  Cognitive and Psychological Sciences on the Internet. A list of resources, maintained by
  Stanford University.
- *Wisconsin/Michigan State Brain Collections*
  http://www.neurophys.wisc.edu/brain/brain.html
- *Yahoo! - Health:Medicine:Neurosciences:Conferences*
  http://www.yahoo.com/Health/Medicine/Neurosciences/Conferences
- *Yahoo! - Health:Diseases and Conditions:Aphasia*
  http://www.yahoo.com/Health/Diseases_and_Conditions/Aphasia/
  → | Health:Diseases and Conditions:Aphasia | Aphasia and Literacy |
  | Language and Communication in Aphasia | National Aphasia Association | On Aphasia |
  What is Aphasia? |
- *Yahoo! - Health:Medicine:Neurosciences*
  http://www.yahoo.com/Health/Medicine/Neurosciences/index.html
  → | Indices | Books | Conferences | Institutes | Journals | Organizations | Paralysis |

REFERENCES

Aaltonen, O., Tuomainen, J., Laine, M., & Niemi, P. (1993). Cortical differences in tonal versus vowel processing as revealed by an ERP components called mismatch negativity (MMN). *Brain and Language, 44,* 139–152.

Abeles, N. (1985). Boston Diagnostic Aphasia Examination. In D. J. Keyser & R. C. Sweetland (Eds.), *Test critiques* (Vol. 1, pp. 117–124). Kansas City, MO: Test Corporation of America.

Aboitiz, F., Scheibel, A. B., Fisher, R. S., & Zaidel, E. (1992a). Fiber composition of the human corpus callosum. *Brain Research, 598,* 143–153.

Aboitiz, F., Scheibel, A. B., Fisher, R. S., & Zaidel, E. (1992b). Individual differences in brain asymmetries and fiber composition in the human corpus callosum. *Brain Research, 598,* 154–161.

Aboitiz, F., Scheibel, A. B., & Zaidel, E. (1992). Morphometry of the human corpus callosum, with emphasis on sex differences. *Brain, 115,* 1521–1541.

Ackerman, H., Hertrich, I., & Ziegler, W. (1993). Prosodische Störungen bei neurologischen Erkrankungen—eine Literaturübersicht [Prosodic disorders in neurological diseases—A review of the literature]. *Fortschritte der Neurologie Psychiatrie, 61,* 241–253.

Ackerman, P. T., Dykman, R. A., Oglesby, D. M., & Newton, J. E. O. (1994). EEG power spectra of children with dyslexia, slow learners, and normally reading children with ADD during verbal processing. *Journal of Learning Disabilities, 27,* 619–630.

Ackley, D. H., Hinton, G. E., & Sejnowski, T. J. (1985). A learning algorithm for Boltzmann machines. *Cognitive Science, 9,* 147–169.

Adamovich, B. L., & Henderson, J. A. (1984). Can we learn more from word fluency measures with aphasic, right brain injured and closed head trauma patients? *Clinical Aphasiology, 14,* 124–131.

Adams, J., Faux, S. F., Nestor, P. G., Shenton, M., Marey, B., Smith, S., & McCarley, R. W. (1993). ERP abnormalities during semantic processing in schizophrenia. *Schizophrenia Research, 10,* 247–257.

Adolphs, R., Tranel, D., Damasio, H., & Damasio, A. (1994). Impaired recognition of emotion in facial expressions following bilateral damage to the human amygdala. *Nature, 372,* 669–672.

Agid, Y., Javoy-Agid, F., Ruberg, M., Pillon, B., Dubois, B., Duyckaerts, C., Hauw, J., Baron, J.-C., & Scatton, B. (1986). Progressive supranuclear palsy: Anatomoclinical and biochemical considerations. *Advances in Neurology, 45,* 191–206.

Aglioti, S., & Fabbro, F. (1993). Paradoxical selective recovery in a bilingual aphasic following a subcortical lesion. *Neuroreport, 4,* 1359–1362.

Ahonen, A. I., Hämäläinen, M. S., Kahola, M. J., Knuutila, J. E. T., Laine, P. P., Lounasmaa, O. V., Parkkonen, L. T., Simola, J. T., & Tesche, C. D. (1993). 122–channel SQUID instrument for investigating the magnetic signals from the human brain. *Physica Scripta, T49,* 198–205.

Aitchison, J. (1987). *Words in the mind: An introduction to the mental lexicon.* Oxford: Basil Blackwell.

Aitchison, J., & Straff, M. (1982). Lexical storage and retrieval: A developing skill (pp. 197–241). In Cutler, A. (Ed.), *Slips of the tongue and language production.* Berlin: Mouton.

Albanese, E., Merlo, A., Albanese, A., & Gómez, E. (1989). Anterior speech region: Asymmetry and weight-surface correlation. *Archives of Neurology, 46,* 307–310.

Albert, M. L., Feldman, R. G., & Willis, A. L. (1974). The subcortical dementia of progressive supranuclear palsy. *Journal of Neurology, Neurosurgery and Psychiatry, 37,* 121–130.

Albert, M. L., & Obler, L. K. (1978). *The bilingual brain.* New York: Academic Press.

Alexander, M. P. (1992). Speech and language deficits after subcortical lesions of the left hemisphere: A clinical, CT, and PET study. In G. Vallar, S. F. Cappa & C.-W. Wallesch (Eds.), *Neuropsychological disorders associated with subcortical lesions* (pp.455–477). New York: Oxford University Press.

Alexander, M. P., Benson, D. F., & Stuss, D. T. (1989). Frontal lobes and language. *Brain and Language, 37,* 656–691.

Alexander, M. P., Hiltbrunner, B., & Fischer R. S. (1989). Distributed anatomy of transcortical sensory aphasia. *Archives of Neurology, 46,* 885–892.

Alexander, M. P., Naeser, M. A., & Palumbo, C. L. (1987). Correlations of subcortical CT lesion sites and aphasia profiles. *Brain, 110,* 961–991.

Ali Cherif, A., Royere, M. L., Gosset, A., Poncet, M., Salamon, G., & Khalil, R. (1984). Troubles du comportement et de l'activité mentale après intoxication oxycarbonée: Lésions pallidales bilatérales. *Revue Neurologique, 140,* 32–40.

Allain, H., Lieury, H., Quemener, V., & Thomas, V. (1995). Procedural memory and Parkinson's disease. *Dementia, 6,* 174–178.

Allen, L. S., Richey, M. F., Chai, Y. M., & Gorski, R. A. (1991). Sex differences in the corpus callosum of the living human being. *Journal of Neuroscience, 11,* 933–942.

Allport, A. (1993). Attention and control: Have we been asking the wrong questions? A critical review of twenty-five years. In D. E. Meyer & S. Kornblum (Eds.), *Attention and Performance 14: Synergies in experimental psychology, artificial intelligence, and cognitive neuroscience* (pp. 183–218). Cambridge: MIT Press.

Almkvist, O. (1994). Neuropsychological deficits in vascular dementia in relation to Alzheimer's disease: Reviewing evidence for functional similarity or divergence. *Dementia, 5,* 203–209.

Almli, C. R., & Finger, S. (1988). Toward a definition of recovery of function. In S. Finger, T. E. Levere, C. R. Almli, & D. G. Stein (Eds.), *Brain injury and recovery: Theoretical and controversial issues* (pp. 1–14). New York: Plenum Press.

Aloia, M. S., Gourovitch, M. L., Weinberger, D. R., & Goldberg, T. E. (1996). An investigation of semantic space in patients with schizophrenia. *Journal of the International Neuropsychological Society, 2,* 267–273.

Alpert, M., Rosen, A., Welkowitz, J., & Lieberman, A. (1990). Interpersonal communication in the context of dementia. *Journal of Communication Disorders, 23,* 337–346.

Alpert, M., Rosen, A., Welkowitz, J., Sobin, C., & Borod, J. C. (1989). Vocal acoustic correlates of flat affect in schizophrenia. *British Journal of Psychiatry, 154,* 51–56.

American Psychiatric Association. (1987). *Diagnostic and statistical manual of mental disorders* (3rd ed.). Washington, DC: Author.

American Psychiatric Association. (1994). *Diagnostic and statistical manual of mental disorders* (4th ed.). Washington, DC: Author.

Anand, A. & Wales, R. J. (1996). Psychotic speech: A neurolinguistic perspective, *Australian and New Zealand Journal of Psychiatry, 28,* 229–238.

Anand, A., Wales, R. J., Jackson, H. J., & Copolov, D. L. (1994). Linguistic impairment in early psychosis. *Journal of Nervous and Mental Disease, 182,* 488–493.

Anderson, J. M., Milner, R. D. G., & Strich, S. J. (1967). Effects of neonatal hypoglycemia on the nervous system: A pathological study. *Journal of Neurology, Neurosurgery, and Psychiatry, 30,* 295–310.

Andreasen, N. C. (1982). The relationship between schizophrenic language and the aphasias. In F. A. Henn & H. A. Nasrallah (Eds.), *Dopamine and psychosis in schizophrenia as a brain disease* (pp. 99–111). New York: Oxford University Press.

Andreasen, N. C., Arndt, S., Swayze, V., Cizadlo, T., Flaum, M., O'Leary, D., Ehrhardt, J. C., & Yuh, W. T. (1994). Thalamic abnormalities in schizophrenia visualized through magnetic resonance image averaging. *Science, 266,* 294–298.

Andrews, R. J. (1991). Transhemispheric diaschisis. A review and comment. *Stroke, 22,* 943–949.

Andy, O. J., & Bhatnager, S. C. (1894). Right-hemispheric language: Evidence from cortical stimulation. *Brain and Language, 23,* 159–166.

Appel, J., Kertesz, A., & Fisman, M. (1982). A study of language functioning in Alzheimer patients. *Brain and Language, 17,* 73–91.

Aram, D., & Eisele, J. A. (1992). Plasticity and recovery of higher cognitive functions following early brain injury. In I. Rapin & S. J. Segalowitz (Eds.), *Handbook of neuropsychology*. Vol. 6, *Child neuropsychology* (pp. 73–92). Amsterdam: Elsevier.

Aram, D. M., & Whitaker, H. A. (1988). Cognitive sequelae of unilateral lesions acquired in early childhood. In D. L. Molfese & S. J. Segalowitz (Eds.), *Brain lateralization in children: Developmental implications* (pp. 417–436). New York: Guilford Press.

Arbib, M. A. (1970). On modelling the nervous system. In H. E. von Gierke, W. D. Keidel, & H. L. Ostreicher (Eds.), *Principles and practice of bionics* (pp. 45–57). Slough, England: Technivision Book Services.

Arbib, M. A., & Caplan, D. (1979). Neurolinguistics must be computational. *Behavioral & Brain Sciences, 2,* 449–483.

Ardila, A., & Ostrosky-Solis, F. (Eds.). (1989). *Brain organization and language and cognitive processes.* New York: Plenum Press.

Ardila, A. (1995). Directions of research in cross-cultural neuropsychology. *Journal of Clinical and Experimental Neuropsychology, 17,* 143–150.

Arguin, M., Bub, D., & Dudek, G. (1996). Shape integration for visual object recognition and its implication in category-specific visual agnosia. *Visual Cognition, 3,* 221–275.

Arias, M., Requena, I., Ventura, M., Pereiro, I., Castro, A., & Alvarez, A. (1995). A case of deaf-mutism as an expression of pure word deafness: Neuroimaging and electrophysiological data. *European Neurology, 2,* 583–585.

Armstrong, S. L., Gleitman, L. R., & Gleitman, H. (1983). What some concepts might not be. *Cognition, 13,* 263–308.

Armus, S. R., Brookshire, R. H., & Nicholas, L. E. (1989). Aphasic and non-brain-damaged adults' knowledge of scripts for common situations. *Brain and Language, 36,* 518–528.

Arnold, R., Yule, W., & Martin, N. (1985). The psychological characteristics of infantile calcemia: A preliminary investigation. *Developmental Medicine and Child Neurology, 27,* 49–59.

Arseni, C., Constantinovici, A., Iliescu, D., Debrota, I., & Gagea, A. (1970). Considerations on posttraumatic aphasia in peacetime. *Psychiatria, Neurologia, Neurochirurgia, 73,* 105–112.

Arvedson, J. C., McNeil, M. R., & West, T. L. (1985). Prediction of revised Token Test overall, subtest, and linguistic unit scores by two shortened versions. *Clinical Aphasiology, 15,* 57–62.

Asperger (1944/1991). Die "Autistischen Psychopathen" im Kindesalter. *Archiv für Psychiatrie und Nervenkrankheiten, 117,* 76–136. ["Autistic psychopathy" in childhood.] In U. Frith (Ed., Trans.), *Autism and Asperger syndrome* (pp. 37–92). Cambridge: Cambridge University Press.

Atkinson, J. M., & Heritage, J. (1984). *Structures of social action: Studies in conversation analysis.* Cambridge, UK: Cambridge University Press.

Aubert, D., & Whitaker, H. A. (1996). David Hartley's model of vibratiuncles seen as a contribution to the localization theory of brain function. *History and Philosophy of Psychology Bulletin, 8,*(1), 14–16.

Auer, R. N., Wieloch, T., Olsson, Y., & Siesjo, B. K. (1984). The distribution of hypoglycemia in brain damage. *Acta Neuropathologica, 64,* 177–191.

Auerbach, S., Allard, T., Naeser, M., Alexander, M., & Albert, M. (1982). Pure word deafness: Analysis of a case with bilateral lesions and a deficit at the pre-phonemic level. *Brain, 105,* 271–300.

Augustine (1952). *The Confessions. Britannica Great Books* (Vol. 18) (E. J. Pusey, Trans.). Chicago: Encyclopedia Britannica. (Original work written in 397.)

Aulanko, R., Hari, R., Lounasmaa, O. V., Näätänen, R., & Sams, M. (1993). Phonetic invariance in the human auditory cortex. *Neuroreport., 4,* 1356–1358.

Auriacombe, S., Grossman, M., Carvell, S., Gollomp, S., Stern, M. B., & Hurtig, H. I. (1993). Verbal fluency deficits in Parkinson's disease. *Neuropsychology, 7,* 182–192.

Awh, E., Smith, E. E., & Jonides, J. (1995). Human rehearsal processes and the frontal lobes: PET evidence. In J. Grafman, K. Holyoak, & F. Boller (Eds.), *Structure and functions of the human prefrontal cortex* (pp. 97–117). New York: *Annals of the New York Academy of Sciences, 769.*

Aynsley-Green, A., Polak, J. M., & Bloom, S. R. (1981). Nesidioblastosis of the pancreas: Definition of the syndrome and the management of the severe hyperinsulinemic hypoglycemia. *Archives of the Diseases of Childhood, 56,* 496–508.

Bachevalier, J., & Mishkin, M. (1986). Visual recognition impairment follows ventromedial but not dorsolateral prefrontal lesions in monkeys. *Behavioural Brain Research, 20,* 249–261.

Bach-y-Rita, P. (1990). Brain plasticity as a basis for recovery of function in humans. *Neuropsychologia, 28,* 547–554.

Bach-y-Rita, P. (1992). Application of principles of brain plasticity and training to restore function. In R. R. Young & P. J. Delwaide (Eds.), *Principles and practice of restorative neurology* (pp. 54–65). London: Butterworth Heinemann.

Baddeley, A.D (1976). *The psychology of memory.* New York: Basic Books.

Baddeley, A. D. (1986). *Working memory.* Oxford: Oxford University Press.

Baddeley, A. D. (1993a). Short-term phonological memory and long-term learning: A single case study. *European Journal of Cognitive Psychology, 5,* 129–148.

Baddeley, A. D. (1993b). A theory of rehabilitation without a model of learning is a vehicle without an engine: A comment on Caramazza and Hillis. *Neuropsychological Rehabilitation, 3,* 235–244.

Baddeley, A. D., Logie, R. H., Bressi, S., Della Sala, S., & Spinnler, H. (1986). Dementia and working memory. *Quarterly Journal of Experimental Psychology, 38A,* 603–618.

Baddeley, A. D., Papagno, C., & Vallar, G. (1988). When long-term memory depends on short-term storage. *Journal of Memory and Language, 27,* 586–595.

Baddeley, A. D., Vallar, G., & Wilson, B. A. (1987). Sentence comprehension and phonological memory: Some neuropsychological evidence. In M. Coltheart (Ed.), *Attention and performance XII* (pp. 509–529). Hillsdale, NJ: Lawrence Erlbaum.

Badecker, W., & Caramazza, A. (1985). On considerations of method and theory governing the use of clinical categories in neurolinguistics and cognitive neuropsychology: The case against agrammatism. *Cognition, 20,* 97–125.

Badecker, W., & Caramazza, A. (1986). A final brief in the case against agrammatism. *Cognition, 24,* 277–282.

Badecker, W., & Caramazza, A. (1987). The analysis of morphological errors in a case of acquired dyslexia. *Brain and Language, 32,* 278–305.

Badecker, W., Hillis, A., & Caramazza, A. (1990). Lexical morphology and its role in the writing process: Evidence from a case of acquired dysgraphia. *Cognition, 35,* 205–234.

Baharav, E. (1990). Agrammatism in Hebrew: Two case studies. In L. Menn & L. K. Obler (Eds.), *Agrammatic aphasia: A Cross-Language Narrative Sourcebook* (Vol. 2, pp. 1087–1190). Amsterdam: John Benjamins.

Bailey, D. L., Jones, T., Friston, K. J., Colebatch, J. G., & Frackowiak, R. S. J. (1991). Physical validation of statistical mapping. *Journal of Cerebral Blood Flow and Metabolism, 11* (Suppl. 2), S150.

Bakker, D. J. (1990). *Neuropsychological treatment of dyslexia.* New York: Oxford University Press.

Bakker, D. J. (1994). Dyslexia and the ecological brain. *Journal of Clinical and Experimental Neuropsychology, 16,* 734–743.

Baldwin, B., & Forstl, H. (1993). 'Pick's disease'—101 years on still there, but in need of reform. *British Journal of Psychiatry, 163,* 100–104.

Baldwin, D. A. (1995). Understanding the link between joint attention and language acquisition. In C. Moore & P. J. Dunham (Eds.), *Joint attention: Its origins and role in development* (pp. 131–158). Hillsdale, NJ: Lawrence Erlbaum.

Ballard, D. H. (1986). Cortical connections and parallel processing: Structure and function. *Behavioral and Brain Sciences, 9,* 67–120.

Balota, D. A., & Chumbley, J. I. (1984). Are lexical decisions a good measure of lexical access? The role of word frequency in the neglected decision stage. *Journal of Experimental Psychology, 10,* 340–357.

Balota, D. A., & Ferraro, F. R. (1993). A dissociation of frequency and regularity effects in pronunciation performance across young adults, older adults, and individuals with senile dementia of the Alzheimer type. *Journal of Memory and Language, 32,* 573–592.

Balota, D. A., & Ferraro, F. R. (1996). Lexical, sublexical, and implicit memory processes in healthy young and healthy older adults and in individuals with dementia of the Alzheimer type. *Neuropsychology, 10,* 82–95.

Bandettini, P. A., & Wong, E. C. (in press). Echo-planar magnetic resonance imaging of human brain activation. In F. Schmitt, M. Stehling, & R. Turner (Eds.), *Echo Planar Imaging*. New York: Springer Verlag.

Bandettini, P. A., Wong, E. C., Hinks, R. S., Tikofsky, R. S., & Hyde, J. S. (1992). Time course EPI of human brain function during task activation. *Magnetic Resonance in Medicine, 25,* 390–397.

Banich, M. T., & Karol, D. (1992). The sum of the parts does not equal the whole: Evidence from bihemispheric processing. *Journal of Experimental Psychology: Human Perception and Performance, 18,* 763–784.

Banich, M. T. (1995a). Interhemispheric interaction: Mechanisms of unified processing. In F. L. Kitterle (Ed.), *Hemispheric communication: Mechanisms and models* (pp. 271–300). Hillsdale, NJ: Lawrence Erlbaum.

Banich, M. T. (1995b). Interhemispheric processing: Theoretical considerations and empirical approaches. In R. J. Davidson & K. Hugdahl (Eds.), *Brain asymmetry* (pp. 427–450). Cambridge: MIT Press.

Banse, R., & Scherer, K. R. (1996). Acoustic profiles in vocal emotion expression. *Journal of Personality and Social Psychology, 70,* 614–636.

Barch, D. M. & Berenbaum, H. (1996). Language production and thought disorder in schizophrenia. *Journal of Abnormal Psychology, 105,* 81–88.

Barch, D. M., Cohen, J. D., Servan-Schreiber, D., Steingard, S., Steinhauer, S. & Van Kammen, D. P. (1995, March). Lexical priming in schizophrenia: The effects of anti-psychotic medication. Poster presented at the Cognitive Neuroscience Annual Meeting, San Francisco, CA.

Barisnikov, K., Van der Linden, M., & Poncelet, M. (1996). Acquisition of new words and phonological working memory in Williams syndrome: A case study. *Neurocase, 2,* 395–404.

Baron, J.-C., D'Antona, R., Pantano, P., Serdaru, M., Samson, Y., & Bousser, M. G. (1986). Effects of thalamic stroke on energy metabolism of the cerebral cortex. *Brain, 109,* 1243–1259.

Baron-Cohen, S., Allen, J., & Gillberg, C. (1992). Can autism be detected at 18 months? The needle, the haystack, and the CHAT. *British Journal of Psychiatry, 161,* 839–843.

Baron-Cohen, S., Tager-Flusberg, H., & Cohen, D. J. (Eds.). (1993). *Understanding other minds: Perspectives from autism.* Oxford: Oxford University Press.

Barr, A., & Geigenbaum, E. A. (Eds.) (1982). *The handbook of artificial intelligence.* Stanford, CA: HeurisTech Press.

Barr, W. B., Bilder, R. M., Goldberg, E., Kaplan, E., & Mukherjee, S. (1989). The neuropsychology of schizophrenic speech. *Journal of Communication Disorders, 22,* 327–349.

Barret, S. E. & Rugg, M. D. (1989). Event-related potentials and the semantic matching of faces. *Brain and Cognition, 14,* 201–212.

Barris, R. W., & Schuman, H. R. (1953). Bilateral anterior cingulate gyrus lesions. *Neurology, 3,* 44–52.

Bartholow, R. (1874). Experimental investigations into functions of the human brain. *American Journal of Medical Sciences, 67,* 305–313.

Bartlett, E. J., Brown, J. W., Wolf, A. P., & Brodie, J. D. (1987). Correlations between glucose metabolic rates in brain regions of healthy male adults at rest and during language stimulation. *Brain and Language, 32,* 1–18.

Bartolomeo, P., Bachoud-Levi, A. -C., Degos, J. -D., & Boller, F. (in press). Right hemisphere contributions to residual reading in pure alexia: Evidence from a patient with consecutive bilateral strokes. In E. Zaidel, M. Iacoboni, and A. Pasqual-Leone, (Eds.), *The human corpus callosum: Anatomy, physiology, and behavior; individual differences and clinical applications.* New York: Plenum.

Basile, L. F. H., Rogers, R. L., Bourbon, W. T., & Papanicolaou, A. C. (1994). Slow magnetic fields from human frontal cortex. *Electroencephalography and Clinical Neurophysiology, 90,* 157–165.

Basile, L. F. H., Simos, P. G., Tarkka, I. M., Brunder, D. G., & Papanicolaou, A. C. (1996). Task-specific magnetic fields from the left human frontal cortex. *Brain Topography, 9,* 1–7.

Basso, A. (1989). Spontaneous recovery and language rehabilitation. In X. Seron & G. Deloche (Eds.), *Cognitive approaches in neuropsychological rehabilitation* (pp. 17–37). Hillsdale, NJ: Lawrence Erlbaum.

Basso, A. (1992). Prognostic factors in aphasia. *Aphasiology, 6,* 337–348.

Basso, A., Capitani, E., & Moraschini, S. (1982). Sex differences in recovery from aphasia. *Cortex, 18,* 469–475.

Basso, A., Capitani, E., & Vignolo, L. A. (1979). Influence of rehabilitation on language skills in aphasic patients. A controlled study. *Archives of Neurology, 36,* 190–196.

Basso, A., Capitani, E., Laiacona, M., & Zanobio, M. E. (1985). Crossed aphasia: One or more syndromes. *Cortex, 21,* 25–46.

Basso, A., Farabola, M., Grassi, M. P., Laiacona, M., & Zanobio, M. E. (1990). Aphasia in left-handers. Comparison of aphasia profiles and language recovery in non-right-handed and matched right-handed patients. *Cortex, 38,* 233–252.

Basso, A., Gardelli, M., Grasso, M. P., & Marriotti, M. (1989). The role of the right hemisphere in the recovery of aphasia: Two case studies. *Cortex, 25,* 555–566.

Basso, A., Lecours, A.-R., Moraschini, S., & Vanier, M. (1985). Anatomo-clinical correlations of the aphasias as defined through computerized tomography: On exceptions. *Brain and Language, 26,* 201–229.

Basso, A., & Scarpa, M. T. (1990). Traumatic aphasia in children and adults: A comparison of clinical features and evolution. *Cortex, 26,* 502–514.

Bastiaanse, R. (1995). Broca's aphasia: A syntactic and/or morphological disorder? A case study. *Brain and Language, 48,* 1–32.

Bastian, H. C. (1869). On the various forms of loss of speech in cerebral disease. *British Foreign Medical and Chirurgical Review, 43,* 209–236, 470–492.

Bastian, H. C. (1898). *Aphasia and other speech defects.* London: .

Bates, E., Appelbaum, M., & Allard, L. (1991). Statistical constraints on the use of single cases in neuropsychological research. *Brain and Language, 40,* 295–329.

Bates, E., Hamby, S., & Zurif, E. (1983). The effects of focal brain damage on pragmatic expression. *Canadian Journal of Psychology, 37,* 59–84.

Bates, E., & Wulfeck, B. (1989). Crosslinguistics studies of aphasia. In B. MacWhinney & E. Bates (Eds.), *The crosslinguistic study of sentence processing* (pp. 328–371). New York: Cambridge University Press.

Bates, E., Wulfeck, B., & MacWhinney, B. (1991a). Cross-linguistic research in aphasia: An overview. *Brain and Language, 40,* 1–15.

Bates, E., Wulfeck, B., & MacWhinney, B. (1991b). Cross-linguistic studies in aphasia: An overview. *Brain and Language, 41,* 123–148.

Battison, R. (1978). *Lexical borrowing in American Sign Language.* Silver Spring, MD: Linstok Press.

Baum, S. (1992). The influence of word length on syllable duration in aphasia: Acoustic analyses. *Aphasiology, 6,* 501–513.

Baum, S. (1993). An acoustic analysis of rate of speech effects on vowel production in aphasia. *Brain and Language, 44,* 414–430.

Baum, S. (1996). Fricative production in aphasia: Effects of speaking rate. *Brain and Language, 52,* 328–341.

Baum, S., & Pell, M. (1997). Production of affective and linguistic prosody by brain-damaged patients. *Aphasiology, 11,* 177–198.

Baum, S., & Ryan, L. (1993). Rate of speech effects in aphasia: Voice onset time. *Brain and Language, 44,* 431–445.

Baum, S., & Slatkovsky, K. (1993). Phonemic false evaluation?: Preliminary data from a conduction aphasia patient. *Clinical Linguistics and Phonetics, 7,* 207–218.

Baum, S. R., & Pell, M. D. (in press). The neural bases of prosody: Insights from lesion studies and neuroimaging. In M. Lynch (Ed.), *The cognitive science of prosody: Interdisciplinary perspectives.* Amsterdam: Elsevier/North Holland Science Publisher.

Baum, S. R., Blumstein, S. E., Naeser, M. A., & Palumbo, C. L. (1990). Temporal dimensions of consonant and vowel production: An acoustic and CT scan analysis of aphasic speech. *Brain and Language, 39,* 33–56.

Bavry, J. L. (1991). *STAT-POWER: Statistical design analysis system* (2nd ed.). Chicago: Scientific Software.

Baxter, D. M., & Warrington, E. K. (1985). Category specific phonological dysgraphia. *Neuropsychologia, 23,* 653–666.

Baxter, L. R., Mazziotta, J. C., Pahl, J. J., George-Hyslop, P. S., Haines, J. L., Gusella, J. F., Szuba, M. P., Selin, C. E., Guze, B. H., & Phelps, M. E. (1992). Psychiatric, genetic and positron emission tomographic evaluation of persons at risk for Huntington's disease. *Archives of General Psychiatry, 49,* 148–154.

Bayles, K. A. (1982). Language function in senile dementia. *Brain and Language, 16,* 265–280.

Bayles, K. A. (1993). Pathology of language behavior in dementia. In G. Blanken, J. Dittman, H. Grimm, J. Marshall, & C.-W. Wallesch, (Eds.), *Linguistic disorders and pathologies.* Berlin: Walter de Gruyter.

Bayles, K. A. (1994). Management of neurogenic communication disorders associated with dementia. In R. Chapey (Ed.), *Language intervention strategies in adult aphasia* (pp. 535–545). Baltimore: Williams & Wilkins.

Bayles, K. A., & Tomoeda, C. K. (1983). Confrontation naming impairment in dementia. *Brain and Language, 19,* 98–114.

Bayles, K. A., Tomoeda, C. K., & Trosset, M. W. (1990). Naming and categorial knowledge in Alzheimer's disease: The process of semantic memory deterioration. *Brain and Language, 39,* 498–510.

Bayles, K. A., Tomoeda, C. K., & Trosset, M. W. (1992). Relation of linguistic communication abilities of Alzheimer's patients to stage of disease. *Brain and Language, 42,* 454–472.

Bayles, K. A., Trosset, M. W., Tomoeda, C. K., Montgomery, E. B., & Wilson, J. (1993). Generative naming in Parkinson's disease patients. *Journal of Clinical and Experimental Neuropsychology, 15,* 547–562.

Bayley, N. (1969). *Bayley scales of infant development.* New York: Psychological Corporation.

Baynes, K., Tramo, M. J., & Gazzaniga, M. S. (1992). Reading with a limited lexicon in the right hemisphere of a callosotomy patient. *Neuropsychologia, 30,* 187–200.

Baynes, K., Wessinger, C. M., Fendrich, R., & Gazzaniga, M. S. (1995). The emergence of the capacity to name left visual field stimuli in a callosotomy patient: Implications for functional plasticity. *Neuropsychologia, 33,* 1225–1242.

Bear, D. M. (1983). Hemispheric specialization and the neurology of emotion. *Archives of Neurology, 40,* 195–202.

Bear, D. M., & Fedio, P. (1977). Quantitative analysis of interictal behavior in temporal lobe epilepsy. *Archives of Neurology, 34,* 454–467.

Beatty, W. W., & Monson, N. (1989). Lexical processing in Parkinson's disease and multiple sclerosis. *Journal of Geriatry, Psychiatry and Neurology, 2,* 145–152.

Beatty, W. W., Monson, N., & Goodkin, D. E. (1989). Access to semantic memory in Parkinson's disease and multiple sclerosis. *Journal of Geriatry, Psychiatry and Neurology, 2,* 153–162.

Beatty, W. W., Staton, R. D., Weir, W. S., Monson, N., & Whitaker, H. A. (1989). Cognitive disturbances in Parkinson's disease. *Journal of Geriatry, Psychiatry and Neurology, 2,* 22–33.

Beaumanoir, A. (1985). The Landau-Kleffner syndrome. In J. B. Roger, C. Dravet, M. Bureau, F. E. Dreifuss, & P. Wolf (Eds.), *Epileptic syndromes in infancy, childhood and adolescence* (pp. 181–191). London: John Libbey.

Bédard, M. A., Massioui, F. E., Pillon, B., & Nandrino, J. L. (1993). Time for reorienting of attention: A premotor hypothesis of the underlying mechanism. *Neuropsychologia, 31,* 241–249.

Bédard, M. A., Pillon, B., Dubois, B., Masson, H., & Agid, Y. (in press). Acute and long-term administration of anticholinergics in Parkinson's disease: Specific effects on the subcortico-frontal syndrome. *Brain and Cognition.*

Beeghly, M., Weiss-Perry, B., & Cicchetti, D. (1990). Beyond sensorimotor functioning: Early communicative and play development of children with Down syndrome. In D. Cicchetti & M. Beeghly (Eds.), *Children with Down syndrome: A developmental perspective* (pp. 329–368). New York: Cambridge University Press.

Behrens, S. J. (1988). The role of the right hemisphere in the production of linguistic stress. *Brain and Language, 33,* 104–127.

Behrens, S. J. (1989). Characterizing sentence intonation in a right hemisphere-damaged population. *Brain and Language, 37,* 181–200.

Behrmann, M. (1987). The rites of righting writing: Homophone remediation in acquired dysgraphia. *Cognitive Neuropsychology, 4,* 365–384.

Behrmann, M., Black, S. E., & Bub, D. (1990). The evolution of pure alexia: A longitudinal study of recovery. *Brain and Language, 39,* 405–427.

Behrmann, M., & Lieberthal, T. (1989). Category-specific treatment of a lexical semantic deficit: A single case study of global aphasia. *British Journal of Communication Disorders, 24,* 281–299.

Behrmann, M., Moscovitch, M., Black, S. E., & Mozer, M. (1990). Perceptual and conceptual mechanisms of neglect: Two contrasting case studies. *Brain, 113,* 1163–1183.

Behrmann, M., Moscovitch, M., & Mozer, M. (1991). Directing attention to words and nonwords in normal subjects and in a computational model: Implications for neglect dyslexia. *Cognitive Neuropsychology, 8,* 213–248.

Behrmann, M., & Shallice, T. (1995). Pure alexia: A nonspatial visual disorder affecting letter activation. *Cognitive Neuropsychology, 12,* 409–454.

Béland, R., & Favreau, Y. (1991). On the special status of coronals in aphasia. In C. Paradis & J.-F. Prunet (Eds.), *Phonetics and phonology: The special status of coronals: Internal and external evidence* (Vol. 2, pp. 201–221). New York: Academic Press.

Béland, R. (1990). Vowel epenthesis in aphasia. In J.-L. Nespoulous & P. Villiard (Eds.), *Morphology, phonology, and aphasia* (pp. 235–252). New York: Springer Verlag.

Béland, R., Caplan, D., & Nespoulous, J. L. (1990). The role of abstract phonological representations in word production: Evidence from phonemic paraphasias. *Journal of Neurolinguistics, 5,* 125–164.

Belleville, S., Peretz, I., & Arguin, H. (1992). Contribution of articulatory rehearsal to short-term memory: Evidence from a selective disruption. *Brain and Language, 43,* 713–746.

Bellugi, U., Bihrle, A., & Corina, D. (1991). Linguistic and spatial development: Dissociations between cognitive domains. In N. A. Krasnegor, D. M. Rumaugh, R. L. Schiefelbusch, & M. Studdert-Kennedy (Eds.), *Biological determinants of language development* (pp. 363–393). Hillsdale, NJ: Lawrence Erlbaum.

Belota, D. A., & Duchek, J. M. (1988). Age-related differences in lexical access, spreading activation and simple pronunciation. *Psychology and Aging, 3,* 84–93.

Benbadis, S. R., Dinner, D. S., Chelune, G. J., Piedmonte, M., & Lüders, H. O. (1995). Objective criteria for reporting language dominance by intracarotid amobarbital procedure. *Journal of Clinical and Experimental Neuropsychology, 17,* 668–690.

Bennett, F. C., LaVeck, B., & Sells, C. J. (1978). The Williams elfin facies syndrome: The psychological profile as an aid in syndrome identification. *Pediatrics, 61,* 303–306.

Benson, D. F., & Geschwind, N. (1975). Psychiatric conditions associated with focal lesions of the central nervous system. In S. Arieti & M. Reiser (Eds.), *American handbook of Psychiatry* (pp. 208–243). New York: Basic Books.

Benson, D. F., & Patten, D. H. (1967). The use of radioactive isotopes in the localization of aphasia-producing lesions. *Cortex, 3,* 258–271.

Benson, D. F. (1973). Psychiatric aspects of aphasia. *British Journal of Psychiatry, 123,* 555–566.

Benson, D. F. (1978). Age, aphasia and stroke localization. *Archives of Neurology, 35,* 619–620.

Benson, D. F. (1984). The neurology of human emotion. *Bulletin of Clinical Neurosciences, 49,* 4923–4942.

Benson, F., & Ardila, A. (1996). *Aphasia: A clinical perspective.* New York: Oxford University Press.

Benson, F. (1979). Neurologic correlates of anomia. In H. Whitaker and H. A. Whitaker (Eds.), *Studies in Neurolinguistics* (Vol. 4, pp. 293–328). San Diego: Academic Press.

Benson, F. D., & Geschwind, N. (1985). Aphasia and related disorders: A clinical approach. In M. M. Mesulam (Ed.), *Principles of behavioral neurology* (pp. 193–238). Philadelphia: F. A. Davis.

Bentin, S., Kutas, M., & Hillyard, S. A. (1995). Semantic processing and memory for attended and unattended words in dichotic listening: Behavioral and electrophysiological evidence. *Journal of Experimental Psychology: Human Perception and Performance, 21,* 54–67.

Bentin, S., Silverberg, R., & Gordon, H. W. (1981). Asymmetrical cognitive deterioration in demented and Parkinson patients. *Cortex, 17,* 533–543.

Benton, A. L. (1994). Neuropsychological assessment. *Annual Review of Psychology, 45,* 1–23.

Benton, A. L., & Hamsher, K. (1989). *Multilingual aphasia examination* (2nd ed.). Iowa City: AJA Association.

Benton, A. L., & Joynt, R. J. (1960). Early descriptions of aphasia. *Archives of Neurology and Psychiatry, 3,* 205–222.

Benton, A. L., & Van Allen, M. W. (1968). Impairment in facial recognition in patients with cerebral disease. *Cortex, 4,* 344–358.

Berger, M. S., Cohen, W. A., & Ojemann, G. A. (1990). Correlation of motor cortex brain mapping data with magnetic resonance imaging. *Journal of Nuerosurgery, 72,* 383–387.

Bergvall, U., Habib, M., Jiddane, M., Rumeau, C., & Salamon, G. (1986). Cortical asymmetry on MRI related to unilateral hemispheric function. *Acta Radiologica, 369,* 205–207.

Berk, R. A. (Ed.). (1980). *Criterion-referenced measurement: The state of the art.* Baltimore: Johns Hopkins University Press.

Berko-Gleason, J., Goodglass, H., Obler, L. K., Green, E., Hyde, M. R., & Weintraub, S. (1980). Narrative strategies of aphasics and normal-speaking subjects. *Journal of Speech and Hearing Research, 23,* 370–382.

Berndt, R., Mitchum, C. C., & Haendiges, A. H. (1996). Comprehension of reversible sentences: A meta-analysis. *Cognition, 58,* 289–308.

Berndt, R. S. (1987). Symptom co-occurrence and dissociation in the interpretation of agrammatism. In M. Coltheart, G. Sartori, & R. Job (Eds.). *The cognitive neuropsychology of language.* London: Lawrence Erlbaum.

Berndt, R. S. (1991). Sentence processing in aphasia. In M. T. Sarno (Ed.), *Acquired aphasia* (pp. 223–270). San Diego: Academic Press.

Berndt, R. S., Mitchum, C. C., & Price, T. R. (1991). Short-term memory and sentence comprehension. An investigation of a patient with crossed aphasia. *Brain, 114,* 263–280.

Bernstein, I. H., Garbin, C. P., & Teng, G. K. (1988). *Applied multivariate analysis.* New York: Springer-Verlag.

Bernstein, I. H., Prather, P. A., & Rey-Casserly, C. (1995). Neuropsychological assessment in preoperative and postoperative evaluation. *Neurosurgery Clinics of North America, 6,* 443–454.

Berry, I., Démonet, J.-F., Warach, S., Viallard, G., Boulanouar, K., Franconi, J.-M., Marc-Vergnes, J.-P., Edelman, R., & Manelfe, C. (1995). Activation of association auditory cortex demonstrated with functional MRI. *NeuroImage, 2,* 215–219.

Berthier, M. L., Starkstein, S. E., Lylyk, P., & Leiguarda, R. (1990). Differential recovery of languages in a bilingual patient: A case study using selective amytal test. *Brain and Language, 38,* 449–453.

Besner, D., Twilley, L., McCann, R., & Seergobin, K. (1990). On the association between connectionism and data: Are a few words necessary? *Psychological Review, 97,* 432–446.

Best, C. T. (1988). The emergence of cerebral asymmetries in early human development: A literature review and a neuroembryological model. In D. L. Molfese & S. J. Segalowitz (Eds.), *Brain lateralization in children: Developmental implications* (pp. 5–34). New York: Guilford Press.

Bhatia, K. P., & Marsden, C. D. (1994). The behavioural and motor consequences of focal lesions of the basal ganglia in man. *Brain, 117,* 859–876.

Bhatnagar, S. C., & Andy, O. J. (1981). Language in the non-dominant right hemisphere. *Archives of Neurology, 40,* 728–731.

Bickerton, D. (1984). The language bioprogram hypothesis. *Behavioral and Brain Sciences, 7,* 173–187.

Bigler, E. D., & Naugle, R. I. (1985). Case studies in cerebral plasticity. *International Journal of Clinical Neuropsychology, 7,* 12–23.

Bihrle, A. M., Brownell, H. H., Powelson, J. A., & Gardner, H. (1986). Comprehension of humorous and nonhumorous materials by left and right brain-damaged patients. *Brain and Cognition, 5,* 399–411.

Binder, J., Rao, S. M., Hammeke, T. A., Yetkin, F. Z., Jesmanowicz, A., Bandettini, P. A., Wong, E. C., Estkowski, L. D., Goldstein, M. D., Haughton, V. M., & Hyde, J. S. (1994). Functional magnetic resonance imaging of human auditory cortex. *Annals of Neurology, 35,* 662–672.

Binder, J. R., Rao, S. M., Hammeke, T. A., Frost J. A., Bandettini, P. A., Jesmanowicz, A., & Hyde, J. S. (1995). Lateralized human brain language systems demonstrated by task subtraction functional magnetic resonance imaging. *Archives of Neurology, 52,* 593–601.

Binder, J. R., Swanson, S. J., Hammeke, T. A., Morris, G. L., Mueller, W. M., Fischer, M., Benbadis, S., Frost, J. A., Rao, S. M., & Haughton, V. M. (1996). Determination of language dominance using functional MRI: A comparison with the Wada test. *Neurology, 46,* 978–984.

Bird, E. K., & Chapman, R. S. (1994). Sequential recall in individuals with Down syndrome. *Journal of Speech and Hearing Research, 37,* 1369–1380.

Bishop, D. V. M. (1983). Linguistic impairment after left hemidecortication for infantile hemiplegia? A reappraisal. *Quarterly Journal of Experimental Psychology, 35A,* 199–207.

Bishop, D. V. M. (1985). Age of onset and outcome in 'acquired aphasia with convulsive disorder' (Landau-Kleffner syndrome). *Developmental Medicine and Child Neurology,* 27, 705–712,

Blanken, G. (1988). Anmerkungen zur Methodologie der kognitiven Neurolinguistik. *Neurolinguistik, 2,* 127–147.

Blanken, G., Dittmann, J., Haas, J. C., & Wallesch, C. W. (1987). Spontaneous speech in senile dementia and aphasia: Implications for a neurolinguistic model of language production. *Cognition, 27,* 247–274.

Blanken, G., Dittmann, J., Haas. J-C., & Wallesch, C-W. (1987). Spontaneous speech in senile dementia and aphasia: implications for a neurolinguistic model of language production. *Cognition, 27,* 219–274.

Bleuler, E. (1911). *Dementia praecox or the group of schizophrenias.* New York: International Universities Press.

Blevins, J. (1995). The syllable in phonological theory. In J. Goldsmith (Ed.), *The handbook of phonological theory* (pp. 206–244). Cambridge, MA: Basil Blackwell.

Blomert, L. (1990). What functional assessment can contribute to setting goals for aphasia threrapy. *Aphasiology, 4,* 307–320.

Blomert, L. (1994). *Assessment and recovery of verbal communication in aphasia.* Published dissertation. University of Nijmegen, The Netherlands.

Blomert, L., & de Roo, E. (1992). Recovery of language functions during the first year: Syntactic and verbal communicative changes. Poster presented at the 10th European Workshop on Cognitive Neuropsychology, Bressanone, Italy.

Blomert, L., & de Roo, E. (1995). Evolution of presumed syntactic deficits. Abstract. *Brain and Language, 51,* 96–98.

Blomert, L., Kean, M.-L., Koster, C., & Schokker, J. (1994). Amsterdam-Nijmegen Everyday Language Test: Construction, reliability and validity. *Aphasiology, 8,* 381–407.

Blonder, L., Pickering, J., Heath, R., Smith, C., & Butler, S. (1995). Prosodic characteristics of speech pre- and post-right hemisphere stroke. *Brain and Language, 51,* 318–335.

Blonder, L. X., Bowers, D., & Heilman, K. M. (1992). The role of the right hemisphere in emotional communication. *Brain, 114,* 1115–1127 [published erratum appeared in *Brain, 115* (1992), 654].

Blonder, L. X., Gur, R., & Gur, R. (1989). The effects of right and left hemiparkinsonism on prosody. *Brain and Language, 36,* 193–207.

Bloom, R. L., Obler, L. K., De Santi, S., & Ehrlich, J. (Eds.). (1994). *Discourse analysis and application: Studies in adult clinical populations.* Hillsdale, NJ: Lawrence Erlbaum.

Blumstein, S. (1990). Phonological deficits in aphasia: Theoretical perspectives. In A. Caramazza (Ed.), *Cognitive neuropsychology and neurolinguistics* (pp. 33–53). Hillsdale, NJ: Lawrence Erlbaum.

Blumstein, S. (1991). Phonological aspects of aphasia. In M. T. Sarno (Ed.), *Acquired aphasia* (2nd ed., pp. 151–180). New York: Academic Press.

Blumstein, S. (1994). Impairments of speech production and speech perception in aphasia. *Philosophical Transactions of the Royal Society of London B, 346,* 29–36.

Blumstein, S., Alexander, M., Ryalls, J., Katz, W., & Dworetzky, B. (1987). On the nature of the foreign accent syndrome: A case study. *Brain and Language, 31,* 215–244.

Blumstein, S., Baker, E., & Goodglass, H. (1977). Phonological factors in auditory comprehension in aphasia. *Neuropsychologia, 15,* 19–30.

Blumstein, S., & Baum, S. (1987). Consonant production deficits in aphasia. In J. Ryalls (Ed.), *Phonetic approaches to speech production in aphasia and related disorders* (pp. 3–21). Boston: Little, Brown.

Blumstein, S., Burton, M., Baum, S., Waldstein, R., Katz, D. (1994). The role of lexical status on the phonetic categorization of speech in aphasia. *Brain and Language, 46,* 181–197.

Blumstein, S., Cooper, W. E., Goodglass, H., Statlender, S., & Gottlieb, J. (1980). Production deficits in aphasia: A voice-onset time analysis. *Brain and Language, 9,* 153–170.

Blumstein, S. E. (1973). *A phonological investigation of aphasic speech.* The Hague: Mouton.

Boatman, D., Crone, N. E., Lesser, R. P., Nathan, S., Hart, J., Schwerdt, P., Sieracki, J. M., Poon, P., Webber, R., Uematsu, S., & Gordon, B. (1992). The localization of speech perception processes using direct cortical electrical interference and electrocorticography. *Epilepsia, 33,* 119.

Bobes, M. A., Valdés-Sosa, M., & Olivares, E. (1994). An ERP study of expectancy violation in face perception. *Brain and Cognition, 26,* 1–22.

Bogen, J. E. (1969). The other side of the brain II: An appositional mind. *Bulletin of the Los Angeles Neurological Societies, 324,* 191–219.

Bogen, J. E. (1979). A systematic quantitative study of anomia, tactile cross-retrieval and verbal cross-cueing in the long term following complete cerebral commissurotomy. Invited address, Academy of Aphasia, San Diego, 1979.

Bogen, J. E. (1993). The callosal syndrome. In K. M. Heilman & E. Valenstein (Eds.), *Clinical Neuropsychology* (3rd ed., pp. 337–407). New York: Oxford University Press.

Bogen, J. E. (1997). Does cognition in the disconnected right hemisphere require right hemisphere possession of language. In D. Van Lancker (Ed.), special issue of *Brain and Language,* "Current studies of right hemisphere function," *57,* 12–21.

Boldton, J. S. (1911). A contribution to the localization of cerebral function, based on the clinico-pathological study of mental disease: Chapter 1. *Brain, 33,* 26–37.

Boles, D. B., & Douglas, C. E.(1983). Visual and phonetic codes and the process of generation in letter matching. *Journal of Experimental Psychology: Human Perception and Performance, 9,* 657–674.

Boliek, C. A., & Obrzut, J. E. (1994). Perceptual laterality in developmental learning disabilities. In R. J. Davidson & K. Hugdahl (Eds.), *Brain asymmetry* (pp. 637–658). Cambridge: MIT Press.

Bolinger, D. (1964). Around the edge of language: Intonation. In D. Bolinger (Ed.), *Intonation* (pp. 19–29). Harmondsworth, UK: Penguin Books.

Boller, F. (1977). Johann Baptist Schmidt: A pioneer in the history of aphasia. *Archives of Neurology, 34,* 306–307.

Boller, F., Becker, J. T., Holland, A. L., Forbes, M. M., Hood, P. C., & McGonigle-Gibson, K. L. (1991). Predictors of decline in Alzheimer's disease. *Cortex, 27,* 9–17.

Bolter, J. D. (1984). *Turing's man: Western culture in the computer age.* Chapel Hill: University of North Carolina Press.

Borod, J. C. (1993). Cerebral mechanisms underlying facial, prosodic, and lexical emotional expression: A review of neuropsychological studies and methodological issues. *Neuropsychology, 7,* 445–463.

Bottini G., Corcoran, R., Sterzi, R., Paulesu, E., Schenone, P., Scarpa, P., & Frackowiak, R. S. J. (1994). The role of the right hemisphere in the interpretation of figurative aspects of language. *Brain, 117,* 1241–1253

Bouillaud, J. B. (1825). Recherches cliniques propre à démontrer que la perte de la parole correspond à la lésion des lobules antérieurs du cerveau, et à confirmer l'opinion de M. Gall, sur le siège de l'organe du langage articulé. (Mémoire lu à l'Académie royale de Médecine, le 21 février 1825.) *Archives Générales de Médecine, 3*(8), 25–45.

Bouton, C. (1991). *Neurolinguistics: Historical and theoretical perspectives.* New York: Plenum Press.

Bowers, D., Bauer, R. M., & Heilman, K. M. (1993). The nonverbal affect lexicon: Theoretical perspectives from neurological studies of affect perception. *Neuropsychology, 7,* 433–444.

Bowers, D., Coslett, H. B., Bauer, R. M., Speedie, L. J., & Heilman, K. M. (1987). Comprehension of emotional prosody following unilateral hemisphere lesions: Processing defect versus distraction defect. *Neuropsychologia, 25,* 317–328.

Braak, H., & Braak, E. (1991). Neuropathological stageing of Alzheimer-related changes. *Acta Neuropathologica, 82,* 239–259.

Braak, H., & Braak, E. (1992). Cortical morphological changes in dementia. In I. Kostovic, S. Knezevic, H. M. Wisneiwsi, & G. J. Spilich (Eds.), *Neurodevelopment, aging and cognition* (pp. 215–226). Boston: Birkhauser.

Bracco, L., Gallato, R., Grigoletto, F., Lippi, A., Pedone, D., Bino, G., Lazzaro, M. P., Carella, M., Pozzilli, C., Giometto, B., & Amaducci, L. (1994). Factors affecting course and survival in Alzheimer's disease: A 9–year longitudinal study. *Archives of Neurology, 51,* 1213–1219.

Bradley, L., & Bryant, P. E. (1983). Categorising sounds and learning to read: A causal connection. *Nature, 301,* 419–421.

Bradshaw, J. L., & Nettleton, N. C. (1988). Monaural asymmetries. In K. Hugdahl (Ed.), *Handbook of dichotic listening: Theory, methods and research* (pp. 45–69). Chichester, UK: John Wiley.

Branch, C., Milner, B., & Rasmussen, T. (1964). Intracarotid sodium amytal for the lateralization of cerebral speech dominance: Observations in 123 patients. *Journal of Neurosurgery, 21,* 399–405.

Branch, W. B., Cohen, M. J., & Hynd, G. W. (1995). Academic achievement and attention-deficit/hyperactivity disorder in children with left- or right-hemisphere dysfunction. *Journal of Learning Disabilities, 28,* 35–43.

Brandt, J., Folstein, S. E., & Folstein, M. F. (1988). Differential cognitive impairment in Alzheimer's and Huntington's disease. *Annals of Neurology, 23,* 555–561.

Brauer, D., McNeil, M. R., Duffy, J. R., Keith, R. L., & Collins, M. J. (1989). The differentiation of normal from aphasic performance using PICA discriminant function scores. *Clinical Aphasiology, 18,* 117–129.

Breedin, S., Saffran, E., & Coslett, H. (1994). Reversal of the concreteness effect in a patient with semantic dementia. *Cognitive Neuropsychology, 11,* 617–660.

Breedin, S. D., & Martin, R. C. (1996). Patterns of verbs impairment in aphasia: An analysis of four cases. *Cognitive Neuropsychology, 13,* 51–91.

Breitmeyer, B. G. (1993). Sustained (P) and transient (M) channels in vision: A review and implications for reading. In D. M. Willows, R. S. Kruk, & E. Corcos (Eds.), *Visual processes in reading and reading disabilities* (pp. 95–110). Hillsdale, NJ: Lawrence Erlbaum.

Brenneise-Sarshad, R., Nicholas, L. E., & Brookshire, R. H. (1991). Effects of apparent listener knowledge and picture stimuli on aphasic and non-brain-damaged speakers' narrative discourse. *Journal of Speech and Hearing Research, 34,* 168–176.

Brentari, D., Poizner, H., & Kegl, J. (1995). Aphasic and Parkinsonian signing: Differences in phonological disruption. *Brain and Language, 48,* 69–105.

Broadbent, G. (1872). *On the cerebral mechanisms of speech and thought.* London (cited by Wallesch & Papagno, 1988).

Broca, P. P. (1861a). Nouvelle observation d'aphémie produite par une lésion de la troisième circonvolution frontale. *Bulletin de la Société d'anatomie, 36,* 398–407.

Broca, P. P. (1861b). Remarques sur le siège de la faculté du langage articulé; suivies d'une observation d'aphémie (Perte de la parole). *Bulletin de la Société anatomie de Paris, 36,* 330–357.

Broca, P. P. (1863). Localisation des fonctions cérébrales. Siège du langage articulé. *Bulletin de la Société d'Anthropologie, 4,* 200–203.

Broca, P. P. (1865a). Du siège de la faculté du langage articulé. *Bulletin de la Société d'anthropologie, 7,* 377.

Broca, P. P. (1865b). Sur le siège de la faculté du langage articulé. *Bulletin de la Société d'Anthropologie, 6,* 337–393.

Brodin, J., & Magnusson, M. (Eds.) (1993). Virtual reality and disability: Proceedings of the first Nordic conference on virtual reality and disability (Report No. 9). Stockholm: Department of Education, Stockholm University, Sweden.

Brody, J. E. (1992). Brain injuries cause aphasia, difficulty in speaking, writing and understanding words. *New York Times,* June 10, B7, C13.

Brookshire, R. H., & Nicholas, L. E. (1993). *Discourse comprehension test.* Tucson, AZ: Communication Skill Builders.

Brookshire, R. H., & Nicholas, L. E. (1994). Speech sample size and test–retest stability of connected speech measures for adults with aphasia. *Journal of Speech and Hearing Research, 37,* 399–407.

Brookshire, R. H., & Nicholas, L. E. (1994). Test–retest stability of measures of connected speech in aphasia. *Clinical Aphasiology, 22,* 119–133.

Broom, Y. M., & Doctor, E. A. (1995a). Developmental phonological dyslexia: A case study of the efficacy of a remediation programme. *Cognitive Neuropsychology, 12,* 725–766.

Broom, Y. M., & Doctor, E. A. (1995b). Developmental surface dyslexia: A case study of the efficacy of a remediation programme. *Cognitive Neuropsychology, 12,* 69–110.

Brown, G. D. A. (in press). Connectionism, phonology, reading and regularity in developmental dyslexia. *Brain and Language.*

Brown, J. W. (1973). *Aphasia by Arnold Pick.* Springfield, IL: C. C. Thomas.

Brown, J. W. (1988). *The life of the mind: Selected papers.* Hillsdale, NJ: Lawrence Erlbaum.

Brown, J. W., & Grober, E. (1983). Age, sex, and aphasia type: Evidence for a regional cerebral growth process underlying lateralization. *Journal of Nervous and Mental Disease, 171,* 431–434.

Brown, J. W., & Jaffe, J. (1975). Hypothesis on cerebral dominance. *Neuropsychologia, 13,* 107–110.

Brown, P., & Levinson, S. (1987). *Politeness: Some universals in language use.* Cambridge: Cambridge University Press.

Brown, R. G., & Marsden, C. D. (1988). Internal versus external cues and the control of attention in Parkinson's disease. *Brain, 111,* 323–345.

Brown, R. M., Crane, A. M., & Goldman, P. S. (1979). Regional distribution of monoamines in the cerebral cortex and subcortical structures of the rhesus monkey: Concentrations and in vivo synthesis rates. *Brain Research, 168,* 133–150.

Brown, W. S., Larson, E. B., & Jeeves, M. A. (1994). Directional asymmetries in interhemispheric transmission time: Evidence from visual evoked potentials. *Neuropsychologia, 32,* 439–448.

Brownell, H., & Martin, G. (in press). In M. Beeman & C. Chiarello (Eds.), *Getting it right: The cognitive neuroscience of right hemisphere language comprehension.* Hillsdale, NJ: Lawrence Erlbaum.

Brownell, H. H., Carroll, J. J., Rehak, A., & Wingfield, A. (1992). The use of pronoun anaphora and speaker mood in the interpretation of conversational utterances by right hemisphere brain-damaged patients. *Brain and Language, 43,* 121–147.

Brownell, H. H., Gardner, H., Prather, P., & Martino, G. (1995). Language, communication and the right hemisphere. In H. S. Kirshner (Ed.), *Handbook of neurological speech and language disorders* (pp. 325–349). New York: Marcel Dekker.

Brownell, H. H., & Joanette, Y. (Eds.). (1993). *Narrative discourse in neurological impaired and normal aging adults.* San Diego: Singular Publishing.

Brownell, H. H., Michel, D., Powelson, J., & Gardner, H. (1983). Surprise but not coherence: Sensitivity to verbal humor in right-hemisphere patients. *Brain and Language, 18,* 20–27.

Brownell, H. H., Potter, H. H., Bihrle, A. M., & Gardner, H. (1986). Inference deficits in right brain-damaged patients. *Brain and Language, 27,* 310–321.

Brownell, H. H., Simpson, T. L., Bihrle, A. M., Potter, H. H., & Gardner, H. (1990). Appreciation of metaphoric alternative word meanings by left and right brain-damaged patients. *Neuropsychologia, 28,* 375–384.

Bruder, G. E., Quitkin, F. M., Stewart, J. W., Martin, C., Voglmaier, M., & Harrison, W. M. (1989). Cerebral laterality and depression: Differences in perceptual asymmetry among diagnostic subtypes. *Journal of Abnormal Psychology, 98,* 177–186.

Brumfitt, S. (1995). Psychotherapy in aphasia. In C. Code & D. Müller (Eds.), *Treatment of aphasia: From theory to practice* (pp. 18–28). London: Whurr Publishers.

Brun, A. (1993). Frontal lobe dementia of the non-Alzheimer type revisited. *Dementia, 4,* 126–131.

Brundert, S., Elger, C. E., Solymosi, L., Kurthen, M., & Linke, D. (1993). Der selektive Amobarbitaltest in der Epileptologie. *Radiologe, 33,* 213–218.

Brunn, J. L., & Farah, M. J. (1991). The relation between spatial attention and reading: Evidence from the neglect syndrome. *Cognitive Neuropsychology, 8,* 59–75.

Bryan, K. (1989). Language prosody and the right hemisphere. *Aphasiology, 3,* 285–299.

Bryden, M., & Ley, R. (1983). Right-hemisphere involvement in the perception and expression of emotion in normal humans. In K. Heilman & P. Satz (Eds.), *Neuropsychology of human emotion* (pp. 6–44). New York: Guilford Press.

Bryden, M. P. (1988). An overview of the dichotic listening procedure and its relation to cerebral organization. In K. Hugdahl (Ed.), *Handbook of dichotic listening: Theory, methods and research* (pp. 1–43). Chichester, UK: John Wiley.

Bub, D. N., & Arguin, M. (1995). Visual word activation in pure alexia. *Brain and Language, 49,* 77–103.

Bub, J., & Bub, D. (1988). On the methodology of single-case studies in cognitive neuropsychology. *Cognitive Neuropsychology, 5,* 565–582.

Buckingham, H. (1992). The mechanisms of phonemic paraphasia. *Clinical Linguistics and Phonetics, 6,* 41–63.

Buckner, R. L., Corbetta, M., Schatz, J., Raichle, M. E., & Petersen, S. E. (1996). Preserved speech abilities and compensation following prefrontal damage. *Proceedings of the National Academy of Sciences, USA, 93,* 1249–1253.

Buckner, R. L., Raichle, M. E., & Petersen, S. E. (1995). Dissociation of human prefrontal cortical areas across different speech production tasks and gender groups. *Journal of Neurophysiology, 74,* 2163–2173.

Buhlman-Fleming, M. B., & Bryden, M. P. (1994). Simultaneous verbal and affective laterality effects. *Neuropsychologia, 32,* 787–797.

Burke, H. L., Yeo, R. A., Delaney, H. D., & Conner, L. (1993). CT scan cerebral hemispheric asymmetries: Predictors of recovery from aphasia. *Journal of Clinical and Experimental Neuropsychology, 15,* 191–204.

Burklund, C. W., & Smith, A. (1977). Language and the cerebral hemispheres. *Neurology, 27,* 627–633.

Burr, C. W. (1905). Loss of the signed language in a deaf-mute from cerebral tumor and softening. *New York Medical Journal, 81,* 1106–1108.

Butters, N., Sax, D., Montgomery, K., & Tarlow, S. (1978). Comparison of the neuropsychological deficits associated with early and advanced Huntington's disease. *Archives of Neurology, 35,* 585–589.

Butterworth, B. (1983). Lexical representation. In B. Butterworth (Ed.), *Language Production* (Vol. 2, pp. 257–294). New York: Academic Press.

Butterworth, B. (1992). Disorders of phonological encoding. *Cognition, 42,* 261–286.

Butterworth, B., Campbell, R., & Howard, D. (1986). The uses of short-term memory: A case study. *Quarterly Journal of Experimental Psychology, 38A,* 705–737.

Butterworth, B., Howard, D., & McLoughlin, P. (1984). The semantic deficit in aphasia: the relationship between semantic errors in auditory comprehension and picture naming. *Neuropsychologia, 22,* 409–426.

Buxbaum, D. J., & Coslett, H. B. (in press). Deep dyslexic phenomena in a letter-by-letter reader. *Brain and Language.*

Buytenhuijs, E. L., Berger, H. J., Van Spaendonck, K. P., Horstink, M. W., Borm, G. F., & Cools, A. R. (1994). Memory and learning strategies in patients with Parkinson's disease. *Neuropsychologia, 32,* 335–342.

Byng, S. (1988). Sentence processing deficits: Theory and therapy. *Cognitive Neuropsychology, 5,* 629–676.

Byng, S. (1995). What is aphasia therapy? In C. Code & D. Muller (Eds.), *The treatment of aphasia: From theory to practice.* London: Whurr Publishers.

Byng, S., & Black, M. (1995). What makes a therapy? Some parameters of therapeutic intervention in aphasia. *European Journal of Disorders of Communication, 30,* 303–316.

Byrne, J., Dywan, C., & Connolly, J. (1995a). An innovative method to assess the receptive vocabulary of children with cerebral palsy using event-related brain potentials. *Journal of Clinical and Experimental Neuropsychology, 17,* 9–19.

Byrne, J., Dywan, C., & Connolly, J. (1995b). Assessment of children's receptive vocabulary using brain event-related potentials: Development of a clinically valid test. *Child Neuropsychology, 1,* 211–223

Bzoch, K., & League, R. (1978). *Assessing language skills in infancy.* Baltimore: University Park Press.

Cambier, J., Elghozi, D., Signoret, J. L., & Henin, D. (1983). Contribution de l'hémisphère droit au langage des aphasiques: Disparition de ce langage après lésion droite. *Revue Neurologique, 139,* 55–63.

Cambier, J., Masson, M., Viader, F., Limodier, J., & Strube, A. (1985). Le syndrome frontal de la maladie de Steele–Richardson–Olszewski. *Revue Neurologie, 141,* 528–536.

Campbell, J. I. D., & Clark, J. M. (1988). An encoding complex view of cognitive number processing: Comment on McCloskey, Sokol, & Goodman (1986). *Journal of Experimental Psychology: General, 117,* 204–214.

Campbell, S., & Whitaker, H. A. (1986). Cortical maturation and developmental neurolinguistics. In J. E. Obrzut & G. W. Hynd (Eds.), *Child neuropsychology.* Vol. I, *Theory and research* (pp. 55–72). New York: Academic Press.

Canadian study of health and aging working group. (1994). Canadian study of health and aging: Study methods and prevalence of dementia. *Canadian Medical Association Journal, 150*, 899–913.

Cancelliere, A., & Kertesz, A. (1990). Lesion localization in acquired deficits of emotional expression and comprehension. *Brain and Cognition, 13*, 133–147.

Caplan, D. (1986). In defense of agrammatism. *Cognition, 24*, 263–276.

Caplan, D. (1987). *Neurolinguistics and linguistic aphasiology.* Cambridge, UK: Cambridge University Press.

Caplan, D. (1988). On the role of group studies in neuropsychology and pathopsychological research. *Cognitive Neuropsychology, 5*, 535–548.

Caplan, D. (1991). Agrammatism is a theoretically coherent aphasic category. *Brain and Language, 110*, 274–281.

Caplan, D. (1992). *Language: Structure, processing, and disorders.* Cambridge: MIT Press.

Caplan, D. (1995). Language disorders. In R. L. Mapou, & J. Spector (Eds.), *Clinical neuropsychological assessment: A cognitive approach.* New York: Plenum Press.

Caplan, D., Baker, C., & Dehaut, F. (1985). Syntactic determinants of sentence comprehension in aphasia. *Cognition, 21*, 117–175.

Caplan, D., & Hildebrandt, N. (1988). *Disorders of syntactic comprehension.* Cambridge: MIT Press.

Caplan, D., Hildebrandt, N., & Makris, N. (1996). Location of lesions in stroke patients with deficits in syntactic processing in sentence comprehension. *Brain, 119*, 933–949.

Caplan, D., & Waters, G. S. (1990). Short-term memory and language comprehension: A critical review of the neuropsychological literature. In G. Vallar & T. Shallice (Eds.), *Neuropsychological impairments of short-term memory* (pp. 337–389). Cambridge: Cambridge University Press.

Caplan, D., & Waters, G. S. (1995). Aphasic disorders of syntactic comprehension and working memory capacity. *Cognitive Neuropsychology, 12*, 637–649.

Cappa, S., Cavalotti, G., Vignolo, L. A. (1981). Phonemic and lexical errors in fluent aphasia: correlation with lesion site. *Neuropsychologia, 19*, 171–177.

Cappa, S., & Vallar, G. (1992). The role of the left and right hemispheres in the recovery from aphasia. *Aphasiology, 6*, 354–372.

Cappa, S. F., Miozzo, A., & Frugoni, M. (1994). Glossolalic jargon after a right hemispheric stroke in a patient with Wernicke's aphasia. *Aphasiology, 8*, 83–87.

Cappa, S. F., Perani, D., Bressi, S., Paulesu, E., Franceschi, M., & Fazio, F. (1993). Crossed aphasia: A positron emission tomography follow-up study of two cases. *Journal of Neurology Neurosurgery and Psychiatry, 56*, 665–671.

Cappa, S. F., Perani, D., Grassi, F., Bressi, S., Alberoni, M., Franceschi, M., Bettinardi, V., Todde, S., Fazio, F. (1997). A PET follow-up study of recovery after stroke in acute aphasics. *Brain and Language, 56*, 55–67.

Cappa, S. F., Perani, D., Vallar, G., Paulesu, E., Bressi, S., Alberoni, M., Franceschi, M., Bettinardi, V., Rizzo, G., Matarrese, M., Lenzi, G. L., Canal, N., & Fazio, F. (1991). Recovery from aphasia after acute stroke: A PET metabolic flow up study. *Neurology, 41* (Suppl. 1), 885S.

Caprez, G. (1992). Sinn und Unsinn des Computereinsatzes in der neuropsychologischen Rehabilitation. In V. M. Roth (Ed.), *Computer in der Sprachtherapie* (pp. 17–25). Tübingen: Gunter Narr Verlag.

Caramazza, A. (1984). The logic of neuropsychological research and the problem of patient classification in aphasia. *Brain and Language, 21*, 9–20.

Caramazza, A. (1986). On drawing inferences about the structure of normal cognitive systems from the analysis of patterns of impaired performance: The case for single-patient studies. *Brain and Cognition, 5*, 41–66.

Caramazza, A. (1989). Cognitive neuropsychology and rehabilitation: An unfulfilled promise? In X. Seron & G. Deloche (Eds.), *Cognitive approaches in neuropsychological rehabilitation* (pp. 383–398). Hillsdale, NJ: Lawrence Erlbaum.

Caramazza, A. (1992). Is cognitive neuropsychology possible? *Journal of Cognitive Neuroscience, 4*, 80–95.

Caramazza, A. (1996). The brain's dictionary. *Nature, 380*, 485–486.

Caramazza, A., & Badecker, W. (1989). Patient classification in neuropsychological research. *Brain and Cognition, 10,* 256–295.

Caramazza, A., Berndt, R. S., & Basili, A. G. (1983). The selective impairment of phonological processing: a Case study. *Brain and Language, 18,* 128–174.

Caramazza, A., Berndt, S., Basili, A., & Koller, J. (1981). Syntactic processing deficits in aphasia. *Cortex, 17,* 333–348.

Caramazza, A., & Hillis, A. (1989a). The disruption of sentence production: A case of selected deficit to positional level processing. *Brain and Language, 35,* 625–650.

Caramazza, A., & Hillis, A. (1989b). The disruption of sentence production: Some dissociations. *Brain and Language, 36,* 625–650.

Caramazza, A., Hillis, A., Rapp, B., & Romani, C. (1990). The multiple semantics hypothesis: Multiple confusions? *Cognitive Neuropsychology, 7,* 161–189.

Caramazza, A., & Hillis, A. E. (1990a). Levels of representations, co-ordinate frames, and unilateral neglect. *Cognitive Neuropsychology, 7,* 391–445.

Caramazza, A., & Hillis, A. E. (1990b). Where do semantic errors come from? *Cortex, 26,* 95–122.

Caramazza, A., & Hillis, A. E. (1991). Lexical organization of nouns and verbs in the brain. *Nature, 349,* 788–790.

Caramazza, A., Laudanna, A., & Romani, C. (1988). Lexical access and inflectional morphology. *Cognition, 28,* 297–332.

Caramazza, A., & McCloskey, M. (1987). Dissociations of calculation processes. In G. Deloche & X. Seron (Eds.), *Mathematical disabilities: A cognitive neuropsychological perspective* (pp. 221–234). Hillsdale, NJ: Lawrence Erlbaum.

Caramazza, A., & McCloskey, M. (1988). The case for single-patient studies. *Cognitive Neuropsychology, 5,* 517–528.

Caramazza, A., & Zurif, E. B. (1976). Dissociation of algorithmic and heuristic processes in language comprehension: Evidence from aphasia. *Brain and Language, 3,* 572–582.

Cardebat, D. (1987). Incohérence narrative: Analyse comparée de récits de patients aphasiques et de patients déments. *Cahiers du Centre Interdisciplinaire des Sciences du Langage, 6,* 151–175.

Cardebat, D., Démonet, J.-F., Celsis, P., Puel, M., Viallard, G., & Marc-Vergnes, J.-P. (1994). Right temporal compensatory mechanisms in a deep dysphasic patient: A case report with activation study by SPECT. *Neuropsychologia, 32,* 97–103.

Cardebat, D., Démonet, J.-F., Puel, M., Nespoulous, J.-L., & Rascol, A. (1991). Langage et Démences [Language and dementias]. In M. Habib, Y. Joanette, & M. Puel (Eds.), Démences et syndromes démentiels. Approche neuropsychologique (pp. 153–164). Paris: Masson.

Carlomagno, S., & Parlato, V. (1989). Writing rehabilitation in brain-damaged adult patients: A cognitive approach. In X. Seron & G. Deloche (Eds.), *Cognitive approaches in neuropsychological rehabilitation.* London: Lawrence Erlbaum.

Carpenter, P. A., Miyake, A., & Just, M. A. (1994). Working memory constraints in comprehension: Evidence from individual differences, aphasia and aging. In M. A. Gernsbacher (Ed.), *Handbook of psycholinguistics* (pp. 1075–1122). New York: Academic Press.

Carpenter, P. A., Miyake, A., & Just, M. A. (1995). Language comprehension: Sentence and discourse processing. *Annual Review of Psychology, 46,* 91–120.

Caselli, R. J. (1995). Progresssive aphasia in patients with motor neuron disease-dementia complex. *Brain and Language, 51,* 71–72.

Castles, A., & Coltheart, M. (1993). Varieties of developmental dyslexia. *Cognition, 47,* 149–180.

Castro-Caldas, A., & Bothello, M. (1980). Dichotic listening in the recovery of aphasia after stroke. *Brain and Language, 10,* 145–151.

Chaika, E. (1974). A linguist looks at "schizophrenic" language. *Brain and Language, 1,* 257–276.

Chaika, E. (1982). A unified explanation for the diverse structural deviations reported for adult schizophrenics with disrupted speech. *Journal of Communication Disorders, 15,* 167–189.

Chapey, R. (Ed.) (1994). *Language intervention strategies in adult aphasia* (3rd ed.). Baltimore: Williams & Wilkins.

Chapman, R. M., Ilmoniemi, R., Barbanera, S., & Romani, G. L. (1984). Selective localization of alpha brain activity with neuromagnetic measurements. *Electroencephalography and Clinical Neurophysiology, 58,* 569–572.

Chapman, R. S., Schwartz, S. E., & Bird, E. K. (1991). Language skills of children and adolescents with Down syndrome: I. Comprehension. *Journal of Speech and Hearing Research, 34,* 1106–1120.

Chapman, S. B., Culhane, K. A., Levin, H. S., Harward, H., Mendelsohn, D., Ewing-Cobbs, L., Fletcher, J. M., & Bruce, D. (1992). Narrative discourse after closed head injury in children and adolescents. *Brain and Language, 43,* 42–65.

Chapman, S. B., & Ulatowska, H. A. (1989). Discourse in aphasia: Integration deficits in processing reference. *Brain and Language, 36,* 651–668.

Chapman, S. B., & Ulatowska, H. K. (1994). Differential diagnosis in aphasia. In R. Chapey (Ed.), *Language intervention in adult aphasia* (pp. 121–131). Baltimore: Williams & Wilkins.

Charolles, M. (1986). Grammaire de texte—Théorie du discours—Narrativité. *Pratiques, 11–12,* 133–154.

Chary, P. (1986). Aphasia in a multilingual society: A preliminary study. In J. Vaid (Ed.), *Language processing in bilinguals* (pp. 183–197). Hillsdale, NJ: Lawrence Erlbaum.

Chawluk, J. B., Grossman, M., Calcano-Perez, J. A., Alavi, A., Hurtig, H. I., & Reivich, M. (1990). Positron emission tomographic studies of cerebral metabolism in Alzheimer's disease. In M. F. Schwartz (Ed.), *Modular deficits in Alzheimer-type dementia* (pp. 101–141). Cambridge: MIT Press.

Chenery, H. J. (1996). Semantic priming in Alzheimer's dementia. *Aphasiology, 10,* 1–20.

Chenery, H. J., & Murdoch, B. E. (1994). The production of narrative discourse in response to animations in persons with dementia of the Alzheimer's type: Preliminary findings. *Aphasiology, 8,* 159–171.

Cherry, B., Hellige, J. B., & McDowd, J. M. (1995). Age differences and similarities in patterns of cerebral hemispheric asymmetry. *Psychology and Aging, 10,* 191–203.

Chertkow, H., & Bub, D. (1990). Semantic memory loss in Alzheimer-type dementia. In M. F. Schwartz (Ed.), *Modular deficits in Alzheimer-type dementia* (pp. 207–244). Cambridge: MIT Press.

Chertkow, H., & Bub, D. (1990). Semantic memory loss in dementia of Alzheimer's type. What do various measures measure. *Brain, 113,* 397–417.

Chertkow, H., Bub, D., Bergman, H., Bruemmer, A., Merling, A., & Rothfleisch, J. (1994). Increased semantic priming in patients with dementia of the Alzheimer type. *Journal of Clinical and Experimental Neuropsychology, 16,* 608–622.

Chertkow, H., Bub, D., & Seidenberg, M. (1989). Priming and semantic memory loss in Alzheimer's disease. *Brain and Language, 36,* 420–446.

Chi, J. G., Dooling, E. C., & Gilles, F. H. (1977). Gyral development of the human brain. *Annals of Neurology, 1,* 86–93.

Chiarello, C. (1988). *Right hemisphere contributions to lexical semantics.* Berlin: Springer Verlag.

Chiarello, C. (1991). Interpretation of word meanings in the cerebral hemispheres: One is not enough. In P. J. Schwanenflugel (Ed.), *The psychology of word meanings* (pp. 251–278). Hillsdale, NJ: Lawrence Erlbaum.

Chiarello, C. (1995). Does the corpus callosum play a role in the activation and suppression of ambiguous word meanings? In F. L. Kitterle (Ed.), *Hemispheric communication: mechanisms and models* (pp. 177–188). Hillsdale, NJ: Lawrence Erlbaum.

Chiarello, C., Knight, R., & Mandel, M. (1982). Aphasia in a prelingually deaf woman. *Brain, 105,* 29–51.

Chiarello, C., & Maxfield, L. (1996) Varieties of interhemispheric inhibition, or how to keep a good hemisphere down. *Brain and Cognition, 30,* 81–109.

Chiarello, C., Maxfield, L., Richards, L., & Kahan, T. (1995). Activation of lexical codes for simultaneously presented words: Modulation by attention and pathway strength. *Journal of Experimental Psychology: Human Perception and Performance, 21,* 776–808.

Chiarello, C., & Nuding, S. (1987). Visual field effects for processing function and content words. *Neuropsychologia, 25,* 539–548.

Chiron, C., Raynaud, C., Tzourio, N., Diebler, C., Dulac, O., Zilbovicius, M., & Syrota, A. (1989). Regional cerebral blood flow by SPECT imaging in Sturge-Weber disease: An aid for diagnosis. *Journal of Neurology, Neurosurgery, and Psychiatry, 52,* 1402–1409.

Chollet, F., Di Piero, V., Wise, R. S. J., Brooks, D. J., Dolan, R. J., & Frackowiak, R. S. J. (1991). The functional anatomy of motor recovery after stroke in humans: A study with positron emission tomography. *Annals of Neurology, 29,* 63–71.

Chomsky, N. (1965). *Aspects of the theory of syntax.* Cambridge: MIT Press.

Chomsky, N. (1981). *Lectures on government and binding.* Dordrecht, The Netherlands: Foris.

Chomsky, N. (1993). A minimalist program for linguistic theory. In K. Hale & J. Keyser (Eds.), *The view from building 20* (pp. 1–52). Cambridge: MIT Press.

Chomsky, N. (1994). *Bare phrase structure.* MIT Occasional Papers in Linguistics. Cambridge: MIT Press.

Christiansen, J. A. (1995). Coherence violations and propositional usage in the narratives of fluent aphasics. *Brain and Language, 51,* 291–317.

Christman, S. (1994). Target-related neologism formation in jargonaphasia. *Brain and Language, 46,* 109–128.

Chugani, H. T., Mazziotta, J. C., & Phelps, M. E. (1989). Sturge-Weber syndrome: A study of cerebral glucose utilization with positron emission tomography. *Journal of Pediatrics, 114,* 244–253.

Chui, H. C., Teng, E. L., Henderson, V. W., & Moy, A. C. (1985). Clinical subtypes of dementia of the Alzheimer type. *Neurology, 35,* 1544–1550.

Chui, H. C., Victoroff, J. I., Margolin, D., Jagust, W., Shankle, R., & Katzman, R. (1992). Criteria for the diagnosis of ischemic vascular dementia proposed by the state of California Alzheimer's Disease Diagnostic and Treatment Centers. *Neurology, 42,* 473–480.

Cicone, M., Wapner, W., & Gardner, H. (1980). Sensitivity to emotional expressions and situations in organic patients. *Cortex, 16,* 145–158.

Cimino, C., Verfaellie, M., Bowers, D., & Heilman, K. (1991). Autobiographical memory: Influence of right hemisphere damage on emotionality and specificity. *Brain and Language, 15,* 106–118.

Cipolotti, L., & Butterworth, B. (1995). Toward a multiroute model of number processing: Impaired number transcoding with preserved calculation skills. *Journal of Experimental Psychology: General, 124,* 375–390.

Clark, A., Harvey, P., & Alpert, M. (1994). Medication effects on referent communication in schizophrenic patients: An evaluation with a structured task. *Brain and Language, 46,* 392–401.

Clark, H. H. (1973). The language-as-fixed-effect fallacy: A critique of language statistics in psychological research. *Journal of Verbal Learning and Verbal Behavior, 12,* 335–369.

Clark, H. H. (1985). Language use and language users. In G. Lindzey & E. Aronson (Eds.), *Handbook of social psychology* (3rd ed., Vol 2, pp. 179–231). New York: Random House.

Clark, L. E., & Knowles, J. B. (1973). Age differences in dichotic listening performance. *Journal of Gerontology, 28,* 173–178.

Clark, R. G. (1995). Fear and loathing in the amygdala. *Current Biology, 5,* 246–248.

Clarke, E., & Jacyna, L. S. (1987). *Nineteenth-century origins of neuroscientific concepts.* Berkeley: University of California Press.

Clarke, E., & O'Malley, C. D. (1968). *The human brain and spinal cord. A historical study.* Berkeley: University of California Press.

Clarke, J. M., Lufkin, R. B., & Zaidel, E. (1993). Corpus callosum morphometry and dichotic listening performance: Individual differences in functional interhemispheric inhibition? *Neuropsychologia, 31,* 547–557.

Clarke, J. M., & Zaidel, E. (1994). Anatomical-behavioral relationships: corpus callosum morphometry and hemispheric specialization. *Behavioral Brain Research, 64,* 185–202.

Clarke, S., Kraftsik, R., van der Loos, H., & Innocenti, G.M. (1989). Forms and measures in adult and developing human corpus callosum: Is there sexual dimorphism? *Journal of Comparative Neurology, 280,* 213–230.

Code, C. (1987). *Language, aphasia, and the right hemisphere.* Chichester, UK: John Wiley.

Code, C. (1997). "Can the right hemisphere speak?" In D. Van Lancker (Ed.), special issue of *Brain and Language,* "Current studies of right hemisphere function," *57,* 38–59.

Coelho, C. A., Liles, B. Z., & Duffy, R. J. (1991a). Analysis of conversational discourse in head-injured adults. *Journal of Head Trauma Rehabilitation, 6,* 92–98.

Coelho, C. A., Liles, B. Z., & Duffy, R. J. (1991b). Discourse analyses with closed head injured adults: Evidence for differing patterns of deficits. *Archives of Physical and Medical Rehabilitation, 72,* 465–468.

Coelho, C. A., Liles, B. Z., & Duffy, R. J. (1991c). The use of discourse analyses for the evaluation of higher level traumatically brain-injured adults. *Brain Injury, 5,* 381–392.

Coelho, C. A., Liles, B. Z., & Duffy, R. J. (1995). Impairments of discourse abilities and executive functions in traumatically brain injured adults. *Brain Injury, 9,* 471–477.

Cohen, B. (1978). Referent communication disturbances in schizophrenia. In S. Schwartz (Ed.), *Language and cognition in schizophrenia* (pp. 1–34). Hillsdale, NJ: Lawrence Erlbaum.

Cohen, D. (1972). Magnetoencephalography: Detection of the brain's electrical activity with a superconducting magnetometer. *Science, 175,* 664–666.

Cohen, G., & Freeman, R. (1978). Individual differences in reading strategies in relation to cerebral asymmetry. In J. Requin (Ed.), *Attention and performance VII* (pp. 411–426). Hillsdale, NJ: Lawrence Erlbaum.

Cohen, H., Bouchard, S., Scherzer, P., & Whitaker, H. A. (1994). Language and verbal reasoning in Parkinson's disease. *Neuropsychiatry, Neuropsychology and Behavioral Neurology, 7,* 166–175.

Cohen, H., Laframboise, M., Labelle, A., & Bouchard, S. (1993). Speech timing deficits in Parkinson's disease. *Journal of Clinical and Experimental Neuropsychology, 15,* 102–103.

Cohen, J. (1988). *Statistical power analysis of the behavioral sciences* (2nd ed.). Hillsdale, NJ: Lawrence Erlbaum.

Cohen, J. D., Dunbar, K., & McClelland, J. L. (1990). On the control of automatic processes: A parallel distributed account of the Stroop effect. *Psychological Review, 97,* 332–361.

Cohen, L., & Dehaene, S. (1991). Neglect dyslexia for numbers? A case report. *Cognitive Neuropsychology, 8,* 39–58.

Cohen, L., & Dehaene, S. (1995). Number processing in pure alexia: The effect of hemispheric asymmetries and task demands. *NeuroCase, 1,* 121–137.

Cohen, L., & Dehaene, S. (1996). Cerebral networks for number processing: Evidence from a case of posterior callosal lesion. *NeuroCase, 2,* 155–174.

Cohen, L., Dehaene, S., & Verstichel, P. (1994). Number words and number non-words: A case of deep dyslexia extending to Arabic numerals. *Brain, 117,* 267–279.

Cohen, M. J., Riccio, C. A., & Flannery, A. M. (1994). Expressive aprosodia following stroke to the right basal ganglia: A case report. *Neuropsychology, 8,* 242–245.

Colby, K. M., Christinaz, D., Parkison, R. C., Graham, S., & Karpf, C. (1981). A word-finding computer program with a dynamic lexical–semantic memory for patients with anomia using an intelligent speech prosthesis. *Brain and Language, 14,* 272–281.

Collins, A. M., & Loftus, E. F. (1975). A spreading-activation theory of semantic processing. *Psychological Review, 82,* 407–428.

Collins, A. M., & Quillian, M. R. (1969). Retrieval time from semantic memory. *Journal of Verbal Learning and Verbal Behavior, 8,* 240–247.

Coltheart, M. (1980). Deep dyslexia: A right hemisphere hypothesis. In M. Coltheart, K. Patterson, & J. C. Marshall (Eds.), *Deep dyslexia.* London: Routledge & Kegan Paul.

Coltheart, M. (1983). The right hemisphere and disorders of reading. In A. W. Young (Ed.), *Functions of the right cerebral hemisphere* (pp. 171–120). London: Academic Press.

Coltheart, M. (1985). Cognitive neuropsychology and the study of reading. In M. I. Posner & O. S. M. Marin (Eds.), *Attention and performance XI* (pp. 3–37). Hillsdale, NJ: Lawrence Erlbaum.

Coltheart, M., Bates, A., & Castles, A. (1994). Cognitive neuropsychology and rehabilitation. In M. J. Riddoch & G. W. Humphreys (Eds.), *Cognitive neuropsychology and cognitive rehabilitation* (pp. 17–37). Hove, UK: Lawrence Erlbaum.

Coltheart, M., & Byng, S. (1989). A treatment for surface dyslexia. In X. Seron & G. Deloche (Eds.), *Cognitive approaches in neuropsychological rehabilitation.* London: Lawrence Erlbaum.

Coltheart, M., Curtis, B., Atkins, P., & Haller, M. (1993). Models of reading aloud: Dual-route and parallel-distributed processing approaches. *Psychological Review, 100,* 589–608.

Coltheart, M., Patterson, K. E., & Marshall, J. C. (1987). *Deep dyslexia* (2nd ed.). London: Routledge & Kegan Paul.

Coltheart, M., Sartori, G., & Job, R. (Eds.). (1987). *The cognitive neuropsychology of language.* London: Lawrence Erlbaum.

Connolly, J. F., & Phillips, N. A. (1994). Event-related potential components reflect phonological and semantic processing of the terminal word of spoken sentences. *Journal of Cognitive Neuroscience, 63,* 256–266.

Conrad, R. (1971). The chronology of the development of covert speech in children. *Developmental Psychology, 24,* 505–514.

Conrad, R., & Hull, A. J. (1964). Information, acoustic confusion and memory span. *British Journal of Psychology, 55,* 429–432.

Cooper, I. S., Riklan, M., Stellar, S., Waltz, J. M., Levita, E., Ribera, V. A., & Zimmerman, J. (1968). A multidisciplinary investigation of neurosurgical rehabilitation in bilateral parkinsonism. *Journal of the American Geriatrics Society,* 16, 1177–1306.

Cooper, J. A., Sagar, H. J., & Sullivan, E. V. (1993). Short-term memory and temporal ordering in early Parkinson's disease: Effects of disease chronicity and medication. *Neuropsychologia, 31,* 933–949.

Cooper, W. (1983). Brain cartography: Electrical stimulation of processing sites or transmission lines? *The Behavioral and Brain Sciences, 6,* 212–213.

Cooper, W., & Klouda, G. (1987). Intonation in aphasic and right-hemisphere–damaged patients. In J. Ryalls (Ed.), *Phonetic approaches to speech production in aphasia and related disorders* (pp. 45–57). Boston: College Hill Press.

Cooper, W., Soares, C., Nicol, J., Michelow, D., & Goloskie, S. (1984). Clausal intonation after unilateral brain damage. *Language and Speech, 27,* 17–24.

Corballis, M. C., & Morgan, M. J. (1978). On the biological basis of human laterality: I. Evidence for a maturational left-right gradient. *Behavioral and Brain Sciences, 2,* 261–336.

Corballis, M. C., & Trudel, C. I. (1993). The role of the forebrain commissures in interhemispheric integration. *Neuropsychology, 7,* 1–19.

Corbett, A. J., McCusker, E. A., & Davidson, O. R. (1988). Acalculia following a dominant-hemisphere subcortical infarct. *Archives of Neurology,* 43, 964–966.

Corina, D. P. (1989). Recognition of affective and noncanonical linguistic facial expressions in hearing and deaf subjects. *Brain and Cognition, 9,* 227–237.

Corina, D. P., Bellugi, U., Kritchevsky, M., O'Grady-Batch, L., & Norman, F. (1990). *Spatial relations in signed versus spoken language: Cluses to right parietal functions.* Academy of Aphasia.

Corina, D. P., Kritchevsky, M., & Bellugi, U. (1992). Linguistic permeability of unilateral neglect: Evidence from American Sign Language. *Proceedings of the 14th Annual Conference of the Cognitive Science Society* (pp. 384–389). Hillsdale, NJ: Lawrence Erlbaum.

Corina, D. P., Kritchevsky, M., & Bellugi, U. (1996). Visual language processing and unilateral neglect: Evidence from American Sign Language. *Cognitive Neuropsychology, 13,* 321–351.

Corina, D. P., Poizner, H. P., Feinberg, T., Dowd, D., & O'Grady, L. (1992). Dissociation between linguistic and non-linguistic gestural systems: A case for compositionality. *Brain and Language, 43,* 414–447.

Corina, D. P., & Sandler, W. (1993). On the nature of phonological structure in sign language. *Phonology, 10,* 165–207.

Cornell, T. L. (1995). On the relation between representational and processing models of asyntactic comprehension. *Brain and Language, 50,* 304–324.

Coryell, J. (1985). Infant rightward asymmetries predict right-handedness in childhood. *Neuropsychologia, 23,* 269–271.

Coslett, H. B., & Monsul, N. (1994). Reading with the right hemisphere: Evidence from transcranial magnetic stimulation. *Brain and Language, 46,* 198–211.

Coslett, H. B., & Saffran, E. M. (1989). Evidence for preserved reading in "pure alexia." *Brain, 112* (April), 327–359.

Coslett, H. B., & Saffran, E. M. (1994). Mechanisms of implicit reading in alexia. In M. J. Farah & G. Ratcliff (Eds.), *The neuropsychology of high level vision* (pp. 299–330). Hillsdale, NJ: Lawrence Erlbaum.

Coslett, H. B., Saffran, E. M., Greenbaum, S., & Schwartz, H. (1993). Reading in pure alexia. *Brain, 116,* 21–37.

Cotman, C. W., & Nieto-Sampedro, M. (1985). Progress in facilitating the recovery of function after central nervous system trauma. *Annals of the New York Academy of Sciences, 457,* 83–104.

Cotman, C. W., Cummings, B. J., & Pike, C. J. (1993). Molecular cascades in adaptive versus pathological plasticity. In A. Gario (Ed.), *Neuroregeneration* (pp. 217–240). New York: Raven Press.

Cottrell, G. W. (1985). Connectionist parsing. In *Proceedings of the 7th Annual Cognitive Science Society* (pp. 216–225). Irvine, CA: Cognitive Science Society.

Coubes, P., Baldy-Moulinier, M., Zanca, M., Boire, J. Y., Child, R., Bourbotte, G., & Frerebeau, P. (1995). Monitoring sodium methohexital distribution with [99m Tc] HMPAO with single photon emission computed tomography during Wada test. *Epilepsia, 36,* 1041–1049.

Coulter, G. R., & Anderson, S. R. (1993). Introduction. In G. Coulter (Ed.), *Phonetics and phonology* (pp. 1–16). San Diego: Academic Press.

Courchesne, E., Townsend, J., Akshoomoff, N. A., Saitoh, O., Yeung-Courchesne, R., Lincoln, A. J., James, H. E., Haas, R. H., Schreibman, L., & Lau, L. (1994). Impairment in shifting attention in autistic and cerebellar patients. *Behavioural Neuroscience, 108,* 848–865.

Crerar, M. A., & Ellis, A. W. (1995). Computer-based therapy for aphasia: Towards second generation clinical tools. In C. Code & D. Müller (Eds.), *Treatment of aphasia: From theory to practice* (pp. 223–250). London: Whurr Publishers.

Crerar, M. A., Ellis, A. W., & Dean, E. C. (1996). Remediation of sentence processing deficits in aphasia using a computer-based microworld. *Brain and Language, 52,* 229–275.

Critchley, M. (1964). The neurology of psychotic speech. *British Journal of Psychiatry, 110,* 353–364.

Critchley, M. (1970). *The dyslexic child.* London: Heinemann.

Critchley, M. D. (1938). Aphasia in a partial deaf-mute. *Brain, 61,* 163–169.

Crocker, L., & Algina, J. (1986). *Introduction to classical and modern test theory.* Fort Worth, TX: Harcourt Brace Jovanovich.

Croisile, B., Adeleine, P., Carmoi, T., Aimard, G., & Trillet, M. (1995). Évaluation de l'orthographie dans la maladie d'Alzheimer [Evaluation of orthography in Alzheimer's disease]. *Revue de Neuropsychologie, 5,* 23–51.

Crone, N. E., Hart, J., Boatman, D., Lesser, R. P., & Gordon, B. (1994). Regional cortical activation during language and related tasks identified by direct cortical electrical recording. *Brain and Language, 47,* 466–468.

Crossley, M., & Hiscock, M. (1992). Age-related differences in concurrent-task performance of normal adults: Evidence for a decline in processing resources. *Psychology and Aging, 7,* 499–506.

Crosson, B. (1984). The role of the dominant thalamus in language: A review. *Psychological Bulletin, 96,* 491–517.

Crosson, B. (1985). Subcortical aphasia: A working model. *Brain and Language, 25,* 257–292.

Crosson, B. (1992). *Subcortical functions in language and memory.* New York: Guilford Press.

Crosson, B., & Hughes, C. W. (1987). Role of the thalamus in language: Is it related to schizophrenic thought disorder? *Schizophrenia Bulletin, 13,* 605–621.

Crosson, B., Moberg D. J., Boone, J. R., Rothi, L. J. G., Raymer, A. (in press). Category-specific naming deficit for medical terms after dominant thalamic/capsular hemorrhage. *Brain and Language.*

Crosson, B., Parker, J. C., Warren, R. L., Kepes, J. J., Kim, A. K., & Tulley, R. C. (1986). A case of thalamic aphasia with postmortem verification. *Brain and Language, 29,* 301–314.

Crosson, B., Rao, S. M., Woodley, S. J., Rosen, A. C., Hammeke, T. A., Bobholz, J. A., Cunningham, J. M., Fuller, S. A., Binder, J. R, & Cox, R. W. (1996). Mapping of semantic versus phonological versus orthographic verbal working memory in normal adults with *fMRI. NeuroImage, 3,* S538.

Crow, B. (1983). Topic shifts in couples' conversations. In R. T. Craig & K. Tracy (Eds.), *Conversational coherence: Form, structure and strategy* (pp. 136–156). Beverly Hills, CA: Sage.

Crystal, D. (1982). *Profiling linguistic disability.* London: Edward Arnold.

Crystal, D., Fletcher, P., & Garman, M. (1989). *The grammatical analysis of language disability* (2nd ed.). London: Cole and Whurr.

Cuenod, C. A., Bookheimer, S. Y., Hertz-Pannier, L., Zeffiro, T. A., Theodore, W. H., & Le Bihan, D. (1995). Functional MRI during word generation, using conventional equipment: A potential tool for language localization in the clinical environment. *Neurology, 45,* 1821–1827.

Culler, F. L., Jones, K. L., & Deftos, L. J. (1985). Impaired calcitonic secretion in patients with Williams syndrome. *Journal of Pediatrics, 107,* 720–723.

Cummings, J. L. (1985). *Clinical neuropsychiatry.* Orlando, FL: Grune & Stratton.

Cummings, J. L. (1986). Subcortical dementia: Neuropsychology, neuropsychiatry, and pathophysiology. *British Journal of Psychiatry, 149,* 682–697.

Cummings, J. L. (1988). Depression in vascular dementia. *Hillside Journal of Clinical Psychiatry, 10*(2), 209–231.

Cummings, J. L. (1993). Frontal-subcortical circuits and human behavior. *Archives of Neurology, 50,* 873–880.

Cummings, J. L., & Benson, D. F. (1992). *Dementia: A clinical approach* (2nd ed.). Boston: Butterworth-Heinemann.

Cummings, J. L., Benson, D. F., Hill, M. A., & Read, S. (1985). Aphasia in dementia of the Alzheimer type. *Neurology, 35,* 394–397.

Cummings, J. L., Benson, D. F., Walsh, M., & Levine, H. L. (1979). Left to right transfer of language dominance: A case study. *Neurology, 29,* 1547–1550.

Cummings, J. L., Darkins, A., Mendez, M., Hill, M. A., & Benson, D. F. (1988). Alzheimer's disease and Parkinson's disease: Comparison of speech and language alterations. *Neurology, 38,* 680–684.

Cummings, J. L., & Duchen, L. W. (1981). Klüver-Bucy syndrome in Pick's disease: Clinical and pathologic correlations. *Neurology, 31,* 1415–1422.

Cunningham, R., Farrow, V., Davies, C., & Lincoln, N. (1995). Reliability of the assessment of communicative effectiveness in severe aphasia. *European Journal of Disorders of Communication, 30,* 1–16.

Curtiss, S., Jackson C. A., Kempler D., Hanson R., & Metter E. J. (1986). Length vs. structural complexity in sentence comprehension in aphasia. *Clinical Aphasiology, 16,* 45–53.

Cushing, H. (1909). A note upon the faradic stimulation of the postcentral gyrus in conscious patients. *Brain, 32,* 44–54.

Cutler, A. (Ed.). (1982). *Slips of the tongue and language production.* Amsterdam: Walter de Gruyter/ Mouton.

Cutting, John. (1990). *The right cerebral hemisphere and psychiatric disorders.* Oxford: Oxford University Press.

Czopf, J. (1972). Über die Rolle der nicht dominanten Hemisphäre in der Restitution der Sprache der Aphasischen. *Archiv für Psychiatrie und Nervenkrankheiten, 216,* 162–171.

Dale, P., Bates, E., Reznick, S., & Morisset, C. (1989). The validity of a parent report instrument. *Journal of Child Language, 16,* 239–249.

Dallal, G. E. (1988). PITMAN: A FORTRAN program for exact randomization tests. *Computers in Biomedical Research, 21,* 9–15.

Damasio, A. (1994). *Decartes' error: Emotion, reason and the human brain.* New York: Avon Books.

Damasio, A., Bellugi, U., Damasio, H., Poizner, H., & VanGilder, J. (1986). Sign language aphasia during left-hemisphere amytal injection. *Nature, 322,* 363–365.

Damasio, A., & Van Hoesen, G. (1983). Emotional disturbances associated with focal lesions of the frontal lobe. In K. Heilman & P. Satz (Eds.), *The neuropsychology of human emotion: Recent advances* (pp. 85–100). New York: Guilford Press

Damasio, A. R. (1989). Time-locked multi-regional retroactivation: A systems-level proposal for the neural substrates of recall and recognition. *Cognition, 33,* 25–62

Damasio, A. R. (1992). Aphasia. *New England Journal of Medicine, 326,* 531–539.

Damasio, A. R., & Damasio, H. (1983). The anatomic basis of pure alexia. *Neurology, 33,* 1573–1583.

Damasio, A. R., & Damasio, H. (1992). Brain and language. *Scientific American, 267,* 89–95.

Damasio, A. R., & Geschwind, N. (1984). The neural basis of language. *Annual Review of Neuroscience, 7,* 127–147.

Damasio, A. R., & Tranel, D. (1993). Nouns and verbs are retrieved with differently distributed neural systems. *Proceedings of the National Academic of Sciences USA, 90,* 4957–4960.

Damasio, A. R., Tranel, D., & Damasio, H. (1990). Individuals with sociopathic behavior caused by frontal damage fail to respond autonomically to social stimuli. *Behavioural Brain Research, 41,* 81–94.

Damasio, H. (1991). Neuroanatomical correlates of the aphasias. In M. Taylor Sarno (Ed.), *Acquired aphasia* (2nd ed., pp. 45–71). San Diego: Academic Press.

Damasio, H. (1995). *Human brain anatomy in computerized images.* New York: Oxford University Press.

Damasio, H., & Damasio, A. R. (1989). *Lesion analysis in neuropsychology.* New York: Oxford University Press.

Damasio, H., Grabowski, T. J., Tranel, D., Hichwa, R. D., & Damasio, A. R. (1996). A neural basis for lexical retrieval. *Nature, 380,* 499–505.

Daneman, M., & Carpenter, P. A. (1980). Individual differences in working memory and reading. *Journal of Verbal Learning and Verbal Behavior, 19,* 450–466.

Danesi, M. (1994). The neuroscientific perspective in second language acquisition research: A critical synopsis. *International Review of Applied Linguistics, 32,* 201–228.

Daniele, A., Giustolisi, L., Silveri, M. C., Colosimo, C., & Gainotti, G. (1994). Evidence for a possible neuroanatomical basis for lexical processing of nouns and verbs. *Neuropsychologia, 32,* 1325–1341.

Danly, M., Cooper, W., & Shapiro, B. (1983). Fundamental frequency, language processing, and linguistic structure in Wernicke's aphasia. *Brain and Language, 19,* 1–24.

Danly, M., & Shapiro, B. (1982). Speech prosody in Broca's aphasia. *Brain and Language, 16,* 171–190.

D'Antona, R., Baron, J. C., Samson, Y., Serdaru, M., Viader, F., Agid, Y., & Cambier, J. (1985). Subcortical dementia: Frontal cortex hypometabolism detected by positron tomography in patients with progressive supranuclear palsy. *Brain, 108,* 785–799.

Darby, D. G. (1993). Sensory aprosodia: A clinical clue to lesions of the inferior division of the right middle cerebral artery? *Neurology, 43,* 567–572.

Darkins, A. W., Fromkin, V. A., & Benson, D. F. (1988). A characterization of the prosodic loss in Parkinson's disease. *Brain and Language, 34,* 315–327.

Darley, F. L., Aronson, A. E., & Brown, J. R. (1975). *Motor speech disorders.* Philadelphia: W. B. Saunders.

Darwin, C. (1877). A biographical sketch of an infant. *Mind, 2,* 292–294.

David, N. J., Mackey, E. A., & Smith, J. L. (1968). Further observations in progressive supranuclear palsy. *Neurology, 18,* 349–356.

David, R. M., & Skilbeck, C. E. (1984). Raven IQ and language recovery following stroke. *Journal of Clinical Neuropsychology, 6,* 302–308.

Davidson, R. (1994). Asymmetric brain function, affective style, and psychopathology: The role of early experience and plasticity. *Development and Psychopathology, 6,* 741–758.

Davidson, R. J., & Tomarken, A. J. (1989). Laterality and emotions: An electrophysiological approach. In F. Boller & J. Grafman (Eds.), *Handbook of neurospsychology* (pp. 419–441). Amsterdam: Elsevier.

Davidson, R., Leslie, S., & Saron, C. (1990). Reaction time measures of interhemispheric transfer time in reading disabled and normal children. *Neuropsychologia, 28,* 471–485.

Davies, K. G., Maxwell, R. E., Beniak, T. E., Destafney, E., & Fiol, M. E. (1995). Language function after temporal lobectomy without stimulation mapping of cortical function. *Epilepsia, 36,* 130–136.

DeArmond, S. J., Fusco, M., & Dewey, M. (1976). *Structure of the human brain* (2nd ed.). New York: Oxford University Press.

De Bleser, R., Bayer, J., & Luzzatti, C. (1996). Linguistic theory and morphosyntactic impairments in German and Italian aphasics. *Journal of Neurolinguistics, 9,* 175–185.

DeFries, J. C. (1991). Genetics and dyslexia: An overview. In M. Snowling & M. Thomson (Eds.), *Dyslexia: Integrating theory and practice* (pp. 3–20). London: Whurr Publishers.

de Groot, J. M., Rodin, G., & Olmstead, M. P. (1995). Alexithymia, depression and treatment outcome in bulimia nervosa. *Comprehensive Psychiatry, 36,* 53–60.

Dehaene, S. (1992). Varieties of numerical abilities. *Cognition, 44,* 1–42.

Dehaene, S., & Akhavein, R. (1995). Attention, automaticity, and levels of representation in number processing. *Journal of Experimental Psychology: Learning, Memory and Cognition, 21,* 314–326.

Dehaene, S., & Cohen, L. (1991). Two mental calculation systems: A case study of severe acalculia with preserved approximation. *Neuropsychologia, 29,* 1045–1074.

Dehaene, S., & Cohen, L. (1995). Towards an anatomical and functional model of number processing. *Mathematical Cognition, 1,* 83–120.

Dehaene, S., & Cohen, L. (in press). Contrasting patterns of acalculia following inferior parietal and sub-cortical lesions. *Cortex, 38.*

Dehaene, S., Dupoux, E., & Mehler, J. (1990). Is numerical comparison digital: Analogical and symbolic effects in two-digit number comparison. *Journal of Experimental Psychology: Human Perception and Performance, 16,* 626–641.

Dejerine, J. J. (1901). *Anatomie des centres nerveux.* Paris: Rueff.

De la Monte, S. M., Wells, S. E., Hedley-Whyte, E. T., & Growdon, J. H. (1989). Neuropathological distinction between Parkinson's dementia and Parkinson's plus Alzheimer's disease. *Annals of Neurology, 26,* 309–320.

Delay, J., & Brion, S. (1962). *Les démences tardives* [The senile dementias]. Paris: Masson.

Delis, D. C., Wapner, W., Gardner, H., & Moses, J. A. (1983). The contribution of the right hemisphere to the organization of paragraphs. *Cortex, 19,* 43–50.

Delis, D. E., Kramer, J. H., Kaplan, E., & Ober, B. A. (1987). *California Verbal Learning Test: Manual.* New York: The Psychological Corporation.

Dell, G. S. (1986). A spreading activation theory of lexical retrieval in sentence production. *Psychological Review, 93,* 283–321.

Dell, G. S. (1986). A spreading-activation theory of retrieval in sentence production. *Psychological Review, 93,* 283–321.

Deloche, G., & Willmes, K. (in press). On two crucial assumptions of the McCloskey et al. calculation and number processing model. *Cortex.*

Deloche, G., Dordain, M., & Kremin, H. (1993). Rehabilitation of confrontational naming in aphasia: Relations between oral and written modalities. *Aphasiology, 7,* 201–216.

Delwaide, P. J., & Young, R. R. (1992). Introduction—why restorative neurology? In R. R. Young & P. J. Delwaide (Eds.), *Principles and practice of restorative neurology* (pp. 1–4). London: Butterworth Heinemann.

Demb, J. B., Desmond, J. E., Wagner, A. D., Vaidya, C. J., Glover, G. H., & Gabrieli, J. D. E. (1995). Semantic encoding and retrieval in the left inferior prefrontal cortex: A functional MRI study of task difficulty and process specificity. *Journal of Neuroscience, 15,* 5870–5878.

Demeurisse, G., & Capon, A. (1991). Brain activation during a linguistic task in conduction aphasia. *Cortex, 27,* 285–294.

Demeurisse, G., & Capon, A. (1987). Language recovery in aphasic stroke patients: Clinical, CT and CBF studies. *Aphasiology, 1,* 301–315.

Démonet, J.-F. (1987). *Les aphasies "sous-corticales": Étude linguistique, radiologique et hémodynamique de 31 observations.* Thèse pour le doctorat d'état en médecine, Université Paul Sabatier, Toulouse, France.

Démonet, J.-F. (1995). Studies of language processes using positron emission tomography. In F. Boller & J. Grafman (Series Eds.) & R. Johnson Jr. & J.-C. Baron (Volume Eds.), *Handbook of Neuropsychology* (Vol 10, pp. 423–437). Amsterdam: Elsevier.

Démonet, J.-F., Celsis, P., Nespoulous, J.-L., Viallard, G., Marc-Vergnes, J.-P., & Rascol, A. (1992). Cerebral blood flow correlates of word monitoring in sentences: Influence of semantic incoherence. A SPECT study in normals. *Neuropsychologia, 30,* 1–11.

Démonet, J.-F., Chollet, F., Ramsay, S., Cardebat, D., Nespoulous, J.-L., Wise, R., Rascol, A., & Frackowiak, R. S. J. (1992). The anatomy of phonological and semantic processing in normal subjects. *Brain, 115,* 1753–1768.

Démonet, J.-F., Fiez, J. A., Paulesu, E., Petersen, S. E., & Zatorre, R. J. (1996). PET studies of phonological processing. A critical reply to Poeppel. *Brain and Language, 55.* 352–379.

Démonet, J.-F., Price, C., Wise, R., & Frackowiak, R. S. J. (1994a). A PET study of cognitive strategies in normal subjects during language tasks: Influence of phonetic ambiguity and sequence processing on phoneme monitoring. *Brain, 117,* 671–682.

Démonet, J.-F., Price, C., Wise, R., & Frackowiak, R. S. J. (1994b). Differential activation of right and left posterior sylvian regions by semantic and phonological tasks: A positron-emission tomography study in normal human subjects. *Neuroscience Letters, 182,* 25–28.

Dempster, F. N., & Brainerd, C. J. (1995). *Interference and inhibition in cognition.* San Diego: Academic Press.

Den Heyer, K., Goring, A., Gorgichuk, S., Richards, L., & Landry, M. (1988). Are lexical decisions a good measure of lexical access? Repetition blocking suggests the affirmative. *Canadian Journal of Psychology, 42,* 274–296.

Denes, G., Balliello, S., Volterra, V., & Pellegrini, A. (1986). Oral and written language in a case of childhood phonemic deafness. *Brain and Language, 29,* 252–267.

Denes, G., & Dalla Barba, G. (1995). Vico, precursor of cognitive neuropsychology? The first reported cases of noun-verb dissociation following brain damage. *Brain and Language, 51,* 7–8.

Denes, G., Semenza, C., & Bisiacchi, P. (Eds.). (1988). *Perspectives on cognitive neuropsychology.* Hove, UK: Lawrence Erlbaum.

Dennis, M. (1976). Dissociated naming and locating of body parts after left anterior temporal lobe resection: An experimental case study. *Brain and Language, 3,* 147–163.

Dennis, M., & Barnes, M. A. (1990). Knowing the meaning, getting the point, bridging the gap and carrying the message: Aspects of discourse following closed head injury in childhood and adolescence. *Brain and Language, 39,* 428–446.

Dennis, M., Lovett, M., & Wiegel-Crump, C.A. (1981). Written language acquisition after left or right hemidecortication in infancy. *Brain and Language, 12,* 54–91.

Dennis, M., & Whitaker, H. A. (1976). Language acquisition following hemidecortication: Linguistic superiority of the left over the right hemisphere. *Brain and Language, 3,* 404–433.

Dennis, M., & Whitaker, H. A. (1977). Hemispheric equipotentiality and language acquisition. In S. J. Segalowitz & F. A. Gruber (Eds.), *Language development and neurological theory* (pp. 93–106). New York: Academic Press.

Deonna, T., Beaumanoir, A., Gaillard, F., & Assal G. (1977). Acquired aphasia in childhood with seizure disorder: A heterogeneous syndrome. *Neuropediatrics, 8,* 263–273.

de Partz, M.-P. (1986). Re-education of a deep dyslexic patient: Rationale of the method and results. *Cognitive Neuropsychology, 3,* 149–178.

de Partz, M.-P., Seron, X., & Van der Linden, M. (1992). Re-education of a surface dysgraphia with a visual imagery strategy. *Cognitive Neuropsychology, 9,* 369–401.

de Renzi, E., Colombo, A., & Scarpa, M. (1991). A clinical-CT scan study of a particularly severe subgroup of global aphasics. *Brain, 114,* 1719–1730.

de Renzi, E., & Vignolo, L. A. (1962). The token test: A sensitive test to detect receptive disturbances in aphasia. *Brain, 85,* 665–678.

Desmond, J. E., Sum, J. M., Wagner, A. D., Demb, J. B., Shear, P. K., Glover, G. H., Gabrieli, J. D. E., & Morrell, M. J. (1995). Functional MRI measurement of language lateralization in Wada-tested patients. *Brain, 118,* 1411–1419.

Deutsch, A., & Bentin, S. (1994). Attention mechanisms mediate the syntactic priming effects in auditory word identification. *Journal of Experimental Psychology: Learning, Memory, & Cognition, 20,* 595–607.

DeVilliers, J. G. (1974). Quantitative aspects of agrammatism in aphasia. *Cortex, 10,* 36–54.

Devinsky, O., Morrell, M. J., & Vogt, B. A. (1995). Contributions of anterior cingulate cortex to behavior. *Brain, 118,* 279–306.

De Vreese, L. P., Motta, M., & Toschi A. (1988). Compulsive and paradoxical translation behaviour in a case of presenile dementia of the Alzheimer type. *Journal of Neurolinguistics, 3,* 233–259.

Dijkstra, T., & de Smedt, K. (Eds.) (1996). *Computational psycholinguistics: Artificial intelligence and connectionist models of human language processing.* London: Taylor & Francis.

Di Sciullo, A. M., & Williams, E. (1987). *On the definition of word.* Cambridge: MIT Press.

Dix, M. R., Harrison, M. J. G., & Lewis, P. D. (1971). Progressive supranuclear palsy: A report of 9 cases with particular reference to the oculomotor disorder. *Journal of the Neurological Sciences, 13,* 237–256.

Docherty, N. M., DeRosa, M., & Andreasen, N. C. (1996). Communication disturbances in schizophrenia and mania, *Archives of General Psychiatry, 53,* 358–364.

Docherty, N. M., Evans, I. M., Sledge, W. H., Seibyl, J. P., & Krystal, J. H. (1994). Affective reactivity of language in schizophrenia. *Journal of Nervous and Mental Disorders, 182,* 98–102.

Docherty, N. M., Hawkins, K. A., Hoffman, R. E., Quinlan, D. M., Rakfeldt, J., & Sledge W. (1996). Working memory, attention, and communication disturbances in schizophrenia, *Journal of Abnormal Psychology, 105,* 212–219.

Docherty, N. M., Sledge, W. H., & Wexler, B. E. (1994). Affective reactivity of language in stable schizophrenic outpatients and their parents. *Journal of Nervous and Mental Disease, 182,* 313–318.

Dodrill, C. B. (1993). Preoperative criteria for identifying eloquent brain: Intracarotid amytal for language and memory testing. *Neurosurgery Clinics of North America, 4,* 211–216.

Dogil, G. (1989). The phonological and acoustic form of neologistic jargon aphasia. *Clinical Linguistics and Phonetics, 3,* 265–279.

Dogil, G., Hildebrandt, G., & Schürmeier, K. (1990). The communicative function of prosody in a semantic jargon aphasia. *Journal of Neurolinguistics, 5,* 353–369.

Dooling, E. C., Chi, J. G., & Gilles, F. H. (1983). Telencephalic development: Changing gyral patterns. In F. H. Gilles, A. Leviton, & E. C. Dooling (Eds.), *The developing human brain: Growth and epidemiologic neuropathy* (pp. 94–104). Boston: John Wright, PSG.

Dos Santos, G., Nespoulous, J.-L., & Whitaker, H. A. (in preparation). Grasset's Polygon.

Douglass, E., & Richardson, J. (1959). Aphasia in a congenital deaf-mute. *Brain, 82,* 68–80.

Downes, J. J., Sharp, H. M., Costall, B. M., Sagar, H. J., & Howe, J. (1993). Alternating fluency in Parkinson's disease: An evaluation of the attentional control theory of cognitive impairment. *Brain, 116,* 887–902.

Dressler, W. U., & Pléh, C. (1988). On text disturbances in aphasia. In W. U. Dressler & J. A. Stark (Eds.), *Linguistic analyses of aphasic language* (pp. 151–178). New York: Springer Verlag.

Drobny, M., Krajnak, V., Svalekova, A., & Pithova, B. (1991). Creutzfeldt-Jakob disease: Clinical picture analysis. *European Journal of Epidemiology, 7,* 511–516.

Dronkers, N. F. (1996). A new brain region for coordinating speech articulation. *Nature, 384,* 159–161.

Dronkers, N. F., & Pinker, S. (in press). Language and the aphasias. In E. Kandel, J. Schwartz, & T. Jessell (Eds.), *Principles in neural science.* New York: Elsevier, North Holland.

Dronkers, N. F., Redfern, B. B., & Ludy, C. A. (1995). Lesion localization in chronic Wernicke's aphasia. *Brain and Language, 51*(1), 62–65.

Dronkers, N. F., Shapiro, J. K., Redfern, B., & Knight, R. T. (1992). The role of Broca's area in Broca's aphasia. *Journal of Clinical and Experimental Neuropsychology, 14,* 52–53.

Dronkers, N. F., Wilkins, D. P., Van Valin, R., Jr., Redfern, B., & Jaeger, J. (1994). A reconsideration of the brain areas involved in the disruption of morphosyntactic comprehension. *Brain and Language, 47,* 461–463.

Druks, J., & Marshall, J. C. (1991). Agrammatism: An analysis and critique, with new evidence from four Hebrew-speaking aphasic patients. *Cognitive Neuropsychology, 8,* 415–433.

Druks, J., & Marshall, J. C. (1995). When passives are easier than actives: Two case studies of agrammatic comprehension. *Cognition, 55,* 311–331.

Druks, J., & Marshall, J. C. (1996). Syntax, strategies, and the single case: A reply to Zurif. *Cognition, 58,* 281–287.

Duara, R., Grady, C., Haxby, J., Ingvar, D., Sokoloff, L., Margolin, R., Manning, R., Cutler, N., & Rapoport, S. (1984). Human brain glucose utilization and cognitive function in relation to age. *Annals of Neurology, 16,* 702–713.

Duara, R., Lopez-Alberola, R. F., Barker, W. W., Loewenstein, D. A., Zatinsky, M., Eisdorfer, C. E., & Weinberg, G. B. (1993). A comparison of familial and sporadic Alzheimer's disease. *Neurology, 43,* 1377–1384.

Dubois, B., & Pillon, B. (1992). Biochemical correlates of cognitive changes and dementia in Parkinson's disease. In S. J. Huber & J. L. Cummings (Eds.), *Parkinson's disease: Neurobehavioral aspects* (pp. 178–198). Oxford: Oxford University Press.

Duchovny, M., Jayakar, P., Harvey, A. S., Resnick, T., Alvarez, L., Dean, P., & Levin, B. (1996). Language cortex representation: Effects of developmental versus acquired pathology. *Annals of Neurology, 40,* 31–38.

Duffy, J. R. (1994). Schuell's stimulation approach to rehabilitation. In R. Chapey (Ed.), *Language intervention strategies in adult aphasia* (3rd ed.; pp. 146–174). Baltimore: Williams & Wilkins.

Duffy, J. R. (1995). *Motor speech disorders.* St. Louis: Mosby.

Dugas, M., Masson, M., Le Henzey, M. F., Regnier, N. (1982). Aphasie "acquise" de l'enfant avec épilepsie (syndrome de Landau et Kleffner). Douze observations personnelles. *Revue Neurologique, 138,* 755–780.

Dulac, O., Billard, C., & Arthuis, M. (1983). Aspects électro-cliniques et évolutifs de l'épilepsie dans le syndrome aphasie-épilepsie. *Archives Français de Pédiatrie,* 40, 299–308.

Duncan, C., Rumsey, J. M., Wilkniss, S. M., Denkla, M. B., Hamburger, S. D., & Odou-Potkin, M. (1994). Developmental dyslexia and attention dysfunction in adults: Brain potential indices of information processing. *Psychophysiology, 31,* 386–401.

Dunn, J. C., & Kirsner, K. (1988). Discovering functionally independent mental processes: The principle of reversed association. *Psychological Review, 95,* 91–101.

Dunn, L. M. (1959). *Peabody picture vocabulary test.* Circle Pines, MN: American Guidance Service.

Edgington, E. S. (1995). *Randomization tests* (3rd ed.). New York: Marcel Dekker.

Edmundson, A., & McIntosh, J. (1995). Cognitive neuropsychology and aphasia therapy: Putting the theory into practice. In C. Code & D. Müller (Eds.), *The treatment of aphasia: From theory to practice.* London: Whurr Publishers.

Eggert, G. (Ed.) (1977). *Wernicke's works on aphasia.* The Hague: Mouton.

Ehri, L. C. (1992). Reconceptualising the development of sight word reading and its relationship to recoding. In P. B. Gough, L. C. Ehri, & R. Treiman (Eds.), *Reading acquisition* (pp. 107–143). Hillsdale, NJ: Lawrence Erlbaum.

Ehrlich, J., & Barry, P. (1989). Rating communication behaviours in the head injured. *Brain Injury, 3,* 193–198.

Ehrlich, J. S. (1988). Selective characteristics of narrative discourse in head-injured and normal adults. *Journal of Communication Disorders, 21,* 1–9.

Ehrlich, M.-F., & Tardieu, H. (1993). Modèles mentaux, modèles de situation et compréhension de textes. In M.-F. Ehrlich, H. Tardieu, & M. Cavazza (Eds.), *Les modèles mentaux: Une approche cognitive des représentations* (pp. 47–77). Paris: Masson.

Eimas, P. D. (1975). Speech perception in early infancy. In B. Cohen & P. Salapatec (Eds.), *Infant perception: From sensation to cognition* (Vol. 2, pp. 193–231). New York: Academic Press.

Eisenson, J. (1973). *Adult aphasia.* New York: Appleton-Century-Crofts.

Ekman, P. (1992). An argument for basic emotions. *Cognition and Emotion, 6,* 169–200.

Ekman, P., & Davidson, R. J. (1993). Voluntary smiling changes regional brain activity. *Psychological Science, 4,* 342–345.

Ekman, P., & Friesen, W. V. (1978). *Facial action coding system.* Palo Alto, CA: Consulting Psychologists Press.

Elbert, J. C. (1993). Occurrence and pattern of impaired reading and written language in children with attention deficit disorders. *Annals of Dyslexia, 43,* 26–43.

Eling, P. (Ed.) (1994). *Reader in the history of aphasia.* Amsterdam: John Benjamins.

Ellis, A. W. (1984). *Reading, writing and dyslexia: A cognitive analysis.* Hillsdale, NJ: Lawrence Erlbaum.

Ellis, A. W., Flude, B. M., & Young, A. W. (1987). "Neglect dyslexia" and the early visual processing of letters in words and nonwords. *Cognitive Neuropsychology, 4,* 439–464.

Ellis, A. W., & Young, A. W. (1988). *Human cognitive neuropsychology.* London: Lawrence Erlbaum.

Ellis, H., & Young, A. (1990) Accounting for delusional misidentifications. *British Journal of Psychiatry, 157,* 239–248.

Ellis, H. D., Ellis, D. M., Fraser, W., & Deb, S. (1994). A preliminary study of right hemisphere cognitive deficits and impaired social judgements among young people with Asperger's syndrome. *European Child and Adolescent Psychiatry, 3,* 255–266.

Elman, J. L. (1991). Distributed representations, simple recurrent networks, and grammatical structure. *Machine Learning, 7,* 195–225.

Embretson, S. E. (Ed.). (1985). *Test design: Developments in psychology and psychometrics.* New York: Academic Press.

Emmorey, K. (1987). The neurological substrates for prosodic aspects of speech. *Brain and Language, 30,* 305–320.

Emmorey, K. (1996). The confluence of space and language in signed languages. In P. Bloom, M. Peterson, L. Nadel, & M. Garrett (Eds.), *Language and space* (pp. 171–209). Cambridge: MIT Press.

Emmorey, K., Corina, D. P., & Bellugi, U. (1995). Differential processing of topographic and referential functions of space. In K. Emmorey & J. Reilly (Eds.), *Language, gesture and space* (pp. 43–62). Hillsdale, NJ: Lawrence Erlbaum.

Eng, N., Obler, L., Harris, K., & Abramson, A. (in press). Tone perception deficits in Chinese-speaking Broca's aphasics. *Aphasiology.*

Engelien, A., Silbersweig, D., Stern, E., Huber, W., Döring, W., Frith, C. D., & Frackowiak, R. S.J. (1995). The functional anatomy of recovery from auditory agnosia: A PET study of sound categorization in a neurological patient and normal controls. *Brain, 118,* 1395–1409.

Engle, R. W., Conway, A. R. A., Tuholski, S. T., & Shisler, R. J. (1995). A resource account of inhibition. *Psychological Science, 6,* 122–125.

Entus, A. K. (1977). Hemispheric asymmetry in processing of dichotically presented speech and nonspeech stimuli by infants. In S. J. Segalowitz & F. A. Gruber (Eds.), *Language development and neurological theory* (pp. 63–73). New York: Academic Press.

Erickson, B., Lind, E. A., Johnson, B. C., & O'Barr, W. M. (1978). Speech style and impression formation in a court setting: The effects of "Powerful" and "Powerless" speech. *Journal of Experimental Social Psychology, 14,* 266–279.

Ericsson, K. A., & Kintsch, W. (1995). Long-term working memory. *Psychological Review, 102,* 211–245.

Esiri, M. M. (1994). Dementia and normal aging: Neuropathology. In F. A. Huppert, C. Brayne, & D. W. O'Connor (Eds.), *Dementia and normal aging* (pp. 385–436). Cambridge, UK: Cambridge University Press.

Eslinger, P. J., & Damasio, A. R. (1985). Severe disturbance of higher cognition after bilateral frontal lobe ablation. *Neurology, 35,* 1731–1741.

Eslinger, P. J., & Grattan, L. M. (1993). Frontal lobe and frontal-striatal substrates for different forms of human cognitive flexibility. *Neuropsychologia, 31,* 17–28.

Eulitz, C., Diesch, E., Pantev, C., Hampson, S., & Elbert, T. (1995). Magnetic and electric brain activity evoked by the processing of tone and vowel stimuli. *Journal of Neuroscience, 15,* 2748–2755.

Eulitz, C., Elbert, T., Bartenstein, P., Weiler, C., Müller, S. P., & Pantev, C. (1994). Comparison of magnetic and metabolic brain activity during a verb generation task. *NeuroReport, 6,* 97–100.

Eviatar, Z. (1995). Reading direction and attention: Effects of lateralized ignoring. *Brain and Cognition, 29,* 137–150.

Eviatar, Z. (1997). Language experience and right hemisphere tasks: The effects of scanning habits and multilingualism. *Brain and Language, 58,* 157–173.

Eviatar, Z., Menn, L., & Zaidel, E. (1990). Concreteness: Nouns, verbs and hemispheres. *Cortex, 26,* 611–624.

Eviatar, Z., & Zaidel, E. (1991). The effects of word length and emotionality on hemispheric contribution to lexical decision. *Neuropsychologia, 29,* 415–428.

Eviatar, Z., & Zaidel, E. (1994). Shape and name: Letter matching within and between the disconnected hemispheres. *Brain and Cognition, 25,* 128–137.

Fabbro, F., & Paradis, M. (1995). Differential impairments in four multilingual patients with subcortical lesions. In M. Paradis (Ed.), *Aspects of bilingual aphasia* (pp. 139–176). Oxford: Pergamon Press.

Faber, R., Abrams, R., Taylor, M. A., Kasprisin, S., Morris, C., & Weiss, R. (1983). A comparison of schizophrenic patients with formal thought disorder and neurologically impaired patients with aphasia. *American Journal of Psychiatry, 140,* 1348–1351.

Faber-Langendoen, K., Morris, J. C., Knesevich, J. W., LaBarge, E., Miller, J. P., & Berg, L. (1988). Aphasia in senile dementia of the Alzheimer type. *Annals of Neurology, 23,* 365–370.

Farah, M., & McClelland, J. (1992). Neural network models and cognitive neuropsychology. *Psychiatric Annals, 22,* 148–153.

Farah, M. J. (1994). Neuropsychological inference with an interactive brain: A critique of the locality assumption. *Behavioral and Brain Sciences, 17,* 43–104.

Farah, M. J., & McClelland, J. L. (1991). A computational model of semantic memory impairment: Modality specificity and emergent category specificity. *Journal of Experimental Psychology: General, 120,* 339–357.

Farah, M. J., & Wallace, M. A. (1992). Semantically-bounded anomia: Implications for the neural implementation of naming. *Neuropsychologia, 30,* 609–621.

Faure, S., & Blanc-Garin, J. (1994). Right hemisphere semantic performance and competence in a case of partial interhemispheric disconnection. *Brain and Language, 47,* 557–581.

Faust, M., Kravetz, S., & Babkoff, H. (1993). Hemispheric specialization or reading habits: Evidence from lexical decision research with Hebrew words and sentences. *Brain and Language, 44,* 254–263.

Fay, W. H., & Schuler, A. L. (1980). *Emerging language in autistic children.* Baltimore: University Park Press.

Fayol, M., & Lemaire, P. (1993). Levels of approach to discourse. In H. H. Brownell & Y. Joanette (Eds.), *Narrative discourse in neurologically impaired and normal aging adults* (pp. 3–21). San Diego: Singular Publishing.

Feeney, D. M., & Baron, J. C. (1986). Diaschisis. *Stroke, 17,* 817–830.

Feldman, J. A., & Ballard, D. H. (1982). Connectionist models and their properties. *Cognitive Science, 6,* 205–254.

Feldman, J. A., Fanty, M. A., Goddard, N. H., & Lynne, K. J. (1988). Computing with structured connectionist networks. *Communications of the ACM, 31,* 170–187.

Fergusson, A. (1994). The influence of aphasia, familiarity and activity on conversational repair. *Aphasiology, 8,* 143–157.

Fernald, A. (1985). Four-month-old infants prefer to listen to motherese. *Infant Behavior and Development, 8,* 181–195.

Fiez, J. A., Raichle, M. E., Balota, D. A., Tallal, P., & Petersen, S. E. (1996). PET activation of posterior temporal regions during auditory word presentation and verb generation. *Cerebral Cortex, 6,* 1–10.

Fiez, J. A., Raichle, M. E., Miezin, F. M., Petersen, S. E., Tallal, P., & Katz, W. F. (1995). PET studies of auditory and phonological processing: Effects of stimulus characteristics and task demands. *Journal of Cognitive Neuroscience, 7,* 357–375.

Fine, J., Bartolucci, G., Ginsberg, G. & Szatmari, P. (1991). The use of intonation to communicate in pervasive developmental disorders. *Journal of Child Psychology and Psychiatry, 32,* 771–782.

Finger, S. (1994). *Origins of neuroscience.* New York: Oxford University Press.

Finger, S., & Almli, C. R. (1985). Brain damage and neuroplasticity: Mechanisms of recovery or development? *Brain Research Reviews, 10,* 177–186.

Finger, S., LeVere, T. E., Almli, C. R., & Stein, D. G. (1988). Recovery of function: Sources of controversy. In S. Finger, T. E. Levere, C. R. Almli, & D. G. Stein (Eds.), *Brain injury and recovery: Theoretical and controversial issues* (pp. 351–361). New York: Plenum Press.

Finger, S., & Stein, D. (1982). *Brain damage and recovery: Research and clinical perspectives.* New York: Academic Press.

Fischer, B., Biscaldi, M., & Otto, P. (1993). Saccadic eye movements of dyslexic adult subjects. *Neuropsychologia, 31,* 887–906.

Fischer, B., & Weber, H. (1990). Saccadic reaction times of dyslexic and age-matched normal subjects. *Perception, 19,* 805–818.

Fischer, J. M., Kennedy, J. L., Caine, E. D., & Shoulson, I. (1983). Dementia in Huntington's disease: A cross-sectional analysis of intellectual decline. In R. Mayeux & W. G. Rosen (Eds.), *The dementias* (pp. 229–238). New York: Raven Press.

Fisher, P., Gatterer, G., & Danielczyk, W. (1988). *Functional Neurology, 3,* 301–307.

Fisher, R. A. (1935). *Design of experiments.* Edinburgh: Oliver and Boyd.

Fitch, R. H., & Denenberg, V. H. (1995). A role for ovarian hormones in sexual differentiation of the brain. *Psycoloquy [On-line], 6.* Available: psyc@princeton.edu; http://www.princeton.edu/~harnad/psyc.html.

Flanagan, S., McDonald, S., & Togher, L. (1995). Evaluation of the BRISS as a measure of social skills in the traumatically brain injured. *Brain Injury, 9,* 321–338.

Flechsig, P. (1901). Developmental (myelogenetic) localization of the cerebral cortex in the human subject. *Lancet, 2,* 1027.

Fleming, R. E., Hubbard, D. J., Schinsky, L., Datta, K. (1982). Consideration of PICA subtest variability in cases of aphasia secondary to blunt head trauma. *Clinical Aphasiology, 12,* 85–91.

Flicker, C., Ferris, S. H., Crook, T., Bartus, R., & Reisberg, B. (1986). Cognitive decline in advanced age: Future directions for the psychometric differentiation of normal and pathological age changes in cognitive functions. *Developmental Neuropsychology, 2,* 309–322.

Flor-Henry, P. (1969). Psychosis and temporal lobe epilepsy: A controlled investigation. *Epilepsia, 10,* 363–395.

Fodor, J. A. (1983). *The modularity of mind.* Cambridge: MIT Press.

Fodor, J. A., Bever, T. G., & Garrett, M. F. (1974). *The psychology of language.* New York: McGraw-Hill.

Fodor, J. A. & Pylyshyn, Z. W. (1988). Connectionism and cognitive architecture: A critical analysis. In J. M. Steven Pinker (Ed.), *Connections and symbols* (pp. 3–71). Cambridge: MIT Press.

Foldi, N. S. (1987). Appreciation of pragmatic interpretation of indirect commands: Comparison of right and left hemisphere brain-damaged patients. *Brain and Language, 31,* 88–108.

Folstein M. F., & Breitner, J. C. S. (1981). Language disorder predicts familial Alzheimer's disease. *Johns Hopkins Medical Journal, 149,* 145–147.

Foss, D. (1988). Experimental psycholinguistics. *Annual Review of Psychology, 39,* 301–348.

Foundas, A. L., Leonard, C. M., Gilmore, R., Fennell, E., & Heilman, K. M. (1994). Planum temporale asymmetry and language dominance. *Neuropsychologia, 32,* 1225–1231.

Foundas, A. L., Leonard, C. M., Gilmore, R. L., Fennell, E. B., & Heilman, K. M. (1996). Pars triangularis asymmetry and language dominance. *Proceedings of the National Academy of Science (USA), 93,* 719–722.

Foundas, A. L., Leonard, C. M., & Heilman, K. M. (1995). Morphologic cerebral asymmetries and handedness: The pars triangularis and planum temporale. *Archives of Neurology, 52,* 501–508.

Fowler, C., Napps, S. E., & Feldman, L. (1985). Relations among regular and irregular morphologically related words in the lexicon as revealed by repetition priming. *Memory and Cognition, 13,* 241–255.

Fox, P. T., & Raichle, M. E. (1986). Focal physiological uncoupling of cerebral blood flow and oxydative metabolism during somatosensory stimulation in human subjects. *Proceedings of the National Academy of Sciences of the USA, 83,* 1140–1144.

Fraser W. I., King, K., Thomas, P., & Kendell, R. E. (1986). The diagnosis of schizophrenia by language analysis. *British Journal of Psychiatry, 156,* 211–215.

Frauenfelder, U. H., & Schreuder, R. (1992). Constraining psycholinguistic models of morphological processing and representation: The role of productivity. In G. E. Booij & J. van Marle (Eds.), *Yearbook of morphology 1991* (pp. 165–183). Dordrecht, The Netherlands: Kluwer.

Frazen, M. D., Tishelman, A. C., Seime, R. J., & Friedman, A. (1986). Neuropsychological evaluation of an individual with congenital left hemisphere porencephaly. *International Journal of Clinical Neuropsychology, 8,* 156–163.

Frazier, L. (1987). *Sentence processing: A tutorial review.* London: Lawrence Erlbaum.

Frazier, L., & Macnamara, P. (1995). Favor referential representations. *Brain and Language, 49,* 224–240.

Frederiksen, C. H., & Stemmer, B. (1993). Conceptual processing of discourse by a right hemisphere brain-damaged patient. In H. H. Brownell & Y. Joanette (Eds.), *Narrative discourse in neurologically impaired and normal aging adults* (pp. 239–278). San Diego: Singular Publishing.

Frederiksen, C. H., Bracewell, R. J., Breuleux, A., & Renaud, A. (1990). The cognitive representation and processing of discourse: Function and dysfunction. In Y. Joanette & H. H. Brownell (Eds.), *Discourse ability and brain damage: Theoretical and empirical explanations* (pp. 69–110). New York: Springer Verlag.

Freedman, M., Alexander, M. P., & Naeser, M. A. (1984). Anatomic basis of transcortical motor aphasia. *Neurology, 34,* 409–417.

Freedman, M., Rivoria, P., Butters, N., Sax, D., & Feldman, R. G. (1984). Retrograde amnesia in Parkinson's disease. *Canadian Journal of Neurology, 11,* 297–301.

Freeman, R. D. & Thibos, L. N. (1973). Electrophysiological evidence that abnormal early visual experience can modify the human brain. *Science, 180,* 876–878.

Freud, S. (1960). *Jokes and their relation to the unconscious.* New York: W. W. Norton.

Frey, R. T., Woods, D. L., Knight, R. T., Scabini, D., & Clayworth, C. (1987). Defining functional areas with averaged CT scans. *Society for Neuroscience, 13,* 1266.

Friederici, A., (1988). Agrammatic comprehension: Picture of a computational mismatch. *Aphasiology, 2,* 279–282.

Friederici, A. (1994). Arnold Pick. In P. Eling (Ed.), *Reader in the history of aphasia* (pp. 252–280). Amsterdam: John Benjamins.

Friederici, A. (1995). The time course of syntactic activation during language processing: A model based on neuropsychological and neurophysiological data. *Brain and Language, 50,* 259–281.

Friederici, A., & Kilborn, K. (1989). Temporal constraints on language processing: Syntactic priming in Broca's aphasia. *Journal of Cognitive Neuroscience, 1,* 262–272.

Friederici, A., Pfeifer, E., & Hahne, A. (1993). Event-related brain potentials during natural speech processing: Effects of semantic, morphological and syntactic violations. *Cognitive Brain Research, 1,* 183–192.

Friederici, A. D., & Frazier, L. (1992). Thematic analysis in agrammatic comprehension: Thematic structures and task demands. *Brain and Language, 42,* 1–29.

Friederici, A. D., Wessels, J., Emmorey, K., & Bellugi, U. (1992). Sensitivity of inflectional morphology in aphasia: A real-time processing perspective. *Brain and Language, 43,* 774–763.

Friedman, A., Polson, M. C., & Dafoe, C. G. (1988). Dividing attention between the hands and the head: Performance trade-offs between rapid finger tapping and verbal memory. *Journal of Experimental Psychology: Human Perception and Performance, 7,* 1031–1058.

Friedman, D., Hamberger, M. J., Stern, Y. , & Marder, K. (1992). Event-related potentials (ERPs) during repetition priming in Alzheimer's patients and young and older controls. *Journal of Clinical and Experimental Neuropsychology, 14,* 448–462.

Friedman, R. B. (1996). Recovery from deep alexia to phonological alexia: Points on a continuum. *Brain and Language, 52,* 114–128.

Friedman, R. B., & Alexander, M. P. (1984). Pictures, images and pure alexia: A case study. *Cognitive Neuropsychology, 1,* 9–23.

Friedman, W. J., & Wilkins, A. J. (1985). Scale effects in memory for the time of events. *Memory and Cognition, 13,* 168–175.

Friedmann, N., & Grodzinsky, Y. (1994). Verb inflection in agrammatism: A dissociation between tense and agreement. *Brain and Language, 47,* 402–405.

Friedmann, N., & Grodzinsky, Y. (1997). Tense and agreement in agrammatic production: Pruning the syntactic tree. *Brain and Language, 56,* 397–425.

Friston, K. J., Frith, C. D., Liddle, P. F., & Frackowiak R. S. J. (1993). Functional connectivity: The principal-component analysis of large (PET) data sets. *Journal of Cerebral Blood Flow and Metabolism, 13,* 5–14.

Frith, C. D., Friston, K. J., Liddle, P. F., & Frackowiak R. S. J. (1991). A PET study of word finding. *Neuropsychologia, 29,* 1137–1148.

Frith, U., & Snowling, M. (1983). Reading for meaning and reading for sound in autistic and dyslexic children. *Journal of Developmental Psychology, 1,* 329–342.

Frith, U. (1989a). A new look at language and communication in autism. *British Journal of Disorders of Communication, 24,* 123–150.

Frith, U. (1989b). *Autism: Explaining the enigma.* Oxford: Basil Blackwell.

Frith, U. (Ed.). (1991). *Autism and Asperger syndrome.* Cambridge, UK: Cambridge University Press.

Fromkin, V. (1975). A linguist looks at "A linguist looks at 'schizophrenic' language." *Brain and Language, 2,* 498–503.

Fromkin, V. A. (1991). Language and brain: Redefining the goals and methodology of linguistics. In A. Kasher (Ed.), *The Chomskyan turn* (pp. 78–103). Cambridge: Basil Blackwell.

Fromm, D., & Holland, A. L. (1989). Functional communication in Alzheimer's disease. *Journal of Speech and Hearing Disorders, 54,* 535–540.

Frumkin, B., & Rapin, I. (1980). Perception of vowels and consonant-vowels of varying duration in language impaired children. *Neuropsychologia, 18,* 443–454.

Fuentes, L. J., Carmona, E., Agis, I. F., & Catena, A. (1994). The role of the anterior attention system in semantic processing of both foveal and parafoveal words. *Journal of Cognitive Neuroscience, 6,* 17–25.

Funnell, E., & Sheridan, J. (1992). Categories of knowledge? Unfamiliar aspects of living and nonliving things. *Cognitive Neuropsychology, 9,* 135–153.

Furlan, M., Marchal, G., Viader, F., Derlon, J.-M., & Baron, J.-C. (1996). Spontaneous neurological recovery after stroke and the fate of the ischemic penumbra. *Annals of Neurology, 40,* 216–226.

Fuster, J. M. (1989). *The prefrontal cortex* (2nd ed., pp. 51–192). New York: Raven Press.

Gadler, H. P., & Zechner, K. (1992). AUSWEGE—WEGE in Österreich. In V. M. Roth (Ed.), *Computer in der Sprachtherapie* (pp. 147–160). Tübingen: Gunter Narr Verlag.

Gagnepain, J. (1982). Du vouloir dire. Vl. 1. Oxford: Pergamon Press.

Gainotti, G. (1976). The relationships between semantic impairment in comprehension and naming in aphasic patients. *The British Journal of Disorders of Communication, 11,* 77–81.

Gainotti, G. (1990). The categorical organization of semantic and lexical knowledge in the brain. *Behavioural Neurology, 3,* 109–115.

Gainotti, G. (1993). Riddle of the right hemisphere's contribution to the recovery of language. *Journal of Disorders of Communication, 28,* 227–246.

Gainotti, G., Caltagirone, C., & Zoccolotti, P. (1993). Left/right and cortical/subcortical dichotomies in the neuropsychological study of human emotions. In F. N Watts (Ed.), *Cognition and Emotion* (special issue), *7,* 71–93.

Gainotti, G., Silveri, M. C., Daniele, A., & Giustolisi, L. (1995). Neuroanatomical correlates of category-specific semantic disorders: A critical survey. *Memory, 3,* 247–264.

Galaburda, A. M. (1984). Anatomical asymmetries. In N. Geschwind & A. M. Galaburda (Eds.), *Cerebral dominance: The biological foundations* (pp. 11–25). Cambridge: Harvard University Press.

Galaburda, A. M. (1990). Introduction. *Neuropsychologia, 28,* 515–516.

Galaburda, A. M. (1993). Neuroanatomic basis of developmental dyslexia. *Neurological Clinics (Behavioral Neurology), 11,* 161–173.

Galaburda, A. M. (1994). Anatomic basis of cerebral dominance. In R. J. Davidson & K. Hugdahl (Eds.), Brain asymmetry (pp. 51–73). Cambridge: MIT Press.

Galaburda, A. M. (1995). Anatomic basis of cerebral dominance. In R. J. Davidson & K. Hugdahl (Eds.), *Brain asymmetry* (pp. 51–73). Cambridge: MIT Press.

Galaburda, A. M., Le May, M., Kemper, T., & Geschwind, N. (1978). Left-right asymmetries in the brain. *Science, 199,* 852–856.

Galaburda, A. M., Menard, M. T., & Rosen, G. D. (1994). Evidence of aberrant auditory anatomy in developmental dyslexia. *Proceedings of the National Academy of Sciences, USA, 91,* 8010–8013.

Galaburda, A. M., Rosen, G. D., & Sherman, G. F. (1990) Individual variability in cortical organization: Its relationship to brain laterality and implications to function. *Neuropsychologia, 28,* 529–543.

Galaburda, A. M., & Sanides, F. (1980). Cytoarchitectonic organization of the human auditory cortex. *Journal of Comparative Neurology, 190,* 597–610.

Galaburda, A. M., Sherman, G. F., Rosen, G. D., Aboitiz, F., & Geschwind, N. (1985). Developmental dyslexia: Four consecutive patients with cortical anomalies. *Annals of Neurology, 18,* 222–233.

Gallen, C. C., Schwartz, B., Rieke, K., Pantev, C., Sobel, D., Hirschkoff, E., & Bloom, F. E. (1994). Intrasubject reliability and validity of somatosensory source localization using a large array biomagnetometer. *Electroencephalography and Clinical Neurophysiology, 90,* 145–156.

Gamsu, C. V. (1986). Confrontation naming in Parkinsonian patients: Post-operative anomia revisited. *Neuropsychologia, 24,* 727–729.

Gandour, J., Akamanon, C., Dechongkit, S., Khunadorn, F., & Boonklam, R. (1994). Sequences of phonemic approximations in a Thai conduction aphasic. *Brain and Language, 46,* 69–95.

Gandour, J. (1987). Tone production in aphasia. In J. Ryalls (Ed.), *Phonetic approaches to speech production in aphasia and related disorders* (pp. 45–57). Boston: Little, Brown.

Gandour, J. (1995). Tone and intonation in Thai after unilateral brain damage [abstract]. *Brain and Language, 51,* 191–193.

Gandour, J. (in press). Aphasia in tone languages. In P. Coppens, A. Basso, & Y. Lebrun (Eds.), *Aphasia in atypical populations.* Hillsdale, NJ: Lawrence Erlbaum.

Gandour, J., & Dardarananda, R. (1983). Identification of tonal contrasts in Thai aphasic patients. *Brain and Language, 18,* 98–114.

Gandour, J., Dechongkit, S., Ponglorpisit, S., & Khunadorn, F. (1994). Speech timing at the sentence level in Thai after unilateral brain damage. *Brain and Language, 46,* 419–438.

Gandour, J., Dechongkit, S., Ponglorpisit, S., Khunadorn, F., & Boongird, P. (1993). Intraword timing relations in Thai after unilateral brain damage. *Brain and Language, 45,* 160–179.

Gandour, J., Holasuit Petty, S., & Dardarananda, R. (1989). Dysprosody in Broca's aphasia: A case study. Brain and Language, 37, 232–257.

Gandour, J., & Ponglorpisit, S. (1990). Disruption of tone space in a Thai-speaking patient with subcortical aphasia. *Journal of Neurolinguistics, 5,* 333–351.

Gandour, J., Ponglorpisit, S, Dechongkit, S., Khunadorn, F., Boongird, P., & Potisuk, S. (1993). Anticipatory tonal coarticulation in Thai noun compounds after unilateral brain damage. *Brain and Language, 45,* 1–20.

Gandour, J., Ponglorpisit, S., Khunadorn, F., Dechongkit, S., Boongird, P., Boonklam, R., & Potisuk, S. (1992a). Lexical tones in Thai after unilateral brain damage. *Brain and Language, 43,* 275–307.

Gandour, J., Ponglorpisit, S., Khunadorn, F., Dechongkit, S., Boongird, P., & Boonklam, R. (1992b). Stop voicing in Thai after unilateral brain damage. *Aphasiology, 6,* 535–547.

Gandour, J., Ponglorpisit, S., Khunadorn, F., Dechongkit, S., Boongird, P., & Boonklam, R. (1992c). Timing characteristics of speech after brain damage: Vowel length in Thai. *Brain and Language, 42,* 337–345.

Gandour, J., Potisuk, S., Ponglorpisit, S., Dechongkit, S., Khunadorn, F., & Boongird, P. (1996). Tonal coarticulation in Thai after unilateral brain damage. *Brain and Language, 52,* 505–535.

Gandour, J., Potisuk, S., Ponglorpisit, S., Khunadorn, F., Boongird, P., & Dechongkit, S. (in press). Interaction between tone and intonation in Thai after unilateral brain damage. *Brain and Language.*

Ganis, G., Kutas, M., & Sereno, M. I. (1996). The search for "common sense": An electrophysiological study of the comprehension of words and pictures in reading. *Journal of Cognitive Neuroscience, 8,* 89–106.

Garcia, L., & Joanette, Y. (1994) Conversational topic-shifting analysis in dementia. In R. L. Bloom, L. K. Obler, S. De Santi, & J. Ehrlich (Eds.), *Discourse analysis and application: Studies in adult clinical populations* (pp. 161–183). Hillsdale, NJ: Lawrence Erlbaum.

Gardner, H. (1985). *The mind's new science.* New York: Basic Books.

Gardner, H, Brownell, H. H., Wapner, W., & Michelow, D. (1983). Missing the point: The role of the right hemisphere in the processing of complex linguistic material. In E. Perecman (Ed.), *Cognitive processing in the right hemisphere* (pp. 169–192). New York: Academic Press.

Garnham, A., Oakhill, J. V., & Johnson-Laird, P. N. (1982). Referential continuity and the coherence of discourse. *Cognition, 11,* 29–46.

Garnsey, S. M. (1993). Event-related potentials in the study of language. *Language and Cognitive Processes, 8,* 337–356.

Garnsey, S. M., Tanenhaus, M. K., , & Chapman, R. M. (1989). Evoked potentials and the study of sentence comprehension. *Journal of Psycholinguistic Research, 18,* 51–60.

Garret, M. F. (1980). Levels of processing in sentence production. In B. Butterworth (Ed.), *Language production.* Vol. 1, *Speech and talk* (pp. 177–220). London: Academic Press.

Garrett, M. F. (1982). Production of speech: Observations from normal and pathological language use. In A. W. Ellis (Ed.), *Normality and pathology in cognitive functions* (pp. 19–76). London: Academic Press.

Garrison, F. H. (1929). *An introduction to the history of medicine* (4th ed.). Philadelphia: W. B. Saunders.

Gascon, G., Victor, D., Lombroso, C. T., & Goodglass, H. (1973). Language convulsive disorders and electroencephalographic abnormalities. *Archives of Neurology, 28,* 156–162.

Gathercole, S. E., & Baddeley, A. D. (1993). *Working memory and language.* Hillsdale, NJ: Lawrence Erlbaum.

Gazzaniga, M. S., & Hillyard, S. A. (1971). Language and speech capacity of the right hemisphere. *Neuropsychologia, 9,* 273–280.

Gazzaniga, M. S., & Smylie, C. E (1984). Dissociation of language and cognition: A psychological profile of two disconnected right hemispheres. *Brain, 107,* 145–153.

Gazzaniga M. S., Smylie, C. S., Baynes, K., McCleary, C., & Hurst, W. (1984). Profiles of right hemisphere language and speech following brain bisection. *Brian and Language, 22,* 206–220.

Geffen, G., & Wale, J. (1979). Development of selective listening and hemispheric asymmetry. *Developmental Psychology, 15,* 138–146.

Geffner, D. S., & Hochberg, I. (1971). Ear laterality performance of children from low and middle socio-economic levels on a verbal dichotic listening task. *Cortex, 8,* 193–203.

Geiselman, R., & Crawley, J. (1983). Incidental processing of speaker characteristics: Voice as connotative information. *Journal of Verbal Learning and Verbal Behavior, 22,* 15–23.

Gerard, C. L., Dugas, M., Valdois, S., Franc, S., & Lecendraux M. (1993). Landau-Kleffner syndrome diagnosed after 9 years of age: Another Landau-Kleffner syndrome. *Aphasiology, 7,* 463–473.

Gernsbacher, M. (Ed.). (1994). *Handbook of psycholinguistics.* San Diego: Academic Press.

Gerson, S. N., Benson, F., & Frazier, S. (1977). Diagnosis: Schizophrenia versus posterior aphasia. *American Journal of Psychiatry, 134,* 966–969.

Geschwind, N. (1965). Disconnection syndromes in animals and man. *Brain, 88,* 237–294 and 585–644.

Geschwind, N. (1967). The varieties of naming errors. *Cortex, 3,* 97–112.

Geschwind, N. (1970). The organization of language and the brain. *Science, 170,* 940–944.

Geschwind, N. (1971). Aphasia. *New England Journal of Medicine, 284,* 654–656.

Geschwind, N. (1974). Late changes in the nervous system. In D. Stein, J. Rosen, & N. Butters (Eds.), *Plasticity and recovery of function in the central nervous system.* New York: Academic Press.

Geschwind, N. (1975). The apraxias: Neural mechanisms of disorders of learned movement. *American Scientist, 63,* 188–195.

Geschwind, N. (1985). Mechanisms of change after brain lesions. In F. Nottebohm (Ed.), *Annals of the New York Academy of Sciences.* Vol. 457, *Hope for a new neurology* (pp. 1–11). New York: New York Academy of Sciences.

Geschwind, N., & A. M. Galaburda (1987). *Cerebral lateralization: Biological mechanisms, associations, and pathology.* Cambridge: MIT Press.

Geschwind, N., & Galaburda, A. M. (1985). Cerebral lateralization: Biological mechanisms, associations and pathology: A hypothesis and a program for research. I. II. III. *Archives of Neurology, 42,* 426–457, 521–552, 634–654.

Geschwind, N., & Levitsky, W. (1968). Human brain: Left-right asymmetries in temporal speech region. *Science, 161,* 186–187.

Gevins, A., Leong, H., Smith, M. E., Le, J., & Du, R. (1995). Mapping cognitive brain function with modern high-resolution electroencephalography. *Trends in Neurosciences, 18,* 429–436.

Giard, M. H., Collet, L., Bouchet, P., & Pernier, J. (1994). Auditory selective attention in the human cochlea. *Brain Research, 633,* 353–356.

Gibb, W. R. G., Scott, T., & Lees, A. J. (1991). Neuronal inclusions of Parkinson's disease. *Movement Disorders, 6,* 2–11.

Gigley, H. (1983). HOPE-AI and the dynamic process of language behavior. *Cognition and Brain Theory, 6,* 39–88.

Gleason, J. B., Goodglass, H., Ackerman, N., Green, E., & Hyde, M. R. (1975). The retrieval of syntax in Broca's aphasia. *Brain and Language, 2,* 451–471.

Glick, S. D. (Ed.). (1985). *Cerebral lateralization in nonhuman species.* Orlando, FL: Academic Press.

Glisky, E. L., Schlacter, D. L., & Tulving, E. (1986). Learning and retention of computer-related vocabulary in memory-impaired patients: Method of vanishing cues. *Journal of Clinical and Experimental Neuropsychology, 8,* 292–312.

Glosser, G., & Deser, T. (1990). Patterns of discourse production among neurological patients with fluent language disorders. *Brain and Language, 40,* 67–88.

Glosser, G., & Friedman, R. B. (1990). The continuum of deep/phonological dyslexia. *Cortex, 26,* 343–359.

Glosser, G., & Goodglass, H. (1990). Disorders in executive control functions among aphasics and other brain damaged patients. *Journal of Clinical and Experimental Neuropsychology, 12,* 485–501.

Godfrey, H. P. D., Knight, R. G., & Bishara, S. N. (1991). The relationship between social skill and family problem solving following very severe closed head injury. *Brain Injury, 5,* 207–211.

Godfrey, H. P. D., Knight, R. G., Marsh, N. V., Moroney, B., & Bishara, S. M. (1989). Social interaction and speed of information processing following severe head injury. *Psychological Medicine, 19,* 175–182.

Gold, J. M., Randolph, C., Coppola, R., Carpenter, C. J., Goldberg, T. E., & Weinberg, D. R. (1992). Visual orienting in schizophrenia. *Schizophrenia Research, 7,* 203–209.

Goldberg, G. (1985). Supplementary motor area structure and function: Review and hypotheses. *Behavioral and Brain Sciences, 8,* 567–615.

Goldenberg, G., Podreka, I., Steiner, M., & Willmes, K. (1987). Patterns of regional cerebral blood flow related to memorizing of high and low imagery words—an emission computer tomography study. *Neuropsychologia, 25,* 473–485.

Goldenberg, G., & Spatt, J. (1994). Influence of size and site of cerebral lesions on spontaneous recovery of aphasia and on success of language therapy. *Brain and Language, 47,* 684–698.

Goldfarb, R., Stocker, B., Eisenson, J., & DeSanti, S. (1994). Communicative responsibility and semantic task in aphasia and "schizophasia." *Perceptual and Motor Skills, 79,* 1027–1039.

Goldman-Rakic, P. S. (1987). Circuitry of primate prefrontal cortex and regulation of behaviour by representational knowledge. In F. Plum & V. Mountcastle (Eds.), *Handbook of physiology* (Vol. 5; pp. 373–417). Bethesda, MD: American Physiological Society.

Goleman, D. (1995). *Emotional intelligence.* New York: Bantam.

Gomez, M. R., & Bebin, E. M. (1987). Sturge-Weber Syndrome. In M. R. Gomez (Ed.), *Neurocutaneous diseases. A practical approach* (pp. 356–367). Boston: Butterworths.

Gomez-Tortosa, E., Martin, E. M., Gaviria, M., Charbel, F. & Ausman, J. I. (1995). Selective deficit of one language in a bilingual patient following surgery in the left perisylvian area. *Brain and Language, 48,* 320–325.

Goodglass, H. (1976). Agrammatism. In H. Whitaker & H. A. Whitaker (Eds.), *Studies in neurolinguistics* (pp. 237–260). New York: Academic Press.

Goodglass, H. (1993). *Understanding aphasia.* San Diego: Academic Press.

Goodglass, H., & Berko, (1960). Agrammatism and inflectional morphology in English. *Journal of Speech and Hearing Research, 3,* 257–267.

Goodglass, H., Christiansen, J. A., & Gallagher, R. E. (1994). Syntactic constructions used by agrammatic speakers: Comparison with conduction aphasics and normals. *Neuropsychology, 8,* 598–613.

Goodglass, H., & Hunt, J. (1958). Grammatical complexity and aphasic speech. *Word, 14,* 197–207.

Goodglass, H., & Kaplan, E. (1972). *The assessment of aphasia and related disorders.* Philadelphia: Lea & Febiger.

Goodglass, H., & Kaplan, E. (1983). *The assessment of aphasia and related disorders* (2nd ed.). Philadelphia: Lea & Febiger.

Goodglass, H., Klein, B., Carey, P., & Jones, K. (1966). Specific semantic word categories in aphasia. *Cortex, 2,* 74–89.

Goodglass, H., Wingfield, A., Hyde, M. R., & Theurkauf, J. C. (1986). Category-specific dissociations in naming and recognition by aphasic patients. *Cortex, 22,* 87–102.

Goodyear, P., & Hynd, G. W. (1992). Attention deficit disorder with (ADD/H) and without (ADD/WO) hyperactivity: Behavioral and neuropsychological differentiation. *Journal of Clinical Child Psychology, 21,* 273–305.

Gordon, B. (1997). Models of naming. In H. Goodglass & A. Wingfield (Eds.). San Diego: Academic Press.

Gordon, B., Hart, J., Boatman, D., Crone, N., Nathan, S., Uematsu, S., Holcomb, H., Krauss, G., Selnes, O., & Lesser, R, P. (1994). Language and brain organization from the perspectives of cortical electrical interference, direct cortical recording, PET scanning, and acute lesion studies. In D. C. Gajdusek & G. M. McKhann (Eds.), *Evolution and neurology of language.* FESN Series Vol. 10, Nos. 182. Amsterdam: Elsevier.

Gordon, B., Hart, J., Lesser, R. P., & Arroyo, S. (1996). Mapping cerebral sites for emotion and emotional expression with direct cortical electrical stimulation and seizure discharges. In G. Holstege, R. Bandler, & C. B. Sapre (Eds.), *Progress in brain research* (Vol. 107, pp. 617–622) Amsterdam: Elsevier.

Gordon, B., Hart, J., Lesser, R., Schwerdt, P., Bare, M., Fisher, R., Krauss, G., Uematsu, S., & Selnes, O. (1990). Individual variations in perisylvian language representation. *Neurology, 40* (Suppl. 1), 172.

Gordon, B., Hart, J., Lesser, R., Schwerdt, P., Bare, M., Selnes, O., Fisher, R., & Uematsu, S. (1991). Visual confrontation naming mapped with direct cortical electrical stimulation. *Neurology, 41* (Supp. 1), 186–187.

Gordon, B., Lesser, R. P., & Hart, J. (1989). Fractionation of language functions by direct electrical cortical stimulation: Functional and neuroanatomical aspects. *Neurology, 39,* 175.

Gordon, H. (1970). Hemispheric asymmetries for the perception of musical chords. *Cortex, 6,* 387–398.

Gordon, H., & Bogen, J. E. (1981). Hemispheric lateralization of singing after intracarotid sodium amylobartitone. *Journal of Neurology, Neurosurgery, and Psychiatry, 37,* 727–738.

Gordon, H. W. (1980). Right hemisphere comprehension of verbs in patients with complete forebrain commissurotomy: Use of the dichotic method and manual performance. *Brain and Language, 11,* 76–86.

Gordon, J., & Baum, S. (1994). Rhyme priming in aphasia: The role of phonology in lexical access. *Brain and Language, 47,* 661–683.

Gordon, N. (1990). Acquired aphasia in childhood: The Landau-Kleffner syndrome. *Developmental Medicine and Child Neurology, 32,* 270–274.

Gordon, W. P., & Illes, J. (1987). Neurolinguistic characteristics of language production in Huntington's disease. *Brain and Language, 31,* 1–10.

Goswami, U. (1993). Toward an interactive analogy model of reading development: Decoding vowel graphemes in beginning reading. *Journal of Experimental Child Psychology, 56,* 443–475.

Goswami, U., & Bryant, P. (1990). *Phonological skills and learning to read.* Hove, UK: Lawrence Erlbaum.

Gotman, J., Bouwer, M., & Jones-Gotman, M. (1992). Intracranial EEG study of brain structures affected by internal carotid injection of amobarbital. *Neurology, 42,* 2136–2143.

Gott, F. S. (1973). Language after dominant hemispherectomy. *Journal of Neurology, Neurosurgery and Psychiatry, 36,* 1062–1068.

Gough, M. H. (1984). The surgical treatment of hyperinsulinism in infancy and childhood. *British Journal of Surgery, 71,* 75–78.

Gough, P. B., & Hillinger, M. L. (1980). Learning to read: An unnatural act. *Bulletin of the Orton Society, 30,* 179–196.

Gowers, W. (1895). *Malattie del sistema nervoso.* Milan: Vallardi.

Gowers, W. R. (1887). *Lectures in the diagnosis of diseases of the brain.* Philadelphia: Blakiston.

Grabow, J. D., Aronson, A. E., Rose, D. E., & Greene, K. L. (1980). Summated potentials evoked by speech sounds for determining cerebral dominance for language. *Electroencephalography and Clinical Neurophysiology, 49,* 38–47.

Grafman, J., Kampen, D., Rosenberg, J., Salazar, A., & Boller, F. (1989). Calculation abilities in a patient with a virtual left hemispherectomy. *Behavioural Neurology, 2,* 183–194.

Grafman, J., Thompson, K., Weingartner, H., Martinez, R., Lawlor, B., & Sunderland, T. (1991). Script generation as an indicator of knowledge representation in patients with Alzheimer's disease. *Brain and Language, 40,* 344–358.

Graves, R., & Landis, T. (1985). Hemispheric control of speech expression in aphasia. *Archives of Neurology, 42,* 249–251.

Gray, J. A., Feldon, J., Rawlins, J. N., Hemsley, D. R., & Smith, A. D. (1991). The neuropsychology of schizophrenia. *Behavioral and Brain Sciences, 14,* 1–84.

Greden, J. F., Albala, A. A., Smokler, I. A., Gardner, R., & Carroll, B. J. (1981). Speech pause time: A marker of psychomotor retardation among endogenous depressives. *Biological Psychiatry, 16,* 851–859.

Green, D. W. (1986). Control, activation, and resource: A framework and a model for the control of speech in bilinguals. *Brain and Language, 27,* 210–223.

Green, M. F., Hugdahl, K., & Mitchell. S. (1994). Dichotic listening during auditory hallucinations in patients with schizophrenia. *American Journal of Psychiatry, 151,* 357–362.

Greenblatt, S. H. (1973). Alexia without agraphia or hemianopsia: Anatomical analysis of an autopsied case. *Brain, 96,* 307–316.

Greenfield, P. M. (1991). Language, tools and brain: The ontogeny and phylogeny of hierarchically organized sequential behavior. *Behavioral and Brain Sciences, 14,* 531–595.

Grice, H. P. (1975). Logic and conversation. In P. Cole & J. L. Morgan (Eds.), *Syntax and semantics.* Vol. 3, *Speech acts* (pp. 41–58). New York: Academic Press.

Grice, P. (1978). Further notes on logic and conversation. In P. Cole (Ed.), *Syntax and semantics.* Vol. 9, *Pragmatics* (pp. 113–127). New York: Academic Press.

Grinvald, A., Lieke, E., Frostig, R. D., Gilbert, C. D., & Wiesel, T. N. (1986). Functional architecture of cortex revealed by optical imaging of intrinsic signals. *Nature, 324,* 361–364.

Grodzinsky, J. (1990). *Theoretical perspectives on language deficits.* Cambridge: MIT/Batsford.

Grodzinsky, Y. (1984). The syntactic characterization of agrammatism. *Cognition, 16,* 99–120.

Grodzinsky, Y. (1986). Language deficits and the theory of syntax. *Brain and Language, 27,* 135–159.

Grodzinsky, Y. (1989). Agrammatic comprehension of relative clauses. *Brain and Language, 37,* 480–499.

Grodzinsky, Y. (1990). *Theoretical perspectives on language deficits.* Cambridge: MIT Press.

Gropen, J., Pinker, S., Hollander, M., & Goldberg, R. (1991). Syntax and semantics in the acquisition of locative verbs. *Journal of Child Language, 18,* 115–151.

Grosjean, F. (1982). *Life with two languages: An introduction to bilingualism.* Cambridge: Harvard University Press.

Grossman, M., Carvell, S., Gollomp, S., Stern, M. B., Vernon, G., & Hurtig, H. I. (1991). Sentence comprehension and praxis deficits in Parkinson's disease. *Neurology, 41,* 1620–1626.

Grossman, M., Carvell, S., & Peltzer, L. (1993). The sum and substance of it: the appreciation of mass and count quantifiers in Parkinson's disease. *Brain and Language, 44,* 351–384.

Grossman, M., Carvell, S., Stern, M. B., Gollomp, S., & Hurtig, H. I. (1992). Sentence comprehension in Parkinson's disease: The role of attention and memory. *Brain and Language, 42,* 347–384.

Grossman, M., Onishi, K., Hughes, E., D'Esposito, M., Biassou, N., Seidl, A., White, H., Devine, T. W., & Robinson, K. M. (1995). Comprehension deficits in Alzheimer's disease, vascular dementia, and frontal dementia. *Brain and Language, 51,* 144–146.

Grossman, M., Stern, M. B., Gollomp, S., Vernon, G., et al. (1994). Verb learning in Parkinson's disease. *Neuropsychology, 8,* 413–423.

Gross-Tsur, V., Shalev, R. S., Manor, O., & Amir, N. (1995). Developmental right hemisphere syndrome: Clinical spectrum of the nonverbal learning disability. *Journal of Learning Disabilities, 28,* 80–86.

Gross-Tsur, V., Shalev, R. S., Wertman-Elad, R., Landau, H., & Amir, N. (1994). Neurobehavioral profile of children with persistent hyperinsulinemic hypoglycemia of infancy. *Developmental Neuropsychology, 10,* 153–163.

Grosswasser, Z., Mendelson, L., Stern, M. J., Schecter, I., & Najenson, T. (1977). Re-evaluation of prognostic factors in rehabilitation after severe head injury. *Scandinavian Journal of Rehabilitation Medicine, 9,* 147–149.

Grou, C., & Whitaker, H. A. (1992). "Le cerveau: Petite histoire de la localisation des fonctions. *Interface, 13*(5), 14–21.

Gruen, A. K., Frankle, B. C., & Schwartz, R. (1990). Word fluency generation skills of head-injured patients in an acute trauma center. *Journal of Communication Disorders, 23,* 163–170.

Gunter, T., Jackson, J., & Mulder, G. (1995). Language, memory, and aging: An electrophysiological exploration of the N400 during reading of memory-demanding sentences. *Psychophysiology, 32,* 215–229.

Gupta, P., & Touretzky, D. (1994). Connectionist models and linguistic theory: Investigations of stress systems in language. *Cognitive Science, 18,* 1–50.

Gurd, J. M., & Ward, C. D. (1989). Retrieval from semantic and letter-initial categories in patients with Parkinson's disease. *Neuropsychologia, 27,* 743–746.

Gurd, J. M., Ward, C. D., & Hodges, J. (1990). Parkinson's disease and the frontal hypothesis: Task alternation in verbal fluency. *Advances in Neurology, 53,* 321–325.

Gustafson, L. (1993). Clinical picture of frontal lobe degeneration of non-Alzheimer type. *Dementia, 4,* 143–148.

Gustafson, L., Brun, A., & Risberg, J. (1990). Frontal lobe dementia of the non-Alzheimer type. *Advances in Neurology, 51,* 65–71.

Guyard, H., Masson, V., & Quiniou, R. (1990). Computer-based aphasia treatment meets artificial intelligence. *Aphasiology, 4,* 599–613.

Haarmann, H. J., & Kolk, H. H. J. (1991a). A computer model of the temporal course of agrammatic sentence understanding: The effects of variation in severity and sentence complexity. *Cognitive Science, 15,* 49–87.

Haarmann, H. J., & Kolk, H. H. J. (1991b). Syntactic priming in Broca's aphasics: Evidence for slow activation. *Aphasiology, 5,* 247–263.

Haarmann, H. J., & Kolk, H. H. J. (1992). The production of grammatical morphology in Broca's and Wernicke's aphasics: Speed and accuracy factors. *Cortex, 28,* 97–112.

Haarmann, H. J., & Kolk, H. H. J., (1994). On-line sensitivity to subject–verb agreement violations in Broca's aphasics: The role of syntactic complexity and time. *Brain and Language, 46,* 493–516.

Haarmann, H. J., Just, M. A., & Carpenter, P. A. (in press). Aphasic sentence comprehension as a resource deficit: A computational approach. *Brain and Language.*

Haber, L. R. (1988). Porch Index of Communicative Ability. In D. J. Keyser & R. C. Sweetland (Eds.), *Test critiques* (Vol. 7, pp. 446–455). Kansas City, MO: Test Corporation of America.

Habib, M., & Démonet, J.-F. (1996). Cognitive neuroanatomy of language: The contribution of functional neuroimaging. *Aphasiology, 10,* 217–234.

Habib, M., Gayraud, D., Oliva, A., Regis, J., Salamon, G., & Khalil, R. (1991). Effects of handedness and sex on the morphology of the corpus callosum: A study with magnetic resonance imaging. *Brain and Cognition, 16,* 41–61.

Habib, M., Robichon, F., Lévrier, O., Khalil, R., & Salamon, G. (1995). Diverging asymmetries of temporo-parietal cortical areas: A reappraisal of Geschwind/Galaburda theory. *Brain and Language, 48,* 238–258.

Hachinski, V. C., Lassen, N. A., & Marshall, J. (1974). Multi-infarct dementia a cause of mental deterioration. *Lancet, 2,* 207–209.

Hagan, C. (1984). Language disorders in head trauma. In A. Holland (Ed.), *Language disorders in adults* (pp. 245–282). San Diego: College Hill Press.

Hagiwara, H. (1994). Functional categories and language breakdown. Paper presented at a symposium on agrammatism, 10th Tennett conference, Montreal, May 1994.

Hagiwara, H. (1995). The breakdown of functional categories and the economy of derivation. *Brain and Language, 50,* 92–116.

Haglund, M. M., Ojemann, G. A., & Hochman, D. W. (1992). Optical imaging of epileptiform and functional activity in human cerebral cortex. *Nature, 358,* 668–671.

Hagoort, P. (1989). Processing of lexical ambiguities: A comment on Milberg, Blumstein, and Dworetzky, 1987. *Brain and Language, 36,* 335–348.

Hagoort, P. (1991). Tracking the time course of language understanding in aphasia. Dissertation, University of Nijmegen, The Netherlands.

Hagoort, P. (1993). Impairments of lexical-semantic processing in aphasia: Evidence from the processing of lexical ambiguities. *Brain and Language, 45,* 189–232.

Hagoort, P. (1997). Semantic priming in Broca's aphasics at a short SOA: No support for an automatic access deficit. *Brain and Language, 56,* 287–300.

Hagoort, P., Brown, C., & Groothusen, J. (1993). The Syntactic Positive Shift (SPS) as an ERP measure of syntactic processing. *Language and Cognitive Processes, 8,* 439–483.

Hagoort, P., Brown, C. M., & Swaab, T. Y. (1996). Lexical-semantic event-related potential effects in patients with left hemisphere lesions and aphasia, and patients with right hemisphere lesions without aphasia. *Brain, 119,* 627–649.

Hagoort, P., & Kutas, M. (1995). Electrophysiological insights into language deficits. In F. Boller & J. Grafman (Eds.), *Handbook of neuropsychology* (Vol. 1, pp. 105–134). Amsterdam: Elsevier.

Halgren, E. (1990). Insights from evoked potentials into the neuropsychological mechanisms of reading. In A. B. Scheibel & A. F. Wechsler (Eds.), *Neurobiology of higher cognitive function* (pp. 103–150). New York: Guilford Press.

Halle, M. (1992). Phonological features. In W. Bright (Ed.), *International encyclopedia of linguistics* (Vol. 3, pp. 207–212). Oxford: Oxford University Press.

Halliday, M. A. K. (1985). *An introduction to functional grammar.* Sydney: Edward Arnold.

Halliday, M. A. K., & Hasan, R. (1985). *Language, context, and text: Aspects of language in a social-semiotic perspective.* Victoria: Deakin University Press.

Hamberger, M. J., Friedman, D., Ritter, W., & Rosen, J. (1995). Event-related potential and behavioral correlates of semantic processing in Alzheimer's patients and normal controls. *Brain and Language, 48,* 33–68.

Hambleton, R. K., & Novick, M. R. (1973). Toward an integration of theory and method for criterion-referenced tests. *Journal of Educational Measurement, 10,* 159–170.

Hambleton, R. K., Swaminathan, H., Algina, J., & Coulson, D. B. (1978). Criterion-referenced testing and measurement: A review of technical issues and developments. *Review of Educational Research, 48,* 1–47.

Hamilton, C. R., Nargeot, M.-C., Vermeire, B. A., & Bogen, J. E. (1986). Comprehension of language by the right hemisphere. *Society for Neuroscience Abstracts, 12,* 721.

Hamilton, H. E. (1994). Requests for clarification as evidence of pragmatic comprehension difficulty: The case of Alzheimer's disease. In R. L. Bloom, L. K. Obler, S. De Santi, & J. Ehrlich (Eds.), *Discourse analysis and application: Studies in adult clinical populations* (pp. 185–199). Hillsdale, NJ: Lawrence Erlbaum.

Hammer, M. (1977). Lateral differences in the newborn infant's response to speech and noise stimuli. *Dissertation Abstracts International, 38,* 1493B.

Hanley, J. R., Dewick, H. C., Davies, A. D.M., Playfer, J., & Turnbull, C. (1990). Verbal fluency in Parkinson's disease. *Neuropsychologia, 28,* 737–741.

Hanley, J. R., Young, A. W., & Pearson, N. A. (1991). Impairment of the visuo-spatial sketchpad. *Quarterly Journal of Experimental Psychology, 43A,* 101–125.

Hanson, W. R., Metter, J. F., & Riege, W. H. (1989). The course of chronic aphasia. *Aphasiology, 3,* 19–29.

Happé, F. G. E. (1993). Communicative competence and theory of mind in autism: A test of Relevance theory. *Cognition, 48,* 101–119.

Happé, F. G. E. (1994). *Autism: An introduction to psychological theory.* London: UCL Press.

Happé, F. G. E. (1995). The role of verbal ability in the theory of mind task performance of subjects with autism. *Child Development, 66,* 843–855.

Hardyck, C. (1991). Shadow and substance: Attentional irrelevancies and perceptual constraints in hemispheric processing of language stimuli. In F. L. Kitterle (Ed.), *Cerebral laterality: Theory and research* (pp. 133–154). Hillsdale, NJ: Lawrence Erlbaum.

Hari, R., Hämäläinen, M., Kaukoranta, E., Mäkelä, J., Joutsiniemi, S. L., & Tiihonen, J. (1989). Selective listening modifies activity in the human auditory cortex. *Experimental Brain Research, 74,* 463–470.

Harley, T. (1993). Connectionist approaches to language disorders. *Aphasiology, 7,* 221–249.

Harrington, A. (1987). *Medicine, mind and the double brain.* Princeton, NJ: Princeton University Press.

Hart, J. & Gordon, B. (1990). Delineation of single-word semantic comprehension deficits in aphasia, with anatomical correlation. *Annals of Neurology, 27,* 226–231.

Hart, J., & Gordon, B. (1992). Neural subsystems for object knowledge. *Nature, 359,* 60–64.

Hart, J., Jr., Berndt, R. S., & Caramazza, A. (1985). Category-specific naming deficit following cerebral infarction. *Nature, 316,* 439–440.

Hart, J., Jr., Lesser, R. P., Fisher, R. S., Schwerdt, P., Bryan, R. N., & Gordon, B. (1991). Dominant-side intracarotid amobarbital spares comprehension of word meaning. *Archives of Neurology, 48,* 55–58.

Hart, J., Jr., Lewis, P. J., Lesser, R. P., Fisher, R. S., Monsein, L. H., Schwerdt, P., Bandeen-Roche, K., & Gordon, B. (1993). Anatomic correlates of memory from intracarotid amobarbital injections with Technetium Tc 99m Hexamethylpropyleneamine Oxime SPECT. *Archives of Neurology, 50,* 745–750.

Hart S. (1988). Language and dementia: a review. *Psychological Medicine, 18,* 99–112.

Hartley, L. L., & Jensen, P. J. (1991). Narrative and procedural discourse after closed head injury. *Brain Injury, 5,* 267–285.

Hartley, L. L., & Jensen, P. J. (1992). Three discourse profiles of closed-head-injured speakers: Theoretical and clinical implications. *Brain Injury, 6,* 271–382.

Harvey, P. (1983). Speech competence in manic and schizophrenic psychoses: The association between clinically rated thought disorder and cohesion and reference performance. *Journal of Abnormal Psychology, 92,* 368–377.

Harvey, P. D., & Serper, M. R. (1990). Linguistic and cognitive failures in schizophrenia. A multivariate analysis. *Journal of Nervous and Mental Disease, 178,* 487–494.

Hasan, R. (1985). The texture of a text. In M. A. K. Halliday & R. Hasan (Eds.), *Language, context, and text: Aspects of language in a social-semiotic perspective* (pp. 43–59) Victoria: Deakin University Press.

Hatcher, P., Hulme, C., & Ellis, A. W. (1994). Ameliorating early reading failure by integrating the teaching of reading and phonological skills: The phonological linkage hypothesis. *Child Development, 65,* 41–57.

Hatfield, F. M. (1983). Aspects of acquired dysgraphia and implications for reeducation. In C. Code & D. J. Müller (Eds.), *Aphasia therapy* (pp. 157–169). London: Edward Arnold.

Haug, H. (1987). Brain sizes, surfaces, and neuronal sizes of the cortex cerebri: A stereological investigation of man and his variability and a comparison with some mammals. *American Journal of Anatomy, 180,* 126–142.

Hawkins, H. L., Hillyard, S. A., Luck, S. J., Mouloua, M., Downing, C. J., & Woodward, D. P. (1990). Visual attention modulates signal detectability. *Journal of Experimental Psychology: Human Perception and Performance, 16,* 802–811.

Head, H. (1926). *Aphasia and kindred disorders of speech.* (Vols. 1–2). Cambridge: Cambridge University Press.

Hebb, D. O. (1949). *The organization of behavior.* New York: John Wiley.

Hecaen, H., De Agostini, M., & Monzon-Montes, A. (1981). Cerebral organization in left-handers. *Brain and Language, 12,* 261–284.

Hecker, M. (1971). Speaker recognition: An interpretative survey of the literature. *ASHA Monographs, 16.*

Heeschen, C. (1985). Agrammatism versus paragrammatism: A fictitious opposition. In M.-L. Kean (Ed.), *Agrammatism* (pp. 207–248). New York: Academic Press.

Heilman, K. M., Rothi, L., McFarling, D., & Rottmann, A. L. (1981). Transcortical sensory aphasia with relatively spared spontaneous speech and naming. *Archives of Neurology, 38,* 236–239.

Heilman, K. M., Safran, A., & Geschwind, N. (1971). Closed head trauma and aphasia. *Journal of Neurology, Neurosurgery and Psychiatry, 34,* 265–269.

Heilman, K. M., Scholes, R. & Watson, R. T. (1975). Auditory affective agnosia. *Journal of Neurology, Neurosurgery, and Psychiatry, 38,* 69–72.

Heiss, W.-D., Kessler, J., Karbe, H., Fink, G. R., & Pawlik, G. (1993). Cerebral glucose metabolism as a predictor of recovery from aphasia in ischemic stroke. *Archives of Neurology, 50,* 958–964.

Hellige, J. B. (1993a). *Hemispheric asymmetry: What's right and what's left.* Cambridge: Harvard University Press.

Hellige, J. B. (1993b). Unity of thought and action: Varieties of interaction between the left and right cerebral hemispheres. *Current Directions in Psychological Science, 2,* 21–25.

Hellige, J. B. (1994). Babies, bath water and the chicken's way out. *Brain and Cognition, 26,* 66–73.

Hellige, J. B. (1995). Hemispheric asymmetry for components of visual information processing. In R. J. Davidson, & K. Hugdahl (Eds.), *Brain asymmetry* (pp. 99–122). Cambridge: MIT Press.

Hellige, J. B. (1996). Hemispheric asymmetry for visual information processing. *Acta Neurobiologiae Experimentalis, 56,* 485–497.

Hellige, J. B., & Cowin, E. L. (1996). Effects of stimulus arrangement on hemispheric differences and interhemispheric interaction for processing letter trigrams. *Neuropsychology, 10,* 247–253.

Hellige, J. B., Cowin, E. L., & Eng, T. (1995). Recognition of CVC syllables from LVF, RVF and central locations: Hemispheric differences and interhemispheric interaction. *Journal of Cognitive Neuroscience, 7,* 258–266.

Hellige, J. B., & Longstreth, L. E. (1981). Effects of concurrent hemisphere-specific activity on unimanual tapping rate. *Neuropsychologia, 19,* 395–405.

Hellige, J. B., Taylor, A. K., & Eng, T. L. (1989). Interhemispheric interaction when both hemispheres have access to the same stimulus information. *Journal of Experimental Psychology: Human Perception and Performance, 15,* 711–722.

Helm-Estabrooks, N. (1991). *Aphasia diagnostic profiles.* San Antonio, TX: Special Press.

Helm-Estabrooks, N., Ramsberger, G., Morgan, A. R., & Nicholas, M. (1989). Boston assessment of severe aphasia. Chicago: Riverside Publishing.

Henderson, V. W., Buckwalter, J. G., Sobel, E., Freed, D. M., & Diz, M. M. (1992). The agraphia of Alzheimer's disease. *Neurology, 42,* 777–784.

Henderson, V. W., Mack, W., Freed, D. M., Kempler, D., & Andersen, E. S. (1990). Naming consistency in Alzheimer's disease. *Brain and Language, 39,* 530–538.

Herholz, K., Thiel, A., Wienhard, K., Pietrzyk, U., von Stockhausen, H.-M., Karbe, H., Kessler, J., Bruckbauer, T., Halber, M., & Heiss, W.-D. (1996). Individual functional anatomy of verb generation. *NeuroImage, 3,* 185–194.

Hermann, B. P., & Wyler, A. R. (1988). Comparative results of dominant temporal lobectomy under general or local anesthesia: Language outcome. *Journal of Epilepsy, 1,* 127–134.

Hermann, B. P., Wyler, A. R., Somes, G., & Clement, L. (1994). Dysnomia after left anterior temporal lobectomy without functional mapping: Frequency and correlates. *Neurosurgery, 35,* 52–57.

Herrick, C. J. (1948). *The brain of the tiger salamander.* Chicago: University of Chicago Press.

Hesketh, A., & Bishop, D. V. M. (1996). Agrammatism and adaptation theory. *Aphasiology, 10,* 49–80.

Hickock, G., & Avrutin, S. (1996). Comprehension of Wh-questions in two Broca's aphasics. *Brain and Language, 52,* 314–327.

Hickock, G., Bellugi, U., & Klima, E. S. (1996). The neurobiology of sign language and its implications for the neural basis of language. *Nature, 381,* 699–702.

Hickock, G., Clark, K., Erhard, P., Helms-Tillery, K., Naeve-Velguth, S., Adriany, G., Hu X., Tomaso, H., Bellugi, U., Strick, P. L., & Ugurbil, K. (1995). Effect of modality on the neural organization for language: An fMRI study of sign language production. *Society for Neuroscience Abstracts, 21,* 694.

Hickock, G., Klima, E., Kritchevsky, M., & Bellugi, U. (1995). A case of "sign blindness" following left occipital damage in a deaf signer. *Neuropsychologia, 33*(12), 1597–1606.

Hickok, G. (1992). *Agrammatic comprehension and the trace-deletion hypothesis.* MIT Occasional Paper. Cambridge: MIT Press.

Hickok, G., Zurif, E., & Canseco-Gonzales, E. (1993). Structural description of agrammatic comprehension. *Brain and Language, 45,* 371–395.

Hier, D. B., Hagenlocker, K., & Shindler, A. G. (1985). Language disintegration in dementia: Effects of etiology and severity. *Brain and Language, 25,* 117–133.

Higginson, R., & MacWhinney, B. (1990). *CHILDES/BIB: An annotated bibliography of child language and language disorders.* Hillsdale, NJ: Lawrence Erlbaum.

Higginson, R., & MacWhinney, B. (1994). *CHILDES/BIB 1994* (Suppl.). Hillsdale, NJ: Lawrence Erlbaum.

Hill, P. M., & McCune-Nicolich, L. (1981). Pretend play and patterns of cognition in Down's syndrome children. *Child Development, 52,* 611–617.

Hillis, A. E. (1994). Contributions from cognitive analyses. Roberta Chapey (Ed.), *Language intervention strategies in adult aphasia* (3rd Ed. pp. 207–219). Baltimore: Williams and Wilkins.

Hillis, A. E. (1993). The role of models of language processing in rehabilitation of language impairments. *Aphasiology, 7,* 5–26.

Hillis, A. E., & Caramazza, A. (1990). The effects of attentional deficits on reading and spelling. In A. Caramazza (Ed.), *Cognitive neuropsychology and neurolinguistics: Advances in models of cognitive function and impairment* (pp. 211–275). Hillsdale, NJ: Lawrence Erlbaum.

Hillis, A. E., & Caramazza, A. (1991). Mechanisms for accessing lexical representations for output: Evidence from a cetegory specific semantic deficit. *Brain and Language, 40,* 106–144.

Hillis, A. E., & Caramazza, A. (1994). Aspects of semantic processing revealed by optic aphasia: Additional insights provided by new evidence. *Brain and Language, 47,* 380–383.

Hillis, A. E., & Caramazza, A. (1995). Representation of grammatical categories of words in the brain. *Journal of Cognitive Neuroscience, 7,* 396–407.

Hillis, A. E., Rapp, B., Romani, C., & Caramazza, A. (1990). Selective impairment of semantics in lexical processing. *Cognitive Neuropsychology, 7,* 191–243.

Hines, M., Chiu, L., McAdams, L. A., Bentler, P. M., & Lipcamon, J. (1992). Cognition and the corpus callosum: Verbal fluency, visuospatial ability, and language lateralization related to midsagittal surface areas of callosal subregions. *Behavioral Neuroscience, 106,* 3–14.

Hines, T. M., & Volpe, B. T. (1985). Semantic activation in patients with Parkinson's disease. *Experimental Aging Research, 11,* 105–107.

Hinton, G. E., & Shallice, T. (1991). Lesioning an attractor network: Investigations of acquired dyslexia. *Psychological Review, 98,* 74–95.

Hirsch, E., Marescaux, C., Maquet, P., Metz-Lutz, M. N., Kiesmann, M., Salmon, E., Franck, G., & Kurtz, D. (1990). Landau-Kleffner syndrome: A clinical and EEG study of five cases. *Epilepsia, 31,* 756–767.

Hirsch, H. J., Loo, S., Evans, N., Grigler, J. F., Filler, R. M., & Gabbay, K. H. (1977). Hypoglycemia of infancy and nesidioblastosis: Studies with somatostatin. *New England Journal of Medicine, 296,* 1323–1326.

Hiscock, M. (1988). Behavioral asymmetries in normal children. In D. L. Molfese & S. J. Segalowitz (Eds.), *Brain lateralization in children: Developmental implications* (pp. 85–169). New York: Guilford Press.

Hiscock, M., & Beckie, J. L. (1993). Overcoming the right-ear advantage: A study of focused attention in children. *Journal of Clinical and Experimental Neuropsychology, 15,* 754–772.

Hiscock, M., & Kinsbourne, M. (1987). Specialization of the cerebral hemispheres: Implications for learning. *Journal of Learning Disabilities, 20,* 130–143.

Hiscock, M., & Kinsbourne, M. (1994). Phylogeny and ontogeny of cerebral lateralization. In R. J. Davidson & K. Hugdahl (Eds.), *Brain asymmetry* (pp. 535–578). Cambridge: MIT Press.

Hittmair-Delazer, M., Sailer, U., & Benke, T. (1995). Impaired arithmetic facts but intact conceptual knowledge—a single case study of dyscalculia. *Cortex, 31,* 139–147.

Hittmair-Delazer, M., Semenza, C., & Denes, G. (1994). Concepts and facts in calculation. *Brain, 117,* 715–728.

Hochheimer, W. (1936). Zur Psychologie des Choreatiker. *Zeitschrift für Psychologie und Neurologie, 47,* 49–115.

Hochman, D. W., Whitaker, H. A., Haglund, M., & Ojemann, G. A. (1995). Is there concordance between functional brain mapping techniques [comparing intraoperative electrical stimulation and optical imaging]? *Human Brain Mapping* (Supp. 1), 85.

Hodges, J. R., Graham, N., & Patterson, K. (1995). Charting the progression in semantic dementia: Implications for the organisation of semantic memory. *Memory, 3,* 463–495.

Hodges, J. R., & Gurd, J. M. (1994). Remote memory and lexical retrieval in a case of frontal Pick's disease. *Archives of Neurology, 51*(8), 821–827.

Hodges, J. R., Patterson, K., Oxbury, S., & Funnell, E. (1992). Semantic dementia. *Brain, 115,* 1783–1806.

Hodges, J. R., Salmon, D. P., & Butters, N. (1991). The nature of the naming deficit in Alzheimer's and Huntington's disease. *Brain, 114,* 1547–1558.

Hodges, J. R., Salmon, D. P., & Butters N. (1992). Semantic memory impairment in Alzheimer's disease: Failure of access or degraded knowledge? *Neuropsychologia, 30,* 301–314.

Hoffman, J. E., & Subramaniam, B. (1995). The role of visual attention in saccadic eye movements. *Perception & Psychophysics, 57,* 787–795.

Hoffman, R. E. (1986). Verbal hallucinations and language production processes in schizophrenia. *Brain and Behavioral Sciences, 9,* 503–548.

Hoffman, R. E., Hogben, G. L., Smith, H., & Calhoun W. F. (1985). Message disruption during syntactic processing in schizophrenia. *Journal of Communication Disorders, 18,* 183–202.

Hoffman, R. E., & Sledge, W. (1988). An analysis of grammatical deviance occurring in spontaneous schizophrenic speech. Journal of Neurolinguistics, 3, 189–191.

Hoffman, R. E., Stopek, S., & Andreasen, N. D. (1986). A comparative study of manic vs. schizophrenic speech disorganization. *Archives of General Psychiatry, 43,* 831–838.

Hofstede, B. T. M., & Kolk, H. H. J. (1994). The effects of task variation on the production of grammatical morphology in Broca's aphasia: A multiple case-study. *Brain and Language, 46,* 278–328.

Holcomb, P. J., & Neville, H. J. (1990). Auditory and visual semantic priming in lexical decision: A comparison using event-related brain potentials. *Language and Cognitive Processes, 5,* 281–312.

Holcomb, P. J. , & Neville, H. J. (1991). Natural speech processing: An analysis using event-related brain potentials. *Psychobiology, 19,* 286–300.

Holland, A. L. (1980). *Communicative abilities in daily living.* Baltimore: University Park Press.

Holland, A. L. (1984). When is aphasia aphasia? The problem of closed head injury. In R. W. Brookshire (Ed.), *Clinical aphasiology* (Vol. 14, pp. 345–349). Minneapolis: BRK Publishers.

Holland, A. L. (1989). Recovery in aphasia. In F. Boller & J. Grafman (Eds.), *Handbook of neuropsychology* (Vol. 2; pp. 83–90). Amsterdam: Elsevier Science.

Holland, A. L. (1994). Cognitive neuropsychological theory and treatment for aphasia: Exploring the strengths and limitations. *Clinical Aphasiology, 22,* 275–282.

Holland, A. L., & Forbes, M. M. (1993). *Aphasia treatment: World perspectives.* San Diego: Singular Publishing.

Holland, A. L., McBurney, D. H., Moossy, J., & Reinmuth, O. M. (1985). The dissolution of language in Pick's disease with neurofibrillary tangles: A case study. *Brain and Language, 24,* 36–58.

Homan, R. W., Criswell, E., Wada, J. A., & Ross, E. D. (1982). Hemispheric contributions to manual communication (signing and finger spelling). *Neurology, 32,* 1020–1023.

Hood, A. (1824). Case 4th—July 28, 1824 (Mr. Hood's cases of injuries of the brain. *Phrenological Journal and Miscellany.* Vol. 2, 82–94.

Hopfield, J. J. (1982). Neural networks and physical systems with emergent collective computational abilities. *Proceedings of the National Academy of Sciences, 79,* 2554–2558.

Hoptman, M. J., Davidson, R. J., Gudmundsson, A., Schreiber, R. T., & Ershler, W. B. (1996). Age differences in visual evoked potential estimates of interhemispheric transfer. *Neuropsychology, 10,* 263–271.

Horenstein, S., Chung, G., & Brenner, S. (1978). Aphasia in two verified cases of left thalamic hemorrhage. *Annals of Neurology, 4,* 177.

Horev, Z., Ipp, M., Levey, P., & Daneman, D. (1991). Familial hyperinsulinism: Successful conservative management. *Journal of Pediatrics, 119,* 717–720.

Horn, J. L., & Cattell, R. B. (1967). Age differences in fluid and crystallized intelligence. *Acta Psychologica, 26,* 107–129.

Horner, J., Dawson, D. V., Eller, M. A., Buoyer, F. G., Crowder, J. L., & Reus, C. M. (1995). Prognosis for improvement during acute rehabilitation as measured by the Western Aphasia Battery. *Clinical Aphasiology, 23,* 141–154.

Horner, J., Heyman, A., Dawson, D., & Rogers H. (1988). The relationship of agraphia to the severity of dementia in Alzheimer's disease. *Archives of Neurology, 45,* 760–763.

Horner, J., & Loverso, F. L. (1991). Models of aphasia treatment. In T. E. Prescott (Ed.), *Clinical aphasiology* (Vol. 20; pp. 61–75). Austin, TX: Pro-Ed.

Horner, J., Loverso, F. L., & Gonzalez Rothi, L. (1994). Models of aphasia treatment. In R. Chapey (Ed.), *Language intervention strategies in adult aphasia* (3rd ed.; pp. 135–145). Baltimore: Williams & Wilkins.

Horwitz, B., Duara, R., & Rappoport, S. I. (1984). Intercorrelations of glucose rates between brain regions: Application to healthy males in a reduced state of sensory input. *Journal of Cerebral Blood Flow and Metabolism, 4,* 484–499.

Horwitz, B., Maisog, J., Kirschner, P., Haxby, J. V., McIntosh, A. R., Schapiro, M. B., Friston, K. J., Ungerleider, L. G., & Grady C. L. (1993). Functional pathways in the brain during object and spatial visual processing: An rCBF PET/correlation analysis. *Journal of Cerebral Blood Flow and Metabolism, 13* (Suppl. 1), S527.

Hough, M. (1990). Narrative comprehension in adults with right and left hemisphere brain-damage: Theme organization. *Brain and Language, 38,* 253–277.

House, A., Dermot, R., & Standen, P. J. (1987). Affective prosody in the reading voice of stroke patients. *Journal of Neurology, Neurosurgery, and Psychiatry, 50,* 910–912.

Howard, D. (1995). Short-term recall without short-term memory. In R. Campbell & M. A. Conway (Eds.), *Broken memories: Case studies in memory impairment* (pp. 285–301). Oxford: Basil Blackwell.

Howard, D., & Franklin, S. (1988). *Missing the meaning?: A cognitive neuropsychological study of the processing of words by an aphasic patient.* Cambridge: MIT Press

Howard, D., & Hatfield, F. M. (1987). *Aphasia therapy: Historical and contemporary issues.* Hillsdale, NJ: Lawrence Erlbaum.

Howard, D., & Patterson, K. (1989). Models for therapy. In X. Seron & G. Deloche (Eds.), *Cognitive approaches in neuropsychological rehabilitation* (pp. 39–64). Hillsdale, NJ: Lawrence Erlbaum.

Howard, D., Patterson, K., Franklin, S., Orchard-Lisle, V., & Morton, J. (1985). The treatment of word retrieval deficits in aphasia: A comparison of two therapy methods. *Brain, 108,* 817–829.

Howard, D., Patterson, K., Wise, R., Brown, W. D., Friston, K., Weiller, C., & Frackowiak, R. S. J. (1992). The cortical localisation of the lexicons: PET evidence. *Brain, 115,* 1769–1782.

Huber, S. J., Shuttleworth, E. C., & Freidenberg, D. L. (1989). Neuropsychological differences between the dementias of Alzheimer's and Parkinson's diseases. *Archives of Neurology, 46,* 1287–1291.

Huber, W. (1990). Text comprehension and production in aphasia: Analysis in terms of micro- and macrostructure. In. Y. Joanette & H. H. Brownell (Eds.), *Discourse ability and brain damage: Theoretical and empirical perspectives* (pp. 154–179). New York: Springer Verlag.

Huber, W., Poeck, K., & Willmes, K. (1984). The Aachen Aphasia Test (AAT). *Advances in Neurology, 42,* 291–303.

Huber, W., Willmes, K., & Göddenhenrich, S. (1988). Die Diagnose von aphasischen Leistungsdissoziationen beim lexikalischen Diskriminieren. In K.-B. Günther (Ed.), *Sprachstörungen: Probleme ihrer Diagnostik bei mentalen Retardierungen, Entwicklungsdysphasien und Aphasien* (pp. 306–329). Heidelberg: Edition Schindele.

Huber, W., Willmes, K., Poeck, K., Van Vleymen, B., & Deberdt, W. (1997). Piracetam as an adjuvant to language therapy for aphasia: A randomised double-blind placebo-controlled pilot study. *Archives of Physical Medicine and Rehabilitation, 78,* 245–250.

Huff, F. J., Corkin, S., & Growdon, J. H. (1986). Semantic impairment and anomia in Alzheimer's disease. *Brain and Language, 28,* 235–249.

Hugdahl, K., & Andersson, L. (1986). The "forced-attention paradigm" in dichotic listening to CV-syllables: A comparison between adults and children. *Cortex, 22,* 417–422.

Hugdahl, K., & Nordby, H. (1994). Electrophysiological correlates to cued attentional shifts in the visual and auditory modalities. *Behavioral and Neural Biology, 62,* 21–32.

Hughes, C. F., Uhlmann, C., & Pennebaker, J. W. (1994). The body's response to processing emotional trauma: Linking verbal text with autonomic activity. *Journal of Personality, 62,* 565–585.

Hughes, C. P., Chan, J.-L., & Ming, S.-S. (1983). Aprosodia in Chinese patients with right cerebral hemisphere lesion. *Archives of Neurology, 40,* 732–736.

Hulme, C., Maugham, S., & Brown, G. (1991). Memory for familiar and unfamiliar words: Evidence for a long-term contribution to short-term span. *Journal of Memory and Language, 30,* 685–701.

Humphreys, G. W., & Riddoch, M. J. (Eds.). (1994). *Cognitive neuropsychology and cognitive rehabilitation.* Hove, UK: Lawrence Erlbaum.

Hunter, R., & Macalpine, I. (Eds.). (1982). *Three hundred years of psychiatry 1535–1860.* Hartsdale: Carlisle.

Hupet, M., & Nef, F. (1994). Vieillissement cognitif et langage. In M. Van der Linden & M. Hupet (Ed.), *Le vieillissement cognitif* (pp. 37–85). Paris: Presses Universitaires de France.

Iacoboni, M., & Zaidel, E. (1996). Hemispheric independence in word recognition: Evidence from bilateral presentations. *Brain and Language, 53,* 121–140.

Ide, A., & Aboitiz, F. (1996). Bifurcation patterns in the human sylvian fissure: Hemispheric and sex differences. *Cerebral Cortex, 6,* 717–725.

Ide, A., Rodríguez, E., Zaidel, E., & Aboitiz, F. (1996). Bifurcation patterns in the human Sylvian fissure: Hemispheric and sex differences. *Cerebral Cortex, 6,* 717–725.

Illes, J. (1989). Neurolinguistic features of spontaneous language production dissociate three forms of neurodegenerative disease: Alzheimer's, Huntington's, and Parkinson's. *Brain and Language, 37,* 628–642.

Inhoff, A. W., Pollatsek, A., Posner, M. I., & Rayner, K. (1989). Covert attention and eye movements during reading. *Quarterly Journal of Experimental Psychology, 41A,* 63–89.

Innocenti, G. M. (1986). General organization of callosal connections in the cerebral cortex. In E. G. Jones & A. Peters (Eds.), *Cerebral cortex* (Vol. 5, pp. 291–354). New York: Plenum Press.

Innocenti, G. M. (1995). Cellular aspects of callosal connections and their development. *Neuropsychologia, 33,* 961–988.

Intriligator, J., Hanaff, M. A., & Michel, F. (1995). A patient suffering from damage to the posterior portion of the corpus callosum can name items in both visual fields but cannot report whether they are the same or different. *Society for Neuroscience Abstracts, 36,* 5470.

Irigaray, L. (1973). *Le langage des déments* [The language of the demented]. The Hague and Paris: Mouton.

Isen, A. M., Niedenthal, P. M., & Cantor, N. (1992). An influence of positive affect on social categorization. *Motivation and Emotion, 16,* 65–78.

Ishii, N., Nishihara, Y., & Imamura, T. (1986). Why do frontal lobe symptoms predominate in vascular dementia with lacunes? *Neurology, 36,* 340–345.

Ito, M. (1993). Movement and thought: Identical control mechanisms by the cerebellum. *Trends in Neuroscience, 16,* 448–450.

Ivan, T. M., & Franco, K. (1994). Poststroke pathological laughing and crying. *American Journal of Psychiatry, 151,* 290–291.

Ivry, R. B., & Robertson, L. C. (in press). *The two sides of perception.* Cambridge: MIT Press.

Jack, C. R., Nichols, D. A., Sharbrough, F. W., Marsh, W. R., Petersen, R. C., Hinkeldey, N. S., Ivnik, R. J., Cascino, G. D., & Ilstrup, D. M. (1989). Selective posterior cerebral artery injection of amytal: New method of preoperative memory testing. *Mayo Clinic Proceedings, 64,* 965–975.

Jackendoff, R. (1975). Morphological and semantic regularities in the lexicon. *Language, 51,* 639–671.

Jackendoff, R. (1983). *Semantics and cognition.* Cambridge: MIT Press.

Jackendoff, R. (1987). On beyond zebra: The relation of linguistic and visual information. *Cognition, 26,* 89–114.

Jackendoff, R. (1990). *Semantic structures.* Cambridge: MIT Press.

Jackendoff, R. (1992). *Languages of the mind: Essays on mental representation.* Cambridge: MIT Press.

Jackson, J. H. (1868). Notes on the physiology and pathology of the nervous system. *Medical Times and Gazette, 2,* 208, 358, 526, 696; (1869) *1,* 245, 600; *2,* 481. Reprinted in Taylor (1958).

Jackson, J. H. (1874). On the nature of the duality of the brain. In J. Taylor (Ed.), *Selected writings of John Hughlings Jackson* (Vol 2. [1932], pp. 129–145). London: Hodder & Stoughton.

Jackson, J. H. (1878). *Brain, 1,* 64. [Reprinted in J. H. Jackson (1932), *Selected writings of Hughlings Jackson.* London: Taylor].

Jackson, J. H. (1878–1879). On affections of speech from disease of the brain. *Brain, 1,* 304–330; (1879–1880) *Brain, 2,* 203–222 and 323–356. Reprinted in Taylor (1958).

Jackson, J. H. (1879–1880). On affections of speech from disease of the brain. *Brain, 2,* 203–222.

Jackson, J. H. (1893). Words and other symbols in mentation. *Medical Press and Circular, 2,* 205. Reprinted in Taylor (1958).

Jackson, S. T. , & Tompkins, C. A. (1991). Supplemental aphasia tests: Frequency of use and psychometric properties. *Clinical Aphasiology, 20,* 91–99.

Jacobs, B., & Schumann, J. H. (1992). Language acquisition and the neurosciences: Toward a more integrative perspective. *Applied Linguistics, 13,* 283–301.

Jacobs, B. (1988). Neurobiological differentiation of primary and secondary language acquisition. *Studies in Second Language Acquisition, 10,* 303–337.

Jakobson, R. (1956). Two aspects of language and two types of aphasic disturbance. In R. Jakobson & M. Halle (Eds.), *Fundamentals of language.* The Hague: Mouton.

James, W. (1980). *The principles of psychology.* New York: Holt (first published in 1890 by Holt and Company).

Jampala, V. C., Taylor, M. A., & Abrams, R. (1989) The diagnostic implications of formal thought disorder in mania and schizophrenia: A reassessment. *American Journal of Psychiatry, 146,* 459–463.

Janati, A., & Appel, A. R. (1984). Psychiatric aspects of progressive supranuclear palsy. *Journal of Nervous and Mental Disease, 172,* 85–89.

Jäncke, L., Schlaug, G., Huang, Y., & Steinmetz, H. (1995). Asymmetry of the planum parietale. *NeuroReport, 5,* 1161–1163.

Jarema, G. (1993). In sensu non in situ: the prodromic cognitivism of Kussmaul. *Brain and Language, 45,* 495–510.

Jarema, G., & Kadzielawa, D. (1990). Agrammatism in Polish: A case study. In L. Menn & L. K. Obler (Eds.), *Agrammatic aphasia: A cross-language narrative sourcebook* (Vol. 2, pp. 817–893). Amsterdam: John Benjamins.

Jarema, G., & Kehayia, E. (1992). Impairment of inflectional morphology and lexical storage. *Brain and Language, 43,* 541–564.

Jay, T. (1992). *Cursing in America.* Philadelphia: John Benjamins.

Jeannerod, M. (1994). The representing brain: Neural correlates of motor intention and imagery. *Behavioral and Brain Sciences, 17,* 187–245.

Jeffery, P. J., Monsein, L. H., Szabo, Z., Hart, J., Fisher, R. S., Lesser, R. P., Debrun, G. M., Gordon, B., Wagner, H. N., & Camargo, E. E. (1991). Mapping the distribution of amobarbital sodium in the intracarotid Wada test by use of Tc-99m HMPAO with SPECT. *Radiology, 178,* 847–850.

Jellinger, K. (1986). Pathology of parkinsonism. In S. Fahn, C. D. Marsden, P. Jenner, & P. Teychenne (Eds.), *Recent developments in Parkinson's disease* (pp. 33–66). New York: Raven Press.

Jenkins, W. M., Merzenich, M. M., & Recanzone, G. M. (1990). Neocortical representational dynamics in adult primates: Implications for neuropsychology. *Neuropsychologia, 28,* 573–584.

Jennings, J. R., & Coles, M. G. H. (Eds.). (1991). *Handbook of cognitive psychophysiology.* New York: John Wiley.

Jerison, H. J. (1991). *Brain size and the evolution of mind.* (Fifty-ninth James Arthur Lecture on the Evolution of the Human Brain). New York: American Museum of Natural History.

Jernigan, T. L., Bellugi, U., Sowell, E., Doherty, S., & Hesselink, J. R. (1993). Cerebral morphological distinctions between Williams and Down syndromes. *Archives of Neurology, 50,* 186–191.

Jespersen, O. (1969). *Essentials of English grammar.* London: George Allen & Unwin.

Joanette, Y., Ali-Cherif, A., Delpuech, F., Habib, M., Pelissier, J. F., & Poncet, M. (1983). Évolution de la sémiologie aphasique avec l'âge. *Revue Neurologique, 139,* 657–664.

Joanette, Y., & Brownell, H. (Eds.). (1990). *Discourse ability and brain damage: Theoretical and empirical perspectives.* New York: Springer Verlag.

Joanette, Y., & Goulet, P. (1990). Narrative discourse in right-brain-damaged right-handers. In Y. Joanette & H. H. Brownell (Eds.), *Discourse ability and brain damage: Theoretical and empirical perspectives* (pp. 131–153). New York: Springer Verlag.

Joanette, Y., Goulet, P., & Daoust, H. (1991). Incidence et profils des troubles de la communication verbale chez les cérébrolésés droits. *Revue de Neuropsychologie, 1,* 3–27.

Joanette, Y., Goulet, P., & Hannequin, D. (1990). *Right hemisphere and verbal communication.* New York: Springer Verlag.

Joanette, Y., Goulet, P., Ska, B., & Nespoulous, J.-L. (1986). Informative content of narrative discourse in right-brain-damaged right-handers. *Brain and Language, 29,* 81–105.

Joanette, Y., Lecours, A. R., Lepage, Y., & Lamoureux, M. (1983). Language in right-handers with right hemisphere lesions: A preliminary study including anatomical, genetic, and social factors. *Brain and Language, 20,* 329–339.

Job, R., & Sartori, G. (1984). Morphological decomposition: Evidence from crossed phonological dyslexia. *Quarterly Journal of Experimental Psychology, 36A,* 435–458.

Johanson, A., & Hagberg, B. (1989). Psychometric characteristics in patients with frontal lobe degenerations of non-Alzheimer type. *Archives of Gerontology and Geriatrics, 8,* 129–137.

Johnson, J. P., Summers, R. K., & Weidner, W. E. (1977). Dichotic ear preference in aphasia. *Journal of Speech and Hearing Research, 20,* 116–129.

Johnson, L. E. (1984). Vocal responses to left visual field stimuli following forebrain commissurotomy. *Neuropsychologia, 22,* 153–166.

Johnson, R. C., Cole, R. E., Bowers, J. K., Foiles, S. V., Nikaido, A. M., Patrick, J. W., & Woliver, R. E. (1979). Hemisphere efficiency in middle and later adulthood. *Cortex, 15,* 109–119.

Johnson, R., Jr. (1989). Auditory and visual P300s in temporal lobectomy patients: Evidence for modality dependent generators. *Psychophysiology, 26,* 633–650.

Johnson, W. F., Emde, R. N., Scherer, K. R. & Klinnert, M. D. (1986). Recognition of emotion from vocal cues. *Archives of General Psychiatry, 43,* 280–283.

Jones, E. (1986). Building the foundations for sentence production in a non-fluent aphasic. *British Journal of Disorders of Communication, 21,* 63–82.

Jones, E. G. (1985). The thalamus. New York: Plenum Press.

Jones-Gotman, M. (1987). Commentary: Psychological evaluation-testing hippocampal function. In J. Engel Jr. (Ed.), *Surgical treatment of the epilepsies* (pp. 203–211). New York: Raven Press.

Jones-Gotman, M., Barr, W. B., Dodrill, C. B., Gotman, J., Meador, K. J., Rausch, R., Sass, K. J., Sharbrough, F. W., Silfvenius, H., & Wieser, H.-G. (1993). Postscript: Controversies concerning the use of intraarterial amobarbital procedures. In J. Engel Jr. (Ed.), *Surgical treatment of the epilepsies* (2nd ed., pp. 445–449). New York: Raven Press.

Jordan, F. M., Murdoch, B. E., & Buttsworth, D. L. (1991). Closed head injured childrens' performance on narrative tasks. *Journal of Speech and Hearing Research, 34,* 572–582.

Jorm, A. F. (1986). Controlled and automatic information processes in senile dementia: A review. *Psychological Medicine, 16,* 77–88.

Josiassen, R. C., Curry, L. M., & Mancall, E. L. (1983). Development of neuropsychological deficits in Huntington's disease. *Archives of Neurology, 40,* 791–796.

Junqué, C., Vendrell, P., & Vendrell, J. (1995). Differential impairments and specific phenomena in 50 Catalan–Spanish bilingual aphasic patients. In M. Paradis (Ed.), *Aspects of bilingual aphasia* (pp. 177–209). Oxford: Pergamon Press.

Just, M., & Carpenter, P. (1993). The intensity dimension of thought: Pupillometric indices of sentence processing. *Canadian Journal of Experimental Psychology, 47,* 310–339.

Just, M. A., & Carpenter, P. A. (1992). A capacity theory of comprehension: Individual differences in working memory. *Psychological Review, 99,* 122–149.

Kaas, J. H. (1995). The reorganisation of sensory and motor maps in adult mammals. In M. S. Gazzaniga (Ed.), *The cognitive neurosciences* (pp. 51–71). Cambridge: MIT Press.

Kaasgaard, K., & Lauritsen, P. (1995). The use of computers in cognitive rehabilitation in Denmark. *American Journal of Speech-Language Pathology, 4,* 5–8.

Kaes, T. (1907). *Die Grosshirnrinde des Menschen in ihren Massen und in ihrem Fasergehalt.* Jéna: Fischer.

Kahn, H. J., Joanette, Y., Ska, B., & Goulet, P. (1990). Discourse analysis in neuropsychology: Comment on Chapman and Ulatowska. *Brain and Language, 38,* 454–461.

Kandel, E. R., Schwartz, J. H., and Jessell, T. M. (1991). *Principles of neural science* (3rd ed.). Norwalk: Appleton & Lange.

Kanner, L. (1943). Autistic disturbances of affective contact. *Nervous Child, 2,* 217–250.

Kaplan, E., Goodglass, H., & Weintraub, S. (1976). *The Boston Naming Test* (Experimental Edition). Philadelphia: Lea & Febiger.

Kaplan, E., Goodglass, H., & Weintraub, S. (1978, 1983). *The Boston Naming Test.* Philadelphia: Lea & Febiger.

Kaplan, J. A., Brownell, H. H., Jacobs, J. R., & Gardner, H. (1990). The effects of right hemisphere damage on the pragmatic interpretation of conversational remarks. *Brain and Language, 38,* 315–333.

Karamat, E., Ilmberger, J., Poewe, W., & Gerstenbrand, F. (1991). Memory dysfunction in Parkinson patients: an analysis of verbal learning processes. *Journal of Neural Transmission, 33,* 93–97.

Karanth, P., & Rangamani, G. N. (1988). Crossed aphasia in multilinguals. *Brain and Language, 34,* 169–180.

Karbe, H., Herholz, K., Szelies, B., Pawlik, G., Wienhard, K., & Heiss, W. D. (1989). Regional metabolic correlates of Token Test results in cortical and subcortical left hemispheric infarction. *Neurology, 39,* 1083–1088.

Karbe, H., Kertesz, A., & Polk, M. (1993). Profiles of language impairment in primary progressive aphasia. *Archives of Neurology, 50,* 193–201.

Karbe, H., Kessler, J., Herholz, K., Fink, G. R., & Heiss, W.-D. (1995). Long-term prognosis of post-stroke aphasia studied with positron emission tomography. *Archives of Neurology, 52,* 186–190.

Kasari, C., Sigman, M., Baumgartner, P., & Stipek, D. (1993). Pride and mastery in children with autism. *Journal of Child Psychology and Psychiatry, 34,* 353–362.

Kase, C. S., Wolf, P. A., Backman, D. L., Linn, R. T., & Cupples, L. A. (1989). Dementia and stroke: The Framingham study. In M. D. Ginsberg & W. D. Dietrich (Eds.), *Cerebrovascular diseases: 16th research (Princeton) conference* (pp. 193–198). New York: Raven Press.

Kataria, S., Goldstein, D. J., & Kushnik, T. (1984). Developmental delays in Williams (elfin facies) syndrome. *Applied Research in Mental Retardation, 5,* 419–423.

Katayama, J., & Yagi, A. (1992). Negative brain potentials elicited by an unexpected color patch or word. *Electroencephalography and Clinical Neurophysiology, 83,* 248–253.

Katz, L., Rexer, K., & Lukatela, G. (1991). The processing of inflected words. *Psychological Research, 53,* 25–32.

Katz, R. C. (1990). Intelligent computerized treatment or artificial aphasia therapy? *Aphasiology, 4,* 621–624.

Katz, R. C. (1995). Aphasia treatment and computer technology. In C. Code & D. Müller (Eds.), *Treatment of aphasia: From theory to practice* (pp. 253–285). London: Whurr Publishers.

Katz, R. C., & Wertz, R. T. (1992). Computerized hierarchical reading treatment in aphasia. *Aphasiology, 6,* 165–177.

Katz, R. C., & Wertz, R. T. (1997). The efficacy of computer-provided reading treatment for chronic aphasic adults. *Journal of Speech and Hearing Research.*

Katz, R. C., Wertz, R. T., Davidoff, M., Shubitowski, Y. D., & Devitt, E. W. (1989). A computer program to improve written confrontation naming in aphasia. In T. E. Prescott (Ed.), *Clinical aphasiology: 1988 conference proceedings* (pp. 321–338). Austin, TX: Pro-Ed.

Katz, W. (1988). Anticipatory coarticulation in aphasia: Acoustic and perceptual data. *Brain and Language, 35,* 340–368.

Katz, W., Machetanz, J., Orth, U., & Schönle, P. (1990a). A kinematic analysis of anticipatory coarticulation in the speech of anterior aphasic subjects using electromagnetic articulography. *Brain and Language, 38,* 555–575.

Katz, W., Machetanz, J., Orth, U., & Schönle, P. (1990b). Anticipatory labial coarticulation in the speech of German-speaking anterior aphasic subjects: Acoustic analyses. *Journal of Neurolinguistics, 5,* 295–320.

Kaukoranta, E., Hari, R., & Lounasmaa, O. V. (1987). Response of the human auditory cortex to vowel onset after fricative consonants. *Experimental Brain Research, 69,* 19–23.

Kawashima, R., Itoh, M., Hatazawa, J., Miyazawa, H., Yamada, K., Matsuzawa, T., & Fukuda, H. 1993. Changes of regional cerebral blood flow during listening to an unfamiliar spoken language. *Neuroscience Letters, 161,* 69–72.

Kay, J., & Franklin, S. (1995). Cognitive neuropsychology and assessment. In R. L. Mapou & J. Spector (Eds.), *Clinical neuropsychological assessment: A cognitive approach.* New York: Plenum Press.

Kay, J., Lesser, R., & Coltheart, M. (1992). *PALPA: Psycholinguistic assessments of language processing in aphasia.* Hove, UK: Lawrence Erlbaum.

Kaye, J., Lowenstamm, J., & Vergnaud, J.-R. (1985). The internal structure of phonological elements: A theory of charm and government. *Phonology, 2,* 305–328.

Kean, M.-L. (1977). The linguistic interpretation of aphasic syndromes: Agrammatism in Broca's aphasia. An example. *Cognition, 5,* 9–46.

Kean, M.-L. (1995). The elusive character of agrammatism. *Brain and Language, 50,* 369–384.

Kegl, J. (1995). Levels of representation and units of access relevant to agrammatism. *Brain and Language, 50,* 151–200.

Kegl, J., & Poizner, H. (1991). The interplay between linguistic and spatial processing in a right-lesioned signer. *Journal of Clinical and Experimental Neuropsychology, 13,* 38–39.

Kehayia, E. (1990). Morphological deficits in agrammatic aphasia: A comparative linguistic study. Doctoral dissertation. Montreal: McGill University.

Kehayia, E., & Jarema, G. (1994). Morphological priming (or prim#ing?) of inflected verb forms: A comparative study. *Journal of Neurolinguistics, 8,* 83–94.

Kehayia, E., Jarema, G., & Kadzielawa, D. (1990). A cross-linguistic study of morphological errors in aphasia: Evidence from English, Greek and Polish. In J.-L. Nespoulous & P. Villiard (Eds.), *Morphology phonology and aphasia* (pp. 140–155). New York: Springer Verlag.

Kemper, S., Anagnopoulos, C., Lyons, K., & Heberlein, W. (1994). Speech accommodations to dementia. *Journal of Gerontology: Psychological Sciences, 49,* 223–229.

Kempler, D., Curtiss, S., & Jackson, C. (1987). Syntactic preservation in Alzheimer's disease. *Journal of Speech and Hearing Research, 30,* 343–350.

Kempler, D., Van Lancker, D., & Read, S. (1988). Proverb and idiom comprehension in Alzheimer disease. *Alzheimer Disease and Associated Disorders, 2,* 38–49.

Kennedy, M., & Murdoch, B. E. (1993). Chronic aphasia subsequent to striato-capsular and thalamic lesions in the left hemisphere. *Brain and Language, 44,* 284–295.

Kennedy, M. R. T., Strand, E. A., Burton, W., & Peterson, C. (1994). Analysis of first-encounter conversations of right-hemisphere-damaged adults. In M. L. Lemme (Ed.), *Clinical aphasiology* (Vol. 22, pp. 67–80). Austin, TX: Pro-Ed.

Kent, R., & Rosenbek, J. (1982). Prosodic disturbance and neurologic lesion. *Brain and Language, 15,* 259–291.

Kertesz, A. (1982). *Western Aphasia Battery.* New York: Grune & Stratton, Inc.

Kertesz, A. (Ed.). (1983). *Localization in neuropsychology.* New York: Academic Press.

Kertesz, A. (1989). Anatomical and physiological correlations and neuroimaging techniques in language disorders. In A. Ardila & F. Ostrosky-Solis (Eds.), *Brain organization of language and cognitive processes* (pp. 37–59). New York: Plenum Press.

Kertesz, A. (1993). Clinical forms of aphasia. *Acta Neurochirurgica Supplementum, 56,* 52–58.

Kertesz, A. (1993). Neurobiological foundations of aphasia rehabilitation. In M. Paradis (Ed.), *Foundations of aphasia rehabilitation* (pp. 365–377). Oxford: Pergamon Press.

Kertesz, A. (1994a). Localization and function: Old issues revisited and new developments. In A. Kertesz (Ed.), *Localization and neuroimaging in neuropsychology* (pp. 1–33). San Diego: Academic Press.

Kertesz, A. (Ed.). (1994b). *Localization and neuroimaging in neuropsychology.* San Diego: Academic Press.

Kertesz, A., Appell, J., & Fisman, M. (1986). The dissolution of language in Alzheimer's disease. *Canadian Journal of Neurological Sciences, 13,* 415–418.

Kertesz, A., Caselli, R., Graff-Radford, N., & Miller, B. (1995). Primary progressive aphasia: Linguistic and biological aspects. *Brain and Language, 51,* 69–70.

Kertesz, A., & Clydesdale, S. (1994). Neuropsychological deficits in vascular dementia vs. Alzheimer's disease: Frontal lobe deficits prominent in vascular dementia. *Archives of Neurology, 51,* 1226–1231.

Kertesz, A., Harlock, W., & Coates, R. (1979). Computer tomographic localization, lesion size, and prognosis in aphasia and nonverbal impairment. *Brain and Language, 8,* 34–50.

Kertesz, A., Lau, W. K., & Polk, M. (1993). The structural determinants of recovery in Wernicke's aphasia. *Brain and Language, 44,* 153–164.

Kertesz, A., Lesk, D., & McCabe, P. (1977). Isotope localization of infarcts in aphasia. *Archives of Neurology, 34,* 590–601.

Kertesz, A., & McCabe, P. (1977). Recovery patterns and prognosis in aphasia. *Brain, 100,* 1–18.

Kertesz, A., Polk, M., Black, S. E., & Howell, J. (1992). Anatomical asymmetries and functional laterality. *Brain, 115,* 589–605.

Kertesz, A., Polk M., Carr, T. (1990). Cognition and white matter changes on magnet resonance imaging in dementia. *Archives of Neurology, 47,* 387–391.

Kettrick, K., & Hatfield, N. (1986). Bilingualism in a visuo-gestural mode. In J. Vaid (Ed.), *Language processing in bilinguals* (pp. 253–269). Hillsdale, NJ: Lawrence Erlbaum.

Kimberg, D. Y., & Farah, M. J. (1993). A unified account of cognitive impairment following frontal lobe damage: The role of working memory in complex, organised behaviour. *Journal of Experimental Psychology: General, 122,* 411–428.

Kimura, D. (1979). Neuromotor mechanisms in the evolution of human communication. In I. H. D. Steklis & M. J. Raleigh (Eds.), *Neurobiology of social communication in primates,* (pp. 197–219). New York: Academic Press.

Kimura, D. (1981). Neural mechanisms in manual signing. *Sign Language Studies, 33,* 291–312.

Kimura, D. (1993). *Neuromotor mechanisms in human communication.* Oxford: Oxford University Press.

Kimura, D., Barnett, J. M., & Burkhart, G. (1981). The psychological test pattern in progressive supranuclear palsy. *Neuropsychologia, 19,* 301–306.

Kimura, D., Battison, R., & Lubert, B. (1976). Impairment of non-linguistic hand movements in a deaf aphasic. *Brain and Language, 3,* 566–571.

King, J., & Kutas, M. (1995). Who did what and when? Using word- and clause-level ERPs to monitor working memory usage in reading. *Journal of Cognitive Neuroscience, 7,* 376–395.

King, K., Fraser, W. I., Thomas, P., & Kendell, R. E. (1990). Reexamination of the language of psychotic subjects. *British Journal of Psychiatry, 156,* 211–215.

Kinkel, W., & Jacobs, L. (1976). Computerized axial transverse tomography in cerebrovascular disease. *Neurology, 26,* 924–930.

Kinsbourne, M. (1970a). The cerebral basis of lateral asymmetries in attention. *Acta Psychologica, 33,* 193–201.

Kinsbourne, M. (1970b). A model for the mechanism of unilateral neglect of space. *Transaction of the American Neurological Association, 95,* 143–145.

Kinsbourne, M. (1971). The minor hemisphere as a source of aphasic speech. *Archives of Neurology, 25,* 303–306.

Kinsbourne, M. (1974). Mechanisms of hemispheric interaction in man. In M. Kinsbourne & W. L. Smith (Eds.), *Hemispheric disconnection and cerebral function* (pp. 260–285). Springfield, IL: C. C. Thomas.

Kinsbourne, M. (1982). Hemisphere specialization and the growth of human understanding. *American Psychologist, 37,* 411–420.

Kinsbourne, M. (1984). Why is neuropsychology progressing so slowly? *Contemporary Psychology, 29,* 793–794.

Kinsbourne, M., & Hiscock, M. (1983). Asymmetries of dual-task performance. In J. B. Hellige (Ed.), *Cerebral hemisphere asymmetry: Method, theory and application.* (pp. 255–334). New York: Praeger.

Kinsbourne, M., & Hiscock, M. (1983). The normal and deviant development of functional lateralization of the brain. In P. H. Mussen (Series Ed.) & M. M. Haith & J. J. Campos (Vol. Eds.), *Handbook of child psychology:* Vol. 2, *Infancy and developmental psychobiology* (4th ed., pp. 157–280). New York: John Wiley.

Kinsbourne, M., & Hiscock, M. (1987). Language lateralization and disordered language development. In S. Rosenberg (Ed.), *Advances in applied psycholinguistics.* Vol. 1, *Disorders of first-language development* (pp. 220–263). Cambridge, UK: Cambridge University Press.

Kintsch, W., & van Dijk, T. A. (1978). Toward a model of text comprehension and production. *Psychological Review, 85,* 363–394.

Kintsch, W. (1988). The role of knowledge in discourse comprehension: A construction–integration model. *Psychological Review, 95,* 163–182.

Kintsch, W. (1993). Information accretion and reduction in text processing: Inferences. *Discourse Processes, 16,* 193–202.

Kitterle, F. L., Christman, S., & Conesa, J. (1995). Spatial-frequency selectivity in hemispheric transfer. In F. L. Kitterle (Ed.), *Hemispheric communication: Mechanisms and models* (pp. 319–346). Hillsdale, NJ: Lawrence Erlbaum.

Klauer, K. J. (1987). *Kriteriumsorientierte Tests.* Göttingen: Hogrefe.

Klein, D., Behrmann, M., & Doctor, E. (1994). The evolution of deep dyslexia: Evidence for the spontaneous recovery of the semantic reading route. *Cognitive Neuropsychology, 11,* 579–611.

Klein, D., Zatorre, R. T., Milner, B., Meyer, E., & Evans, A. C. (1995). The neural substrates of bilingual language processing: Evidence from positron emission tomography. In M. Paradis (Ed.), *Aspects of bilingual aphasia* (pp. 23–36). Oxford: Pergamon Press.

Klein, S. K., Kurtzberg, D., Brattson, A., Kreuzer, J. A., Stappels, D. R., Dunn, M. A., Rapin, I., & Vaughan, H. G. (1995). Electrophysiologic manifestations of impaired temporal lobe auditory processing in verbal auditory agnosia. *Brain and Language, 51,* 383–405.

Klima, E., & Bellugi, U. (1979). *The signs of language.* Cambridge: Harvard University Press.

Klin, A. (1991). Young autistic children's listening preferences in regard to speech: A possible characterisation of the symptom of social withdrawal. *Journal of Autism and Developmental Disorders, 21,* 29–42.

Klin, A., Volkmar, F. R., Sparrow, S. S., Cicchetti, D. V., & Rourke, B. P. (1995). Validity and neuropsychological characterisation of Asperger syndrome: Convergence with nonverbal learning disabilities syndrome. *Journal of Child Psychology and Psychiatry, 36,* 1127–1140.

Klisz, D. (1978). Neuropsychological evaluation in older persons. In M. Storandt, I. C. Siegler, & M. F. Elias (Eds.), *The clinical psychology of aging* (pp. 71–95). New York: Plenum Press.

Kluender, R. , & Kutas, M. (1993a). Bridging the gap: Evidence from ERPs on the processing of unbounded dependencies. *Journal of Cognitive Neuroscience, 5,* 196–214.

Kluender, R., & Kutas, M. (1993b). Subjacency as a processing phenomenon. *Language and Cognitive Processes, 8,* 573–633.

Knight, R. T., Scabini, D., Woods, D. L. & Clayworth, C. C. (1989). Contributions of temporal-parietal junction to the human auditory P3. *Brain Research, 502,* 109–116.

Knopman, D. S., Christensen, K. J., Schut, L. J., Harbaugh, R. E., Reeder, T., Ngo, T., & Frey, W. (1989). The spectrum of imaging and neuropsychological findings in Pick's disease. *Neurology, 39,* 362–368.

Knopman, D. S., Rubens, A. B., Selnes, O. A., Klassen, A. D., & Meyer, M. W. (1984). Mechanisms of recovery from aphasia: Evidence from serial xenon 133 cerebral blood flow studies. *Annals of Neurology, 15,* 530–535.

Knopman, D. S., Selnes, O. A., Niccum, M., & Rubens, A. B. (1984). Recovery of naming in aphasia: Relationship to fluency, comprehension and CT findings. *Neurology, 34,* 1461–1471.

Knopman, D. S., Selnes, O. A., Niccum, N., Rubens, A. B., Yock, D., & Larson, D. (1983). A longitudinal study of speech fluency in aphasia: CT correlates of recovery and persistent nonfluency. *Neurology, 33,* 1170–1178.

Kohn, S. (1989). The nature of the phonemic string deficit in conduction aphasia. *Aphasiology, 3,* 209–239.

Kohn, S. (1993). Segmental disorders in aphasia. In G. Blanken, J. Dittman, H. Grimm, J. C. Marshall, & C.-W. Wallesch (Eds.), *Linguistic disorders and pathologies: An international handbook* (pp. 197–209). New York: Walter de Gruyter.

Kohonen, T. (1982). Self-organized formation of topologically correct feature maps. *Biological Cybernetics, 43,* 59–69.

Kolb, B. (1992). Mechanisms underlying recovery from cortical injury: Reflections on progress and directions for the future. In F. D. Rose & D. A. Johnson (Eds.), *Recovery from brain damage* (pp. 169–186). New York: Plenum Press.

Kolb, B. (1995). *Brian development, plasticity and behavior.* Hove, UK: Lawrence Erlbaum.

Kolb, B., & Taylor, L. (1981). Affective behavior in patients with localized cortical excisions: Role of lesion site and side. *Science, 214,* 89–91.

Kolb, B., & Whishaw, I. Q. (1983). Performance of schizophrenic patients on tests sensitive to left or right frontal, temporal or parietal function in neurological patients. *Journal of Mental and Nervous Disease, 171,* 435–443.

Kolk, H. H. J. (1987). A theory of grammatical impairment in aphasia. In G. Kempen (Ed.), *Natural language generation: New results in artificial intelligence, psychology, and linguistics* (pp. 377–391). Dordrecht, The Netherlands: Martinus Nijhoff.

Kolk, H. H. J. (1995). A time-based approach to agrammatic production. *Brain and Language, 50,* 282–303.

Kolk, H. H. J., & Friederici, A. D. (1985). Strategy and impairment in sentence understanding by Broca's and Wernicke's aphasics. *Cortex, 21,* 47–67.

Kolk, H. H. J., & Heeschen, C. (1990). Adaptation symptoms and impairment symptoms in Broca's aphasia. *Aphasiology, 4,* 221–231.

Kolk, H. H. J., & Heeschen, C. (1992). Agrammatism, paragrammatism and the management of language. *Language and Cognitive Processes, 7,* 89–129.

Kolk, H. H. J., & Heeschen, C. (1996). The malleability of agrammatic symptoms: A reply to Hesketh and Bishop. *Aphasiology, 10,* 81–96.

Kolk, H. H. J., & Hofstede, B. M. T. (1994). The choice for ellipsis: A case study of stylistic shifts in an agrammatic speaker. *Brain and Language, 47,* 507–509.

Kolk, H. H. J., & van Grunsven, M. J. F. (1985). Agrammatism as a variable phenomenon. *Cognitive Neuropsychology, 2,* 347–384.

Kolk, H. H. J., van Grunsven, M. J. F., & Keyser, A. (1985). On parallelism between production and comprehension in agrammatism. In M.-L. Kean (Ed.), *Agrammatism* (pp. 165–206). New York: Academic Press.

Kolk, H. H. J., & Weijts, M. (1996). Judgements of semantic anomaly in agrammatic patients: Argument movement, syntactic complexity and the use of heuristics. *Brain and Language, 54,* 86–135.

Koller, W. C., & Hubble, J. P. (1992). Classification of parkinsonism. In W. C. Koller (Ed.), *Handbook of Parkinsons's disease* (pp. 59–103). New York: Marcel Dekker.

Kontiola, P., Laaksonen, R., Sulkava, R., & Erkinjuntti, T. (1990). Pattern of language impairment is different in Alzheimer's disease and multi-infarct dementia. *Brain and Language, 38,* 364–383.

Kosa, U. (Ed.) (1994). Sprechende Computer in der pädagogischen Praxis. Weinheim: Deutscher Studien Verlag.

Koss, E., Edland, S., Fillenbaum, G., Mohs, R., Clark, C., Galasko, D., & Morris, J. C. (1996). Clinical and neuropsychological differences between patients with earlier and later onset of Alzheimer's disease: A CERAD analysis, Part 12. *Neurology, 46,* 136–141.

Kosslyn, S. M., & Intriligator, J. M. (1992). Is cognitive neuropsychology plausible? The perils of sitting on a one-legged stool. *Journal of Cognitive Neuroscience, 4,* 96–106.

Kosslyn, S. M., & Koenig, O. (1992). *Wet mind: The new cognitive neuroscience.* New York: Free Press.

Kosslyn, S. M., & Van Kleek, M. H. (1990). Broken brains and normal minds: Why Humpty Dumpty needs a skeleton. In E. L. Schwartz (Ed.), *Computational Neuroscience.* Cambridge, MA: MIT Press.

Kotten, A. (1989). Aphasia treatment: A multidimensional process. In E. Perecman (Ed.), *Integrating theory and practice in neuropsychology* (pp. 293–315). Hillsdale, NJ: Lawrence Erlbaum.

Kounios, J., & Holcomb, P. (1994). Concreteness effects in semantic processing: ERP evidence supporting dual-coding theory. *Journal of Experimental Psychology: Learning, Memory, and Cognition, 20,* 804–823.

Kraepelin, E. (1971/1919). *Dementia praecox and paraphrenia.* Facsimile 1919 edition. R. E. Huntington, NY: Krieger Publishing.

Kraus, N., McGee, T., Micco, A., Sharma, A., Carrell, T., & Nicol, T. (1993). Mismatch negativity in school-age children to speech stimuli that are just perceptibly different. *Electroencephalography and Clinical Neurophysiology: Evoked Potentials, 88,* 123–130.

Kreiman, J. (1997). Listening to voices: Theory and practice in voice perception. In K. Johnson & J. W. Mullenis (Eds.), *Talker variability in speech processing* (pp. 85–108). New York: Academic Press.

Kremin, H. (1990). Naming and its disorders. In F. Boller & J. Grafman (Eds.), *Handbook of neuropsychology:* (Vol. 1, pp. 307–328). Amsterdam: Elsevier Science Publishers.

Kulynych, J. J., Vladar, K., Jones, D. W., & Weinberger, D. R. (1994). Gender differences in the normal lateralization of the supratemporal cortex: MRI surface-rendering morphometry of Heschl's gyrus and the planum temporale. *Cerebral Cortex, 4,* 107–118.

Kumin, L., Councill, C., & Goodman, M. (1994). A longitudinal study of the emergence of phonemes in children with Down syndrome. *Journal of Communication Disorders, 27,* 293–303.

Kurowski, K., Blumstein, S., & Alexander, M. (1996). The foreign accent syndrome: A reconsideration. *Brain and Language, 54,* 1–25.

Kurthen, M. (1992). Der intrakarotidale Armobarbitaltest: Indikation, Durchführung, Ergebnisse. *Nervenarzt, 63,* 713–724.

Kurthen, M., Biersack, H. J., Limke, D. B., Brassel, F., Wappenschmidt, J., Durwen, H. F., Reuter, B. M., Solymosi, L., & Reichmann, K. (1988). Präoperative Diagnostik: Wada-Test mit SPECT-Kontrolle zur Hirnfunktionslokalisation bei therapieresistenter Epilepsie. *Neurochirurgia, 31,* 96–98.

Kurthen, M., Helmstaedter, C., Linke, D. B., Hufnagel, A., Elger, C. E., & Schramm, J. (1994). Quantitative and qualitative evaluation of patterns of cerebral language dominance. *Brain and Language, 46,* 536–564.

Kurthen, M., Solymosi, L., & Linke, D. (1993). Der intrakarotidale amobarbital Test: Neuroradiologische und neuropsychologische Aspekte. *Radiologe, 33,* 204–212.

Kussmaul, A. (1877). *Die Störungen der Sprache.* Leipzig: Vogel. (cited by Wallesch & Papagno, 1988).

Kutas, M. (1993). In the company of other words: Electrophysiological evidence for single-word and sentence context effects. *Language and Cognitive Processes, 8,* 533–572.

Kutas, M., & Hillyard, S. A. (1980a). Event-related brain potentials to semantically inappropriate and surprisingly large words. *Biological Psychology, 11,* 99–116.

Kutas, M., & Hillyard, S. A. (1980b). Reading senseless sentences: Brain potentials reflect semantic incongruity. *Science, 207,* 203–205.

Kutas, M., & Hillyard, S. A. (1983). Event-related brain potentials to grammatical errors and semantic anomalies. *Memory and Cognition, 11,* 539–550.

Kutas, M., & Hillyard, S. A. (1984). Brain potentials during reading reflect word expectancy and semantic association. *Nature, 307,* 161–163.

Kutas, M., & Hillyard, S. A. (1989). An electrophysiological probe of incidental semantic association. *Journal of Cognitive Neuroscience, 1,* 38–49.

Kutas, M., & Van Petten, C. K. (1988). Event-related potential studies of language. In P. Ackles, J. R. Jennings, & M. G. H. Coles (Eds.), *Advances in psychophysiology* (pp. 139–187). Greenwich, CT: JAI Press.

Kutas, M., & Van Petten, C. K. (1994). Psycholinguistics electrified: Event-related potential investigations. In M. A. Gernsbacher (Ed.), *Handbook of psycholinguistics* (pp. 83–143). San Diego: Academic Press.

Kwapil, T. R., Hegley, D. C., Chapman, L. J., & Chapman, J. P. (1990). Facilitation of word recognition by semantic priming in schizophrenia. *Journal of Abnormal Psychology, 99,* 215–221.

Kwong See, S. T., & Ryan, E. B. (1995). Cognitive mediation of adult age differences in language performance. *Psychology and Aging, 10,* 458–468.

LaBerge, D., & Buchsbaum, M. S. (1990). Positron emission tomographic measurements of pulvinar activity during an attention task. *Journal of Neuroscience, 10,* 613–619.

LaBerge, D., Brown, V., Carter, M., Bash, D., & Hartley, A. (1991). Reducing the effects of adjacent distractors by narrowing attention. *Journal of Experimental Psychology: Human Perception and Performance, 17,* 65–76.

Lackner, J., & Shattuck-Hufnagel, S. (1982). Alterations in speech shadowing ability after cerebral injury in man. *Neuropsychologia, 20,* 709–714

Lacroix, J., Joanette, Y., & Bois, M. (1994). Un nouveau regard sur la notion de validité écologique: Apport du cadre conceptuel de la CIDIH. *Revue de neuropsychologie, 4,* 115–141.

Ladavas, E., Paladini, R., & Cubelli, R. (1993). Implicit associative priming in a patient with left visual neglect. *Neuropsychologia, 31,* 1307–1320.

Laffal, J. (1965). *Pathological and normal language.* New York: Atherton.

Lagrèze, H. L., Hartmann, A., Anzinger, G., Schaub, A., & Deister, A. (1993). Functional cortical interaction patterns in visual perception and visuospatial problem solving. *Journal of the Neurological Sciences, 114,* 25–35.

Laine, M., Niemi, J., Koivuselkä-Sallinen, P., & Hyönä, J. (1995). Morphological processing of polymorphemic nouns in a highly inflected language. *Cognitive Neuropsychology, 12.* 457–502.

Laine, M., Niemi, J., & Marttila, R. (1990). Changing error patterns during reading recovery: A case study. *Journal of Neurolinguistics, 5,* 75–81.

Laland, S., Braun, C. M., Charlebois, N., & Whitaker, H. A. (1992). Effects of right and left hemisphere cerebrovascular lesions on discrimination of prosodic and semantic aspects of affect in sentences. *Brain and Language, 42,* 165–186.

Lallemand, C.-F. (1820, 1823). Recherches anatomo-pathologiques sur l'encéphale et ses dépendances (Vols. 1–2). Paris: Baudouin fils and Bechet jeune.

Lambert, A. J. (1991). Interhemispheric interaction in the split brain. *Neuropsychologia, 29,* 941–948.

Lambert, A. J. (1993). Attentional interaction in the split brain: Evidence from negative priming. *Neuropsychologia, 31,* 313–324.

Lambrecht, K. J., & Marshall, R. C. (1983). Comprehension in severe aphasia. *Clinical Aphasiology, 13,* 186–192.

Landau, B., & Jackendoff, R. (1993). "What" and "where" in spatial language and spatial cognition. *Behavioral and Brain Sciences, 16,* 217–265.

Landau, W. M., & Kleffner, F. R. (1957). Syndrome of acquired aphasia with convulsive disorder in children. *Neurology, 7,* 523–530.

Landis, T., Regard, M., & Serrat, A. (1980). Iconic reading in a case of alexia without agraphia caused by a brain tumor: A tachistoscopic study. *Brain and Language, 11,* 45–53.

Landre, N. A., Taylor, M. A., & Kearns, K. P. (1992). Language functioning in schizophrenic and aphasic patients. *Neuropsychiatry, Neuropsychology and Behavioral Neurology, 5,* 7–14.

Lane, H. (1984). *When the mind hears: A history of the deaf.* Cambridge: Harvard University Press.

Langley, J., & Zaidel, E. (1990). The role of the right hemisphere in deep dyslexia and in the generation of semantic errors. *Brain and Language, 39,* 613.

Lanoe, Y., Fabry, B., Lanoe, A., Pedetti, L., Fahed, M., & Benoit, T. (1992). Aphasie croisée chez un adulte: Représentation du langage dans les deux hémisphères. *Revue de Neuropsychologie, 2,* 373–392.

Lapointe, S. (1985). A theory of verb form use in the speech of agrammatic aphasics. *Brain and Language, 24,* 100–155.

LaRue, A. (1992). *Aging and neuropsychological assessment.* New York: Plenum Press.

Lassen, N. A., & Ingvar, D. H. (1961). The blood flow of the cerebral cortex determined by radioactive Krypton-85. *Experientia, 17,* 42–43.

Lassonde, M., Bryden, M. P., & Demers, P. (1990). The corpus callosum and cerebral speech lateralization. *Brain and Language, 38,* 195–206.

Laubstein, A. S. (1993). Inconsistency and ambiguity in Lichtheim's model. *Brain and Language, 45,* 588–603.

Lawlor, B. A. , Ryan, T. M., Schmeidler, J., Mohs, R. C., & Davis, K. L. (1994). Clinical symptoms associated with age at onset in Alzheimer's disease. *American Journal of Psychiatry, 151,* 1646–1649.

Lecours, A. R. (1989). Literacy and acquired aphasia. In A. M. Galaburda (Ed.), *From reading to neurons* (pp. 27–39). Cambridge: MIT Press.

Lecours, A. R. (Ed.). (1996). *Écriture: Histoire, théorie et maladies*. Isbergues: Ortho.

Lecours, A. R., & Joanette, Y. (1980). Linguistic and other psychological aspects of paroxysmal aphasia. *Brain and Language, 10* 1–23.

Lecours, A. R., Lhermitte, F., & Bryans, B. (1983). *Aphasiology*. London: Bailliere Tindall.

LeDoux, J. (1995). Emotion: Clues from the brain. *Annual Reviews of Psychology, 46,* 209–235.

Lee, G. P., Loring, D. W., Smith, J. R., & Flanigin, H. F. (1995). Intraoperative hippocampal cooling and Wada memory testing in the evaluation of amnesia risk following anterior temporal lobectomy. *Archives of Neurology, 52,* 857–861.

Lee, H., Nakada, T., Deal, J. L., Lynn, S., & Kwee, I. L. (1984). Transfer of language dominance. *Annals of Neurology, 15,* 304–307.

Lee, L. (1974). *Developmental Sentence Analysis*. Evanston, IL: Northwestern University Press.

Leech, G. N. (1983). *Principles of pragmatics*. London: Longman.

Lehmkuhle, S. (1993). Neurological basis of visual processes in reading. In D. M. Willows, R. S. Kruk, & E. Corcos (Eds.), *Visual processes in reading and reading disabilities* (pp. 77–94). Hillsdale, NJ: Lawrence Erlbaum.

Leifer, J., & Lewis, M. (1984). Acquisition of conversational response skills by young Down syndrome and nonretarded children. *American Journal of Mental Deficiency, 88,* 610–618.

Leiner, H. C., Leiner, A. L., & Dow, R. S. (1993). Cognitive and language functions of the human cerebellum. *Trends in Neuroscience, 16,* 444–447.

Leischner, A. (1943). Die "Aphasie" der Taubstummen. *Archiv für Psychiatrie und Nervenheilkunde, 115,* 469–548.

LeMay, M. (1984). Radiological, developmental, and fossil asymmetries. In N. Geschwind & A. M. Galaburda (Eds.), *Cerebral dominance: The biological foundations* (pp. 26–42). Cambridge: Harvard University Press.

Le Moal, M., & Simon, H. (1991). Mesocorticolimbic dopaminergic network: Functional and regulatory roles. *Physiological Reviews, 71,* 155–234.

Lenneberg, E. H. (1967). *Biological foundations of language*. New York: John Wiley.

Lerman, P., Lerman-Sagie, T., & Kivity S. (1991). Effect of early corticosteroid therapy for Landau-Kleffner syndrome. *Developmental Medicine and Child Neurology, 33,* 257–260.

Lesser, R. (1978). *Linguistic investigations of aphasia*. London: Arnold.

Lesser, R., & Milroy, L. (1993). *Linguistics and aphasia: Psycholinguistic and pragmatic aspects of intervention*. New York: Longman Publishing.

Lesser, R. P., Dinner, D. S., Lüders, H., & Morris, H. H. (1986). Memory for objects presented soon after intracarotid amobarbital sodium injections in patients with medically intractable complex partial seizures. *Neurology, 36,* 895–899.

Lesser, R. P., Lueders, H., Klem, G., Dinner, D. S., Morris, H. H., Hahn, J., & Wyllie, E. (1987). Extraoperative cortical functional localization in patients with epilepsy. *Journal of Clincial Neurophysiology, 4,* 27–53.

Levelt, W. (1989). *Speaking: From intention to articulation*. Cambridge: MIT Press.

Levelt, W. J. M. (1992). Accessing words in speech production. Stages, processes and representations. *Cognition, 42,* 1–22.

Levin, B. E., & Katzen, H. L. (1995). Early cognitive changes and nondementing behavioral abnormalities in Parkinson's disease. *Advances in Neurology, 65,* 85–95.

Levin, B. E., Llabre, M. M., & Weiner, W. J. (1989). Cognitive impairment associated with early Parkinson's disease. *Neurology, 39,* 557–561.

Levin, B. E., & Tomer, R. (1992). A prospective study of language abilities in Parkinson's disease. *Journal of Clinical and Experimental Neuropsychology, 14,* 34.

Levin, H. S., Cantrell, D. T. C., Soukup, V., Crow, W., & Bartha, M. C. (1994). Preliminary results of an incremental intracarotid amobarbital procedure: Evaluation of language and memory without sedation. *Journal of Epilepsy, 7,* 11–17.

Levin, H. S., Grossman, R. G., & Kelly, P. J. (1976). Aphasic disorders in patients with closed head injury. *Journal of Neurology, Neurosurgery and Psychiatry, 39,* 1062–1070.

Levin, H. S., Grossman, R. G., Rose, S. E., & Teasedale, G. (1979). Long term neuropsychological outcome of closed head injury. *Journal of Neurosurgery, 50,* 412–422.

Levin, H. S., Grossman, R. G., Sarwar, M.,& Meyers, C.A. (1981). Linguistic recovery after closed head injury. *Brain and Language, 12,* 360–374.

Levine, D., Fisher, M., & Caveness, V. (1974). Porencephaly with microgyria: A pathological study. *Acta Neuropsychologia, 12,* 505–512.

Levine, D. N., & Sweet, E. (1982). The neurological basis of Broca's aphasia and its implications for the cerebral control of speech. In M. Arbib, D. Caplan, & J. C. Marshall (Eds.), *Neural models of language processes* (pp. 299–325). San Diego: Academic Press.

Levinson, S. (1983). *Pragmatics.* Cambridge, UK: Cambridge University Press.

Levitt, P., Rakič, P., & Goldman-Rakič, P. (1984). Region-specific distribution of catecholamine afferents in primate cerebral cortex: A fluorescence histochemical analysis. *Journal of Comparative Neurology, 227,* 23–36.

Levy, J., & Trevarthen, C. (1977). Perceptual, semantic and phonetic aspects of elementary language processes in split-brain patients. Brain, 100, 105–118.

Lewandowski, L., Costenbader, V., & Richman, R. (1984). Neuropsychological aspects of Turner syndrome. *International Journal of Clinical Neuropsychology, 7,* 144–147.

Lewine, J. D. (1990). Neuromagnetic techniques for the noninvasive analysis of brain function. In S. E. Freeman, E. Fukushima, & E. R. Greene (Eds.), *Noninvasive techniques in Biology and Medicine* (pp. 33–74). San Francisco: San Francisco Press.

Lezak, M. D. (1995). *Neuropsychological assessment* (3rd ed.) New York: Oxford University Press.

Libben, G. (1990). Morphological representations and morphological deficits in aphasia. In J.-L. Nespoulous & P. Villiard (Eds.), *Morphology, phonology, and aphasia* (pp. 20–31). New York: Springer Verlag.

Libben, G. (1994). The role of hierarchical morphological structure: A case study. *Journal of Neurolinguistics, 8,* 49–55.

Liberman, A. M., Cooper, F. S., Shankweiler, D., & Studdert-Kennedy, M. (1967). Perception of the speech code. *Psychological Review, 74,* 431–461.

Lichtheim, L. (1885). On aphasia. *Brain, 7,* 433–484.

Liddell, S. K. (1990). Four functions of a locus: Reexamining the structure of space in ASL. In C. Lucas (Ed.), *Sign language research: Theoretical issues* (pp. 176–198). Washington, DC: Gallaudet University Press.

Liddell, S. K. (1994). Real, surrogate and token space: Grammatical consequences in ASL. In K. Emmorey & J. Reilly (Eds.), *Language, gesture and space* (pp. 19–41). Hillsdale, NJ: Lawrence Erlbaum.

Lieber, R. (1980). On the organization of the lexicon. Doctoral dissertation. Cambridge: Massachusetts Institute of Technology.

Lieberman, P. (1984). *The biology and evolution of language.* Cambridge: Harvard University Press.

Lieberman, P., Kako, E., Friedman, J., Tajchman, G., Feldman, L. S., & Jiminez, E. B. (1992). Speech production, syntax comprehension and cognitive deficits in Parkinson's disease. *Brain and Language, 43,* 169–189.

Liederman, J. (1983). Mechanisms underlying instability in the development of hand preference. In G. Young, S. Segalowitz, C. Corter, & S. Trehub (Eds.), *Manual specialization and the developing brain* (pp. 71–92). New York: Academic Press.

Liederman, J. (1987). Neonates show an asymmetric degree of head rotation but lack an asymmetric tonic neck reflex asymmetry: Neuropsychological implications. *Developmental Neuropsychology, 3,* 101–112.

Liederman, J. (1988). Misconceptions and new conceptions about early brain damage, functional asymmetry, and behavioral outcome. In D. L. Molfese, & S. J. Segalowitz (Eds.), *Brain lateralization in children: Developmental implications* (pp. 375–400). New York: Guilford Press.

Liederman, J. (1995). A reinterpretation of the split-brain syndrome: Implications for the function of corticocortical fibers. In R. J. Davidson & K. Hugdahl (Eds.), *Brain asymmetry* (pp. 451–490). Cambridge: MIT Press.

Liederman, J., & Kinsbourne, M. (1980). Rightward bias in neonates depends upon parental right handedness. *Neuropsychologia, 18,* 579–584.

Light, L. L., Singh, A., & Capps, J. L. (1986). Dissociation of memory and awareness in young and older adults. *Journal of Clinical and Experimental Neuropsychology, 8,* 594–610.

Liles, B. Z., Coelho, C. A., Duffy, R. J., & Zalagens, M. R. (1989). Effects of elicitation procedures on the narratives of normal and closed head injured adults. *Journal of Speech and Hearing Disorders, 54,* 356–366.

Lillo-Martin, D. C. (1991). *Universal grammar and American Sign Language: Setting the null argument parameters.* Dordrecht, The Netherlands: Kluwer.

Lillo-Martin, D. C., & Klima, E. S. (1990). Pointing out differences: ASL pronouns in syntactic theory. In S. Fischer, D. Siple, P. DeCaro, & J. James (Eds.), *Theoretical issues in sign language research, 1: Linguistics* (pp. 191–210). Chicago: University of Chicago Press.

Lilly, R., Cummings, J. L., Benson, D. F., & Frankl, M. (1983). The human Klüver-Bucy syndrome. *Neurology, 33,* 1141–1145.

Linebarger, M. C. (1989). Neuropsychological evidence for linguistic modularity. In G. M. Carlson & M. K. Tanenhaus (Eds.), *Linguistic structure in language processing* (pp. 129–189). Dordrecht, The Netherlands: Kluwer Academic Publishers.

Linebarger, M. C. (1990). Neuropsychology of sentence parsing. In A. Caramazza (Ed.), *Cognitive neuropsychology and neurolinguistics* (pp. 55–122). Hillsdale, NJ: Lawrence Erlbaum.

Linebarger, M. C. (1995). Agrammatism as evidence about grammar. *Brain and Language, 50,* 52–91.

Linebarger, M. C., Schwartz, M., & Saffran, E. (1983). Sensitivity to grammatical structure in so-called agrammatic aphasics. *Cognition, 13,* 361–392.

Lingraphica: The language prothesis for aphasia [Portable computer device]. (1992). Mountain View, CA: Tolfa Corporation.

Liotti, M., & Tucker, D. M. (1994). Emotion in asymmetric corticolimbic networks. In R. J. Davidson & K. Hugdahl (Eds.), *Human brain laterality* (pp. 389–424). New York: Oxford University Press.

Livingston, K. E., & Escobar, A. (1973). Tentative limbic system models for certain patterns of psychiatric disorders. In L. V. Laitinen & K. E. Livingston (Eds.), *Surgical approaches in psychiatry* (pp. 245–252). Baltimore: University Park Press.

Livingstone, M. S., Rosen, G. D., Drislane, F., & Galaburda, A. M. (1991). Physiological and anatomical evidence for a magnocellular defect in developmental dyslexia. *Proceedings of the National Academy of Sciences, USA, 88,* 7943–7947.

Loeb, C., & Gandolfo, C. (1988). Intellectual impairment and cerebral lesions in multiple cerebral infarcts: A clinical computed tomography study. *Stroke, 19,* 560–565.

Logie, R. H. (1993). Working memory in everyday cognition. In G. M. Davies & R. H. Logie (Eds.), *Memory in everyday life* (pp. 173–218). Amsterdam: Elsevier.

Logie, R. H. (1995). *Visuo-spatial working memory.* Hillsdale, NJ: Lawrence Erlbaum.

Lojek-Osiejuk, E. (1996). Knowledge of scripts reflected in discourse of aphasics and right-brain-damaged patients. *Brain and Language, 53,* 58–80.

Lord, F. M., & Novick, M. R. (1968). *Statistical theories of mental test scores.* Reading, MA: Addison-Wesley.

Loring, D. W., Lee, G. P., & Meador, K. J. (1994). Intracarotid amobarbital (Wada) assessment. In A. R. Wyler & B. P. Hermann (Eds.), *The surgical management of epilepsy* (pp. 97–110). Boston: Butterworth-Heinemann.

Loring, D. W., Meador, K. J., & Lee, G. P. (1992a). Amobarbital dose effects on Wada memory testing. *Journal of Epilepsy, 5,* 171–174.

Loring, D. W., Meador, K. J., & Lee, G. P. (1992b). Criteria and validity issues in Wada assessment. In T. L. Bennett (Ed.), *The neuropsychology of epilepsy* (pp. 233–245). New York: Plenum Press.

Loring, D. W., Meador, K. J., Lee, G. P. & King, D. W. (1992). *Amobarbital effect and lateralized brain function: The Wada test.* New York: Springer Verlag.

Loring, D. W., Meador, K. J., Lee, G. P., Flanigin, H. F., King, D. W., & Smith, J. R. (1990). Crossed aphasia in a patient with complex partial seizures: Evidence from intracarotid amobarbital testing, functional cortical mapping, and neuropsychological assessment. *Journal of Clinical and Experimental Neuropsychology, 12,* 340–354.

Loring, D. W., Meador, K. J., Lee, G. P., King, D. W., Gallagher, B. B., Murro, A. M., & Smith, J. R. (1994). Stimulus timing effects on Wada memory testing. *Archives of Neurology, 51,* 806–810.

Loring, D. W., Meador, K. J., Lee, G. P., Murro, A. M., Smith, J. R., Flanigin, H. F., Gallagher, B. B., & King, D. W. (1990). Cerebral lateralization: Evidence from intracarotid amobarbital testing. *Neuropsychologia, 28,* 831–838.

Lovegrove, W., Martin, F., & Slaghuis, W. (1986). A theoretical and experimental case for a visual deficit in specific reading disability. *Cognitive Neuropsychology, 3,* 225–267.

Lovegrove, W. J., & Williams, M. C. (1993). Visual temporal processing deficits in specific reading disability. In D. M. Willows, R. S. Kruk, & E. Corcos (Eds.), *Visual processes in reading and reading disabilities* (pp. 311–329). Hillsdale, NJ: Lawrence Erlbaum.

Loverso, F. L. (1987). Unfounded expectations: Computers in rehabilitation. *Aphasiology, 1,* 157–160.

Loverso, F. L., Prescott, T. E., & Selinger, M. (1992). Microcomputer treatment applications in aphasiology. *Aphasiology, 6*(2), 155–163.

Loverso, F. L., Prescott, T. E., Selinger, M., Wheeler, K. M., & Smith, R. D. (1985). The application of microcomputers for the treatment of aphasic adults. In R. H. Brookshire (Ed.), *Clinical aphasiology: 1985 conference proceedings* (pp. 189–195). Minneapolis: BRK Publishers.

Lovett, M. W., Dennis, M., & Newman, J. E. (1986). Making reference: The cohesive use of pronouns in the narrative discourse of hemidecorticate adolescents. *Brain and Language, 29,* 224–251.

Low, A. A. (1931). A case of agrammatism in the English language. *Archives of Neurology and Psychiatry, 25,* 556–597.

Löwenstein, D. A., Arguelles, T., Arguelles, S., Linn-Fuentes, P. (1994). Potential cultural bias in the neuropsychological assessment of the older adult. *Journal of Clinical and Experimental Neuropsychology, 16,* 623–629.

Löwenstein, D. A., & Rupert, M. P. (1992). The NINCDS-ADRDA neuropsychological criteria for the assessment of dementia: Limitations of current diagnostic guidelines. *Behavior, Health and Aging, 2,* 113–121.

Lucchelli, F., & De Renzi, E. (1992). Proper name anomia. *Cortex, 28,* 221–230.

Luck, S. J., Hillyard, S. A., Mangun, G. R., & Gazzaniga, M. S. (1994). Independent attentional scanning in the separated hemispheres of split-brain patients. *Journal of Cognitive Neuroscience, 6,* 84–91.

Ludlow, C. (1973). The recovery of syntax in aphasia: An analysis of syntactic structures used in connected speech during the initial recovery period. Dissertation, New York University.

Ludlow, C. L., Rosenberg, J., Fair, C., Buck, D., Schesselman, S., & Salazar, A. (1986). Brain lesions associated with non-fluent aphasia fifteen years following penetrating head injury. *Brain, 109,* 55–80.

Lueders, H., Lesser, R. P., Dinner, D. S., Morris, H. H., Hahn, J. Friedman, L., Skipper, G., Wyllie, E., & Friedman, D. (1987). Chronic intracranial recording and stimulation with subdural electrodes. In J. Engel Jr. (Ed.), *Surgical treatment of the epilepsies* New York: Raven Press. (pp. 297–321).

Lueders, H., Lesser, R. P., Hahn, J., Dinner, D. S., Morris, H. H., Wyllie, E., & Godoy, J. (1991). Basal temporal language area. *Brain, 114,* 743–754.

Luh, K. E., & Levy, J. (1995) Interhemispheric cooperation: Left is left and right is right, but sometimes the twain shall meet. *Journal of Experimental Psychology: Human Perception and Performance, 21,* 1243–1258.

Lukatela, G., Gligorijevic, B., Kostic, A., & Turvey, M. T. (1980). Representation of inflected nouns in the internal lexicon. *Memory and Cognition, 8,* 415–423.

Lukatela, K., Crain, S., & Shankweiler, D. (1988). Sensitivity to inflectional morphology in agrammatism: Investigation of a highly inflected language. *Brain and Language, 33,* 1–15.

Lund & Manchester [research groups]. (1994). Clinical and neuropathological criteria for frontotemporal dementia. *Journal of Neurology, Neurosurgery, and Psychiatry, 57,* 416–418.

Lundberg, I. (1994). Reading difficulties can be predicted and prevented: A Scandinavian perspective on phonological awareness and reading. In C. Hulme & M. Snowling (Eds.), *Reading development and dyslexia* (pp. 180–199). London: Whurr Publishers.

Lundberg, I., Frost, J., & Petersen, O.-P. (1988). Effects of an extensive programme for stimulating phonological awareness in pre-school children. *Reading Research Quarterly, 33,* 263–284.

Luria, A. R. (1963). *Restoration of function after brain injury.* London: Pergamon Press.

Luria, A. R. (1966). *Higher cortical functions in man.* New York: Basic Books.

Luria, A. R. (1973a). *The working brain.* Harmondsworth, England: Penguin Books.

Luria, A. R. (1973b). *The working brain: An introduction to neuropsychology* (B. Haigh, Trans.). New York: Basic Books.

Luria, A. R. (1976). *Basic problems in neurolinguistics.* The Hague: Mouton.

Luria, A. R., Naydin, V. L., Tsvetkova, L. S., & Vinarskaya, E. N. (1969). Restoration of higher cortical function following local brain damage. In P. J. Vinken & G. W. Bruyn (Eds.), *Handbook of clinical neurology* (Vol. 3; pp. 368–433). Amsterdam: North-Holland Publishing Company.

Luzzatti, C., & De Bleser, R. (1996). Morphological processing in Italian agrammatic speakers: Eight experiments in lexical morphology. *Brain and Language, 51,* 26–74.

Luzzatti, C., & Whitaker, H. A. (1996a). Johannes Schenck and Johannes Jakob Wepfer: Clinical and anatomical observations in the prehistory of neurolinguistics and neuropsychology. *Journal of Neurolinguistics, 9,* 157–164.

Luzzatti, C., & Whitaker, H. A. (1996b). Jean Baptiste Bouillaud, François Lallemand and the role of the frontal lobe: Location and mislocation of language in the early 19th century. *Brain and Cognition, 32,* 103–104.

Lynch, G., Granger, R., Ambros-Ingerson, J., Davis, C. M., Schehr, R., & Kessler, M. (1997). Evidence that a positive modulator of AMPA-type glutamate receptor improves delayed recall in aged humans. *Experimental Neurology, 145(1),* 89–92.

Macaruso, P., McCloskey, M., & Aliminosa, D. (1993). The functional architecture of the cognitive numerical-processing system: Evidence from a patient with multiple impairments. *Cognitive Neuropsychology, 10,* 341–376.

MacKain, K., Studdert-Kennedy, M., Spieker, S., & Stern, D. (1983). Infant intermodal speech perception is a left hemisphere function. *Science, 214,* 1347–1349.

MacLean (1987). Triune brain. In G. Adelman (Ed.), *Encyclopedia of neuroscience* (pp. 1235–1237). Boston: Birkhäuser.

MacWhinney, B. (1991a). *The CHILDES database.* Dublin, OH: Discovery Systems.

MacWhinney, B. (1991b). *The CHILDES project: Tools for analyzing talk.* Hillsdale, NJ: Lawrence Erlbaum.

MacWhinney, B. (1993). *The CHILDES database* (2nd ed.). Dublin, OH: Discovery Systems.

MacWhinney, B. (1994a). *The CHILDES database* (3rd ed.). Dublin, OH: Discovery Systems.

MacWhinney, B. (1994b). *The CHILDES Project: Tools for analyzing talk* (2nd ed.). Hillsdale, NJ: Lawrence Erlbaum.

MacWhinney, B. (1994c). New horizons for CHILDES research. In J. Sokolov & C. Snow (Eds.), *Handbook for research in language development using CHILDES* (pp. 408–452). Hillsdale, NJ: Lawrence Erlbaum.

MacWhinney, B. (1995). Computational tools for analyzing language. In P. Fletcher & B. MacWhinney (Eds.), *Handbook of child language research* (pp. 152–175). London: Basil Blackwell.

MacWhinney, B. (in press). The CHILDES system. In W. Ritchie & T. Bhatia (Eds.), *Handbook of language acquisition.* New York: Academic Press.

MacWhinney, B., & Bates, E. (1978). Sentential devices for conveying givenness and newness: A cross-cultural developmental study. *Journal of Verbal Learning and Verbal Behavior, 17,* 539–558.

MacWhinney, B., & Bates, E. (Eds.). (1989). *The crosslinguistic study of sentence processing.* New York: Cambridge University Press.

MacWhinney, B., & Osman-Sagi, J. (1991). Inflectional marking in Hungarian aphasics. *Brain and Language, 41,* 165–183.

MacWhinney, B., & Snow, C. (1985). The Child Language Data Exchange System. *Journal of Child Language, 12,* 271–295.

MacWhinney, B., & Snow, C. (1990). The Child Language Data Exchange System: An update. *Journal of Child Language, 17,* 457–472.

Magnúsdóttir, S., & Thráinsson, H. (1990). Agrammatism in Icelandic: Two case studies. In L. Menn & L. K. Obler (Eds.), *Agrammatic aphasia: A cross-language narrative sourcebook* (Vol. 1, pp. 443–478). Amsterdam: John Benjamins.

Magnusson, M. (1996). *Telematic pilot study: Study of expectations using videotelephony.* Poster session presented at the 7th International Aphasia Rehabilitation Conference, Cambridge, MA (18 August 1996).

Maher, B. (1972). The language of schizophrenia: A review and interpretation. *British Journal of Psychiatry, 120,* 3–17.

Maher, E. R., Smith, E. M., & Lees, A. J. (1985). Cognitive deficits in the Steele–Richardson–Olsewski syndrome. *Journal of Neurology, Neurosurgery, and Psychiatry, 48,* 1234–1239.

Mahoney, G., Glover, A., & Finger, I. (1981). Relationship between language and sensorimotor development of Down's syndrome and nonretarded children. *American Journal of Mental Deficiency, 86,* 21–27.

Mahoney, G., & Snow, K. (1983). The relationship of sensorimotor functioning to children's response to early language training. *Mental Retardation, 21,* 248–254.

Mahurin, R. K., Feher, E. P., Nance, M. L., Levy, J. K., & Pirozzolo, F. J. (1992). Cognition in Parkinson's disease and related disorders. In R. Parks, R. Zec, & R. Wilson (Eds.), *Neuropsychology of Alzheimer's disease and other dementias* (pp. 308–349), New York: Oxford University Press.

Malapani, C., Pillon, B., Dubois, B., & Agid, Y. (1994). Impaired simultaneous cognitive task performance in Parkinson's disease: A dopamine-related dysfunction. *Neurology, 144,* 319–326.

Mandell, A. M., Alexander, M. P., & Carpenter, S. (1989). Creutzfeld-Jakob disease presenting as an isolated aphasia. *Neurology, 39,* 55–58.

Mandler, J. M., & Johnson, N. S. (1977). Remembrance of things parsed: Story structure and recall. *Cognitive Psychology, 9,* 111–151.

Manschreck, T. C., Maher, B., Celada, M. T., Schneyer, M., & Fernandez, R. (1991). Object chaining and thought disorder in schizophrenic speech. *Psychological Medicine, 21,* 443–446.

Mansfield, P. (1977). Multi-planar image formation using spinechoes. *Journal of Solid State Physics, 10,* L55–58.

Mantovani, J. F., & Landau, W. M. (1980). Acquired aphasia with convulsive disorder: Course and prognosis. *Neurology, 30,* 524–529.

Mapou, R. L., & Spector, J. (Eds.). (1995). *Clinical neuropsychological assessment: A cognitive approach.* New York: Plenum Press.

Maquet, P., Hirsch, E., Dive, D., Salmon, E., Marescaux, C., & Franck, G. (1990). Cerebral glucose utilization during sleep in Landau-Kleffner syndrome: A PET study. *Epilepsia, 31,* 778–783.

Marcie, P., Roudier, M., & Boller, F. (1994). Spontaneous language and impairment of communication in Alzheimer's disease. *Linguistische Berichte 6* (special issue), 111–127.

Marescaux, C., Hirsch, E., Finck, S., Maquet, P., Schlumberger, E., Sellal, F., Metz-Lutz, M. N., Alembik, Y., Salmon, E., & Franck, G. (1990). Landau-Kleffner syndrome: A pharmacologic study of five cases. *Epilepsia, 31,* 768–777.

Margolin, D. I. (Ed.). (1992). *Cognitive neuropsychology in clinical practice.* New York: Oxford University Press.

Margolin, D. I., & Goodman-Schulman, R. (1992). Oral and written spelling impairments. In D. Margolin (Ed.), *Cognitive neuropsychology in clinical practice.* Oxford: Oxford University Press.

Marie, P. (1906). Révision de la question de l'aphasie: Que faut-il penser des aphasies sous-courticales (aphasies pures)? *La Semaine Médicale, 42,* 17 October 1906. (cited by Démonet, J.-F. [1987]. Les aphasies "sous-corticales": Étude linguistique, radiologique et hémodynamique de 31 observations. Thèse pour le doctorat d'état en médecine, Université Paul Sabatier, Toulouse, France.

Marien, P., Saerens, J., Verslegers, W., Borggreve, F., & De-Deyn, P. P. (1993). Some controversies about type and natrure of aphasic symptomatology in Landau-Kleffner's syndrome. *Acta Neurologica Belgica, 93,* 183–203.

Marin, O. S. M., Saffran, E. M., & Schwartz, M. F. (1976). Dissociations of language in aphasia: Implications for normal functions. *Annals of New York Academy of Sciences, 280,* 868–884.

Mark, V. W., Thomas, B. E., & Berndt, R. S. (1992). Factors associated with improvement in global aphasia. *Aphasiology, 6,* 121–134.

Markowicz, H. & Woodward, J. (1978). Language and the maintenance of ethnic boundaries in the deaf community. *Communication and Cognition, 2,* 29–38.

Marr, D. (1982). *Vision.* San Francisco: Freeman.

Marsh, N. V., & Knight, R. G. (1991a). Behavioural assessment of social competence following severe head injury. *Journal of Clinical and Experimental Neuropsychology, 13,* 729–740.

Marsh, N. V., & Knight, R. G. (1991b). Relationship between cognitive deficits and social skill after closed head injury. *Neuropsychology, 5,* 107–117.

Marshall, J. (1983). A rose by any other name. *The Behavioral and Brain Sciences, 6,* 216–217.

Marshall, J. (1986). The description and interpretation of aphasic language disorders. *Neuropsychologia, 24,* 5–24.

Marshall, J. C. (1994). Henry Charlton Bastian. In P. Eling (Ed.), *Reader in the history of aphasia* (pp. 99–132). Amsterdam: John Benjamins.

Marshall, J. C., & Newcombe, F. (1973). Patterns of paralexia: A psycholinguistic approach. *Journal of Psycholinguistic Research, 2,* 175–199.

Marshall, J. C., & Newcombe, F. (1984). Putative problems and pure progress in neuropsychological single-case studies. *Journal of Clinical Neuropsychology, 6,* 65–70.

Marshall, J. F. (1985). Neural plasticity and recovery of function after brain injury. *International Review of Neurobiology, 26,* 201–247.

Marshall, R., Gandour, J., & Windsor, J. (1988). Selective impairment of phonation: A case study. *Brain and Language, 35,* 313–339.

Marshall, R. C., Thompkins, C. A., & Phillips, D. S. (1982). Improvement in treated aphasia: Examination of selected prognostic factors. *Folia Phoniatrica, 34,* 304–315.

Marsolek, C. J., Kosslyn, S. M., & Squire, L. R. (1992). Form-specific visual priming in the right cerebral hemisphere. *Journal of Experimental Psychology: Learning, Memory and Cognition, 18,* 492–508.

Martin, A. (1990). Neuropsychology of Alzheimer's disease: The case for subgroups. In M. F. Schwartz (Ed.), *Modular deficits in Alzheimer-type dementia* (pp. 143–175). Cambridge: MIT Press.

Martin, A., Brouwers, P., Lalonde, F., Cox, C., Teleska, P., Fedio, P., Foster, N. L., & Chase, T. N. (1986). Towards a behavioral typology of Alzheimer's patients. *Journal of Clinical and Experimental Neuropsychology, 8,* 594–610.

Martin, A., & Fedio, P. (1983). Word production and comprehension in Alzheimer's disease: The breakdown of semantic knowledge. *Brain and Language, 19,* 124–141.

Martin, A., Haxby, J. V., Lalonde, F. M., Wiggs, C. L., & Ungerleider, L. G. (1995). Discrete cortical regions associated with knowledge of color and knowledge of action. *Science, 270,* 102–105.

Martin, A., Wiggs, C. L., Ungerleider, L. G., & Haxby, J. V. (1996). Neural correlates of category-specific knowledge. *Nature, 379,* 649–652.

Martin, J. B. (1984). Huntington's disease: New approaches to an old problem. *Neurology, 34,* 1059–1072.

Martin, J. H. (1991). The collective electrical behavior of cortical neurons: The electroencephalogram and the mechanisms of epilepsy. In E. R. Kandel, J. H. Schwartz, & T. M. Jessell (Eds.), *Principles of neural science* (3rd ed., pp. 777–791). New York: Elsevier.

Martin, N., & Saffran, E. M. (1992). A computational account of deep dysphasia: Evidence from a single case study. *Brain & Language, 43,* 240–274.

Martin, N., Saffran, E. M., & Dell, G. S. (1996). Recovery in deep dysphasia: Evidence for a relations between auditory–verbal STM capacity and lexical errors in repetition. *Brain and Language, 52,* 83–113.

Martin, N., Saffran, E. M., Dell, G. S., & Schwartz, M. F. (1994). Origins of paraphasias in deep dyslexia: Testing the consequences of a decay impairment to an interactive activation model of lexical retrieval. *Brain and Language, 47,* 609–660.

Martin, R. C. (1993). Short-term memory and sentence processing: Evidence from neuropsychology. *Memory and Cognition, 21,* 176–183.

Martin, R. C., (1995). Working memory doesn't work: A critique of Miyake et al.'s capacity theory of aphasic comprehension deficits. *Cognitive Neuropsychology, 12,* 623–636.

Martin, R. C., & Feher, E. (1990). The consequences of reduced memory span for the comprehension of semantic versus syntactic information. *Brain and Language, 38,* 1–20.

Martin, R. C., Frederick-Wetzel, W., Blossom-Stack, C., & Feher, E. (1989). Syntactic loss versus processing deficit. An assessment of two theories of agrammatism and syntactic comprehension deficits. *Cognition, 32,* 157–191.

Martin, R. C., & Romani, C. (1994). Verbal working memory and sentence comprehension: A multiple components view. *Neuropsychology, 8,* 506–523.

Martin, R. C., & Romani, C. (1995). Remembering stories but not words. In R. Campbell & M. A. Conway (Eds.), *Broken memories: Case studies in memory impairment* (pp. 267–284). Oxford: Basil Blackwell.

Martin, R. C., Shelton, J., & Yaffee, L. (1994). Language processing and working memory: Neuropsychological evidence for separate phonological and semantic capacities. *Journal of Memory and Language, 33,* 83–111.

Martins, I. P., Castro-Caldas, A., Van Dongen, H. R., & Van Hout, A. (1991). *Acquired aphasia in children: Acquisition and breakdown of language in the developing brain* (pp. 171–184). Dordrecht, The Netherlands: Kluwer.

Martins, I. P., & Ferro, J. (1992). Recovery of acquired aphasia in children. *Aphasiology, 6,* 431–438.

Maruff, P., Hay, D., Malone, V., & Currie, J. (1995). Asymmetries in the covert orienting of visual spatial attention in schizophrenia. *Neuropsychologia, 33,* 1205–1223.

Massaro, D. W., & Cowan, N. (1993). Information processing models: Microscopes of the mind. *Annual Review of Psychology, 44,* 383–425.

Massaro, D. W., & Egan, P. B. (1996). Perceiving affect from the voice and face. *Psychonomic Bulletin & Review, 3,* 215–221.

Massman, P. J., & Doody, R. S. (1996). Hemispheric asymmetry in Alzheimer's disease is apparent in motor functioning. *Journal of Clinical and Experimental Neuropsychology, 18,* 110–121.

Masur, D. M., Sliwinski, M., Lipton, R. B., Blau, A. D., & Crystal, H. A. (1994). Neuropsychological prediction of dementia and the absence of dementia in healthy elderly persons. *Neurology, 44,* 1427–1432.

Mateer, C. A., Rapport, R. L., & Kettrick, C. (1984). Cerebral organization of oral and signed language responses: Case study evidence from amytal and cortical stimulation studies. *Brain and Language, 21,* 123–135.

Matison, R., Mayeux, R., Rosen, J., & Fahn, S. (1982). "Tip-of-the-tonge" phenomenon in Parkinson's disease. *Neurology, 32,* 567–570.

Matsui, T., & Hirano, A. (1978). *An atlas of the human brain for computerized tomography.* New York: Fischer.

Mauner, G., Fromkin, V. A., & Cornell, T. L. (1993). Comprehension and acceptability judgements in agrammatism: Disruptions in the syntax of referential dependency. *Brain and Language, 45,* 340–370.

Mayeux, R. (1986). Emotional changes associated with basal ganglia disorders. In K. Heilman & P. Satz (Eds.), *Neuropsychology of human emotion* (pp. 141–164). New York: Guilford Press.

Mazoyer, B. M., Dehaene, S., Tzourio, N., Frak, V., Syrota, A., Murayama, N., Levrier, O., Salamon, G., Cohen, L., & Mehler, J. (1993). The cortical representation of speech. *Journal of Cognitive Neuroscienc, 5,* 467–479.

Mazzoni, M., Vista, M., Pardossi, L., Avila, L., Bianchi, F., & Moretti, P. (1992). Spontaneous evolution of aphasia after ischemic stroke. *Aphasiology, 6,* 387–396.

McCallum, W. C., Farmer, S. F., & Pocock, P. V. (1984). The effects of physical and semantic incongruities on auditory event-related potentials. *Electroencephalography and Clinical Neurophysiology, 59,* 477–488.

McCarthy, G., Blamire, A. M., Rothman, D. L., Gruetter, R., & Shulman, R. G. (1993). Echo-planar magnetic resonance imaging studies of frontal cortex activation during word generation in humans. *Proceedings of the National Academy of Sciences, 90,* 4952–4956.

McCarthy, G., Nobre, A. C., Bentin, S., & Spencer, D. D. (1995). Language-related field potentials in the anterior temporal lobe: I. Intracranial distribution and neural generators. *Journal of Neuroscience, 15,* 1080–1089.

McCarthy, G., Wood, C. C., Williamson, P. D., & Spencer, D. D. (1989). Task-dependent field potentials in human hippocampal formation. *Journal of Neuroscience, 9,* 4253–4268.

McCarthy, J. J. (1981). A prosodic theory of nonconcatenative morphology. *Linguistic Inquiry, 12,* 373–418.

McCarthy, R., & Warrington, E. K. (1984). A two-route model of speech production: Evidence from aphasia. *Brain, 107,* 463–485.

McCarthy, R. A., & Warrington, E. K. (1985). Category specificity in an agrammatic patient: The relative impairment of verbs retrieval and comprehension. *Neuropsychologia, 23,* 709–727.

McCarthy, R. A., & Warrington, E. K. (1988). Evidence for modality-specific meaning systems in the brain. *Nature, 334,* 428–430.

McCarthy, R. A., & Warrington, E. K. (1990a). Cognitive neuropsychology: A clinical introduction. San Diego: Academic Press.

McCarthy, R. A., & Warrington, E. K. (1990b). The dissolution of semantics. *Nature, 343,* 599.

McClelland, J. L. (1993). Toward a theory of information processing in graded, random, interactive networks. In D. E. Meyer & S. Kornblum (Eds.), *Attention and performance XIV: Synergies in experimental psychology, artificial intelligence, and cognitive neuroscience* (pp. 655–688). Cambridge: MIT Press.

McClelland, J. L., & Rumelhart, D. E. (1981). An interactive activation model of context effects in letter perception: Part I. An account of basic findings. *Psychological Review, 88,* 375–405.

McCloskey, M. (1992). Cognitive mechanisms in numerical processing: Evidence from acquired dyscalculia. *Cognition, 44,* 107–157.

McCloskey, M. (1993). Theory and evidence in cognitive neuropsychology: A "radical" response to Robertson, Knight, Rafal, & Shimamura. *Journal of Experimental Psychology: Learning, Memory, and Cognition, 19,* 718–734.

McCloskey, M., & Caramazza, A. (1987). Cognitive mechanisms in normal and impaired number processing. In G. Deloche & X. Seron (Eds.), *Mathematical disabilities: A cognitive neuropsychological perspective* (pp. 201–219). Hillsdale, NJ: Lawrence Erlbaum.

McCloskey, M., & Caramazza, A. (1988). Theory and methodology in cognitive neuropsychology: A response to our critics. *Cognitive Neuropsychology, 5,* 534–574.

McCloskey, M., Caramazza, A., & Basili, A. (1985). Cognitive mechanisms in number processing and calculation: Evidence from dyscalculia. *Brain and Cognition, 4,* 171–196.

McCloskey, M., Sokol, S. M., & Goodman, R. A. (1986). Cognitive processes in verbal-number production: Inferences from the performance of brain-damaged subjects. *Journal of Experimental Psychology: General, 115,* 307–330.

McCormick, D. A., & Feeser, H. R. (1990). Functional implications of burst firing and single spike activity in lateral geniculate relay neurons. *Neuroscience, 39,* 103–113.

McDonald, S., (1992a). Communication disorders following closed head injury: New approaches to assessment and rehabilitation. *Brain Injury, 6,* 283–292.

McDonald, S., (1992b). Differential pragmatic language loss following closed head injury: Ability to comprehend conversational implicature. *Applied Psycholinguistics, 13,* 295–312.

McDonald, S. (1993a). Major review. Viewing the brain sideways? Right hemisphere versus anterior models of non-aphasic language disorders. *Aphasiology, 7,* 535–549.

McDonald, S., (1993b). Pragmatic language loss following closed head injury: Inability to meet the informational needs of the listener. *Brain and Language, 44,* 28–46.

McDonald, S., & Pearce, S. (1995). The dice game: A new test of organisational skills in language. *Brain Injury, 9,* 255–271.

McDonald, S., & Pearce, S. (1996). Clinical insights into pragmatic language theory: The case of sarcasm. *Brain and Language 53,* 81–104.

McDonald, S., & Pearce, S. (in press). *Executive impairment and communication skills: The ability to make requests that overcome listener reluctance.* Proceedings of the 5th Conference of the International Association for the Study of Traumatic Brain Injury and the 20th Conference of the Australian Society for the Study of Brain Impairment. Bowen Hills, Queensland: Australian Academic Press.

McDonald, S., & van Sommers, P. (1993). Differential pragmatic language loss following closed head injury: Ability to negotiate requests. *Cognitive Neuropsychology, 10,* 297–315.

McFarling, D., Rothi, L. J., & Heilman, K. M. (1982). Transcortical aphasia from ischaemic infarcts of the thalamus: A report of two cases. *Journal of Neurology, Neurosurgery, and Psychiatry, 45,* 107–112.

McGlone, J. (1977). Sex differences in the cerebral organization of verbal function and cognitive impairment in stroke. Age, sex, aphasia type and laterality differences. *Brain, 100,* 775–793.

McGlone, J. (1980). Sex differences in human brain asymmetry: A critical survey. *Behavioral and Brain Sciences, 3,* 215–264.

McGlynn, E. A. (1996). Domains of study and methodological challenges. In L. I. Sederer & B. Dickey (Eds.), *Outcomes assessment in clinical practice* (pp. 19–24). Baltimore: Williams & Wilkins.

McHenry, L. C. (1969). *Garrison's history of neurology* (revised and enlarged). Springfield, IL: C. C. Thomas.

McKelvey, J. R., Lambert, R., Mottron, L., & Shevell, M. I. (1995). Right-hemisphere dysfunction in Asperger's syndrome. *Journal of Child Neurology, 10,* 310–314.

McKenna, P., & Warrington, E. K. (1978). Category-specific preservation: A single case study. *Journal of Neurology, Neurosurgery and Psychiatry, 41,* 571–574.

McKenna, P., & Warrington, E. K. (1993). The neuropsychology of semantic memory. In F. Boller & J. Grafman (Eds.), *Handbook of neuropsychology* (Vol. 8, pp. 193–213). Amsterdam: Elsevier.

McNeil, J. E., & Warrington, E. K. (1994). A dissociation between addition and subtraction within written calculation. *Neuropsychologia, 32,* 717–728.

McNeil, M., Hashi, M., & Southwood, H. (1994). Acoustically derived perceptual evidence for coarticulatory errors in apraxic and conduction aphasic speech production. *Clinical Aphasiology, 22,* 203–218.

McNeil, M. R., Liss, J. M., Tseng, C., & Kent, R. D. (1990). Effects of speech rate on the absolute and relative timing of apraxic and conduction aphasic sentence production. *Brain and Language, 38,* 135–158.

McNeil, M. R., & Prescott, T. E. (1978). *Revised Token Test.* Baltimore: University Park Press.

McPherson, S. E., Kuratani, J. D., Cummings, J. L., Shih, J., Mischel, P. S., & Vinters, H. V. (1994). Creutzfeldt-Jakob disease with mixed transcortical aphasia: Insights into echolalia. *Behavioral Neurology, 7,* 197–203.

Meadows, K. (1980). *Deafness and child development.* Berkeley: University of California Press.

Meckler, R. J., Mack, J. L., & Bennett, R. (1979). Sign language aphasia in a non-deaf-mute. *Neurology, 29*(7), 1037–1040.

Mecklinger, A., Schriefers, H., Steinhauer, K., & Friederici, A. D. (1995). Processing relative clauses varying on syntactic and semantic dimensions: An analysis with event-related potentials. *Memory and Cognition, 23,* 477–494.

Meerson, Y. A., & Tarkhan, A. U. (1988). Role of structures of the left and right hemispheres in perception of the prosodic characteristics of speech: A combined clinical and experimental study. *Human Physiology, 14,* 339–346.

Mehler, J., & Fox, R. (1985). *Neonate cognition: Beyond the blooming buzzing confusion.* Hillsdale, NJ: Lawrence Erlbaum.

Mehrabian, A. (1968). Communication without words. *Psychology Today, 2,* 52–55.

Mehta, C., & Patel, N. (1995). *User manual: StatXact 3 for Windows.* Cambridge, MA: Cytel Software.

Meier, R. P. (1991). Language acquisition by deaf children. *American Scientist, 79,* 60–70.

Mendez, M., Adams, N. L., & Lewandowski, K. S. (1989). Neurobehavioral changes associated with caudate lesions. *Neurology, 39,* 349–354.

Mendez, M., & Mendez, M. A. (1991). Differences between multi-infarct dementia and Alzheimer's disease on unstructured neuropsychological tasks. *Journal of Clinical and Experimental Neuropsychology, 13,* 923–932.

Mendez, M. F. (1994). Huntington's disease: Update and review of neuropsychiatric aspects. *International Journal of Psychiatry in Medicine, 24,* 189–208.

Mendez, M. F., Selwood, A., Mastri, A. R., & Frey, W. H. (1993). Pick's disease versus Alzheimer's disease: A comparison of clinical characteristics. *Neurology, 43,* 289–292.

Menn, L., & Obler, L. K. (1990). *Agrammatic aphasia: A cross-language narrative sourcebook.* Amsterdam: John Benjamins.

Menn, L., Ramsberger, G., & Helm-Estabrooks, N. (1994). A linguistic communication measure for aphasic narratives. *Aphasiology, 8,* 343–359.

Mentis, M., & Prutting, C. A. (1987). Cohesion in the discourse of normal and head injured adults. *Journal of Speech and Hearing Research, 30,* 88–98.

Mentis, M., & Prutting, C. A. (1991). Analysis of topic as illustrated in a head-injured and a normal adult. *Journal of Speech and Hearing Research, 34,* 583–595.

Mentis, M., Briggs-Whittaker, J., & Gramigna, G. D. (1995). Discourse topic management in senile dementia of the Alzheimer's type. *Journal of Speech and Hearing Research, 38,* 1054–1066.

Mervis, C. B., & Bertrand, J. (1995). Early lexical development of children with Williams syndrome. In U. Bellugi & C. A. Morris (Eds.), Williams syndrome: From cognition to gene. Abstracts from the Williams Syndrome Association National/International Professional Conference. *Genetic Counseling, 6,* 134–135.

Messert, B., & Nuis, C. V. (1966). A syndrome of paralysis of downward gaze, dysarthria, pseudobulbar palsy, axial rigidity of the neck and trunk dementia. *Journal of Nervous and Mental Disease, 143,* 47–54.

Messick, S. (1980). Test validation and the ethics of assessment. *American Psychologist, 35,* 1012–1027.

Messmer, D., & Roth, V. M. (1996). WinWEGE (FAX +49 7531 15583).

Mesulam, M. M. (1982). Slowly progressive aphasia without generalized dementia. *Annals of Neurology, 11,* 592–598.

Mesulam, M. M. (1990). Large-scale neurocognitive networks and distributed processing for attention, language and memory. *Annals of Neurology, 28,* 597–613.

Méthe, S., Huber, W., & Paradis, M. (1993). Inventory and classification of rehabilitation methods. In M. Paradis (Ed.), *Foundations of aphasia rehabilitation* (pp. 3–60). Oxford: Pergamon Press.

Metter, E. J. (1987). Neuroanatomy and physiology of aphasia: Evidence from positron emission tomography. *Aphasiology, 1,* 3–33.

Metter, E. J., & Hanson, W. R. (1994). Use of positron emission tomography to study aphasia. In A. Kertesz (Ed.), *Localization and neuroimaging in neuropsychology* (pp. 123–148). San Diego: Academic Press.

Metter, E. J., Hanson, W. R., Jackson, C. A., Kempler, D., van Lancker, D., Mazziotta, J. C., & Phelps, M. E. (1986). Temporoparietal cortex in aphasia. Evidence from positron emission tomography. *Archives of Neurology, 47,* 1235–1238.

Metter, E. J., Jackson, C. A., Kempler, D., & Hanson, W. R. (1992). Temporoparietal cortex and the recovery of language comprehension in aphasia. *Aphasiology, 6,* 349–358.

Metter, E. J., Jackson, C. A., Kempler, D., Riege, W. H., Hanson, W. R., Mazziotta, J. C., & Phelps, M. E. (1986). Left hemisphere intracerebral hemorrages studied by F-18–Deoxyglucose PET. *Neurology, 36,* 1155–1162.

Metter, E. J., Riege, W. H., Hanson, W. R., Camras, L. R., Phelps, M. E., & Khul, D. E. 1984. Correlation of glucose metabolism and structural damage to language function in aphasia. *Brain and Language, 21,* 187–207.

Metter, E. J., Riege, W. H., Hanson, W. R., Kuhl, D. E., & Phelps, M. E. (1984). Commonality and differences in aphasia: Evidence from BDAE and PICA. *Clinical Aphasiology, 14,* 70–77.

Metter, E. J., Riege, W. H., Hanson, W. R., Kuhl, D. E., Phelps, M. E., Squire, L. R., Wasterlain, C. G., & Benson, D. F. (1983). Comparison of metabolic rates, language and memory in subcortical aphasias. *Brain and Language, 19,* 33–47.

Metter, E. J., Riege, W. H., Kuhl, D. E., & Phelps, M. E. (1984). Cerebral metabolic relationships for selected brain regions in healthy adults. *Journal of Cerebral Blood Flow and Metabolism, 4,* 1–7.

Metter, E. J., Wasterlain, C. G., Kuhl, D. E., Hanson, W. R., & Phelps, M. E. (1981). 18 FDG positron emission computed tomography in a study of aphasia. *Annals of Neurology, 10,* 173–183.

Metz-Lutz, M.-N. (1995). Les méthodes d'exploration en temps réel en aphasiologie. *Revue de Neuropsychologie, 5,* 225–252.

Meyer, D. E., & Schvaneveldt, R. W. (1971). Facilitation in recognizing pairs of words: Evidence of a dependence between retrieval operations. *Journal of Experimental Psychology, 90,* 227–234.

Meyer, J. S., Sakai, F., Yamaguchi, F., Yamamoto, M., & Shaw, T. (1980). Regional changes in cerebral blood flow during standard behavioral activation in patients with disorders of speech and mentation compared to normal volunteers. *Brian and Language, 9,* 61–77.

Miceli, G., Caltagirone, C., Gainotti, G. Masullo, C., Silveri, M. C., & Villa, G. (1981). Influence of age, sex, literacy and pathologic lesion on incidence, severity and type of aphasia. *Acta Neurologica Scandinavica, 64,* 370–382.

Miceli, G., & Caramazza, A. (1988). Dissociation of inflectional and derivational morphology. *Brain and Language, 35,* 24–65.

Miceli, G., Mazzucchi, A., Menn, L., & Goodglass, H. (1983). Contrasting cases of Italian agrammatic aphasia without comprehension disorder. *Brain and Language, 19,* 65–97.

Miceli, G., Silveri, M. C., Nocentini, U., & Caramazza, A. (1988). Patterns of dissociation in comprehension and production of nouns and verbs. *Aphasiology, 2,* 351–358.

Miceli, G., Silveri, M. C., Romani, C., & Caramazza, A. (1989). Variation in the pattern of omissions and substitutions of grammatical morphemes in the spontaneous speech of so-called agrammatic patients. *Brain and Language, 26,* 447–492.

Miceli, G., Silveri, M. C., Villa, G., & Caramazza, A. (1984). On the basis for the agrammatic's difficulty in producing main verbs. *Cortex, 20,* 207–220.

Michel, G. F. (1983). Development of hand-use preference during infancy. In G. Young, S. J. Segalowitz, C. M. Corter, & S. E. Trehub (Eds.), *Manual specialization and the developing brain* (pp. 33–70). New York: Academic Press.

Miikulainen, R. (in press). Dyslexic and category-specific impairments in a self-organizing feature map model of the lexicon. *Brain and Language.*

Miklowitz, D. J., Velligan, D. I., Goldstein, M. J., Nuechterlein, K. H., Gitlin, M. J., Ranlett, G., & Doane, J. A. (1991). Communication deviance in families of schizophrenic and manic patients. *Journal of Abnormal Psychology, 100,* 153–173.

Milberg, W., Blumstein, S., & Dworetzky, B. (1988). Phonological processing and lexical access in aphasia. *Brain and Language, 34,* 279–293.

Miles, T. (1983). *Dyslexia: The pattern of difficulties.* London: Granada.

Miller, B. L., Cummings, J. L., Villanueva-Meyer, J., Boone, K., Meringer, C. M., Lesser, I. M., & Mena, I. (1991). Frontal lobe degeneration: Clinical, neuropsychological, and SPECT characteristics. *Neurology, 41,* 1374–1382.

Miller, E. (1985). Possible frontal impairments in Parkinson's disease: A test using a measure of verbal fluency. *British Journal of Clinical Psychology, 24,* 211–212.

Mills, D., Coffey-Corina, S., & Neville, H. (1993). Language acquisition and cerebral specialization in 20–month-old infants. *Journal of Cognitive Neuroscience, 5,* 317–334.

Mills, R. H. (1982). Microcomputerized auditory comprehension training. In R. H. Brookshire (Ed.), *Clinical aphasiology: 1982 conference proceedings* (pp. 147–152). Minneapolis: BRK Publishers.

Milton, S. B., Prutting, C. A., & Binder, G. M. (1984). Appraisal of communication in head-injured adults. *Journal of Speech and Hearing Research, 30,* 88–98.

Milton, S. B., & Wertz, R. T. (1986). Management of persisting communication deficits in patients with traumatic brain injury. In B. P. Uzzell & Y. Gross (Eds.), *Clinical neuropsychology of intervention* (pp. 223–255). Boston: Martinus Nijhoff.

Mimouni, Z., & Jarema, G. (1995). *Agrammatic aphasia in Arabic.* CLASNET Working Papers No. 3.

Mimouni, Z., Kehayia, E., & Jarema, G. (1992). Morphological priming of inflected nouns in a nonconcatenative language: Implications for lexical storage and access. Academy of Aphasia 30th Annual Meeting, Toronto.

Minkowski, M. (1928). Sur un cas d'aphasie chez un polyglotte. *Revue Neurologique, 49,* 361–366. (Trans. M. Paradis [Ed.] 1993. *Readings on aphasia in bilinguals and polyglots* [pp. 274–279]. Montreal: Didier.)

Miozzo, A., Soardi, M., & Cappa, S. F. (1994). Pure anomia with spared action naming due to a left temporal lesion. *Neuropsychologia, 32,* 1101–1109.

Mirsky, A. F., Anthony, B. J., Duncan, C. C., Ahearn, M. B., & Kellam, S. G. (1991). Analysis of the elements of attention: A neuropsychological approach. *Neuropsychology Review, 2,* 109–145.

Mirsky, A. F., Fantie, B. D., & Tatman, J. E. (1995). Assessment of attention across the lifespan. In R. L. Mapou & J. Spector (Eds.), *Clinical neuropsychological assessment: A cognitive approach* (pp. 17–48). New York: Plenum Press.

Mishkin, M. (1982). A memory system in the monkey. *Philosophical Transactions of the Royal Society of London: Series B, 298,* 85–95.

Mishkin, M., & Murray, E. A. (1994). Stimulus recognition. *Current Opinion in Neurobiology, 4,* 200–206.

Mishkin, M., & Phillips, R. R. (1990). A corticolimbic memory path revealed through its disconnection. In C. Trevarthen (Ed.), *Brain circuits and functions of the mind: Essays in honor of Roger W. Sperry* (pp. 196–210). New York: Cambridge University Press.

Mitchum, C. C., & Berndt, R. S. (1995). The cognitive neuropsychological approach to treatment of language disorders. *Neuropsychological Rehabilitation, 5,* 1–16.

Miyake, A., Carpenter, P. A., & Just, M. A. (1994). A capacity approach to syntactic comprehension disorders: Making normal adults perform like aphasic patients. *Cognitive Neuropsychology, 11,* 671–717.

Miyake, A., Carpenter, P. A., & Just, M. A. (1995). Reduced resources and specific impairments in normal and aphasic sentence comprehension. *Cognitive Neuropsychology, 12,* 651–679.

Moeller, J. R., Struther, S. C., Sidtis, J. J., & Rottenberg, D. A. (1987). Scaled subprofile model: A statistical approach to the analysis of functional patterns in emission tomographic data. *Journal of Cerebral Blood Flow and Metabolism, 7,* 649–658.

Moen, I. (1991). Functional lateralisation of pitch accents and intonation in Norwegian: Monrad-Krohn's study of an aphasic patient with altered "melody of speech." *Brain and Language, 41,* 538–554.

Moen, I. (1993). Functional lateralization of the perception of Norwegian word tones—evidence from a dichotic listening experiment. *Brain and Language, 44,* 400–413.

Moen, I., & Sundet, K. (1996). Production and perception of word tones (pitch accents) in patients with left and right hemisphere damage. *Brain and Language, 53,* 267–281.

Mohr, B., Pulvermüller, F., Rayman, J., & Zaidel, E. (1994). Interhemispheric cooperation during lexical processing is mediated by the corpus callosum: Evidence from a split-brain patient. *NeuroReports, 181,* 17–21.

Mohr, B., Pulvermüller, F., & Zaidel, E. (1994). Lexical decision after left, right, and bilateral presentation of words and non-words: Evidence for interhemispheric interaction. *Neuropsychologia, 32,* 105–124.

Mohr, J. P., Biller, J., Hilal, S. K., Yuh, W. T. C., Tatemichi, T. K., Hedges, S., Tali, E., Nguyen, H., Mun, I., Adams, H. P., Grimsman, K., & Marler, J. R. (1995). Magnetic resonance versus computed tomographic imaging in acute stroke. *Stroke, 26,* 807–812.

Mohr, J. P., Pessin, M. S., Finkelstein, S., Funkenstein, H. H., Duncan, G. W., & David, K. R. (1978). Broca's aphasia: Pathological and clinical. *Neurology, 28,* 311–324.

Molfese, D. L. (1978). Neuroelectrical correlates of categorical speech perception in adults. *Brain and Language, 5,* 25–35.

Molfese, D. L. (1980a). Hemispheric specialization for temporal information: Implications for the perception of voicing cues during speech perception. *Brain and Language, 11,* 285–299.

Molfese, D. L. (1980b). The phoneme and the engram: Electrophysiological evidence for the acoustic invariant in stop consonants. *Brain and Language, 9,* 372–376.

Molfese, D. L. (1989). Electrophysiological correlates of word meaning in 14-month-old infants. *Developmental Neuropsychology, 5,* 79–103.

Molfese, D. L. (1990). Auditory evoked response recorded from 16-month-old human infants to words they did and did not know. *Brain and Language, 38,* 345–363.

Molfese, D. L., & Betz, J. C. (1988). Electrophysiological indices of the early development of lateralization for language and cognition, and their implications for predicting later development. In D. L. Molfese & S. J. Segalowitz (Eds.), *Brain lateralization in children: Developmental implications* (pp. 171–190). New York: Guilford Press.

Molfese, D. L., Burger-Judisch, L. M., & Hans, L. L. (1991). Consonant discrimination by newborn infants: Electrophysiological differences. *Developmental Neuropsychology, 7,* 177–195.

Molfese, D. L., Freeman, R. B., & Palermo, D. S. (1975). The ontogeny of brain lateralization for speech and non-speech stimuli. *Brain and Language, 2,* 356–368.

Molfese, D. L., & Molfese, V. J. (1985). Electrophysiological indices of auditory discrimination in newborn infants: The bases for predicting later language development? *Infant Behavior and Development, 8,* 197–211.

Molfese, D. L., & Molfese, V. J. (1986). Psychophysical indices of early cognitive processes and their relationship to language. In J. E. Obrzut & G. W. Hynd (Eds.), *Child neuropsychology.* Vol. 1, *Theory and research* (pp. 95–116). New York: Academic Press.

Molfese, D. L. & Molfese, V. J. (1988). Right hemisphere responses from preschool children to temporal cues contained in speech and nonspeech materials: Electrophysiological correlates. *Brain and Language, 33,* 245–249.

Molfese, D. L., & Molfese, V. J. (1994). Short-term and long-term developmental outcomes: The use of behavioral and electrophysiological measures in early infancy as predictors. In G. Dawson & K. Fischer (Eds.), *Human behavior and the developing brain* (pp. 493–517). New York: Guilford Press.

Molfese, D. L., Morse, P. A., & Peters, C. J. (1990). Auditory evoked responses to names for different objects: Cross-modal processing as a basis for infant language acquisition. *Developmental Psychology, 26,* 780–795.

Molfese, D. L., & Wetzel, W. F. (1992). Short- and long-term auditory recognition memory in 14-month-old human infants: Electrophysiological correlates. *Developmental Neuropsychology, 8,* 135–160.

Molfese, D. L., Wetzel, W. F., & Gill, L. A. (1993). Known versus unknown word discriminations in 12-month-old human infants: Electrophysiological correlates. *Developmental Neuropsychology, 9,* 241–258.

Molloy, R., Brownell, H. H., & Gardner, H. (1990). Discourse comprehension by right hemisphere stroke patients: Deficits in prediction and revision. In Y. Joanette & H. H. Brownell (Eds.), *Discourse ability and brain damage: Theoretical and empirical perspectives* (pp. 113–130). New York: Springer Verlag.

Mondor, T. A. (1994). Interaction between handedness and the attentional bias during tests of dichotic listening performance. *Journal of Clinical and Experimental Neuropsychology, 16,* 377–385.

Money, J. (1993). Specific neurocognitional impairments associated with Turner (45, X) and Klinefelter (47, XXY) syndromes: A review. *Social Biology, 40,* 147–151.

Monrad-Krohn, G. H. (1947). Dysprosody or altered "melody of speech." *Brain, 70,* 405–415.

Monroe, N. E. (1985). Communicative abilities in daily living. In D. J. Keyser & R. C. Sweetland (Eds.), *Test critiques* (Vol. 4, pp. 189–194). Kansas City, MO: Test Corporation of America.

Monsch, A. U., Bondi, M. W., Butters, N., Salmon, D. P., Katzman, R., & Thal, L. J. (1992). Comparisons of verbal fluency tests in the detection of dementia of the Alzheimer type. *Archives of Neurology, 49,* 1253–1258.

Monsell, S., Doyle, M., & Haggard, P. (1989). Effects of frequency on visual word recognition tasks: Where are they? *Journal of Experimental Psychology: General, 118,* 43–71.

Monsell, S., Patterson, K. E., Graham, A., Hughs, C., & Milroy, R. (1992). Lexical and sublexical translation of spelling to sound: Strategic anticipation of lexical status. *Journal of Experimental Psychology: Learning, Memory, and Cognition, 18,* 452–467.

Moore, M. K., & Meltzoff, A. N. (1978). Object permanence, imitation, and language development in infancy: Toward a neo-Piagetian perspective on communicative and cognitive development. In F. D. Minifie & L. L. Lloyd (Eds.), *Communicative and cognitive abilities: Early behavioral assessment.* Baltimore: University Park Press.

Moore, W. H. (1989). Language recovery in aphasia: A right hemisphere perspective *Aphasiology, 3,* 101–110.

Moore, W. H., & Weidner, W. (1974). Bilaterial tachistoscopic presentations in aphasic and normal subjects. *Perceptual and Motor Skills, 40,* 379–386.

Moore, W. H., & Weidner, W. (1975). Dichotic word perception of aphasic and normal subjects. *Perceptual and Motor Skills, 40,* 379–386.

Morais, J., Alegria, J., & Content, A. (1987). The relationships between segmental analysis and alphabetic literacy: An interactive view. *European Bulletin of Cognitive Psychology, 7,* 415–438.

Moreaud, O., Pellat, J., Charnallet, A., Carbonnel, S., & Brennen, T. (1995). Déficit de la production et de l'apprentissage des noms propres après lésion tubéro-thalamique gauche. *Revue Neurologique, 151,* 93–99.

Morice, R. D. (1986). Beyond language—speculations on the prefrontal cortex and schizophrenia. *Australian and New Zealand Journal of Psychiatry, 20,* 7–10.

Morice, R. D., & Ingram, J. C. L. (1982). Language analysis in schizophrenia: Diagnostic implications. *Australian and New Zealand Journal of Psychiatry, 16,* 11–21.

Morice, R. D., & McNicol, D. (1986). Language changes in schizophrenia: A limited replication. *Schizophrenia Bulletin, 12,* 239–251.

Morrell, F., Whisler, W. W., Smith, M. C., Hoeppner, T. J., de Toledo-Morrell, L., Pierre-Louis, S. J., Kanner, A. M., Buelow, J. M., Ristanovic, R., Bergen, D. et al. (1995). Landau-Kleffner syndrome: Treatment with subpial intracortical transection. *Brain, 118,* 1529–1546.

Morrell, L. K. & Salamy, J. G. (1971). Hemispheric asymmetry of electrocortical responses to speech stimuli. *Science, 174*, 164–166.

Morris, R., Bakker, D., Satz, P., & Van der Vlugt, H. (1984). Dichotic listening ear asymmetry: Patterns of longitudinal development. *Brain and Language, 22*, 49–66.

Morris, R. G. (1994). Working memory in Alzheimer-type dementia. *Neuropsychology, 8*, 544–554.

Mortensen, E. L., & Kleven, M. (1993). A WAIS longitudinal study of cognitive development during the life span from ages 50 to 70. *Developmental Neuropsychology, 9*, 115–130.

Morton, J. (1980). The logogen model and orthographic structure. In U. Frith (Ed.), *Cognitive approaches in spelling*. London: Academic Press.

Morton, J. (1984). Brain-based and non-brain-based models of language. In D. Caplan, A. R. Lecours, & A. Smith (Eds.), *Biological perspectives on language* (pp. 40–64). Cambridge: MIT Press.

Morton, J., & Jusczyk, P. (1984). On reducing language to biology. *Cognitive Neuropsychology, 1*, 83–116.

Morton, L. L. (1994). Interhemispheric balance patterns detected by selective phonemic dichotic laterality measures in four clinical subtypes of reading-disabled children. *Journal of Clinical and Experimental Neuropsychology, 16*, 556–567.

Moscovitch, M. (1973). Language and the cerebral hemisphere: Reaction-time studies and their implications for models of cerebral dominance. In P. Pliner, L. Krames, & T. Alloway (Eds.), *Communication and affect: Language and thought* (pp. 89–126). New York: Academic Press.

Moscovitch, M. (1977). The development of lateralisation of language and its relation to cognitive and linguistic development. A review and some theoretical speculations. In S. J. Segalowitz & F. A. Gruber (Eds.), *Language development and neurological theory* (pp. 193–211). New York: Academic Press.

Moss, H. E., & Tyler, L. K. (1995). Investigating semantic memory impairments: The contribution of semantic priming. *Memory, 3*, 359–395.

Moutier, F. (1908). L'Aphasie de Broca. Doctoral dissertation, Paris (cited by Wallesch & Papagno, 1988).

Moutier, F. (1908). *L'Aphasie de Broca*. Paris: Steinheil.

Moyer, R. S., & Landauer, T. K. (1967). Time required for judgments of numerical inequality. *Nature, 215*, 1519–1520.

Mozer, M., & Behrmann, M. (1990). On the interaction of selective attention and lexical knowledge: A connectionist account of neglect dyslexia. *Journal of Cognitive Neuroscience, 2*, 96–123.

Mozer, M. C., Zemel, R. S., Behrmann, M., & Williams, C. K. I. (1992). Learning to segment images using dynamic feature binding. *Neural Computation, 4*, 650–665.

Mundy, P., Kasari, C., Sigman, M., & Ruskin, E. (1995). Nonverbal communication and early language acquisition in children with Down syndrome and in normally developing children. *Journal of Speech and Hearing Research, 38*, 157–167.

Mundy, P., Sigman, M., Kasari, C., & Yirmiya, N. (1988). Nonverbal communication skills in Down syndrome children. *Child Development, 59*, 235–249.

Münte, T., & Heinze, H. (1994). Brain potentials reveal deficits of language processing after closed head injury. *Archives of Neurology, 51*, 482–493.

Münte, T. F., Heinze, H.-J., & Mangun, G. R. (1993). Dissociation of brain activity related to syntactic and semantic aspects of language. *Journal of Cognitive Neuroscience, 5*, 335–344.

Murphy, D., & Cutting, J. (1990). Prosodic comprehension and expression in schizophrenia. *Journal of Neurology, Neurosurgery, and Psychiatry, 53*, 727–730.

Mussolino, A., & Dellatolas, G. (1991). Asymétries du cortex cérébral évaluées in vivo par angiographie stéréotaxique-stéréoscopique. *Revue Neurologique, 147*, 35–45.

Myers, P., & Linebaugh, C. (1981). Comprehension of idiomatic expressions by right-hemisphere-damaged adults. In R. H. Brookshire (Ed.), *Clinical aphasiology: Conference proceedings* (pp. 254–261). Minneapolis: BRK Publishers.

Myerson, M. O., & Frank, R. A. (1987). Language, speech & hearing in Williams syndrome: Intervention approaches and research needs. *Developmental Medicine & Child Neurology, 29*, 258–270.

Myerson, R., & Goodglass, H. (1972). Transformational grammars of three agrammatic patients. *Language and Speech, 15*, 40–50.

Näätänen, R., Simson, M., & Loveless, N. E. (1982). Stimulus deviance and evoked potentials. *Biological Psychology, 14*, 53–98.

Nadeau, S. E., & Crosson, B. (1995). A guide to the functional imaging of cognitive processes. *Neuropsychiatry, Neuropsychology, and Behavioral Neurology, 8,* 143–162.

Nadeau, S. E., & Crosson, B. (1997). Subcortical aphasia. *Brain and Language, 58,* 355–402.

Nadel, L. (1992). Multiple memory systems: What and why. *Journal of Cognitive Neuroscience, 4,* 179–188.

Naeser, M., Gaddie, A., Palumbo, C. L., & Stiassny-Eder, D. (1990). Late recovery of auditory comprehension in global aphasia: Improved recovery observed with subcortical temporal isthmus lesion vs Wernicke's cortical area lesion. *Archives of Neurology, 47,* 425–432.

Naeser, M. A., & Hayward, R. W. (1978). Lesion localization in aphasia with cranial computed tomography and the Boston Diagnostic Aphasia Exam. *Neurology, 28,* 545–551.

Naeser, M. A., Helm-Estabrooks, N., Haas, G., Auerbach, S., & Srinivasan, M. (1987). Relationship between lesion extent in "Wernicke's area" on computed tomographic scan and predicting recovery of comprehension in Wernicke's aphasia. *Archives of Neurology, 44,* 73–82.

Naeser, M. A., & Palumbo, C. L. (1994). Neuroimaging and language recovery in stroke. *Journal of Clinical Neurophysiology, 11*(2), 150–174.

Naeser, M. A., & Palumbo, C. L. (1995). How to analyze CT/MRI scan lesion sites to predict potential for long-term recovery in aphasia. In H. S. Gardner (Ed.), *Handbook of neurological speech and language disorders* (pp. 91–148). New York: Marcel Dekker.

Naeser, M. A., Palumbo, C. L., Helm-Estabrooks, N., Stiassny-Eder, D., & Albert, M. L. (1989). Severe nonfluency in aphasia. Role of the medial subcallosal fasciculus and other white matter pathways in the recovery of spontaneous speech. *Brain, 112,* 1–38.

Nagy, W. E., & Herman, P. A. (1987). Breadth and depth of vocabulary knowledge: Implications for acquisition and instruction. In M. G. McKeown & M. E. Curtis (Eds.), *The nature of vocabulary acquisition* (pp. 19–35). Hillsdale, NJ: Lawrence Erlbaum.

Nakasoto, N., Seki, K., Kawamura, T., Fujita, S., Kanno, A., Fujiwara, S., & Yoshimoto, T. (1996). Functional brain mapping using an MRI-linked whole head magnetoencephalography (MEG) system. In: C. Barber, G. Celecia, G. C. Comi, & F. Mauguiere (Eds.) *Functional Neuroscience* (EEG Suppl. 46). Amsterdam: Elsevier, (pp. 119–126)

Natsopoulos, D., Katsarou, Z., Bostantzopoulou, S., Grouios, G., Mentenopoulos, G., & Logothetis, J. (1991). Strategies in comprehension of relative clauses by Parkinsonian patients. *Cortex, 27,* 255–268.

Natsopoulos, D., Mentenopoulos, G., Bostantzopoulou, S., Katsarou, Z., Grouios, G., & Logothetis, J. (1991). Understanding of relational terms *before* and *after* in Parkinsonian patients. *Brain and Language, 40,* 444–458.

Neary, D. (1995). Neuropsychological aspects of frontotemporal degeneration. In J. Grafman, K. Holyoak, & F. Boller (Eds.), *Structure and functions of the human prefrontal cortex* (pp. 15–22). New York: Annals of the New York Academy of Sciences, 769.

Neary, D., Snowden, J. S., Bowen, D. M., Sims, N. R., Mann, D. M. A., Benton, J. S., Northen, B., Yates, P. O., & Davison, A. N. (1986). Neuropsychological syndromes in presenile dementia due to cerebral atrophy. *Journal of Neurology, Neurosurgery, and Psychiatry, 49,* 163–174.

Nebes, R. D., & Sperry, R. W. (1971). Hemispheric disconnection syndrome with cerebral birth injury in the dominant arm area. *Neuropsychologia, 9,* 247–259.

Neely, J. H. (1977). Semantic priming and retrieval from lexical memory: Roles of inhibitionless spreading activation and limited-capacity attention. *Journal of Experimental Psychology: General, 106,* 226–254.

Neely, J. H. (1991). Semantic priming effects in visual word recognition: A selective review of current findings and theories. In D. Besner & G. W. Humphreys (Eds.), *Basic processes in reading: Visual word recognition* (pp. 264–336). Hillsdale, NJ: Lawrence Erlbaum.

Neils, J., Baris, J. M., Carter, C., Dell'Aira, A. L., Nordloh, S. J., Weiler, E., & Weisiger, B. (1995). Effects of age, education, and living environment on Boston Naming Test performance. *Journal of Speech and Hearing Research, 38,* 1143–1149.

Neils, J., Roeltgen, D. P., & Greer, A. (1995). Spelling and attention in early Alzheimer's disease: Evidence for impairment of the graphemic buffer. *Brain and Language, 49,* 241–262.

Nemiah, J. C., Freyberger, H., & Sifneos, P. E. (1976). Alexithymia: A view of the psychosomatic process. In O. W. Hill (Ed.), *Modern trends in psychosomatic medicine* (pp. 430–439). London: Butterworth.

Nenov, V. I., Halgren, E., Smith, M. E., Badier, J.-M., Ropchan, J., Blahd, W. H., & Mandelkern, M. (1991). Localized brain metabolic response correlated with potentials evoked by words. *Behavioral Brain Research, 44,* 101–104.

Nespoulos, J.-L., & Dordain, M. (1991). Variability, attentional factors and the processing of grammatical morphemes in sentence production by an agrammatic patient. In J. Tesak (Ed.), *Grazer Linguistische Studien 35: Neuro- und Patholinguistik* (pp. 33–63). Graz: Institut für Sprachwissenschaft der Universität Graz.

Nespoulous, J.-L., Dordain, M., Perron, C., Ska, B., Bub, D., Caplan, D., Mehler, J., & Lecours, A. R. (1988). Agrammatism in sentence production without comprehension deficits: Reduced availability of syntactic structures and/or of grammatical morphemes. A case study. *Brain and Language, 33,* 273–295.

Netley, C. (1983). Sex chromosome abnormalities and the development of verbal and nonverbal abilities. In C. Cooper & J. Ludlow (Eds.), *Genetic aspects of speech and language disorders* (pp 179–196). New York: Academic Press.

Nettleton, J., & Lesser, R. (1991). Therapy for naming difficulties in aphasia: Application of a cognitive neuropsychological model. *Journal of Neurolinguistics, 6,* 139–157.

Neuberger, M. (1897/1981). *The historical development of experimental brain and spinal cord physiology before Flourens.* Baltimore: Johns Hopkins University Press.

Neville, H. J. (1990). Intermodal competition and compensation in development: Evidence from studies of the visual system in congenitally deaf adults. *Annals of the New York Academy of Sciences, 608,* 71–87.

Neville, H., Coffey, S., Lawson, D., Fischer, A., Emmorey, K., & Bellugi, U. (1997). Neural systems mediating American Sign Language: Effects of sensory experience and age of acquisition. *Brain and Language, 57,* 285–308.

Neville, H., Corina, D., Bavelier, D., Clark, V. P., Jezzard, P., Prinster, A., Padmanhaban, S., Braun, A., Rauschecker, J., & Turner, R. (1995). Effects of early experience on cerebral organization for language: An fMRI study of sentence processing in English and ASL by hearing and deaf subjects. *Human Brain Mapping* (Suppl. 1), 278.

Neville, H., Mills, D., & Lawson, D. (1992). Fractionating language: Different neural subsystems with different sensitive periods. *Cerebral Cortex, 2,* 244–258.

Neville, H., Nicol, J., Barss, A., Forster, K., & Garrett, M. (1991). Syntactically based sentence processing classes: Evidence from event-related brain potentials. *Journal of Cognitive Neuroscience, 3,* 152–165.

Neville, H., Schmidt, A., & Kutas, M. (1983). Altered visual evoked potentials in congenially deaf adults. *Brain Research, 266,* 127–132.

Newcombe, F., & Marshall, J. C. (1988). Idealisation meets psychometrics: The case for the right groups and the right individuals. *Cognitive Neuropsychology, 5,* 549–564.

Newport, E. L., & Meier, R. P. (1985). The acquisition of American Sign Language. In D. I. Slobin (Ed.), *The crosslinguistic study of language acquisition. Vol. 1, The data* (pp. 881–938). Hillsdale, NJ: Lawrence Erlbaum.

Newport, E. L., & Supalla, T. (1980). Clues from the acquisition of signed and spoken language. In U. Bellugi & M. Studdert-Kennedy (Eds.), *Signed and spoken language: Biological constraints on linguistic form* (pp. 187–211). Weinheim/Deerfield Beach, FL: Verlag Chemie.

Niccum, N., & Speaks, C. (1986). Longitudinal dichotic listening patterns for aphasics: Description of recovery curves. *Brian and Language, 28,* 273–288.

Niccum, N., & Speaks, C. (1991). Interpretation of outcome of dichotic listening tests following stroke. *Journal of Clinical and Experimental Neuropsychology, 13,* 614–628.

Nicholas, L. E. & Brookshire, R. H. (1992). A system for scoring main concepts in the discourse of non-brain-damaged and aphasic speakers. In M. L. Lemme (Ed.), *Clinical aphasiology* (Vol. 21, pp. 87–100). Austin, TX: Pro-Ed.

Nicholas, L. E., Brookshire, R. H., MacLennan, D. L., Schumacher, J. G., & Porrazzo, S. A. (1989). The Boston Naming Test: Revised administration and scoring procedures and normative information for non-brain-damaged adults. *Clinical Aphasiology, 18,* 103–115.

Nicholas, M., Obler, L. K., Albert, M. L., & Helm-Estabrooks, N. (1985). Empty speech in Alzheimer's disease and fluent aphasia. *Journal of Speech and Hearing Research, 28,* 405–410.

Nicholas, M. L., Helm-Eastabrooks, N., Ward-Lonergan, J., & Morgan, A. R. (1993). Evolution of severe aphasia in the first two years post-onset. *Archives of Physical Medicine and Rehabilitation, 74,* 830–836.

Nickels, L. (1992). The autocue? Self-generated phonemic cues in the treatment of reading and naming. *Cognitive Neuropsychology, 9,* 155–182.

Niemi, J., Laine, M., Hänninen, R., & Koivuselkä-Sallinen, P. (1990). Agrammatism in Finnish: Two case studies. In L. Menn & L. K. Obler (Eds.), *Agrammatic aphasia. A cross-language narrative sourcebook* (Vol. 2, pp. 1013–1085). Amsterdam: John Benjamins.

Nigam, A., Hoffman, J., & Simons, R. (1992). N400 to semantically anomalous pictures and words. *Journal of Cognitive Neuroscience, 4,* 15–22.

Nilipour, R. (1988). Bilingual aphasia in Iran: A preliminary report. *Journal of Neurolinguistics, 3,* 185–232.

Nilipour, R., & Ashayeri, H. (1989). Alternating antagonism between two languages with successive recovery of a third in a trilingual patient. *Brain and Language, 36* 23–48.

Nippold, M. A., Cuyler, J. S., & Braunbeck-Price, R. (1988). Explanation of ambiguous advertisements: A developmental study with children and adolescents. *Journal of Speech and Hearing Research, 31,* 466–474.

Nitrini, R., Lefèvre, B. H., Mathias, S. C., Caramelli, P., Carrilho, P. E., Sauaia, N., Massad, E., Takiguti, C., Silva, I. O., Porto, C. S., Magila, M. C., & Scaff, M. (1994). Testes neuropsicológicos de aplicação simples para o diagnóstico de demência [Neuropsychological tests of simple application for diagnosing dementia]. *Arquivos de Neuropsiquiatria, 52,* 457–465.

Nitrini, R., Mathias, S. C., Caramelli, P., Carrilho, P. E., Lefèvre, B.H., Porto, C. S., Magila, M. C., Buchpiguel, C., Barros, N. G., Gualandro, S., Bacheschi, L. A., & Scaff, M. (1995). Evaluation of 100 patients with dementia in São Paulo, Brazil: Correlations with socioeconomic status and education. *Alzheimer Disease and Associated Disorders, 9,* 146–151.

Niznikiewicz, M. A., Nestor, P. G., O'Donnell, B. F., Allard, J. E., Shenton, M. E., & McCarley, R. W. (1996). Working memory as a factor in language dysfunction in schizophrenia. *Biological Psychiatry, 39,* 569.

Nobre, A., & McCarthy, G. (1994). Language-related ERPs: Scalp distributions and modulation by word type and semantic priming. *Journal of Cognitive Neuroscience, 6,* 233–255.

Nobre, A. C., & McCarthy, G. (1995). Language-related field potentials in the anterior-medial temporal lobe: II. Effects of word type and semantic priming. *Journal of Neuroscience, 15,* 1090–1099.

Norman, D. A., & Shallice, T. (1986). Attention to action: Willed and automatic control of behavior. In R. J. Davidson, G. E. Schwartz, & D. Shapiro (Eds.), *Consciousness and self-regulation: Advances in Research* (Vol. 4; pp. 1–18). New York: Plenum Press.

Novak, G. P., Kurtzberg, D., Kreuzer, J. A., & Vaughan, H. G. (1989). Cortical responses to speech sounds and their formants in normal infants: Maturational sequence and spatiotemporal analysis. *Electroencephalography and Clinical Neuropsychology, 73,* 295–305.

Novosel, T., & Roth, V. M. (1997). MODAKT97 (FAX +49 7533 7814).

Ober, B. A., Shenaut, G. K., & Vinogradov, S. (1995, March). Semantic priming in paranoid versus non-paranoid schizophrenics. Poster presented at Cognitive Neuroscience Society Annual Meeting, San Francisco, CA.

Obler, L., Fein, D., Nicholas, M., & Albert, M. (1991). Auditory comprehension and aging: Decline in syntactic processing. *Applied Psycholinguistics, 12,* 433–452.

Obler, L. K., & Albert, M. (1984). Language in aging. In M. Albert (Ed.), *Clinical neurology of aging* (pp. 245–253). New York: Oxford University Press.

Obler, L. K., Albert, M., Goodglass, H., & Benson, F. (1978). Aphasia type and aging. *Brain and Language, 6,* 318–322.

Obrzut, J. E., Obrzut, A., Bryden, M. P., & Bartels, S. G. (1985). Information processing and speech lateralization in learning-disabled children. *Brain and Language, 25,* 87–101.

Odor, J. P. (1988). Student models in machine-mediated learning. *Journal of Mental Deficiency Research, 32,* 247–256.

Ogawa, S., Lee, T. M., Kay, A. R., & Tank, D. W. (1990). Brain magnetic resonance imaging with contrast dependent on blood oxygenation. *Proceedings of the National Academy of Sciences of the U.S.A., 87,* 9868–9872.

Ogden, J. A. (1988). Language and memory functions after language recovery periods in left-hemispherectomized subjects. *Neuropsychologia, 26,* 645–659.

Ojemann, G. A. (1975). Language and the thalamus: Object naming and recall during and after thalamic stimulation. *Brain and Language, 2,* 101–120.

Ojemann, G. A. (1983). Brain organization for language from the perspective of electrical stimulation mapping. *The Behavioral and Brain Sciences, 6,* 189–230.

Ojemann, G. A., Creutzfeldt, O., Lettich, E., & Haglund, M. M. (1988). Neuronal activity in human lateral temporal cortex related to short-term verbal memory, naming and reading. *Brain, 111,* 1383–1403.

Ojemann, G. A., Fedio, P., & Van Buren, J. (1968). Anomia from pulvinar and subcortical parietal stimulation. *Brain, 91,* 99–116.

Ojemann, G. A., & Mateer, C. (1979). Human language cortex: Localization of memory, syntax and sequential motor-phoneme identification systems. *Science, 205,* 1401–1403.

Ojemann, G. A., Ojemann, J., Lettich, E., & Berger, M. (1989). Cortical language localization in left, dominant hemisphere: An electrical stimulation mapping investigation in 117 patients. *Journal of Neurosurgery, 71,*316–326.

Ojemann, G. A., & Whitaker, H. A. (1978a). The bilingual brain. *Archives of Neurology, 35,* 409–412.

Ojemann, G. A., & Whitaker, H. A. (1978b). Language localization and variability. *Brain and Language, 6,* 239–260.

Olivares, E., Bobes, M. A., Aubert, E., & Valdés-Sosa, M. (1994). Associative effects with memories of artificial faces. *Cognitive Brain Research, 2,* 39–48.

Olson, A., & Caramazza, A. (1991). The role of cognitive theory in neuropsychological research. In F. Boller and J. Grafman (Eds.), *Handbook of neuropsychology* (Vol. 5, pp. 287–309). Amsterdam: Elsevier.

O'Neill, Y. (1980). *Speech and speech disorders in Western thought before 1600.* Westport, CT: Greenwood Press.

Onghena, P., & Edgington, E. S. (1994). Randomization tests for restricted alternating treatment designs. *Behavioral Research and Therapy, 32,* 783–786.

Osgood, C. E. (1980). *Lectures on language performance.* New York: Springer Verlag.

Osherson, D. N., & Smith, E. E. (1981). On the adequacy of prototype theory as a theory of concepts. *Cognition, 9,* 35–58.

Osterhout, L. (1994). Event-related brain potentials as tools for comprehending language comprehension. In C. Clifton & L. Frazier (Eds.), *Perspectives on sentence processing* (pp. 15–44). Hillsdale, NJ: Lawrence Erlbaum.

Osterhout, L. (in press). On the brain response to syntactic anomalies: Manipulations of word position and word class reveal individual differences. *Brain and Language.*

Osterhout, L., & Holcomb, P. (1992). Event-related potentials elicited by syntactic anomaly. *Journal of Memory and Language, 31,* 785–806.

Osterhout, L., & Holcomb, P. (1993). Event-related potentials and syntactic anomaly: Evidence of anomaly detection during the perception of continuous speech. *Language and Cognitive Processes, 8,* 413–437.

Osterhout, L., & Holcomb, P. (1995). Event-related potentials and language comprehension. In M. D. Rugg & M. G. H. Coles (Eds.), *Electrophysiology of mind* (pp. 171–215). Oxford: Oxford University Press.

Osterhout, L., Holcomb, P., & Swinney, D. (1994). Brain potentials elicited by garden-path sentences: Evidence of the application of verb information during parsing. *Journal of Experimental Psychology: Learning, Memory, and Cognition, 20,* 786–803.

Ostrin, R. K., & Tyler, L. K. (1993). Automatic access to lexical semantics in aphasia: Evidence from semantic and associative priming. *Brain and Language, 45,* 147–159.

Otero, E., Cordova, S., Diaz, F., Garcia-Teruel, I., & Del Brutto, O. H. (1989). Acquired epileptic aphasia (the Landau-Kleffner syndrome) due to neurocysticercosis. *Epilepsia, 30,* 569–572.

Ouellette, G., & Baum, S. (1994). Acoustic analysis of prosodic cues in left- and right-hemisphere damaged patients. *Aphasiology, 8,* 257–283.

Ozonoff, S., Rogers, S. J., & Pennington, B. F. (1991). Asperger's syndrome: Evidence of an empirical distinction from high-functioning autism. *Journal of Child Psychology and Psychiatry, 32,* 1107–1122.

Packard, J. (1986). Tone production deficits in non-fluent aphasic Chinese speech. *Brain and Language, 29,* 212–223.

Packard, J. (1993). *A linguistic investigation of aphasic Chinese speech.* Boston: Kluwer Academic Publishers.

Padden, C., & Humphries, T. (1988). *Deaf in America: Voices from a culture.* Cambridge: Harvard University Press.

Paetau, R., Kajola, M., Korkman, M., Hamalainen, M., Granstrom, M., & Hari, R. (1991). Landau-Kleffner syndrome: Epileptic activity in the auditory cortex. *Neuroreport, 2,* 201–204.

Paivio, A. (1991). Dual coding theory: Retrospect and current status. *Canadian Journal of Psychology, 45,* 255–287.

Pandya, D. N., Seltzer, B., & Barbas, H. (1988). Input-output organization of the primate cerebral cortex. *Comparative Primate Biology, 4,* 39–80.

Pandya, D. N., & Yeterian, E. H. (1990). Prefrontal cortex in relation to other cortical areas in rhesus monkey: Architecture and connections. *Progress in Brain Research, 85,* 63–94.

Pantev, C., Hoke, M., Lehnertz, K., Lütkenhönen, B., Anogianakis, G., & Wittkowski, W. (1988). Tonotopic organization of the human auditory cortex revealed by transient auditory evoked fields. *Electroencephalography and Clinical Neurophysiology, 69,* 160–170.

Papagno, C., & Basso, A. (1993). Impairment of written language and mathematical skills in a case of Landau-Kleffner syndrome. *Aphasiology, 7,* 451–462.

Papanicolaou, A. C., DiScenna, A., Gillespie, L., & Aram, D. (1990). Probe-evoked potential findings following unilateral left-hemisphere lesions in children. *Archives of Neurology, 47,* 562–566.

Papanicolaou, A. C., & Johnstone, J. (1984). Probe evoked potentials: Theory, method and applications. *International Journal of Neuroscience, 24,* 107–131.

Papanicolaou, A. C., Moore, B. D., Deutsch, G., Levin, H. S., & Eisenberg, H. M. (1988). Evidence for right-hemisphere involvement in recovery from aphasia. *Archives of Neurology, 45,* 1025–1029.

Papanicolaou, A. C., Rogers, R. L., Baumann, S. B., Saydjari, C., & Eisenberg, H. M. (1990). Source localization of two evoked magnetic field components using two alternative procedures. *Experimental Brain Research, 80,* 44–48.

Papez, J. W. (1937). A proposed mechanism of emotion. *Archives of Neurology and Psychiatry, 38,* 725–743.

Paquier, P. F., Van Dongen, H. R., & Loonen, M. C. B. (1992). The Landau-Kleffner syndrome or 'Acquired Aphasia with Convulsive Disorder': Long-term follow-up of six children and a review of the recent literature. *Archives of Neurology, 49,* 354–359.

Paradis, M. (1977). Bilingualism and aphasia. In H. Whitaker and H. A. Whitaker (Eds.), *Studies in neurolinguistics* (Vol. 3, pp. 65–121). New York: Academic Press.

Paradis, M. (1987a). Neurofunctional modularity of cognitive skills: Evidence from Japanese alexia and polyglot aphasia. In E. Keller & M. Gopnik (Eds.), *Motor and sensory processes of language* (pp. 277–289). Hillsdale: NJ Lawrence Erlbaum.

Paradis, M. (1987b). Neurolinguistic perspectives on bilingualism. In M. Paradis and G. Libben, *The assessment of bilingual aphasia* (pp. 1–17). Hillsdale: NJ Lawrence Erlbaum.

Paradis, M. (1993a). Acquired organic pathologies of language behavior: Neurolinguistic disorders. In G. Blanken, J. Dittmann, H. Grimm, J. C, Marshall & C. W. Wallesch (Eds.), *Linguisitic Disorders and Pathologies,* (Vol. 9, pp. 278–288). Berlin: Walter de Gruyter.

Paradis, M. (1993b). Linguistic, psycholinguistic, and neurolinguistic aspects of "interference" in bilingual speakers: The Activation Threshold Hypothesis. *International Journal of Psycholinguistics, 9,* 133–145.

Paradis, M. (1994). Neurolinguistic aspects of implicit and explicit memory: Implications for bilingualism and SLA. In N. Ellis (Ed.), *Implicit and explicit learning of languages* (pp. 393–419). London: Academic Press.

Paradis, M., & Goldblum, M.-C (1989). Alternate antagonism with paradoxical translation behavior in two bilingual aphasic patients. *Brain and Language, 15,* 55–69.

Paradis, M., Goldblum, M.-C., & Abidi, R. (1982). Selective crossed aphasia in a trilingual aphasic patient followed by reciprocal antagonism. *Brain and Language, 36,* 62–75.

Paradis, M., & Gopnik, M. (1994). Compensatory strategies in familial language impairment. *McGill Working Papers in Linguistics, 10,* 143–149.

Paradis, M., & Lebben, G. (1987). *The assessment of bilingual aphasia.* Hillsdale, NJ: Lawrence Erlbaum.

Pardo, P. J. & Sams, M. (1993). Human auditory cortex responses to rising versus falling glides. *Neuroscience Letters, 159,* 43–45.

Parisi, D. (1985). A procedural approach to the study of aphasia. *Brain and Language, 26,* 1–15.

Parkinson, J. (1817). *An essay on the shaking palsy.* London: Whittingham and Rowland for Sherwood, Neely and Jones.

Pascual-Castroviejo, I., Lopez-Martin, V., Martinez-Bermejo, A., & Perez-Higueras, A. (1992). Is cerebral arteritis the cause of the Landau-Kleffner syndrome? Four cases in childhood with angiographic study. *Canadian Journal of Neurological Sciences, 19,* 46–52.

Pashek, G. V., & Holland, A. L. (1988). Evolution of aphasia in the first year post-onset. *Cortex, 24,* 411–423.

Pasquier, F., Lebert, F., Grymonprez, L., & Petit, H. (1995). Verbal fluency in dementia of frontal lobe type and dementia of the Alzheimer type. *Journal of Neurology, Neurosurgery, and Psychiatry, 58,* 81–84.

Patel, P. G., & Satz, P. (1994). The language production system and senile dementia of Alzheimer's type: Neuropathological implications. *Aphasiology, 8,* 1–18.

Patry, G., Lyagoubi, S., & Tassinari, C. A. (1971). Subclinical 'electrical status epilepticus' induced by sleep in children: A clinical and electroencephalographic study of six cases. *Archives of Neurology, 24,* 242–252.

Patry, R., & Nespoulous, J.-L. (1990). Discourse analysis in linguistics: Historical and theoretical background. In Y. Joanette & H. H. Brownell (Eds.), *Discourse ability and brain damage: Theoretical and empirical perspectives* (pp. 3–27). New York: Springer Verlag.

Patterson, K. (1986). Lexical but non-semantic spelling? *Cognitive Neuropsychology, 3,* 341–367.

Patterson, K. E. (1982). The relation between reading and phonological coding: Further neuropsychological observations. In A. W. Ellis (Ed.), *Normality and pathology in cognitive functions* (pp. 77–111). London: Academic Press.

Patterson, K. E. (1994). Reading, writing and rehabilitation: A reckoning. In M. J. Riddoch & G. W. Humphreys (Eds.), *Cognitive neuropsychology and cognitive rehabilitation.* Hillsdale, NJ: Lawrence Erlbaum.

Patterson, K. E., Marshall, J. C., & Coltheart, M. (Eds.). (1985). *Surface dyslexia.* London: Lawrence Erlbaum.

Paulesu, E., Connelly, A., Frith, C. D., Friston, K. J., Heather, J., Meyers, R., Gadian, D. G., & Frackowiak, R. S. J. (1995). Functional MRI correlations with positron emission tomography. Initial experience using a cognitive activation paradigms on verbal working memory. *Neuroimaging Clinics of North America, 5,* 207–225.

Paulesu, E., Frith, C. D., & Frackowiak, R. S. J. (1993). The neural correlates of the verbal component of working memory. *Nature, 362,* 342–344.

Pearce, S., McDonald, S., & Coltheart, M. (1995). Interpreting ambiguous advertisements: The effect of frontal lobe damage. Abstracts of the 2nd INS Pacific Rim Conference, Cairns, Australia. *Journal of the International Neuropsychological Society, 1,* 321.

Pedersen, P. M., Joergensen, H. S., Nakayama, H., Raaschou, H. O., & Skyhoj Olsen, T. (1995). Aphasia in acute stroke: Incidence, determinants and recovery. *Annals of Neurology, 38,* 659–666.

Pell, M. D., & Baum, S. R. (1997). The ability to perceive and comprehend intonation in linguistic and affective contexts by brain-damaged adults. In D. Van Lancker (Ed.), Special issue of *Brain and Language,* "Current studies of right hemisphere function." Vol. 57. pp 80–99.

Penfield, W. (1967). *The excitable cortex in conscious man.* Liverpool: Liverpool University Press.

Penfield, W., & Erickson, T. (1941). *Epilepsy and cerebral localization.* Springfield, IL: C. C. Thomas.

Penfield, W., & Jasper, H. (1954). *Epilepsy and the functional anatomy of the human brain.* Boston: Little, Brown.

Penfield, W., & Perot, P. (1963). The brain's record of auditory and visual experience—a final summary and discussion. *Brain, 86,* 595–696.

Penfield, W., & Roberts, L. (1959). *Speech and brain mechanisms.* Princeton, NJ: Princeton University Press.

Penn, C., & Cleary, J. (1988). Compensatory strategies in the language of closed head injured patients. *Brain Injury, 2,* 3–17.

Pennington, B. F. (1991). *Diagnosing learning disorders: A neuropsychological framework.* New York: Guilford Press.

Peper, M., & Irle, E. (1997). Categorial and dimensional decoding of emotional intonations in patients with focal brain lesions. *Brain and Language, 58,* 233–264.

Perani, D., Papagno, C., Cappa, S. F., Gerundini, P., & Fazio, F. (1988). Crossed aphasia: An investigation with single photon emission computerized tomography. *Cortex, 24,* 171–178.

Perecman, E. (Ed.) (1983). *Cognitive processing in the right hemisphere.* New York: Academic Press.

Perez-Abalo, M. C., Rodriguez, R., Bobes, M. A., Gutierrez, J., & Valdés-Sosa, M. (1994). Brain potentials and the availability of semantic and phonological codes over time. *NeuroReport, 5,* 2173–2177.

Perkins, J., Baran, J., & Gandour, J. (1996). Hemispheric specialization in processing intonation contours. *Aphasiology, 10,* 343–362.

Perkins, L. (1995). Applying conversation analysis to aphasia: Clinical implications and analytic issues. *European Journal of Disorders of Communication, 30,* 372–383.

Perniola, T., Margari, L., Buttiglione, M., Andreula, C., Simone, I. L., & Santostasi, R. (1993). A case of Landau-Kleffner syndrome secondary to inflammatory demyelinating disease. *Epilepsia, 34,* 551–556.

Peters, M. (1988). The size of the corpus callosum in males and females: Implications of a lack of allometry. *Canadian Journal of Psychology, 42,* 313–324.

Petersen, E., Fox, P. T., Posner, M. I., Mintun, M., & Raichle, M. E. (1988). Positron emission tomographic studies of the cortical anatomy of single-word processing. *Nature, 331,* 585–589.

Petersen, S. E., Fox, P. T., Posner, M. I., Mintun, M., & Raichle, M. E. (1988). Positron emission tomographic studies of the cortical anatomy of single-word processing. *Nature, 331,* 585–589.

Petersen, S. E., Fox, P. T., Posner, M. I., Mintun, M., & Raichle, M. E. (1989). Positron emission tomographic studies of the processing of single words. *Journal of Cognitive Neurosciences, 1,* 153–170.

Petersen, S. E., Fox, P. T., Snyder, A. Z., & Raichle, M. E. (1990). Activation of extrastriate and frontal cortical areas by visual words and word-like stimuli. *Science, 249,* 1041–1044.

Petheram, B. (1996). Exploring the home-based use of microcomputers in aphasia therapy. *Aphasiology, 10,* 267–282.

Petit, J. M., & Noll, J. D. (1979). Cerebral dominance in aphasia recovery. *Brian and Language, 7,* 191–200.

Petry, M. C., Crosson, B., Gonzalez-Rothi, L. J., Bauer, R. M., & Schauer, C. A. (1994). Selective attention and aphasia in adults: Preliminary findings. *Neuropsychologia, 32,* 1397–1408.

Phaf, R. H., Van der Heijden, A. H. C., & Hudson, P. T. (1990). SLAM: A connectionist model for attention in visual selection tasks. *Cognitive Neuropsychology, 22,* 273–341.

Piaget, J., & Inhelder, B. (1969). *The psychology of the child.* New York: Basic Books.

Pick, A. (1913). *Die agrammatischen Sprachstörungen.* Berlin: Springer Verlag.

Pick, A. (1971). *Aphasia* (Trans. J. W. Brown). Springfield, IL: C. C. Thomas. (Original German edition published in 1931.)

Pickersgill, M. J., & Lincoln, N. B. (1983). Prognostic indicators and the pattern of recovery of communication in aphasic stroke patients. *Journal of Neurology, Neurosurgery, and Psychiatry, 46,* 130–139.

Picton, T., & Hillyard, S. (1988). Endogenous event-related potentials. In T. W. Picton (Ed.), *Handbook of EEG and clinical neurophysiology.* Vol. 3., *Human event-related potentials* (pp. 361–426). New York: Elsevier Science.

Picton, T., Lins, O., & Sherg, M. (1995). The recording and analysis of event-related potentials. In F. Boller & J. Grafman (Eds.), *Handbook of neuropsychology* (Vol. 10, pp. 3–73). Amsterdam: Elsevier.

Pillon, B., Blin, J., Vidailhet, M., Deweer, B., Sirigu, A., Dubois, B., & Agid, Y. (1995). The neuropsychological pattern of corticobasal degeneration: Comparison with progressive supranuclear palsy and Alzheimer's disease. *Neurology, 45,* 1477–1483.

Pillon, B., Dubois, B., Lhermitte, F., & Agid, Y. (1986). Heterogeneity of cognitive impairment in progressive supranuclear palsy, Parkison's disease, and Alzheimer's disease. *Neurology, 36,* 1179–1185.

Pinker, S. (1989). *Learnability and cognition: The acquisition of argument structure.* Cambridge: MIT Press.

Pinker, S. (1994). *The language instinct: How the mind creates language.* New York: Morrow and Company.

Pirozzolo, F. J., Hansch, E. C., Mortimer, J. A., Webster, D. D., & Kuskowski, M. A. (1982). Dementia in Parkinson's disease: A neuropsychological analysis. *Brain and Cognition, 1,* 71–83.

Pitman, E. J. G. (1937). Significance tests which may be applied to samples from any populations. *Journal of the Royal Statistical Society, Series B 4,* 119–130.

Pitman, E. J. G. (1938). Significance tests which may be applied to samples from any populations. III. The analysis of variance test. *Biometrika, 29,* 322–335.

Pitres, A. (1895). Étude de l'aphasie chez les polyglottes. *Revue de Médecine, 15,* 873–899.

Pitres, A. (1898a). L'aphasie amnésique et ses variétés cliniques. *Le Progrès Médical, 21,* 321–324.

Pitres, A. (1898b). L'aphasie amnésique et ses variétés cliniques. *Le Progrès Médical, 28,* 18–23.

Pittam, J., & Scherer, K. R. (1993). Vocal expression and communication of emotion. In M. Lewis & J. M. Haviland (Eds.), *Handbook of emotions* (pp. 185–198). New York: Guilford Press.

Pizzamiglio, L., Mammucari, A., & Razzano, C. (1985). Evidence for sex differences in brain organization in recovery from aphasia. *Brain and Language, 25,* 213–223.

Plaut, D. (1996). Relearning after damage in connectionist networks: Toward a theory of rehabilitation. *Brain and Language, 52,* 25–82.

Plaut, D. C., & Shallice, T. (1993a). Deep dyslexia: A case study of connectionist neuropsychology. *Cognitive Neuropsychology, 10,* 377–500.

Plaut, D. C., & Shallice, T. (1993b). Perseverative and semantic influences on visual object naming errors in optic aphasia: A connectionist account. *Journal of Cognitive Neuroscience, 5,* 89–117.

Ploog, D. W. (1981). Neurobiology of primate audio-vocal behavior. *Brain Research Reviews, 3,* 35–61.

Ploog, D. W. (1992). Neuroethological perspectives on the human brain: From the expression of emotions to intentional signing and speech. In A. Harrington (Ed.), *So human a brain: Knowledge and values in the neurosciences* (pp. 3–13). Boston: Birkhauser.

Podoll, K., Caspary, P., Lange, H. W., & Noth, J. (1988). Language functions in Huntington's disease. *Brain, 111,* 1475–1503.

Podoll, K., Schwarz, M., & Noth, J. (1991). Language functions in progressive supranuclear palsy. *Brain, 114,* 1457–1472.

Poeck, K. (1982). *Klinische Neuropsychologie.* Stuttgart: Georg Thieme Verlag.

Poeck, K. (1983). What do we mean by "aphasic syndromes"? A neurologist's view. *Brain and Language, 20,* 79–89.

Poeck, K., de Bleser, R., & von Keyserlingk, D. G. (1984). Computed tomography localization of standard aphasic syndromes. In F. C. Rose (Ed.), *Advances in neurology.* Vol. 42, *Progress in aphasiology* (pp. 71–89). New York: Raven Press.

Poeck, K., & Luzzatti, C. (1988). Slowly progressive aphasia in three patients. *Brain, 111,* 151–168.

Poirier, J., Gray, F., & Escourolle, R. (1990). *Manual of basic neuropathology* (3rd ed.) (Trans. L. J. Rubinstein). Philadelphia: W. B. Saunders. (Original work published 1971.)

Poizner, H., & Battison, R. (1980). Cerebral asymmetry for sign language: Clinical and experimental evidence. In H. Lane & F. Grosjean (Eds.), *Recent perspectives on American Sign Language* (pp. 79–101). Hillsdale, NJ: Lawrence Erlbaum.

Poizner, H., & Kegl, J. (1992). Neural basis of language and motor behavior: Perspectives from American Sign Language. *Aphasiology, 6,* 219–256.

Poizner, H., Klima, E. S., & Bellugi, U. (1987). *What the hands reveal about the brain.* Cambridge: MIT Press.

Polich, J. (1985). Semantic categorization and event-related potentials. *Brain and Language, 26,* 304–321.

Polich, J. (1993). Cognitive brain potentials. *Current Directions in Psychological Science, 2,* 175–179.

Poline, J.-B., & Mazoyer, B. M. (1993). Analysis of individual positron emission tomography activation maps by detection of high signal-to-noise-ratio pixel clusters. *Journal of Cerebral Blood Flow and Metabolism, 13,* 425–437.

Pollatsek, A., Bolozky, S., Well, A. D., & Rayner, K. (1981). Asymmetries in perceptual span for Israeli readers. *Brain and Language, 14,* 174–180.

Pollatsek, A., Raney, G. E., Lagasse, L., & Rayner, K. (1993). The use of information below fixation in reading and visual search. *Canadian Journal of Experimental Psychology, 47,* 179–200.

Pollock, J.-Y. (1989). Verb movement, universal grammar and the structure of IP. *Linguistic Inquiry, 20,* 365–424.

Poncet, M., & Habib, M. (1994). Atteinte isolée des comportements motivés et lésions des noyaux gris centraux. *Revue Neurologique, 150,* 588–593.

Porch, B. E. (1971). *Porch Index of Communicative Ability. Vol. 1, Theory and development.* Palo Alto, CA: Consulting Psychologists Press.

Porch, B. E. (1981). *Porch Index of Communicative Ability. Vol. 2, Administration, scoring, and interpretation* (3rd ed.). Palo Alto, CA. Consulting Psychologists Press.

Porter, R. J., Jr., & Berlin, C. I. (1975). On interpreting developmental changes in the dichotic right-ear advantage. *Brain and Language, 2,* 186–200.

Posner, M. I. (1978). Chronometric explorations of mind. Hillsdale, NJ: Lawrence Erlbaum.

Posner, M. I. (1995). Attention in cognitive neuroscience: An overview. In M. S. Gazzaniga (Ed.), *The cognitive neurosciences* (pp. 615–624). Cambridge: MIT Press.

Posner, M. I., Early, T. S., Reiman, E., Pardo, P. J., & Dhawan, M. (1988). Asymmetries in hemispheric control of attention in schizophrenia. *Archives of General Psychiatry, 45,* 814–821.

Posner, M. I., & Mitchell, R. F. (1967). Chronometric analysis of classification. *Psychological Review, 74,* 392–409.

Posner, M. I., & Peterson, S. E. (1990). The attention system of the human brain. *Annual Review of Neuroscience, 13,* 25–42.

Posner, M. I., Petersen, S. E., Fox, P. T., & Raichle, M. E. (1988). Localization of cognitive operations in the human brain. *Science, 240,* 1627–1631.

Posner, M. I., & Raichle, M. E. (1994). *Images of mind.* New York: Scientific American Library.

Posner, M. I., Sandson, J., Dhawan, M., & Shulman, G. L. (1989). Is word recognition automatic? A cognitive-anatomical approach. *Journal of Cognitive Neuroscience, 1,* 50–60.

Powell, A. L., Cummings, J. L., Hill, M. A., & Benson, D. F. (1988). Speech and language alterations in multi-infarct dementia. *Neurology, 38,* 717–719.

Praamstra, P., Meyer, A., & Levelt, W. (1994). Neurophysiological manifestations of phonological processing: Latency variation of a negative ERP component timelocked to phonological mismatch. *Journal of Cognitive Neuroscience, 6,* 204–219.

Praamstra, P., & Stegeman, D. G. (1993). Phonological effects on the auditory N400 event-related brain potential. *Cognitive Brain Research, 1,* 73–86.

Preissl, H., Pulvermüller, F., Lutzenberger, W., & Birbaumer, N. (1995). Evoked potentials distinguish between nouns and verbs. *Neuroscience Letters, 197,* 81–83.

Previc, F. (1991). A general theory concerning the prenatal origins of cerebral lateralization in humans. *Psychological Review, 98,* 299–334.

Previc, F. (1994). Assessing the legacy of the GBG model. *Brain and Cognition, 26,* 174–180.

Pribram, K. H. (1960). A review of theory in physiological psychology. *Annual Review of Psychology, 11,* 1–40.

Pribram, K. H. (1991). *Brain and perception: Holonomy and structure in figural processing.* Hillsdale, NJ: Lawrence Erlbaum.

Pribram, K. H., & McGuinness, D. (1975). Arousal, activation, and effort in the control of attention. *Psychological Review, 82,* 6–149.

Price, C., Wise, R. J., Ramsay, S., Friston, K., Howard, D., Patterson, K., & Frackowiak, R. S. J. (1992). Regional response differences within the human auditory cortex when listening to words. *Neuroscience Letters, 146,* 179–182.

Price, C., Wise, R. J. S., & Frackowiak, R. S. J. (1996). Demonstrating the implicit processing of visually presented words and pseudowords. *Cerebral Cortex, 6,* 62–70

Price, C. J., & Humphreys, G. W. (1993). Attentional dyslexia: The effects of co-occurring deficits. *Cognitive Neuropsychology, 10*, 569–592.

Price, C. J., Moore, C. J., & Frackowiak R. S. J. (1996). The effect of varying stimulus rate and duration on brain activity during reading. *NeuroImage, 3*, 40–52.

Price, C. J., Wise, R. J. S., Watson, J. D. G., Patterson, K., Howard, D., & Frackowiak R. S. J. (1994). Brain activity during reading: The effects of exposure duration and task. *Brain, 117*, 1255–1269.

Prigatano, G. P., Roueche, J. R., & Fordyce, D. J. (1986). *Neuropsychological rehabilitation after brain injury*. Baltimore: John Hopkins University Press.

Provins, K. A. (1992). Early infant motor asymmetries and handedness: A critical evaluation of the evidence. *Developmental Neuropsychology, 8*, 325–365.

Pruess, J. B., Vadasy, P. F., & Fewell, R. R. (1987). Language development in children with Down syndrome: An overview of recent research. *Education and Training in Mental Retardation, 22*, 44–55.

Prusiner, S. B. (1987). Prions and neurodegenerative diseases. *New England Journal of Medicine, 317*, 1571–1581.

Pulvermüller, F. (1996). Hebb's concept of cell assemblies and the psychophysiology of word processing. *Psychophysiology, 33*, 317–333.

Pulvermüller, F., Lutzenberger, W., & Birbaumer, N. (1995). Electrocortical distinction of vocabulary types. *Electroencephalography and Clinical Neurophysiology, 94*, 357–370.

Pulvermüller, F., & Preibl, H. (1991). A cell assembly model of language. *Network, 2*, 455–468.

Pulvermüller, F., Preissl, H., Lutzenberger, W., & Birbaumer, N. (1996). Brain rhythms of language: Nouns versus verbs. *European Journal of Neuroscience, 8*, 937–941.

Pulvermüller, F., & Roth, V. M. (1991). Communicative aphasia treatment as a further development of PACE therapy. *Aphasiology, 5*, 39–50.

Pulvermüller, F., & Roth, V. M. (1993). Integrative und computerunterstützte Aphasietherapie [Integrative and computer-assisted aphasia therapy]. In M. Grohnfeldt (Ed.), *Handbuch der Sprachtherapie: Zentrale Sprach- und Sprechstörungen 6* (pp. 230–250). Berlin: Wissenschaftsverlag Volker Spiess.

Pulvermüller, F., & Schumann, J. H. (1994). Neurobiological mechanisms of language acquisition. *Language Learning, 44*, 681–734.

Rafal, R., & Robertson, L. (1995). The neurology of visual attention. In M. S. Gazzaniga (Ed.), *The cognitive neurosciences* (pp. 625–648). Cambridge: MIT Press.

Rafal, R. D., & Grimm, R. K. (1981). Progressive supranuclear palsy: Functional analysis of the response to methysergide and antiparkinsonian agents. *Neurology, 31*, 1507–1518.

Ragin, A. B., & Oltmanns, T. F. (1986). Lexical cohesion and formal thought disorder during and after schizophrenic episodes. *Journal of Abnormal Psychology, 95*, 181–183.

Raichle, M. E., Fiez, J. A., Videen, T. O., MacLeod, A. K., Pardo, J. V., Fox, P. T., & Petersen, S. E. (1994). Practice-related changes in human brain functional anatomy during non-motor learning. *Cerebral Cortex, 4*, 8–26.

Ramier, A. M., & Hecaen, H. (1970). Rôle respectif des atteintes frontales et de la latéralisation fonctionnelle dans les déficits de la fluence verbale. *Revue Neurologique (Paris), 123*, 17–22.

Ramsay, D. S. (1983). Unimanual hand preference and duplicated syllable babbling in infants. In G. Young, S. J. Segalowitz, C. M. Corter, & S. E. Trehub (Eds.), *Manual specialization and the developing brain* (pp. 161–176). New York: Academic Press.

Ramsey, C. (1989). Language planning in deaf education. In U C. Lucas (Ed.), *Sociolinguistics of the deaf community* (pp. 123–146). San Diego: Academic Press.

Randolph, C., Goldberg, T. E., & Weinberger, D. R. (1993). In K. M. Heilman & E. Valenstein (Eds.), *Clinical neuropsychology* (3rd ed.; pp 499–522). New York: Oxford University Press.

Randolph, C., Lansing, A., Ivnik, R. J., Cullum, C. M., & Hermann, B. P. (1996). Determinants of confrontation naming performance. *Journal of the International Neuropsychological Society, 2*, 6.

Rangamani, G. N. (1989). Clinical evidence towards the cerebral organization of language. Ph.D. thesis, University of Mysore, India.

Raoul, P., Lieury, A., Decombe, R., Chauvel, P., & Allain, H. (1992). Déficit mnésique au cours de la maladie de Parkinson. Vieillissement accéléré des processus de rappel. *Presse Médicale, 21*, 69–73.

Rapcsak, S. Z., Arthur, S. A., Bliklen, D. A., & Rubens, A. B. (1989). Lexical agraphia in Alzheimer's disease. *Archives of Neurology, 46,* 65–68.

Rapcsak, S. Z., Comer, J. F., & Rubens, A. B. (1993). Anomia for facial expressions: Neuropsychological mechanisms and anatomical correlates. *Brain and Language, 45,* 233–252.

Rapcsak, S. Z., Kaszniak, A. W., & Rubens, A. B. (1989). Anomia for facial expressions: Evidence for a category specific visual-verbal disconnection syndrome. *Neuropsychologia, 27,* 1031–1041.

Rapcsak, S. Z., Krupp, L. B., Rubens, A. B., & Reim, J. (1990). Mixed transcortical aphasia without anatomic isolation of the speech area. *Stroke, 21,* 953–956.

Rapcsak, S. Z., & Rubens, A. B. (1990). Disruption of semantic influence on writing following a left prefrontal lesion. *Brain and Language, 38,* 334–344.

Rapin, I., Mattis, S., Rowan, A. J., & Golden, J. J. (1977). Verbal auditory agnosia in children. *Developmental Medicine and Child Neurology, 19,* 192.

Rapp, B., Benzing, L., & Caramazza, A. (1995). The representation of grammatical category at the level of phonological and orthographic lexical form. *Brain and Language, 51,* 46–49.

Rapp, B. C., & Caramazza, A. (1991). Spatially determined deficits in letter and word processing. *Cognitive Neuropsychology, 8,* 275–311.

Rapp, B. C., & Caramazza, A. (1993). On the distinction between deficits of access and deficits of storage: A question of theory. *Cognitive Neuropsychology, 10,* 113–141.

Rapp, B. C., Hillis, A. E., & Caramazza, A. (1993). The role of representations in cognitive theory: More on multiple semantics and the agnosias. *Cognitive Neuropsychology, 10,* 235–249.

Rapport, R., Tan, C. T., & Whitaker, H. A. (1983). Language function and dysfunction among Chinese and English speaking polyglots: Cortical stimulation, Wada testing and clinical studies. *Brain and Language, 18,* 342–366.

Raskin, S. A., Sliwinski, M., & Borod, J. C. (1992). Clustering strategies on tasks of verbal fluency in Parkinson's disease. *Neuropsychologia, 30,* 95–99.

Rasmussen T., & Milner, B. (1977). The role of early left-brain injury in determining lateralization of cerebral speech functions. *Annals of the New York Academy of Sciences, 299,* 355–369.

Rausch, R., & Risinger, M. (1990). Intracarotid sodium amobarbital procedure. In A. A. Boulton, G. B. Baker, & M. Hiscock (Eds.), *Neuromethods:* Vol. 17, *Neuropsychology* (pp. 127–146). Clifton, NJ: Humana Press.

Rauch, R. A., & Jinkins, J. R. (1994). Analysis of cross-sectional area measurements of the corpus callosum adjusted for brain size in male and female subjects from childhood to adulthood. *Behavioral Brain Research, 64,* 65–78.

Rausch, R., Silfvenius, H., Wieser, H.-G., Dodrill, C. B., Meador, K. J., & Jones-Gotman, M. (1993). Intraarterial amobarbital procedures. In J. Engel Jr. (Ed.), *Surgical treatment of the epilepsies* (2nd ed., pp. 341–357). New York: Raven Press.

Rayman, J., & Zaidel, E. (1991). Rhyming and the right hemisphere. *Brain and Language, 40,* 89–105.

Raymer, A. M., Moberg, P. J., Crosson, B., Nadeau, S. E., & Rothi, L. G. J. (1997). Lexical-semantic deficits in two cases of thalamic lesion. *Neuropsychologia 35,* 211–219.

Rayner, K., Sereno, S. C., Lesch, M. F., & Pollatsek, A. (1995). Phonological codes are automatically activated during reading: Evidence from an eye movement priming paradigm. *Psychological Science, 6,* 26–32.

Recanzone, G. H., Schreiner, C. E., & Merzenich, M. M. (1993). Plasticity in the frequency representation of primary auditory cortex following discrimination training in adult owl monkeys. *Journal of Neurophysiology, 67,* 1071–1091.

Rehak, A., Kaplan, J. A., & Gardner, H. (1992). Sensitivity to conversational deviance in right-hemisphere-damaged patients. *Brain and Language, 42,* 203–217.

Rehak, A., Kaplan, J. A., Weylman, S. T., Kelly, B., Brownell, H. H., & Gardner, H. (1992). Story processing in right-hemisphere brain-damaged patients. *Brain and Language, 42,* 320–336.

Reicher, G. M. (1969). Perceptual recognition as a function of meaningfulness of stimulus material. *Journal of Experimental Psychology, 81,* 274–280.

Reilly, J., Klima, E. S., & Bellugi, U. (1991). Once more with feeling: Affect and language in atypical populations. *Development and Psychopathology, 2,* 367–391.

Reilly, R., & Sharkey, N. (1992). *Connectionist approaches to natural language processing.* Hove, UK: Lawrence Erlbaum.

Reuter-Lorenz, P. A., & Baynes, K. (1992). Modes of lexical access in the callosotomized brain. *Journal of Cognitive Neuroscience, 4,* 155–164.

Revonsuo, A. (1995). Words interact with colors in a globally aphasic patient: Evidence from a stroop-like task. *Cortex, 31,* 377–386.

Revonsuo, A., & Laine, M. (1996). Semantic processing without conscious understanding in a global aphasic: Evidence from auditory event-related brain potentials. *Cortex, 32,* 29–48.

Rey, M., Dellatolas, G., Bancaud, C., & Talairach, J. (1988). Hemispheric lateralization of motor and speech functions after early brain lesion: Study of 73 epileptic patients with intracarotid amytal test. *Neuropsychologia, 26,* 167–172.

Rheingold, H. (1991). *Virtual reality.* New York: Summit Books.

Ribot, T. (1891). *Les maladies de la mémoire.* Paris: Baillère.

Ribot, T. (1904). *Les maladies de la mémoire.* Paris: Alcan.

Richardson, B., & Wuillemin, D. (1995). Reply [to "Another sighting of differential language laterality in multilinguals, this time in Loch Tok Pisin" by M. Paradis]. *Brain and Language, 49,* 187.

Richer, F., Martinez, M., Robert M., Bouvier, G., & Saint-Hillaire, J. M. (1993). Stimulation of human somatosensory cortex: Tactile and body displacement perceptions in medical regions. *Experimental Brain Research, 93,* 1173–1176.

Ricks, D. M., & Wing, L. (1976). Language, communication and the use of symbols in normal and autistic children. In L. Wing (Ed.), *Early childhood autism: Clinical, educational and social aspects* (2nd ed.; pp. 93–134). Oxford: Pergamon Press.

Riddoch, M. J., & Humphreys, G. (1994). *Cognitive neuropsychology and cognitive rehabilitation.* Hove, UK: Lawrence Erlbaum.

Rieber, R. W., & Vetter, H. (1994). The problem of language and thought in schizophrenia: A review. *Journal of Psycholinguistic Research, 23,* 149–195.

Riklan, M., & Cooper, I. S. (1975). Psychometric studies of verbal functions following thalamic lesions in humans. *Brain and Language, 2,* 45–64.

Ringo, J. L., Doty, R. W., Demeter, S., & Simard, P. Y. (1994). Time is of the essence: A conjecture that hemispheric specialization arises from interhemispheric conduction delay. *Cerebral Cortex, 4,* 331–343.

Rizzolatti, G., Riggio, L., Dascola, I., & Umiltà, C. (1987). Reorienting attention across the horizontal and vertical meridians: Evidence in favor of a premotor theory of attention. *Neuropsychologia, 25,* 31–40.

Roberts, G. W. (1991). Schizophrenia: A neuropathological perspective. *British Journal of Psychiatry, 158,* 8–17.

Robertson, L. C. (1995). Hemispheric specialization and cooperation in processing complex visual patterns. In F. L. Kitterle (Ed.), *Hemispheric communication: Mechanisms and models* (pp. 301–318). Hillsdale, NJ: Lawrence Erlbaum.

Robertson, L. C., Knight, R. T., Rafal, R., & Shimamura, A. P. (1993). Cognitive neuropsychology is more than single-case studies. *Journal of Experimental Psychology: Learning, Memory, and Cognition, 19,* 710–717.

Robin, D., Klouda, G., & Hug, L. (1991). Neurogenic disorders of prosody. In M. Cannito & D. Vogel (Eds.), *Treating disordered speech motor control: For clinicians by clinicians* (pp. 241–271). Austin, TX: Pro-Ed.

Robin, D., Tranel, D., & Damasio, H. (1990). Auditory perception of temporal and spectral events in patients with focal left and right cerebral lesions. *Brain and Language, 39,* 539–555.

Robinson, B. W. (1976). Limbic influences on human speech. *Annals of the New York Academy of Sciences, 280,* 761–771.

Robinson, I. (1990). Does computerized cognitive rehabilitation work? A review. *Aphasiology, 4,* 381–405.

Robinson, R., Kubos, K., Starr, L., Rao, K., & Price, T. (1984). Mood disorders in stroke patients: Importance of location of lesion. *Brain, 197,* 81–93.

Robinson, R. G., Starr, L. B., & Price, T. R. (1984). A two-year longitudinal study of mood disorders following stroke. *British Journal of Psychiatry, 144,* 256–262.

Rochester, S., & Martin, J. (1979). *Crazy talk: A study of the discourse of schizophrenic speakers.* New York: Plenum Press.

Rochon, E., Waters, G. S., & Caplan, D. (1994). Sentence comprehension in patients with Alzheimer's disease. *Brain and Language, 46,* 329–349.

Rodel, M., Cook, H. D., Regard, M., & Landis, T. (1992). Hemispheric dissociation in judging semantic relations: Complementarity for close and distant associates. *Brain and Language, 43,* 448–459.

Roeltgen, D., & Rapcsak, S. Z. (1993). Acquired disorders of writing and spelling. In G. Blanken, J. Dittman, & C. Wallesch (Eds.), *Linguistic disorders and pathologies* (pp. 262–278). Berlin: Walter de Gruyter.

Rogers, R. L., Basile, L. F. H., Papanicolaou, A. C., & Eisenberg, H. M. (1993). Magnetoencephalography reveals two distinct sources associated with late positive evoked potentials during visual oddball task. *Cerebral Cortex, 3,* 163–169.

Rogers, R. L., Baumann, S. B., Papanicolaou, A. C., Bourbon, T. W., Alagarsamy, S., & Eisenberg, H. M. (1991). Localization of the P3 sources using magnetoencephalography and magnetic resonance imaging. *Electroencephalography and Clinical Neurophysiology, 79,* 308–321.

Rogers, R. L., Papanicolaou, A. C., Baumann, S. B., & Eisenberg, H. M. (1992). Late magnetic fields and positive evoked potentials following infrequent and unpredictable omissions of visual stimuli. *Electroencephalography and Clinical Neurophysiology, 83,* 146–152.

Roman, G. C., Tatemichi, T. K., Erkinjuntti, T., Cummings, J. L., Masdeu, J. C., Garcia, J. H., Amaducci, L., Orgogozo, J.M., Brun, A., Hofman, A., Moody, D.M., O'Brian, M.D., Yamaguchi, T., Grafman, J., Drayer, B.P., Bennet, D. A., Fisher, M., Ogata, J., Kokmen, E., Bermejo, F., Wolf, P. A., Gorelick, P. B., Bick, K. L., Pajeau, A. K., Bell, M. A., DeCarli, C., Culebras, A., Korczyn, A.D., Bogousslavsky, J., Hartmann, A., & Scheinberg, P. (1993). Vascular dementia: Diagnostic criteria for research studies. *Neurology, 43,* 250–260.

Romani, G. L., Williamson, S. J., & Kaufman, L. (1982). Tonotopic organization of the human auditory cortex. *Science, 216,* 1339–1340.

Rosenbek, J. C. (1979). Wrinkled feet. In R. H. Brookshire (Ed.), *Clinical aphasiology: 1979 conference proceedings* (pp. 163–176). Minneapolis: BRK Publishers.

Rosenbek, J. C. (1995). Efficacy in dysphagia. *Dysphagia, 10,* 263–267.

Rosin, M., Swift, E., Bless, D., & Kluppel Vetter, D. (1988). Communication profiles of adolescents with Down syndrome. *Journal of Childhood Communication Disorders, 12,* 49–64.

Rösler, F., Pütz, P., Friederici, A., & Hahne, A. (1993). Event-related brain potentials while encountering semantic and syntactic constraint violations. *Journal of Cognitive Neuroscience, 5,* 345–362.

Ross, E. (1980). The aprosodias: Functional-anatomical organization of the affective components of language in the right hemisphere. *Archives of Neurology, 38,* 561–569.

Ross, E., Edmondson, J., Seibert, G., & Homan, R. (1988). Acoustic analysis of affective prosody during right-sided Wada test: A within-subjects verification of the right hemisphere's role in language. *Brain and Language, 33,* 128–145.

Rosser, A. E., & Hodges, J. R. (1994a). Initial letter and semantic category fluency in Alzheimer's disease, Huntington's disease, and progressive supranuclear palsy. *Journal of Neurology, Neurosurgery, and Psychiatry, 57,* 1389–1394.

Rosser, A. E., & Hodges, J. R. (1994b). The Dementia Rating Scale in Alzheimer's disease, Huntington's disease, and progressive supranuclear palsy. *Journal of Neurology, 241,* 531–536.

Rossing, H. (1975). Zur sprachlichen Leistung der rechten (nicht-dominanten) Hemisphäre. *Zeitschrift Dialektologie und Linguistik, 13,* 172–197.

Roth, V. M. (1989). PAKT und STACH. Therapeutisches Gespräch in der Aphasikerfamilie und Sprachtraining am Computer. Zwei sich ergänzende Erweiterungen der Aphasietherapie? In V. M. Roth (Ed.), *Kommunikation trotz gestörter Sprache* (pp. 101–118). Tübingen: Gunter Narr Verlag.

Roth, V. M. (Ed.) (1992a). *Computer in der Sprachtherapie.* Tübingen: Gunter Narr Verlag.

Roth, V. M. (1992b). SICH-ÄUSSERNDES Verstehen. NeueWEGE. In V. M. Roth (Ed.), *Computer in der Sprachtherapie* (pp. 187–214). Tübingen: Gunter Narr Verlag.

Roth, V. M. (1992c). WEGE: „Infotainment" und Klinische Linguistik. Computer in der Sprachtherapie. In G. Rickheit, R. Mellies, & A. Winnecken (Eds.), *Linguistische Aspekte der Sprachtherapie. Forschung und Intervention bei Sprachstörungen* (pp. 277–297). Opladen: Westdeutscher Verlag.

Roth, V. M. (1994). *Ich war in der Sprache wie tot* [Video]. Available from Medien- und Verlagswerkstatt, Macairestr. 3, D-78467 Konstanz, Germany, Fax: (+49) 7531–54929.

Roth, V. M. (1995). Neue Wege wagen? Wandel und Variation von Sozialformen beim sprachheilpädagogischen „Multimedia"-Einsatz. *Sprachheilarbeit, 40,* 171–177.

Roth, V. M. (1996). Sprechende Bilder—BOX—NeueWEGE in der Sprachtherapie. Im Verbund. In B. Simons (Ed.), *Gruppentherapie bei Aphasie. Probleme und Lösungen* (pp. 9–33). Bern: Peter Lang. Europäischer Verlag der Wissenschaften.

Roth, V. M., and Schönle, P. W. (1992). Sprachtraining für Aphasiker mit Computerhilfe (STACH und WEGE) in einer Selbsthilfegruppe. *Rehabilitation, 31,* 91–97.

Rothi, L. J. G., Goldstein, L. P., Teas, E., Schoenfeld, D., Moss, S., & Ochipa, C. (1985). Treatment of alexia without agraphia: A case report. *Journal of Clinical and Experimental Neuropsychology, 7,* 607.

Rueckl, J. G., & Dror, I. E. (1994). The effect of orthographic-semantic systematicity on the acquisition of new words. In C. Umiltà & M. Moscovitch (Eds.), *Attention and performance XV: Conscious and nonconscious information processing* (pp. 571–588). Cambridge: MIT Press.

Rugg, M. D. (1995). ERP studies of memory. In M. D. Rugg & M. G.H. Coles (Eds.), *Electrophysiology of mind* (pp.132–170). Oxford: Oxford University Press.

Rugg, M. D., Doyle, M. C., & Melan, C. (1993). An event-related potential study of the effects of within- and across-modality word repetition. *Language and Cognitive Processes, 8,* 357–377.

Rumelhart, D. E., Hinton, G. E., & Williams, R. J. (1986). Learning internal representations by error propagation. In D. E. Rumelhart & J. L. McClelland & the PDP Research Group (Eds.), *Parallel distributed processing: Explorations in the microstructures of cognition* (pp. 318–362). Cambridge: MIT Press.

Rumelhart, D. E., McClelland, J. L., & the PDP Research Group (1986). *Parallel distributed processing: Explorations in the microstructure of cognition.* Vol. 1, *Foundations.* Cambridge: MIT Press.

Russell, R. W. (1992). Interactions among neurotransmitters: Their importance to the "integrated organism." In E. D. Levin, M. W. Decher, & L. L. Butcher (Eds.), *Neurotransmitter interactions and cognitive function* (pp. 183–195). Boston: Birkhauser.

Russell, R. W., Smith, C. A., Booth, R. A., Jenden, D. J., & Waite, J. J., (1986). Behavioral and physiological effects associated with changes in muscarinic receptors following administration of an irreversible cholinergic agonist (BM123). *Psychopharmacology, 90,* 308–315.

Russell, W. R., & Espir, M. L. E. (1961). *Traumatic aphasia.* London: Oxford University Press.

Ryalls, J. (Ed.). (1987a). *Phonetic approaches to speech production in aphasia and related disorders.* Boston: Little, Brown.

Ryalls, J. (1987b). Vowel production in aphasia: Towards an account of the consonant–vowel dissociation. In J. Ryalls (Ed.), *Phonetic approaches to speech production in aphasia and related disorders* (pp. 23–43). Boston: Little, Brown.

Ryalls, J. (1988). Concerning right-hemisphere dominance for affective language. *Archives of Neurology, 45,* 337–338.

Ryalls, J., & Behrens, S. (1988). An overview of changes in fundamental frequency associated with cortical insult. *Aphasiology, 2,* 107–115.

Ryalls, J., & Reinvang, I. (1986). Functional lateralization of linguistic tones: Acoustic evidence from Norwegian. *Language and Speech, 29,* 389–398.

Sacchett, C., & Humphreys, G. W. (1992). Calling a squirrel a squirrel but a canoe a wigwam: A category-specific deficit for artefactual objects and body parts. *Cognitive Neuropsychology, 9,* 73–86.

Sackeim, H. A., Greenberg, M. S., Weiman, M. A., Gur, R. C., Hungerbuhler, J. P., & Geschwind, N. (1982). Hemispheric asymmetry in the expression of positive and negative emotions. *Archives of Neurology, 39,* 210–218.

Sacks, H., Schegloff, E., & Jefferson, G. (1974). A Simplest systematic for the organisation of turn taking in conversation. *Language, 50,* 696–735.

Saddy, J. D. (1995). Variables and events in the syntax of agrammatic speech. *Brain and Language, 50,* 135–150.

Saffran, E. M., Berndt, R. S., & Schwartz, M. F. (1989). The quantitative analysis of agrammatic production: Procedure and data. *Brain and Language, 37,* 440–479.

Saffran, E. M., Bogyo, L. C., Schwartz, M. F., & Marin, O. S. M. (1980). Does deep dyslexia reflect right-hemisphere reading? In M. Coltheart, K. Patterson, & J. C. Marshall (Eds.), *Deep Dyslexia.* London: Routledge & Kegan Paul.

Saffran, E. M., & Marin, O. S. M. (1975). Immediate memory for word lists and sentences in a patient with deficient auditory short-term memory. *Brain and Language, 2,* 420–433.

Saffran E. M., & Schwartz, M. F. (1994). Of cabbages and things: Semantic memory from a neuropsychological perspective: A tutorial review. In C. Umiltà & M. Moscovitch (Eds.), *Attention and performance XV: Conscious and nonconscious information processing* (pp. 507–536). Cambridge: MIT Press.

Sagar, H. J., Sullivan, E. V., Cooper, J. A., & Jordan, N. (1991). Normal release from proactive interference in untreated patients with Parkinson's disease. *Neuropsychologia, 29,* 1033–1044.

Sagar, H. J., Sullivan, E. V., Gabrieli, J. D., Corkin, S., & Growdon, J. H. (1988). Temporal ordering and short-term deficits in Parkinson's disease. *Brain, 111,* 525–539.

Saint-Cyr, J. A., Taylor, A. E., & Lang, A. E. (1988). Procedural learning and neostriatal dysfunction in man. *Brain, 111,* 949–959.

Sakurai, Y., Momose, T., Iwata, M., Watanabe, T., Ishikawa, T., & Kanazawa, I. (1993). Semantic process in Kana word reading: Activation studies with positron emission tomography. *NeuroReport, 4,* 327–330.

Sakurai, Y., Momose, T., Iwata, M., Watanabe, T., Ishikawa, T., Takeda, K., & Kanazawa, I. (1992). Kanji word reading process analysed by positron emission tomography. *NeuroReport, 3,* 445–448.

Salmelin, R., Hari, R., Lounasmaa, O. V., & Sams, M. (1994). Dynamics of brain activation during picture naming. *Nature, 368,* 463–465.

Salthouse, T. A. (1988). The role of processing resources in cognitive aging. In M. L. Howe & C. J. Brainerd (Eds.), *Cognitive development in adulthood* (pp. 185–239). New York: Springer Verlag.

Salthouse, T. A., Mitchell, D. R. D., Skovronek, E., & Babcock, R. L. (1989). Effects of adult age and working memory on reasoning and spatial abilities. *Journal of Experimental Psychology, Learning, Memory and Cognition, 15,* 507–516.

Samar, V. J., & Berent, G. P. (1986). The syntactic priming effect: Evoked response evidence for a prelexical locus. *Brain and Language, 28,* 250–272.

Samra, K., Riklan, M., Levita, E., Zimmerman, J., Waltz, J. M., Bergmann, L., & Cooper, I. S. (1969). Language and speech correlates of anatomically verified lesions in thalamic surgery for parkinsonism. *Journal of Speech and Hearing Research, 12,* 510–540.

Sams, M., Aulanko, R., Aaltonen, O., & Näätänen, R. (1990). Event-related potentials to infrequent changes in synthesized phonetic stimuli. *Journal of Cognitive Neuroscience, 2,* 344–357.

Sanides, F. (1970). Functional architecture of motor and sensory cortices in primates in the light of a new concept of neocortex evolution. In C. R. Noback & W. Montagna (Eds.), *The primate brain: Advances in primatology* (Vol. 1, pp. 137–208). New York: Appleton-Century-Crofts.

Sapir, E. (1926–1927). Speech as a personality trait. *American Journal of Sociology, 32,* 892–905.

Sarno, J., Swisher, L., & Sarno, M. (1969). Aphasia in a congenitally deaf man. *Cortex, 5,* 398–414.

Sarno, M. T. (1988). Head injury: Language and speech defects. *Scandinavian Journal of Rehabilitation, 17* (Med. Suppl.), 55–64.

Sarno, M. T. (1991). Recovery and rehabilitation in aphasia. In M. T. Sarno (Ed.), *Acquired aphasia* (2nd ed.; pp. 521–582). San Diego: Academic Press.

Sarno, M. T., & Levita, E. (1981). Some observations on the nature of recovery in global aphasia after stroke. *Brain and Language, 13,* 1–12.

Sartori, G., & Job, R. (1988). The oyster with four legs: A neuropsychological study on the interaction of visual and semantic information. *Cognitive Neuropsychology, 5,* 105–132.

Sartori, G., Job, R., Miozzo, M., Zago, S., & Marchiori, G. (1993). Category-specific form knowledge deficit in a patient with herpes simplex. *Journal of Clinical and Experimental Neuropsychology, 15,* 280–299.

Sasanuma, S., Kamio, A., & Kubota, M. (1990). Agrammatism in Japanese: Two case studies. In L. Menn & L. K. Obler (Eds.), *Agrammatic aphasia: A cross-language narrative sourcebook* (Vol. 2, pp. 1225–1307). Amsterdam: John Benjamins.

Sasanuma, S., & Park, H. S. (1995). Patterns of language deficits in two Korean-Japanese aphasic patients. In M. Paradis (Ed.), *Aspects of bilingual aphasia* (pp. 111–122). Oxford: Pergamon Press.

Sass, K. J., Sass, A., Westerveld, M., Lencz, T., Novelly, R. A., Kim, J. H., & Spencer, D. D. (1992). Specificity in the correlation of verbal memory and hippocampal neuron loss: Dissociation of memory, language, and verbal intellectual ability. *Journal of Clinical and Experimental Neuropsychology, 14,* 662–672.

Sass, L. A.(1992) *Madness and modernism: Insanity in the light of modern art, literature, and thought.* New York: Basic Books.

Satz, P. (1977). Laterality tests: An inferential problem. *Cortex, 13,* 208–212.

Satz, P., & Bullard-Bates, C. (1981). Acquired aphasia in children. In M. T. Sarno (Ed.), *Acquired aphasia* (pp. 399–426). New York: Academic Press.

Satz, P., Strauss, E., & Whitaker, H. (1990). The ontogeny of hemispheric specialization: Some old hypotheses revisited. *Brain and Language, 38,* 596–614.

Satz, P., Strauss, E., Wada, J., & Orsini, D. L. (1988). Some correlates of intra- and interhemispheric speech organization after left focal brain injury. *Neuropsychologia, 26,* 345–350.

Sawhney, I. M., Suresh, N., Dhand, U. K., & Chopra, J. S. (1988). Acquired aphasia with epilepsy: Landau-Kleffner syndrome. *Epilepsia, 29,* 283–287.

Scalise, S. (1984). *Generative morphology.* Dordrecht, The Netherlands: Foris.

Scarborough, H. S. (1990). Index of productive syntax. *Applied Psycholinguistics, 11,* 1–22.

Schacter, S., & Singer, J. (1962). Cognitive, social and physiological determinants of emotional state. *Psychological Review, 69,* 379–399.

Schaie, K. W. (1993). The Seattle longitudinal studies of adult intelligence. *Current Directions in Psychological Science, 2,* 171–175.

Schank, R. C., & Abelson, R. P. (1977). *Scripts, goals, plans and understanding.* Hillsdale, NJ: Lawrence Erlbaum.

Schenck, J. (1584). *Observationes medicae de capite humano.* Basel: Frobeniana.

Scherer, K. R. (1979). Personality markers in speech. In H. Giles (Ed.), *Social markers in speech* (pp. 147–201). Cambridge: Cambridge University Press.

Schieffelin, B., & Ochs, E. (1987). *Language acquisition across cultures.* New York: Cambridge University Press.

Schlaug, G., Jäncke, L., Huang, Y., & Steinmetz, H. (1995). In vivo evidence of structural brain asymmetry in musicians. *Science, 267,* 699–701.

Schlenck, C., Schlenck, K. J., & Springer, L. (1995). *Die Behandlung des schweren Agrammatismus: Reduzierte Syntaxtherapie (REST).* Stuttgart: Georg Thieme Verlag.

Schlenk, K. J., & Springer, L. (1989). Agrammatismus in der Rückbildung: Eine Fallstudie. *Neurolinguistik, 1,* 57–72.

Schmidt, A. L., Arthur, D. L., Kutas, M., & Flynn, E. (1989). Neuromagnetic responses during reading meaningful and nonmeaningful sentences. *Psychophysiology Abstracts, 26,* S6.

Schnider, A., Benson, D. F., & Scharre, D. W. (1994). Visual agnosia and optic aphasia: Are they anatomically distinct? *Cortex, 30,* 445–457.

Scholten, I. M., Kneebone, A. C., Denson, L. A., Fields, C. D., & Blumbergs, P. (1995). Primary progressive aphasia: Serial linguistic, neuropsychological findings with neuropathological results. *Aphasiology, 9,* 495–516.

Schuell, H. (1965). *The Minnesota Test for Differential Diagnosis of Aphasia.* Minneapolis: University of Minnesota Press.

Schuell, H., Jenkins, J. J., & Jimenez-Pabon, E. (1965). *Aphasia in adults.* New York: Harper & Row.

Schulman-Galambos, C. (1977). Dichotic listening performance in elementary and college students. *Neuropsychologia, 15,* 577–584.

Schumann, J. H. (1994). Emotion and cognition in second language acquisition. *Studies in Second Language Acquisition, 16,* 231–242.

Schwanenflugel, P. J. (Ed.). (1991). *The psychology of word meanings.* Hillsdale, NJ: Lawrence Erlbaum.

Schwartz, M. F. (1984). What the classical aphasia categories can't do for us and why. *Brain and Language, 21,* 3–8.

Schwartz, M. F., Dell, G. S., Martin, N., & Saffran, E. M. (1994). Normal and aphasic naming in an interactive spreading activation model. *Brain and Language, 47,* 391–394.

Schwartz, M. F., Fink, R. B., & Saffran, E. M. (1995). The modular treatment of agrammatism. *Neuropsychological Rehabilitation, 5,* 93–127.

Schwartz, M. F., Linebarger, M. C., Saffran, E., & Pate, D. (1987). Syntactic transparency and sentence interpretation in aphasia. *Language and Cognitive Processes, 2,* 85–113.

Schwartz, M. F., Saffran, E. M., & Marin, O. S. M. (1980). Fractionating the reading process in dementia: Evidence for word-specific print-to-sound associations. In M. Coltheart, K. Patterson, & J. C. Marshall (Eds.), *Deep dyslexia.* London: Routledge & Kegan Paul.

Schwartz, M. F., Saffran, E. M., Fink, R. B., Myers, J. L., & Martin, N. (1994). Mapping therapy: A treatment program for agrammatism. *Aphasiology, 8,* 19–54.

Schwartz, M. F., Saffran, E., & Marin, O. (1980). The word order problem in agrammatism: I. Comprehension. *Brain and Language, 10,* 249–262.

Schwartz, S. (1982). Is there a schizophrenic language? *Behavioral and Brain Sciences, 5,* 579–626.

Schweiger, A., Zaidel, E., Field, T., & Dobkin, B. (1989). Right hemisphere contribution to lexical access in an aphasic with deep dyslexia. *Brain and Language, 37,* 73–89.

Scott, C., & Byng, S. (1989). Computer assisted remediation of a homophone comprehension disorder in surface dyslexia. *Aphasiology, 3,* 301–320.

Scott, S., Caird, F., & Williams, B. (1984). Evidence for an apparent sensory speech disorder in Parkinson's disease. *Journal of Neurology, Neurosurgery, and Psychiatry, 47,* 840–843.

Searle, J. R. (1976). The classification of illocutionary acts. Language in Society, 5, 1–24. (Reprinted in Searle (1979). Expression and meaning. Cambridge: Cambridge University Press).

Segalowitz, S. J. (1989). ERPs and research in neurolinguistics. In A. Ardila & F. Ostrosky (Eds.) *Brain organization of language and cognitive processes* (pp. 61–82). New York: Plenum Press.

Segalowitz, S. J., & Berge, B. E. (1995). Functional asymmetries in infancy and early childhood: A review of electrophysiological studies and their implications. In R. J. Davidson & K. Hugdahl (Eds.), *Brain asymmetry* (pp. 579–616). Cambridge: MIT Press.

Segalowitz, S. J., & Berge, B. E. M. (1994). Functional asymmetries in infancy and early childhood: A review of electrophysiologic studies and their implications. In R. J. Davidson & K. Hugdahl (Eds.), *Brain asymmetry* (pp. 579–615). Cambridge: MIT Press.

Segalowitz, S. J., & Chapman, J. S. (1980). Cerebral asymmetry for speech in neonates: A behavioral measure. *Brain and Language, 9,* 281–288.

Segalowitz, S. J., & Cohen, H. (1989). Right hemisphere EEG sensitivity to speech. *Brain and Language, 37,* 220–231.

Segalowitz, S. J., Menna, R., & MacGregor, L. (1987). Left and right hemisphere participation in reading: Evidence from ERPs. *Journal of Clinical and Experimental Neuropsychology, 9,* 274.

Segalowitz, S. J., Wagner, W. J., & Menna, R. (1992). Lateral versus frontal ERP predictors of reading skill. Special issue: The role of frontal lobe maturation in cognitive and social development. *Brain and Cognition, 20,* 85–103.

Segui, J., Dupoux, E., & Mehler, J. (1990). The role of the syllable in speech segmentation, phoneme identification, and lexical access. In G. M. T. Altman (Ed.), *Cognitive models of speech processing* (pp. 263–280). Cambridge: MIT Press.

Seidenberg, M. (1993). Connectionist models and cognitive theory. *Psychological Science, 4,* 228–235.

Seidenberg, M. (1994). Language and connectionism: The developing interface. Cognition, 50, 385–401.

Seidenberg, M., & McClelland, J. L. (1989). A distributed, developmental model of word recognition and naming. *Psychological Review, 96,* 523–568.

Seidenberg, M., Waters, G., Barnes, M., & Tanenhaus, M. (1984). When does irregular spelling or pronunciation influence word recognition? *Journal of Verbal Learning and Verbal Behavior, 23,* 383–404.

Seidman, L. J., Cassens, G. P., Kremens, W. S., & Pepple, J. R. (1992). Neuropsychology of schizophrenia. In R. F. White (Ed.), *Clinical syndromes in adult neuropsychology: The practitioner's handbook* (pp. 381–449). New York: Elsevier Science.

Sejnowski, T., Koch, C., & Churchland, P. (1988). Computational neuroscience. *Science, 241,* 1299–1306.

Selinger, M., Prescott, T. E., & Katz, R. C. (1987). Handwritten versus typed responses on PICA graphic subtests. In R. H. Brookshire (Ed.), *Clinical aphasiology: 1987 conference proceedings* (pp. 136–142). Minneapolis: BRK Publishers.

Selnes, O. A., Knopman, D. S., Niccum, N., & Rubens, A. B. (1983). CT-scan correlates of auditory comprehension deficits in aphasia: A prospective recovery study. *Annals of Neurology, 13,* 558–566.

Selnes, O. A., Knopman, D. S., Niccum, N., & Rubens, A. B. (1985). The critical role of Wernicke's area in sentence repetition. *Annals of Neurology, 17,* 549–557.

Selnes, O. A., Niccum, N., Knopman, D. S., & Rubens, A. B. (1984). Recovery of single-word comprehension: CT scan correlates. *Brain and Language, 21,* 72–84.

Semenza, C., & Zettin, M. (1988). Generating proper names: A case of selective inability. *Cognitive Neuropsychology, 5,* 711–724.

Semenza, C., & Zettin, M. (1989). Evidence from aphasia for the role of proper names as pure referring expressions. *Nature, 342,* 678–679.

Serafetinides, E. A., & Falconer, M. A. (1963). The effects of temporal lobectomy in epileptic patients with psychosis. *Journal of Mental Science, 108,* 584–593.

Sergent, J. (1986). Prolegomena to the use of the tachistoscope in neuropsychological research. *Brain and Cognition, 5,* 127–130.

Sergent, J. (1987). A new look at the human split brain. *Brain, 110,* 1375–1392.

Sergent, J. (1988). Some theoretical and methodological issues in neuropsychological research. In F. Boller and J. Grafman (Eds.), Handbook of Neuropsychology (Vol. 1, pp. 69–81). Amsterdam: Elsevier.

Sergent, J. (1990). Furtive incursions into bicameral minds: Integrative and coordinating role of subcortical structures. *Brain, 113,* 537–568.

Sergent, J. (1991). Processing of spatial relations within and between the disconnected cerebral hemispheres. *Brain, 114,* 1025–1043.

Sergent, J., Ohta, S., & Mac Donald, B. (1992). Functional neuroanatomy of face and object processing. *Brain, 115,* 15–36.

Sergent, J., Zuck, E., Levesque, M., & Mac Donald B. (1992). Positron emission tomography study of letter and object processing: empirical findings and methodological considerations. *Cerebral Cortex, 2,* 68–80.

Seron, X., & Deloche, G. (1989). *Cognitive approaches in neuropsychological rehabilitation.* Hillsdale, NJ: Lawrence Erlbaum.

Seron, X., Deloche, G., Moulard, G., & Rouselle, M. (1980). A computer-based therapy for the treatment of aphasic subjects with writing disorders. *Journal of Speech and Hearing Disorders, 45,* 45–58.

Sethi, P. K., & Rao, S. (1976). Gelastic, quiritarian, and cursive epilepsy. *Journal of Neurology, Neurosurgery, and Psychiatry, 39,* 823–828.

Seymour, P. H. K. (1986). *Cognitive analysis of dyslexia.* London: Routledge & Kegan Paul.

Seymour, P. H. K. (in press). Foundations of orthographic development. In C. Perfetti, L. Rieben, & M. Fayol (Eds.), *Learning to spell.* Hillsdale, NJ: Lawrence Erlbaum.

Seymour, P. H. K., & Bunce, F. (1994). Application of cognitive models to remediation in cases of developmental dyslexia. In M. J. Riddoch & G. W. Humphreys (Eds.), *Cognitive neuropsychology and cognitive rehabilitation* (pp. 349–377). Hove, UK: Lawrence Erlbaum.

Seymour, P. H. K., & Evans, H. M. (1993). The visual (orthographic) processor and developmental dyslexia. In D. M. Willows, R. S. Kruk, & E. Corcos (Eds.), *Visual processes in reading and reading disabilities* (pp. 347–376). Hillsdale, NJ: Lawrence Erlbaum.

Seymour, S. E., Reuter-Lorenz, P. A., & Gazzaniga, M. S. (1994). The disconnection syndrome: Basic findings reaffirmed. *Brain, 117,* 105–115.

Shallice, T. (1979). Case study approach in neuropsychological research. *Journal of Clinical Neuro-Psychology, 1,* 183–211.

Shallice, T. (1988). *From neuropsychology to mental structure.* Cambridge: Cambridge University Press.

Shallice, T. (1993). Multiple semantics: Whose confusions? *Cognitive Neuropsychology, 10,* 251–261.

Shallice, T., & Burgess, P. W. (1991). Deficits in strategy application following frontal lobe damage in man. *Brain, 114,* 727–741.

Shallice, T., & Butterworth, B. (1977). Auditory-verbal short-term memory impairment and spontaneous speech. *Neuropsychologia, 15,* 729–735.

Shallice, T., & Saffran, E. M. (1986). Lexical processing in the absence of explicit word identification: Evidence from a letter-by-letter reader. *Cognitive Neuropsychology, 3,* 429–458.

Shallice, T., & Warrington, E. K. (1977). The possible role of selective attention in acquired dyslexia. *Neuropsychologia, 15,* 31–41.

Shankweiler, D., & Crain, S. (1987). Language mechanisms and reading disorder: A modular approach. In P. Bertelson (Ed.), *The onset of literacy* (pp. 139–168). Cambridge: MIT Press.

Shankweiler, D., Crain, S., Gorrell, P., & Tuller, B. (1989). Reception of language in Broca's aphasia. *Language and Cognitive Processes, 4,* 1–33.

Shapiro, D. (1965). *Neurotic styles.* New York: Basic Books.

Shapiro, L. P., & Danly, M. (1985). The role of the right hemisphere in the control of speech prosody in propositional and affective contexts. *Brain and Language, 25,* 19–36.

Shapiro, L. P., Gordon, B., Hack, N., & Killackey, J. (1993). Verb–argument structure processing in complex sentences in Broca's and Wernicke's aphasia. *Brain and Language, 42,* 423–447.

Shapiro, L. P., & Levine, B. A. (1990). Real-time sentence processing in aphasia. *Brain and Language, 38,* 21–47.

Sharma, A., Kraus, N., McGee, T., Carrell, T., & Nicol, T. (1993). Acoustic versus phonetic representation of speech as reflected by the mismatch negativity event-related potential. *Electroencephalography and Clinical Neurophysiology: Evoked Potentials, 88,* 64–71.

Shastri, L. & Ajjanagadde, V. (1993). From simple associations to systematic reasoning: A connectionist representation of rules, variables, and dynamic bindings using temporal synchrony. *Behavioral and Brain Sciences, 16,* 417–494.

Shaywitz, B. A., Shaywitz, S. E., Pugh, K. R., Constable, R. T., Skudlarski, P., Fulbright, R. K., Bronen, R. A., Fletcher, J. M., Shankweiler, D. P., Katz, L., & Gore, J.C. (1995). Sex differences in the functional organization of the brain for language. *Nature, 373,* 607–609.

Sherratt, S., & Penn, C. (1990). Discourse in a right-hemisphere brain-damaged subject. *Aphasiology, 4,* 539–560.

Shindler, A. G., Caplan, L. R., & Hier, D. B. (1984). Intrusions and perseverations. *Brain and Language, 23,* 148–158.

Shipley, T. (1961). *Classics in psychology.* New York: Philosophical Library.

Shipley-Brown, F., Dingwall, W., Berlin, C., Yeni-Komshian, G., & Gordon-Salant, S. (1988). Hemispheric processing of affective and linguistic intonation contours in normal sujbects. *Brain and Language, 33,* 16–26.

Shucard, D. W., Shucard, J. L., Clopper, R. R., & Schacter, M. (1992). Electrophysiological and neuro-physiological indices of cognitive processing deficits in Turner syndrome. *Developmental Neuropsychology, 8,* 399–323.

Shui, L. P., & Pashler, H. (1995). Spatial attention and vernier acuity. *Vision Research, 35,* 337–343.

Shuttleworth, E. C., & Huber, S. J. (1988). The naming disorder of dementia of the Alzheimer type. *Brain and Language, 34,* 222–234.

Shye, S., & Elizur, D. (1995). *Introduction to facet theory.* Newbury Park, CA: Sage.

Sidtis, J. J. (1980). On the nature of the cortical function underlying right hemisphere auditory perception. *Neuropsychologia, 18,* 321–330.

Sidtis, J. J. (1984). Music, pitch perception, and the mechanisms of cortical hearing. In M. S. Gazzaniga (Ed.), *Handbook of cognitive neuroscience* (pp. 91–114). New York: Plenum Press.

Sidtis, J. J., & Volpe, T. B. (1988). Selective loss of complex-pitch or speech discrimination after unilateral cerebral lesion. *Brain and Language, 34,* 235–245.

Sidtis, J. J., Volpe, T. B., Holtzman, J., Wilson, D., & Gazzaniga, M. (1981). Cognitive interaction after staged callosal section: Evidence for transfer of semantic activation. *Science, 212,* 344–346.

Sieroff, E., Pollatsek, A., & Posner, M. I. (1988). Recognition of visual letter strings following injury to the posterior visual spatial attention system. *Cognitive Neuropsychology, 5,* 427–449.

Silbersweig, D. A., Stern, E., Frith, C. D., Cahill, C., Grootoonk, S., Spinks, T. J., Clark, J., Blasberg, R. G., Plum, F., Frackowiak, R. S. J., & Jones, T. (1993). Detection of thirty-second cognitive activations in single subjects with positron emission tomography: A new low-dose $H_2^{15}O$ regional cerebral blood flow three-dimensional imaging technique. *Journal of Cerebral Blood Flow and Metabolism, 13,* 617–629.

Silbersweig, D. A., Stern, E., Frith, C. D., Seward, J., Holmes, A., Schnorr, L., Cahill, C., McKenna, P., Chua, S., Jones, T., & Frackowiak, R. S. J. (1995). Mapping the neural correlates of auditory hallucinations in schizophrenia: Involuntary perceptions in the absence of external stimuli (abstract). *Human Brain Mapping* (Suppl. 1), 422.

Silverberg, R., & Gordon, H. W. (1979). Differential aphasia in bilinguals. *Neurology, 29,* 51–55.

Silveri, M. C., & Di Betta, A. M. (in press). Nouns–verbs dissociation in brain-damaged patients: Further evidence, further consideration. *Neurocase.*

Silvestrini, M., Troisi, E., Matteis, M., Cupini, L. M., & Caltagirone, C. (1995). Involvement of the healthy hemisphere in recovery from aphasia and motor deficit in patients with cortical ischemic infarction: A transcranial Doppler study. *Neurology, 45,* 1815–1820.

Simmons, N. N., & Buckingham, H. W. (1992). Recovery in jargonaphasia. *Aphasiology, 6,* 403–414.

Simos, P. G., Basile, L. F. H., & Papanicolaou, A. C. (in press). Source localization of the N400 response in a sentence-reading paradigm using evoked magnetic fields and magnetic resonance imaging. *Brain Research.*

Ska, B., & Guénard, D. (1993). Narrative schema in dementia of the Alzheimer's type. In H. H. Brownell & Y. Joanette (Eds.), *Narrative discourse in neurologically impaired and normal aging adults* (pp. 299–316). San Diego: Singular Publishing.

Skinner, J. E., & Yingling, C. D. (1977). Central gating mechanisms that regulate event-related potentials and behavior. In J. E. Desmedt (Ed.), *Attention, voluntary contraction and event-related cerebral potentials* (Vol. 1, pp. 30–69). Basel: Karger.

Slavin, M. D., Laurence, S., & Stein, D. G. (1988). Another look at vicariation. In S. Finger, T. E. Levere, C. R. Almli, & D. G. Stein (Eds.), *Brain injury and recovery: Theoretical and controversial issues* (pp. 165–179). New York: Plenum Press.

Slobin, D. (1985). Crosslinguistic evidence for the language-making capacity. In D. Slobin (Ed.), *The crosslinguistic study of language acquisition.* Vol. 2, *Theoretical issues* (pp. 1157–1256). Hillsdale, NJ: Lawrence Erlbaum.

Slobin, D. I. (1991). Aphasia in Turkish: Speech production in Broca's and Wernicke's patients. *Brain and Language, 41,* 149–164.

Small, S. L. (1990). Learning lexical knowledge in context: experiments with recurrent feedforward networks. In *Proceedings of the Twelfth Annual Conference of the Cognitive Science Society* (pp. 479–483), Massachusetts Institute of Technology, Cambridge, MA: Lawrence Erlbaum.

Small, S. L., Hart, J., Nguyen, T., & Gordon, B. (1995). Distributed representations of semantic knowledge in the brain. *Brain, 118,* 441–453.

Small, S. L., Noll, D. C., Perfetti, C. A., Xu, B., & Schneider, W. (1994). Activation of left frontal operculum and motor cortex with functional MRI of language processing. *Society for Neuroscience Abstracts, 20,* 6.

Smit, A., Hand, L., Freilinger, J., Bernthal, J., & Bird, A. (1990). The Iowa articulation norms project and its Nebraska replication. *Journal of Speech and Hearing Disorders, 55,* 779–798.

Smith, A. (1966). Speech and other functions after left (dominant) hemispherectomy. *Journal of Neurology, Neurosurgery, and Psychiatry, 29,* 467–471.

Smith, A. D., & Fullerton, A. M. (1981). Age differences in episodic and semantic memory: Implications for language and cognition. In D. S. Beasly & G. A. Davis (Eds.), *Aging, communication process and disorders* (pp. 139–156). New York: Grune & Stratton.

Smith, C. U. M. (in press). Descartes pineal neuropsychology. *Brain and Cognition.*

Smith, D. W., & Jones, K. L. (1982). *Recognizable patterns of human malformation: Genetic, embryologic and clinical aspects* (3rd ed.). Philadelphia: W. B. Saunders.

Smith, L., & von Tetzchner, S. (1986). Communicative, sensorimotor, and language skills of young children with Down syndrome. *American Journal of Mental Deficiency, 91,* 57–66.

Smith, M. E., Stapleton, J. M., & Halgren, E. (1986). Human medial temporal lobe potentials evoked in memory and language tasks. *Electroencephalography and Clinical Neurophysiology, 63,* 145–159.

Smith, S., Butters, N., White, R., Lyon, L., & Granholm, E. (1988). Priming semantic relations in patients with Huntington's disease. *Brain and Language, 33,* 27–40.

Snow, P., Lambier, J., Parson, C., Mooney, L., Couch, D., & Russell, J. (1987). Conversational skills following closed head injury: Some preliminary findings. In C. Field, A. Kneebone, & M. W. Reid

(Eds.), *Brain impairment: Proceedings of the eleventh annual brain impairment conference* (pp. 87–97). Richmond, Victoria: Australian Society for the study of Brain Impairment.

Snowden, J. S., Goulding, P. J., & Neary, D. (1989). Semantic dementia: A form of circumscribed cerebral atrophy. *Behavioural Neurology, 2,* 167–182.

Snowdon, D. A., Kemper, S. J., Mortimer, J. A., Greiner, L. H., Wekstein, D. R., & Markesbery, W. R. (1996). Linguistic ability in early life and cognitive function in Alzheimers disease in late life. Findings from the nun study. *Journal of the American Medical Association, 275,* 528–532.

Snyder, P. J., Novelly, R. A., & Harris, L. J. (1990). Mixed speech dominance in the intracarotid sodium amytal procedure: Validity and criteria issues. *Journal of Clinical and Experimental Neuropsychology, 12,* 629–643.

Soderfeldt, B., Ronnberg, & J. Risberg, J. (1994). Regional cerebral blood flow in sign language users. *Brain and Language, 46,* 59–68.

Sohlberg, M. M., & Mateer, C. A. (1989). *Introduction to cognitive rehabilitation.* New York: Guilford Press.

Sokol, S. M., Goodman-Schulman, R., & McCloskey, M. (1989). In defense of a modular architecture for the number-processing system: Reply to Campbell and Clark. *Journal of Experimental Psychology: General, 118,* 105–110.

Soury, J. (1899). *Le système nerveux central: structure et fonctions.* Paris: Carré et Naud.

Speedie, L. J., Brake, N., Folstein, S., Bowers, D., & Heilman, K. (1990). Comprehension of prosody in Huntington's disease. *Journal of Neurology, Neurosurgery, and Psychiatry, 53,* 607–610.

Speedie, L. J., Coslett, H. B., & Heilman, K. M. (1984). Repetition of affective prosody in mixed transcortical aphasia. *Archives of Neurology, 41,* 268–270.

Speedie, L., Wertman, E., Verfaellie, M., Zilberman, N., Lichtenstein, M., & Heilman, K. M. (1995). Contralateral neglect and reading directionality. *Journal of the International Neuropsychological Society, 1,* 142.

Spence, S. E., Godfrey, H. P. D., Knight, R. G., & Bishara, S. N. (1993). First impressions count: A controlled investigation of social skill following closed head injury. *British Journal of Clinical Psychology, 32,* 309–318.

Sperber, D., & Wilson, D. (1986). *Relevance: Communication and cognition.* London: Basil Blackwell.

Sperry, R. W., Zaidel, E., & Zaidel, D. (1979). Self-recognition and social awareness in the disconnected minor hemisphere. *Neuropsychologia, 17,* 153–166.

Spitzer, M., Weisker, I., Winter, M., Maier, S., Hermle, L., & Maher, B. (1994). Semantic and phonological priming in schizophrenia. *Journal of Abnormal Psychology, 103,* 485–494.

Spreen, O. (1973). Psycholinguistics and aphasia: The contribution of Arnold Pick. In H. Goodglass & S. Blumstein (Eds.), *Psycholinguistics and aphasia* (pp. 141–170). Baltimore: Johns Hopkins University Press.

Spreen, O., & Benton, A. L. (1977). *Neurosensory Center Comprehensive Examination for Aphasia.* Victoria, BC: Neuropsychology Laboratory, University of Victoria.

Spreen, O., Risser, A. H., & Edgell, D. (1995). *Developmental neuropsychology.* New York: Oxford University Press.

Spreen, O., & Strauss, E. (1991). *A compendium of neuropsychological tests: Administration, norms, and commentary.* Oxford: Oxford University Press.

Springer, L., & Willmes, K. (1993). Efficacy of language systematic learning approaches to treatment. In M. Paradis (Ed.), *Foundations of aphasia rehabilitation* (pp. 77–97). Oxford: Pergamon Press.

Springer, L., Willmes, K., & Haag, E. (1993). Training in the use of wh-questions and prepositions in dialogues: A comparison of two different approaches in aphasia therapy. *Aphasiology, 7,* 251–270.

Squire, L. R. (1986). Mechanisms of memory. *Science, 232,* 1612–1619.

Squire, L. R., & Zola-Morgan, S. (1991). The medial temporal lobe memory system. *Science, 253,* 1380–1386.

St. James-Roberts, I. (1981). A reinterpretation of hemispherectomy data without functional plasticity of the brain. *Brain and Language, 13,* 31–53.

Stachowiak, F. J. (1992). Was leisten Computer in der Aphasietherapie? In W. Widdig, I. Ohlendorf, T. A. Pollow, & J. P. Malin (Eds.), *Sprache und Sprechen aus neurolinguistischer und medizinischer Sicht* (pp. 85–139). Bochum: Universitätsklinik Bergmannsheil.

Stachowiak, F. J. (1993). Computer-based aphasia therapy with the Lingware/STACH system. In F. J. Stachowiak (Ed.), *Developments in the assessment and rehabilitation of brain-damaged patients* (pp. 353–380). Tübingen: Gunter Narr Verlag.

Stadie, N., Cholewa, J., De Bleser, R., & Tabatabaie, S. (1994). Das neurolinguistische Expertensystem LeMo I. Theoretischer Rahmen und Konstruktionsmerkmale des Testteils LEXIKON. *Neurolinguistik, 8,* 1–25.

Stanovich, K. E. (1988). Explaining the difference between dyslexic and garden variety poor readers: The phonological-core variable-difference model. *Journal of Learning Disabilities, 21,* 590–604.

Stark, H., & Stark, J. (1990). Syllable structure in Wernicke's aphasia. In J.-L. Nespoulous & P. Villiard (Eds.), *Morphology, phonology, and aphasia* (pp. 213–234). New York: Springer Verlag.

Starkstein, S., Federoff, J., Price, T., Leiguarda, R., & Robinson, R. (1994). Neuropsychological and neuroradiologic correlates of emotional prosody comprehension. *Neurology, 44,* 515–522.

Starkstein, S. E., Preziosi, T. J., Berthier, M. L., Bolduc, P. L., Mayberg, H. S., & Robinson, R. G. (1989). Depression and cognitive impairment in Parkinson's disease. *Brain, 112,* 1141–1153.

Staubli, U., Perez, Y., Xu, F., Rogers, G., Ingvar, M., Stone-Elander, S., & Lynch, G. (1994). Centrally active modulators of glutamate (AMPA) receptors facilitate the induction of LTP in vivo. *Proceedings of the National Acadamy of Sciences, USA, 91,* 11158–11162.

Steele, J. C. (1972). Progressive supranuclear palsy. *Brain, 95,* 693–704.

Steele, J. C., Richardson, J. C., & Olszewski, J. (1964). Progressive supranuclear palsy: A heterogeneous degeneration involving the brain stem, basal ganglia and cerebellum, with vertical gaze and pseudobulbar palsy, nuclear dystonia and dementia. *Archives of Neurology, 10,* 333–359.

Steele, R. D., Weinrich, M., Kleczewska, M. K., Wertz, R. T., & Carlson, G. S. (1987). Evaluating performance of severely aphasic patients on a computer-aided visual communication system. In R. H. Brookshire (Ed.), *Clinical aphasiology: 1987 conference proceedings* (pp. 46–54). Minneapolis: BRK Publishers.

Stein, D. G., Brailowsky, S., & Will, B. (1995). *Brain repair.* New York: Oxford University Press.

Stein, D. G., & Glazier, M. M. (1992). An overview of developments in research on recovery from brain injury. *Advances in Experimental and Medical Biology, 325,* 1–22.

Steiner, J. (1993). Ganzheitliche Aphasiebehandlung und -forschung. In M. Grohnfeldt (Ed.), *Handbuch der Sprachtherapie: Zentrale Sprach- und Sprechstörungen 6* (pp. 300–326). Berlin: Wissenschaftsverlag Volker Spiess.

Steinmetz, H., Ebeling, U., Huang, Y., & Kahn, T. (1990). Sulcus topography of the parietal opercular region: An anatomic and MR study. *Brain and Language, 38,* 515–533.

Steinmetz, H, Jäncke, L., Kleindschmidt, A., Schlaug, G., Volkman, J., & Huang, Y. (1992). Sex but no hand difference in the isthmus of the corpus callosum. *Neurology, 42,* 749–752

Steinmetz, H. & Seitz, R. J. (1991). Functional anatomy of language processing: Neuroimaging and the problem of individual variability. *Neuropsychologia, 29,* 1149–1161.

Steinmetz, H., Staiger, J. F., Schlaug, G., Huang, Y., & Jäncke, L. (1995). Corpus callosum and brain volume in women and men. *Neuroreport, 6,* 1002–1004.

Steinmetz, H., Volkmann, J., Jäncke, L., & Freund, H. J. (1991). Anatomical left-right asymmetry of language-related temporal cortex is different in left- and right-handers. *Annals of Neurology, 29,* 315–319.

Stemberger, J. P. (1984). Structural errors in normal and agrammatic speech. *Cognitive Neuropsychology, 1,* 281–313.

Stemmer, B. (1994). A pragmatic approach to neurolinguistics: Requests (re)considered. *Brain and Language, 46,* 565–591.

Stemmer, B., Giroux, F., & Joanette, Y. (1994). Production and evaluation of requests by right hemisphere brain-damaged individuals. *Brain and Language, 47,* 1–31.

Stemmer, B., & Joanette, Y. (in press). The interpretation of narrative discourse of brain-damaged individuals within the framework of a multi-level discourse model. In M. Beeman & C. Chiarello (Eds.), *Right hemisphere language comprehension: Perspectives from cognitive neuroscience.* Hove, UK: Lawrence Erlbaum.

Steno, N. (1699/1965). *Nicolas Steno's lecture on the anatomy of the brain.* Copenhagen: Nyt Nordisk Forlag.

Steriade, D. (1995). Underspecification and markedness. In J. Goldsmith (Ed.), *The handbook of phonological theory* (pp. 114–174). Cambridge, MA: Basil Blackwell.

Steriade, M., Domich, L., & Oakson, G. (1986). Reticularis thalami neurons revisited: Activity changes during shifts in states of vigilance. *Journal of Neuroscience, 6,* 68–81.

Steriade, M., Jones, E. G., & Llinas, R. R. (1990). *Thalamic oscillations and signaling.* New York: John Wiley.

Sternberg, S. (1969). The discovery of processing stages: Extensions of Donder's method. *Acta Psychologica, 30,* 276–315.

Stewart, F., Parkin, A. J., & Hunkin, N. M. (1992). Naming impairments following recovery from herpes simplex encephalitis: Category specific? *Quarterly Journal of Experimental Psychology, 44A,* 261–284.

Stokoe, W., Casterline, D., & Corneberg. C. (1965). *A dictionary of American Sign Language on linguistic principles.* Silver Spring, MD: Linstok Press.

Strang, J. D., & Rourke, B. P. (1985). Adaptive behavior of children who exhibit specific arithmetic disabilities and associated neuropsychological abilities and deficits. In B. P. Rourke (Ed.), *Neuropsychology of learning disabilities: Essentials of subtype analysis* (pp. 302–328). New York: Guilford Press.

Strauss, E., Satz, P., & Wada, J. (1990). An examination of the crowding hypothesis in epileptic patients who have undergone the carotid amytal test. *Neuropsychologia, 28,* 1221–1227.

Strome, M. (1981). Down's syndrome: A modern otorhinolaryngological perspective. *Laryngoscope, 91,* 1581–1594.

Strong, C. J., & Shaver, J. P. (1991). Stability of cohesion in the spoken narratives of language impaired and normally developing school-aged children. *Journal of Speech and Hearing Research, 34,* 95–111.

Strub, R. L., & Black, F. W. (1988). *Neurobehavioral disorders.* Philadelphia: F. A. Davis.

Studdert-Kennedy, M. (1983). Mapping speech: More analysis, less synthesis, please. *The Behavioral and Brain Sciences, 6,* 218–219.

Suen, H. K. (1990). *Principles of test theories.* Hillsdale, NJ: Lawrence Erlbaum.

Sultzer, D. L., Levin, H. S., Mahler, M. E., High, W. M., & Cummings, J. L. (1993). A comparison of psychiatric symptoms in vascular dementia and Alzheimer's disease. *American Journal of Psychiatry, 150*(12), 1806–1812.

Supalla, T. (1986). The classifier system in American Sign Language. In C. Craig (Ed.), *Noun classes and categorization* (pp. 181–214). Amsterdam: John Benjamins.

Swaab, T., Brown, C., & Hagoort, P. (1997). Spoken sentence comprehension in aphasia: Event-related potential evidence for a lexical integration deficit. *Journal of Cognitive Neuroscience, 39*–66.

Swearer, J. M., O'Donnell, B. F., Drachman, D. A., & Woodward, B. M. (1992). Neuropsychological features of familial Alzheimer's disease. *Annals of Neurology, 32,* 687–694.

Swindell, C. S., Holland, A. L., & Fromm, D. (1984). Classification of aphasia: WAB type versus clinical impression. *Clinical Aphasiology, 14,* 48–54.

Swinney, D., & Zurif, E. B. (1995). Syntactic processing in aphasia. *Brain and Language, 50,* 225–239.

Swinney, D., Zurif, E., Prather, P., & Love, T. (1994, March). Localization of elemental linguistic resources underlying structural processing: Evidence from aphasia. Poster presented at the Cognitive Neuroscience Society Annual Meeting, San Francisco, CA.

Swinney, D., Zurif, E., Prather, P., & Love, T. (1996). Neurological distribution of processing resources underlying language comprehension. *Journal of Cognitive Neuroscience, 8,* 174–184.

Szabo, C. A., & Wyllie, E. (1993). Intracarotid amobarbital testing for language and memory dominance in children. *Epilepsy Research, 15,* 239–246.

Tabossi, P., & Zardon, F. (1993). Processing ambiguous words in context. *Journal of Memory and Language, 32,* 359–372.

Taft, M. (1979). Recognition of affixed words and the word frequency effect. *Memory and Cognition, 7,* 263–272.

Taft, M. (1988). A morphological-decomposition model lexical representation. *Linguistics, 26,* 657–667.

Taft, M., & Forster, K. (1975). Lexical storage and retrieval of prefixed words. *Journal of Verbal Learning and Verbal Behavior, 14*, 635–647.

Tager-Flusberg, H. (1993). What language reveals about the understanding of minds in children with autism. In S. Baron-Cohen, H. Tager-Flusberg, & D. J. Cohen (Eds.), *Understanding other minds: Perspectives from autism* (pp. 138–157). Oxford: Oxford University Press.

Tager-Flusberg, H., & Sullivan, K. (1994). Predicting and explaining behavior: A comparison of autistic, mentally retarded and normal children. *Journal of Child Psychology and Psychiatry, 35*, 1059–1074.

Talairach, J., & Tournoux, P. (1988). Co-planar stereotaxic atlas of the human brain. 3-dimensional proportional system: An approach to cerebral imaging. (Trans. Mark Rayport). New York: Thieme.

Tallal, P. (1980). Auditory temporal perception, phonics and reading disabilities in children. *Brain and Language, 9*, 182–198.

Tallal, P., Galaburda, A., Llinas, R., & von Euler, C. (1993). *Temporal information processing in the nervous system.* New York: Annals of the New York Academy of Science, 682.

Tallal, P., Miller, S., Bedi, G., Byma, G., Wang, X., Nagarajan, S., Schriner, C., Jenkins, W., & Merzenich, M. (1996). Language comprehension in language-learning impaired children improved with acoustically modified speech. *Science, 271*, 81–84.

Tallal, P., & Piercy, M. (1974). Developmental aphasia: Rate auditory processing and selective impairment of consonant perception. *Neuropsychologia, 12*, 83–94.

Tallal, P., & Piercy, M. (1975). Developmental aphasia: The perception of brief vowels and extended stop consonants. *Neuropsychologia, 13*, 69–74.

Tallal, P., Sainburg, R. L., & Jernigan, T. (1991). The neuropathology of developmental dysphasia—behavioral, morphological and physiological evidence for a pervasive temporal processing disorder. *Reading and Writing, 3*, 363–377.

Tallal, P., & Stark R. E. (1987). Speech acoustic–cue discrimination abilities of normally developing and language impaired children. *Journal of the Acoustical Society of America, 69*, 568–574.

Tartter, V. C. (1980). Happy talk: Perceptual and acoustic effects of smiling on speech. *Perception and Psychophysics, 27*, 24–27.

Tassinari, C. A., Bureau, M., Dravet, C., Dalla Bernardina, D., & Roger, J. (1985). Epilepsy with continuous spikes and waves during slow sleep, otherwise described as ESES (epilepsy with electrical status epilepticus during slow sleep). In J. Roger, C. Dravet, M. Bureau, F. E. Dreifuss, & P. Wolf (Eds.), *Epileptic syndromes in infancy, childhood and adolescence* (pp. 194–204). London: John Libbey.

Tate, R. L., Fenelon, B., Manning, M. L., & Hunter, M. (1991). Patterns of neuropsychological impairment after severe blunt head injury. *Journal of Nervous and Mental Disease, 179*, 117–126.

Tate, R. L., Lulham, J. M., Broe, G. A., Strettles, B., & Pfaff, A. (1989). Psychosocial outcome for the survivors of severe blunt head injury: The results from a consecutive series of 100 patients. *Journal of Neurology, Neurosurgery and Psychiatry, 52*, 1128–1134.

Taylor, A. E., Saint-Cyr, J. A., & Lang, A. E. (1986). Frontal lobe dysfunction in Parkinson's disease: the cortical focus of neostrital flow. *Brain, 109*, 845–883.

Taylor, A. E., Saint-Cyr, J. A., & Lang, A. E. (1987). Parkinson's disease: Cognitive changes in relation to treatment response. *Brain, 110*, 35–51.

Taylor, A. E., Saint-Cyr, J. A., & Lang, A. E. (1990). Memory and learning in early Parkinson's disease: Evidence for a "frontal lobe syndrome". *Brain and Cognition, 13*, 211–232.

Taylor, J. (1958). *Selected writings of John Hughlings Jackson.* London: Staples Press.

Taylor, M. A., Reed, R., & Berenbaum, S. (1994). Patterns of speech disorders in schizophrenia and mania. *Journal of Nervous and Mental Disease, 182*, 319–326.

Tebay, T. E. (1848). Case report. *The Lancet I*, 260.

Tegner, R., & Levander, M. (1993). Word length coding in neglect dyslexia. *Neuropsychologia, 31*, 1217–1223.

Temple, C. M. (1986). Anomia for animals in a child. *Brain and Language, 109*, 1225–1242.

TenHouten, W. D., Hoppe, K. D., Bogen, J. E., & Walter, D. O. (1986). Alexithymia: An experimental study of cerebral commissurotomy patients and normal control subjects. *American Journal of Psychiatry, 143*, 312–316.

Terman, L., & Merrill, M. (1973). *Stanford-Binet intelligence scale.* Boston: Houghton Mifflin.

Teuber, H. L. (1955). Physiological psychology. *Annual Review of Psychology, 6,* 267–296.

Thal, L. J., Grundman, M., & Klauber, M. R. (1988). Dementia: Characteristics of a referral population and factors associated with progression. *Neurology, 38,* 1083–1090.

Thomas, P., King, K., Fraser, I., & Kendell, R. E. (1990). Linguistic performance in schizophrenia: A comparison of acute and chronic patients. *British Journal of Psychiatry, 156,* 204–210.

Thomas, R., Leudar, I., Newby D., & Johnston, M. (1993). Syntactic processing and written language output in first onset psychosis. *Journal of Communication Disorders, 26,* 209–230.

Thompson, C. K., & McReynolds, L. V. (1986). Wh-interrogative production in agrammatic aphasia: An experimental analysis of auditory-visual stimulation and direct-production treatment. *Journal of Speech and Hearing Research, 29,* 193–206.

Thompson, C. K., & Shapiro, L. P. (1995). Training sentence production in agrammatism: Implications for normal and disordered language. *Brain and Language, 50,* 201–224.

Thomsen, I. V. (1975). Evaluation and outcome of aphasia in patients with severe closed head trauma. *Journal of Neurology, Neurosurgery and Psychiatry, 38,* 713–718.

Thomsen, I. V. (1984). Late outcome of very severe blunt head trauma: A 10–15 years second follow up. *Journal of Neurology, Neurosurgery and Psychiatry, 47,* 260–268.

Tinbergen, N. (1951). *The study of instinct.* Oxford: Oxford University Press.

Tissot, R. J., Mounin, G., & Lhermitte, F. (1973). *L'agrammatisme.* Brussels: Dessart.

Toga, A. W., & Mazziotta, J. C. (Eds.). (1996). *Brain mapping: The methods.* San Diego: Academic Press.

Togher, L., Hand, L., & Code, C. (1995). The issue of power in relationships: Evidence from the analysis of systemic functional grammar. In A. Ferguson (Ed.), *Proceedings of the aphasiology symposium of Australia* (pp. 101–116). Newcastle: Coop Bookshop.

Tollkühn, S. (in prep.). *Zur Akzeptanz des Computertrainings in Selbsthilfegruppen.* Leipzig: Universität, Förderpädagogik (Prof. Stachowiak).

Tomoeda, C. K., & Bayles, K. A. (1993). Longitudinal effects of Alzheimer disease on discourse production. *Alzheimer Disease and Associated Disorders, 7,* 223–236.

Tompkins, C. A. (1991a). Automatic and effortful processing of emotional intonation after right or left hemisphere brain damage. *Journal of Speech and Hearing Research, 34,* 820–830.

Tompkins, C. A. (1991b). Redundancy enhances emotional inferencing by right- and left-hemisphere-damaged adults. *Journal of Speech and Hearing Research, 34,* 1142–1149.

Tompkins, C. A. (1995). *Right hemisphere communication disorders: Theory and management.* San Diego: Singular Press.

Tompkins, C. A., Baumgaertner, A., Lehman, M. L., & Fossett, T. R. D. (1995). Suppression and discourse comprehension in right-brain-damaged adults. *Brain and Language, 51,* 181–183.

Tompkins, C. A., Boada, R., McGarry, K., Jones, J., Rahn, A. E., & Ranier, S. (1992). Connected speech characteristics of right-hemisphere-damaged adults: a re-examination. In M. L. Lemme (Ed.) *Clinical Aphasiology* (Vol. 21, pp. 113–122).

Toso, V., Moschini, M., Cagnini, G., & Antoni, D. (1981). Aphasie acquise de l'enfant avec épilepsie. *Revue Neurologique, 137,* 425–434.

Townsend, D. W., Geissbuhler, A., Defrise, M., Hoffman, E. J., Spinks, T. J., Bailey, D. L., Gilardi, M.-C., & Jones, T. (1991). Fully three-dimensional reconstruction for a PET camera with retractable septa. *IEEE Transactions in Medical Imaging, 10,* 505–512.

Tracy, J. Glosser, G., & DellaPietra, L. (1996). A cognitive/linguistic model of single word production abnormalities in schizophrenia: Data from two case reports. *Brain and Cognition, 30,* 311–315.

Tramo, M. J., Baynes, K., Fendrich, R., Mangun, G. R., Phelps, E. A., Reuter-Lorenz, P. A., & Gazzaniga, M. S. (1995). Hemispheric specialization and interhemispheric integration: Insights from experiments with commissurotomy patients. In A. G. Reeves & D. W. Roberts (Eds.), *Epilepsy and the corpus callosum 2* (pp. 263–295). New York: Plenum Press.

Tramo, M. J., & Bharucha, J. J. (1991). Musical priming by the right hemisphere post-callosotomy. *Neuropsychologia, 29,* 313–325.

Tranel, D. (1992). Neurology of language. *Current opinion in neurology and neurosurgery, 5,* 77–82.

Tröster, A. I., Osorio, I., Fields, J. A., Lai, C.-W., Eckard, D. A., Preston, D. F., & Whitaker, H. A. (1996). Transient alexia (letter-by-letter reading) after single photon emission computed tomography-verified selective left posterior cerebral artery sodium amytal injection. *Neuropsychiatry, Neuropsychology, and Behavioral Neurology, 9,* 209–217.

Treiman, R. (1992). The role of intra-syllabic units in learning to read and spell. In P. B. Gough, L. C. Ehri, & R. Treiman (Eds.), *Reading acquisition* (pp. 65–106). Hillsdale, NJ: Lawrence Erlbaum.

Treisman, A., & Sato, S. (1990). Conjunction search revisited. *Journal of Experimental Psychology: Human Perception and Performance, 16,* 459–478.

Trojano, L., Balbi, P., Russo, G., & Elefante, R. (1994). Patterns of recovery and change in verbal and nonverbal functions in a case of crossed aphasia: Implications for models of functional brain lateralization and localization. *Brain and Language, 46,* 637–661.

Trupe, E. H. (1984). Reliability of rating spontaneous speech in the Western Aphasia Battery: Implications for classification. *Clinical Aphasiology, 14,* 55–69.

Tseng, C. H., McNeil, M. R., & Milenkovic, P. (1993). An investigation of attention allocation deficits in aphasia. *Brain and Language, 45,* 276–296.

Tsvjetkova, L. S., & Glozman, Z. M. (1978). *Agrammatism pri afasji.* Moscow: University of Moscow Press.

Tucker, D. M. (1981). Lateral brain function, emotion, and conceptualization. *Psychological Bulletin, 89,* 19–46.

Tucker, D. M. (1992). Developing emotions and cortical networks. In M. Gunnar & C. Nelson (Eds.), *Developmental behavioral neuroscience: Minnesota symposium on child psychology* (Vol. 24, pp. 75–127). Hillsdale, NJ Lawrence: Erlbaum.

Tucker, D. M. (1993). Emotional experience and the problem of vertical integration: Discussion of the special section on emotion. *Neuropsychology, 7,* 500–509.

Tucker, D. M., Liotti, M., Potts, G. F., Russell, G. S., & Posner, M. I. (1994). Spatiotemporal analysis of brain electrical fields. Human Brain Mapping, 1, 134–152.

Tucker, D. M., Luu, P., & Pribram, K. H. (1995). Social and emotional self-regulation. *Annals of the New York Academy of Sciences, 769,* 213–239.

Tucker, D. M., & Williamson, P. A. (1984). Asymmetric neural control systems in human self-regulation. *Psychological Review, 91,* 185–215.

Tulving, E. (1983). *Elements of episodic memory.* Oxford: Oxford University Press.

Tulving, E., Kapur, S., Craik, F. I. M., Moscovitch, M., & Houle, S. (1994). Hemispheric encoding/retrieval asymmetry in episodic memory: Positron emission tomography findings. *Proceedings of the National Academy of Sciences, 91,* 2016–2020.

Tuomainen, J., & Laine, M. (1991). Multiple oral re-reading technique in rehabilitation of pure alexia. *Aphasiology, 5,* 401–409.

Tureen, L., Smolik, E., & Tritt, J. (1951). Aphasia in a deaf mute. *Neurology, 1,* 237–244.

Turkewitz, G. (1988). A prenatal source for the development of hemispheric specialization. In D. L. Molfese & S. J. Segalowitz (Eds.), *Brain lateralization in children: Developmental implications* (pp. 73–81). New York: Guilford Press.

Turkstra, L., McDonald, S., & Kaufman, P. (1995). A test of pragmatic language function in traumatically brain-injured adolescents. *Brain Injury, 10,* 329–345.

Tweedy, J. R., Langer, K. G., & McDowell, F. H. (1982). The effect of semantic relations on the memory deficit associated with Parkinson's disease. *Journal of Clinical Neuropsychology, 4,* 235–247.

Twist, D., Squires, N., Spielholz, N., & Silverglide, R. (1991). Event-related potentials in disorders of prosodic and semantic linguistic processing 1. *Neuropsychiatry, Neuropsychology, and Behavioral Neurology, 4,* 281–304.

Tyler, L. K., Behrens, S., Cobb, H., & Marslen-Wilson, W. (1990). Processing distinctions between stems and affixes: Evidence from a non-fluent aphasic patient. *Cognition, 36,* 129–153.

Tyler, L. K., & Cobb, H. (1987). Processing bound grammatical morphemes in context: The case of an aphasic patient. *Language and Cognitive Processes, 2,* 245–262.

Tyler, S. K., & Tucker, D. M. (1982). Anxiety and perceptual structure: Individual differences in neuropsychological function. *Journal of Abnormal Psychology, 91,* 210–220.

Udwin, O., & Yule, W. (1991). A cognitive and behavioral phenotype in Williams syndrome. *Journal of Clinical and Experimental Neuropsychology, 13,* 232–244.

Ulatowska, H. K., Allard, L., Donnell, A., Bristow, J., Macaluso-Haynes, S., Flower, A., & North, A. J. (1988). Discourse performance in subjects with dementia of the Alzheimer type. In H. A. Whitaker (Ed.), *Neuropsychological studies of non focal brain damage: Trauma and dementia* (pp. 108–131). New York: Springer Verlag.

Ulatowska, H. K., Allard, L., Reyes, B. A., Ford, J., & Chapman, S. (1992). Conversational discourse in aphasia. *Aphasiology, 6,* 325–331.

Ulatowska, H. K., & Chapman, S. B. (1994). Discourse macrostructure in aphasia. In R. L. Bloom, L. K. Obler, S. De Santi, & J. Ehrlich (Eds.), *Discourse analysis and application: Studies in adult clinical populations* (pp. 29–46). Hillsdale, NJ: Lawrence Erlbaum.

Ulatowska, H. K., Doyel, A. W., Freedman-Stern, R. F., Macaluso-Haynes, S., & North, A. J. (1983). Production of procedural discourse in aphasia. *Brain and Language, 18,* 315–341.

Ulatowska, H. K., Freedman-Stern, R. F., Doyel, A. W., Macaluso-Haynes, S., & North, A. J. (1983). Production of narrative discourse in aphasia. *Brain and Language, 19,* 317–334.

Ulatowska, H. K., North, A. J., & Macaluso-Haynes, S. (1981). Production of narrative and procedural discourse in aphasia. *Brain and Language, 13,* 345–371.

Underwood, J. K., & Paulson, C. J. (1981). Aphasia and congenital deafness: A case study. *Brain and Language, 12*(2), 285–291.

Ungerleider, L. G. (1995). Functional brain imaging studies of cortical mechanisms for memory. *Science, 270,* 769–775.

Ungerleider, L. G., & Mishkin, M. (1982). Two cortical visual systems. In D. J. Ingle, M. A. Goodale, & R. W. J. Mansfield (Eds.), *The analysis of visual behavior* (pp. 594–586). Cambridge: MIT Press.

Vaid, J., Bellugi, U., & Poizner, H. (1989). Hand dominance for signing: Clues to brain lateralization of language. *Neuropsychologia, 27,* 949–960.

Vaid, J., & Hall, D. G. (1991). Neuropsychological perspectives on bilingualism: Right, left, and center. In A. Reynolds (Ed.), *Bilingualism, multiculturalism and second language learning* (pp. 81–112). Hillsdale, NJ: Lawrence Erlbaum.

Vaid, J., & Singh, M. (1989). Asymmetries in the perception of facial affect: Is there an influence of reading habits? *Neuropsychologia, 27,* 1277–1287.

Valdés-Sosa, M., Gonzalez, A., Xiang, L., Yi, H., & Bobes, M. A. (1993). Brain potentials in a phonological matching task using Chinese characters. *Neuropsychologia, 8,* 853–864.

Valdois, S. (1990). Internal structure of two consonant clusters. In J.-L. Nespoulous & P. Villiard (Eds.), *Morphology, phonology, and aphasia* (pp. 253–269). New York: Springer Verlag.

Valdois, S., Joanette, Y., & Nespoulous, J. L. (1989). Intrinsic organization of sequences of phonemic approximations: A preliminary study. *Aphasiology, 3,* 55–73.

Valenstein, E. S. (1990). The prefrontal area and psychosurgery. *Progress in Brain Research, 85,* 539–554.

Valentin, M. (1988). *François Broussais: Empereur de la médecine.* Cesson-Sévigné: La Presse de Bretagne.

Vallar, G., & Baddeley, A. D. (1984). Fractionation of working memory: Neuropsychological evidence for a phonological short-term store. *Journal of Verbal Learning and Verbal Behavior, 23,* 151–161.

Vallar, G., & Baddeley, A. D. (1987). Phonological short-term store and sentence processing. *Cognitive Neuropsychology, 4,* 417–438.

Vallar, G., & Baddeley, A. D. (1989). Developmental disorders of verbal short-term memory and their relation to sentence comprehension: A reply to Howard and Butterworth. *Cognitive Neuropsychology, 6,* 465–473.

Vallar, G., & Papagno, C. (1993). Preserved vocabulary acquisition in Down's syndrome: The role of phonological short-term memory. *Cortex, 29,* 467–483.

Vallar, G., Perani, D., Cappa, S. F., Messa, C., Lenzi, G. L., & Fazio, F. (1988). Recovery from aphasia and neglect after subcortical stroke: Neuropsychological and cerebral perfusion study. *Journal of Neurology, Neurosurgery, and Psychiatry, 51,* 1269–1276.

Valsiner, J. (Ed.). (1986). *The individual subject and scientific psychology.* New York: Plenum Press.

Van Bourgondien, M. E., & Mesibov, G. B. (1987). Humour in high-functioning autistic adults. *Journal of Autism and Developmental Disorders, 17,* 417–424.

Van De Sandt-Koenderman, M. (1994). Multicue, a computer program for word finding in aphasia. In Forschungsprojekt Auswege (Eds.), *Sprache-Therapie-Computer. 1. Internationaler Kongress* (abstracts). Graz, Austria: University of Graz.

Van De Sandt-Koenderman, W. M. E., Smit, I. A.C., Van Dongen, H. R., & Van Hest, J. B. C. (1986). A case of acquired aphasia and convulsive disorder: Some linguistic aspects of recovery and breakdown. *Brain and Language, 21,* 174–183.

Van Demark, A .A. (1982). Predicting post-treatment scores on the Boston Diagnostic Aphasia Examination. *Clinical Aphasiology, 12,* 103–110.

Van der Linden, M. (1994). Neuropsychologie de la mémoire. In X. Seron & M. Jeannerod (Eds.), *Traité de neuropsychologie humaine* (pp. 282–316). Brussels: Mardaga.

Van der Linden, M., Coyette, F., & Seron, X. (1992). Selective impairment of the central executive component of working memory: A single case. *Cognitive Neuropsychology, 9,* 301–326.

van Dijk, T. A. (1985). *Handbook of discourse analysis* (Vols. 1–4). London: Academic Press.

van Dijk, T. A., & Kintsch, W. (1983). *Strategies of discourse comprehension.* New York: Academic Press.

Van Gorp, W. G., Satz, P., & Mitrushina, M. (1990). Neuropsychological processes associated with normal aging. *Developmental Neuropsychology, 6,* 279–290.

Van Hout, A. (1992). Acquired aphasia in children. In F. Boller & J. Grafman (Series Eds.) & S. J. Segalowitz & I. Rapin (Vol. Eds.), *Handbook of neuropsychology.* Vol. 7. *Child neuropsychology* (pp. 139–161). Amsterdam: Elsevier.

Van Lancker, D. (1980). Cerebral lateralization of pitch cues in the linguistic signal. *International Journal of Human Communication, 13,* 201–227.

Van Lancker, D. (1988). Nonpropositional speech: Neurolinguistic studies. In A. Ellis (Ed.), *Progress in the psychology of language (Vol. 3,* pp. 49–118). London: Lawrence Erlbaum.

Van Lancker, D. (1990). The neurology of proverbs. *Behavorial Neurology, 3,* 169–187.

Van Lancker, D., & Canter, G. J. (1982). Impairment of voice and face recognition in patients with hemispheric damage. *Brain and Cognition, 1,* 185–195.

Van Lancker, D., & Fromkin, V. A. (1973). Hemispheric specialization for pitch and tone: Evidence from Thai. *Journal of Phonetics, 1,* 101–109.

Van Lancker, D., & Kempler, D. (1987). Comprehension of familiar phrases by left- but not by right-hemisphere damaged patients. *Brain and Language, 32,* 265–277.

Van Lancker, D., Kreiman, J., & Cummings, J. (1989). Voice perception deficits: Neuroanatomical correlates of phonagnosia. *Journal of Clinical and Experimental Neuropsychology, 11,* 665–674.

Van Lancker, D., & Nicklay, C. (1992). Comprehension of personally relevant (PERL) versus novel language in two globally aphasic patients. *Aphasiology, 6,* 37–61.

Van Lancker, D., Pachana, N. A., Cummings, J. L., Sidtis, J., & Erickson, C. (1996). Dysprosodic speech following basal ganglia stroke: Role of frontosubcortical circuits. *Journal of the International Neuropsychological Society, 2,* 5.

Van Lancker, D., & Sidtis, J. (1992). The identification of affective-prosodic stimuli by left- and right-hemisphere-ddamaged subjects: All errors are not created equal. *Journal of Speech and Hearing Research, 35,* 963–970.

Van Petten, C. (1993). A comparison of lexical and sentence-level context effects in event-related potentials. *Language and Cognitive Processes, 8,* 458–531.

Van Petten, C. (1995). Words and sentences: Event-related potential measures. *Psychophysiology, 32,* 511–525.

Van Petten, C., & Kutas, M. (1990). Interactions between sentence context and word frequency in event-related brain potentials. *Memory and Cognition, 18,* 380–393.

Van Petten, C., & Kutas, M. (1991). Influences of semantic and syntactic context on open- and closed-class words. *Memory and Cognition, 19,* 95–112.

Van Petten, C., Kutas, M., Kluender, R., Mitchener, M., & McIsaac, H. (1991). Fractionating the word repetition effect with event-related potentials. *Journal of Cognitive Neuroscience, 3,* 131–150.

Van Rossum, A. (1972). Spastic pseudosclerosis (Creutzfeldt-Jakob disease). In P. J. Vinken & G. W. Bruyn (Eds.), *Diseases of the basal ganglia* (Vol. 6, pp. 726–760). Amsterdam: North-Holland Publishing Company.

Vanier, M., & Caplan, D. (1990). CT-scan correlates of agrammatism. In L. Menn & L. Obler (Eds.), *Agrammatic aphasia: A cross-linguistic narrative sourcebook* (Vol. 1, pp. 37–114). Amsterdam: John Benjamins.

Varga-Khadem, F., O'Gorman, A. M., & Walters, G. V. (1985). Aphasia and handedness in relation to hemispheric side, age at injury and severity of cerebral lesion during childhood. *Brain, 108,* 677–696.

Vico, G. B. (1744). Principi di scienza nuova. terza impressione. Naples: Stamperia Muziana.

Vignolo, L. A. (1964). Evolution of aphasia and language rehabilitation: A retrospective exploratory study. *Cortex, 1,* 344–367.

Vijayan, A., & Gandour, J. (1995). On the notion of a "subtle phonetic deficit" in fluent/posterior aphasia. *Brain and Language, 48,* 106–119.

Villardita, C. (1993). Alzheimers disease compared with cerebrovascular dementia: Neuropsychological similarities and differences. *Acta Neurologica Scandinavica, 87,* 299–308.

Vitu, F., O'Regan, J. K., Inhoff, A. W., & Topolski, R. (1995). Mindless reading: Eye movement characteristics are similar in scanning letter strings and reading texts. *Perception & Psychophysics, 57,* 352–364.

Voeller, K. K. S., & Heilman, K. M. (1988). Attention deficit disorder in children: A neglect syndrome? *Neurology, 38,* 806–808.

Volden, J., & Lord, C. (1991). Neologisms and idiosyncratic language in autistic speakers. *Journal of Autism and Developmental Disorders, 21,* 109–130.

Von Monakow, C. (1914). Diaschisis. In K. H. Pribram (Ed.), *Brain and behaviour I: Moods, states and mind* (Trans. G. Harris, 1969) (pp. 27–36). Baltimore: Penguin.

Von Monakow, C. (1914). *Die Lokalisation im Grosshirn und der Abbau der Funktion durch kortikale Herde.* Wiesbaden: Bergmann.

Vriezen, E. R., & Moscovitch, M. (1990). Memory for temporal order and conditional associative-learning in patients with Parkinson's disease. *Neuropsychologia, 28,* 1283–1293.

Vroomen, J., & de Gelder, B. (1995). Metrical segmentation and lexical inhibition in spoken word recognition. *Journal of Experimental Psychology: Human Perception and Performance, 21,* 89–108.

Wada, J. (1949). A new method for determining the side of cerebral dominance: A preliminary report on the intra-carotid injection of sodium amytal in man. *Igaku to Seibutsugaku* (Medicine and Biology), *14,* 221–222.

Wada, J., & Rasmussen, T. (1960). Intracarotid injection of sodium amytal for the lateralization of cerebral speech dominance: Experimental and clinical observations. *Journal of Neurosurgery, 17,* 266–282.

Wade, D. T., Hewer, R. L., David, R. M., & Enderby, P. M. (1986). Aphasia after stroke: Natural history and associated deficits. *Journal of Neurology, Neurosurgery, and Psychiatry, 49,* 11–16.

Wainer, H., & Brown, H. P. (1988). *Test validity.* Hillsdale, NJ: Lawrence Erlbaum.

Wallace, G. L., & Canter, G. J. (1985). Effects of personally relevant language materials on the performance of severely aphasic individuals. *Journal of Speech and Hearing Disorders, 50,* 385–390.

Wallesch, C.-W., Bak, T., & Schulte-Mönting, J. (1992). Acute aphasia-patterns and prognosis. *Aphasiology, 6,* 373–385.

Wallesch, C.-W., & Papagno, C. (1988). Subcortical aphasia. In F. C. Rose, R. Whurr, & M. A. Wyke (Eds.), *Aphasia* (pp. 256–287). London: Whurr Publishers.

Walsh, K. W. (1986). *Understanding brain damage: A primer of neuropsychological evaluation.* Edinburgh: Churchill Livingston.

Wang, P. P., & Bellugi, U. (1993). Williams syndrome, Down syndrome, and cognitive neuroscience. *Cognitive Neuroscience, 147,* 1246–1251.

Wang, P. P., & Bellugi, U. (1994). Evidence from two genetic syndromes for a dissociation between verbal and visual-spatial short-term memory. *Journal of Clinical and Experimental Neuropsychology, 16,* 317–322.

Wang, P. P., Doherty, S., Hesselink, J. R., & Bellugi, U. (1992). Callosal morphology concurs with neurobehavioral and neuropathological findings in two neurodevelopmental disorders. *Archives of Neurology, 49,* 407–411.

Wang, P. P., Hesselink, J. R., Jernigan, T. L., Doherty, S., & Bellugi, U. (1992). Specific neurobehavioral profile of Williams syndrome is associated with neocerebellar hemispheric preservation. *Neurology, 42,* 1999–2002.

Warburton, E., Wise, R. J. S., Price, C. J., Weiller, C., Hadar, U., Ramsay, S., & Frackowiak R. S. J. (1996). Noun and verb retrieval by normal subjects: Studies with PET. *Brain, 119,* 159–179.

Warrington, E. K. (1975). The selective impairment of semantic memory. *Quarterly Journal of Experimental Psychology, 27,* 635–657.

Warrington, E. K. (1981a). Concrete word dyslexia. *British Journal of Psychology, 72,* 175–196.

Warrington, E. K. (1981b). Neuropsychological studies of verbal semantic systems. *Philosophical Transactions of the Royal Society of London, B295,* 411–423.

Warrington, E. K., & Cipolotti, L. (1996). Word comprehension: The distinction between refractory and storage impairments. *Brain, 119,* 611–625.

Warrington, E. K., Cipolotti, L., & McNeil, J. (1993). Attentional dyslexia: A single case study. *Neuropsychologia, 31,* 871–885.

Warrington, E. K., & McCarthy, R. (1983). Category-specific access dysphasia. *Brain, 106,* 859–878.

Warrington, E. K., & McCarthy, R. (1987). Categories of knowledge: Further fractionation and an attempted integration. *Brain, 110,* 1273–1296.

Warrington, E. K., & McCarthy, R. A. (1994). Multiple meaning systems in the brain: A case for visual semantics. *Neuropsychologia, 32,* 1465–1473.

Warrington, E. K., & Shallice, T. (1979). Semantic access dyslexia. *Brain, 102,* 43–63.

Warrington, E. K., & Shallice, T. (1980). Word form dyslexia. *Brain, 103,* 99–112.

Warrington, E. K., & Shallice, T. (1984a). Category specific semantic impairments. *Brain, 102,* 43–63.

Warrington, E. K., & Shallice, T. (1984b). Category-specific semantic impairment. *Brain, 107,* 829–854.

Watanabe, M. (1990). Prefrontal unit activity during associative learning in the monkey. *Experimental Brain Research, 80,* 296–309.

Waters, G. S., & Caplan, D. (1995). On the nature of the phonological output planning processes involved in verbal rehearsal: Evidence from aphasia. *Brain and Language, 48,* 191–220.

Waters, G. S., & Caplan, D. (1996). The capacity theory of sentence comprehension: Critique of Just and Carpenter. *Psychological Review,* 761–772.

Waters, G. S., Caplan, D., & Hildebrandt, N. (1991). On the structure of verbal short-term memory and its functional role in sentence comprehension: Evidence from neuropsychology. *Cognitive Neuropsychology, 8,* 82–126.

Waters, G. S., Caplan, D., & Rochon, E. (1995). Processing resources and sentence comprehension in patients with Alzheimer's disease. *Cognitive Neuropsychology, 12,* 1–30.

Webb, W. G. (1995). Language batteries in aphasia. In H. Kirshner (Ed.), *Handbook of neurological speech and language disorder,* (Vol. 33, pp. 431–441). New York: M. Kekker.

Weber-Fox, C., & Neville, H. (1996). Maturational constraints of functional specializations for language processing: ERP and behavioral evidence in bilingual speakers. *Journal of Cognitive Neuroscience, 8* 231–256.

Wechsler, A. (1973). The effect of organic brain disease on recall of emotionally charged vs. neutral narrative texts. *Neurology, 23,* 130–135.

Wechsler, A., Verity, A., Rosenchein, S., Fried, I., & Schrebel, A. (1982). A clinical, computed tomographic, and histologic study with Golgi impregnation observations. *Archives of Neurology, 39,* 287–290.

Weigl, E. (1979). Neuropsychologische und neurolinguistische Grundlagen eines Programms zur Rehabilitation aphasischer Störungen. In G. Peuser (Ed.), *Studien zur Sprachtherapie.* München: Fink.

Weigl, E., & Bierwisch, M. (1970). Neuropsychology and linguistics: Topics of common research. *Language, 6,* 1–18.

Weiller, C., Chollet, F., Friston, K. J., Wise, R. J. S., & Frackowiak, R. S. J. (1992). Functional reorganization of the brain in recovery from striatocapsular infarction in man. *Annals of Neurology, 31,* 463–472.

Weiller, C., Isensee, C., Rijintjes, M., Huber, W., Müller, S., Bier, D., Dutschka, K., Woods, R. P., Noth, J., & Diener, H. C. (1995). Recovery from Wernicke's aphasia—a PET study. *Annals of Neurology, 37,* 723–732.

Weiller, C., Willmes, K., Reiche, W., Thron, A., Insensee, C., Buell, U., & Ringelstein, E. B. (1993). The case of aphasia or neglect after striatocapsular infarction. *Brain, 116,* 1509–1525.

Weinberger, D. R. (1987). Implications of normal brain development for the pathogenesis of schizophrenia. *Archives of General Psychiatry, 44,* 660–669.

Weinberger, N. M. (1995). Dynamic regulation of receptive fields and maps in the adult sensory cortex. *Annual Review of Neuroscience, 18,* 129–158.

Weinberger, N. M., Ashe, J. H., Metherate, R., McKenna, T. M., Diamond, D. M., & Bakin, J. (1990). Retuning auditory cortex by learning: A preliminary model of receptive field plasticity. *Concepts in Neuroscience, 1,* 91–132.

Weinrich, M., McCall, D., Weber, C., Thomas, K., & Thornburg, L. (1995). Training on an iconic communication system for severe aphasia can improve natural language production. *Aphasiology, 9,* 343–364.

Weinrich, M., Steele, R. D., Kleczewska, M., Carlson, G. S., Baker, E., & Wertz, R. T. (1989). Representation of verbs in a computerized visual communication system, *Aphasiology 3,* 501–512.

Weinstein, E. A., & Kahn, R. B. (1955). *Denial of illness* (pp. 62–63). New York: Charles Thomas.

Weisenburg, T., & McBride, K. (1935). *Aphasia* (2nd reprint, 1973). New York: Hafner.

Weniger, D. (1993). Disorders of prosody in aphasia. In G. Blanken, J. Dittman, H. Grimm, J. Marshall, & C. Wallesch (Eds.), *Linguistic disorders and pathologies: An international handbook* (pp. 209–215). Berlin: Walter de Gruyter.

Weniger, D., & Bertoni, B. (1993). Which route to aphasia therapy? In A. L. Holland & M. M. Forbes (Eds.), *Aphasia treatment: World perspectives* (pp. 291–318). San Diego: Singular Publishing.

Weniger, D., & Sarno, M. T. (1990). The future of aphasia therapy: More than just new wine in old bottles? *Aphasiology, 4,* 301–306.

Weniger, D., Springer, L., & Poeck, K. (1987). The efficacy of deficit-specific therapy materials. *Aphasiology, 1,* 215–222.

Wenzel, R., Bartenstein, P., Dieterich, M., Danek, A., Weindl, A., Minoshima, S., Ziegler, S., Schwaiger, M., & Brandt, T. (1996). Deactivation of human visual cortex during involuntary ocular oscillations. *Brain, 119,* 101–110.

Wepfer, J. J. (1727). *Observationes medico-practicae de affectionis capitis internis & externis.* Schaffhausen: Ziegler.

Wepman, J., Bock, R. D., Jones, L. V., & Van Pelt, D. (1956). Psycholinguistic study of aphasia: A revision of the concept of anomia. *Journal of Speech and Hearing Disorders, 21,* 468–474.

Werner, H. (1957). *The comparative psychology of mental development.* New York: Harper.

Wernicke, C. (1874). *Der aphasische Symptomenkomplex.* Breslau: Cohn and Weigert.

Wertz, R. T. (1979). Word fluency measures. In F. L. Darley (Ed.), *Evaluation of appraisal techniques in speech and language pathology* (pp. 243–246). Reading, MA: Addison-Wesley.

Wertz, R. T. (1995). Efficacy. In C. Code & D. Müller (Eds.), *Treatment of aphasia: From theory to practice* (pp. 309–339). London: Whurr Publishers.

Wertz, R. T., Deal, J. L., & Robinson, A. J. (1984). Classifying the aphasias: A comparison of the Boston Diagnostic Aphasia Examination and the Western Aphasia Battery. *Clinical Aphasiology, 14,* 40–47.

Wertz, R. T., Dronkers, N. F., & Shubitowski, Y. (1986). Discriminant function analysis of performance by normals and left hemisphere, right hemisphere, and bilaterally brain damaged patients on a word fluency measure. *Clinical Aphasiology, 16,* 257–266.

Wertz, R. T., Dronkers, N. F., and Hume, J. L. (1992). PICA intrasubtest variability and prognosis for improvement in aphasia. *Clinical Aphasiology, 21,* 207–211

Westermann, R., & Hager, W. (1983). On severe tests of trend hypotheses in psychology. *Psychological Record, 33,* 201–211.

Westermann, R., & Hager, W. (1986). Error probabilities in educational and psychological research. *Journal of Educational Statistics, 11,* 117–146.

Westerveld, M., Zawacki, T., Sass, K. J., Spencer, S., Novelly, R. A., & Spencer, D. D. (1994). Intracarotid amytal procedure evaluation of hemispheric speech and memory function in children and adolescents. *Journal of Epilepsy, 7,* 295–302.

Wetzel, W. F., & Molfese, D. L. (1992). The processing of presuppositional information contained in sentences: Electrophysiological correlates. *Brain and Language, 42,* 286–307.

Weylman, S. T., Brownell, H. H., Roman, M., & Gardner, H. (1989). Appreciation of indirect requests by left and right brain-damaged patients: The effects of verbal context and conventionality of wording. *Brain and Language, 36,* 580–591.

Wheeler, R., Davidson, R., & Tomarken, A. (1993). Frontal brain asymmetry and emotional reactivity: a biological substrate of affective style. *Psychophysiology, 30,* 82–89.

Whitaker, H. A. (1979). Electrical stimulation mapping of language cortex. *Experiemental Brain Research* (Suppl. 2, *Hearing mechanisms and speech),* 193–204.

Whitaker, H. A. (1988). William Elder (1864–1931): Diagram Maker and Experimentalist. In L. Hyman and C. Li (Eds.), *Language, speech and mind* (pp. 163–174). London: Routledge.

Whitaker, H. A. (1997). Some comments on pre-modern agrammatism research. In H. A. Whitaker (Ed.), *Agrammatism.* San Diego: Singular Press.

Whitaker, H. A., (Ed.). (1988). *Phonological processes and brain mechanisms.* New York: Springer Verlag.

Whitaker, H. A., & Etlinger, S. C. (1993). Theodor Meynert's contribution to classical 19th century aphasia studies. *Brain and Language, 45,* 560–571.

Whitaker, H., Habiger, J., & Ivers, R. (1985). Acalculia from a lenticular-caudate lesion. *Neurology, 35* (Suppl. 1), 161.

Whitaker, H. A., Markovits, H., Savary, F., Grou, C., & Braun, C.J. (1991). Inference deficits after brain damage. *Journal of Clinical and Experimental Neuropsychology, 13,* 38.

Whitaker, H. A., & Ojemann, G. (1977a). Graded localization of naming from electrical stimulation mapping of left cerebral cortex. *Nature, 270,* 50–51.

Whitaker, H. A., & Ojemann, G. (1977b). Lateralization of higher cortical functions: A critique. *Annals of the New York Academy of Sciences, 299,* 459–473.

Whitaker, H. A., & Selnes, O. A. (1976). Anatomic variations in the cortex: Individual differences and the problem of the localization of language functions. In S. Harnad, H. Steklis, & J. Lancaster (Eds.), *Origin and evolution of language and speech* (pp. 844–854). New York: New York Academy of Sciences, 280.

Whitaker, H. A., & Slotnick, H. (1988). Comments on "The case for single patient studies": Is (neuro)psychology possible? *Cognitive Neuropsychology, 5,* 529–534.

White, H., & Sreenivasan, V. (1987). Epilepsy-aphasia syndrome in children: An unusual presentation to psychiatry. *Canadian Journal of Psychiatry, 32,* 599–601.

Whiteman, B. C., Simpson, G. B., & Compton, W. C. (1986). Relationship of otitis media and language impairment in adolescents with Down syndrome. *Mental Retardation, 24,* 353–356.

Whorf, B. L. (1961). Linguistic relativity and the relation of linguistic processes to perception and cognition. In S. Saporta (Ed.), *Psycholinguistics: A book of readings* (pp. 460–467). New York: Holt, Rinehart & Winston.

Whyte, J. (1994). Attentional processes and dyslexia. *Cognitive Neuropsychology, 11,* 99–116.

Wiig, E. H., & Secord, W. (1989). *Test of language competence: Expanded edition.* San Antonio, TX: Psychological Corporation.

Wilcox, S. (1992). *The phonetics of fingerspelling.* (Series: Studies in speech pathology and clinical linguistics) (pp. 1–108). Amsterdam: John Benjamins.

Williams, J., & Rausch, R. (1992). Factors in children that predict performance on the intracarotid amobarbital procedure. *Epilepsia, 33,* 1036–1041.

Williams, J. C. P., Barratt-Boyes, B. G., & Lowe, J. B. (1961). Supravalvular aortic stenosis. *Circulation, 24,* 1311.

Williams, J. K. (1994). Behavioral characteristics of children with Turner syndrome and children with learning disabilities. *Western Journal of Nursing Research, 16,* 26–39.

Williams, J. K., Richman, L. C., & Yarbrough, D. B. (1992). Comparison of visual-spatial performance strategy training in children with Turner syndrome and learning disabilities. *Journal of Learning Disabilities, 25,* 658–664.

Williams, S. E., & Canter, G. J. (1987). Action naming performance in four syndromes of aphasia. *Brain and Language, 32,* 124–136.

Willmes, K. (1990). Statistical methods for a single-case study approach to aphasia therapy research. *Aphasiology, 4,* 415–436.

Willmes, K. (1993). Diagnostic methods in aphasiology. In G. Blanken, J. Dittmann, H. Grimm, J. C. Marshall, & C.-W. Wallesch (Eds.), *Linguistic disorders and pathologies: An international handbook* (pp. 137–153). Berlin: Walter de Gruyter.

Willmes, K. (1995). Aphasia therapy research: Some psychometric considerations and statistical methods for the single-case study approach. In C. Code and D. Müller (Eds.), *Treatment of aphasia: From theory to practice* (pp. 286–308). London: Whurr Publishers.

Willmes, K., & Poeck, K. (1993). To what extent can aphasic syndromes be localized? *Brain, 116,* 1527–1540.

Willmes, K., Poeck, K., Weniger, D., & Huber, W. (1983). Facet theory applied to the construction and validation of the Aachen Aphasia Test. *Brain and Language, 18,* 259–276.

Willmes, K., & Ratajczak, H. (1987). The design and application of a data- and methodbase system for the Aachen Aphasia Test. *Neuropsychologia, 25,* 725–733.

Wilson, B. A. (1987). *Rehabilitation of memory.* New York: Guilford Press.

Wilson, B. A., & Baddeley, A. D. (1993). Spontaneous recovery of impaired memory span: Does comprehension recover? *Cortex, 29,* 153–159.

Wilson, B. A., & Patterson, K. E. (1990). Rehabilitation for cognitive impairment: Does cognitive psychology apply? *Journal of Applied Cognitive Psychology, 4,* 247–260.

Wilson, R. S., & Gilley, D. W. (1992). Ideational fluency in Parkinson's disease. *Brain and Cognition, 20,* 236–244.

Winner, E., & Gardner, H. (1977). The comprehension of metaphor in brain-damaged patients. *Brain, 100,* 717–729.

Wischik, C. M., Harrington, C. R., & Mukaetova-Ladinska, E. B. (1994). Molecular characterization of the neurodegenerative changes which distinguish normal aging from Alzheimer's disease. In F. A. Huppert, C. Brayne, & D. W. O'Connor (Eds.), *Dementia and normal aging* (pp. 470–491). Cambridge, UK: Cambridge University Press.

Wise, R. J., Chollet, F., Hadar, U., Friston, K., Hoffner, E., & Frackowiak, R. (1991). Distribution of cortical neural networks involved in word comprehension and word retrieval. *Brain, 114,* 1803–1817.

Witelson, S. F. (1983). Bumps on the brain: Right-left anatomic asymmetry as a key to functional lateralization. In S. J. Segalowitz (Ed.), *Language functions and brain organization* (pp. 117–144). New York: Academic Press.

Witelson, S. F. (1985). The brain connection: The corpus callosum is larger in left-handers. *Science, 229,* 665–668.

Witelson, S. F. (1989). Handedness and sex differences in the isthmus and genu of the corpus callosum in humans. *Brain, 112,* 799–835.

Witelson, S. F. (1995). Neuroanatomical bases of hemispheric functional specialization in the human brain: possible developmental factors. In F. L. Kitterle (Ed.), *Hemispheric communication: Mechanisms and models* (pp. 61–84). Hillsdale, NJ: Lawrence Erlbaum.

Witelson, S. F., Glezer, I. I., & Kigar, D. L. (1995). Women have greater density of neurons in posterior temporal cortex. *Journal of Neuroscience, 15,* 3418–3428.

Witelson, S. F., & Goldsmith, C. H. (1991). The relationship of hand preference to anatomy of the corpus callosum in men. *Brain Research, 545,* 175–182

Witelson, S. F., & Kigar, D. L. (1992). Sylvian fissure morphology and asymmetry in men and women: Bilateral differences in relation to handedness in men. *Journal of Comparative Neurology, 323,* 326–340.

Witelson, S. F., & Nowakowski, R. S. (1991). Left out axons make men right: A hypothesis for the origin of handedness and functional asymmetry. *Neuropsychologia, 29,* 327–333.

Wolf, D., Moreton, J., & Camp, L. (1994). Children's acquisition of different kinds of narrative discourse: Genres and lines of talk. In J. Sokolov & C. Snow (Eds.), *Handbook of research in language development using CHILDES* (pp. 174–209). Hillsdale, NJ: Lawrence Erlbaum.

Wolfe, G. R. (1987). Microcomputers and treatment of aphasia. *Aphasiology, 1,* 165–170.

Wolfe, N., Linn, R., Babikian, V. L., Knoefel, J. E., & Albert, M. L. (1990). Frontal systems impairment following multiple lacunar infarcts. *Archives of Neurology, 47,* 129–132.

Wolff, A. B., Sass, K. J., & Keidan, J. (1994). Case report of an intracarotid amobarbital procedure performed for a deaf patient. *Journal of Clinical and Experimental Neuropsychology, 16,* 15–20.

Woods, B. T. (1980). The restricted effects of right-hemisphere lesions after age one: Wechsler test data. *Neuropsychologia, 18,* 65–70.

Woods, B. T., & Teuber, H.-L. (1978). Changing patterns of childhood aphasia. *Annals of Neurology, 32,* 239–246.

Woods, R. P., Dodrill, C. B., & Ojemann, G. A. (1988). Brain injury, handedness, and speech lateralization in a series of amobarbital studies. *Annals of Neurology, 23,* 510–518.

World Health Organization. (1980). *The international classification of impairments, disabilities and handicaps—a manual of classification relating to the consequences of disease.* Geneva: Author.

Wulfeck, B. (1987). Grammaticality judgments and sentence comprehension in agrammatic aphasia. *Journal of Speech and Hearing Research, 31,* 72–81.

Wychoff, L. H. (1984). Narrative and procedural discourse following closed head injury. Doctoral dissertation, University of Florida, Gainesville.

Yakovlev, P. I. (1948). Motility, behavior and the brain. Stereodynamic organization and neural coordinates of behavior. *Journal of Nervous Mental Diseases, 107,* 313–335.

Yakovlev, P. I. (1963). Telokinesis and handedness (an empirical generalization). In J. Wortis (Ed.), *Recent advances in biological psychiatry* (pp. 21–30). New York: Plenum Press.

Yakovlev, P. I. (1968). Telencephalon "impar", "semipar" and "totopar" (Morphogenetic, tectogenetic and architectonic definitions). *International Journal of Neurology, 6,* 245.

Yakovlev, P. I. (1970). The structural and functional "trinity" of the body, brain and behavior. *Topics and Problems in Psychiatry and Neurology, 10,* 197.

Yakovlev, P. I., & Lecours, A. R. (1967). The myelinogenetic cycles or regional maturation of the brain. In A. Minkowski (Ed.), *Regional development of the brain in early life* (pp. 3–70). Oxford: Basil Blackwell.

Yamadori, A., & Albert, M. L. (1973). Word category aphasia. *Cortex, 9,* 112–125.

Yamaguchi, F., Meyer, J. S., Sakai, F., & Yamamoto, M. (1980). Case reports of three dysphasic patients to illustrate rCBF responses during behavioral activation. *Brain and Language, 9,* 145–148.

Yantis, S., & Johnson, D. N. (1990). Mechanisms of attentional priority. *Journal of Experimental Psychology: Human Perception and Performance, 16,* 812–825.

Yazgan, M. Y., Wexler, B. E., Kinsbourne, M., Peterson, B., & Leckman, J. F. (1995). Functional significance of individual variations in callosal area. *Neuropsychologia, 33,* 769–779.

Yingling, C. D., & Skinner, J. E. (1977). Gating of thalamic input to cerebral cortex by nucleus reticularis thalami. In J. E. Desmedt (Ed.), *Attention, voluntary vontraction and event-related cerebral potentials* (Vol.1, pp. 70–96). Basel: Karger.

Yiu, E., & Fok, A. (1995). Lexical tone disruption in Cantonese aphasic speakers. *Clinical Linguistics and Phonetics, 9,* 79–92.

York, G. K., & Steinberg, D. A. (1995). Hughlings Jackson's theory of recovery. *Neurology, 45,* 834–838.

Young, G., & Gagnon, M. (1990). Neonatal laterality, birth stress, familial sinistrality, and left-brain inhibition. *Developmental Neuropsychology, 6,* 127–150.

Young, G., Segalowitz, S. J., Corter, C. M., & Trehub, S. E. (Eds.). (1983). *Manual specialization and the developing brain.* New York: Academic Press.

Young, G., Segalowitz, S. J., Misek, P., Alp, I. E., & Boulet, R. (1983). Is early reaching left-handed? Review of manual specialization research. In G. Young, S. J. Segalowitz, C. M. Corter, & S. E. Trehub, (Eds.) (1983). *Manual specialization and the developing brain* (pp. 13–32). New York: Academic Press.

Zacks, R. T., & Hasher, L. (1994). Directed ignoring: Inhibitory regulation of working memory. In D. Dagenbach & T. H. Carr (Eds.), *Inhibitory processes in attention, memory and language* (pp. 241–264). San Diego: Academic Press.

Zaidel, D. W. (1988). Observations on right hemisphere language function. In F. C. Rose, R. Whurr, & M. A. Wyke (Eds.), *Aphasia,* (pp. 170–187). London: Whurr Publishers.

Zaidel, E. (1977). Lexical organization in the right hemisphere. In P. A. Buser & E. Rouguel-Buser (Eds.), *Cerebral correlates of conscious experience* (pp. 177–197). Amsterdam: Elsevier.

Zaidel, E. (1978). Lexical organization in the right hemisphere. In P. A. Buser, & Rougeul-Buser, A. (Eds.), *Cerebral correlates of conscious experience,* (pp. 177–197) Amsterdam: Elsevier.

Zaidel, E. (1982). Reading in the disconnected right hemisphere: An aphasiological perspective. In Y. Zotterman (Ed.), *Dyslexia: Neuronal, cognitive and linguistics aspects* (pp. 67–91). Oxford: Pergamon Press.

Zaidel, E. (1983a). A response to Gazzaniga: Language in the right hemisphere, an empirical perspective. *American Psychologist, 38,* 542–546.

Zaidel, E. (1983b). Disconnection syndrome as a model for laterality effects in the normal brain. In J. B. Hellige (Ed.), *Cerebral hemisphere asymmetry: Method, theory, and application* (pp. 95–151). New York: Praeger.

Zaidel, E. (1983c). On multiple representations of the lexicon in the brain: the case of the two hemispheres. In M. Studdert-Kennedy (Ed.), *Psychobiology of Language,* (pp. 105–125). Cambridge: MIT Press.

Zaidel, E. (1990). Language functions in the two hemispheres following cerebral commissurotomy and hemispherectomy. In F. Boller & J. Grafman (Eds.), *Handbook of neuropsychology* (Vol. 4 pp. 115–150). Amsterdam: Elsevier.

Zaidel, E. (1994). Interhemispheric transfer in the split brain: Long-term status following complete cerebral commissurotomy. In R. H. Davidson & K. Hugdahl (Eds.), *Human laterality* (pp. 491–532). Cambridge: MIT Press.

Zaidel, E., Aboitiz, F., Clarke, J., Kaiser, D., & Matteson, R. (1995). Sex differences in interhemispheric relations for language. In F. L. Kitterle (Ed.), *Hemispheric communication: Mechanisms and models* (pp. 85–175). Hillsdale, NJ: Lawrence Erlbaum.

Zaidel, E., & Peters, A. M. (1981). Phonological encoding and ideographic reading by the disconnected right hemisphere: Two case studies. *Brain and Language, 14,* 205–234.

Zajonc, R. B. (1960). The process of cognitively tuning in communication. *Journal of Abnormal and Social Psychology, 61,* 159–167.

Zatorre, R. J. (1988). Pitch perception of complex tones and human temporal lobe function. *Journal of the Acoustical Society of America, 84,* 566–572.

Zatorre, R. J., Evans, A. C., Meyer, E., & Gjedde, A. (1992). Lateralization of phonetic and pitch discrimination in speech processing. *Science, 256,* 846–849.

Zatorre, R. J., Meyer, E., Gjedde, A., & Evans, A. C. (1996). PET studies of phonetic processing of speech: Review, replication and reanalysis. *Cerebral Cortex, 6,* 21–30.

Zechner, K., & Roth, V. M. (1996). NeueWEGE/MODAKT. In D. Novosel & V. M. Roth (Eds.), *VALMOD & MODAKT zur computerunterstützten systemischen Therapie von Sprech-/Sprachstörungen* (pp. 35–71). Konstanz: Querblick.

Zei, B., & Sikic, N. (1990). Agrammatism in Serbo-Croatian: Two case studies. In L. Menn & L. K. Obler (Eds.), *Agrammatic aphasia: A cross-language narrative sourcebook* (Vol. 2, pp. 895–974). Amsterdam: John Benjamins.

Zihl, J., & von Cramon, D. (1979). Restitution of visual function in patients with cerebral blindness. *Journal of Neurology, Neurosurgery, and Psychiatry, 42,* 312–322.

Zingeser, L. B., & Berndt, R. S. (1988). Grammatical class and context effects in a case of pure anomia: Implications for models of language production. *Cognitive Neuropsychology, 5,* 473–516.

Zingeser, L. B., & Berndt, R. S. (1990). Retrieval of nouns and verbs in agrammatism and anomia. *Brain and Language, 39,* 14–32.

Zivi, A., Broussard, G., Daymas, S., Hazard, J., & Sicard, C. (1990). Syndrome aphasie acquise-épilepsie avec psychose: à propos d'une observation. *Annales de Pédiatrie, 37,* 391–394.

Zivin, J. A., & Choi, D. W. (1991). Stroke therapy. *Scientific American, 265,* 56–63.

Zurif, E., Gardner, H., & Brownell, H. (1989). The case against the case against group studies. *Brain and Cognition, 10,* 237–255.

Zurif, E., Swinney, D., & Fodor, J. A. (1991). An evaluation of assumptions underlying the single-patient-only position in neuropsychological research: A reply. *Brain and Cognition, 16,* 198–210.

Zurif, E. B. (1996). Grammatical theory and the study of sentence processing in aphasia: Comments on Druks and Marshall. *Cognition, 58,* 271–279.

Zurif, E. B., Caramazza, A., & Myerson, R. (1972). Grammatical judgments of agrammatic aphasics. *Neuropsychologia, 10,* 405–417.

Zurif, E. B., Swinney, D., Prather, P., Solomon, J., & Bushell, C. (1993). An on-line analysis of syntactic processing in Broca's and Wernicke's aphasia. *Brain and Language, 45,* 448–464.

INDEX

ISBN 0-12-666055-7

90018

9 780126 660555